Core Principles of Special and General Relativity

Core Principles of Special and General Relativity

James H. Luscombe

CRC Press
Taylor & Francis Group
Boca Raton London New York

CRC Press is an imprint of the
Taylor & Francis Group, an **informa** business

CRC Press
Taylor & Francis Group
6000 Broken Sound Parkway NW, Suite 300
Boca Raton, FL 33487-2742

First issued in paperback 2021

ISBN 13: 978-0-367-78067-8 (pbk)
ISBN 13: 978-1-138-54294-5 (hbk)

Visit the Taylor & Francis Web site at
http://www.taylorandfrancis.com

and the CRC Press Web site at
http://www.crcpress.com

Contents

Preface

THE theory of relativity is a core component of physics curricula, yet the level at which it's taught can differ widely, from minimal coverage of special relativity (SR) in modern physics courses, to treatments using four-vectors in mechanics courses, to covariant treatments of electrodynamics, to graduate courses on general relativity (GR). I have sought to create a text aimed at advanced undergraduate/first-year graduate students, which starts with the foundations of SR and continues through to GR, at roughly the same level of sophistication. What makes that a challenge is the mathematics involved toward the end of the journey. General relativity requires the mathematics of curved spaces, the province of differential geometry. If linear algebra comprises the mathematics of quantum mechanics, differential geometry is the lingua franca of GR, *and most physics students learn this branch of mathematics in courses on GR.* We start at the beginning developing the mathematics as required with the goal of providing in one voice, hopefully in an accessible style, the full picture of the subject. I assume students have had, or are taking, the standard courses in undergraduate physics curricula—analytical mechanics, quantum mechanics, electrodynamics, and mathematical methods—but not dedicated courses in relativity beyond what one encounters in a modern physics course. I assume familiarity with the Michelson-Morley experiment (MM). I do not presuppose a mastery of tensors; we supply a reasonably in-depth treatment of tensors, on flat and curved spaces. There are numerous texts on relativity available, of varying degrees of rigor. I have sought a middle ground between treatments that are qualitative and lacking in mathematical details and works written by experts for experts.

Here are some points of note.

Minus signs: Minus-sign ambiguities arise at several places in relativity. The first is the Lorentz metric. We choose $(-+++)$; this seems best (to me)—it singles out time as the quantity warranting special treatment, so true in relativity, and it leaves alone the Euclidean metric for spatial variables. Students must learn from the outset that relativity mostly *is* about time. The perennial debate over the Lorentz metric will not be settled here. Another source of minus sign confusion is in the Riemann curvature tensor $R^{\alpha}{}_{\beta\gamma\delta}$; I have put the indices associated with derivatives in the third and fourth places, i.e., γ and δ. We take the Ricci tensor as the contraction over the first and third indices of the Riemann tensor, $R_{\mu\nu} = R^{\alpha}{}_{\mu\alpha\nu}$. Finally, the energy-momentum tensor is defined so that $T^{00} \geq 0$.

Notation: An attempt has been made at being consistent. Scalar quantities are indicated in italic font: the speed of light, c. Vector quantities are indicated with boldface italic font: force $\boldsymbol{F}$. Tensors considered as geometric objects are indicated with boldface Roman font: $\mathbf{T}$ (this notation doesn't appear until Chapter 5). Components of tensors are indicated in italic font with indices: $T_{\mu\nu}$. Tensor densities are indicated with Gothic symbols, $\mathfrak{T}$; that notation is sparingly used.

Units: I have kept all the factors of c, G, and $\hbar$ in formulas. There is a certain *panache* in advanced physics of working in units where $c = G = 1$, etc. The aim of this practice is to: 1) avoid repetitively writing the same old factors, and 2) gain insight into the geometric meaning of formulas. In a first—and perhaps only—exposure to the subject, I have consistently worked in SI units.

Mathematics: Relativity is a mathematical theory; there's no way around that. Tensors constitute the very *language* of relativity: An equation of physics expressed as a relation between tensors, if valid in one reference frame, is valid in all reference frames. Yet the mathematical preparation of students in this area is often insufficient for a study of relativity, and the power of the theory cannot be harnessed without knowledge of its mathematical structure. To fill this gap, roughly 25% of the book is devoted to the mathematics of relativity. Chapter 5 is an introduction to tensors on flat spaces. Most courses will not cover all this material; consider the latter half of Chapter 5 reference material (which is used throughout the book). The first half of Chapter 5 comprises a "tensor starter kit"—a foundation for the use of tensors in SR. For GR, a deeper understanding must be developed. To study GR at anything beyond a superficial level requires a working knowledge of tensor fields on curved spaces, which is developed in Chapters 13 and 14. I considered putting the material in Chapter 13 (manifolds) into an appendix, but decided against: It should be part of the main exposition of the subject. Nevertheless, it could be skipped on a first reading. Chapter 14 (curvature) presumes a familiarity with manifolds, but not all their properties in detail. Consult the latter half of Chapter 5 and Chapter 13 as needed. The mathematics contained in Chapters 5, 13, and 14, if encountered for the first time, would be daunting despite my attempts to guide you through the maze. It takes time to become proficient in the theory of relativity, to learn its methods and scope. Physics students tend to learn mathematics on a "need-to-know" basis, and most learn this material in courses on GR. Physicists often find themselves strangers in a strange land of mathematics.

Organization: Chapter 1 presents an overview of SR and GR. Chapters 2–10 develop non-gravitational phenomena (SR), first without, and then with the use of tensors. Chapters 11 and 12 introduce the principle of equivalence (the equivalence of local gravity and acceleration) and the treatment of accelerated motion in SR. Chapters 13 and 14 are where a traditional book on GR would begin. Chapters 15–18 present Einstein's field equation, the standard first topics in GR, and the extent to which they have been tested, mainly on the scale of the solar system. Chapter 19 concludes with a brief introduction to cosmology. Appendices contain specialized topics.

History: I have reproduced passages from the writings of Newton, Einstein, Minkowski, and others. It's instructive for students to see how the luminaries of physics have grappled with the very subject they are encountering. No attempt has been made to offer a history of relativity.

Going outside the box: Relativity is foundational to much of physics. The book is offered against the backdrop of the corpus of physical theory, to which the student is assumed to have had exposure. When instructive I point out parallels with other branches of physics; I do not pretend that other parts of physics don't exist.

Disclaimers: In addition to typos and outright blunders, I welcome comments on what is not *clear*. Invariably, when delving into a subject with sufficient depth you get "hot" on the material, and many conclusions seem obvious. Later, however, they may not be so obvious. I have attempted to give all the details necessary to derive the important equations. If the presentation seems ploddingly slow at times, I've succeeded in bringing you up to speed. It's all relative!

Acknowledgments: I thank my colleague Brett Borden for being my LaTeX guru and differential geometry sounding board. I thank the editorial staff at CRC Press, in particular Francesca McGowan and Rebecca Davies. I thank Evelyn Helminen for making figures. I thank my family, for they have seen me too often buried in a computer. My wife Lisa I thank for her encouragement and consummate advice on how not to mangle the English language. Finally, to the students of NPS, I have learned from you, more than you know. Try to remember that science is a "work in progress"; more is unknown than known.

James H. Luscombe

Monterey, California

CHAPTER 1

Relativity

A theory of space, time, and gravity

RELATIVITY is a theory of space and time that provides the foundation for much of physics. It applies to any branch of physics that makes use of the four variables x, y, z, t, where x, y, z are independent spatial coordinates and t denotes time.[1] While originating from a reasonable premise (see below), the theory of relativity[2] implies conceptions of space, time, matter, and motion vastly different from what our everyday experience of the world leads us to formulate. To understand physics in full, as applied to phenomena beyond ordinary experience, one must study relativity (as well as quantum mechanics); our everyday experience is but a special case of all that's possible in the universe. We'll see that relativity consists of two theories: the special theory of relativity (SR) and the general theory of relativity (GR).

1.1 THE PRINCIPLE OF RELATIVITY
TO VANQUISH COORDINATES, TRANSCEND THEM

In broadest terms, relativity holds that the universe doesn't care what systems of coordinates, or reference frames we use to describe physical phenomena.[3] Such a statement hardly sounds revolutionary, yet its implications are far-reaching because in the theory of relativity time is taken as a *coordinate* in a four-dimensional geometry of space and time, rather than as a parameter in pre-relativistic physics.[4] Coordinates are essential for making measurements and performing calculations, yet they're not fundamental—they don't exist in nature—*they're artifacts of our thinking*, what we as humans impose on the world. Therein lies the rub. We need coordinates for practical purposes, yet the goal of physics is to formulate laws of nature as manifestations of an objective reality, that which occurs independently of human beings.[5] *The laws of physics should be expressed in a way that's independent of coordinate system.* Relativity is an outgrowth of a single idea, *the*

[1] Isn't that all of physics? Classical thermodynamics, for example, utilizes variables that characterize the *state* of thermal equilibrium, which is independent of position and time.

[2] Referring to relativity as a *theory* can give the impression that it's speculative. Relativity has been thoroughly tested and is among the most secure theories in physics. It's up to us to fit our minds to the Procrustean bed of physics.

[3] We use the terms reference frame and coordinate system interchangeably.

[4] *Classical physics* refers to non-quantum physics; relativity belongs to classical physics. *Pre-relativistic* refers to physics developed prior to the advent of relativity, which dates to the year 1905.

[5] We use the term *objective* as it's used in science, to refer to objects that exist, or processes that occur, independently of the presence of human beings. That idea conflicts with the *acausality* of measurement as taught in quantum mechanics. There is a successful marriage of quantum mechanics with SR (the Dirac equation), but not with GR. Progress has been made in incorporating quantum effects into GR, such as Hawking radiation, but there is not presently a consistent theory of quantum GR, what's referred to as quantum gravity.

principle of relativity, that physical laws be independent of the reference frame used to represent them. Relativity is therefore a *law about laws*.[6] Albert Einstein said: "... time and space are modes by which we think, and not conditions in which we live."[2, p81] The program of relativity is to express equations of physics in such a way that, if true in one system of space-time coordinates, are true in *any* coordinate system, and thereby *transcend* coordinates. We will travel far in the theory of relativity in pursuit of this goal, which, as we'll see, is achieved by expressing equations as relations between *tensors*,[7] tensors defined on a four-dimensional geometry where time is a dimension.

1.2 THE LAW OF INERTIA: FOUNDATION OF SPECIAL RELATIVITY

Motion exists ... relatively to things that lack it.—Galileo, 1632[3, p116]

Motion is ubiquitous, yet learning to describe it correctly took a long time to achieve. Galileo taught, for the purposes of formulating laws of motion, that *states of uniform motion are the same as rest*,[8] when observed from reference frames in which the *law of inertia* holds, *inertial reference frames* (IRFs).[9] There are an unlimited number of possible IRFs, which therefore comprise a *class* of frames from which to describe motion. Our first order of business is to examine inertia and IRFs, because *SR is based on the equivalence of IRFs*. That we have singled out a particular type of reference frame is what puts the "special" in SR. There are two aspects to the principle of relativity: The *type* of phenomena that are the same for observers in equivalent reference frames, and the *class* of equivalent frames of references. With SR, Einstein showed that mechanical and electromagnetic phenomena obey the same laws for all inertial observers;[10] with GR, he extended the class of equivalent observers to *all* observers, wherein he provided an explanatory framework for gravitational phenomena. We must understand how relativity is implemented for IRFs (SR) before tackling arbitrary frames of reference (GR).

1.2.1 Inertia

The property of matter known as inertia, so familiar to us today, had a difficult time in becoming established. Pick up a rock and throw it. What makes it move when it leaves your hand? According to Aristotle, "Everything that is in motion must be moved by something," an idea seemingly so compelling, it stood for almost 20 centuries.[11] Galileo refuted that idea with a simple experiment.[12] Drop a stone from the mast of a ship that's at rest; note where it lands. Now repeat the experiment on a ship that's in uniform motion. In the Aristotelian theory, the rock would land at a point *displaced*

[6]The principle of relativity is a different kind of law than other physical principles. It presumes the existence of laws of nature, that there are reproducible manifestations of the workings of nature waiting for us to describe, of which we possess a language rich enough to accurately describe. That language is mathematics, which physics relies on heavily. It's remarkable that mathematics, a human invention, applies so well to the description of nature. To quote Eugene Wigner:[1] "... the mathematical formulation of the physicist's often crude experience leads in an uncanny number of cases to an amazingly accurate description of a large class of phenomena. This shows that the mathematical language has more to commend it than being the only language which we can speak; it shows that it is, in a very real sense, the correct language."

[7]If you're uneasy about tensors, don't worry; students are frequently ill-prepared when it comes to tensors. The mathematics of tensors will be developed as we proceed. Vectors are special cases of tensors.

[8]Galileo did not explicitly isolate the concept of inertial motion as a general principle, yet it's quite clear from his writings that he understood it. Even today, students of physics are well advised to read Galileo's *Dialogue*.[3]

[9]There *are* reference frames in which the law of inertia does not hold, *noninertial reference frames*—see Section 1.6.

[10]We'll refer to *inertial observers* as observers at rest relative to IRFs. The "observer" is essentially the reference frame.

[11]Aristotle classified motion as natural and unnatural. Natural motion occurs among the four elements air, earth, fire, and water, which seek to find their natural places, e.g., heavy objects naturally move toward the center of the earth. Natural motion is unforced, not requiring the action of an external agency. Unnatural motion, however, such as horizontal motion on Earth, is forced and requires a mover. What's the "mover" when the rock leaves your hand? Aristotle argued that air, displaced by the motion of the rock, wraps around the rock and pushes it on. A rock thrown in vacuum would not move!

[12]The history of inertia is more involved than our account here. A succession of investigators in the time between Aristotle and Galileo questioned the Aristotelian theory.

from the mast by the distance the ship had moved during the fall.[13] Galileo maintained there would be no displacement because, first, the *rock shares in the motion of the ship*,[14] and second, *free particles move without movers*, that free particles—those with no forces acting on them—once set in motion, *maintain* that state of motion, termed *inertial motion.*[15] The rock accelerates under the action of gravity, but maintains its constant motion in the direction of the uniform motion of the ship because there is no force acting in that direction (assuming negligible wind resistance).[16]

The *primary* state of motion, that exhibited by free particles, is inertial—in a straight line at constant speed. *Free particles of and by themselves cannot change their states of motion.* That fact is highly important (essential, actually) for SR and GR. The unfolding of the inertia concept mirrors the historical development of physics, from Aristotle to Einstein, at least as far as our understanding of motion is concerned. Galileo's experiment with the ship is a variant of an argument used by Aristotle to prove that Earth is immobile: An object projected straight up from the surface of the earth returns to the same place and thus Earth could not have moved in the meantime. Galileo maintained that *nothing* can be inferred from such an argument about Earth's motion or rest. What Galileo asserted is that, except in relation to other objects, *uniform motion of one's reference frame cannot be detected*—a fundamental tenet of relativity—in this case by mechanical means.[17]

Isaac Newton conceived of inertia not just as the property of free objects to maintain states of uniform motion, but also by what he called the *inherent force*, the property by which matter *resists* changes in motion: "Inherent force of matter is the power of resisting by which every body, so far as it is able, perseveres in its state either of resting or of moving uniformly straight forward."[4, p404] Thus there are two aspects of inertia: *perseverance* and *resistance*. His definition of inertia should be read together with his first law of motion: "Every body perseveres in its state of being at rest or of moving uniformly straight forward, except insofar as it is compelled to change its state by forces impressed."[4, p416] Objects move inertially unless *prevented* from doing so by imposed forces, to which they provide a resistance, the *inertial force*.[18] The inertial force is the reaction by which objects "push back" against forces attempting to prevent states of inertial motion:

> Because of the inertia of matter, every body is only with difficulty put out of its state either of resting or of moving. Consequently, inherent force may also be called by the very significant name of force of inertia. Moreover, a body exerts this force only during a change of its state, caused by another force impressed upon it, and this exercise of force is, depending on the view point,[19] both resistance and impetus:[20] resistance insofar as the body, in order to maintain its state, strives against the impressed force, and impetus insofar as the same body, yielding only with difficulty to the force of a resist-

---

[13] In the Aristotelian theory, once the stone has been released (and no longer has a mover), it can only undergo its "natural" motion toward the center of Earth; where the ship goes after the release of the rock is immaterial.

[14] This point, obvious to us today, was one that Galileo had to take pains to establish, that objects can have a *superposition* of motions, i.e., velocity is a vector quantity. In the Aristotelian theory, objects not subject to movers can only have their natural motions. That objects can have "two motions" (downwards and sideways) was foreign to the Aristotelian worldview.

[15] Galileo based this conclusion on his experiments with inclined planes: Objects accelerate on planes oriented downward, decelerate on those oriented upwards, and have no acceleration on horizontal planes.

[16] Truth in advertising: A particle dropped from a sufficiently high point *would* show a displacement from the Coriolis acceleration. By Earth's rotation, a body dropped from a high elevation has a higher transverse velocity than the ground. Such a displacement actually confirms Galileo's hypothesis that different types of motion can be imparted to particles.

[17] In the Aristotelian theory, the speed of the ship could be inferred from the displacement of the rock. Perhaps one has ridden in a train through a tunnel (or a submarine), where, if the ride is smooth enough, one doesn't have a sense of motion. The MM experiment failed to detect uniform motion by electromagnetic means.

[18] We'll see in GR that your weight is the force which must be supplied to *prevent* you from continuing in a state of inertial motion. What's seen as accelerated motion in three dimensions (under the force of gravity) corresponds to a constant state of motion in four-dimensional *spacetime* (defined on page 5). As shown in GR, gravity is a property of spacetime.

[19] What we refer to as reference frame, Newton called *point of view*.

[20] Impetus is another word for momentum. What we call momentum, Newton called *quantity of motion*, defined [4, p404] as "the velocity and quantity of matter jointly"; hence momentum $p = mv$. In SR, momentum is defined as $p = m\gamma v$ (where $\gamma = (1 - v^2/c^2)^{-1/2}$ and c is the speed of light), an alternative "quantity of motion." For $v \ll c$, $\gamma \approx 1$.

ing obstacle, endeavors to change the state of that obstacle. Resistance is commonly attributed to resting bodies and impetus to moving bodies; but motion and rest ... are distinguished from each other only by point of view, and bodies commonly regarded as being at rest are not always truly at rest.[4, p404]

1.2.2 Inertial reference frames

In IRFs the law of inertia holds true, that *free particles* move in *straight lines* at *constant speed*. In view of the transition to GR, several issues are exposed by this benign statement.

1. *What's a free particle*? The answer is seemingly self-evident: If free particles are unaccelerated, then not-free particles are accelerated, right? Not so fast. Such reasoning doesn't take into account how acceleration is measured. Not all unaccelerated particles are free, and not all free particles are unaccelerated: It depends on the reference frame. *In IRFs, acceleration is caused solely by forces*. No force, no acceleration, and *forces arise from physical interactions*. In *noninertial reference frames* (see Section 1.6), acceleration can be an artifact of the choice of frame and not necessarily the result of forces. Forces can be identified from their physical *sources*. Acceleration—seemingly the quantity most accessible to direct observation—is not unambiguous because to measure it a *standard of rest* must be specified. Consider Earth in the gravitational field of the sun. In a frame with the sun at rest, Earth's acceleration is in the direction of the force produced by the sun; Newton's second law of motion is satisfied. In a frame with Earth at rest, however, it is *not satisfied* because Earth's acceleration is zero. *Newton's second law is not a general law of physics* because we're free to choose reference frames in which it doesn't work.[21] *IRFs are frames in which objects with no forces acting on them have no acceleration*.

2. *What's a straight line*? In a given geometry, the straightest possible line is called a *geodesic curve*, a concept that we'll develop. But what specifies the geometry? In GR, the geometry of spacetime[22] is not something known *a priori*, but is instead *determined* by its energy-momentum content. Spacetime geometry is therefore *physical*, something that *emerges* from the distribution of matter-energy-momentum. Spacetime in GR is not something passive and inert; it *evolves* in response to matter. The version of Newton's first law that survives to GR is that free particles follow geodesic paths in spacetime, those determined by the distribution of energy-momentum. We return to this idea when we take up GR.

3. *What's constant speed*? For speed, we need *time*. But whose time? Newtonian mechanics utilizes an *absolute time* that pervades the universe—see page 8. In relativity, time and space do not have separate existences *and are reference-frame specific*.

1.2.3 Equivalence of inertial reference frames

Once a frame has been found meeting the criteria for an IRF, any other frame moving relative to it with constant velocity also constitutes an IRF.[23] *A natural equivalence among IRFs is established by free particles*: All inertial observers agree that the trajectories of free particles are described by constant velocity; all agree on the law of inertia. The *value* of the speed is reference-frame specific, but all agree on its constancy. Thus, all inertial observers agree on the laws of mechanics: Forces manifest in changes of states of inertial motion. *Different inertial observers can observe the same phenomena and describe them by the same laws*. Transforming from one set of inertial observers to another does not change the laws—the very heart of the principle of relativity.

[21] In a sense, that's the problem GR fixes.

[22] Spacetime is defined on page 5. Is it obvious what the *geometry* of spacetime should be?

[23] The motion of free objects is seen as unaccelerated in both frames.

1.2.4 Coordinate transformations and the principle of covariance

Transformation is central to relativity. Transformations between reference frames are effected mathematically as transformations among the different *coordinates* assigned to the same event by all the different, yet equivalent inertial observers. An *event* is a point in space at a point in time. Anything that happens, or has happened or will happen, comprises an event. The totality of all events is a four-dimensional continuum referred to as *spacetime* (no hyphen). We require that *the mathematical form of the laws of physics be unaffected by changes in reference frames, changes in the coordinates assigned to events*, a theme that accompanies us from Newtonian mechanics to SR to GR, that the laws of physics be expressed in a way that their *form* is invariant under progressively more general coordinate transformations. *Form invariance* of physical laws is called the *principle of covariance*, the requirement that the equations of physics adhere to the principle of relativity by having the same mathematical form in all reference frames.

Coordinate transformations in SR must be linear. All inertial observers agree that the spacetime trajectories (*worldlines*) of free particles are straight (see Section 1.4). Coordinate transformations between IRFs must be such as to *map straight lines in spacetime onto straight lines* so as to preserve the law of inertia. Only homogeneous, linear transformations map straight lines onto straight lines, where both lines pass through the same origin of the coordinate system. We'll work through some examples to see how inertial frames can differ and yet be equivalent.

1.2.4.1 Boosts

Figure 1.1 shows frames S and S' with origins displaced by vector $\boldsymbol{R}$, where the coordinate axes

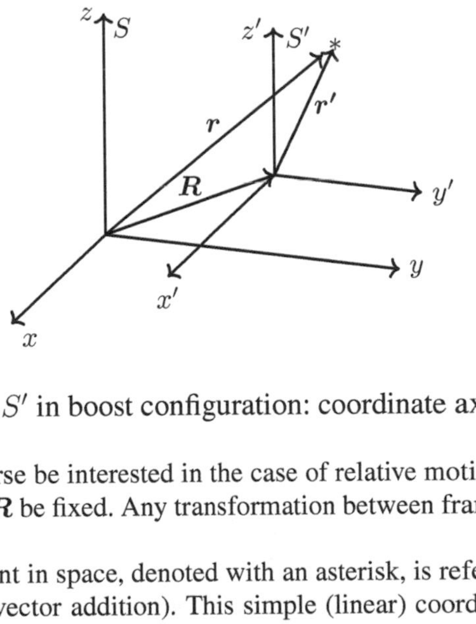

Figure 1.1 Frames S and S' in boost configuration: coordinate axes are parallel.

are parallel. We will of course be interested in the case of relative motion where $\boldsymbol{R} = \boldsymbol{R}(t)$ is time dependent, but for now let $\boldsymbol{R}$ be fixed. Any transformation between frames with parallel axes (as in Fig. 1.1) is called a *boost.*

In Fig. 1.1 the same point in space, denoted with an asterisk, is referenced by vectors $\boldsymbol{r}$ and $\boldsymbol{r}'$, with $\boldsymbol{r}' = \boldsymbol{r} - \boldsymbol{R}$ (law of vector addition). This simple (linear) coordinate transformation can be "inverted" by interchanging primed and unprimed quantities and letting $\boldsymbol{R} \to -\boldsymbol{R}$, $\boldsymbol{r} = \boldsymbol{r}' + \boldsymbol{R}$. That rule will stand us in good stead with linear coordinate transformations: Interchange primed and unprimed quantities and reverse the transformation parameter (velocity, angle, etc.). Suppose S is an IRF, i.e., a frame in which a free particle is unaccelerated, $\ddot{\boldsymbol{r}} = 0$. By differentiating the transformation equation we conclude that $\ddot{\boldsymbol{r}}' = 0$. If S is an IRF, so is S' when it's connected to S by a displacement. *There is no unique origin for IRFs.*

1.2.4.2 Rotations

A more complicated example of a linear transformation is a *rotation*. Figure 1.2 shows frames S and

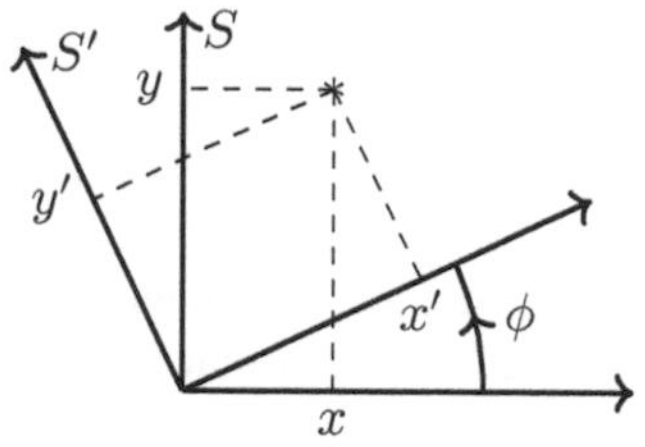

Figure 1.2 Frames having a common origin with axes rotated through a fixed angle ϕ.

S' having a common origin but with coordinate axes rigidly rotated relative to each other by a fixed angle ϕ. How are the coordinates assigned to the same point related? It's an exercise in trigonometry to show that

$$\begin{pmatrix} x' \\ y' \end{pmatrix} = \begin{pmatrix} \cos\phi & \sin\phi \\ -\sin\phi & \cos\phi \end{pmatrix} \begin{pmatrix} x \\ y \end{pmatrix} \equiv R_z(\phi) \begin{pmatrix} x \\ y \end{pmatrix} , \tag{1.1}$$

where we've introduced the rotation operator, $R_z(\phi)$, which effects a rotation about the z-axis (coming out of the paper, not shown) through an angle ϕ. The inverse transformation is obtained by interchanging primed and unprimed quantities and by letting $\phi \to -\phi$. If in S a free particle is observed to be unaccelerated, with $\ddot{x} = 0$ and $\ddot{y} = 0$, then because ϕ is constant, $\ddot{x}' = 0$ and $\ddot{y}' = 0$. A frame rotated relative to an IRF is also an IRF.[24] *There is no unique orientation of IRFs*. General linear transformations involving both boosts and rotations are covered in Chapter 6.

1.2.4.3 Galilean transformations

Now let $\boldsymbol{R}$ in Fig. 1.1 vary linearly with time, $\boldsymbol{R} = \boldsymbol{v}t$, where $\boldsymbol{v}$ is a constant vector. Both observers carry identical clocks, which are synchronized when the origins coincide. By "common sense" reasoning, $\boldsymbol{r}$ and $\boldsymbol{r}'$ are related by $\boldsymbol{r}' = \boldsymbol{r} - \boldsymbol{v}t$. Implicit is the assumption that time in S', t', is the same as that in S, $t' = t$ (absolute time, see page 8). This "obvious" assumption was rarely made explicit in pre-relativistic physics. By differentiating the transformation formula, we have the *Galilean velocity addition formula*[25] $\boldsymbol{u}' = \boldsymbol{u} - \boldsymbol{v}$, where $\boldsymbol{u} \equiv \mathrm{d}\boldsymbol{r}/\mathrm{d}t$ and $\boldsymbol{u}' \equiv \mathrm{d}\boldsymbol{r}'/\mathrm{d}t'$. If in S a free particle is described by $\ddot{\boldsymbol{r}} = 0$, then $\ddot{\boldsymbol{r}}' = 0$ as well. If S is an IRF, then so is S' if it's moving uniformly relative to S. It's difficult to appreciate at first the deep implications of this result!

The transformation $\boldsymbol{r}' = \boldsymbol{r} - \boldsymbol{v}t$ can be written in terms of its vector components:

$$\begin{pmatrix} x' \\ y' \\ z' \end{pmatrix} = \begin{pmatrix} x \\ y \\ z \end{pmatrix} - t \begin{pmatrix} v_x \\ v_y \\ v_z \end{pmatrix} . \tag{1.2}$$

Equation (1.2) underscores the pre-relativistic concept that we live in a three-dimensional world with time as a universal parameter ($t' = t$). If time is included as a separate dimension, however, $\boldsymbol{r}' = \boldsymbol{r} - \boldsymbol{v}t$ and $t' = t$ can be expressed as a linear transformation in four-dimensional spacetime:

$$\begin{pmatrix} t' \\ x' \\ y' \\ z' \end{pmatrix} = \begin{pmatrix} 1 & 0 & 0 & 0 \\ -v_x & 1 & 0 & 0 \\ -v_y & 0 & 1 & 0 \\ -v_z & 0 & 0 & 1 \end{pmatrix} \begin{pmatrix} t \\ x \\ y \\ z \end{pmatrix} . \qquad \text{(Galilean transformation)} \tag{1.3}$$

[24]The frames S and S' are related through a *fixed* angle. A rotating reference frame, with $\phi = \phi(t)$ is not an IRF.
[25]How velocities transform between IRFs in SR is treated in the next two chapters.

Let's get in the habit of listing the time "coordinate" first,[26] as in Eq. (1.3). Equation (1.3) is the *Galilean transformation* (GT), the form of relativity based on everyday experience. Despite its common-sense appeal, the GT does not lead to predictions in agreement with experiment;[27] it will be replaced by another linear transformation of spacetime coordinates that *does* lead to agreement with experiment—the *Lorentz transformation* (LT).[28]

1.2.4.4 Form invariance

The idea of form invariance can be illustrated using the GT, because acceleration is invariant under that transformation: $\boldsymbol{a}' \equiv (\mathrm{d}^2/\mathrm{d}t'^2)\boldsymbol{r}' = (\mathrm{d}^2/\mathrm{d}t^2)(\boldsymbol{r} - \boldsymbol{v}t) = (\mathrm{d}^2/\mathrm{d}t^2)\boldsymbol{r} = \boldsymbol{a}$. Observers in S and S' agree on the form of Newton's second law: $\boldsymbol{F}' = m\boldsymbol{a}' = m\boldsymbol{a} = \boldsymbol{F}$, where mass is the same in all IRFs.[29] The laws of mechanics are invariant under the GT. What about electromagnetism?

Maxwell's equations predict the existence of electromagnetic waves that propagate with a speed given in terms of electromagnetic parameters, $c = 1/\sqrt{\epsilon_0 \mu_0}$. It's shown in Appendix A that the wave equation transforms under the GT for frames in relative motion along a common x-axis as:

$$\frac{\partial^2}{\partial x^2} - \frac{1}{c^2}\frac{\partial^2}{\partial t^2} = \left(1 - v^2/c^2\right)\frac{\partial^2}{\partial x'^2} - \frac{1}{c^2}\frac{\partial^2}{\partial t'^2} + \frac{2v}{c^2}\frac{\partial^2}{\partial x' \partial t'} \ . \tag{A.3}$$

Form invariance therefore does *not* hold for the wave equation under the GT, *implying a crack in the foundation of physics.* The inconsistency is that Maxwell's equations are fundamental laws of physics, yet a prediction of those equations is not invariant under the GT, while the laws of mechanics are. Let's consider the three possible explanations for this inconsistency:

1. The principle of relativity applies to mechanics, but not to electromagnetism. Maxwell's equations predict a speed of electromagnetic waves, but don't specify a reference frame. Perhaps there is only one reference frame in which the speed of light is c? If so, one could detect that frame by electromagnetic means—the MM experiment.

2. The principle of relativity applies to mechanics and electromagnetism but Maxwell's equations are incorrect. If so, one should find discrepancies between the predictions of Maxwell's equations and experimental results. Such discrepancies have yet to be found.

3. The principle of relativity applies to mechanics and electromagnetism, but Newton's laws are incorrect. If so, one should find discrepancies between the predictions of Newton's laws and experimental results—something routinely done at particle accelerators which produce speeds $v \lesssim c$. If Newton's laws are incorrect, so is the GT, and we're back to square one.

Einstein opted for the third explanation. He asserted that the principle of relativity applies to all of physics, not just to mechanics. He then took that idea to its logical extreme. The speed of light is a law of physics, not merely something that we measure. Einstein took the bold step of asserting that the speed of light is the same for all inertial observers, which experiment has shown to be true!

1.3 SPACE, TIME, AND SPACETIME

1.3.1 Newtonian space and time

Relativity is concerned with space and time and how the two are related through motion. It's useful to state Newton's conceptions of space and time, which, while not satisfactory by today's standards, continue to frame the discussion:[4, p408]

[26] In pre-relativistic physics, time is a parameter, not a coordinate.

[27] What's wrong with common sense? If you had to put your finger on it, it would be the assumption that $t' = t$, the notion of absolute simultaneity.

[28] The properties of Lorentz transformations are developed throughout this book.

[29] Mass is the same in all IRFs, wherein all observers claim themselves at rest.

- Absolute space, of its own nature without reference to anything external, always remains homogeneous and immovable.
- Absolute, true, and mathematical time, in and of itself and of its own nature, without reference to anything external, flows uniformly and by another name is called duration.

What's meant by *absolute*? Einstein gave a good definition:[5, p55] "...absolute means not only physically real, but also independent in its physical properties, having a physical effect, but not itself influenced by physical conditions." We'll use absolute in Einstein's sense—physically existing, but not influenced by physical conditions. Newton's space and time are absolute in that sense: They exist—by definition—independent of anything else. These notions unravel in relativity. *Space and time are not independent of each other*, but are two aspects of a single entity: spacetime.

It's understandable that space would be conceived as absolute. Look out at the night sky. Space appears as a vast, fixed arena containing the objects of the universe. Already we're up against cosmological questions. Does space exist independently of the objects in the universe (as Newton would have it), passively containing them, or do the properties of space manifest *because* of the objects in the universe (the picture afforded by GR)? Is the universe separate from the objects it contains? Is it a vast collection of independent objects, or is it a single entity? GR will weigh in on these questions.

1.3.2 Simultaneity—the death knell of absolute time

Snap your fingers. In the Newtonian framework you've just specified "now" *at every point of the universe*, no matter how distant, because time exists independently of space. That notion is indicated in Fig. 1.3. Two points in space having the same time are said to be *simultaneous*. An instant of time

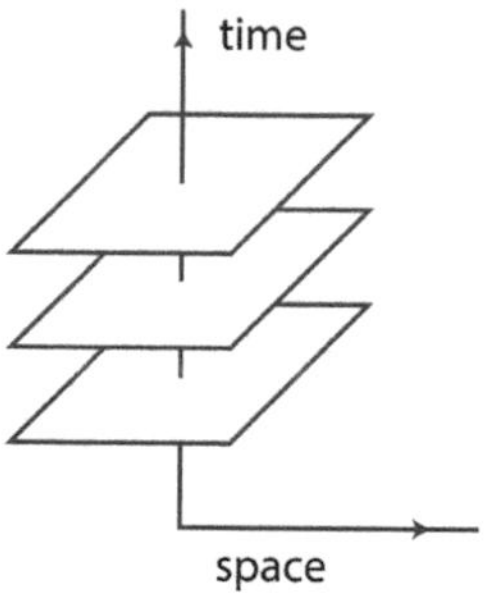

Figure 1.3 Surfaces of simultaneity in Newtonian spacetime.

thus determines a three-dimensional *surface of simultaneity*,[30] extending throughout all of space. Simultaneity is therefore absolute in pre-relativistic physics, existing independently of anything else. In relativity, *simultaneity is not absolute*—two events simultaneous in one IRF, are not in another. Sit equidistant between two friends, and have them snap their fingers at the same time; you hear both simultaneously. To someone walking past you at a constant rate, however, the same finger-snaps would not be simultaneous.[31] The finger-snap would be heard first from the sound source that the walker is moving toward. Whose description of these events is "right"? *Relativity shows there is no absolute meaning to the "same time."* Absolute time does not exist—it's not true that time exists independently of anything else. Time is not a parameter provided by the universe, as it is in pre-relativistic physics; relativity shows that *time exists locally*, relative to a given reference frame.

[30] Actually a three-dimensional *hypersurface*. Our familiar notion of surface (such as the surface of an apple) is a two-dimensional set of points, or manifold, embedded in three-dimensional space. A hypersurface is an $(n-1)$-dimensional manifold embedded in n-dimensional space. Manifolds and hypersurfaces will be systematically introduced in later chapters.

[31] The *relativity of simultaneity* is illustrated in Fig. 1.6.

The term *relativity* is misleading. Relativity does not claim that "everything is relative" (as is sometimes falsely stated), only that *some* things are relative, such as simultaneity. *Relative* refers to measurements made relative to a given reference frame, the results of which may not be the same in all reference frames. The purpose of relativity is to discover what is *not* relative, that what is the same for all observers is a law of physics. Relativity shows that *simultaneity is not a law of physics*.

1.3.3 Absolute space—is it real?

Absolute space, "homogeneous and immovable," would be the ultimate reference frame from which it could be decided whether objects are "really" at rest. How would we recognize an object absolutely at rest? The answer is, we can't.[32] Rest cannot be ascertained against a backdrop of "nothingness" (absolute space); there must be other objects around to compare with—*rest exists only in relation to other objects*, which can be considered reference frames. The same is true of motion. We cannot perceive motion in itself (relative to absolute space); motion is perceived only in relation to objects—*all motion is relative*.[33] Nevertheless, if a reference frame could exist from which all motion is relative to, yet which is itself absolutely at rest, let yourself be at rest in that frame. Someone drifting by in a rocket ship would say you're in motion! *Everything moves with respect to everything else*, and *every inertial observer claims they are at rest*.

Absolute space is thus an empty concept because only relative motion can be observed. Perhaps that's why it went largely unchallenged in the 200 years between the time of Newton and the late 19th century, because it has no observable consequences.[34] The concept of absolute space received support, however, from Maxwellian electrodynamics. Maxwell's equations predict a speed of electromagnetic waves, but they don't specify a reference frame—what better evidence for a preferred frame like absolute space? Physicists of the late 19th century inferred there must be only *one* reference frame in which the speed of light is c (called the *ether frame*, presumably absolute space). Einstein, however, reached the opposite conclusion: If Maxwell's equation don't specify a reference frame, *all inertial observers measure the same speed of light*.

1.3.4 Spacetime coordinates and notational conventions

In the theory of relativity time is taken as a *coordinate* in the specification of physical phenomena, in addition to spatial coordinates. Ask a friend to meet you for coffee. You must specify a point in space, three coordinates (on the surface of Earth usually two suffice), at a point in time, making four numbers in all. Thus, you're asking to meet your friend at a specified spacetime point, i.e., event.

The "gist" of relativity is that different observers assign different coordinates to the same events, underscoring that coordinates are without fundamental significance. *Events are physical and exist independently of the coordinates assigned to them*.[35] The procedure in SR by which coordinates are assigned to events, the *coordinization* of spacetime, is discussed below. In GR, the assignment of spacetime coordinates is associated with its mathematical structure as a *manifold*. In SR, spacetime is *flat*, while in GR spacetime is *curved*. Flat geometries can be *covered* by a single system of coordinates, whereas curved geometries require overlapping coordinate systems. Curved geome-

---

[32]The *unobservability* of absolute space underscores a lesson from the history of physics: Physics is based on what can be measured. Notions of what might or could exist "anyway," but that we can't detect, like absolute space, tend to get excised from physics. "Excess" theoretical structures imply that alternative theories are possible.

[33]Recall Galileo's words (page 2): "Motion exists relative to things that lack it".

[34]There were objections to absolute space most notably from George Berkeley and Ernst Mach. Berkeley's 1721 essay *On Motion* objected to absolute space because it's not observable; see [6], paragraphs 58, 59, and 64. Mach's *Science of Mechanics* [7] (published in 1883) provided the most incisive and influential critique of Newtonian mechanics. Mach contended we're not *allowed* to invent concepts like absolute space. In the world we know of, motion is relative. We should not invent concepts that contravene that fact. "No one is competent to predicate things about absolute space and absolute motion; they are things of thought, pure mental concepts, that cannot be produced in experience."

[35]Spacetime in SR is *absolute*—existing, but not influenced by physical conditions.

tries, however, are *locally flat*—what we learn about coordinatizing spacetime in SR applies to limited regions of spacetime in GR. To locate an object in three-dimensional space, three numbers, or coordinates, must be specified. In the Cartesian coordinate system, the numbers are traditionally denoted (x, y, z). But there are other coordinate systems, e.g., spherical coordinates, (r, θ, ϕ). We'll denote spatial coordinates in a way that doesn't commit to a particular coordinate system with the notation (x^1, x^2, x^3), or simply x^i, where it's understood that $i = 1, 2, 3$. The use of superscripts takes some getting used to, but it's standard notation in tensor analysis.[36] When there's a possibility for confusion, we'll denote the square of x as $(x)^2$ to avoid mistaking it with the coordinate x^2; contrary to what you might think, problems of that sort do not occur often. The time coordinate will be parameterized, for reasons explained in Section 1.4, as $x^0 \equiv ct$. An event thus has coordinates x^0, x^1, x^2, x^3. To save writing, spacetime coordinates are conventionally denoted x^μ, where it's understood that $\mu = 0, 1, 2, 3$. Greek letters denote spacetime coordinates, x^ρ, while Roman letters denote spatial coordinates, x^k. The indices ρ and k are *dummy indices* having no absolute meaning. Thus, $\sum_{\nu=0}^{3} x^\nu = x^0 + \sum_{j=1}^{3} x^j$. As we'll see, two *types* of coordinates arise in *non-orthogonal* coordinate systems: *contravariant*, denoted with superscripts, x^μ, and *covariant*, denoted with subscripts, x_ν. Because GR seeks to work in arbitrary coordinate systems—not necessarily orthogonal—both types of coordinates, x^ν and x_ν, will be used.

Sidebar discussion: In 1908 Hermann Minkowski delivered a seminal presentation, *Space and Time*,[37] in which he showed that the results of SR, as derived algebraically by Einstein in 1905, have a natural and intelligible explanation when space and time are conceived *geometrically* as belonging to a four-dimensional continuum with a non-Euclidean geometry.

> The views of space and time which I wish to lay before you have sprung from the soil of experimental physics, and therein lies their strength. They are radical. Henceforth space by itself, and time by itself, are doomed to fade away into mere shadows, and only a kind of union of the two will preserve as an independent entity.

Many of the terms we use in relativity are due to Minkowski: Proper time, spacelike vector, timelike vector. He didn't use the term *lightcone*, but he did speak of "front" and "back" cones, which we will call future and past lightcones. It's clear that Minkowski had worked out much concerning the geometry of spacetime, what today we call Minkowski space (see Chapter 5). Minkowski died suddenly in 1909 at age 44; one can only wonder what additional contributions he might have made. What we call spacetime, Minkowski called *the world*: "A point of space at a point of time, that is, a system of values x, y, z, t, I will call a *world-point*. The multiplicity of all thinkable x, y, z, t systems we will christen the *world*." The term *worldline* is due to Minkowski:

> We fix our attention on the substantial point which is at the world-point x, y, z, t, and imagine that we are to recognize this substantial point at any other time. Let the variations $\mathrm{d}x, \mathrm{d}y, \mathrm{d}z$ of the space coordinates of this substantial point correspond to a time element $\mathrm{d}t$. Then we obtain, as an image, so to speak, of the everlasting career of the substantial point, a curve in the world, a *worldline*, the points of which can be referred unequivocally to the parameter t from $-\infty$ to $+\infty$. The whole universe is seen to resolve itself into similar worldlines, and ... in my opinion physical laws might find their most perfect expression as reciprocal relations between these worldlines.

---

[36] To quote O. Veblen (from 1927), [8, p1] "Recent advances in the theory of differential invariants and the wide use of this theory in physical investigations have brought about a rather general acceptance of a particular type of notation, the essential feature of which is the systematic use of subscripts and superscripts ...." The use of subscripts and superscripts is not as arbitrary as it might first appear; the way the two types of indices are used in calculations is quite logical and consistent.

[37] Reprinted in *The Principle of Relativity* [9, p73], an important collection of articles by Einstein, Lorentz, Minkowski, and Weyl. A chance to read the original literature in English translation.

1.4 SPACETIME DIAGRAMS

Comprehending relativity is greatly facilitated through the use of *spacetime diagrams*, also called Minkowski diagrams, and we'll use them freely. On such diagrams, time is displayed along the vertical axis, with spatial dimensions displayed on horizontal axes (see Fig. 1.4). It's simplest to

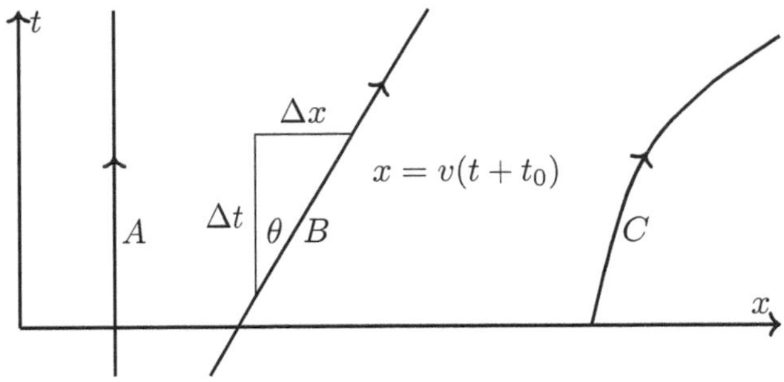

Figure 1.4 Particle worldlines: A is at rest, B is in uniform motion, and C is accelerated.

take time as orthogonal[38] to the three-dimensional space of spatial variables, as in Fig. 1.3. While we employ an orthogonal spacetime coordinate system, the geometry of spacetime is non-Euclidean (as we'll show); don't be fooled into thinking that an orthogonal set of axes implies a Euclidean geometry. Many ingrained habits must be unlearned in "doing" geometry on spacetime diagrams, particularly in calculating distances. Particle A is at rest in the reference frame of Fig. 1.4. The "motion" (history) in spacetime of a stationary object is a line parallel to the time axis. Particle B has constant velocity; its worldline is straight, with speed $v = \Delta x/\Delta t = \tan\theta$ m s^{-1}. Particle C is accelerating; its worldline is curved. In IRFs the worldlines of free particles are *straight*.

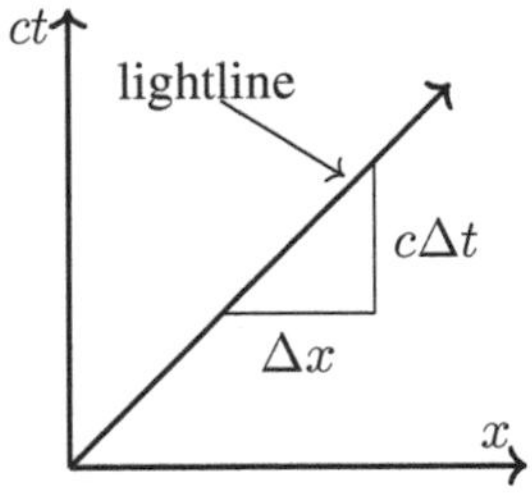

Figure 1.5 Lightline (photon worldline) on a spacetime diagram.

Of particular interest are the worldlines of photons; see Fig. 1.5. Using meters and seconds as the units of length and time, the worldline of a photon would be almost parallel to the spatial axis, with $\theta \approx \pi/2$. There's nothing fundamental about units, however; one size doesn't fit all and it's common to adopt units that are suited to the problem at hand (e.g., the electron volt). It's convenient to scale times by $1/c \approx 3.3$ ns m^{-1}, the time for light to travel one meter. With $t \to t/(1/c) = ct$, the worldline of a photon—the *lightline*—is at the angle $\pi/4$ with respect to the ct *and* x-axes. We'll draw lightlines at 45° angles relative to the space and time axes.

The relativity of simultaneity can be illustrated on a spacetime diagram. Consider a photon source C in a train car[39] situated equidistant between detectors A and B; see Fig. 1.6. The source emits photons back to back. In a reference frame at rest with respect to the train car, photons arrive at the detectors simultaneously. In a reference frame at rest relative to the train station, however,

[38] In rotating reference frames, the orthogonality between space and time breaks down.

[39] Einstein's relativity tends to get done in train stations and in elevators.

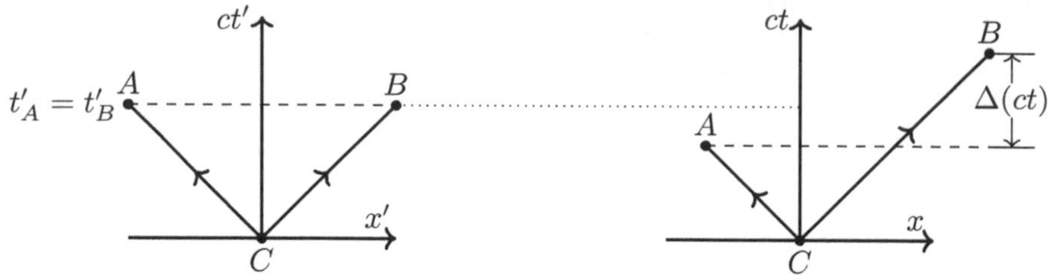

Figure 1.6 Relativity of simultaneity. Photons are received simultaneously in the frame of the emitter, but not in a frame in which emitter and receiver are moving to the right.

which the train is assumed passing through, the photon source and the detectors are in motion with speed v from left to right. The two frames synchronize their identical clocks when the origins of their coordinate systems coincide, whereupon the photons are emitted. Seen from the frame of the station, event A happens *before* B; the photon first encounters detector A moving toward it. Simultaneity is not absolute: What's observed as simultaneous in one IRF, is not in another.

There's a fundamental reason to use ct as the temporal coordinate. The fusion of space and time into spacetime requires that spacetime coordinates *all have the same dimension.* The coordinates of an event in one IRF are, under the LT, a *linear combination* of the coordinates in another IRF, which can be accomplished only if space and time coordinates have the same dimension. We require a conversion factor between spatial and temporal measures, *which must be the same for all IRFs.* We'll show (Chapter 3) that a LT followed by a LT, is itself a LT—what's required by the principle of relativity that all IRFs be equivalent. Such universality is possible only if the conversion factor is universal. *The principle of relativity requires a universal speed.* Experiment shows that speed is the speed of light. For frames in relative motion along a common x-axis (see Fig. 3.1), the spacetime coordinates transform under the LT (Eq. (3.17))

$$\begin{pmatrix} ct' \\ x' \\ y' \\ z' \end{pmatrix} = \begin{pmatrix} \gamma & -\beta\gamma & 0 & 0 \\ -\beta\gamma & \gamma & 0 & 0 \\ 0 & 0 & 1 & 0 \\ 0 & 0 & 0 & 1 \end{pmatrix} \begin{pmatrix} ct \\ x \\ y \\ z \end{pmatrix} , \tag{1.4}$$

where $\gamma \equiv (1 - \beta^2)^{-1/2}$ is the *Lorentz factor*, with $\beta \equiv v/c$. Under the LT, the time coordinate in S' is a mixture of the time *and space* coordinates in S.[40] For that reason, the time coordinate $x^0 = ct$ must have the dimension of length.

The worldline of an object at rest in a given IRF is parallel to the time axis in that frame, e.g., worldline A in Fig. 1.4. The worldline of A might just as well *be* the time axis in that frame, what we'll assume from now on. Let observer B be at rest relative to A—see Fig. 1.7. At time t_1, A emits a photon toward B that's reflected by a mirror attached to B, with the return of the photon recorded at time t_2. A concludes that B has the spatial coordinate $x_p = c(t_2 - t_1)/2$, *half the time difference between emission and reception*, and that the reflection event occurred at time $t_p = (t_2 + t_1)/2$, *the mean of the two times*. This procedure is called the *radar method* of coordinatizing spacetime; it assigns spacetime coordinates to events,

$$(ct_p, x_p) = \left(\tfrac{1}{2}c(t_2 + t_1), \tfrac{1}{2}c(t_2 - t_1)\right) \tag{1.5}$$

[40]It is sometimes said (erroneously) that the GT is the version of the LT for low speeds, $v \ll c$, yet that cannot be true—the GT does not mix in the spatial coordinate for the new time coordinate, at any speed.

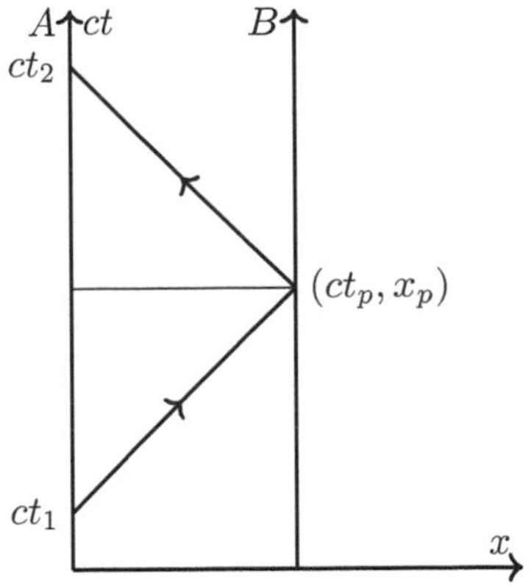

Figure 1.7 Radar method of assigning coordinates to events.

based on measurements made by A using light signals.[41] The radar method builds in the isotropy of the speed of light (established in the MM experiment). The "outbound" speed of light is the same as that for the photon's return journey, and we are free to orient A and B in any direction.

We can redraw Fig. 1.7 as the left portion of Fig. 1.8. A photon emitted at time t_1 is reflected

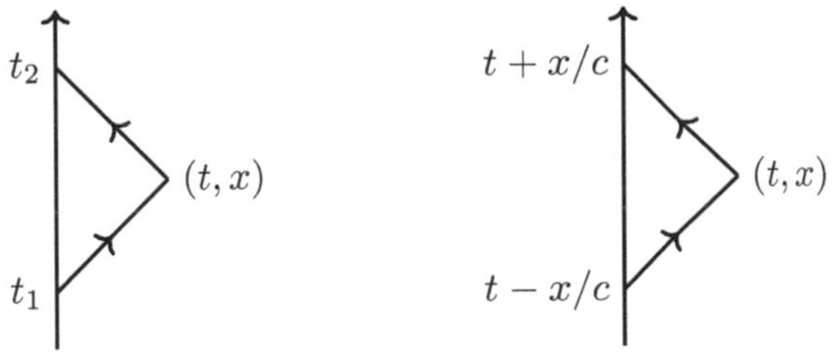

Figure 1.8 Photon emitted at $t - x/c$ is reflected at (t, x) and received at $t + x/c$.

from spacetime point (t, x) and received at time t_2. We haven't drawn observer B in Fig. 1.8, whose only role in Fig. 1.7 was to hold a reflector. Using Eq. (1.5), we can solve for t_1 and t_2 in terms of (t, x), $t_1 = t - x/c$ and $t_2 = t + x/c$; these times are shown in the right portion of Fig. 1.8.

1.5 RELATIVITY OF CAUSALITY: SPACELIKE AND TIMELIKE

While spacetime coordinates are reference-frame dependent, there is an *invariant* involving the *squares* of coordinates that's the same for all inertial observers,[42] the *spacetime separation*

$$s^2 \equiv -(x^0)^2 + \sum_{i=1}^{3}(x^i)^2 \,. \tag{1.6}$$

For an event with coordinates x^μ in one IRF, the coordinates of the same event in another IRF, $x^{\mu'}$, are such that[43]

$$-(ct')^2 + (x')^2 + (y')^2 + (z')^2 = s^2 = -(ct)^2 + (x)^2 + (y)^2 + (z)^2 \,. \tag{1.7}$$

[41]Note that the radar method does not call upon us to compare times as measured in different reference frames; it uses measurements made in a single reference frame.

[42]That there *is* an invariant quantity among coordinates assigned by different inertial observers to the same event implies that spacetime possesses an intrinsic geometry—the subject matter of the rest of the book.

[43]Equation (1.7) applies to IRFs having a common spacetime origin. Note the prime placed on the index, $x^{\mu'}$.

Equation (1.7) can be verified using the special case of a LT given by Eq. (1.4); it's true, however, for any LT. In Chapter 4 we *define* a LT as any linear transformation that leaves s^2 invariant. The spacetime separation is an example of a quantity that's *not relative*—it's observer independent. A way to motivate the invariance of s^2 is to consider two IRFs in relative motion along their common x-axis. At the instant their origins coincide, a flash of light is emitted. Both see an expanding wavefront that in their coordinates is described by $-(ct)^2 + x^2 = 0$. In whatever way the coordinates transform between IRFs, by the principle of relativity both must conclude that $x^2 - (ct)^2 = 0 = (x')^2 - (ct')^2$. While a wavefront of light is described by $s^2 = 0$, Eq. (1.7) holds for any value of s^2.

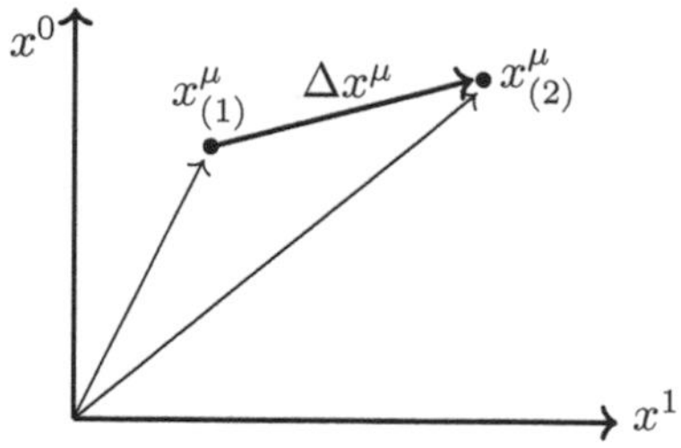

Figure 1.9 Spacetime separation vector between distinct events.

The separation *between* events is defined analogously.[44] Consider two events in a given IRF that have coordinates $x^\mu_{(1)}$ and $x^\mu_{(2)}$. Define the difference vector[45] Δx^μ (see Fig. 1.9)

$$\Delta x^\mu \equiv x^\mu_{(2)} - x^\mu_{(1)} = \begin{pmatrix} x^0 \\ x^1 \\ x^2 \\ x^3 \end{pmatrix}_{(2)} - \begin{pmatrix} x^0 \\ x^1 \\ x^2 \\ x^3 \end{pmatrix}_{(1)} \equiv \begin{pmatrix} \Delta x^0 \\ \Delta x^1 \\ \Delta x^2 \\ \Delta x^3 \end{pmatrix} .$$

The spacetime separation between these events is defined in the same way as in Eq. (1.6):

$$(\Delta s)^2 = -(\Delta x^0)^2 + \sum_{i=1}^{3} (\Delta x^i)^2 . \tag{1.8}$$

Even though the separation has been defined as the *square* of the quantity Δs, the value of $(\Delta s)^2$ can be, depending on the events, positive, zero, *or negative*. There is the temptation to define Δs itself as an imaginary quantity when $(\Delta s)^2 < 0$, a temptation we will resist.[46]

The three possible signs of $(\Delta s)^2$ provide an *absolute* way of characterizing spacetime separations. Because $(\Delta s)^2$ is an invariant, *no LT can change its sign.*

$$(\Delta s)^2 \text{ is called: } \begin{cases} \text{spacelike} & \text{if } (\Delta s)^2 > 0 ; \\ \text{lightlike} & \text{if } (\Delta s)^2 = 0 ; \\ \text{timelike} & \text{if } (\Delta s)^2 < 0 . \end{cases}$$

Figure 1.10 shows examples of the three types of spacetime separations. Timelike separations do not have to be "above" the lightline, nor spacelike separations "below." It's the slope of the lines that counts, not their location in a spacetime diagram.

---

[44] Because coordinates are defined relative to an origin, s^2 is the separation between the event with coordinates x^μ and the event at the origin.

[45] Such a vector is called a *four-vector*; see Chapter 5.

[46] One could either work with a Euclidean geometry that allowed pure imaginary distances, or one could work with a non-Euclidean geometry from the outset. The latter is more in keeping with the requirements of GR; we will not venture down the path of "ict"—as was done in the early days of relativity.

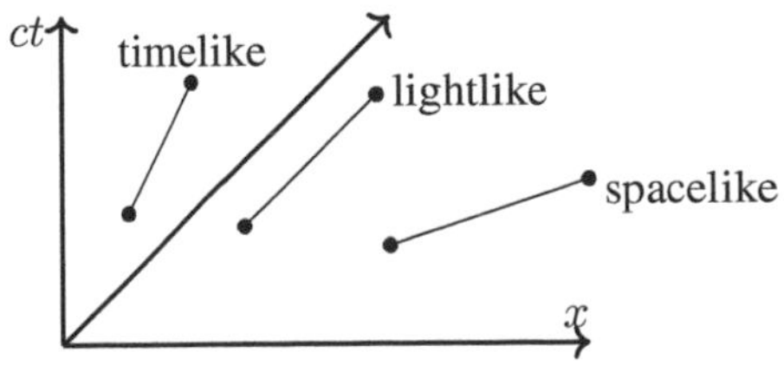

Figure 1.10 Timelike, lightlike, and spacelike separations of spacetime events.

We can always find a reference frame in which spacelike-separated events are simultaneous: For $\Delta t = 0$, it is automatically the case that $(\Delta s)^2 > 0$. However, for frames in which $(\Delta s)^2 > 0$, Δt can be of either sign or zero. Thus, *one cannot speak of a causal relation between spacelike-separated events*. For A and B spacelike-separated events, one can find frames in which *the events occur in either order*,[47] in which A precedes B or B precedes A. *This fact is a major departure from pre-relativistic physics*, in which the time order of events is absolute.[48] Timelike-separated events, on the other hand, *can never be simultaneous*: No reference frame can be found for which $\Delta t = 0$ as it would violate $(\Delta s)^2 < 0$. The temporal order in which timelike-separated events occur is therefore absolute because we can't find a frame in which Δt vanishes. *Only for timelike-separated events can we speak of causality.*

1.6 SEGUE TO GENERAL RELATIVITY: NONINERTIAL FRAMES

Newton's laws work in IRFs, which as we have seen, are frames of reference in which Newton's laws work! What saves us from a circular trap is the ability to identify physical sources of force; *only* in IRFs is the acceleration of objects solely due to forces—only in IRFs does the Newtonian paradigm apply ($\boldsymbol{F} = m\boldsymbol{a}$). From the point of view of fundamental physics, Newton's second law is *limited* by its specialization to IRFs. GR provides equations of motion valid in arbitrary frames of reference. *To what extent do noninertial reference frames find use in Newtonian dynamics*, despite nominally being excluded from the framework of pre-relativistic mechanics? Such a question might appear off topic, but given that SR is based on the equivalence of IRFs, and that GR frees itself from IRFs, it's useful to look at pre-relativistic uses of noninertial frames.

1.6.1 Linear acceleration

Referring to Figure 1.1, $\boldsymbol{r} = \boldsymbol{R} + \boldsymbol{r}'$ where now we allow all quantities to be time dependent. Differentiating twice with respect to absolute time, $\ddot{\boldsymbol{r}} = \boldsymbol{A} + \ddot{\boldsymbol{r}}'$ where $\boldsymbol{A} \equiv \ddot{\boldsymbol{R}}$ is the relative acceleration between frames.[49] Let S be an IRF in which the observed acceleration of a particle of mass m is associated with a force, $\ddot{\boldsymbol{r}} = \boldsymbol{F}/m$. We therefore have an equation similar to Newton's second law:

$$\ddot{\boldsymbol{r}}' = \frac{1}{m}(\boldsymbol{F} - m\boldsymbol{A}) \ . \tag{1.9}$$

The acceleration observed in the accelerated frame ($\ddot{\boldsymbol{r}}'$) is due to forces ($\boldsymbol{F}$) and the *force-like* quantity $-m\boldsymbol{A}$, termed the *fictitious force*, so named because, while it has the dimension of force, is not a force; genuine forces can be traced to physical interactions. Acceleration and force have the same values in all IRFs; they are *absolute*, observer-independent quantities. In noninertial frames, $\ddot{\boldsymbol{r}}'$ is an *apparent acceleration*: It's not absolute, it's reference-frame dependent; from Eq. (1.9) $\ddot{\boldsymbol{r}}'$ is

[47] Demonstrated in Section 2.4.

[48] For the most part the predictions of SR smoothly go over to those of Newtonian mechanics as $v/c \to 0$. Certain conclusions, however, have no counterpart in pre-relativistic physics, such as the *acausality* of spacelike-separated events.

[49] The transformation of acceleration under the LT, which does not assume absolute time, is covered in Chapter 3.

offset from the acceleration due to forces $\boldsymbol{F}/m$ by the acceleration of the reference frame, $\boldsymbol{A}$. The acceleration $\boldsymbol{A}$ in the fictitious force is the acceleration of the *frame*, not that of the *particle*.

Figure 1.11 shows a noninertial frame N, an elevator accelerating relative to IRF I.[50] In I, a

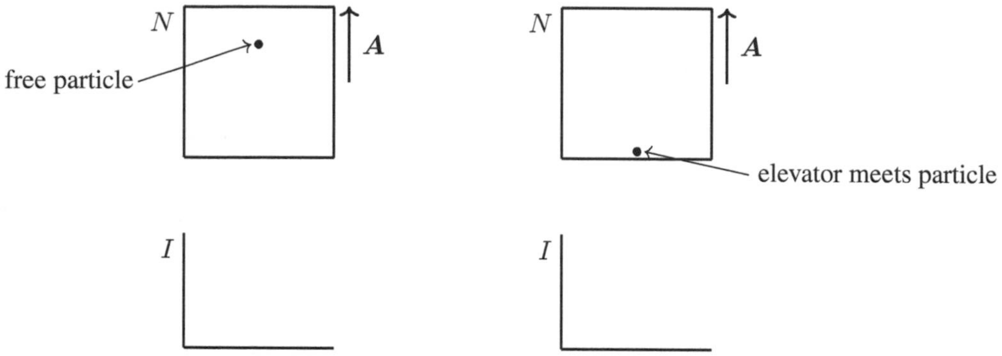

Figure 1.11 Left: In I $\ddot{\boldsymbol{r}} = 0$ (free particle); in N, apparent acceleration $\ddot{\boldsymbol{r}}' = -\boldsymbol{A}$. Right: In N, $\ddot{\boldsymbol{r}}' = 0$ (at rest); in I, $\boldsymbol{F} = m\boldsymbol{A}$.

free particle has inertial motion, $\ddot{\boldsymbol{r}} = 0$, whereas in N it has acceleration $\ddot{\boldsymbol{r}}' = -\boldsymbol{A}$. An observer in N concludes a force produces the observed acceleration, yet there is no force, no physical agency acting on the particle, which is why $-m\boldsymbol{A}$ is called fictitious. When the elevator floor meets the particle, however, *the fictitious force becomes real*.[51] At this point, the elevator *prevents* the particle from continuing ("persevering") in its inertial motion. An inertial observer concludes there is a force on the object, $\boldsymbol{F} = m\boldsymbol{A}$, which follows from Eq. (1.9) with $\ddot{\boldsymbol{r}}' = 0$. The object *resists* changes in its inertial state and exerts a force back on the elevator, $-m\boldsymbol{A} \equiv \boldsymbol{F}_i$ ("endeavors to change the state of that obstacle").[52] When observed from a noninertial frame, a particle moving by inertia appears to accelerate in the direction opposite to the acceleration of the frame; we can still apply Newton's second law in this case by regarding the apparent acceleration as caused by a fictitious force. When, however, the particle is *prevented* from moving by inertia and made to move with the acceleration of the frame, the particle resists acceleration through a real force, the inertial force.

1.6.2 Rotating reference frame

Inertial forces arise in rotating reference frames. Consider a frame (x', y', z') rotating at a constant rate Ω relative to an IRF (x, y, z) about the common z, z' axis. As is well known,[53] the acceleration observed in a rotating frame is related to the force $\boldsymbol{F}$ through an equation analogous to Eq. (1.9),

$$\ddot{\boldsymbol{r}}' = \frac{1}{m}\left(\boldsymbol{F} - 2m\boldsymbol{\Omega} \times \dot{\boldsymbol{r}}' - m\boldsymbol{\Omega} \times \boldsymbol{\Omega} \times \boldsymbol{r}\right) . \tag{1.10}$$

The inertial force $\boldsymbol{F}_i = -2m\boldsymbol{\Omega} \times \dot{\boldsymbol{r}}' - m\boldsymbol{\Omega} \times \boldsymbol{\Omega} \times \boldsymbol{r}$ involves the *Coriolis force* and the *centrifugal force*. These forces are quite real, as anyone who has ridden a merry-go-round can attest.

[50] Assume the elevator is sufficiently outside the gravitational field of Earth that gravity can be ignored.

[51] It's sometimes said, incorrectly, that *any* force observed in a noninertial frame is a fictitious force. Forces in noninertial frames are quite real, as anyone who's ridden in an automobile can attest. "Real fictitious forces" are best given another name—inertial force—because they arise from the inertia of matter.

[52] We're referring to Newton's words cited in Section 1.2.

[53] Any standard text on classical mechanics will have a derivation of the centrifugal and Coriolis forces.

1.6.3 D'Alembert's principle

By *d'Alembert's principle*,[54] Newton's second law is written in a seemingly trivial way $\boldsymbol{F}-m\boldsymbol{a}=0$, equivalently $\boldsymbol{F}+\boldsymbol{F}_i=0$, so that an object *in motion* can be treated *as if in static equilibrium* between impressed forces $\boldsymbol{F}$ and the inertial force $\boldsymbol{F}_i$ (produced by the mass in response to the changes in inertial motion brought about by $\boldsymbol{F}$).[10, p88] That a mass in motion can be treated *as if* at rest underscores the relativity of motion. An object appears at rest *in a frame moving with an object* ($\ddot{\boldsymbol{r}}'=0$), and in such a frame we have from either Eq. (1.9) or Eq. (1.10) equilibrium between impressed and inertial forces. D'Alembert's principle could be considered a precursor to GR—it gives insight into how an equation of motion might appear in an arbitrary frame of reference.

An example from elementary mechanics illustrates these ideas. The left portion of Fig. 1.12 schematically shows a car undergoing acceleration $\boldsymbol{A}$ as seen from an IRF. Attached to the car is

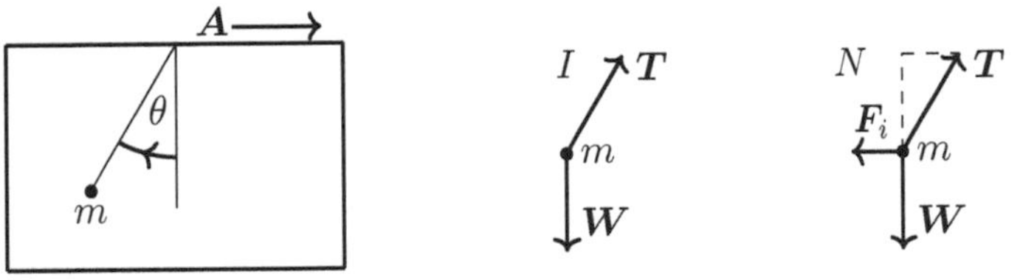

Figure 1.12 Ball hanging from the ceiling of an accelerating car. Forces as seen from an inertial frame, I, and a noninertial frame, N.

a ball of mass m hanging from a string. The forces "impressed" on the ball are the tension $\boldsymbol{T}$ in the string and its weight $\boldsymbol{W}$. As shown in the middle portion of Fig. 1.12, these are the forces that cause the ball to undergo the acceleration observed in inertial frame I, $\boldsymbol{T}+\boldsymbol{W}=m\boldsymbol{A}$. In the usual coordinate system involving horizontal and vertical components, where vertical is defined by the direction of gravity, Newton's second law separates into two scalar equations $T\sin\theta=mA$ and $T\cos\theta-mg=0$, from which we find $\tan\theta=A/g$ and $T=m\sqrt{g^2+A^2}$. In the noninertial frame N of the car, no acceleration is observed and Eq. (1.9) gives the same equation of force balance: $0=\boldsymbol{T}+\boldsymbol{W}-m\boldsymbol{A}$. When the car is *not* accelerated, the ball hangs "straight down" in the direction of gravity with the tension equal to the weight, $T=mg$. With the car accelerated, we can view the tension as balancing the resultant of $\boldsymbol{W}$ and the inertial force $\boldsymbol{F}_i=-m\boldsymbol{A}$, $\boldsymbol{T}=-(\boldsymbol{W}+\boldsymbol{F}_i)$ (right portion of Fig. 1.12). Alternatively, the inertial force is the opposite of (reaction to) the resultant of the physical forces, $\boldsymbol{T}$ and $\boldsymbol{W}$, $\boldsymbol{F}_i=-(\boldsymbol{T}+\boldsymbol{W})$.

1.7 GENERAL RELATIVITY: A THEORY OF GRAVITATION

1.7.1 Newtonian gravitation—consistent with the theory of relativity?

In elementary physics one first learns about Newton's law of motion $F=ma$, which applies for any force F, and, second, Newton's law of gravitation—an expression for a force law—that masses m_1 and m_2 at locations $\boldsymbol{r}_1$ and $\boldsymbol{r}_2$ experience an attractive force of magnitude

$$F=G\frac{m_1m_2}{|\boldsymbol{r}_1-\boldsymbol{r}_2|^2},$$

where G is the gravitational constant. Newton's law of gravity works well in explaining many phenomena, from predicting solar eclipses to sending satellites to distant planets. Despite its successes, however, *Newtonian gravitation is not consistent with relativity*, for two main reasons.

[54]D'Alembert's principle is a formulation of classical mechanics equivalent to Hamilton's principle (a more well-known formulation of classical mechanics.) See Appendix D.

1. The locations specified by $\boldsymbol{r}_1$ and $\boldsymbol{r}_2$ in Newton's formula are implicitly assumed to occur *at the same time*. Relativity shows there is no absolute meaning to the "same time."
2. What *mediates* gravity? If m_1 were to move suddenly, Newton's formula would have that the force on m_2 would change *instantly*, yet instantaneous interactions are not physical, the bugaboo of *action at a distance*.[55]

Newton wrote in 1693:[11, p217] "... so that one body may act upon another at a distance through a vacuum without the mediation of anything else, by and through which their action and force may be conveyed from one to another, is to me so great an absurdity, that I believe no man who has in philosophic matters a competent faculty of thinking could ever fall into it." In 1713 he wrote,[4, p943] "I have not as yet been able to deduce from phenomena the reason for these properties of gravity, and I do not feign hypotheses ... it is enough that gravity really exists and acts according to the laws that we have set forth, and is sufficient to explain all the motions of the heavenly bodies ...". Newton appeals to pragmatism: Even though he can't explain the workings of gravity, his law of gravity works and works well and, as he tells it, explains "all" the motions of celestial bodies. Or *does* it?

1.7.2 Do we need a relativistic theory of gravitation?

Under what conditions do relativistic effects become important in gravitational physics? We know that modifications to Newtonian dynamics manifest as speeds become comparable with the speed of light, $v \lesssim c$. What relativistic effects are specifically associated with *gravity*? Consider the energy of the gravitational field. In Newtonian theory, the energy stored in the gravitational field of a mass M of radius R with uniform mass density is given by the expression[56]

$$E_{\text{grav}} = \frac{3}{5}\frac{GM^2}{R} .$$

Let's ignore the numerical factor and take as a measure of gravitational energy the terms GM^2/R. The *rest energy* is another kind of energy, $E_{\text{rest}} = Mc^2$. By forming the ratio $E_{\text{grav}}/E_{\text{rest}}$ we obtain a characteristic dimensionless number specifying the gravitational energy relative to the rest energy,

$$\frac{E_{\text{grav}}}{E_{\text{rest}}} = \frac{GM}{Rc^2} \equiv \frac{\Phi}{c^2} , \tag{1.11}$$

where Φ is the *gravitational potential*—the gravitational potential energy per mass—which has the dimension of speed squared.[57] Newton's law of gravity, like Coulomb's law, is a $1/r^2$ law. Any result obtained in electrostatics has an analog in Newtonian gravity. For future reference, Table 1.1 compares the properties of the Newtonian gravitational field with those of the electrostatic field.

[55] Newton's law of gravitation was controversial when it was introduced. Aristotle taught that heavenly objects (stars and planets) *by their nature* move in circles at constant speed, while on Earth heavy objects move toward the center of Earth. Stones fall, but planets don't. Descartes, in an attempt to *explain* planetary orbits, proposed that the sun sets up a whirlpool motion to keep planets moving in circular motion. Kepler (at roughly the same time) discovered that planets move in elliptical orbits, not circular. It's against this backdrop that Newton's law of gravity is startling. Newton offered *no explanation* of how the sun could exert an influence on Earth over vast distances—action at a distance. He "merely" offered a formula that predicts the motion of objects subject to gravity. With his inverse-square law, Newton could account for Kepler's three laws of planetary motion; he also showed that Descartes's whirlpool hypothesis contradicts Kepler's third law. To illustrate the difficulty inherent with action at a distance, what would you think of a theory purporting that radiant energy disappears from the sun and eight minutes later appears on Earth *without accounting for how it happens*? Newton's law of gravity is an effective, phenomenological description that provides no explanation for the mechanism of gravity. As we'll see, GR holds that spacetime itself is the underlying substrate that mass couples to.

[56] A similar expression holds for the energy stored in the electric field associated with a uniform ball of charge, a calculation you've probably already done.

[57] The electrostatic potential is the energy per charge, which is given a special unit—a volt is a joule per coulomb. Gravitational energy per mass (the gravitational potential) has the dimension of speed squared; just think of kinetic energy $\propto mv^2$.

Table 1.1 Comparison of Newtonian gravitation theory with electrostatics.

	Newtonian gravitation	Electrostatics
Force between point objects	$\boldsymbol{F}_{\text{grav}} = -G\dfrac{Mm}{r^2}\hat{\boldsymbol{r}}$	$\boldsymbol{F}_{\text{elec}} = \dfrac{Qq}{4\pi\epsilon_0 r^2}\hat{\boldsymbol{r}}$
Field vector of a point source	$\boldsymbol{g} = -\dfrac{GM}{r^2}\hat{\boldsymbol{r}}$	$\boldsymbol{E} = \dfrac{Q}{4\pi\epsilon_0 r^2}\hat{\boldsymbol{r}}$
Gauss's law	$\boldsymbol{\nabla}\cdot\boldsymbol{g} = -4\pi G\rho$	$\boldsymbol{\nabla}\cdot\boldsymbol{E} = \rho_{\text{elec}}/\epsilon_0$
Irrotational field (of point source)	$\boldsymbol{\nabla}\times\boldsymbol{g} = 0$	$\boldsymbol{\nabla}\times\boldsymbol{E} = 0$
Potential energy of point objects	$U(r) = -\dfrac{GMm}{r}$	$U_{\text{elec}}(r) = \dfrac{Qq}{4\pi\epsilon_0 r}$
Potential of a point source	$\Phi(r) = -\dfrac{GM}{r}$	$\Phi_{\text{elec}}(r) = \dfrac{Q}{4\pi\epsilon_0 r}$
Poisson equation	$\nabla^2\Phi = 4\pi G\rho$	$\nabla^2\Phi_{\text{elec}} = -\rho_{\text{elec}}/\epsilon_0$
Potential of an extended source	$\Phi(\boldsymbol{r}) = -G\displaystyle\int\frac{\rho(\boldsymbol{r}')}{\lvert\boldsymbol{r}-\boldsymbol{r}'\rvert}\mathrm{d}^3r'$	$\Phi_{\text{elec}}(\boldsymbol{r}) = \dfrac{1}{4\pi\epsilon_0}\displaystyle\int\frac{\rho_{\text{elec}}(\boldsymbol{r}')}{\lvert\boldsymbol{r}-\boldsymbol{r}'\rvert}\mathrm{d}^3r'$
Local energy density	$-\dfrac{\lvert g(r)\rvert^2}{8\pi G}$	$\dfrac{\epsilon_0}{2}\lvert E(r)\rvert^2$

A large value of the ratio in Eq. (1.11) (of order unity) would indicate an object for which the gravitational energy is comparable to Mc^2. The dimensionless quantity in Eq. (1.11) occurs in GR as a measure of the significance of relativistic effects in gravity; be on the lookout for it. While $v \ll c$ is an indicator that Newtonian dynamics provides an accurate description, $\Phi \ll c^2$ is an indicator that Newtonian gravitation should suffice. Numerical values of this ratio are listed in Table 1.2 for various systems.

Table 1.2 Ratio of gravitational to rest-mass energy.

System	$GM/(Rc^2) \equiv \Phi/c^2$	Comment
Earth	10^{-9}	GPS system inoperable without relativistic corrections
Sun	10^{-6}	Precession of planetary orbits unaccounted by Newtonian mechanics
Black hole	0.5	As relativistic as it gets
Universe	0.5	Ditto for the universe!

The gravitational energy of the Earth is seemingly a negligible fraction (10^{-9}) of its rest energy and thus we would conclude that the Newtonian theory of gravity should suffice. While largely true, there are nonetheless small effects due to time dilation in a gravitational field that must be taken into account if the global positioning system (GPS) is to operate properly. *Gravitational time dilation* is not the *special relativistic* time dilation ("moving clocks run slow"), but rather is an effect associated with gravity, that clocks run slower the deeper they are in a gravitational potential well. The GPS system would go wrong in a matter of minutes if relativistic effects were not taken into account. *Even for weak gravity there are important effects that Newtonian theory cannot describe.*

For the sun, with $E_{grav} \approx 10^{-6}Mc^2$, the orbit of Mercury *precesses* at a small but measurable rate that cannot be accounted for in Newtonian mechanics, yet which is explained precisely by GR. The *precession of orbits* is one of the classic tests of GR.[58]

Newton's law of gravity contains no characteristic length scale over which it applies: It's intended to apply for *any* distance. GR, however, features an intrinsic length associated with a spherically symmetric mass M, the *Schwarzschild radius* $r_S \equiv 2GM/c^2$. (Remarkably, the Schwarzschild radius can be obtained from Newtonian mechanics as the radius of an object for which the escape velocity $v_{esc} = \sqrt{2GM/R} = c$.) If M lies within the Schwarzschild radius, then $r = r_S$ defines an *event horizon* for external observers: Signals emitted cannot reach outside observers and we have a *black hole*. Black holes are regions of spacetime from which nothing, not even light, can escape. For black holes, $E_{grav}/E_{rel} = \frac{1}{2}$. Clearly implicit in the description of a black hole is the prediction that gravity affects the propagation of light. *Gravitational lensing*, the deflection of light by gravity, is an experimental tool for investigating *dark matter*, a hypothesized form of matter that, while not luminous, can nevertheless be inferred from its gravitational influences.

For the universe, GM/Rc^2 can be estimated from its mean mass density ρ and size R: $M = \frac{4}{3}\pi R^3\rho$. Let ρ be the *critical density* obtained from cosmological theory,[59] $\rho_c \equiv 3H_0^2/(8\pi G) \approx 10^{-29}$ g cm^{-3}, where H_0 is the *Hubble constant*. Thus, $GM/Rc^2 = \frac{1}{2}(RH_0/c)^2$. Take the size of the universe to be $R = ct_H$ where $t_H \equiv H_0^{-1}$ is the *Hubble time*, the approximate age of the universe. With these substitutions, $GM/Rc^2 = \frac{1}{2}$. While one can question any of these assumptions, the larger point is that the universe is "just" as relativistic as a black hole!

Because gravity is always attractive, why doesn't the universe collapse? Newton concluded that the universe must be infinite in extent to avoid such a collapse. GR, however, predicts an *expanding* universe! To preclude this possibility,[60] Einstein introduced an adjustable constant, the *cosmological constant* Λ, with the purpose of producing a static, finite-sized universe. It was later shown (in 1922, by Alexander Friedmann) that GR predicts an expanding universe no matter what the value[61] of Λ. The "standard model" of cosmology, the Friedmann-Robertson-Walker model, is derived from GR, including Λ, a term now thought to be associated with *dark energy*, a proposed form of energy that leads to a universe that's not only expanding, but is *accelerating* in its expansion. In 1998 an acceleration to the expansion of the universe was discovered, and Λ was invoked as an explanation.[62]

Thus, astrophysical and cosmological phenomena[63] *require for their explanation a relativistic theory of gravitation.* We need a theoretical framework that can handle arbitrary gravitational fields, from the environment near planets and stars, to that of black holes, and ultimately the universe. GR is a theoretical tool for describing spacetime that incorporates the effects of gravity.

1.7.3 Thinking about relativistic gravity

Can Newtonian gravity be "fixed up" so as to be relativistically correct? The short answer is *no*. No "tweak" of Newton's formula has ever been found, perhaps with factors of γ here and there; it takes a *major* revamping of our concepts of space and time.

It's instructive to ask, given that action at a distance is a flaw of Newtonian gravitation, how is that problem sidestepped with Coulomb's law, which has the same structure as Newton's law of

---

[58] The three classic tests of GR are the precession of orbits, the bending of light by gravity, and the gravitational redshift.

[59] The mean density of the universe ρ is thought to be quite close to the critical value, ρ_c. Knowledge of ρ is of crucial importance to cosmology, as it determines whether the universe is open or closed. It's found that $\rho/\rho_c = 1.0023 \pm 0.005$. When contemplating a number like 10^{-29} g cm^{-3}, it's helpful to keep in mind the density of Earth (~ 5.5 g cm^{-3}) or the density of the sun (~ 1.4 g cm^{-3}). The universe as a gravitating system can perhaps be considered another state of matter, that which is governed by an incomprehensibly small density.

[60] In 1929, it was deduced that the universe *is* expanding from the redshift in spectral lines observed from distant galaxies. To Einstein in 1917, it was obvious that the universe must be static.

[61] As shown by Friedmann, Einstein's static solution of the equations of GR is not stable against small perturbations.

[62] The 2011 Nobel Prize in Physics was awarded for the discovery of the accelerating expansion of the universe.

[63] And even the terrestrial GPS system.

gravity? The force between charges q_1 and q_2 has magnitude

$$F = k \frac{q_1 q_2}{|\boldsymbol{r}_1 - \boldsymbol{r}_2|^2} ,$$

where k is a proportionality factor that depends on the unit of charge adopted. Coulomb's law suffers from the same disease as Newton's law—action at a distance and instantaneous interactions. What saves the day is the *field* concept. Charge q_1 sets up a condition in space—the *electric field*—that q_2 interacts with at its location, which can be symbolized: $\text{Charge}_1 \longleftrightarrow \text{Field} \longleftrightarrow \text{Charge}_2$. We obtain an expression for the static electric field simply by rewriting Coulomb's law,

$$F = q_2 \left(k \frac{q_1}{|\boldsymbol{r}_1 - \boldsymbol{r}_2|^2} \right) \equiv q_2 E .$$

Now, merely writing $F = qE$ would be a change of variables if we didn't ascribe physical reality to the field. And we *do* ascribe reality to the field because we discover—using Maxwell's equations—that the electromagnetic field is a *dynamical* quantity that propagates at the speed of light and transports energy and momentum. Through Maxwell's equations, we discover that the electromagnetic field satisfies a wave equation. Thus, electromagnetism is not transmitted instantaneously as Coulomb's law would lead us to suspect, but is instead a propagating field at finite speed. Is the same true of gravity? *The concept of a field, one that has dynamical properties, answers the problem posed by action at a distance*: It's the field that mediates the interaction between particles, and the field propagates at finite speed.

Physics thrives on analogies. The paradigm of propagating fields leads us to ask: *Can we formulate a field theory of gravity*? Start by rewriting the force law:

$$F = m_2 \left(G \frac{m_1}{|\boldsymbol{r}_1 - \boldsymbol{r}_2|^2} \right) \equiv m_2 g ,$$

where g signifies the gravitational field. The Newtonian gravitational field satisfies Gauss's law $\boldsymbol{\nabla} \cdot \boldsymbol{g} = -4\pi G \rho$, where ρ is the local mass density. Note that the divergence of $\boldsymbol{g}$ is *negative*—there's a negative flux of field lines through any closed surface; gravity is always attractive. This seems like a promising start, but what are the other "Maxwell equations" for gravity? Recall the crucial discoveries in electromagnetism: Charges in motion (currents) produce magnetic fields, time-varying electric fields induce magnetic fields, and time-varying magnetic fields induce electric fields. Are there analogous phenomena in gravity? Does matter in motion lead to new phenomena, akin to a magnetic field, that affect the motion of nearby masses?

There are no "Maxwell equations" for the gravitational field that have been discovered through experiments, akin to Faraday induction. Thus there is no way, based on analogies with the electromagnetic field, to develop a field theory of gravity. Yet that's what GR accomplishes—*a relativistic field theory of gravity* distinct from the theory of the electromagnetic field. Once the machinery of GR has been developed, we'll discover analogs between gravity and electromagnetism *in limiting cases*, that the gravitational field satisfies a set of equations analogous to the equations of electrostatics and magnetostatics. GR predicts *frame dragging*, a gravitational analog of the Lorentz force in electromagnetism—that spacetime is altered by objects in motion, "dragging" nearby objects out of position compared to the predictions of Newtonian physics. While the frame-dragging effect is small, experimental confirmation was reported in 2011. GR also predicts a propagating disturbance in spacetime, *gravitational waves*, which were detected in 2016.[64]

[64]The 2017 Nobel Prize in Physics was awarded for the observation of gravitational waves.

1.7.4 How does GR work?

The central content of GR is the Einstein field equation which schematically has the form

$$\text{Local curvature of spacetime} = \frac{8\pi G}{c^4}\,(\text{Local energy-momentum density})\,.$$

The *curvature* of spacetime, or equivalently, the *geometry* of spacetime, is determined by the energy-momentum contained in that spacetime. Spacetime curvature in turn completely determines the trajectories of particles. Mathematically, the Einstein field equation is a relation between second-rank tensor fields[65]

$$G_{\mu\nu} = \frac{8\pi G}{c^4} T_{\mu\nu}\,. \tag{1.12}$$

(You'll know what this all means soon enough: $G_{\mu\nu}$ is the curvature tensor and $T_{\mu\nu}$ is the energy-momentum tensor that describes the density and flux of energy-momentum in spacetime.) Just as Maxwell's equations relate the electromagnetic field to its *sources* (charge and current densities), the Einstein equation relates spacetime curvature to *its* source: energy-momentum density. In Maxwell's equations, the electromagnetic field is *on* spacetime; in Einstein's equation, spacetime *itself* is the field! Gravity is not a force in the usual sense; gravity *is* spacetime!

The spacetime separation, Eq. (1.8) can be written

$$(\Delta s)^2 = \sum_{\mu=0}^{3}\sum_{\nu=0}^{3} \eta_{\mu\nu}\Delta x^\mu \Delta x^\nu\,, \tag{1.13}$$

where the quantity $\eta_{\mu\nu}$ is the *Lorentz metric tensor*

$$\eta_{\mu\nu} = \begin{pmatrix} -1 & 0 & 0 & 0 \\ 0 & 1 & 0 & 0 \\ 0 & 0 & 1 & 0 \\ 0 & 0 & 0 & 1 \end{pmatrix}. \tag{1.14}$$

The metric tensor[66] contains the information required to calculate the separation between spacetime points with coordinate differences Δx^μ. Note the metric "signature" in Eq. (1.14), the terms on the diagonal, $(-+++)$. This pattern holds for all inertial observers; the metric tensor in SR is *fixed.* Because of the minus sign for the time coordinate, the geometry is not Euclidean.[67]

The worldlines of truly free particles would be straight *throughout all of spacetime*. When one contemplates gravitation, one realizes that *global* inertial frames (holding for all of spacetime) are an idealization: We can't avoid the rest of the matter of the universe! Force-free motion can therefore have only approximate validity. In GR the separation between spacetime points, which in Eq. (1.13) applies for *finite* coordinate differences Δx^μ, is replaced by *infinitesimally separated* spacetime coordinates $\mathrm{d}x^\mu$:

$$(\mathrm{d}s)^2 = \sum_{\mu=0}^{3}\sum_{\nu=0}^{3} g_{\mu\nu}(x)\mathrm{d}x^\mu \mathrm{d}x^\nu\,, \tag{1.15}$$

---

[65] The Einstein field equations are a set of 10 equations between the elements of second-rank symmetric tensors in the four-dimensional geometry of spacetime. These equations are variously referred to in both the singular (Einstein's field equation), because it's one equation between two tensors, and in the plural (field equations), because there are 10 independent equations between the components of the curvature tensor and the energy-momentum tensor. When you refer to "Einstein's equation," people assume you're referring to $E = mc^2$. The Einstein field equations are *vastly* richer in content than $E = mc^2$.

[66] All things tensor will be explained in upcoming chapters.

[67] The spacetime geometry of SR is called *semi-Euclidean* because while not strictly a Euclidean geometry, which would have metric signature $(++++)$, is nevertheless a flat geometry.

where the quantities $g_{\mu\nu}(x)$ are not constant, but vary throughout spacetime—a metric tensor *field*.[68]

The curvature tensor $G_{\mu\nu}$ in Eq. (1.12) is, as we'll see, a complicated expression involving derivatives of the metric tensor field $g_{\mu\nu}(x)$. The Einstein field equation implies a set of ten *non-linear* partial differential equations for $g_{\mu\nu}(x)$. Once the tensor components $g_{\mu\nu}(x)$ are known, the equations of motion for particles and photons are known. Particles and photons in free fall (subject to no forces other than gravity) follow *geodesic curves*, the shortest possible paths in spacetime, determined through a variational principle $\delta \int \mathrm{d}s = 0$. Motion, however, determines the energy-momentum tensor $T_{\mu\nu}$. There is thus a feedback mechanism; see Fig. 1.13. Motion determines the

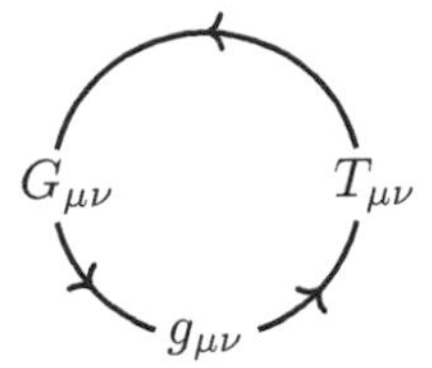

Figure 1.13 Motion determines spacetime curvature, which determines motion.

energy-momentum tensor $T_{\mu\nu}$, which determines the curvature tensor through the Einstein equation, the solution of which determines the metric tensor $g_{\mu\nu}$, which controls motion. GR explains gravity in terms of a varying metrical relation between neighboring spacetime points, $g_{\mu\nu}(x)$, wherein particles get closer together in the future than they are now. Gravity is a manifestation of the curvature of spacetime, that determined by the distribution of energy-momentum.

1.7.5 Gravity is spacetime

That last statement requires elaboration, which we'll do in a roundabout way. What do we need to understand GR? For one, we have to enlarge our mathematical toolbox. The mathematics of curvature is the province of differential geometry, the theory of tensor fields on curved manifolds. The requirement imposed by the principle of relativity, that laws of physics be independent of the reference frame used to represent them, leads to a program, the principle of covariance, of expressing equations of physics as relations among tensors because, if a tensor equation is true in one reference frame, it's true in all reference frames. Once the mathematical foundation of tensor fields has been laid, the Einstein field equation can be introduced forthwith. It's important to recognize that a truly *new* equation of physics cannot be derived from something more fundamental. Once the field equation has been written down (however it was conceived), there isn't a lot of wiggle room: Either its predictions agree with experimental measurements or they don't, and so far GR has passed every test put to it. Is that all we need, more math, in particular the mathematics of tensors? (That and the physical insight of Albert Einstein.) What about SR? In a sense GR doesn't require SR, implausible as that might sound. The thesis of GR is that energy-momentum causes spacetime curvature, where spacetime is modeled as a four-dimensional manifold. The surface of Earth is curved yet we know locally it can be approximated as flat. So too with spacetime: Curved spacetime is locally flat. Manifolds are locally flat at any point, where the condition for flatness is that the derivatives of the metric tensor vanish in a neighborhood of that point. That leaves open the question of what the metric tensor should be for small regions of spacetime. Einstein's answer is that it should correspond to the metric of SR. In 1911, in the time between the development of SR and that of GR, Einstein proposed the *equivalence principle* that the effects of gravitation are *eliminated* in a reference frame *of limited spatial extent* that's freely falling in gravity. In a freely falling frame, objects not subject to forces (other than gravity) remain either at rest or in a state of uniform motion;[69] a freely falling

[68] We study tensor fields in Chapter 13.

[69] Einstein called this the happiest thought of his life.

frame is therefore an IRF, where SR holds sway. SR therefore becomes a theoretical "boundary condition" on GR: the theory of spacetime that results for vanishing gravity. GR must give rise to two *incompatible* limits as shown in Fig. 1.14: SR as $G \to 0$ and Newtonian gravity for $v \ll c$. GR thus

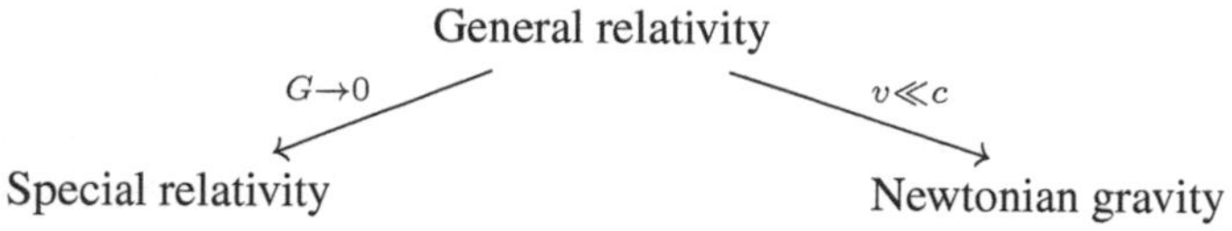

Figure 1.14 General relativity supersedes special relativity and Newtonian gravity

supersedes both theories.[70] It's only in this sense that GR needs SR; GR is the more comprehensive theory. Even though it serves as but a limiting case of GR, it's important to develop SR to understand what is discarded from the Newtonian framework. We do this first without tensors, and then, once tensors have been introduced, we develop special-relativistic physics in tensor form. Getting back to gravity, freely falling particles move along geodesic curves in four-dimensional spacetime at a constant rate—no acceleration, no force required.[71] In *three-dimensional*, space-only geometry, such particles appear to accelerate, which the Newtonian paradigm associates with a force. It's from this perspective we say that gravity is not a force in the usual sense but rather is a manifestation of the properties of spacetime. This point of view is fully developed in the book.

1.8 HASTA LA VISTA, GRAVITY

In the next chapter we begin a systematic exposition of SR, first without tensors, and then once tensors have been introduced (Chapter 5), we cover special-relativistic physics using tensors, following through with Einstein's program of the principle of covariance. Only in Chapter 15 is gravity taken up as a manifestation of spacetime curvature. At this point we say *hasta la vista* gravity, knowing that we'll catch up with you further on down the trail.

SUMMARY

We have presented an overview of the special and general theories of relativity without delving into specifics. Many definitions have been introduced which form the basic vocabulary of the subject.

- The theory of relativity is an outgrowth of a single idea, the principle of relativity, that the laws of physics be expressed in a way that's independent of the reference frame used to represent them. The principle of covariance is the requirement that equations of physics adhere to the principle of relativity by having the same mathematical form in all reference frames, a goal achieved by expressing equations as relations between tensors defined on a four-dimensional geometry where time is a dimension.

- A primitive concept in relativity is an *event*—a point in space at a point in time. The totality of all events is a four-dimensional continuum: spacetime. The theory of relativity is the study of the geometry of spacetime, the relation between points of space and points of time—which in broad terms is what physics is about. In SR spacetime is absolute—physically existing, but not influenced by physical conditions. In GR, spacetime geometry is determined by the distribution of energy and momentum. GR achieves a symmetry between spacetime acting on matter (particles follow geodesic curves established by the curvature of spacetime) and

[70]Newtonian mechanics is correct for phenomena with $v \ll c$ and for which Planck's constant can be ignored. To the diagram in Fig. 1.14 we should add another axis with $\hbar \neq 0$, a theory that has yet to be formulated.

[71]See Section 14.3.5.

matter acting on spacetime (curvature determined by energy-momentum through the Einstein field equation). Such a symmetry, which might be expected generally—Newton's third law, implies that spacetime is physical.

- SR is based on the equivalence of IRFs established by free particles; SR is the law of inertia expressed in spacetime. All inertial observers see the worldlines of free particles as straight, and all inertial observers can claim themselves at rest. The geometry of spacetime in SR is flat. The conditions for flatness were not specified in this chapter, but a hallmark of a flat geometry is that the metric tensor consists of constant elements, as in the Lorentz metric, Eq. (1.14). SR shows that the laws of mechanics and electromagnetism are equivalent for all inertial observers, but not gravitational phenomena, which requires the machinery of GR—the transition from global to local IRFs.

EXERCISES

1.1 Are the events A and B in Fig. 1.6 timelike, lightlike, or spacelike separated? That they are simultaneous in one frame, but not in another suggests what type of spacetime separation? What if in the right portion Fig. 1.6, the train was traveling from right to left, would event A still occur before B?

Note: The remainder of the exercises for Chapter 1 require no relativity. Work them using nonrelativistic physics.

1.2 Objects O_1, O_2, separated by a distance L, move along the x-axis with speed $v < c$. O_1 emits a photon toward O_2, reflecting at event E_1, which O_1 absorbs at event E_2. See Fig. 1.15. Calculate the times t_1 and t_2 in terms of L, v, and c. It may be helpful to draw the "space only" version of the diagram.

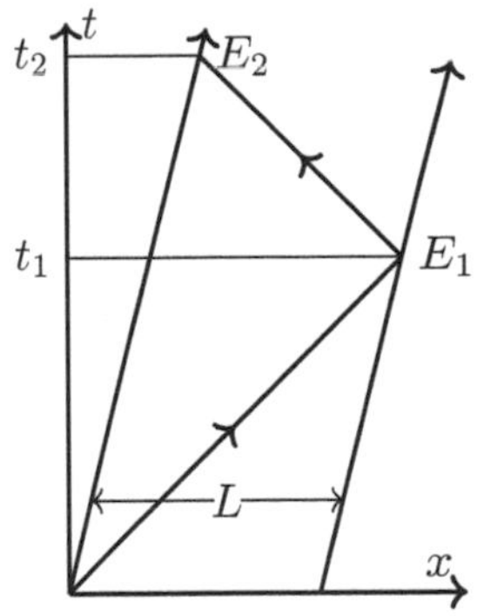

Figure 1.15 Figure for Exercise 1.2.

1.3 A river of width L flows with speed v_r with respect to its banks. Two swimmers can swim relative to still water at speed c, where $c > v_r$. The swimmers decide to have a contest. One will swim across the river and back. Call the time to accomplish this task $T_\perp$. The other will swim up the river the same distance L and back. Call the time to accomplish this task $T_\parallel$.

a. Show that

$$T_\parallel = \frac{2L/c}{1-\beta^2} \equiv \frac{2L}{c}\gamma^2 \qquad T_\perp = \frac{2L/c}{\sqrt{1-\beta^2}} \equiv \frac{2L}{c}\gamma\,, \tag{P1.1}$$

where $\beta \equiv v_r/c$. Thus, $T_{\|} > T_{\perp}$. For the across-the-river swim to arrive at the point on the other bank directly across from the starting point, the swim must be "aimed" upstream at an angle $\tan\theta = v_r T_{\perp}/(2L)$ relative to the line joining the points directly across the river from each other.

b. Show that for small β:

$$T_{\|} - T_{\perp} = \frac{L}{c}\beta^2 + O(\beta^4)\,. \tag{P1.2}$$

1.4 Consider Fig. 1.16. In the same river, a swimmer swims out to a distance L_1 and back at a constant angle ϕ relative to the bank. Call $T(\phi)$ the time to accomplish this task.

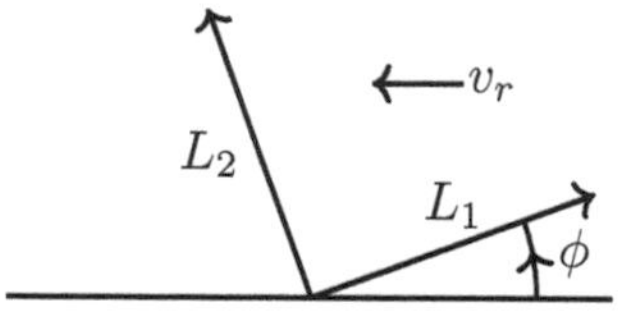

Figure 1.16 A swimmer swims to a distance L_1 in the river, and back, at an angle ϕ relative to the bank. A second swimmer swims to a distance L_2 and back, at an angle $\phi + \pi/2$.

a. Derive a formula for $T(\phi)$. It's instructive to use the Galilean velocity transformation. Let $\boldsymbol{c}$ denote the velocity of the swimmer in the frame of the water. Relative to *still* water, the swimmers swim at speed c. The velocity of the swimmer as observed from the bank is $\boldsymbol{v} = \boldsymbol{v}_r + \boldsymbol{c}$, or $\boldsymbol{c} = \boldsymbol{v} - \boldsymbol{v}_r$. "Dot" this vector into itself to find

$$c^2 = v^2 + v_r^2 - 2vv_r\cos(v, v_r)\,,$$

where (v, v_r) denotes the angle between the vectors $\boldsymbol{v}$ and $\boldsymbol{v}_r$. Show that

$$T(\phi) = \frac{2L_1/c}{1-\beta^2}\sqrt{1-\beta^2\sin^2\phi}\,, \tag{P1.3}$$

where $\beta = v_r/c$.

b. Show that Eq. (P1.3) reduces to Eq. (P1.1) in the appropriate cases.

c. A second swimmer swims out to a distance L_2 and back at a constant angle $\phi + \pi/2$ relative to the bank. Let $T(\phi + \pi/2)$ be the time to accomplish this task. Write down a formula for $T(\phi + \pi/2)$. Take Eq. (P1.3) and let $L_1 \to L_2$ and $\phi \to \phi + \pi/2$.

d. Calculate the *difference* in time for the swimmers to accomplish their tasks, $\Delta T(\phi) \equiv T(\phi + \pi/2) - T(\phi)$. Show that

$$\Delta T(\phi) = \frac{2L_2/c}{1-\beta^2}\sqrt{1-\beta^2\cos^2\phi} - \frac{2L_1/c}{1-\beta^2}\sqrt{1-\beta^2\sin^2\phi}\,. \tag{P1.4}$$

e. Using Eq. (P1.4), show that for small β

$$\Delta T(\pi/2) - \Delta T(0) = \frac{\beta^2}{c}(L_1 + L_2) + O(\beta^4)\,. \tag{P1.5}$$

f. Now imagine that we continuously change the angle ϕ in Fig. 1.16. Using Eq. (P1.4), show that, to leading order in small β, the time difference between the two legs changes with the angle according to

$$\frac{\mathrm{d}}{\mathrm{d}\phi}\Delta T(\phi) = \frac{\beta^2(L_1 + L_2)}{c}\sin 2\phi + O(\beta^4)\,. \tag{P1.6}$$

CHAPTER 2

Basic special relativity

THE basics of special relativity (SR) are presented in this chapter using spacetime diagrams.

2.1 COMPARISON OF TIME INTERVALS: THE BONDI K-FACTOR

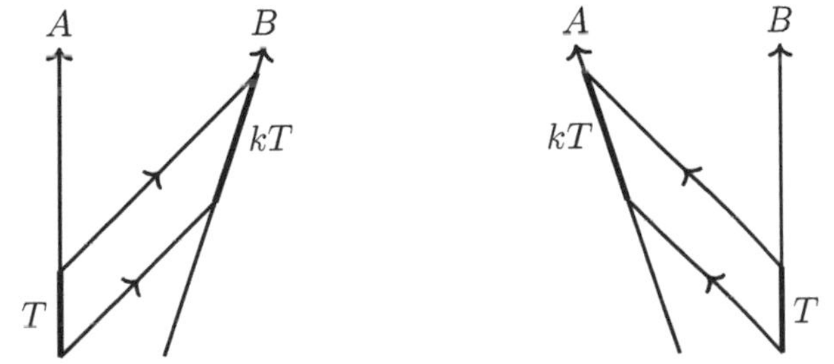

Figure 2.1 Inertial observers A and B emit photons separated by time T. Each sees the other move away at the same speed, v. Photons in the moving frames are received separated in time by kT, where $k = k(v)$.

Let inertial observers A and B in relative motion carry identical clocks, the worldlines of which are shown in Fig. 2.1.[1] A sends two flashes of light to B, a time T apart. What time separation does B measure? Not T—the second photon has further to travel. Perhaps if we knew the relative speed between A and B, the time in B could be calculated? That presupposes, however, the Newtonian conception that time is the same everywhere. SR shows that time is specified by a clock *in a reference frame*. We know that worldlines of free particles are straight in IRFs, and that spacetime coordinates in different IRFs are related by a linear mapping (Section 1.4). The time difference measured in B must therefore be *proportional*[2] to T, call it kT. The *Bondi k-factor* [12] is a function of the relative speed between frames, $k = k(v)$. In particular, as $v \to 0$, $k(v) \to 1$. Both observers see the *other* moving away at the same speed, and thus *the k-factor must be the same for both observers* (equivalence of IRFs). Photons emitted by B separated by time T are measured by A to have a time separation kT (see Fig. 2.1). The rabbit is in the hat.

A and B synchronize their clocks when their worldlines cross (see Fig. 2.2). After time T, A emits a photon toward B, which is reflected back to A. On B's clock, the photon arrives at time kT.

[1]The worldline of an observer *is* the time axis in its reference frame. Imagine yourself holding a clock in a room: *You* define the time axis for your reference frame. We don't show the spatial axes in spacetime diagrams except when necessary. The worldline is the location of $x = 0$ in that frame.

[2]This one assumption is really the whole show. The time measurements involved are at the same spatial locations in each frame, $\Delta x = 0$. Thus, Δt in one frame is linearly related to Δt in another.

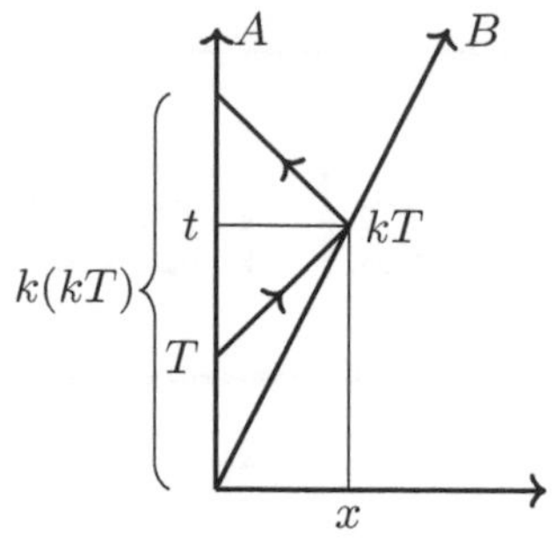

Figure 2.2 Photon emitted at time T, reflected at time kT, and received at time k^2T.

A records the arrival of the photon at a time that's a multiple, k, of the time at which B reflected the photon, kT. Thus, A records the arrival of the photon at time k^2T. Using Eq. (1.5), A assigns coordinates to the photon reflection from B:

$$(ct, x) = \left(\tfrac{1}{2}c(k^2T + T), \tfrac{1}{2}c(k^2T - T)\right) . \tag{2.1}$$

What was never in doubt is that A sees B moving at speed v, which in A's coordinates is expressed as $x = vt$. Thus, using Eq. (2.1),

$$\beta = \frac{v}{c} = \frac{x}{ct} = \frac{cT(k^2-1)/2}{cT(k^2+1)/2} = \frac{k^2-1}{k^2+1} . \tag{2.2}$$

Solve Eq. (2.2) for k:

$$k = \sqrt{\frac{1+\beta}{1-\beta}} . \tag{2.3}$$

Voila! We see that $k(0) = 1$ as expected. There's a sign convention implicit in Eq. (2.3): $\beta > 0$ corresponds to the "receiver" moving *away* from the photon source. Thus, $k > 1$ for $\beta > 0$. For the source *approaching* the receiver, let $\beta \to -\beta$, $0 < k < 1$. From Eq. (2.3), $k(-v) = k^{-1}(v)$.

The k-factor, which relates time intervals, is the *inverse* of the relativistic Doppler factor, which relates frequencies[3] (derived in Appendix B, Eq. (B.3)). It seems that we've arrived at a fundamental result of SR *without invoking any relativity*! If we examine the argument, however, we see that it uses the principle of relativity, that all inertial observers can claim themselves at rest, and the isotropy of the speed of light (through the use of the radar method). We motivated the k-factor by appealing to *linearity*, that all inertial observers see straight worldlines of free particles. The k-factor is thus firmly rooted in the fundamentals of relativity. As we now show, *all the standard results of SR can be derived using the k-factor.*

We can see the connection with the Doppler effect by referring to Fig. 2.3. An emitter emits signals regularly with time separation Δt; it thus emits at the frequency $f_e \equiv (\Delta t)^{-1}$. The receiver receives signals separated by time Δt_{rec}; hence the received frequency is $f_{rec} = (\Delta t_{rec})^{-1}$. The reception time is related to the emission time through the k-factor, $\Delta t_{rec} = k\Delta t$. We therefore have the relativistic Doppler effect, in agreement with Eq. (B.3),

$$f_{rec} = \frac{1}{k} f_e . \tag{2.4}$$

While the receiver is approaching the emitter, $k < 1$, a blueshift, and after, $k > 1$, a redshift.

[3]The relativistic Doppler effect is the classical Doppler effect combined with time dilation. Time dilation is not something we "officially" know about yet; it's derived in Section 2.2.

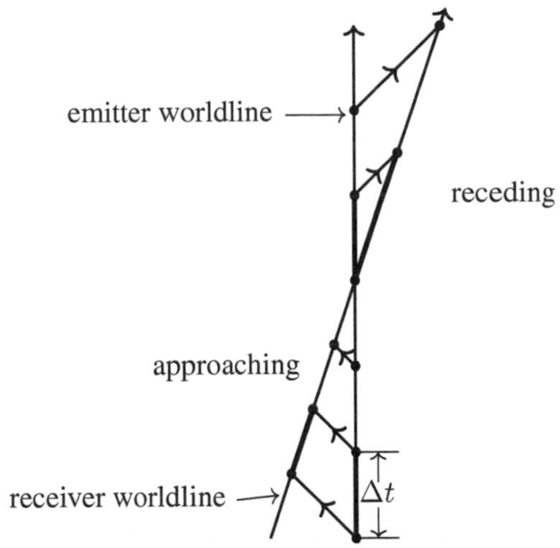

Figure 2.3 Doppler effect. Approaching observer receives photons at a blueshifted frequency; receding observer receives photons at a redshifted frequency.

2.2 TIME DILATION

Figure 2.4 shows the worldlines of inertial observers A and B who have synchronized their clocks.

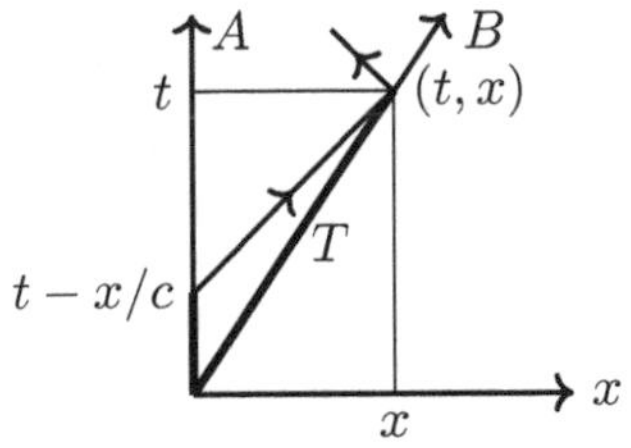

Figure 2.4 Time dilation. Proper time T occurs at time $t = \gamma T$ in reference frame A.

A emits a photon at time $t - x/c$ that's reflected by B. A assigns coordinates (t, x) to the reflection event. B records the arrival of the photon at time T using its clock. *B is at rest relative to the clock*; time measured in that frame is the *proper time*. The time T is related to the time $t - x/c$ through the k-factor:

$$T = k(t - x/c) \,. \tag{2.5}$$

In A, $x = vt$, implying

$$T = k(t - \beta t) = kt(1 - \beta) = t\sqrt{1 - \beta^2} \,, \tag{2.6}$$

where we've used Eq. (2.3). Equation (2.6) is usually written

$$t = \frac{1}{\sqrt{1 - \beta^2}} T = \gamma T \,. \tag{2.7}$$

Equations (2.6) or (2.7) are referred to as *time dilation*—"moving clocks run slow." Suppose B measures T to be one hour. A will measure a time *longer* than one hour, and conclude that the moving clock runs slower. We show in Chapter 4 that the effect is symmetrical: Both observers claim a moving clock runs slow.

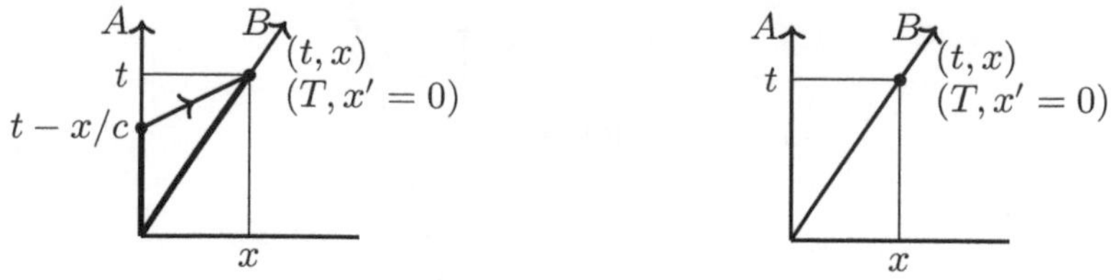

Figure 2.5 Left: k-factor relates time for two events (black dots), $T = k(t - x/c)$ Right: Time dilation relates the time coordinates assigned to the same event, $t = \gamma T$.

Let's be clear on *what* times are being compared. The k-factor relates the time coordinates of *two distinct events*—emission and reception of a photon, shown as black dots in Fig. 2.5. Emission occurs at time $t - x/c$ in A, and reception occurs at event $(T, x' = 0)$ in B. The worldline of the clock, B, defines the line $x' = 0$, just as the time axis in A is the line $x = 0$. The k-factor relates the reception time in B to the emission time in A, with $T = k(t - x/c)$. The time dilation factor on the other hand relates the time coordinates assigned to *the same event* by two observers, with $t = \gamma T$.

2.3 VELOCITY ADDITION

Figure 2.6 shows the worldlines of three inertial observers, A, B, and C, which synchronize their clocks as they pass at the origin. A emits two photons with a time separation T. The two photons

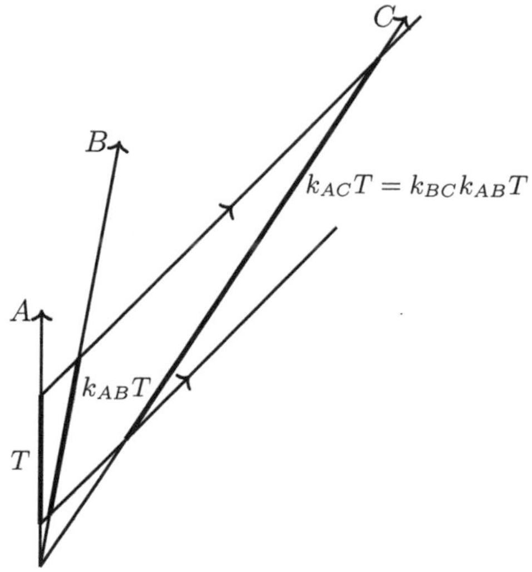

Figure 2.6 Composition rule for k-factors.

arrive in B separated by time $k_{AB}T$, where k_{AB} is the k-factor associated with the relative speed between A and B. The photons arrive in C separated by time $k_{AC}T$. Alternatively, the photons leave frame B separated in time $k_{AB}T$, which arrive in frame C with time separation $k_{BC}k_{AB}T$. Thus, $k_{AC}T = k_{BC}k_{AB}T$, implying the k-factors satisfy a multiplicative composition law:

$$k_{AC} = k_{AB}k_{BC} \,. \tag{2.8}$$

The k-factor is a proxy for speed: the larger is β, the larger is k. The projection of the interval T (along the time axis in the rest frame) onto the time axes of frames in motion is a measure of speed. The way the geometry of spacetime works, k-factors are multiplicative between frames.

Combining Eq. (2.2) with Eq. (2.5),

$$\beta_{AC} = \frac{k_{AC}^2 - 1}{k_{AC}^2 + 1} = \frac{k_{AB}^2 k_{BC}^2 - 1}{k_{AB}^2 k_{BC}^2 + 1} . \tag{2.9}$$

Using Eq. (2.3) for each of the k-factors, it follows that (show this)

$$\beta_{AC} = \frac{\beta_{AB} + \beta_{BC}}{1 + \beta_{AB}\beta_{BC}} , \tag{2.10}$$

the *Einstein velocity addition formula.* For low speeds, $\beta_{AB} \ll 1$, $\beta_{BC} \ll 1$, and $\beta_{AC} \approx \beta_{AB} + \beta_{BC}$, the Galilean velocity addition formula. If we substitute $\beta_{BC} = 1$ in Eq. (2.10), we obtain $\beta_{AC} = 1$ for *any* β_{AB}. There is an *invariant limiting speed* implied by the theory, $\beta = 1$.

Example. Apply the velocity addition formula to the speed of light in a moving medium. The speed of light *in the rest frame* of a medium of index of refraction n, is c/n. What is the speed of light u when the medium has speed v relative to the lab frame? Using Eq. (2.10),

$$u = \frac{v + c/n}{1 + v/(cn)} = \frac{c}{n} + v\left(1 - \frac{1}{n^2}\right)\frac{1}{1 + v/(nc)} .$$

The coefficient multiplying v, $(1 - n^{-2})$, the *Fresnel drag coefficient*, was confirmed in the Fizeau water tube experiment of 1851, where $v \ll c$. Relativistic velocity addition thus has an observable consequence at relatively slow speeds—the speed of the flow of water in the Fizeau experiment.[4]

Example. Particles A and C have velocities $\beta_A = 0.95$ and $\beta_C = -0.95$ relative to a linear accelerator. What is the velocity of C relative to A? It's helpful to draw a spacetime diagram—see Fig. 2.7. B is a laboratory observer, situated between A and C; the precise location of B is

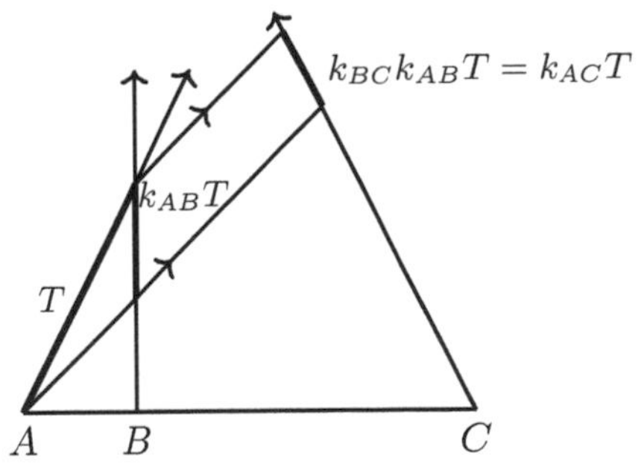

Figure 2.7 Spacetime diagram for particles A and C approaching each other.

unimportant. Because A and C are both approaching B, we can set $\beta_{AB} = \beta_{BC} = -0.95$ and use Eq. (2.10) to conclude that $\beta_{AC} = -0.9987$. In Fig. 2.7, particle A emits two photons separated by time T. Because A is approaching B, $k_{AB} < 1$; similarly for k_{BC}.

---

[4]The Fizeau experiment is worth learning about—an ingenious interferometric experiment not unlike the MM experiment. It uses a tube constructed so that water can flow in opposite directions, through which beams of light pass in such a way that each beam propagates in the direction of water flow. The light beams are then brought together to interfere, where the change in phase is correlated with the speed of the water. In the Fizeau experiment, the flow of water can simply be turned off, something that Michelson and Morley couldn't do—turn off the motion of the earth!

2.4 LORENTZ TRANSFORMATION

Figure 2.8 shows the worldlines of observers A and B in relative motion (what we often call frames S and S'), which synchronize their clocks as they pass. Each uses the radar method to assign coordinates to the same event, P: (t, x) and (t', x'). A photon emitted by A at time $t - x/c$ reflects from event P and is received at time $t + x/c$. B emits a photon at time $t' - x'/c$ which reflects from the same event, and is recorded at time $t' + x'/c$. Note the symmetry: Both observers claim to be at rest; both emit a photon at the "same time" using their respective coordinates, $t - x/c$ and $t' - x'/c$.

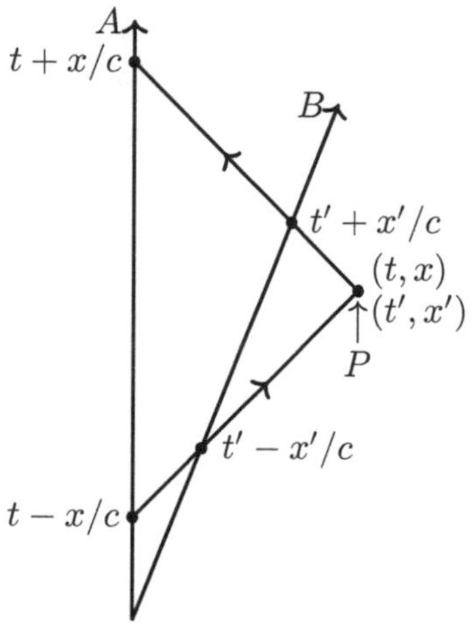

Figure 2.8 Inertial observers A and B use the radar method to assign coordinates to the same event, (t, x) and (t', x').

The emission and reception times in the two frames are naturally related through the k-factor:

$$\begin{aligned} t' - x'/c &= k(t - x/c) \\ t + x/c &= k(t' + x'/c) \,. \end{aligned} \tag{2.11}$$

Solve Eq. (2.11) for (t', x'):

$$\begin{aligned} ct' &= \tfrac{1}{2}\left(k^{-1} + k\right) ct + \tfrac{1}{2}\left(k^{-1} - k\right) x \\ x' &= \tfrac{1}{2}\left(k^{-1} - k\right) ct + \tfrac{1}{2}\left(k^{-1} + k\right) x \,. \end{aligned}$$

Using Eq. (2.3), we have the matrix equation (show this)

$$\begin{pmatrix} ct' \\ x' \end{pmatrix} = \gamma \begin{pmatrix} 1 & -\beta \\ -\beta & 1 \end{pmatrix} \begin{pmatrix} ct \\ x \end{pmatrix} , \tag{2.12}$$

the same as Eq. (A.6).

Location of x'-axis: Lines of simultaneity

With the LT, we can find the location of the x'-axis—the spatial axis of S'—in relation to the space and time axes of S. The t'-axis is the worldline seen in S, $x = vt$, or $ct = \beta^{-1}x$. The same result follows from Eq. (2.12) as the locus of points with $x' = 0$ (check it!). What about the x'-axis? Answer: The locus of points associated with $t' = 0$, which from Eq. (2.12) is $ct = \beta x$. Figure 2.9 shows the x' and t' axes both situated at the angle ϕ with respect to the x, t axes, where $\tan\phi = \beta$. As $\beta \to 1$, $\phi \to \pi/4$. The coordinates assigned to the same event in each reference

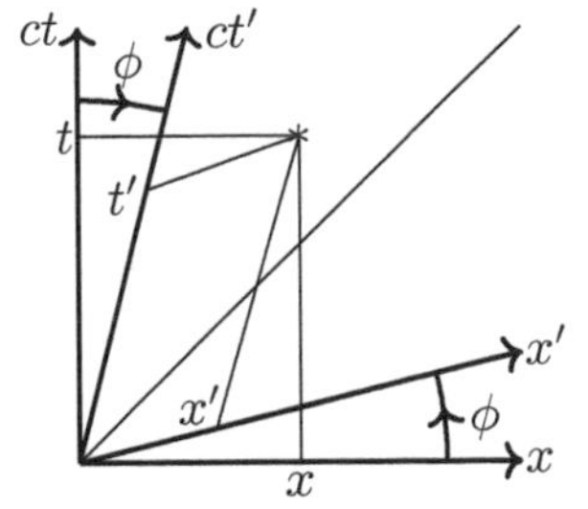

Figure 2.9 Coordinates assigned to the same event—the asterisk—in reference frames in relative motion with speed β: (t, x) and (t', x'). The t' and x'-axes form the same angle ϕ with respect to the t and x-axes, with $\tan\phi = \beta$.

frame are found by projecting onto the respective space and time-axes, as shown. Knowing the x'-axis provides a way to test for the simultaneity of events (in S'): If two events can be connected by a line parallel to the x' axis, they have the same time *in that frame*. Lines parallel to the x-axis are *lines of simultaneity*.[5]

We began our discussion of spacetime diagrams by agreeing to take the time axis as orthogonal to the space of spatial variables. All IRFs are equivalent, yet the space and time axes of S' do not appear orthogonal in Fig. 2.9. We don't know yet how to form the inner product of spacetime vectors. We'll see that the t' and x'-axes are indeed orthogonal in S'.

Example. Particle B moves away from A at speed $\beta = 0.25$, from left to right. What are the coordinates *in the moving frame* assigned to an event that in the rest frame occurs at $ct = 2.25$ and $x = 1.5$? For $\beta = 0.25$, $\gamma = 4/\sqrt{15} = 1.03$. Use the LT, Eq. (2.12):

$$\begin{pmatrix} ct' \\ x' \end{pmatrix} = 1.03 \begin{pmatrix} 1 & -0.25 \\ -0.25 & 1 \end{pmatrix} \begin{pmatrix} 2.25 \\ 1.5 \end{pmatrix} = \begin{pmatrix} 1.94 \\ 0.97 \end{pmatrix} .$$

These are the coordinates shown in Fig. 2.9. What if the speed is negative (particle moves from right to left)? Figure 2.10 shows the spacetime diagram for $\beta = -0.25$.

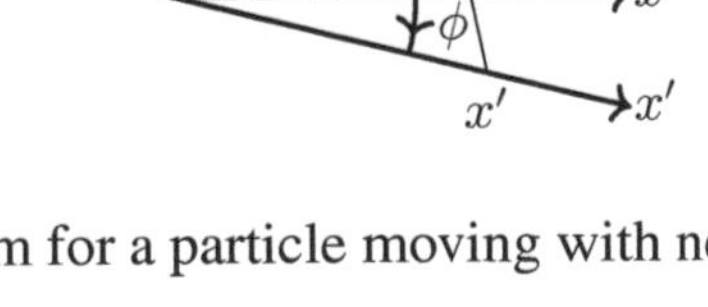

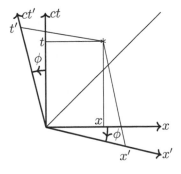

Figure 2.10 Spacetime diagram for a particle moving with negative velocity, right to left.

---

[5]Lines parallel to the t-axis are lines of *co-locality*; between timelike separated events one can always find a frame of reference where the events occur at the same location in space.

Example. Relativity of causality
Spacelike-separated events can occur simultaneously or in *either* time order, depending on the reference frame (mentioned in Section 1.5). The left portion of Fig. 2.11 shows spacelike-separated events A and B as seen from a reference frame moving with speed $\beta = 0.26$ relative to the unprimed frame. A precedes B in both frames. In the right portion of Fig. 2.11 the same events occur in the opposite order in a frame moving with speed $\beta = 0.71$.

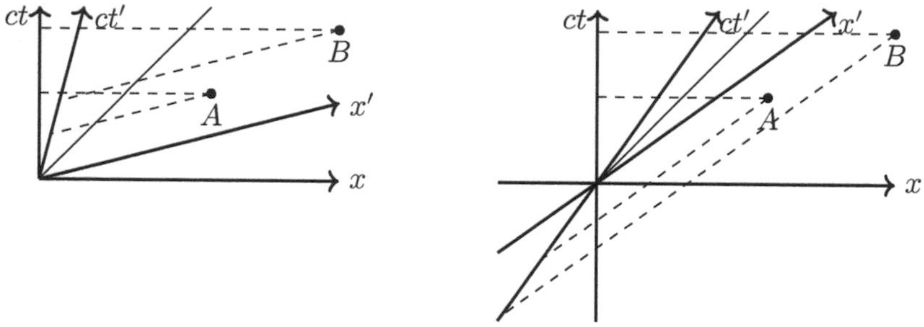

Figure 2.11 Time order of spacelike-separated events is reference-frame dependent.

2.5 LENGTH CONTRACTION

We now discuss, from several points of view, *length contraction*, the converse of time dilation, a phenomenon students (and others) tend to find more confusing than time dilation.

2.5.1 Using the k-factor

A rod of rest length D moves along the x-axis with speed v from left to right. Figure 2.12 shows the worldlines of the front and back ends of the moving rod as B_{front} and B_{back} from the perspective of reference frame A. Clocks are synchronized when the *front* edge of the rod passes the origin, O. You may find it helpful to visualize how you would use a radar gun to measure the distance to an approaching rod, traveling straight at you. As we now show, the length of the rod as measured in A is $d = D/\gamma$, the phenomenon of length contraction.

A emits a photon at time $-d/c$ (negative time, relative to the origin O), which reflects from the back end of the rod, event E. The reflected photon arrives in A at time d/c. The coordinate of the back end of the rod is, from the radar method, d. In B the emitted photon passes the front end of the rod at point P in Fig. 2.12. It's as if in frame B a photon was emitted from the front end of the rod toward the back end. Such a photon would have been emitted at time $-d/(kc)$ in the B-frame. We've used that $k(-v) = k^{-1}(v)$; the receiver (front end of the rod) is moving toward the source—negative speed. The reflected photon encounters B_{front} at point Q in Fig. 2.12, which occurs at time kd/c (use the k-factor together with the time d/c in the A-frame). By the radar method, the coordinates for event E in frame B are

$$(ct', x') = \left(\frac{d}{2}(k - k^{-1}), \frac{d}{2}(k + k^{-1})\right) = (d\beta\gamma, d\gamma) \,. \tag{2.13}$$

Thus $D = \gamma d$, or $d = D/\gamma$: Rods in motion appear shorter than the length measured in its rest frame. We'll show in Chapter 4 that the relation is symmetrical: Both observers claim that a rod in motion has a length shorter than its rest length. The line labeled D in Fig. 2.12 is where the x'-axis in B intersects event E (a line of simultaneity). In both reference frames, *the spatial coordinates of the two ends of the rod have been obtained at the same time.*

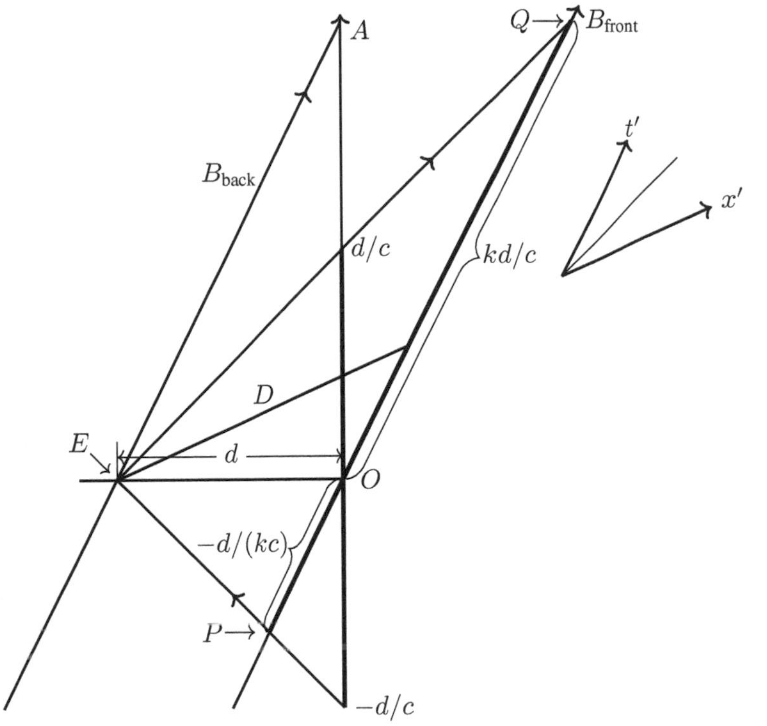

Figure 2.12 Length contraction. Rod of rest length D has length D/γ measured in A.

2.5.2 Using the Lorentz transformation

Length contraction can be more readily demonstrated using the Lorentz transformation (LT). Figure 2.13 shows the worldlines of the front and back ends of the moving rod as B_{front} and B_{back} as seen in the frame of observer A. Both observers want to measure the length of the rod, and both are careful to measure the two ends of the rod at the same time in their reference frames.[6] But of course, what's simultaneous in one frame is not in another. Observer B, at rest relative to the rod, measures $\Delta x'$ at $t' = 0$ as the rest length of the rod. Observer A records the locations of the two ends of the rod at time t, measuring the length as Δx. The events used to measure length in frame A are not simultaneous in frame B, and vice versa; see Fig. 2.13. Referring to the events with $\Delta t = 0$, we have from Eq. (2.12),

$$\begin{pmatrix} c\Delta t' \\ \Delta x' \end{pmatrix} = \gamma \begin{pmatrix} 1 & -\beta \\ -\beta & 1 \end{pmatrix} \begin{pmatrix} 0 \\ \Delta x \end{pmatrix} .$$

Thus, $\Delta x' = \gamma \Delta x$ (length contraction) and $c\Delta t' = -\beta\gamma\Delta x$ (relativity of simultaneity).

2.5.3 Pole and barn

A paradox is a conflict between reality and your feeling of what reality ought to be.
—Richard Feynman [13, p18-9]

One of the more well-known of the supposed paradoxes associated with SR is the pole and barn problem. In this thought experiment, a runner carries a pole that in its rest frame is 20 m long (a

[6] To measure the length of a stick moving past you, you wouldn't first measure the location of one end of the stick, and then only an hour later measure the location of the other end.

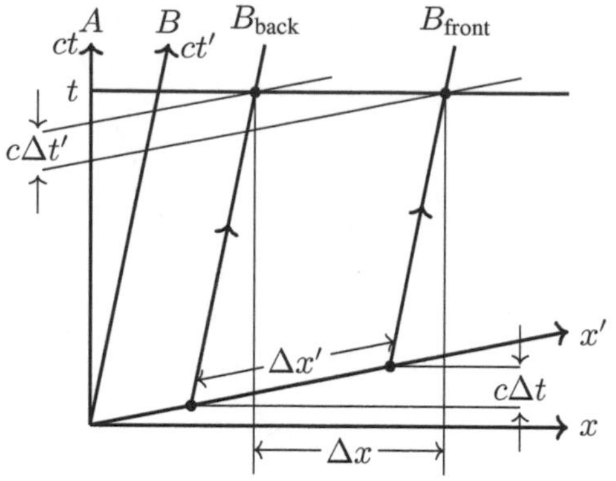

Figure 2.13 Length contraction $\Delta x = \Delta x'/\gamma$ where $\Delta x'$ is the proper length.

long pole!). The runner is headed toward a barn that in *its* rest frame is 10 m long. The speed of the runner is such that $\gamma = 2$ ($\beta = \sqrt{3}/2$). In the frame of the barn, the pole appears 10 m long because of relativistic length contraction. Thus, the pole fits entirely within the barn at one instant of time. In the frame of the pole, however, the barn appears 5 m long, and the pole cannot fit entirely within the barn. Can these descriptions be reconciled?

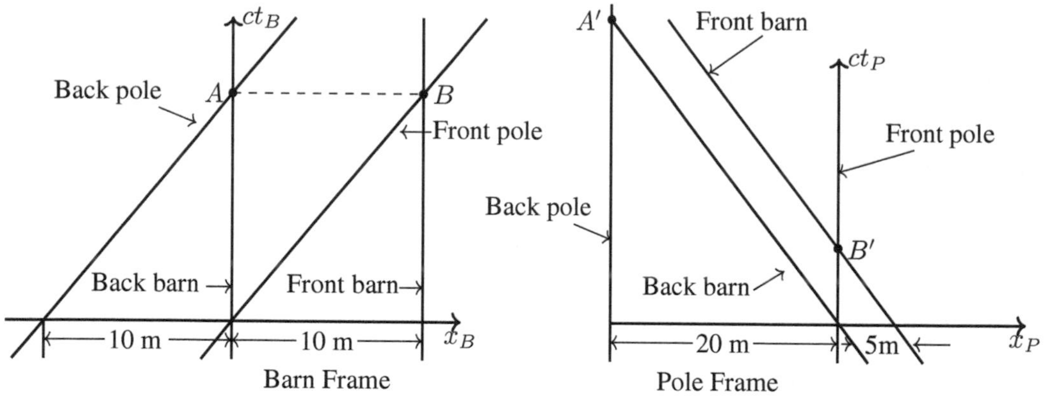

Figure 2.14 Reference frames of barn and pole on spacetime diagrams (not to scale).

The left portion of Fig. 2.14 shows the events in the reference frame of the barn, with the pole approaching from left to right. We consider the front of the pole and the front of the barn to be on the right, with the back of the pole and the back of the barn to the left. The front of the pole first encounters the back of the barn. As the pole passes through the barn there is an instant of time when the pole fits entirely within the barn. These events are labeled in Fig. 2.14 as A (back of the pole encountering the back of the barn) and B (front of the pole encountering the front of the barn).

The same events are shown in the reference frame of the pole in the right portion of Fig. 2.14. We can place the origin of spacetime coordinates wherever we want, but to use the LT formula *the origins of the two systems of spacetime coordinates must coincide.*[7] In both diagrams the origin is the event in which the front of the pole encounters the back of the barn. Events A and B (from the left diagram) are shown in the right diagram as events A' and B'. The *events* are the same, but the coordinates assigned to them are different in the two frames.

[7]Because the Lorentz transformation is a linear, homogeneous coordinate transformation.

In the barn frame, A has coordinates $(20/\sqrt{3}, 0)$—the time to cover 10 m at speed $\beta = \sqrt{3}/2$. In the pole frame, the same event has coordinates

$$\begin{pmatrix} ct_P \\ x_P \end{pmatrix} = \gamma \begin{pmatrix} 1 & -\beta \\ -\beta & 1 \end{pmatrix} \begin{pmatrix} ct_B \\ x_B \end{pmatrix} = \begin{pmatrix} 2 & -\sqrt{3} \\ -\sqrt{3} & 2 \end{pmatrix} \begin{pmatrix} 20/\sqrt{3} \\ 0 \end{pmatrix} = \begin{pmatrix} 40\sqrt{3} \\ -20 \end{pmatrix} .$$

B has coordinates in the barn frame $(20/\sqrt{3}, 10)$ and coordinates $(10/\sqrt{3}, 0)$ in the pole frame.

Where's the paradox? Well, there isn't one, save for our intuition-derived expectation that the pole fitting entirely within the barn should be an objective fact, the same for all observers. That's the point of the exercise: What's simultaneous in one frame (events A and B in the barn frame) is not in another (A' and B' in the pole frame).

2.5.4 Length contraction, Minkowski, and the fourth dimension

Objects in motion have a length L contracted in the direction of motion, $L \le L_0$, where L_0 is the rest length. Does that mean objects in motion shrink? Consider an analogy from crystallography. Through a crystal lattice, various planes may be drawn containing lattice points, *lattice planes*, labeled by Miller indices[8] (hkl), e.g., the (100) plane or the (110) plane—see Fig. 2.15. A plane is

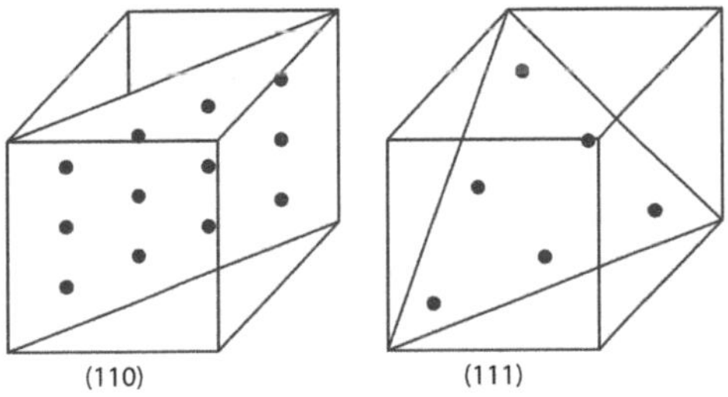

Figure 2.15 Separation between atoms in a crystal is lattice-plane dependent.

a two-dimensional *slice* of a three-dimensional geometry. Restricting your attention to one of these planes, what is the distance between lattice points? The answer depends on the plane. Does it make sense to ask what is the *real* distance between atoms? There's no definitive answer if all we see is a two-dimensional sampling of the points that have an arrangement in three dimensions.

The same reasoning applies to objects we see at an instant of time in our three-dimensional world, objects that exist in four-dimensional spacetime. Figure 2.16 shows the worldlines of the

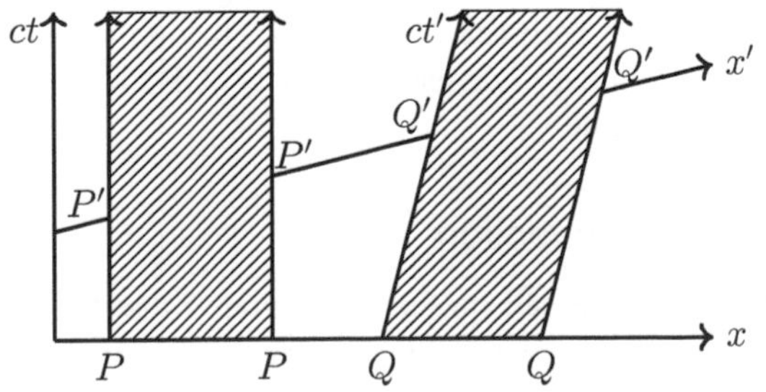

Figure 2.16 Worldtubes intersecting lines of simultaneity for frames in relative motion.

ends of two identical rods, much as in the previous figures. The points of an extended object (such

[8] This nomenclature is discussed in any book on solid-state physics.

as a rod), considered in spacetime as a collection of worldlines, comprise a *worldtube*, shown in crosshatch in Fig. 2.16. In the (t, x) coordinate system the rest length, or *proper length*, is the line PP. In the (t', x') system, attached to an identical rod moving along the x-axis, the rest length is the line $Q'Q'$. The line PP, as seen from the (t', x') system is shown as $P'P'$, while $Q'Q'$ is shown as QQ in the (t, x) system.[9]

Minkowski argued that the apparent deformation of a moving object can be understood as arising from a *three-dimensional slice* (surface of simultaneity) of a four-dimensional entity. The length depends on the intersection of the worldtube with an observer's *space*—surface of simultaneity. The way the non-Euclidean geometry of spacetime works, the intersection with the *rest-space* of an observer produces the largest length.[10] By considering spacetime as a whole, by taking a *geometric* point of view, Minkowski found that the perplexing results of SR can be given an intelligible explanation. His most far-reaching conclusion is that *observers in relative motion have different spaces as well as times*. One must arrive at this conclusion if surfaces of simultaneity (observer-dependent slices) in a four-dimensional spacetime are three-dimensional spaces.

The usual length contraction hypothesis, according to Minkowski

> ...sounds extremely fantastical, for the contraction is not to be looked upon as a consequence of resistances in the ether, or anything of that kind, but simply as a gift from above—as an accompanying circumstance of the circumstance of motion.

Minkowski held that the idea of a four-dimensional world *explains* the principle of relativity:

> ...the word *relativity-postulate* ... seems to me very feeble. Since the postulate comes to mean that only the four-dimensional world of space and time is given by phenomena, but that the projection in space and in time may still be undertaken with a certain degree of freedom, I prefer to call it the *postulate of the absolute world* (or briefly, the world-postulate).
>
> We should then have in the world no longer *space*, but an infinite number of spaces, analogously as there is in three-dimensional space an infinite number of planes. Three-dimensional geometry becomes a chapter in four-dimensional physics.

Basically, because in a four-dimensional world observers in relative motion have their own spaces and times, all inertial observers describe phenomena the same way because all are at rest in their respective frames. Thus, every inertial observer measures the same speed of light using its own rest-space coordinates and time. There cannot be absolute motion in the sense of Newton because there is not just *one* space and *one* time.

Minkowski's explanation of length contraction—the same four-dimensional worldtube intersected by spaces of different observers—makes a compelling case for the reality of four-dimensional spacetime. Einstein did not at first embrace Minkowski's theory, but soon started to make use of tensor methods in spacetime geometry. General relativity (GR) would not be possible without a geometrical perspective on spacetime.

2.5.5 FitzGerald-Lorentz contraction

In 1888 Oliver Heaviside showed (based on the ether model) that the electric field surrounding a spherical charge would cease to have spherical symmetry if the charge was in motion relative to the ether. In the Heaviside model, the *longitudinal* component of the electric field (in the direction of motion) is affected by motion, but not the transverse components.[11] In 1889, G.F. FitzGerald took

[9] We're using the notation employed by Minkowski.[9, p78] Why? To encourage you to read the original literature!

[10] In a crystal, the lattice constant is reported as the *shortest* distance between atoms.

[11] This is of course exactly the opposite from SR, where the longitudinal field component is invariant and the transverse components transform between IRFs. See Chapter 8.

Heaviside's result and suggested *ad hoc* that the shape of an object would be altered *in the direction of motion*. As is well known, if the length L of the arm in a Michelson interferometer is distorted in the direction of motion such that $L \to L\sqrt{1-\beta^2}$, it would explain the null result of the MM experiment while preserving the notion of the ether. In 1892, Lorentz independently published the same idea, although Lorentz attempted to work through detailed models of inter-molecular forces that would demonstrate the effect. The idea came to be known as the *FitzGerald-Lorentz contraction hypothesis* (FL).

Einstein's hypothesis that the speed of light is the same in all IRFs also accounts for the null result of the MM experiment, without making assumptions about the internal constitution of matter. As we've seen, an identical formula $L = L_0\sqrt{1-\beta^2}$ is derived in SR, and it's important to understand the difference between relativistic length contraction and the FL contraction. Length contraction in SR is a *coordinate effect*, the difference in spatial coordinates of something that should be seen in its totality in four-dimensional spacetime. There is not implied an *actual* contraction, in distinction to the FL contraction. Asking whether a stick "really" contracts is tantamount to asking whether it's "really" moving, which can be answered only from absolute space. If SR is correct and there is no ether *and* length contraction is a "real" contraction, the MM experiment would show a *positive* result, because the FL contraction would *introduce* a time difference between the arms of the interferometer. A real length contraction is not compatible with the null result of the MM experiment *and* isotropy of the speed of light.

2.5.6 Experimental status

There is no *direct* experimental confirmation of relativistic length contraction. Elementary particles can be made to move rapidly (speeds comparable to c), but their linear dimensions cannot be measured directly, while macroscopic objects, the dimensions of which can be measured, cannot be made to move at relativistic speeds. The predictions of SR all emerge from the principle of relativity, and length contraction is one of its consequences. It's often said that SR rests on two postulates (the way Einstein presented it): the principle of relativity and the invariance of the speed of light in IRFs. The principle of relativity alone predicts a universal speed, which experiment shows to be the speed of light.[12] Time dilation has been confirmed through the measurement of the relativistic Doppler effect (Ives-Stillwell experiment [14]). The MM experiment [15] showed that the speed of light is isotropic, used in the derivation of the radar method. The Kennedy-Thorndike experiment [16] showed that the speed of light is independent of the velocity of the source, implicitly used in the derivation of each of the effects of SR (Doppler effect, time dilation, Lorentz transformation, length contraction). While there is no direct confirmation of length contraction, we show in Chapter 8 that the Lorentz force $q\,(\boldsymbol{E} + \boldsymbol{v} \times \boldsymbol{B})$ can be derived without approximation as a frame transformation, a derivation that relies heavily on length contraction.

2.5.7 Length contraction in one frame is time dilation in another

Time dilation and length contraction are each a consequence of the relativity of simultaneity. Both effects emerge from a comparison of events measured from reference frames in relative motion. In a frame at rest relative to clocks and rods, measurements taken at the same location (proper time) and at the same time (proper length), are different from measurements made in a frame in which clocks and rods are in motion, the measurements of which occur at different locations (time dilation) and at different times (length contraction). As we now show, what can be interpreted as time dilation in one frame can be interpreted as length contraction in another.

[12] Thus, Einstein's second postulate is not strictly necessary. We argue in Chapter 3 that Einstein's second postulate is the assertion that photons have no rest mass.

Referring to Fig. 2.7, particles A and C are launched at time $t = 0$ with speeds of $0.95c$, a distance $L = 3.2$ km apart (the length of the Stanford Linear Accelerator). At what time do the particles collide, in the lab frame and in the frame of one of the particles? In lab-frame coordinates (see Fig. 2.17), the particles collide at half their initial separation, $t_L = L/(2v) = 1.6$ km$/0.95c = 5.61\mu$s. The *proper time*, the time that particle A experiences before the collision is, from Eq. (2.7), $T = t/\gamma = 1.75\mu$s, where $\gamma = 3.2$ for $\beta = 0.95$. Moving clocks run slow.

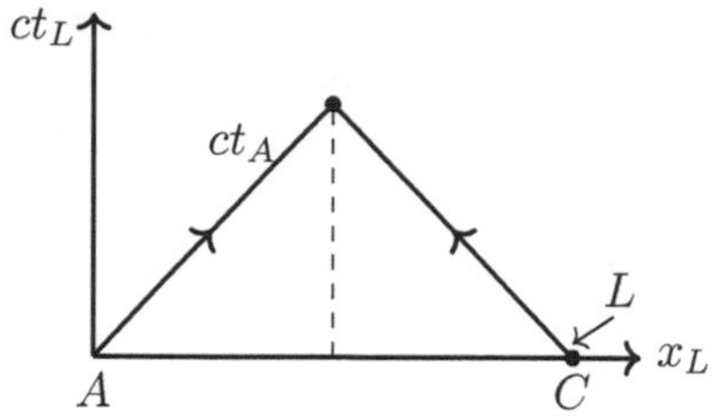

Figure 2.17 In laboratory coordinates, C starts at L and collides with A at $L/2$

Let's calculate that time using length contraction, knowing (page 31) that in the rest frame of A, C approaches with speed $\beta_r = 0.9987$. The Lorentz factor associated with β_r is $\gamma_r = 19.51$. One might think that C "sees" a contracted length $L/\gamma_r = 3200$ m$/19.51 = 164$ m (for the starting separation $L = 3.2$ km). A would then suffer a collision after a time $L/(\gamma_r\beta_r c) = 0.55\mu$s, not the same as our previous calculation of 1.75μs. What's wrong with this apparently too-facile argument? As is often the case, the problem lies in simultaneity. The starting separation is specified in the *lab* frame; *only* in that frame can we say the particles are 3.2 km apart at $t = 0$. What length *should* we use?

Let's write down the LT between the lab frame and the IRF associated with A. The laboratory is rushing from right to left in frame A—negative velocity. From Eq. (2.12) with $\beta \to -\beta$,

$$\begin{pmatrix} ct'_L \\ x'_L \end{pmatrix} = \gamma \begin{pmatrix} 1 & \beta \\ \beta & 1 \end{pmatrix} \begin{pmatrix} ct_A \\ x_A \end{pmatrix} . \tag{2.14}$$

The *inverse* of Eq. (2.14) is found by reversing the speed, $\beta \to -\beta$ (see Exercise 2.3). Thus, we have the equivalent LT

$$\begin{pmatrix} ct_A \\ x_A \end{pmatrix} = \gamma \begin{pmatrix} 1 & -\beta \\ -\beta & 1 \end{pmatrix} \begin{pmatrix} ct'_L \\ x'_L \end{pmatrix} . \tag{2.15}$$

The location of the t_L axis (in the (t_A, x_A) coordinate system) is found from Eq. (2.14) by setting $x_L = 0$ ($ct_A = -\beta^{-1}x_A$); the location of the x_L axis is found by setting $t_L = 0$ in Eq. (2.14) ($ct_A = -\beta x_A$). These axes are shown in Fig. 2.18.

In the lab frame, particle C starts at $(0, L)$. The coordinates in A associated with that event are found from Eq. (2.15):

$$\begin{pmatrix} ct_A \\ x_A \end{pmatrix} = \gamma \begin{pmatrix} 1 & -\beta \\ -\beta & 1 \end{pmatrix} \begin{pmatrix} 0 \\ L \end{pmatrix} .$$

In A, C starts at coordinates $(ct_A = -\beta\gamma L, x_A = \gamma L)$, i.e., in A the *rest length* is γL, because that's what gets contracted to the length L specified in the lab frame, a frame that's now in motion with respect to A. The time coordinate $cT_A = -\beta\gamma L$ is the time difference in A between events that are simultaneous in the lab frame.

The worldline of C in Fig. 2.18 connects the event with coordinates $(-\beta\gamma L, \gamma L)$ with the collision with A at $(cT, 0)$. We can compute T from the known relative velocity between A and C:

$$\beta_r = \frac{\gamma L}{cT + \beta\gamma L} ,$$

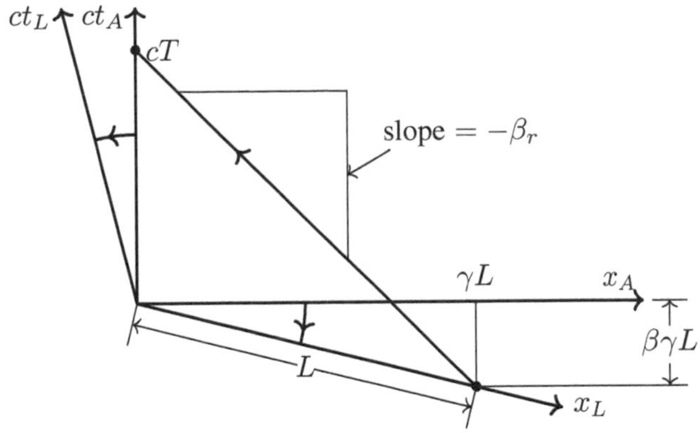

Figure 2.18 Events of Fig. 2.17 shown in the frame of particle A. Not drawn to scale.

or

$$cT = \frac{\gamma L}{\beta_r}(1 - \beta_r \beta) . \tag{2.16}$$

Using Eq. (2.16), $T = 1.75\mu$s, the same as we found in the lab frame using time dilation.

We can ask, what equivalent length, call it D, did particle C traverse at speed $\beta_r c$? That is, express cT in terms of an equivalent length,

$$cT = \frac{\gamma L}{\beta_r}(1 - \beta\beta_r) \equiv \frac{D}{\beta_r} .$$

It is not difficult to show that

$$1 - \beta\beta_r = \frac{1}{\gamma_r} , \tag{2.17}$$

where $\gamma_r \equiv (1 - \beta_r)^{-1/2}$ (use $\beta_r = 2\beta/(1+\beta^2)$). Thus, particle C sees the rest length γL in the A frame contracted to $\gamma L/\gamma_r$.

2.6 FOUNDATIONAL EXPERIMENTS

2.6.1 The Michelson-Morley experiment: Isotropy of c

Exercise 1.4 is modeled on the MM experiment. The "swimmers" are beams of light, and the river is the ether, streaming past us in our reference frame at speed v_r. Our speed relative to the ether is unknown, but can be estimated to be on the order of the speed of Earth in its orbit around the sun, $v_r \approx 3 \times 10^4$ m s^{-1}. (Thus, $\beta_r \approx 10^{-4}$.) Associated with the time difference ΔT between the arms of the interferometer is a shift of $N = f\Delta T$ fringes, where f is the frequency of the light. *Fringes* are alternating bands of light and dark seen when the beams of light in the arms of the interferometer are brought together, and correspond to alternating conditions of constructive and destructive interference. A single fringe represents one wavelength of the light source. Associated with the time difference given by Eq. (P1.2), we should expect to see $N = (L/\lambda)\beta^2$ fringe shifts implied by the motion of Earth relative to the ether. The trouble is, we can't stop the earth to count fringe shifts. Staring through the telescope in the interferometer, there's no "zero" marking on the fringes, implying that $N = (L/\lambda)\beta^2$ can't be tested. Michelson came up with the idea of watching the fringe pattern as the apparatus is rotated; in that way one could actually *observe* the number of fringe shifts ΔN. In rotating the apparatus through $\pi/2$ radians, one should, using Eq. (P1.5), expect to see ΔN fringe shifts, where

$$\Delta N = \frac{c}{\lambda}\left(\Delta T(\pi/2) - \Delta T(0)\right) \approx \frac{\beta^2}{\lambda}\left(L_1 + L_2\right) .$$

As the interferometer is rotated continuously, Eq. (P1.6) gives the expected change in fringe shift per radian

$$\frac{\mathrm{d}}{\mathrm{d}\phi}\Delta N(\phi) = \frac{c}{\lambda}\frac{\mathrm{d}}{\mathrm{d}\phi}\Delta T(\phi) = \beta^2 \frac{L_1 + L_2}{\lambda}\sin 2\phi + O(\beta^4) \,. \tag{2.18}$$

In the MM experiment $L_1 + L_2 \approx 22$ m and the yellow light of sodium was used, $\lambda = 589$ nm. Michelson expected to see 0.4 fringe shift and he estimated he would have been able to observe 0.01 fringe shift. Figure 2.19 shows the fringe-shift measurements from the MM experiment (solid

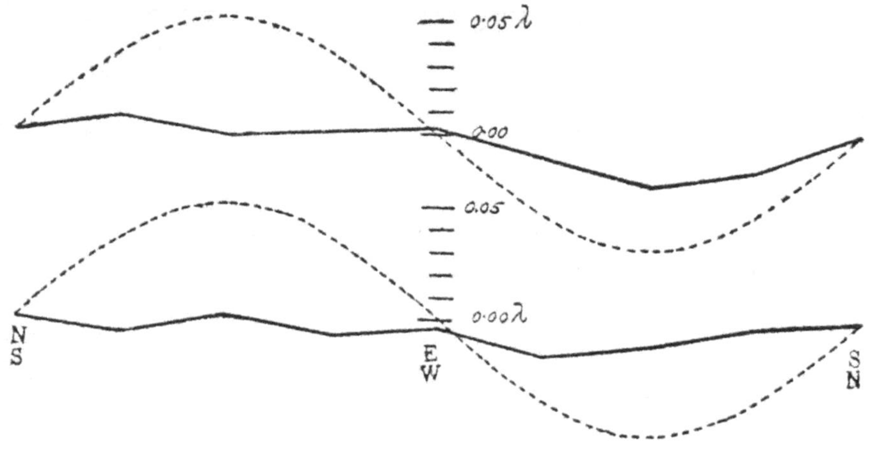

Figure 2.19 Fringe-shift measurements from the MM experiment (solid lines) versus orientation of the interferometer.[15] Dashed curve is Eq. (2.18) divided by eight. Upper data taken at noon, lower data taken in the evening. Reprinted by permission of the *American Journal of Science.*

lines), reported as fractions of a wavelength, for various orientations of the interferometer—upper data taken at noon, lower data taken in the evening.[15] The dashed curve is the result of Eq. (2.18) *divided by eight*. The number 0.05λ in Fig. 2.19 represents 0.4 fringe shift (what they expected to observe) divided by eight. Just to be clear: What they expected to observe would have been *eight times as large* as the dashed curve in Fig. 2.19. The area under the dashed curve, say from N (north) to E (east) is $\int_0^{\pi/2} \sin 2\phi \mathrm{d}\phi = 1$.

The data shown in Fig. 2.19 represents the first experiment in support of SR. Michelson and Morley stated about their results: "It seems fair to conclude from the figure that if there is any displacement due to the relative motion of the earth and the luminiferous ether, this amount cannot be much greater than 0.01 of the distance between the fringes" (which we note is in the noise of their measurements). They concluded: "It appears, from all that precedes, reasonably certain that if there be any relative motion between earth and the luminiferous ether, it must be small . . . ."

The MM experiment can be called the most successful failed experiment.[13] The experiment has been repeated with ever-increasing precision, but always with the same negative result. In 1979 a group reported [17] a fringe shift of $(1.5 \pm 2.5) \times 10^{-15}$, consistent with no effect at all. In 2009, a group reported [18] a measurement of the isotropy of the speed of light, $\Delta c/c \sim 10^{-17}$. It appears we're unable to detect motion relative to the ether, and if it can't be measured, does it exist?

[13]For which Albert Michelson was awarded the 1907 Nobel Prize in Physics!

2.6.2 The Kennedy-Thorndike experiment: $c \neq c(v)$

The Kennedy-Thorndike (KT) experiment [16] is a modification of the MM experiment where the arms of the interferometer are intentionally made as different as possible.[14] Let the longitudinal arm oriented in the direction of the earth's motion have length L_L, and let the transverse arm be oriented perpendicular to the longitudinal arm with length $L_T \neq L_L$.

To analyze the KT experiment, assume 1) the ether frame exists and 2) the FL contraction is real. We know these assumptions can be used to explain the results of the MM experiment; let's see how we do with the KT experiment. Take the lengths L_L and L_T to be measured in the ether frame.[15] The interferometer travels with speed v in the ether frame. We can use Eq. (P1.1) for the round-trip times, except now the longitudinal arm has the contracted length $L_L/\gamma(v)$. The traversal time in the longitudinal direction is then $T_L = 2L_L\gamma(v)/c$, and that for the transverse arm is $T_T = 2L_T\gamma(v)/c$ (no contraction in the transverse direction—Chapter 3). The number of fringe shifts produced as a consequence of the earth's motion relative to the ether is

$$N = f\left(T_L - T_T\right) = \frac{2\Delta L}{c} f\gamma(v) \,, \tag{2.19}$$

where f is the frequency of the light. Equation (2.19) has in common with the MM experiment that it can't be tested directly because we can't stop the earth. In the MM experiment the apparatus was rotated to detect a shift in fringe pattern. In the KT experiment, the apparatus was firmly fixed in the laboratory; the "rotation" is provided by Earth itself, either from its daily rotation or its annual orbit around the sun. The fringe shift between Earth having velocity $\boldsymbol{v}$ and $\boldsymbol{v}'$ is, from Eq. (2.19),

$$\Delta N = \frac{2\Delta L}{c} f\left(\gamma(v') - \gamma(v)\right) \,. \tag{2.20}$$

Like the MM experiment, the KT experiment produced a null result within experimental uncertainties, $\Delta N \approx 0$, which from Eq. (2.20) would not seem possible because $\boldsymbol{v}' \neq \boldsymbol{v}$. The experimental finding can be reconciled with the prediction of Eq. (2.20) *if we attribute to the ether a new ability*, that of altering the frequency of light in a velocity-dependent manner. For Eq. (2.20) to produce $\Delta N = 0$, it must be true that $f'\gamma(v') = f\gamma(v)$, where f' and f are frequencies that have been *altered* (by motion through the ether) relative to the frequency measured in the ether frame, call it f_0. That is, for Eq. (2.20) to produce a null result, it must be the case that $f'\gamma(v') = f\gamma(v) = f_0$.

On the basis of the ether model we require, to explain the MM experiment, the ability of the ether to contract objects in the direction of motion, and, to explain the KT experiment, the ability of the ether to modify the vibrational frequencies of systems in motion, all so that it (the ether) can evade detection! On the other hand, Einstein's hypothesis, that the speed of light is the same in all IRFs, naturally accounts for the MM experiment because there's no preferred orientation of IRFs, and the KT experiment because all IRFs in relative uniform motion are equivalent, *in addition to SR making numerous other predictions*. The MM experiment shows that the round-trip time required for light to cover a distance in free space is independent of direction, i.e., *the speed of light is isotropic*, while the KT experiment shows that the round-trip time for light to cover a distance is independent of the velocity of the source. Thus, *the speed of light is independent of the velocity of the source.* The ether model requires length contraction and time dilation for motion with respect to a *unique* reference frame, the ether, whereas in SR these relations are *symmetric* between IRFs: All inertial observers see rods in motion contracted and moving clocks run slow—Chapter 4.

[14] The MM experiment was carefully crafted to have equal-length arms.

[15] Such an assumption renders these quantities unknowable—that's the problem with absolute space that "makes no impression on our senses"—but we're assuming the ether frame to exist; no conclusion will depend on the value of ΔL.

SUMMARY

Spacetime diagrams were used to illustrate the basic effects of SR, the "ingredients" used in descriptions of processes in space and time: time dilation, length contraction, simultaneity, the Doppler effect, and the velocity addition formula. These phenomena are interrelated—time dilation in one frame can be explained as length contraction in another. It's not always clear which effect is most appropriate to use in analyzing a given problem. Can these effects be viewed from a unified perspective? Yes, and there are two ways of going about that. The first is to consider how *all* the spacetime coordinates change between IRFs; this is accomplished using the LT. With the LT at our disposal, we can focus on the relevant *events* involved in a given problem. We present a systematic derivation of the LT in the next chapter. The second way is to exploit relativistic *invariance*, to focus on what does *not* change between frames, the subject of Chapter 4.

EXERCISES

2.1 Derive Eq. (2.10) from Eq. (2.9) using Eq. (2.3) for each of the k-factors.

2.2 Derive Eq. (2.17). Use that $\beta_r = 2\beta/(1+\beta^2)$.

2.3 The *inverse* of Eq. (2.12) is found by setting $\beta \to -\beta$. Show that

$$\gamma^2 \begin{pmatrix} 1 & \beta \\ \beta & 1 \end{pmatrix} \begin{pmatrix} 1 & -\beta \\ -\beta & 1 \end{pmatrix} = \begin{pmatrix} 1 & 0 \\ 0 & 1 \end{pmatrix} .$$

2.4 Derive the k-factor from the Lorentz transformation. Referring to Fig. 2.20, a photon is emitted at time T and received in a frame moving away at time kT. The Lorentz transformation relates the coordinates assigned to the same event:

$$\begin{pmatrix} ct \\ x \end{pmatrix} = \gamma \begin{pmatrix} 1 & \beta \\ \beta & 1 \end{pmatrix} \begin{pmatrix} kcT \\ 0 \end{pmatrix} .$$

We've used the inverse transformation. The time t is the time T plus the time for the photon to travel the distance x, $t = T + x/c$. Show that k is given by Eq. (2.3).

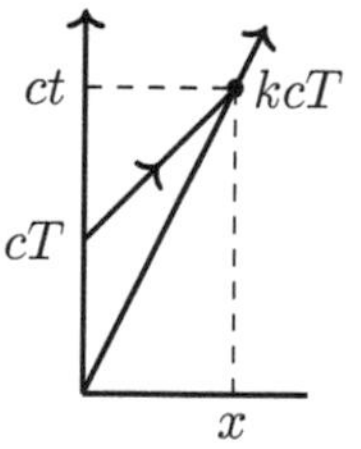

Figure 2.20 Derive the k-factor using the Lorentz transformation.

CHAPTER 3

Lorentz transformation, I

WE provide a systematic derivation of the Lorentz transformation[1] (LT) and examine its kinematical consequences, what can be said without taking into account the *causes* of motion. (Relativistic *dynamics* is taken up in Chapter 7.) We derive the LT first for frames in standard configuration (defined below) and then for frames not in standard configuration. Along the way, the velocity addition formula emerges as a bonus.

We make two assumptions about space and time (appropriate for SR), that space is *isotropic* (all directions are equivalent) and that spacetime is *homogeneous* (no location or instant of time is preferred).[2] These concepts are distinct: Isotropy does not necessarily imply homogeneity, nor does homogeneity necessarily imply isotropy. One could have homogeneous, anisotropic spaces (a crystalline environment, for example, where one direction is preferred over the others), and one could have inhomogeneous, isotropic spaces (all spatial directions equivalent, yet a special location of the origin—a set of concentric spheres about a specified origin). Remarkably, *these two assumptions together with the principle of relativity suffice to determine the LT.*

3.1 FRAMES IN STANDARD CONFIGURATION

Let IRFs S and S' be in relative motion with velocity $\boldsymbol{v}$. Whatever is the direction of $\boldsymbol{v}$, it is by assumption constant (IRF); let $\boldsymbol{v}$ *define* the direction of the x-axis. The observers synchronize their clocks when the origins of their coordinate systems coincide, i.e., where and when they have a common *spacetime origin*. Frames in *standard configuration* move along their common x-axis with their y and z-axes parallel, as shown in Fig. 3.1.

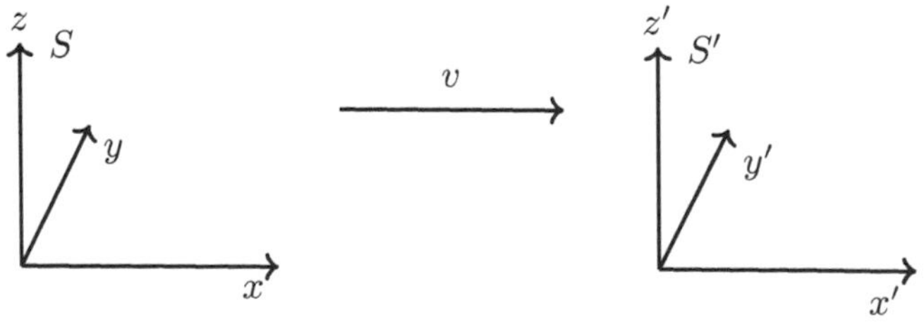

Figure 3.1 Frames in standard configuration. S' moves to the right along the common x-axis with y and z-axes parallel.

[1]The LT is derived in Appendix A as the linear transformation that preserves the form of the wave equation. The same is derived in Section 2.4 using the k-factor method. The more ways you have of looking at something, the better.

[2]In GR, spacetime is neither homogeneous nor isotropic; the gravitational field results from the curvature of spacetime.

3.1.1 General form of the Lorentz transformation

The LT, symbolized $L(v)$, is a linear mapping between the spacetime coordinates assigned to events by different inertial observers. All inertial observers see straight worldlines for free particles, and straight lines are preserved under homogeneous, linear mappings.[3] The most general linear homogeneous mapping between four-dimensional spaces has 16 parameters. For frames in standard configuration, that number can be reduced considerably by invoking homogeneity and isotropy. Write $L(v)$ as a 4×4 matrix containing four unknown functions of v, $\alpha(v), \delta(v), \gamma(v), \eta(v)$:

$$\begin{pmatrix} \theta t' \\ x' \\ y' \\ z' \end{pmatrix} = \left(\begin{array}{cc|cc} \alpha(v) & \delta(v)\theta & 0 & 0 \\ -v\gamma(v)/\theta & \gamma(v) & 0 & 0 \\ \hline 0 & 0 & \eta(v) & 0 \\ 0 & 0 & 0 & \eta(v) \end{array}\right) \begin{pmatrix} \theta t \\ x \\ y \\ z \end{pmatrix} \equiv L(v) \begin{pmatrix} \theta t \\ x \\ y \\ z \end{pmatrix} . \tag{3.1}$$

We've introduced in Eq. (3.1) an unknown parameter θ having the dimension of speed. We argued in Section 1.4 that the principle of relativity *requires* a universal speed, the same in all IRFs.[4] Let θ represent that speed; experiment will show that $\theta = c$.

We indicate mathematically that L is a mapping from the coordinates of S to those of S' with the notation $L : S \to S'$. That is, L associates an element of S with an element[5] of S'. Denote the matrix in Eq. (3.1) in block form: $\left(\begin{array}{c|c} A & B \\ \hline C & D \end{array}\right)$. Block $B = 0$ because of isotropy: We're free to orient the y and z-axes however we choose; the assignment of x' and t' can depend only on the relative speed and not on the orientation of y and z, otherwise clocks situated differently around the x-axis would show different times in violation of the assumption of isotropy. Block $C = 0$ because of homogeneity: The assignment of y' and z' can't depend on the choice of spacetime origin. Block D is diagonal because frames in standard configuration have parallel y and z-axes. The coefficients $\eta(v)$ are the same for y and z because of isotropy; we'll show that $\eta(v) = 1$. In block A there are functions $\alpha(v)$ and $\delta(v)$ in the equation for t', but only one independent function $\gamma(v)$ in the equation for x', because the location of $x' = 0$ (in S) is described by $x = vt$.

3.1.2 What if S' moves to the left?

We could equally well consider the motion of S' along the *negative* x-axis (Fig. 3.2), in which case

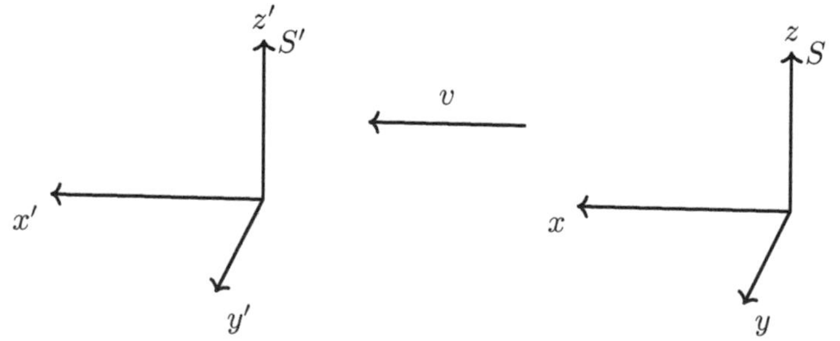

Figure 3.2 Motion of S' along negative x-axis of S.

the LT would follow from Eq. (3.1) by letting $t \to t$, $x \to -x$, $y \to -y$, $z \to z$, $t' \to t'$, $x' \to -x'$,

[3]That is, lines that pass through the origin.

[4]This is because, as we'll show, LTs have the property that a LT followed by a LT is itself a LT.

[5]The level of mathematical maturity is only going to increase from here on. Don't fight it; mathematics is in your future lightcone.

$y' \to -y'$, and $z' \to z$:

$$\begin{pmatrix} \theta t' \\ -x' \\ -y' \\ z' \end{pmatrix} = \begin{pmatrix} \alpha(-v) & \delta(-v)\theta & 0 & 0 \\ v\gamma(-v)/\theta & \gamma(-v) & 0 & 0 \\ 0 & 0 & \eta(-v) & 0 \\ 0 & 0 & 0 & \eta(-v) \end{pmatrix} \begin{pmatrix} \theta t \\ -x \\ -y \\ z \end{pmatrix} \equiv L(-v) \begin{pmatrix} \theta t \\ -x \\ -y \\ z \end{pmatrix} . \tag{3.2}$$

By changing the sign of x but keeping the *sense of time* unchanged (the "orientation" of time), $v \to -v$. We've changed the signs of y and y' for cosmetic purposes: to keep S and S' right-handed systems when we reverse the sense of the x and x'-axes. It's as if we've rotated Fig. 3.1 180 degrees about the z-axis in S to produce Fig. 3.2.

Equation (3.2) can be derived from Eq. (3.1) by defining a "reverse" operator,

$$R \equiv \begin{pmatrix} 1 & 0 & 0 & 0 \\ 0 & -1 & 0 & 0 \\ 0 & 0 & -1 & 0 \\ 0 & 0 & 0 & 1 \end{pmatrix} .$$

Apply R to Eq. (3.1):

$$R \begin{pmatrix} \theta t' \\ x' \\ y' \\ z' \end{pmatrix} = RL(v) \begin{pmatrix} \theta t \\ x \\ y \\ z \end{pmatrix} = RL(v)R^{-1}R \begin{pmatrix} \theta t \\ x \\ y \\ z \end{pmatrix} . \tag{3.3}$$

Comparing Eqs. (3.3) and (3.2), we require that $RL(v)R^{-1} = L(-v)$. By working out $RL(v)R^{-1}$ (do it!), we learn that $\alpha(v)$, $\gamma(v)$, and $\eta(v)$ must be *even* functions, whereas $\delta(v)$ is an *odd* function. An odd function of v can be written $f_{\text{odd}}(v) = v f_{\text{even}}(v)$. Let's represent $\delta(v)$ in terms of the even function $\alpha(v)$, $\theta\delta(v) = -v\alpha(v)/f(v)$, where $f(v)$ is an unknown even function having the dimension of speed. The mapping $L(v) : S \to S'$ can now be parameterized

$$\begin{pmatrix} \theta t' \\ x' \\ y' \\ z' \end{pmatrix} = \begin{pmatrix} \alpha(v) & -v\alpha(v)/f(v) & 0 & 0 \\ -v\gamma(v)/\theta & \gamma(v) & 0 & 0 \\ 0 & 0 & \eta(v) & 0 \\ 0 & 0 & 0 & \eta(v) \end{pmatrix} \begin{pmatrix} \theta t \\ x \\ y \\ z \end{pmatrix} = L(v) \begin{pmatrix} \theta t \\ x \\ y \\ z \end{pmatrix} . \tag{3.4}$$

We still have four functions of v to determine: $\alpha(v), f(v), \gamma(v), \eta(v)$.

3.1.3 Inverse transformation

If frame S sees S' moving away with speed v to the right, S' sees S moving away with speed v to the left. We'll call this the *inverse* transformation. By the principle of relativity, the mapping $S' \to S$ must be of the same form as $L(v)$ in Eq. (3.4), except for $v \to -v$ (see Fig. 3.3):

$$\begin{pmatrix} \theta t \\ x \\ y \\ z \end{pmatrix} = L(-v) \begin{pmatrix} \theta t' \\ x' \\ y' \\ z' \end{pmatrix} = L(-v)L(v) \begin{pmatrix} \theta t \\ x \\ y \\ z \end{pmatrix} , \tag{3.5}$$

where we've used Eq. (3.4). From Eq. (3.5), it must be the case that $L(-v)L(v) = I$, the identity mapping, and hence $L^{-1}(v) = L(-v)$. *The inverse LT is the original LT with the sign of the velocity reversed.* Working out $L(-v)L(v)$ (do it!), we find

$$L(-v)L(v) = \begin{pmatrix} \alpha\left[\alpha - v^2\gamma/(\theta f)\right] & (\alpha - \gamma)v\alpha/f & 0 & 0 \\ (v\gamma/\theta)(\gamma - \alpha) & \gamma\left[\gamma - v^2\alpha/(\theta f)\right] & 0 & 0 \\ 0 & 0 & \eta^2 & 0 \\ 0 & 0 & 0 & \eta^2 \end{pmatrix} . \tag{3.6}$$

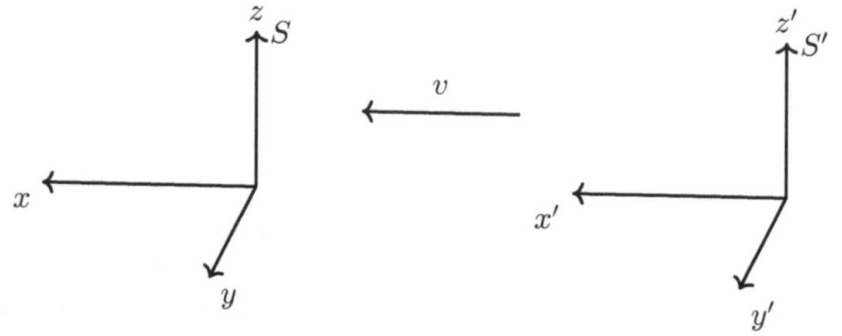

Figure 3.3 Inverse transformation: Motion of S along negative x'-axis.

For the right side of Eq. (3.6) to be the unit matrix we require $\eta(v) = \pm 1$. Because $\eta(0) = 1$, we have $\eta = 1$. Thus, *coordinates transverse to the motion are unaffected*. For the off-diagonal terms in Eq. (3.6) to vanish, we require $\alpha(v) = \gamma(v)$, implying that $\gamma(v) = \left[1 - v^2/(\theta f(v))\right]^{-1/2}$. Thus, the LT for frames in standard configuration has the form

$$L(v) = \begin{pmatrix} \gamma(v) & -v\gamma(v)/f(v) & 0 & 0 \\ -v\gamma(v)/\theta & \gamma(v) & 0 & 0 \\ 0 & 0 & 1 & 0 \\ 0 & 0 & 0 & 1 \end{pmatrix} . \tag{3.7}$$

There's still $f(v)$ and θ to determine.

3.1.4 Group property

All IRFs are equivalent. If we transform from S to S', and then from S' to S'', the net effect must be the same as a single transformation from S to S'', its *group property*. We'll show that the group property requires $f(v)$ to be a constant; it will also establish the Einstein velocity addition theorem. Using Eq. (3.7), transforming from S to S',

$$t' = \gamma(v_1)\left[t - v_1 x/(\theta f(v_1)\right] \qquad x' = \gamma(v_1)(x - v_1 t) , \tag{3.8}$$

where v_1 is the speed of S' as seen from S. Transforming from S' to S'',

$$t'' = \gamma(v_2)\left[t' - v_2 x'/(\theta f(v_2)\right] \qquad x'' = \gamma(v_2)(x' - v_2 t') , \tag{3.9}$$

where v_2 is the speed of S'' as seen from S'. Substitute Eq. (3.8) in Eq. (3.9). We find

$$\begin{aligned} t'' &= \gamma(v_1)\gamma(v_2)\left(1 + \frac{v_1 v_2}{\theta f(v_2)}\right)\left[t - \left(1 + \frac{v_1 v_2}{\theta f(v_2)}\right)^{-1}\left(\frac{v_1}{f(v_1)} + \frac{v_2}{f(v_2)}\right)\frac{x}{\theta}\right] \\ x'' &= \gamma(v_1)\gamma(v_2)\left(1 + \frac{v_1 v_2}{\theta f(v_1)}\right)\left[x - \left(1 + \frac{v_1 v_2}{\theta f(v_1)}\right)^{-1}(v_1 + v_2)t\right] . \end{aligned} \tag{3.10}$$

By the principle of relativity, Eq. (3.10) must be equivalent to a LT from S to S''. Equation (3.10) must therefore have the same form as Eq. (3.8) for some speed w, the speed of S'' as seen from S:

$$t'' = \gamma(w)\left[t - wx/(\theta f(w))\right] \qquad x'' = \gamma(w)(x - wt) . \tag{3.11}$$

The factors multiplying the square brackets in Eq. (3.10) must be identical (so that Eq. (3.10) has the same form as Eq. (3.11)), implying that $f(v_1) = f(v_2)$ or that $f(v)$ is a constant; call it f. With

$f(v) = f$, Eq. (3.10) simplifies:

$$t'' = \gamma(v_1)\gamma(v_2)\left(1 + \frac{v_1 v_2}{\theta f}\right)\left[t - \frac{x}{f\theta}\frac{v_1 + v_2}{1 + v_1 v_2/(\theta f)}\right]$$
$$x'' = \gamma(v_1)\gamma(v_2)\left(1 + \frac{v_1 v_2}{\theta f}\right)\left[x - \frac{v_1 + v_2}{1 + v_1 v_2/(\theta f)}t\right] . \tag{3.12}$$

Comparing Eq. (3.12) with Eq. (3.11) suggests that the compound speed w is given by

$$w = \frac{v_1 + v_2}{1 + v_1 v_2/(\theta f)} . \tag{3.13}$$

Equation (3.13) is more than a suggestion, however; it will be a *requirement* if it can be shown that

$$\gamma(w) = \gamma(v_1)\gamma(v_2)(1 + \frac{v_1 v_2}{\theta f}) \tag{3.14}$$

when w is given by Eq. (3.13). You're going to show (Exercise 3.1) that Eq. (3.14) is an *identity* when the compound speed is given by Eq. (3.13) for any θ and f. We therefore have the form of the velocity addition formula and the LT, except for the constants θ and f.

3.1.5 Existence of a limiting speed

The velocity addition formula Eq. (3.13) implies the existence of a *universal limiting speed*, which we denote for now as ψ. Let v_1 and v_2 both be equal to ψ. We have from Eq. (3.13)

$$w = \frac{2\psi}{1 + \psi^2/(\theta f)} .$$

In order for $w = \psi$, we must have $\psi^2 = \theta f$. If $v_1 = \psi$, Eq. (3.13) (with $\theta f = \psi^2$) implies $w = \psi$ for any v_2. If $v_1 = \psi - \mu_1$ and $v_2 = \psi - \mu_2$, with $\mu_1 \geq 0$ and $\mu_2 \geq 0$, Eq. (3.13) implies that $w \leq \psi$ for any μ_1, μ_2, with equality holding for μ_1 or μ_2 equal to zero or both (see Exercise 3.2).

It might seem that Eq. (3.13) implies three universal speeds, θ, f, and ψ. Simplicity emerges if $\theta = \psi$, which, because $\psi^2 = \theta f$, implies that $f = \theta$. In that case there is a symmetry in the LT—see Eq. (3.7)—the space and time variables transform in an equivalent way.

3.1.6 Value of the limiting speed

The value of the limiting speed must be found experimentally. Figure 3.4 shows four data points for measured speeds β and kinetic energies E_k of electrons.[19] The solid line represents the prediction of SR and the dashed line is the Newtonian prediction. We'll show (in Chapter 7) that kinetic energy is related to speed through the relation $E_k = (\gamma - 1)mc^2$, implying that

$$\beta^2 = 1 - \left(1 + E_k/mc^2\right)^{-2} ,$$

which is plotted in Fig. 3.4 as the solid curve. For low speeds $\beta^2 \approx 2E_k/mc^2$, which is shown as a dashed line. The data clearly show the existence of a limiting speed, in accord with the predictions of SR and completely at odds with Newtonian mechanics, with the limiting speed equal to the speed of light within experimental accuracy. This experiment was repeated at much higher energies, up to 20 GeV, with the limiting speed found to equal c within 2 parts in 10^7.[20] Note that for an energy of 20 GeV, the abscissa in Fig. 3.4 would extend to the right by a factor of 4000.

Taking $\theta = f = c$ as consistent with experiment, we have the LT from Eq. (3.8)

$$t' = \gamma(t - vx/c^2) \qquad x' = \gamma(x - vt) \qquad y' = y \qquad z' = z , \tag{3.15}$$

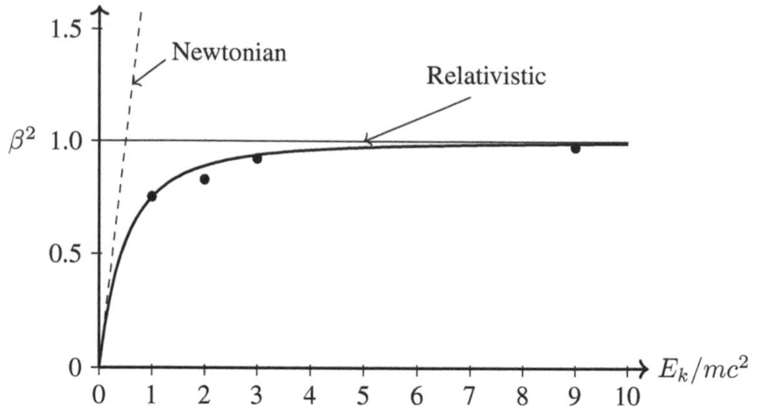

Figure 3.4 Measured speeds [19] (black dots) versus kinetic energy of electrons.

the same as Eq. (A.6) and Eq. (2.12), while the velocity addition formula from Eq. (3.13),

$$w = \frac{v_1 + v_2}{1 + v_1 v_2/c^2} , \tag{3.16}$$

the same as Eq. (2.10). Multiply by c, and the LT for frames in standard configuration is

$$\begin{pmatrix} ct' \\ x' \\ y' \\ z' \end{pmatrix} = \begin{pmatrix} \gamma & -\beta\gamma & 0 & 0 \\ -\beta\gamma & \gamma & 0 & 0 \\ 0 & 0 & 1 & 0 \\ 0 & 0 & 0 & 1 \end{pmatrix} \begin{pmatrix} ct \\ x \\ y \\ z \end{pmatrix} . \tag{3.17}$$

3.1.7 Why c? Do photons have mass?

The question naturally arises why the speed of light is the limiting speed. While there's no definitive answer, the only *particles* that travel at the speed of light are those with zero rest mass. In SR the connection between energy and momentum is, as we'll show, $E^2 = (pc)^2 + (mc^2)^2$. If $m = 0$, then

$$|\boldsymbol{p}| = E/c . \qquad (m = 0) \tag{3.18}$$

As we show in Chapter 7, in SR, $\boldsymbol{p} = \gamma m \boldsymbol{v}$ and $E = \gamma mc^2$. Eliminating γm between these equations, we have the general formula, valid for any m

$$\boldsymbol{p} = E\boldsymbol{v}/c^2 . \qquad (\text{any } m) \tag{3.19}$$

Equation (3.19) is compatible with Eq. (3.18) only if $|\boldsymbol{v}| = c$ for $m = 0$. Does the photon have a rest mass, m_γ? Experiment places an upper bound on a possible photon mass,[6] $m_\gamma < 10^{-18}\text{eV}/c^2$. While extremely small (24 orders of magnitude smaller than the electron mass, and 18 orders of magnitude smaller than the neutrino mass), if $m_\gamma \neq 0$ the speed of light would not be identical with the limiting speed ψ implied by the LT. Photons have momentum because they have energy, even though they have zero mass. That photons act as particles carrying energy and momentum is verified in Compton scattering experiments.

The LT contains a finite universal limiting speed (the same in all IRFs) which we've identified with the speed of light in vacuum. The universality of *c follows* from the principle of relativity; *it does not have to be postulated.* If $c \neq \psi$, photons have a nonzero rest mass, and c would not be universal. *Einstein's postulate of the universality of c is equivalent to the photon having zero mass.*

[6] A good source of information on the properties of elementary particles is the *Particle Data Group*, maintained online.

3.1.8 Discussion

Let's take a moment and review the essentials of the derivation just given. A linear mapping between four-dimensional spaces would have 16 parameters in general. For frames in standard configuration, that number reduces to four independent parameters when homogeneity and isotropy of spacetime are assumed, Eq. (3.4). When the principle of relativity is invoked, leading to $L^{-1}(v) = L(-v)$, that the mapping from $S \to S'$ is the same as that from $S' \to S$ (with $v \to -v$), the LT takes the form of Eq. (3.7) containing $f(v)$ and θ. Invoking the principle of relativity again, that a LT followed by a LT is itself a LT (all IRFs are equivalent), we find that $f(v) = f$, a constant. Comparison with experiment establishes that the limiting speed $\theta = f = c$, leaving us with the LT in the form of Eq. (3.17). That the LT can be derived under such general assumptions lends considerable support to the correctness of SR. In fact, it might lead one to wonder where all the non-intuitive "weirdness" associated with SR comes from; where did we take a "radical" step? It seems that the radical step, if it can be considered such, is in the inclusion of a separate time for each IRF, that time can no longer be considered absolute, that it too is relative to the frame of reference. Once we sign off on the idea that physics is most naturally viewed from the perspective of four-dimensional spacetime, the rest is the equivalence of IRFs, something that has long been known from the law of inertia. *SR is the law of inertia expressed in spacetime.*

3.2 FRAMES NOT IN STANDARD CONFIGURATION

Up to now our picture of reference frames in relative motion has been that of Fig. 3.1. Because the relative velocity $\boldsymbol{v}$ is constant (IRFs), frames in standard configuration suffice for many purposes, where the x-axis is aligned with $\boldsymbol{v}$. There are occasions, however, when we need the LT between reference frames having a more general relationship.

To derive the LT for a general boost (see Fig. 1.1), where $\boldsymbol{v}$ is not aligned with a coordinate axis, express the position vector $\boldsymbol{r}$ as a sum of vectors parallel and perpendicular to $\boldsymbol{v}$, $\boldsymbol{r} = \boldsymbol{r}_{\parallel} + \boldsymbol{r}_{\perp}$ (see Fig. 3.5). The vector $\boldsymbol{r}_{\parallel}$ is the projection of $\boldsymbol{r}$ onto $\boldsymbol{v}$,

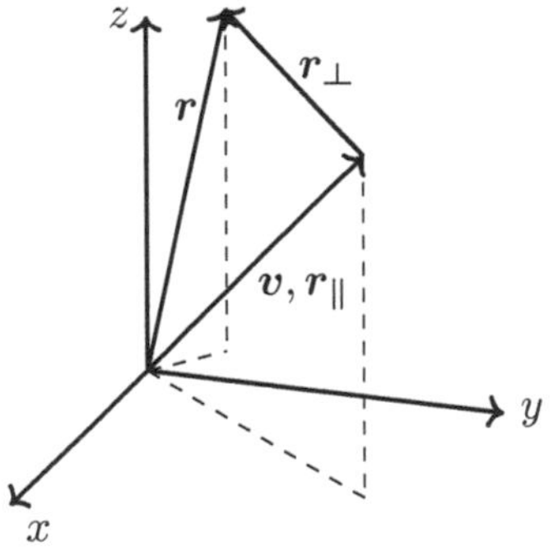

Figure 3.5 Decomposition of $\boldsymbol{r} = \boldsymbol{r}_{\parallel} + \boldsymbol{r}_{\perp}$ into vectors parallel and perpendicular to $\boldsymbol{v}$.

$$\boldsymbol{r}_{\parallel} = (\hat{\boldsymbol{v}} \cdot \boldsymbol{r})\,\hat{\boldsymbol{v}} = (\boldsymbol{v} \cdot \boldsymbol{r})\,\frac{\boldsymbol{v}}{v^2}\,, \tag{3.20}$$

where $\hat{\boldsymbol{v}} \equiv \boldsymbol{v}/v$ is a unit vector. The vector $\boldsymbol{r}_{\perp}$ is by definition $\boldsymbol{r}_{\perp} = \boldsymbol{r} - \boldsymbol{r}_{\parallel}$. For the components of $\boldsymbol{r}_{\parallel}$, the already known LT applies, while the components of $\boldsymbol{r}_{\perp}$ are unchanged. From Eq. (3.15),

$$t' = \gamma\left[t - \boldsymbol{v} \cdot \boldsymbol{r}_{\parallel}/c^2\right] \qquad \boldsymbol{r}'_{\parallel} = \gamma(\boldsymbol{r}_{\parallel} - \boldsymbol{v}t) \qquad \boldsymbol{r}'_{\perp} = \boldsymbol{r}_{\perp}\,. \tag{3.21}$$

We then have, using Eq. (3.21),

$$\boldsymbol{r}' = \boldsymbol{r}'_{\parallel} + \boldsymbol{r}'_{\perp} = \gamma(\boldsymbol{r}_{\parallel} - \boldsymbol{v}t) + \boldsymbol{r}_{\perp} = \gamma(\boldsymbol{r}_{\parallel} - \boldsymbol{v}t) + \boldsymbol{r} - \boldsymbol{r}_{\parallel} = \boldsymbol{r} + (\gamma - 1)\boldsymbol{r}_{\parallel} - \gamma\boldsymbol{v}t\,. \tag{3.22}$$

Combining Eq. (3.20) with Eqs. (3.21) and (3.22), we have the *vector form* of the LT,

$$\begin{aligned} t' =&\gamma\left[t - \boldsymbol{r}\cdot\boldsymbol{v}/c^2\right] \\ \boldsymbol{r}' =&\boldsymbol{r} + (\gamma-1)\frac{(\boldsymbol{r}\cdot\boldsymbol{v})\boldsymbol{v}}{v^2} - \gamma\boldsymbol{v}t = \boldsymbol{r} + \frac{\gamma^2}{c^2(1+\gamma)}(\boldsymbol{r}\cdot\boldsymbol{v})\boldsymbol{v} - \gamma\boldsymbol{v}t \\ =&\gamma(\boldsymbol{r}-\boldsymbol{v}t) + \frac{\gamma^2}{c^2(1+\gamma)}\boldsymbol{v}\times(\boldsymbol{v}\times\boldsymbol{r})\,, \end{aligned} \tag{3.23}$$

where we've used $\boldsymbol{A}\times(\boldsymbol{B}\times\boldsymbol{C}) = \boldsymbol{B}(\boldsymbol{A}\cdot\boldsymbol{C}) - \boldsymbol{C}(\boldsymbol{A}\cdot\boldsymbol{B})$ in the last line.

Referring all vectors to the (x, y, z) basis, Eq. (3.23) can be expressed as a matrix equation,

$$\begin{pmatrix} ct' \\ x' \\ y' \\ z' \end{pmatrix} = \begin{pmatrix} \gamma & -\beta_x\gamma & -\beta_y\gamma & -\beta_z\gamma \\ -\beta_x\gamma & 1+\alpha\beta_x^2 & \alpha\beta_x\beta_y & \alpha\beta_x\beta_z \\ -\beta_y\gamma & \alpha\beta_y\beta_x & 1+\alpha\beta_y^2 & \alpha\beta_y\beta_z \\ -\beta_z\gamma & \alpha\beta_z\beta_x & \alpha\beta_z\beta_y & 1+\alpha\beta_z^2 \end{pmatrix} \begin{pmatrix} ct \\ x \\ y \\ z \end{pmatrix}, \tag{3.24}$$

where $\alpha \equiv \gamma^2/(1+\gamma)$. If $\beta_y = \beta_z = 0$ and $\beta_x = \beta$, Eq. (3.24) reduces to Eq. (3.17). Note the symmetry of the matrix in Eq. (3.24), which arises because boosts connect frames having parallel coordinate axes.

Equation (3.24) is therefore *not* the most general LT, because it prescribes transformations among a particular class of reference frames—those having parallel coordinate axes. We show in Chapter 6 that an arbitrary LT can be represented as a rotation followed by a boost. Rotations are described by three angles and boosts are described by three velocity components. *The most general LT requires six parameters to be completely specified.*

3.3 TRANSFORMATION OF VELOCITY AND ACCELERATION

The LT is a linear mapping between the *coordinates* of IRFs in relative motion. Velocity and acceleration involve *ratios* of differences between space and time coordinates. We can use the LT to "build" the transformation equations for these quantities.[7]

3.3.1 Velocity transformation

Let S' move relative to S with constant velocity $\boldsymbol{v}$. Let $\boldsymbol{u} = \mathrm{d}\boldsymbol{r}/\mathrm{d}t$ be the velocity of a particle as seen in S and let $\boldsymbol{u}' = \mathrm{d}\boldsymbol{r}'/\mathrm{d}t'$ be the velocity of the same particle seen in S'. Form the differentials $\mathrm{d}\boldsymbol{r}'$ and $\mathrm{d}t'$ from Eq. (3.23) holding $\boldsymbol{v}$ constant:

$$\begin{aligned} \mathrm{d}\boldsymbol{r}' =&\gamma\left(\mathrm{d}\boldsymbol{r} - \boldsymbol{v}\mathrm{d}t\right) + \frac{\gamma^2}{c^2(1+\gamma)}\boldsymbol{v}\times(\boldsymbol{v}\times\mathrm{d}\boldsymbol{r}) \\ \mathrm{d}t' =&\gamma\left[\mathrm{d}t - \mathrm{d}\boldsymbol{r}\cdot\boldsymbol{v}/c^2\right] = \gamma\mathrm{d}t\left[1 - \boldsymbol{u}\cdot\boldsymbol{v}/c^2\right]\,. \end{aligned} \tag{3.25}$$

Divide $\mathrm{d}\boldsymbol{r}'$ by $\mathrm{d}t'$ in Eq. (3.25) to obtain the *velocity* transformation equation

$$\boldsymbol{u}' = \frac{\boldsymbol{u}-\boldsymbol{v}}{1-\boldsymbol{v}\cdot\boldsymbol{u}/c^2} + \frac{\gamma}{c^2(1+\gamma)}\frac{\boldsymbol{v}\times(\boldsymbol{v}\times\boldsymbol{u})}{(1-\boldsymbol{v}\cdot\boldsymbol{u}/c^2)}\,. \tag{3.26}$$

By decomposing $\boldsymbol{u} = \boldsymbol{u}_\parallel + \boldsymbol{u}_\perp$ into vectors parallel and perpendicular to $\boldsymbol{v}$, we obtain

$$\boldsymbol{u}'_\parallel = \frac{\boldsymbol{u}_\parallel - \boldsymbol{v}}{1-\boldsymbol{v}\cdot\boldsymbol{u}/c^2}\,, \tag{3.27}$$

[7] Which is to say, the velocity and acceleration vectors do not transform according to the LT because they are not four-vectors. In Chapter 7 we define velocity and acceleration four-vectors by differentiating the four spacetime coordinates with respect to the proper time; these vectors do transform with the LT.

while $\boldsymbol{u}_\perp$ transforms as

$$\boldsymbol{u}'_\perp = \frac{\boldsymbol{u}_\perp}{\gamma\left[1 - \boldsymbol{v}\cdot\boldsymbol{u}/c^2\right]} \,. \tag{3.28}$$

Whereas the *coordinates* transverse to $\boldsymbol{v}$ are left unchanged, $\boldsymbol{r}'_\perp = \boldsymbol{r}_\perp$, the same is not true for the transverse velocity components; time transforms between frames, time is not absolute.

The inverse of Eq. (3.26) provides a clean statement of velocity addition *in vector form*. Switch primes and unprimes, and let $\boldsymbol{v} \to -\boldsymbol{v}$:

$$\boldsymbol{u} = \frac{\boldsymbol{v} + \boldsymbol{u}'}{1 + \boldsymbol{v}\cdot\boldsymbol{u}'/c^2} + \frac{\gamma}{c^2(1+\gamma)} \frac{\boldsymbol{v}\times(\boldsymbol{v}\times\boldsymbol{u}')}{(1 + \boldsymbol{v}\cdot\boldsymbol{u}'/c^2)} \,. \tag{3.29}$$

Equation (3.29) specifies the resultant of adding the velocity of S' relative to S, $\boldsymbol{v}$, to the velocity seen in S', $\boldsymbol{u}'$.

3.3.2 Non-colinear velocities

Equation (3.29) differs from Eq. (3.16), which applies for *colinear* velocities (all in the same line). When $\boldsymbol{v}$ and $\boldsymbol{u}'$ are not colinear, new physical effects manifest.[8] For non-colinear velocities, there's an *asymmetry* in Eq. (3.29): The two velocities do not occur in the formula in a symmetrical manner. We define the *direct sum* of two velocities, which has ordered "slots" for the vectors being added,

$$\boldsymbol{v}_a \oplus \boldsymbol{v}_b \equiv \frac{\boldsymbol{v}_a + \boldsymbol{v}_b}{1 + \boldsymbol{v}_a\cdot\boldsymbol{v}_b/c^2} + \frac{\gamma_a}{c^2(1+\gamma_a)} \frac{\boldsymbol{v}_a\times(\boldsymbol{v}_a\times\boldsymbol{v}_b)}{1 + \boldsymbol{v}_a\cdot\boldsymbol{v}_b/c^2} \,, \tag{3.30}$$

where $\gamma_a \equiv 1/\sqrt{1 - v_a^2/c^2}$ is the Lorentz factor associated with $\boldsymbol{v}_a$. *The relativistic addition of velocities is not associative,*

$$\boldsymbol{v}_1 \oplus \boldsymbol{v}_2 \neq \boldsymbol{v}_2 \oplus \boldsymbol{v}_1 \,. \tag{3.31}$$

Only when the velocities are colinear does $\boldsymbol{v}_1 + \boldsymbol{v}_2 = \boldsymbol{v}_2 + \boldsymbol{v}_1$.

3.3.3 Acceleration transformation

The transformation equation for the acceleration vector $\boldsymbol{a} = \mathrm{d}\boldsymbol{u}/\mathrm{d}t$ can be obtained by differentiating Eq. (3.26) ($\boldsymbol{v}$ is constant),

$$\mathrm{d}\boldsymbol{u}' = \frac{1}{\gamma\left(1 - \boldsymbol{v}\cdot\boldsymbol{u}/c^2\right)^2}\left[\mathrm{d}\boldsymbol{u} - \frac{\gamma}{c^2(1+\gamma)}\left(\boldsymbol{v}\cdot\mathrm{d}\boldsymbol{u}\right)\boldsymbol{v} + \frac{1}{c^2}\boldsymbol{v}\times\boldsymbol{u}\times\mathrm{d}\boldsymbol{u}\right] \,. \tag{3.32}$$

Divide Eq. (3.32) by $\mathrm{d}t'$ in Eq. (3.25) to obtain

$$\boldsymbol{a}' = \frac{1}{\gamma^2\left[1 - \boldsymbol{v}\cdot\boldsymbol{u}/c^2\right]^3}\left[\boldsymbol{a} - \frac{\gamma}{c^2(1+\gamma)}\left(\boldsymbol{v}\cdot\boldsymbol{a}\right)\boldsymbol{v} + \frac{1}{c^2}\boldsymbol{v}\times\boldsymbol{u}\times\boldsymbol{a}\right] \,. \tag{3.33}$$

While Eq. (3.33) is a complicated expression, it suffices to note that $\boldsymbol{a}' \neq \boldsymbol{a}$! That alone tells us that $\boldsymbol{F} = m\boldsymbol{a}$ is not consistent with SR.[9] We show in Chapter 7 how to "fix up" Newton's second law to be relativistically correct. Equation (3.33) can be simplified by decomposing $\boldsymbol{a}$ into vectors parallel and perpendicular to $\boldsymbol{v}$, $\boldsymbol{a} = \boldsymbol{a}_\parallel + \boldsymbol{a}_\perp$. We find:

$$\boldsymbol{a}'_\parallel = \frac{\boldsymbol{a}_\parallel}{\gamma^3\left[1 - \boldsymbol{v}\cdot\boldsymbol{u}/c^2\right]^3} \qquad \boldsymbol{a}'_\perp = \frac{\boldsymbol{a}_\perp + \boldsymbol{v}\times(\boldsymbol{u}\times\boldsymbol{a})/c^2}{\gamma^2\left[1 - \boldsymbol{u}\cdot\boldsymbol{v}/c^2\right]^3} \,. \tag{3.34}$$

We'll use Eq. (3.34) in Chapter 12.

---

[8]The prime example is Thomas precession.

[9]We showed in Chapter 1 that $F = ma$ is invariant under the Galilean transformation.

3.4 RELATIVISTIC ABERRATION AND DOPPLER EFFECT

Let frames S and S' be in standard configuration—see Fig. 3.6. Let an object in S' have velocity $\boldsymbol{u}'$ in the x'-y' plane. Let $\boldsymbol{u}'$ be oriented to the x'-axis at angle θ' so that $u'_x = u' \cos\theta'$ and

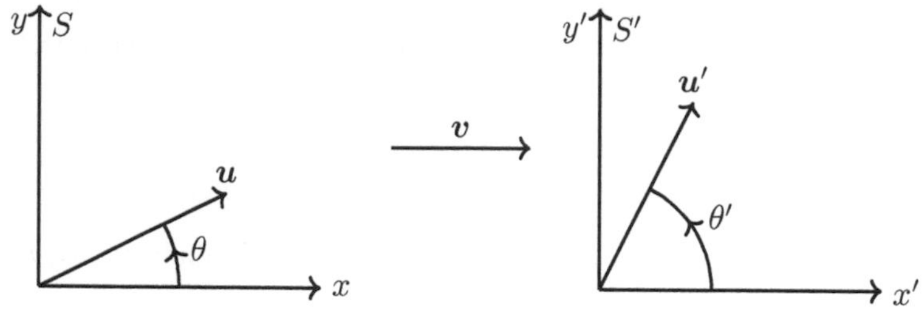

Figure 3.6 Relativistic aberration.

$u'_y = u' \sin\theta'$. What is the angle θ observed in S between the velocity $\boldsymbol{u}$ and the x-axis? That question can be answered using the velocity transformation equations.

Use the inverse transformations of Eqs. (3.27) and (3.28):

$$
\begin{aligned}
u_x = u\cos\theta &= \frac{u'_x + v}{1 + vu'\cos\theta'/c^2} = \frac{u'\cos\theta' + v}{1 + vu'\cos\theta'/c^2} \\
u_y = u\sin\theta &= \frac{u'_y}{\gamma\left(1 + vu'\cos\theta'/c^2\right)} = \frac{u'\sin\theta'}{\gamma\left(1 + vu'\cos\theta'/c^2\right)} .
\end{aligned}
\tag{3.35}
$$

Divide the two equations in Eq. (3.35),

$$
\frac{u_y}{u_x} = \tan\theta = \frac{u'\sin\theta'}{\gamma\left(v + u'\cos\theta'\right)} .
\tag{3.36}
$$

For a light ray in S', set $u' = c$ in which case Eq. (3.36) becomes

$$
\tan\theta = \frac{\sin\theta'}{\gamma(\beta + \cos\theta')} .
\tag{3.37}
$$

Equation (3.37) is the formula for the *relativistic aberration of light*. As $\beta \to 1$, the angle θ gets increasingly compressed into a cone of half-angle $\theta \approx \gamma^{-1}$, a phenomenon known as *relativistic beaming*. To show this, set $\beta = 1$ in Eq. (3.37), in which case we have, approximately,

$$
\tan\theta \approx \frac{1}{\gamma}\tan(\theta'/2) .
\tag{3.38}
$$

Irrespective of the emission angle θ', θ is compressed into γ^{-1} for sufficiently large γ.

Referring now to Fig. 3.7, suppose there's a source of light at rest at the origin of S' that emits signals with frequency f_e (as measured by an observer at rest relative to S'). The source emits signals into the direction θ', measured from the x'-axis. Let E'_1, E'_2 denote the events in S' at which successive light signals are emitted; the first at $t' = 0$ and the second at $\Delta T' \equiv f_e^{-1}$. In S, event E_1 (corresponding to E'_1) occurs at the origin at $t = 0$, for which a light ray is seen to be emitted into the direction θ, measured with respect to the x-axis. (The two frames synchronized their clocks as the origin of S' passed the origin of S.) Event E_2 in S (corresponding to E'_2) occurs at time $t_2 = \gamma\Delta T'$ (time dilation) at position $x_2 = \beta\gamma c\Delta T'$, at which another ray of light is seen to be emitted in S. (Figure 3.7 is not a spacetime diagram.[10]) In S, both rays are detected at a distant

[10]There are no spacetime diagrams in Chapter 3, which works with *three-vectors*, vectors in three spatial dimensions.

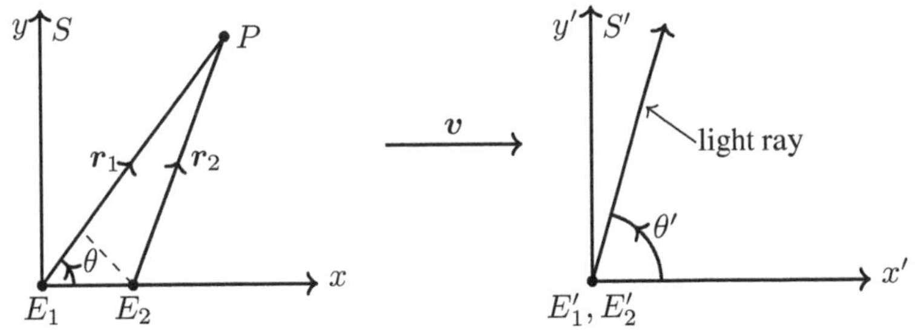

Figure 3.7 Two photons emitted from source at rest in S' in direction θ', observed in S.

location P. The question is, What is the time ΔT in S between the reception of the two signals? The first arrives at P at time $T_1 = r_1/c$, where r_1 is the distance of P to the origin. The second signal arrives at P at time $T_2 = \gamma\Delta T' + r_2/c$, where r_2 is the distance from P to the location x_2. Thus,

$$\Delta T = T_2 - T_1 = \gamma\Delta T' + \frac{1}{c}(r_2 - r_1) . \tag{3.39}$$

Assume that P is sufficiently distant that we can approximate $r_2 \approx r_1 - x_2 \cos\theta = r_1 - \beta\gamma c\Delta T' \cos\theta$. Thus, from Eq. (3.39),

$$\Delta T = \gamma\Delta T' (1 - \beta\cos\theta) . \tag{3.40}$$

Let $f_o \equiv (\Delta T)^{-1}$ denote the frequency observed in S; from Eq. (3.40)

$$f_o = \frac{f_e}{\gamma(1 - \beta\cos\theta)} = \gamma(1 + \beta\cos\theta') f_e , \tag{3.41}$$

where the second equality follows from the aberration formula; see Exercise 3.8b. Equation (3.41) is a general expression for the relativistic Doppler effect.

We'll show in Section 5.3.2 that both Eqs. (3.41) and (3.37) (Doppler shift and aberration) emerge as the result of a single LT involving an appropriately defined *four-vector*, a vector in spacetime. That is, the Doppler effect (involving time) and aberration (involving spatial directions) are two aspects of the same thing when viewed from the perspective of four-dimensional spacetime.

For $\theta' = \pi$ in Eq. (3.41) (radiation emitted against the direction of motion, source receding), we recover our previous result, Eq. (2.4), the *longitudinal Doppler effect*. For $\theta = \pi/2$ in Eq. (3.41) (radiation received in S orthogonal to the direction of motion), we have the *transverse Doppler effect*:

$$f_o = \frac{1}{\gamma} f_e . \tag{3.42}$$

The transverse Doppler effect is a *direct consequence* of the time dilation of a moving clock; there is no analogous effect in pre-relativistic physics. It was first measured in 1979.[21]

SUMMARY

We derived the LT for frames in standard configuration using the homogeneity and isotropy of spacetime, and the principle of relativity, Eq. (3.17). The theory predicts a limiting speed, which experiment shows is the speed of light. We derived the LT for a general boost—where the velocity does not line up with coordinate axes—in Eq. (3.24). The addition of non-colinear velocities $\boldsymbol{v}_1$ and $\boldsymbol{v}_2$ is not associative, $\boldsymbol{v}_1 + \boldsymbol{v}_2 \neq \boldsymbol{v}_2 + \boldsymbol{v}_1$.

EXERCISES

3.1 Show that Eq. (3.14) is an identity when the compound speed is given by Eq. (3.13), for any θ and f. Hint: Square Eq. (3.14) first. Don't forget the definition $\gamma(v) = \left[1 - v^2/(\theta f)\right]^{-1/2}$.

3.2 Show that Eq. (3.13), written in the form $w = \dfrac{v_1 + v_2}{1 + v_1 v_2/\psi^2}$, where ψ is the limiting speed, implies that $w \leq \psi$. Hint: Let $v_1 = \psi - \mu_1$ and $v_2 = \psi - \mu_2$, where $\mu_1 \geq 0$ and $\mu_2 \geq 0$.

3.3 Referring to Fig. 1.1, suppose that the vector $\boldsymbol{R}$ is time dependent, with $\boldsymbol{R} = \boldsymbol{v}t + \frac{1}{2}\boldsymbol{a}t^2$, where $\boldsymbol{v}$ and $\boldsymbol{a}$ are constant vectors. The two observers synchronize their clocks as the origins coincide. Suppose S is an IRF. Show that S' is not an IRF if $\boldsymbol{a} \neq 0$. Assume absolute time.

3.4 Derive Eq. (1.3). Let $\boldsymbol{r}' = \boldsymbol{r} - \boldsymbol{v}t$. Assume absolute time.

3.5 Derive Eq. (1.1). Show that if one frame in Fig 1.2 is an IRF, the other is as well. Thus, *there is no unique orientation to IRFs*. Note: ϕ is fixed here. One system is *rotated* with respect to the other, not *rotating*.

3.6 Write down the inverse transformation to Eq. (1.1). Show that $R_z(\phi)R_z(-\phi) = I_2$, the 2×2 identity matrix.

3.7 Show, under the transformation Eq. (1.1), that $(x')^2 + (y')^2 = x^2 + y^2$. That is, the distance to the origin (axis of rotation) is preserved under a rigid rotation of the coordinate axes.

3.8 a. Referring to Eq. (3.37), show that the aberration formula can be written

$$\cos\theta = \frac{\beta + \cos\theta'}{1 + \beta\cos\theta'} .$$

Hint: $\cos^2\theta = (1 + \tan^2\theta)^{-1}$

b. Show that $\gamma(1 + \beta\cos\theta') = [\gamma(1 - \beta\cos\theta)]^{-1}$.

3.9 Let $\boldsymbol{A} = A\hat{\boldsymbol{z}}$ be a velocity vector in a rectangular (x, y, z) coordinate system. Consider the vector $\boldsymbol{A}+\mathrm{d}\boldsymbol{A}$ where $\mathrm{d}\boldsymbol{A} = \mathrm{d}A_{\|}\hat{\boldsymbol{z}}+\mathrm{d}A_{\perp}\hat{\boldsymbol{y}}$ is a differential velocity with components parallel and perpendicular to $\boldsymbol{A}$. Define another differential velocity $\mathrm{d}\boldsymbol{w} \equiv \gamma_A^2 \mathrm{d}A_{\|}\hat{\boldsymbol{z}} + \gamma_A \mathrm{d}A_{\perp}\hat{\boldsymbol{y}}$, where $\gamma_A \equiv 1/\sqrt{1 - A^2/c^2}$. Using Eq. (3.30) show, to first order in small quantities, that $\boldsymbol{A} \oplus \mathrm{d}\boldsymbol{w} = \boldsymbol{A} + \mathrm{d}\boldsymbol{A}$.

CHAPTER 4

Geometry of Lorentz invariance

WHAT is the essence of SR? If you had to reduce relativity to a one-line description what might it be? The Bondi k-factor was derived in Chapter 2 using the principle of relativity (all inertial observers can claim themselves at rest, isotropy of the speed of light). The standard effects of SR were then derived using the k-factor, including the LT. Can it be said that the k-factor is the essence of SR? In Chapter 3 the LT was derived using the principle of relativity (all IRFs are equivalent), linearity (all inertial observers see straight worldlines for free particles), and isotropy and homogeneity of spacetime. Perhaps the LT is the heart of SR? The invariance of the spacetime separation follows from the LT, but can also be explained using the principle of relativity (all inertial observers claim they are at rest, all measure the same speed of light, Section 1.5). Amongst these interconnected ideas, can one be seen as more fundamental? As we continue in our study of relativity, we will have fewer opportunities to explicitly invoke the principle of relativity, and we'll rely progressively more on the use of Lorentz invariance. A *Lorentz invariant* is a quantity that remains unchanged under the LT, what all inertial observers find to be the same. Lorentz invariance brings to the fore Einstein's program for relativity that what is *not* relative has objective meaning.

In this chapter we look at the geometry of spacetime implied by the Lorentz invariance of the spacetime separation. We will promote invariance to a more fundamental status than the LT. Instead of saying that the invariant separation follows from the LT, the LT will be defined as any linear transformation that preserves the spacetime separation. If we had to come up with a "tagline" for SR, it might be the physics of the invariant separation in absolute spacetime. Such a description presages that for GR, which might be the physics of dynamic spacetime.

4.1 LORENTZ TRANSFORMATIONS AS SPACETIME ROTATIONS

In this section we show that boosts can be considered rotations in spacetime. To develop that idea we first consider rotations in Euclidean space and apply what we learn to LTs.

4.1.1 Active vs. passive transformations

Equation (1.1), which describes how the components of a two-dimensional vector transform upon rigidly rotating the x and y-axes counterclockwise through an angle ϕ (see Fig. 4.1), can be written

$$(\boldsymbol{r})' = R(\phi)\boldsymbol{r} \, . \tag{4.1}$$

Parentheses have been placed around $\boldsymbol{r}$ in Eq. (4.1) to indicate that $\boldsymbol{r}$ is the *same vector* before and after the transformation: only the *coordinates* have changed as a result of changing the coordinate

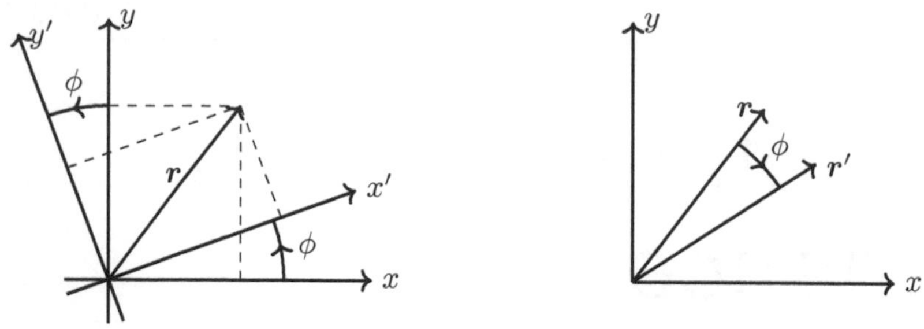

Figure 4.1 Left: Passive transformation—same vector $\boldsymbol{r}$, different coordinate system. Right: Active transformation—different vectors, $\boldsymbol{r}, \boldsymbol{r}'$, same coordinate system.

system (left portion of Fig. 4.1). A rotation can just as well be seen, however, as a transformation of the *vector*, changing not the coordinate system but changing $\boldsymbol{r}$ to a new vector $\boldsymbol{r}'$,

$$\boldsymbol{r}' = R(\phi)\boldsymbol{r} \,, \tag{4.2}$$

where the components of $\boldsymbol{r}$ and $\boldsymbol{r}'$ are expressed with respect to the same coordinate system (right portion of Fig. 4.1). In either case, the coordinates are transformed. The transformation in the form $(\boldsymbol{r})' = R(\phi)\boldsymbol{r}$ is a *passive transformation*, leaving the vector unchanged but changing the coordinate axes, while $\boldsymbol{r}' = R(\phi)\boldsymbol{r}$ is an *active transformation*, changing the vector with respect to the same coordinate system. It's not necessary to make the notational distinction in Eq. (4.1). Rotating the coordinate axes counterclockwise by the angle ϕ is equivalent to rotating the vector $\boldsymbol{r}$ *clockwise* through ϕ. The components of $\boldsymbol{r}'$ in an active rotation are the same as those of $(\boldsymbol{r})'$ in a passive rotation *except for the sign of the angle*: a positive angle in the active transformation is opposite to that for the passive.

4.1.2 Rotational symmetry

Symmetries have two aspects: a transformation and something invariant under the transformation. For rotations the distance to the rotation axis is preserved: $x'^2 + y'^2 = (x\cos\phi + y\sin\phi)^2 + (-x\sin\phi + y\cos\phi)^2 = x^2 + y^2$. The terms $x^2 + y^2$ can be generated from the inner product $\boldsymbol{r}^T\boldsymbol{r}$:

$$\boldsymbol{r}^T\boldsymbol{r} = \begin{pmatrix} x & y \end{pmatrix} \begin{pmatrix} x \\ y \end{pmatrix} = x^2 + y^2 \,. \tag{4.3}$$

Rotational invariance can then be expressed as $\boldsymbol{r}'^T\boldsymbol{r}' = \boldsymbol{r}^T\boldsymbol{r}$. Using Eq. (4.2),

$$\boldsymbol{r}^T\boldsymbol{r} = \boldsymbol{r}'^T\boldsymbol{r}' = \boldsymbol{r}^T R^T(\phi)R(\phi)\boldsymbol{r} \,, \tag{4.4}$$

where we've used for matrices A and B, $(AB)^T = B^T A^T$. Rotational symmetry *requires* of R

$$R^T(\phi)R(\phi) = I \,, \tag{4.5}$$

or that $R^T(\phi) = R^{-1}(\phi) = R(-\phi)$ (an orthogonal matrix). It can be seen explicitly from Eq. (1.1) that $R^T(\phi) = R(-\phi)$.

The inner product in Euclidean space is defined by $\boldsymbol{r}\cdot\boldsymbol{r} \equiv \boldsymbol{r}^T\boldsymbol{r} = x^2 + y^2$. Clearly $\boldsymbol{r}\cdot\boldsymbol{r} \geq 0$, where equality implies $\boldsymbol{r} = 0$. The length (*norm*) of $\boldsymbol{r}$ is defined as $||\boldsymbol{r}|| \equiv \sqrt{\boldsymbol{r}\cdot\boldsymbol{r}} = \sqrt{x^2 + y^2}$. From Eq. (4.4) the inner product is invariant under rotations, as is the norm, $||\boldsymbol{r}'|| = ||R(\phi)\boldsymbol{r}|| = ||\boldsymbol{r}||$. The invariance of the inner product generalizes to different vectors, $\boldsymbol{r}_1'\cdot\boldsymbol{r}_2' = \boldsymbol{r}_1\cdot\boldsymbol{r}_2$. The distance from $\boldsymbol{r}_1$ to $\boldsymbol{r}_2$ is also preserved, $||R(\phi)\boldsymbol{r}_1 - R(\phi)\boldsymbol{r}_2|| = ||R(\phi)(\boldsymbol{r}_1 - \boldsymbol{r}_2)|| = ||\boldsymbol{r}_1 - \boldsymbol{r}_2||$.

Even though the coordinates change under rotation, the norm and the inner product do not. Coordinates are relative to a coordinate system, but the norm and inner product are *absolute* quantities—they have the same value in all frames connected by rotation. Because these quantities are absolute, they have geometric meaning. *A geometry is characterized by its symmetries, its invariants.*

Seen as a passive transformation, the invariance of the inner product under rotations should not come as a surprise—it's the *same* vector before and after the transformation. Seen as an active transformation, rotations map circles onto themselves—an *invariant circle*; see Fig 4.2. Let $x = \cos\alpha$

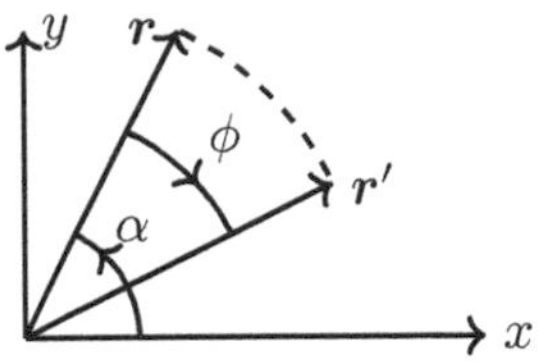

Figure 4.2 Invariant circle under active rotations (dashed line).

and $y = \sin\alpha$ be the coordinates for a point on the unit circle. The rotation matrix $R(\phi)$ maps (x, y) into (x', y'), where

$$\begin{pmatrix} x' \\ y' \end{pmatrix} = \begin{pmatrix} \cos\phi & \sin\phi \\ -\sin\phi & \cos\phi \end{pmatrix} \begin{pmatrix} \cos\alpha \\ \sin\alpha \end{pmatrix} = \begin{pmatrix} \cos\alpha\cos\phi + \sin\alpha\sin\phi \\ \sin\alpha\cos\phi - \cos\alpha\sin\phi \end{pmatrix} = \begin{pmatrix} \cos(\alpha - \phi) \\ \sin(\alpha - \phi) \end{pmatrix} .$$

The transformed point in the active rotation lies on the unit circle at the angle $\alpha - \phi$. *All possible rotations in the Euclidean plane about a fixed axis are represented by the mapping of a point on a circle into another point on the same circle.*

4.1.3 Invariance of the spacetime separation under Lorentz transformations

Under LTs the spacetime separation from the origin is preserved:

$$-(ct')^2 + x'^2 + y'^2 + z'^2 = -(ct)^2 + x^2 + y^2 + z^2 . \tag{1.7}$$

In analogy with rotations, therefore, boost transformations can be considered rotations in spacetime, even though they can't be visualized as such. Rotations in Euclidean space are the result of twisting around an *axis of rotation*. About what axis are we twisting in implementing a LT? It can't be visualized. If we generalize the concept of rotation to be *a mapping affecting pairs of coordinate axes* (x and y, for example), the LT is a rotation in that sense. In four-dimensional spacetime, there are $\binom{4}{2} = 6$ pairs of axes: three pair the time axis with a spatial axis, and the other three involve space-space pairs of axes.[1] We show in Chapter 6 that the most general LT is characterized by six independent parameters pertaining to mappings of the six possible pairs of coordinate axes in four-dimensional spacetime.

Equation (1.7) specifies that for an event with spacetime coordinates (ct, x, y, z) in one frame, then for the coordinates of the *same event* in another IRF, (ct', x', y', z'), the two sets of coordinates are such that Eq. (1.7) is satisfied.[2] For frames in standard configuration, the transverse coordinates are unaffected by the motion, and in that case $-(ct)^2 + x^2$ is an invariant, one that's easier to see the geometric meaning of.

[1] We're using the binomial coefficient, "N choose k," $\binom{N}{k} = N!/(k!(N-k)!)$. In three dimensions there are $\binom{3}{2} = 3$ pairs of coordinate axes, the same number as the dimension of the space, and thus we can associate a three-dimensional vector with a rotation. The case of three dimensions is special, however: $n = 3$ is the non-trivial solution of $\binom{n}{2} = n$. Only in three dimensions can we associate a rotation with a vector in the same space. The axis of rotation in a four-dimensional space would be a vector in a six-dimensional space.

[2] Such a statement is possible only if the two frames have a common spacetime origin.

The hyperbolic form of the LT (for frames in standard configuration) is, from Eq. (A.5):

$$\begin{pmatrix} ct' \\ x' \end{pmatrix} = \begin{pmatrix} \cosh\theta & \sinh\theta \\ \sinh\theta & \cosh\theta \end{pmatrix} \begin{pmatrix} ct \\ x \end{pmatrix} \equiv L(\theta) \begin{pmatrix} ct \\ x \end{pmatrix} , \tag{4.6}$$

where $\tanh\theta = -\beta$. Using Eq. (4.6),

$$\begin{aligned} -(ct')^2 + x'^2 &= -(ct\cosh\theta + x\sinh\theta)^2 + (ct\sinh\theta + x\cosh\theta)^2 \\ &= (\cosh^2\theta - \sinh^2\theta)(-(ct)^2 + x^2) = -(ct)^2 + x^2 . \end{aligned} \tag{4.7}$$

For given coordinates (ct, x), compute the number $k \equiv -(ct)^2 + x^2$, where k can be positive, negative, or zero (contrast with the Euclidean distance, which is always positive). Equation (4.7) shows that $-(ct')^2 + x'^2 = k$ for *all possible* LTs starting with (ct, x). The locus of points such that $-(ct)^2 + x^2 = k$ is a hyperbola; see Fig. 4.3. *The active form of the LT maps hyperbolas onto themselves*, the *invariant hyperbola*. The asymptotes are the lightlines $ct = \pm x$.

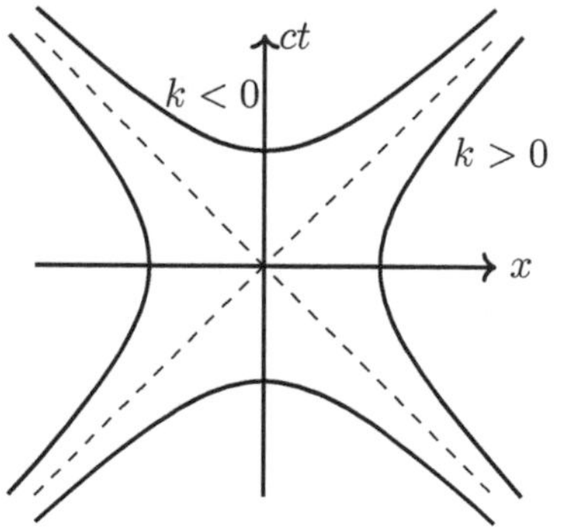

Figure 4.3 Invariant hyperbola $-(ct)^2+x^2 = k$. For $k > 0$ or $k < 0$ there are two branches.

Figure 4.4 shows the LT from the passive and active points of view. As a passive transformation

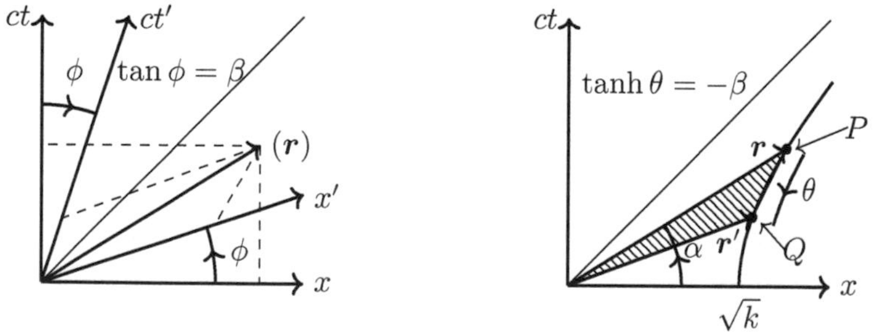

Figure 4.4 Passive and active forms of the Lorentz transformation.

the same spacetime point has coordinates in two reference frames (left portion of Fig. 4.4). To understand the active transformation, consider the hyperbola associated with $k > 0$. As shown in the right portion of Fig. 4.4, the spacetime vector $\boldsymbol{r}$ intersects the hyperbola at point P with coordinates $ct = \sqrt{k}\sinh\alpha$ and $x = \sqrt{k}\cosh\alpha$, where α is a *hyperbolic angle*, the angle between

$\boldsymbol{r}$ and the x-axis.[3] The LT maps the point (ct, x) into (ct', x'), where

$$\begin{pmatrix} ct' \\ x' \end{pmatrix} = \begin{pmatrix} \cosh\theta & \sinh\theta \\ \sinh\theta & \cosh\theta \end{pmatrix} \begin{pmatrix} \sqrt{k}\sinh\alpha \\ \sqrt{k}\cosh\alpha \end{pmatrix} = \sqrt{k} \begin{pmatrix} \sinh(\alpha+\theta) \\ \cosh(\alpha+\theta) \end{pmatrix} .$$

The transformed coordinates (relative to the same coordinate axes) are to be found on the hyperbola with $ct' = \sqrt{k}\sinh(\alpha+\theta)$ and $x' = \sqrt{k}\cosh(\alpha+\theta)$, i.e., at the hyperbolic angle $\alpha+\theta$, the point Q in Fig. 4.4. A boost is therefore a rotation[4] along the invariant hyperbola through the angle θ, where $\tanh\theta = -\beta$.

4.2 KINEMATIC EFFECTS FROM THE INVARIANT HYPERBOLA

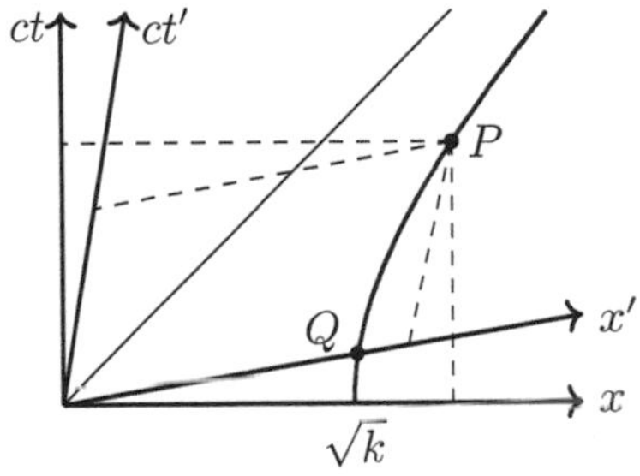

Figure 4.5 P lies on the hyperbola $-(ct)^2 + x^2 = k$ which intersects the x-axis for all IRFs at the same values $x' = \sqrt{k}$.

Event P in Fig. 4.5 has coordinates (ct, x) in one frame and (ct', x') in another. P lies on the hyperbola $-(ct)^2 + x^2 = k$ for some $k > 0$. The hyperbola intersects the x-axis ($t = 0$) at $x = \sqrt{k}$, and the x'-axis ($t' = 0$) *also* at $x' = \sqrt{k}$ (point Q). Because the hyperbola is Lorentz invariant, it intersects the x'-axis associated with *any* LT (starting from (ct, x)) at the *same value*, $x' = \sqrt{k}$.

The invariant hyperbola can be used to illustrate time dilation and length contraction. Referring to Fig. 4.6, IRFs S and S' in relative motion synchronize their clocks when the origins coincide. The worldlines of the clocks consist of the time axes, t and t'. Let the clock in S' show one unit of time at event B' which is where the unit hyperbola ($k = -1$) intersects the t'-axis. The hyperbola intersects the t-axis at event A, which is also one unit of time for the clock in S. In S, event B is simultaneous with B' (draw a line parallel to the x-axis). Because $B > A$, S concludes that the moving clock in S' runs slow. As seen from S', however, event A' is simultaneous with A (draw a line parallel to the x'-axis). Because $A' > B'$, S' concludes that the clock in S runs slow. *Both observers conclude that a clock in motion runs slow.*

Referring again to Fig. 4.6, let there be a rigid rod in S', in motion relative to S. At the instant the back end of the rod is at the origin of both frames, the front end is at Q' in S', where the unit hyperbola ($k = 1$) intersects the x'-axis. (S' measures Q' as the length of the rod at $t' = 0$.) The hyperbola intersects the x-axis at P, also at one unit of length. The worldline of the front end of the rod intersects the x-axis at Q, which is the measured length in S (two ends of the rod at the same time, $t = 0$). Because $Q < P$, S concludes that the moving rod is contracted in length. Now let the rod be at rest in S, in motion relative to S'. S measures the length of the rod as P (both ends at the

[3]Hyperbolic angle is measured in *hyperbolic radians*, twice the area enclosed by the vector $\boldsymbol{r}$, the unit hyperbola, and the x-axis, similar to a circular radian which is twice the area between $\boldsymbol{r}$, the unit circle, and the x-axis.[22, p444] The range of hyperbolic angle is unlimited, as we see from $\tanh\theta = -\beta$.

[4]The non-associativity of velocity addition (Section 3.3) is a consequence of the *non-commutativity* of rotations in four-dimensional spacetime. As is well known, three-dimensional rotations do not commute.

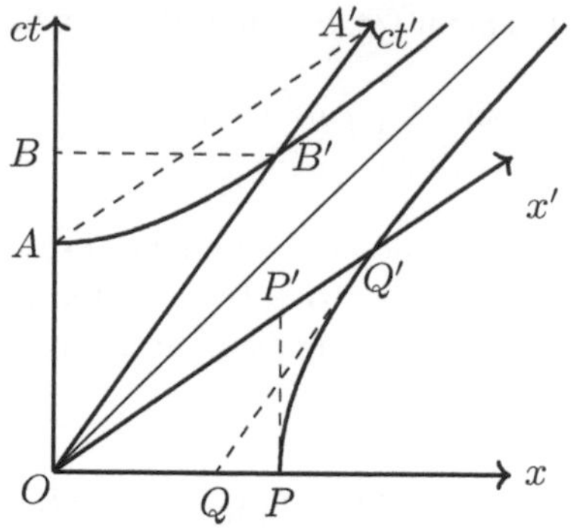

Figure 4.6 Length contraction and time dilation using the invariant hyperbola.

same time, $t = 0$). The worldline of the front edge of the rod intersects the x'-axis at P', which is the length measured by S' (both ends of the rod at the same time, $t' = 0$.) Because $P' < Q'$, S' concludes that a moving rod is contracted in length. *Both observers conclude that a rod in motion is contracted.*

Length contraction and time dilation are thus *symmetric* between IRFs. That's quite possibly the key difference between SR and the ether model (Section 2.6), wherein length contraction and time dilation purportedly occur relative to the ether in order to explain the MM and KT experiments, properties attributed to the ether for the sole purpose of allowing it to evade detection! In SR, length contraction and time dilation are relations between coordinates assigned to events by *any* two IRFs.

4.3 CLASSIFICATION OF LORENTZ TRANSFORMATIONS

We now define a LT to be *any linear transformation that preserves the spacetime separation.* To see how that comes about, define a *time inversion operator* T that maps $t \to -t$ and leaves the spatial coordinates unchanged, a matrix representation of which is:

$$T = \begin{pmatrix} -1 & 0 & 0 & 0 \\ 0 & 1 & 0 & 0 \\ 0 & 0 & 1 & 0 \\ 0 & 0 & 0 & 1 \end{pmatrix} . \tag{4.8}$$

With T, the spacetime separation can be generated from $\boldsymbol{r}^T T \boldsymbol{r} = -(ct)^2 + x^2 + y^2 + z^2$. With the coordinates in another IRF obtained from the LT, $(\boldsymbol{r})' = L(v)\boldsymbol{r}$, we have the invariance of the separation expressed in the form $r'^T T r' = r^T T r$, implying that $r'^T T r' = r^T L^T T L r = r^T T r$. We thus have a requirement on any linear transformation L that preserves the spacetime separation,

$$L^T T L = T \, . \tag{4.9}$$

Equation (4.9) is analogous to the orthogonality condition for Euclidean rotations, Eq. (4.5), which could be written $R^T I R = I$ to make it appear like Eq. (4.9). Instead of the operator T, however, we could just as well define the *parity operator*, which inverts the spatial coordinates and leaves time unchanged:

$$P = \begin{pmatrix} 1 & 0 & 0 & 0 \\ 0 & -1 & 0 & 0 \\ 0 & 0 & -1 & 0 \\ 0 & 0 & 0 & -1 \end{pmatrix} . \tag{4.10}$$

Equation (1.7) would then be generated by the invariance of $r^T Pr$, implying

$$L^T PL = P \,. \tag{4.11}$$

Equations (4.9) and (4.11) each impose a requirement on transformations that preserve the spacetime separation. Which should we use? It's traditional in the relativity literature to write

$$L^T \eta L = \eta \,, \tag{4.12}$$

where η is a diagonal matrix with *either* $(-1,1,1,1)$ on the diagonal, symbolized diag$(-1,1,1,1)$, *or* diag$(1,-1,-1,-1)$. Both conventions are in prevalent use. We will use $\eta = \text{diag}(-1,1,1,1)$, Eq. (1.14).[5] The matrix η is the *Lorentz metric tensor. Any linear transformation satisfying Eq.* (4.12) *is a LT.*

Transformations satisfying Eq. (4.12) have the mathematical property of constituting a *group*, the *Lorentz group*. Without venturing unduly into group theory, the four properties a set of elements must have to be a group are easily demonstrated:[6]

- If L_1 and L_2 are LTs, so is the composition $L \equiv L_1L_2$. From Eq. (4.12), $L_2^T L_1^T \eta L_1 L_2 = L_2^T \eta L_2 = \eta$. A LT followed by a LT is itself a LT, a manifestation of the principle of relativity—all IRFs are equivalent.[7]
- The composition law for LTs (matrix multiplication) is associative: $(L_1L_2)L_3 = L_1(L_2L_3)$.
- There exists an identity element $L = I$, which qualifies as a LT, $I\eta I = \eta$.
- For each L there exists L^{-1} in the group, which is more difficult to prove, although it must be so physically by the principle of relativity. Take the determinant of Eq. (4.12). Using the rules of determinants, including $\det L^T = \det L$, we find $(\det L)^2 = 1$ and hence $\det L = \pm 1$. Because $\det L \neq 0$, L^{-1} exists. To show that it belongs to the group, multiply Eq. (4.12) from the left by $\left(L^T\right)^{-1}$ and from the right by L^{-1}: $\left(L^T\right)^{-1} L^T \eta L L^{-1} = \left(L^T\right)^{-1} \eta L^{-1}$, and thus $\eta = \left(L^{-1}\right)^T \eta L^{-1}$ because for any matrix $\left(L^T\right)^{-1} = \left(L^{-1}\right)^T$. Hence, L^{-1} is a LT.

For later use with tensor analysis, it will be useful to adopt a notation for the elements of the LT matrix, the utility of which will become apparent in the next chapter. Denote the elements of the LT matrix as $L^{\mu}_{\ \nu}$, where the top (bottom) index labels rows (columns), and where Greek letters conventionally have the range $(0,1,2,3)$, with 0 labeling the time coordinate (Section 1.3).

It can be shown (Exercise 4.1) that $\left(L^0_{\ 0}\right)^2 \geq 1$. There are then four possible types of LT: $L^0_{\ 0} \geq 1$ (*orthochronous*), $L^0_{\ 0} \leq -1$ (*non-orthochronous*), $\det L = \pm 1$ (*proper* or *improper*). These constitute four categories of LTs, conventionally denoted as follows.

- $L^{\uparrow}_{+}$: $\det L = +1, L^0_{\ 0} \geq 1$, proper, orthochronous LTs.
  The requirement $\det L = +1$ (proper LTs) excludes the possibility of P or T as LTs. As a result, LTs $\in L^{\uparrow}_{+}$ connect smoothly with the identity transformation as the transformation parameter (speed or rotation angle) continuously goes to zero. The requirement $L^0_{\ 0} \geq 1$ maps positive time onto positive time, "orthochronous."
- $L^{\uparrow}_{-}$: $\det L = -1, L^0_{\ 0} \geq 1$, improper, orthochronous LTs.
  This class allows for the possibility of P as a LT, but excludes T.

[5] I prefer $(-1,1,1,1)$ because it singles out time for special treatment, and relativity is all about time.

[6] I say without *unduly* venturing into group theory because it's a vast subject. Much you may try to resist it, "der Gruppenpest" is an essential part of theoretical physics, which, if studied long enough, will entail picking up at least a nodding acquaintance with group theory. Groups are defined in almost any undergraduate book on algebra.

[7] We showed explicitly in Section 3.1.4 that the group property is satisfied among frames in standard configuration; here we're establishing it for any linear transformation that satisfies Eq. (4.12).

- $L_-^\downarrow$: $\det L = -1, L^0_0 \leq -1$, improper, non-orthochronous LTs.
  This class includes T but excludes P.

- $L_+^\downarrow$: $\det L = +1, L^0_0 \leq -1$, proper, non-orthochronous LTs.
  This class allows for the combined operation PT, inversion of time *and* space.

Only LTs $\in L_+^\uparrow$ are elements of the Lorentz group, because only this class includes the identity transformation. As a result, only LTs $\in L_+^\uparrow$ can be built up out of infinitesimal LTs (Chapter 6). It can be shown that $L \in L_-^\downarrow$ can be written as the product TL', with $L' \in L_+^\uparrow$. Thus we have the mapping $TL_+^\uparrow \to L_-^\downarrow$. Similarly, we have the mappings $PL_+^\uparrow \to L_-^\uparrow$ and $TPL_+^\uparrow \to L_+^\downarrow$. All possible LTs can therefore be obtained from $L \in L_+^\uparrow$ and the discrete transformations T and P.[8]

4.4 SPACETIME GEOMETRY AND CAUSALITY

4.4.1 Vector norm

A geometry involves the ability to specify the distance between points, and a natural way to do that is through the *inner product* between vectors—which allows one to assign a magnitude to vectors. We *define* the inner product between spacetime position vectors[9] using the Lorentz metric:

$$\begin{aligned} \boldsymbol{r}\cdot\boldsymbol{r} \equiv \boldsymbol{r}^T\eta\boldsymbol{r} &= \begin{pmatrix} ct & x & y & z \end{pmatrix} \begin{pmatrix} -1 & 0 & 0 & 0 \\ 0 & 1 & 0 & 0 \\ 0 & 0 & 1 & 0 \\ 0 & 0 & 0 & 1 \end{pmatrix} \begin{pmatrix} ct \\ x \\ y \\ z \end{pmatrix} = \begin{pmatrix} -ct & x & y & z \end{pmatrix} \begin{pmatrix} ct \\ x \\ y \\ z \end{pmatrix} \\ &= -(ct)^2 + x^2 + y^2 + z^2 \,. \end{aligned} \tag{4.13}$$

In this way $\boldsymbol{r}\cdot\boldsymbol{r}$ is a Lorentz invariant: For $(\boldsymbol{r})' = L(v)\boldsymbol{r}$,

$$\boldsymbol{r}'\cdot\boldsymbol{r}' \equiv (L(v)\boldsymbol{r})^T\eta L(v)\boldsymbol{r} = \boldsymbol{r}^T L^T(v)\eta L(v)\boldsymbol{r} = \boldsymbol{r}^T\eta\boldsymbol{r} \equiv \boldsymbol{r}\cdot\boldsymbol{r}\,,$$

where we've used Eq. (4.12). The invariance of the dot product extends to different position vectors, $\boldsymbol{r}_1'\cdot\boldsymbol{r}_2' = \boldsymbol{r}_1\cdot\boldsymbol{r}_2$. The invariant inner product is highly useful in practice: If its value is known in one IRF, it has the same value in any other frame obtained from the first by a LT. Just as with rotations in Euclidean space, the invariance of the inner product among spacetime vectors is a property of the geometry. While coordinates transform between IRFs, there is an intrinsic property of the spacetime geometry, the inner product, which has been defined to generate the spacetime separation.

The *norm* of a spacetime vector can now be defined,

$$||\boldsymbol{r}|| \equiv \begin{cases} \sqrt{\boldsymbol{r}\cdot\boldsymbol{r}} & \text{for } \boldsymbol{r} \text{ spacelike } (\boldsymbol{r}\cdot\boldsymbol{r} > 0) \\ \sqrt{-\boldsymbol{r}\cdot\boldsymbol{r}} & \text{for } \boldsymbol{r} \text{ timelike } (\boldsymbol{r}\cdot\boldsymbol{r} < 0) \\ 0 & \text{for } \boldsymbol{r} \text{ lightlike } (\boldsymbol{r}\cdot\boldsymbol{r} = 0) \end{cases} . \tag{4.14}$$

A vector with unit norm is a *unit vector*. Note that in contrast with Euclidean geometry, where $\boldsymbol{r}\cdot\boldsymbol{r} = 0$ implies $\boldsymbol{r} = 0$, a spacetime vector can have zero norm and be *non-zero*. A vector with zero norm is called a *null vector*. Because the norm has been defined using the inner product, it too is a Lorentz invariant.

Spacetime separations are non-intuitive. The analog of the "3-4-5" right triangle in Euclidean geometry is in spacetime geometry a "3-5-4" triangle (see Fig. 4.7), an instance of time dilation—

[8] In quantum field theory, systems with Lorentz symmetry must also have CPT symmetry—that the physics is unaffected by the combined transformation CPT where C ("charge conjugation") converts a particle into its antiparticle.

[9] The symbol $\boldsymbol{r}$ has been used to denote two-dimensional vectors, as in Eq. (4.3), three-dimensional vectors, as in Eq. (3.23), and now as a four-dimensional vector. Soon we'll refer to a spacetime position vector as a *four-vector*.

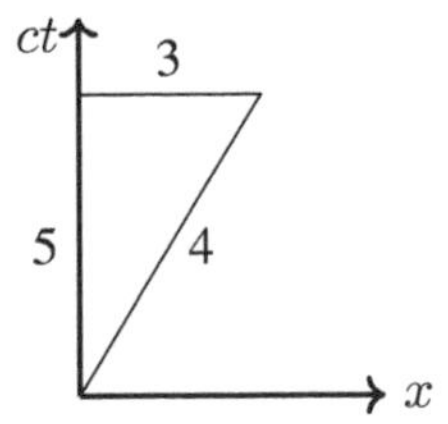

Figure 4.7 Moving clocks run slow: in the non-Euclidean geometry of spacetime.

the length of the hypotenuse (proper time) is shorter than the base of the triangle (time ascribed to the moving clock). The hypotenuse is a timelike vector with norm $\sqrt{-[-5^2+3^2]} = 4$. Does the hypotenuse in Fig. 4.7 *look* shorter than the base of the triangle? Don't bring your geometric expectations, based on a lifetime of Euclidean reasoning, to spacetime diagrams. Figure 4.8 shows

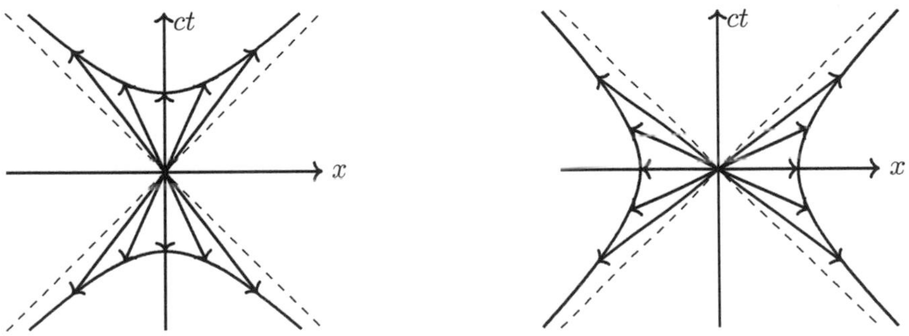

Figure 4.8 Timelike unit vectors (left) and spacelike unit vectors (right).

timelike and spacelike unit vectors: Each vector in Fig. 4.8 connects the origin with a unit hyperbola. These are not all unit vectors in the same frame; each would be a unit vector in some IRF, however.

4.4.2 Orthogonality

Can spacetime vectors be orthogonal? We assumed in setting up spacetime diagrams that the time axis (timelike) is orthogonal to the space axis (spacelike). Vectors can be timelike, lightlike, or spacelike. Is orthogonality possible for each type of vector? For $\boldsymbol{A}$ and $\boldsymbol{B}$ spacetime position vectors, can $\boldsymbol{A}\cdot\boldsymbol{B}=0$, where the inner product is defined in Eq. (4.13)? Denote the time component of $\boldsymbol{A}$ as A^0 and the space components as $\vec{A}$ (a notation we use in Chapter 5). Then, using Eq. (4.13), $\boldsymbol{A}\cdot\boldsymbol{A} = -(A^0)^2 + ||\vec{A}||^2$, where here $||\vec{A}|| = \sqrt{\vec{A}\cdot\vec{A}}$ is the Euclidean norm of the spatial part of the vector. A spacetime vector is spacelike, lightlike, or timelike according to whether $||\vec{A}|| > |A^0|$, $||\vec{A}|| = |A^0|$, or $||\vec{A}|| < |A^0|$. Orthogonality implies that $A^0B^0 = \vec{A}\cdot\vec{B}$. Of the three types of vector (timelike, spacelike, lightlike), which can meet this condition?

- Let $\boldsymbol{A}$ be a timelike vector. Vector $\boldsymbol{B}$ can be orthogonal to $\boldsymbol{A}$ only if $\boldsymbol{B}$ is spacelike.

  Proof: Assume that $A^0B^0 = \vec{A}\cdot\vec{B}$ (orthogonality) and $|A^0| > ||\vec{A}||$ (timelike). Then, $|A^0B^0| = |\vec{A}\cdot\vec{B}| \le ||\vec{A}||\,||\vec{B}||$ (Schwartz inequality). But $|A^0| > ||\vec{A}||$, implying that $|B^0| < ||\vec{B}||$. For orthogonality to hold, $\boldsymbol{B}$ can only be spacelike.

  Timelike vectors cannot be orthogonal, and timelike vectors cannot be orthogonal to lightlike vectors. A timelike vector can only be orthogonal to a spacelike vector.

- Let $\boldsymbol{A}$ be a lightlike vector. Vector $\boldsymbol{B}$ can be orthogonal to $\boldsymbol{A}$ if $\boldsymbol{B}$ is spacelike or lightlike.

Proof: From $A^0B^0 = \vec{A}\cdot\vec{B}$, $|B^0| = |\vec{A}\cdot\vec{B}|/|A^0| \leq ||\vec{A}||\,||\vec{B}||/|A^0|$ (Schwartz). But $|A^0| = ||\vec{A}||$ (lightlike), implying that $||\vec{B}|| \geq |B^0|$. $\boldsymbol{B}$ is either spacelike or lightlike.

Lightlike vectors can be orthogonal to lightlike and spacelike vectors; two lightlike vectors can be orthogonal if and only if they're scalar multiples of each other.

- It is possible for spacelike vectors to be orthogonal.

$\Longrightarrow$ *Two spacetime vectors can be orthogonal if at least one of them is spacelike or both are lightlike, in which case they are proportional to each other.*

4.4.3 Partition of spacetime

Spacetime can be partitioned into five regions (see Fig. 4.9):

$$\begin{aligned}
T_+ &\equiv \text{Future timelike, } (\Delta s)^2 < 0, t > 0;\\
T_- &\equiv \text{Past timelike, } (\Delta s)^2 < 0, t < 0;\\
S &\equiv \text{Spacelike, } (\Delta s)^2 > 0;\\
L_+ &\equiv \text{Future lightlike (future light cone), } (\Delta s)^2 = 0, t > 0;\\
L_- &\equiv \text{Past lightlike (past light cone), } (\Delta s)^2 = 0, t < 0\,.
\end{aligned}$$

The LT maps each of these regions onto itself. Because $(\Delta s)^2$ is Lorentz invariant, the spacelike

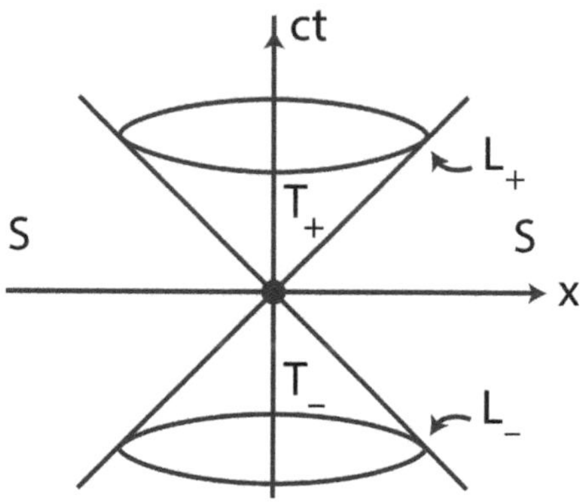

Figure 4.9 Partition of spacetime into spacelike, lightlike, and timelike regions.

region S is mapped onto itself under the LT; likewise with timelike and lightlike regions, where we include future and past sets. The past and future sets, however, are separately preserved under the LT. We prove this for T_+, where for $t > 0$, we must show that $t' > 0$ for $x \in T_+$. From Eq. (4.6),

$$ct' = ct\cosh\theta + x\sinh\theta\,. \tag{4.15}$$

For $x \in T_+$, $-ct < x < ct$. Using this inequality together with Eq. (4.15), it can be shown that $te^{-\theta} < t' < te^{\theta}$, and hence that $t' > 0$ if $t > 0$. The proof for T_- is similar. It's simple to show that the future and past light cones are separately preserved under the LT.

4.4.4 Temporal order and causality

Events in T_+ cannot be simultaneous in any reference frame, and the temporal order in which events occur is the same for all observers (discussed in Section 1.5). Event A in Fig. 4.10 has $t > 0$ and

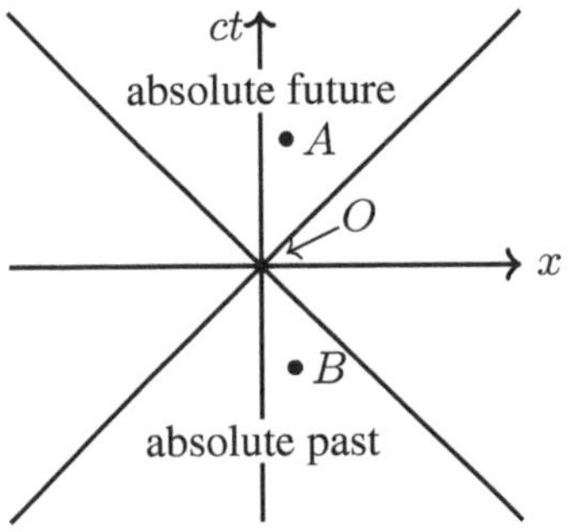

Figure 4.10 Absolute past and absolute future. Events in T_+ occur after the event at O in any reference frame. Events in T_- occur before O in all reference frames.

occurs after the event at the origin, O. No LT can change the time coordinate of A to $t < 0$. Thus A occurs after O in *all* IRFs. For this reason, T_+ is called the *absolute future* because events in this region occur after O in all frames. Likewise, T_- is called the *absolute past* because events in this region occur before O in any frame.

Events on the future light cone, L_+, can be influenced by electromagnetic signals from a source at O. Because physical effects cannot propagate faster than light, any effect originating at O can reach only those points inside T_+ or on L_+. And because the temporal order of events cannot be altered by a LT in this region of spacetime, it's possible to introduce notions of cause and effect in an absolute sense, independent of reference frame. *A causal connection between events can exist only if they are timelike or lightlike separated.* Vectors that are either timelike or null are called *causal* spacetime vectors. At each point in spacetime, there corresponds a light cone with its vertex at that point—see Fig. 4.11. Each event along the worldline of a particle can affect only those events that

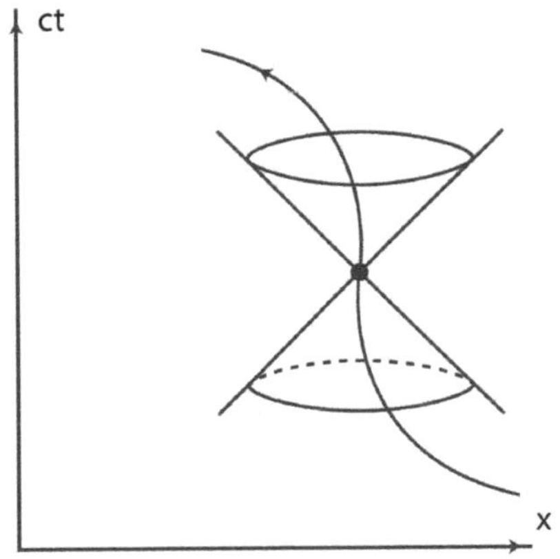

Figure 4.11 A point in spacetime can influence only those events within its future lightcone.

lie in or on its future light cone, and can be affected only by events in or on its past light cone.[10]

[10]In quantum field theory, measurements of a field at the origin and at an event P do not interfere if P is spacelike separated from the origin. Operators must therefore commute for spacelike-separated events.

SUMMARY

- A LT is any linear transformation such that $L^T \eta L = \eta$, where $\eta = \text{diag}(-1, 1, 1, 1)$.
- The inner product between spacetime vectors is Lorentz invariant where $\boldsymbol{r} \cdot \boldsymbol{r} = r^T \eta r$.
- Spacetime can be partitioned into a future timelike region, a past timelike region, a spacelike region, and the future and past light cones. The LT maps each of these regions into themselves.
- There can be a causal connection between events in spacetime, where notions of cause and effect, "earlier" and "later" are independent of reference frame, only if they are timelike or lightlike separated.
- Two spacetime vectors can be orthogonal if one of them is spacelike or both are lightlike.

EXERCISES

4.1 Show that $\left(L^0_0\right)^2 = 1 + \sum_{i=1}^3 \left(L^i_0\right)^2$. Conclude that $\left(L^0_0\right)^2 \geq 1$. Hint: Use Eq. (4.12).

4.2 Show that $L^{-1} = \eta L^T \eta$ and $L^T = \eta L^{-1} \eta$. Hint: $(\eta)^2 = I$.

4.3 Assume a 2D LT, where $\eta = \begin{pmatrix} -1 & 0 \\ 0 & 1 \end{pmatrix}$,

$$L = \begin{pmatrix} L^0_0 & L^0_1 \\ L^1_0 & L^1_1 \end{pmatrix} .$$

a. Use the result of Exercise 4.2 to find L^{-1} in terms of the matrix elements of L. Because we have generally that $L^{-1}(v) = L(-v)$, use your result for L^{-1} to argue that the off-diagonal terms must be odd functions of v, while the diagonal terms are even functions.

b. Use your result for L^{-1} to conclude that L must be a symmetric matrix with $L^1_1 = L^0_0$. Hint: Compare the results of $LL^{-1} = I$ with $L^T \eta L = \eta$.

4.4 Show that the inner product between two spacetime vectors is invariant under the LT, $\boldsymbol{r}'_1 \cdot \boldsymbol{r}'_2 = \boldsymbol{r}_1 \cdot \boldsymbol{r}_2$, where the inner product is defined as $\boldsymbol{r}_1 \cdot \boldsymbol{r}_2 = r_1^T \eta r_2$.

4.5 Events A, B, and C have spacetime coordinates (ct, x) of $(2, 1)$, $(7, 4)$, and $(5, 6)$, respectively. For each pair of events, answer the questions: (1) Are the events timelike, spacelike, or lightlike separated?; (2) Is it possible that one of the events could be caused by the other?

4.6 Show:

a. If two lightlike vectors are orthogonal, they are scalar multiples of each other;

b. That the inner product between a lightlike vector and a timelike vector is negative;

c. That the sum of a lightlike vector and a timelike vector is timelike. Use the result of the previous problem;

d. For two future-pointing timelike vectors, i.e., spacetime vectors $\boldsymbol{A}$, $\boldsymbol{B}$ with $A^0 > 0$ and $B^0 > 0$, that $\boldsymbol{A} \cdot \boldsymbol{B} < 0$;

e. For $\boldsymbol{A}$ a future-pointing timelike vector and $\boldsymbol{B}$ a past-pointing timelike vector, that $\boldsymbol{A} \cdot \boldsymbol{B} > 0$.

CHAPTER 5

Tensors on flat spaces

EINSTEIN'S program for GR, that the laws of physics be independent of coordinate system, is realized by expressing equations as relations between tensors.[1] Tensors are highly useful in SR as well: The time has come to address this important topic.

Tensors are generalizations of vectors. In an n-dimensional space, a vector is specified by n numbers. A second-rank tensor requires n^2 numbers for its specification; a rank-r tensor requires n^r numbers. Physical quantities exist requiring more than n numbers for their specification, the stress tensor for example. We start by defining vectors in spacetime, and then work our way up to tensors. The traditional way of introducing tensors is through their *transformation properties*—how the n^r numbers transform between reference frames (Section 5.1). In Section 5.5 we show that tensors are linear relations between scalars, vectors, and even other tensors.

5.1 TRANSFORMATION PROPERTIES

5.1.1 Spacetime position four-vector

Spacetime is modeled as a four-dimensional continuum[2] obtained from the concatenation of space ($\mathbb{R}^3$) with time ($\mathbb{R}$), $\mathbb{R}^4 = \mathbb{R}^3 \times \mathbb{R}$. Unadorned $\mathbb{R}^4$, however, cannot support the physics of SR; we require a mathematical model having more structure. *Minkowski space* (MS) is a four-dimensional vector space (with points in one-to-one correspondence with those of $\mathbb{R}^4$) spanned by one timelike basis vector, $\boldsymbol{e}_t$, and three spacelike basis vectors, $\boldsymbol{e}_x$, $\boldsymbol{e}_y$, $\boldsymbol{e}_z$, where by convention basis vectors are labeled with subscripts.[3] While any four linearly independent vectors can constitute a basis (known as a *tetrad*), in IRFs we require time to be orthogonal to space. Points in MS (events) are located by a position vector (relative to the origin-event[4]) $\boldsymbol{r} = r^t \boldsymbol{e}_t + r^x \boldsymbol{e}_x + r^y \boldsymbol{e}_y + r^z \boldsymbol{e}_z$, called a *four-vector*, where by convention coordinates are labeled with superscripts. A *change* in reference frame is a change of basis vectors (the passive form of the transformation) in such a way that the components of $\boldsymbol{r}$ transform according to the LT.[5] In Section 4.4 we were careful, in introducing the inner product, to refer to spacetime position vectors, because so far that's the only four-vector we have: a position vector for every event. As we develop SR, a succession of four-vectors will be introduced. The edifice of relativity theory is built on four-vectors and the Lorentz invariants that can be constructed from them. With the understanding that additional four-vectors are forthcoming,

[1] A sizable portion of Einstein's 1916 article is devoted to tensors in a section, "Mathematical Aids to the Formation of Generally Covariant Equations."[9, p111].

[2] Appendix C reviews linear algebra, including the Cartesian product. While the universe is well described as a four-dimensional entity, *string theory* is a proposed framework for quantum gravity that invokes extra spatial dimensions.

[3] Every vector space has a basis, with its dimension the maximum number of linearly independent vectors.

[4] Vector spaces require a zero vector, which we can take as an arbitrarily chosen event for the spacetime origin.

[5] Minkowski space is an inner-product space endowed with a very specific structure; it's not simply a vector space. To speak of timelike and spacelike vectors, an inner product must already have been introduced—Section 4.4.

we refer in this chapter to arbitrary four-vectors $\boldsymbol{A}$:

$$\boldsymbol{A} = A^t \boldsymbol{e}_t + A^x \boldsymbol{e}_x + A^y \boldsymbol{e}_y + A^z \boldsymbol{e}_z \ . \tag{5.1}$$

We'll soon distinguish two *types* of vectors: *contravariant* and *covariant*. Any vector whose provenance can be traced to the position vector (more generally to oriented line elements) is referred to as contravariant. The vector $\boldsymbol{A}$ in Eq. (5.1) has the form of a contravariant vector. Covariant vectors are a geometrically distinct type of vector, related to oriented surface elements.

We adopt a notational convention that allows us to write four-vectors more compactly than Eq. (5.1). We reserve zero to label the time component of four-vectors as well as the associated basis vector, and we use $1, 2, 3$ to label spatial components and basis vectors, instead of x, y, z, or r, θ, ϕ. Thus, Eq. (5.1) can be written $\boldsymbol{A} = A^0\boldsymbol{e}_0 + A^1\boldsymbol{e}_1 + A^2\boldsymbol{e}_2 + A^3\boldsymbol{e}_3$. This convention enables the use of summation notation: $\boldsymbol{A} = \sum_{\alpha=0}^{3} A^\alpha \boldsymbol{e}_\alpha$. Note the Greek letter α as the summation index. A convention in the theory of relativity is that if the sum runs from 0 to 3, use a Greek letter as the summation index; however, if the sum runs from 1 to 3, use a Roman letter. Thus, $\sum_{\alpha=0}^{3} A^\alpha \boldsymbol{e}_\alpha = A^0\boldsymbol{e}_0 + \sum_{i=1}^{3} A^i \boldsymbol{e}_i$. Now, having introduced this convention, much of what we cover in this chapter is general tensor analysis pertaining to *any* space and not specifically to MS. When that's the case there's no need to adopt a notation that singles out the timelike dimension. When we deal with relativity, however, we stick to the convention.

We will encounter expressions involving sums over numerous indices, and writing out the summation symbols becomes cumbersome. The *Einstein summation convention* is that repeated raised and lowered indices imply a sum. Thus, $\sum_{\alpha=0}^{3} A^\alpha \boldsymbol{e}_\alpha \equiv A^\alpha \boldsymbol{e}_\alpha$. Of course, α is a *dummy index* that has no absolute meaning. The expressions $A^\alpha \boldsymbol{e}_\alpha = A^\beta \boldsymbol{e}_\beta = A^\gamma \boldsymbol{e}_\gamma$ are equivalent and imply the same sum. Remember: The rule is that *repeated* upper and lower indices imply a summation. Terms such as $A^\alpha \boldsymbol{e}_\beta$ do not imply a sum. I will gradually work in the summation convention to gain practice with it, but after a point I will simply use it without comment.

We'll use x^μ to denote the components of the *four-position*, $x^\mu = (x^0, x^1, x^2, x^3)$. Not just any collection of four numbers constitutes the components of a four-vector.[6] For example, can we package the components of the electric field vector $\boldsymbol{E}$ into a four-vector, finding something suitable to include as the time component? It turns out the answer is *No*. Likewise there is no four-vector having the components of the orbital angular momentum $\boldsymbol{L}$ as its spatial part.[7]

5.1.2 Metric tensor

We defined the inner product between position four-vectors in Eq. (4.13) so as to produce an invariant under the LT, the spacetime separation. Here we generalize the inner product for arbitrary four-vectors in a way that it generates an invariant under any invertible coordinate transformation, which includes the LT. For vectors defined with respect to the same basis, $\boldsymbol{A} = A^\alpha \boldsymbol{e}_\alpha$, $\boldsymbol{B} = B^\beta \boldsymbol{e}_\beta$ (summation convention), form the inner product by "dotting" them together,

$$\begin{aligned} \boldsymbol{A} \cdot \boldsymbol{B} &= \left(A^0\boldsymbol{e}_0 + A^1\boldsymbol{e}_1 + A^2\boldsymbol{e}_2 + A^3\boldsymbol{e}_3\right) \cdot \left(B^0\boldsymbol{e}_0 + B^1\boldsymbol{e}_1 + B^2\boldsymbol{e}_2 + B^3\boldsymbol{e}_3\right) \\ &= A^\alpha B^\beta \left(\boldsymbol{e}_\alpha \cdot \boldsymbol{e}_\beta\right) \equiv \sum_{\alpha=0}^{3} \sum_{\beta=0}^{3} A^\alpha B^\beta \left(\boldsymbol{e}_\alpha \cdot \boldsymbol{e}_\beta\right) , \end{aligned} \tag{5.2}$$

where a double sum is implied by *two* sets of repeated upper and lower indices. There are 16 terms in Eq. (5.2) when it's expanded out. We've "passed the buck" in defining the inner product between four-vectors to the inner product between basis vectors, $\boldsymbol{e}_\alpha \cdot \boldsymbol{e}_\beta$. We're going to leave these as unspecified for now and represent them with a new symbol labelled by two indices,

$$g_{\alpha\beta} \equiv \boldsymbol{e}_\alpha \cdot \boldsymbol{e}_\beta \ . \tag{5.3}$$

[6] Another way of saying that Minkowski space is not $\mathbb{R}^4$.

[7] Fear not, however: The vectors $\boldsymbol{E}$ and $\boldsymbol{L}$ will find their place as components of tensors.

The 16 quantities $\{g_{\alpha\beta}\}$ are the elements of the *metric tensor*,[8] our first tensor. The metric tensor is one way to define a geometry:[9] Geometric properties such as arc length and surface area can be calculated once the metric has been specified. Said differently, each geometry (including spacetime) has its own metric tensor. Combining Eqs. (5.3) and (5.2),

$$\boldsymbol{A} \cdot \boldsymbol{B} = g_{\alpha\beta} A^\alpha B^\beta \qquad \left(\equiv \sum_{\alpha=0}^{3} \sum_{\beta=0}^{3} g_{\alpha\beta} A^\alpha B^\beta \right) .$$

The components of the metric tensor are *symmetric* in their indices, $g_{\alpha\beta} = g_{\beta\alpha}$; *the metric tensor is always symmetric*. For an n-dimensional space, a symmetric second-rank tensor has $n(n+1)/2$ *independent* elements (show this); in MS there are 10 independent elements of the metric tensor.

To calculate the metric tensor, we must understand what's meant by *basis vector* in this context. Consider the infinitesimal displacement vector[10] in the spherical coordinate system,

$$\begin{aligned} \mathrm{d}\boldsymbol{s} &= \mathrm{d}r\hat{\boldsymbol{r}} + r\mathrm{d}\theta\hat{\boldsymbol{\theta}} + r\sin\theta\mathrm{d}\phi\hat{\boldsymbol{\phi}} \equiv \mathrm{d}r\boldsymbol{e}_r + \mathrm{d}\theta\boldsymbol{e}_\theta + \mathrm{d}\phi\boldsymbol{e}_\phi \\ &= \sum_i (\text{coordinate differential})^i \times (\text{basis vector})_i \equiv \sum_i \mathrm{d}x^i \boldsymbol{e}_i \, . \end{aligned} \tag{5.4}$$

The basis vector $\boldsymbol{e}_i$ is whatever multiplies the coordinate differential $\mathrm{d}x^i$ in the expression for $\mathrm{d}\boldsymbol{s}$. In spherical coordinates, $\boldsymbol{e}_r = \hat{\boldsymbol{r}}$, $\boldsymbol{e}_\theta = r\hat{\boldsymbol{\theta}}$, and $\boldsymbol{e}_\phi = r\sin\theta\hat{\boldsymbol{\phi}}$. Basis vectors are not necessarily unit vectors: *their magnitude and direction generally vary throughout a coordinate system*. The vectors $\{\boldsymbol{e}_i\}$ are *tangent to the coordinate curves that pass through a given point* and point toward increasing values of the coordinate, the *coordinate basis*. Figure 5.1 shows coordinate basis vectors

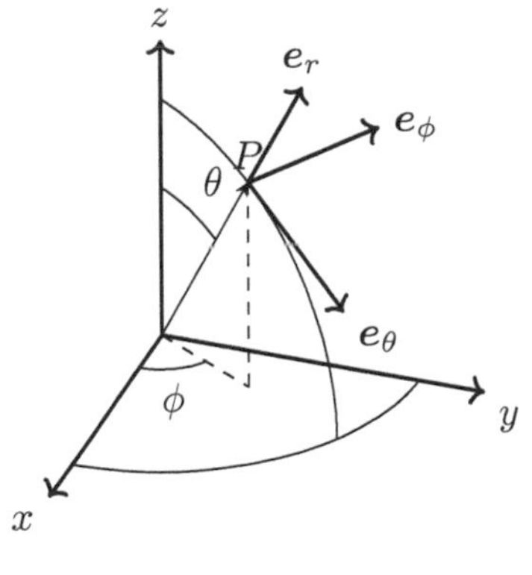

Figure 5.1 Coordinate basis vectors at point P in the spherical coordinate system.

$\boldsymbol{e}_r$, $\boldsymbol{e}_\theta$, and $\boldsymbol{e}_\phi$ "attached" to the point P. Only one coordinate curve is shown in Fig. 5.1, the portion of a semicircle[11] that results for fixed values of r and ϕ, with $0 \le \theta \le \pi/2$. The coordinate curve for the radial coordinate is the ray (for fixed θ and ϕ) $0 \le r < \infty$, while that for the azimuth angle is the circle (for fixed θ and r) $0 \le \phi < 2\pi$.

[8] Actually, Eq. (5.3) specifies the *covariant elements* of the metric tensor, $g_{\alpha\beta}$. We will shortly introduce $g^{\alpha\beta}$, the *contravariant* elements of the metric tensor.

[9] What *is* a geometry? O. Veblen and J.H.C. Whitehead offered:[23, p17] "...a branch of mathematics is called a geometry because the name seems good, on emotional and traditional grounds, to a sufficient number of competent people."

[10] The infinitesimal displacement vector $\mathrm{d}\boldsymbol{s}$ is the *prototype* contravariant vector. Anything called vector (in this case contravariant) must have the properties of the prototype.

[11] Semicircle because the polar angle has the range $0 \le \theta \le \pi$.

We can now calculate the metric tensor for the spherical coordinate system using Eq. (5.3):

$$[g_{ij}] = \begin{pmatrix} g_{rr} & g_{r\theta} & g_{r\phi} \\ g_{\theta r} & g_{\theta\theta} & g_{\theta\phi} \\ g_{\phi r} & g_{\phi\theta} & g_{\phi\phi} \end{pmatrix} = \begin{pmatrix} 1 & 0 & 0 \\ 0 & r^2 & 0 \\ 0 & 0 & r^2\sin^2\theta \end{pmatrix} , \tag{5.5}$$

where $[g_{ij}]$ indicates the tensor components arranged as a matrix. The matrix in Eq. (5.5) is diagonal because the coordinate system is an *orthogonal* coordinate system, with, for example, $\boldsymbol{e}_r \cdot \boldsymbol{e}_\theta = r\hat{\boldsymbol{r}} \cdot \hat{\boldsymbol{\theta}} = 0$. *The metric tensor is always diagonal for orthogonal coordinate systems.* Using Eqs. (5.4) and (5.5), we have the square of the line element in spherical coordinates:

$$(\mathrm{d}s)^2 \equiv \mathrm{d}\boldsymbol{s} \cdot \mathrm{d}\boldsymbol{s} = g_{ij}\mathrm{d}x^i\mathrm{d}x^j = g_{rr}(\mathrm{d}r)^2 + g_{\theta\theta}(\mathrm{d}\theta)^2 + g_{\phi\phi}(\mathrm{d}\phi)^2 . \tag{5.6}$$

The line element $\mathrm{d}s = \sqrt{\mathrm{d}\boldsymbol{s} \cdot \mathrm{d}\boldsymbol{s}}$ represents a physical displacement and must have the dimension of length. *The metric tensor supplies the information required to calculate the distance between points*, the separation of which is characterized by coordinate differentials. If the coordinates do not have a physical dimension, such as angular coordinates, the metric tensor must carry the information so that $g_{ij}\mathrm{d}x^i\mathrm{d}x^j$ has the dimension of length squared (see $g_{\theta\theta}$ and $g_{\phi\phi}$ in Eq. (5.5)).

We can write $(\mathrm{d}s)^2$ in Eq. (5.6) in the following way:

$$\begin{aligned} (\mathrm{d}s)^2 &= \begin{pmatrix} \mathrm{d}r & \mathrm{d}\theta & \mathrm{d}\phi \end{pmatrix} \begin{pmatrix} 1 & 0 & 0 \\ 0 & r^2 & 0 \\ 0 & 0 & r^2\sin^2\theta \end{pmatrix} \begin{pmatrix} \mathrm{d}r \\ \mathrm{d}\theta \\ \mathrm{d}\phi \end{pmatrix} = \begin{pmatrix} \mathrm{d}r & r^2\mathrm{d}\theta & r^2\sin^2\theta\mathrm{d}\phi \end{pmatrix} \begin{pmatrix} \mathrm{d}r \\ \mathrm{d}\theta \\ \mathrm{d}\phi \end{pmatrix} \\ &= (\mathrm{d}r)^2 + r^2(\mathrm{d}\theta)^2 + r^2\sin^2\theta(\mathrm{d}\phi)^2 = g_{ij}\mathrm{d}x^i\mathrm{d}x^j . \end{aligned} \tag{5.7}$$

Equation (5.7) has the form of Eq. (4.13) except with η replaced by $[g_{ij}]$.

Now consider an arbitrary three-dimensional coordinate system where point P is at the intersection of three coordinate curves labeled by (u, v, w) (see Fig. 5.2). For a nearby point Q define the

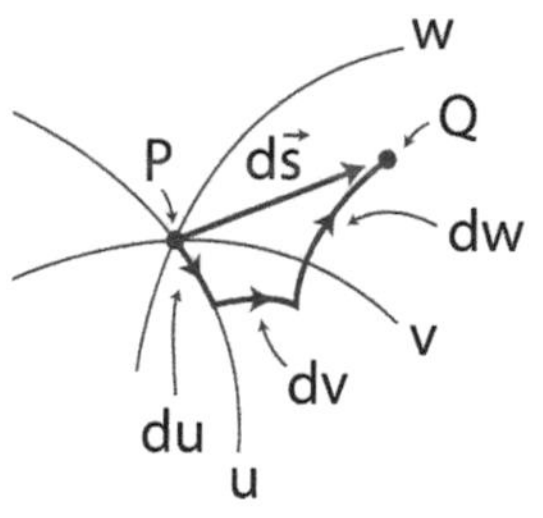

Figure 5.2 General (u, v, w) coordinate system.

vector $\Delta\boldsymbol{s} \equiv \overrightarrow{PQ}$; $\Delta\boldsymbol{s}$ is also the vector $\Delta\boldsymbol{s} \equiv (\boldsymbol{r} + \Delta\boldsymbol{r}) - \boldsymbol{r}$, where $\boldsymbol{r} + \Delta\boldsymbol{r}$ and $\boldsymbol{r}$ are the position vectors for Q and P relative to the origin (not shown). To first order in small quantities,

$$\mathrm{d}\boldsymbol{s} = \frac{\partial \boldsymbol{r}}{\partial u}\mathrm{d}u + \frac{\partial \boldsymbol{r}}{\partial v}\mathrm{d}v + \frac{\partial \boldsymbol{r}}{\partial w}\mathrm{d}w , \tag{5.8}$$

where the derivatives (with respect to coordinates) are evaluated at P. The derivatives

$$\boldsymbol{e}_u \equiv \partial\boldsymbol{r}/\partial u\big|_P \qquad \boldsymbol{e}_v \equiv \partial\boldsymbol{r}/\partial v\big|_P \qquad \boldsymbol{e}_w \equiv \partial\boldsymbol{r}/\partial w\big|_P \tag{5.9}$$

form a local basis—an arbitrary $\mathrm{d}\boldsymbol{s}$ in the neighborhood of P can be expressed as a linear combination of them—and they're tangent to the coordinate curves. *A local set of basis vectors is determined by the local structure of the coordinate system.*

Example. The position vector in spherical coordinates can be written

$$\boldsymbol{r} = r\sin\theta\cos\phi\hat{\boldsymbol{x}} + r\sin\theta\sin\phi\hat{\boldsymbol{y}} + r\cos\theta\hat{\boldsymbol{z}} \ .$$

Applying Eq. (5.9), we have the vectors of the coordinate basis

$$\begin{aligned}
\boldsymbol{e}_r &= \partial\boldsymbol{r}/\partial r = \sin\theta\cos\phi\hat{\boldsymbol{x}} + \sin\theta\sin\phi\hat{\boldsymbol{y}} + \cos\theta\hat{\boldsymbol{z}} \\
\boldsymbol{e}_\theta &= \partial\boldsymbol{r}/\partial\theta = r\cos\theta\cos\phi\hat{\boldsymbol{x}} + r\cos\theta\sin\phi\hat{\boldsymbol{y}} - r\sin\theta\hat{\boldsymbol{z}} \\
\boldsymbol{e}_\phi &= \partial\boldsymbol{r}/\partial\phi = -r\sin\theta\sin\phi\hat{\boldsymbol{x}} + r\sin\theta\cos\phi\hat{\boldsymbol{y}} \ .
\end{aligned}$$

The norms of these vectors are $||\boldsymbol{e}_r|| = 1$, $||\boldsymbol{e}_\theta|| = r$, and $||\boldsymbol{e}_\phi|| = r\sin\theta$ (show this). *The magnitude and direction of the basis vectors are not constant.* The unit vectors are thus

$$\begin{aligned}
\hat{\boldsymbol{e}}_r &= \sin\theta\cos\phi\hat{\boldsymbol{x}} + \sin\theta\sin\phi\hat{\boldsymbol{y}} + \cos\theta\hat{\boldsymbol{z}} \\
\hat{\boldsymbol{e}}_\theta &= \cos\theta\cos\phi\hat{\boldsymbol{x}} + \cos\theta\sin\phi\hat{\boldsymbol{y}} - \sin\theta\hat{\boldsymbol{z}} \\
\hat{\boldsymbol{e}}_\phi &= -\sin\phi\hat{\boldsymbol{x}} + \cos\phi\hat{\boldsymbol{y}} \ .
\end{aligned}$$

Clearly, by definition, $\boldsymbol{e}_r = \hat{\boldsymbol{e}}_r$, $\boldsymbol{e}_\theta = r\hat{\boldsymbol{e}}_\theta$, and $\boldsymbol{e}_\phi = r\sin\theta\hat{\boldsymbol{e}}_\phi$.

What about Minkowski space? For spherical coordinates, the geometry was specified first and then we derived the metric tensor. *Physics* determines the metric of spacetime for us. From Eq. (4.13), $(\mathrm{d}s)^2 = -(\mathrm{d}x^0)^2 + (\mathrm{d}x^1)^2 + (\mathrm{d}x^2)^2 + (\mathrm{d}x^3)^2 \equiv g_{\alpha\beta}\mathrm{d}x^\alpha\mathrm{d}x^\beta$, and thus we have the Lorentz metric tensor, what we previously wrote down in Eq. (1.14):

$$[g_{\alpha\beta}] = \eta = \begin{pmatrix} -1 & 0 & 0 & 0 \\ 0 & 1 & 0 & 0 \\ 0 & 0 & 1 & 0 \\ 0 & 0 & 0 & 1 \end{pmatrix} \ . \tag{1.14}$$

Note that the inner product between the *time* basis vectors, $\boldsymbol{e}_0 \cdot \boldsymbol{e}_0 = -1$, which is non-intuitive but consistent with our definition of timelike unit vector (Section 4.4).

In Euclidean geometry $(\mathrm{d}s)^2 = g_{ij}\mathrm{d}x^i\mathrm{d}x^j > 0$ *for any sign of the differentials* $\mathrm{d}x^i$. A metric is said to be *positive definite*[12] if $(\mathrm{d}s)^2 > 0$ for all $\mathrm{d}x^i$, unless the coordinate differentials vanish, $\mathrm{d}x^i = 0$. Said differently, the distance between two points in a Euclidean geometry vanishes only if the points are coincident. In MS, however, $(\mathrm{d}s)^2$ can be positive, negative, or zero (spacelike, timelike, lightlike), in which case the metric is said to be *indefinite*. With an indefinite metric, two points may be at zero distance $[(\mathrm{d}s)^2 = 0]$ without being coincident ($\mathrm{d}x^i \neq 0$). We show in Section 5.6 that an indefinite metric must have a nonzero null vector.

5.1.3 Dual basis, lowering and raising indices

The basis vectors $\{\boldsymbol{e}_j\}_{j=1}^n$ for an arbitrary coordinate system will not in general be mutually orthogonal. It's highly useful nonetheless to have *some* type of orthogonality relations among basis vectors. To that end, we *define* another set of vectors, $\{\boldsymbol{e}^i\}_{i=1}^n$, the *dual basis*, labeled with *superscripts*, that

[12] A test for positive definiteness is provided by *Sylvester's criterion* that various determinants (principal minors) associated with $[g_{ij}]$ all be positive.[24, p52] The Lorentz metric fails this test: It's not positive definite, and in fact is indefinite.

are orthogonal to each of the vectors $\{e_j\}$, such that[13]

$$e^i \cdot e_j = \delta^i_j \equiv \begin{cases} 1 & \text{if } i = j \\ 0 & \text{if } i \neq j \end{cases}, \qquad (i, j = 1, \cdots, n) \tag{5.10}$$

where we've written the Kronecker delta in a new way.[14] By definition, $e^2 \cdot e_1 = 0$ and $e^1 \cdot e_2 = 0$, but in general $e^1 \cdot e^2 \neq 0$.

Figure 5.3 shows a non-orthogonal basis e_1, e_2 for vectors confined to a plane. Any vector in

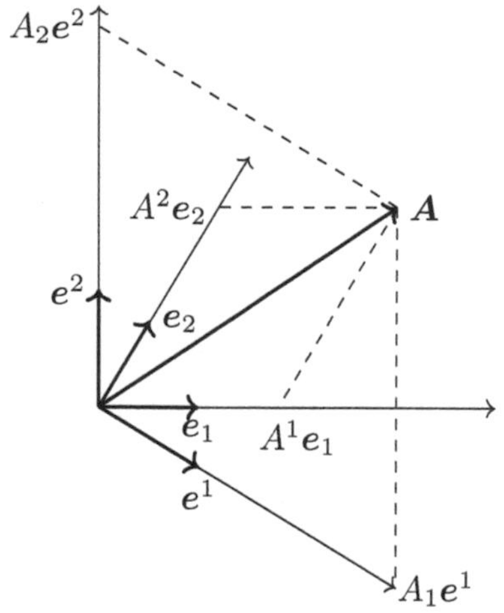

Figure 5.3 Basis vectors e_1, e_2, and dual basis vectors e^1, e^2.

the plane can be expressed as a linear combination $A = A^1 e_1 + A^2 e_2$. The dual basis vectors e^1 and e^2 are shown, constructed so as to satisfy Eq. (5.10): $e^2 \cdot e_1 = 0$, $e^2 \cdot e_2 = 1$, $e^1 \cdot e_2 = 0$, and $e^1 \cdot e_1 = 1$. The same vector can be expressed as a linear combination of the dual basis vectors: $A = A_1 e^1 + A_2 e^2$, where the components of A in the dual basis are labeled with *subscripts*.

We can express a vector in either basis. Using the summation convention,

$$A = A^i e_i = A_k e^k \,. \tag{5.11}$$

There must be a connection between the components A^i and A_k (of the same vector). Take the inner product of both sides of Eq. (5.11) with e_j,

$$e_j \cdot A = A^i e_j \cdot e_i = A^i g_{ji} = A_k e_j \cdot e^k = A_k \delta^k_j = A_j \,,$$

where we've used Eqs. (5.3) and (5.10). Thus,

$$A_j = g_{ji} A^i \,. \tag{5.12}$$

[13]The number of dual basis vectors $\{e^i\}_{i=1}^n$ is the same as that for the original basis $\{e_j\}_{j=1}^n$; the two sets are isomorphic. In crystallography the dual basis is called the reciprocal basis. For the (generally non-orthogonal) directions of crystal axes $\{e_i\}_{i=1}^3$, the dual basis vectors are defined as

$$e^1 = \frac{e_2 \times e_3}{e_1 \cdot (e_2 \times e_3)} \qquad e^2 = \frac{e_3 \times e_1}{e_1 \cdot (e_2 \times e_3)} \qquad e^3 = \frac{e_1 \times e_2}{e_1 \cdot (e_2 \times e_3)} \,.$$

These vectors satisfy $e^i \cdot e_j = \delta^i_j$. Could it be said that one cannot understand solid-state physics without first studying GR?

[14]The dual basis vectors span a logically distinct vector space known as the *dual space*, which plays a fundamental role in tensor analysis. Appendix C contains an introduction to the dual space. From a set of vectors (in general non-orthogonal), a new, orthonormal set of vectors can always be found (Gram-Schmidt process). The dual basis is *not* such a set. The dual basis is in general non-orthogonal; the vectors $\{e^i\}$ are orthonormal to every vector in the set $\{e_j\}$, but not amongst themselves. The Gram-Schmidt process is a basis transformation in a given vector space; the dual basis is the basis of *another* space.

Equation (5.12) is an instance of *lowering an index*: The (covariant) metric tensor connects the components of a vector in the dual basis A_j with its components in the coordinate basis,[15] A^i. Now take the inner product between Eq. (5.11) and $\boldsymbol{e}^j$:

$$\boldsymbol{e}^j \cdot \boldsymbol{A} = A^i \boldsymbol{e}^j \cdot \boldsymbol{e}_i = A^i \delta^j_i = A^j = A_k \boldsymbol{e}^k \cdot \boldsymbol{e}^j \,, \tag{5.13}$$

where we've used Eq. (5.10). We define the *contravariant* elements of the metric tensor as (compare with Eq. (5.3))

$$g^{ij} \equiv \boldsymbol{e}^i \cdot \boldsymbol{e}^j \,, \tag{5.14}$$

where $g^{ij} = g^{ji}$. Combining Eqs. (5.14) and (5.13),

$$A^j = A_k g^{kj} \,. \tag{5.15}$$

Equation (5.15) is an instance of *raising an index*: The contravariant metric tensor connects the components of a vector in the coordinate basis A^j with its components in the dual basis, A_k.

Is there a relation between the contravariant and covariant elements of the metric tensor, g_{ij} and g^{ij}? Combining Eq. (5.15), the raising of an index, $A^i = g^{ij}A_j$, with Eq. (5.12), the lowering of an index, $A_j = g_{jk}A^k$,

$$A^i = g^{ij}A_j = g^{ij}g_{jk}A^k \,. \tag{5.16}$$

Equation (5.16) is equivalent to

$$\left(\delta^i{}_k - g^{ij}g_{jk}\right)A^k = 0 \,. \tag{5.17}$$

But because the $\{A^k\}$ are arbitrary,

$$g^{ij}g_{jk} = \delta^i{}_k \,. \tag{5.18}$$

The two types of metric tensors are inverses of each other. Using Eq. (5.5) we have for spherical coordinates,

$$\left[g^{ij}\right] = \begin{pmatrix} 1 & 0 & 0 \\ 0 & 1/r^2 & 0 \\ 0 & 0 & 1/(r^2\sin^2\theta) \end{pmatrix} \,. \tag{5.19}$$

The representation of vectors in the coordinate and dual bases provides a convenient expression for the inner product,

$$\boldsymbol{A}\cdot\boldsymbol{B} = \left(A^i\boldsymbol{e}_i\right)\cdot\left(B_k\boldsymbol{e}^k\right) = A^iB_k\boldsymbol{e}_i\cdot\boldsymbol{e}^k = A^iB_k\delta^k_i = A^iB_i \,, \tag{5.20}$$

where we've used Eq. (5.10). Likewise, $A^iB_i = g^{ij}A_jg_{ik}B^k = A_jB^k\delta^j{}_k = A_jB^j$, where we've used Eq. (5.18).[16] Note that Eq. (4.13) can be written $\boldsymbol{r}\cdot\boldsymbol{r} = x^\mu x_\mu$.

We now prove a useful result, that we can form an identity *operator* out of the basis vectors, (summation convention)

$$I \equiv \boldsymbol{e}^i\boldsymbol{e}_i \equiv \boldsymbol{e}_i\boldsymbol{e}^i \,, \tag{5.21}$$

where there is no "dot" between the vectors; Eq. (5.21) is an operator.[17] In Cartesian coordinates, Eq. (5.21) reads $I = \boldsymbol{e}^x\boldsymbol{e}_x + \boldsymbol{e}^y\boldsymbol{e}_y + \boldsymbol{e}^z\boldsymbol{e}_z$. Let $I = \boldsymbol{e}^i\boldsymbol{e}_i$ act on a vector defined first with respect to the dual basis, and then with respect to the coordinate basis,

$$\begin{aligned} I\cdot\boldsymbol{A} &= \boldsymbol{e}^i\boldsymbol{e}_i\cdot\left(A_j\boldsymbol{e}^j\right) = \boldsymbol{e}^iA_j\left(\boldsymbol{e}_i\cdot\boldsymbol{e}^j\right) = \boldsymbol{e}^iA_j\delta^j{}_i = \boldsymbol{e}^iA_i = \boldsymbol{A} \\ &= \boldsymbol{e}^i\boldsymbol{e}_i\cdot\left(A^j\boldsymbol{e}_j\right) = \boldsymbol{e}^iA^j\left(\boldsymbol{e}_i\cdot\boldsymbol{e}_j\right) = \boldsymbol{e}^iA^jg_{ji} = \boldsymbol{e}^iA_i = \boldsymbol{A} \,, \end{aligned}$$

[15] Note from Eq. (5.12) that *all* components A^i in the coordinate basis contribute to the components A_j in the dual basis.

[16] Such index manipulations are affectionately known as *index gymnastics.*

[17] The juxtaposition of two vectors without a dot or cross between them is called a *dyad*. A *dyadic* is a sum of dyads. The dyadic identity operator in Eq. (5.21) is analogous to the completeness relation $I = \sum_n |n\rangle\langle n|$ in Dirac notation.

where we've used Eq. (5.10) in the first line and Eq. (5.12) in the second line to lower the index.

Raising and lowering indices applies to basis vectors as well. Expand a dual basis vector in the coordinate basis,

$$\boldsymbol{e}^i = c^{ij}\boldsymbol{e}_j \,, \tag{5.22}$$

where the c^{ij} are unknown expansion coefficients. Take the inner product between $\boldsymbol{e}^k$ and both sides of Eq. (5.22), $\boldsymbol{e}^k \cdot \boldsymbol{e}^i \equiv g^{ik} = c^{ij}\boldsymbol{e}^k \cdot \boldsymbol{e}_j = c^{ij}\delta^k_j = c^{ik}$, where we have used Eq. (5.10). Hence, $c^{ik} = g^{ik}$ and $\boldsymbol{e}^i = g^{ij}\boldsymbol{e}_j$, just like Eq. (5.15). By a similar argument, $\boldsymbol{e}_i = g_{ij}\boldsymbol{e}^j$. We can now establish the identity of the two forms for I in Eq. (5.21), $\boldsymbol{e}_k\boldsymbol{e}^k = g_{kj}\boldsymbol{e}^j g^{kl}\boldsymbol{e}_l = \delta^l{}_j\boldsymbol{e}^j\boldsymbol{e}_l = \boldsymbol{e}^l\boldsymbol{e}_l$.

5.1.4 Coordinate transformations

We now examine how the components of $\mathrm{d}\boldsymbol{s}$ change under invertible coordinate transformations. Dry and technical as this material tends to be, it's highly important for learning about tensors.[18]

Let there be n independent, analytic functions of the coordinates $x^1, \cdots, x^n$, $y^i(x^1, \cdots, x^n)$ $(i = 1, \cdots, n)$, which we can denote $\{y^i(x^j)\}_{i=1}^n$. A set of functions is *independent* if the Jacobian determinant—the determinant of the matrix of partial derivatives $\partial y^i/\partial x^j$ (the *Jacobian matrix*)—does not vanish identically. The functions y^i then provide *another set of coordinates*, a new set of numbers to attach to the *same point in space*,

$$y^i = y^i(x^1, \cdots, x^n) \,, \qquad i = 1, \cdots, n \tag{5.23}$$

and constitute a *transformation of cooordinates*.[19] By assumption (nonvanishing Jacobian determinant), Eq. (5.23) is invertible: $x^j = x^j(y^1, \cdots, y^n)$, $(j = 1, \cdots, n)$.

Consider a point P with coordinates x^i and a neighboring point Q with coordinates $x^i + \mathrm{d}x^i$; see Fig. 5.4. The points (P, Q) define the infinitesimal displacement vector $\mathrm{d}\boldsymbol{s} \equiv \overrightarrow{PQ}$ with components

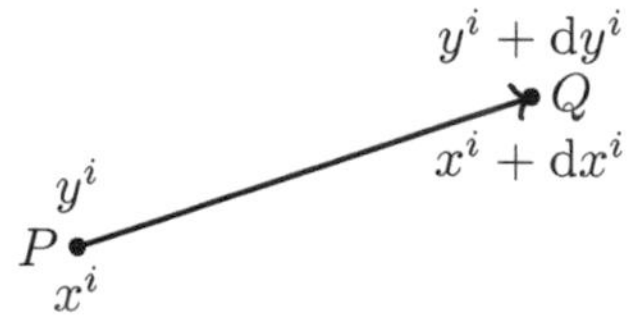

Figure 5.4 Infinitesimally separated points in two coordinate systems.

$\mathrm{d}x^i$. Referring to the same points (P, Q) let there be a different coordinate system, y^j. In this coordinate system the *same vector* $\mathrm{d}\boldsymbol{s}$ has components $\mathrm{d}y^j$. The components of $\mathrm{d}\boldsymbol{s}$ in the two systems of coordinates are related by calculus,

$$\mathrm{d}y^i = \left.\frac{\partial y^i}{\partial x^j}\right|_P \mathrm{d}x^j \equiv A^i_j \mathrm{d}x^j \,, \qquad i = 1, \cdots, n \tag{5.24}$$

where the partial derivatives $A^i_j = \partial y^i/\partial x^j\big|_P$ comprise the elements of the Jacobian matrix associated with the coordinate transformation at point P.[20] The derivatives exist through the analyticity

[18] We should be interested in coordinate transformations for two broad reasons. In SR, a LT is a change in basis vectors in MS. In GR, spacetime cannot be modeled as MS. The curved spacetime of GR requires the mathematical structure of a four-dimensional manifold. A curved manifold cannot be covered by a single coordinate system; it must rely on overlapping "coordinatizations" of spacetime. Overlapping coordinate systems are another way of describing the same spacetime event using different coordinate systems.

[19] For the most part we treat coordinate transformations from the *passive* point of view, where new coordinates are assigned to the *same points*. Coordinate transformations can, however, be viewed in the active sense where new coordinates refer to *different* points in the same coordinate system (same basis vectors), in which case the transformation equations determine a mapping between points.

[20] In writing the elements of the Jacobian matrix as A^i_j, we're using the notation for matrix elements introduced in Section 4.3. The top index labels rows and the bottom index columns.

of the transformation equations, Eq. (5.23). The quantities A^i_j are *constant* for Q in an infinitesimal neighborhood of P. Equation (5.24) then represents a *locally linear* transformation, even though the transformation equations in (5.23) are not necessarily linear.[21]

We adopt a notational device—the *Schouten index convention*—that simplifies tensor transformation equations.[25] Instead of inventing a different symbol for each new coordinate system (y, x', $\bar{x}$, etc), choose x to represent coordinates once and for all. Coordinates in *different coordinate systems* are distinguished by primes attached to *indices*. Thus, Eq. (5.24) is written $\mathrm{d}x^{i'} = A^{i'}_j \mathrm{d}x^j$, with $A^{i'}_j \equiv \partial x^{i'}/\partial x^j$; Eq. (5.23) is written $x^{i'} = x^{i'}(x^j)$.

Coordinate differentials in one coordinate system thus determine the coordinate differentials in another coordinate system. The transformation inverse to Eq. (5.24) is

$$\mathrm{d}x^i = \left.\frac{\partial x^i}{\partial x^{j'}}\right|_P \mathrm{d}x^{j'} \equiv A^i_{j'} \mathrm{d}x^{j'}, \qquad i = 1, \cdots, n \tag{5.25}$$

where $A^i_{j'}$ denotes the partial derivatives $\{\partial x^i/\partial x^{j'}|_P\}$. Combining Eqs. (5.24) and (5.25), $\mathrm{d}x^i = A^i_{j'}\mathrm{d}x^{j'} = A^i_{j'}A^{j'}_k \mathrm{d}x^k$, or

$$\left(\delta^i{}_k - A^i_{j'}A^{j'}_k\right)\mathrm{d}x^k = 0 . \tag{5.26}$$

The *matrices* $\left[A^i_{j'}\right]$ and $\left[A^{j'}_k\right]$ are thus inversely related[22]

$$A^i_{j'}A^{j'}_k = \delta^i{}_k . \tag{5.27}$$

Equation (5.27) is simply the chain rule: $(\partial x^i/\partial x^{j'})(\partial x^{j'}/\partial x^k) = \partial x^i/\partial x^k = \delta^i{}_k$.

Equations (5.24) and (5.25) indicate how the *components* of $\mathrm{d}\boldsymbol{s}$ transform between coordinate systems. How do the basis vectors transform? We now use the key fact that $\mathrm{d}\boldsymbol{s}$ *is the same when expressed in the two basis sets*, $\{\boldsymbol{e}_i\}$ and $\{\boldsymbol{e}_{j'}\}$,[23]

$$\mathrm{d}\boldsymbol{s} = \mathrm{d}x^i\boldsymbol{e}_i = \mathrm{d}x^{j'}\boldsymbol{e}_{j'} . \tag{5.28}$$

Using a familiar strategy, take the inner product between $\mathrm{d}\boldsymbol{s}$ and $\boldsymbol{e}^k$ on both sides of Eq. (5.28),

$$\boldsymbol{e}^k \cdot \mathrm{d}\boldsymbol{s} = \mathrm{d}x^i\boldsymbol{e}^k \cdot \boldsymbol{e}_i = \mathrm{d}x^i\delta^k_i = \mathrm{d}x^k = \mathrm{d}x^{j'}\boldsymbol{e}^k \cdot \boldsymbol{e}_{j'} = A^{j'}_l \mathrm{d}x^l\boldsymbol{e}^k \cdot \boldsymbol{e}_{j'} , \tag{5.29}$$

where we've used Eqs. (5.10) and (5.24). Thus,

$$A^{j'}_l \boldsymbol{e}^k \cdot \boldsymbol{e}_{j'}\mathrm{d}x^l = \mathrm{d}x^k . \tag{5.30}$$

By the reasoning used in Eqs. (5.17) and (5.26), Eq. (5.30) implies that $A^{j'}_l \boldsymbol{e}^k \cdot \boldsymbol{e}_{j'} = \delta^k_l$, in turn implying that $[\boldsymbol{e}^k \cdot \boldsymbol{e}_{j'}]$ is the inverse of the matrix $\left[A^{j'}_l\right]$ (because the inverse of a matrix is unique). Referring to Eq. (5.27), we identify[24]

$$\boldsymbol{e}^k \cdot \boldsymbol{e}_{j'} = A^k_{j'} = \frac{\partial x^k}{\partial x^{j'}} . \tag{5.31}$$

---

[21] We're anticipating the possibility of nonlinear coordinate transformations (which we'll need in GR). At point P, the derivatives $(\partial y^i/\partial x^j)|_P$ are constant only within a small neighborhood of P; a nonlinear transformation can thus be treated as if it's linear within a small region. The LT is strictly linear and the restriction of derivatives to their values at a point is unnecessary.

[22] In other notational schemes one must come up with different symbols for the Jacobian matrix and its inverse; e.g., U and $\overline{U}$, or U and U^{-1}. In either case, one has to remember which matrix applies to which transformation. In the Schouten method there is one symbol with two types of indices, primed and unprimed. Other schemes use the same symbol for the Jacobian matrix and its inverse, but with two ways of writing the indices, $A^i{}_j$ and $A_j{}^i$. Not only does one have to remember which applies to which transformation, such a scheme quickly becomes unintelligible to students at the back of the room.

[23] We have chosen once and for all to represent basis vectors with the symbol $\boldsymbol{e}$.

[24] Note that because of the prime on the index there is (hopefully) no chance of confusing $A^k_{j'} = \boldsymbol{e}^k \cdot \boldsymbol{e}_{j'}$ (from Eq. (5.31)) with $\boldsymbol{e}^k \cdot \boldsymbol{e}_j = \delta^k_j$ (from Eq. (5.10)). In the Schouten method, the symbol A^j_k is *defined* as the Kronecker delta, $A^j_k \equiv \delta^j_k$.

Similarly, in Eq. (5.28) take the inner product between $\mathrm{d}\boldsymbol{s}$ and $\boldsymbol{e}^{k'}$, a dual basis vector in the transformed coordinate system,

$$\boldsymbol{e}^{k'}\cdot\mathrm{d}\boldsymbol{s}=\mathrm{d}x^i\boldsymbol{e}^{k'}\cdot\boldsymbol{e}_i=A^i_{l'}\mathrm{d}x^{l'}\boldsymbol{e}^{k'}\cdot\boldsymbol{e}_i=\mathrm{d}x^{j'}\boldsymbol{e}^{k'}\cdot\boldsymbol{e}_{j'}=\mathrm{d}x^{j'}\delta^{k'}_{j'}=\mathrm{d}x^{k'}\ ,$$

where we've used Eq. (5.25) and the analog of Eq. (5.10) in the transformed coordinate system, $\boldsymbol{e}^{k'}\cdot\boldsymbol{e}_{j'}=\delta^{k'}_{j'}$. We conclude that $A^i_{l'}\boldsymbol{e}^{k'}\cdot\boldsymbol{e}_i=\delta^{k'}_{l'}$ and hence that

$$\boldsymbol{e}^{k'}\cdot\boldsymbol{e}_i=A^{k'}_i=\frac{\partial x^{k'}}{\partial x^i}\ . \tag{5.32}$$

Jacobian matrices therefore connect basis vectors in different coordinate systems (as well as coordinate differentials). From Eqs. (5.31), (5.32), and the identity operator, Eq. (5.21), we obtain the transformation equations between basis vectors,

$$\boldsymbol{e}_i=\boldsymbol{I}\cdot\boldsymbol{e}_i=\boldsymbol{e}_{j'}\boldsymbol{e}^{j'}\cdot\boldsymbol{e}_i=\boldsymbol{e}_{j'}A^{j'}_i \qquad \boldsymbol{e}_{i'}=\boldsymbol{I}\cdot\boldsymbol{e}_{i'}=\boldsymbol{e}_j\boldsymbol{e}^j\cdot\boldsymbol{e}_{i'}=\boldsymbol{e}_jA^j_{i'}\ . \tag{5.33}$$

Likewise, the dual basis vectors transform inversely to Eq. (5.33)

$$\boldsymbol{e}^i=\boldsymbol{I}\cdot\boldsymbol{e}^i=\boldsymbol{e}^{j'}\boldsymbol{e}_{j'}\cdot\boldsymbol{e}^i=A^i_{j'}\boldsymbol{e}^{j'} \qquad \boldsymbol{e}^{i'}=\boldsymbol{I}\cdot\boldsymbol{e}^{i'}=\boldsymbol{e}^j\boldsymbol{e}_j\cdot\boldsymbol{e}^{i'}=A^{i'}_j\boldsymbol{e}^j\ . \tag{5.34}$$

You will refer to these equations more than once; remember where you put them.

5.1.5 Tensor transformation properties: Contravariant, covariant, and all that

5.1.5.1 *Scalar fields: $\phi'(\boldsymbol{r}')=\phi(\boldsymbol{r})$*

An *invariant* is a quantity that does not change under coordinate transformations. The simplest type of invariant is a *scalar*, a number, such as the spacetime separation. A *scalar field* is a function $\phi(\boldsymbol{r})$ that assigns a number to each point in space. *Points* are invariant under passive coordinate transformations, which attach different labels (coordinates) to points, but do nothing to the points themselves. Any *set* of points is therefore invariant, as is any point function. The *value* of a scalar field is invariant under passive coordinate transformations.[25] If a point has coordinates x^i and $x^{j'}$ in two coordinate systems, we require of a scalar field that $\phi(x^i)=\phi(x^i(x^{j'}))\equiv\phi'(x^{j'})$, i.e., the *form* of the function of the transformed coordinates may change, ϕ', but not its value $\phi'(\boldsymbol{r}')$. Scalars do not exhaust the possible types of invariants under coordinate transformations; *invariants other than scalars exist.*

5.1.5.2 *Contravariant tensors*

How do the components of vectors *other* than $\mathrm{d}\boldsymbol{s}$ transform under a change of basis? The question can be answered because we know how basis vectors transform. Calculus was used in Eq. (5.25) to specify the transformation property of the components of $\mathrm{d}\boldsymbol{s}=\mathrm{d}x^i\boldsymbol{e}_i$. We could establish how basis vectors transform (Eqs. (5.33) and (5.34)) by requiring $\mathrm{d}\boldsymbol{s}$ to be the same when represented in different bases, $\mathrm{d}\boldsymbol{s}=\mathrm{d}x^i\boldsymbol{e}_i=\mathrm{d}x^{j'}\boldsymbol{e}_{j'}$, Eq. (5.28). That $\mathrm{d}\boldsymbol{s}$ is the same in different bases implies that *it has an existence independent of coordinate system.* A quantity having such a property is said to be a *geometric object*. The infinitesimal displacement vector $\mathrm{d}\boldsymbol{s}$ is the *prototype* of a *class* of geometric objects referred to as *contravariant vectors.*[26]

A vector $\boldsymbol{T}=T^i\boldsymbol{e}_i$ has contravariant components T^i if they transform as $T^{i'}=A^{i'}_jT^j$, i.e., like Eq. (5.25). In that way $\boldsymbol{T}$ is a geometric object, $\boldsymbol{T}=T^j\boldsymbol{e}_j=T^jA^{i'}_j\boldsymbol{e}_{i'}=T^{i'}\boldsymbol{e}_{i'}$, where we've used

[25] The temperature at a point, for example, doesn't care what coordinates you assign to the point.

[26] By a contravariant vector, we mean a vector with contravariant components. The same terminology applies to contravariant tensors, covariant tensors, etc.

Eq. (5.33). *The contravariant components transform inversely ("contra") to the transformation of basis vectors.* Any set of n quantities $\{T^i\}$ that transform like the components of $\mathrm{d}\boldsymbol{s}$,

$$T^{k'} = A^{k'}_j T^j = \frac{\partial x^{k'}}{\partial x^j} T^j , \tag{5.35}$$

are said to be the contravariant components of a vector. *Any mathematical objects that transform like the components of* $\mathrm{d}\boldsymbol{s}$ *form the contravariant components of a vector.*

A set of $(n)^2$ quantities $\{T^{ij}\}$ are said to be the *contravariant components of a second-rank tensor* if they transform as

$$T^{i'j'} = A^{i'}_k A^{j'}_l T^{kl} = \frac{\partial x^{i'}}{\partial x^k}\frac{\partial x^{j'}}{\partial x^l} T^{kl} . \tag{5.36}$$

One way to create tensors is by multiplying vector components. For A^i and B^j contravariant vector components, $T^{ij} \equiv A^i B^j$ comprise the components of a second-rank tensor because they automatically transform properly. A set of $(n)^r$ objects $\{T^{i_1 \cdots i_r}\}$ that transform as the product of r contravariant vector components, $T^{k_1' k_2' \cdots k_r'} \equiv A^{k_1'}_{m_1} A^{k_2'}_{m_2} \cdots A^{k_r'}_{m_r} T^{m_1 m_2 \cdots m_r}$ are the contravariant components of a *tensor of rank* r.

Example. Show that $\{g^{ij}\}$ are contravariant tensor elements. To do so, we must show that they transform properly. Starting from the definition Eq. (5.14), $g^{i'j'} = \boldsymbol{e}^{i'} \cdot \boldsymbol{e}^{j'} = A^{i'}_k \boldsymbol{e}^k \cdot A^{j'}_l \boldsymbol{e}^l = A^{i'}_k A^{j'}_l g^{kl}$, where we have used Eq. (5.34), in agreement with Eq. (5.36).

Transformation relations such as Eq. (5.36) pertain to the *components* of tensors, but they do not define what a tensor *is*. Like vectors, *tensors are geometric objects.* It's common practice to refer to symbols like T^{ij} as "tensors," but that's not correct. We'll use a special notation to indicate tensors: boldface **Roman** font for tensors, $\mathbf{T}$, as distinguished from boldface ***italic*** font for vectors, $\boldsymbol{A}$. A second-rank tensor $\mathbf{T}$ is a generalization of a vector,[27] $\mathbf{T} \equiv T^{ij}\boldsymbol{e}_i\boldsymbol{e}_j$. A second-rank tensor is independent of coordinate system,

$$\begin{aligned}\mathbf{T}' =& T^{i'j'}\boldsymbol{e}_{i'}\boldsymbol{e}_{j'} = A^{i'}_l A^{j'}_m T^{lm} A^k_{i'} A^n_{j'} \boldsymbol{e}_k\boldsymbol{e}_n = \left(A^k_{i'} A^{i'}_l\right)\left(A^n_{j'} A^{j'}_m\right) T^{lm}\boldsymbol{e}_k\boldsymbol{e}_n \\ =& \delta^k_{\ l}\delta^n_{\ m} T^{lm}\boldsymbol{e}_k\boldsymbol{e}_n = T^{kn}\boldsymbol{e}_k\boldsymbol{e}_n = \mathbf{T} ,\end{aligned}$$

where we've used Eqs. (5.36), (5.33), and (5.27). It's important to distinguish the tensor components T^{ij} from the tensor as a whole, $\mathbf{T}$, just as we distinguish a vector, $\boldsymbol{A}$, from its components, A^i. A rank-r tensor is $\mathbf{T} = T^{k_1 k_2 \cdots k_r}\boldsymbol{e}_{k_1}\boldsymbol{e}_{k_2}\cdots\boldsymbol{e}_{k_r}$. At some point we'll break training and start referring to tensor components as tensors (despite our admonition); continually referring to "the tensor whose components are T^{ij}" becomes cumbersome. The distinction between a tensor and its components should be kept in mind nevertheless.

5.1.5.3 Covariant tensors and mixed tensors

The invariance of scalar fields ($\phi'(\boldsymbol{r}') = \phi(\boldsymbol{r})$) allows us to introduce how derivatives transform between coordinate systems. Using Eq. (5.25),

$$\frac{\partial \phi'}{\partial x^{j'}} = \frac{\partial x^i}{\partial x^{j'}}\frac{\partial \phi}{\partial x^i} = A^i_{j'}\frac{\partial \phi}{\partial x^i} . \tag{5.37}$$

Again, calculus is used to establish a prototype transformation equation. The form of Eq. (5.37) is the inverse of the form of Eq. (5.35) (note the location of the indices); Eq. (5.37), however,

---

[27]The *order* in which the basis vectors is written is important; in general $T^{ij} \neq T^{ji}$.

has the same form as Eq. (5.33). Just as the infinitesimal displacement vector $d\boldsymbol{s}$ is the prototype contravariant vector, the gradient of a scalar field $\boldsymbol{\nabla}\phi$ is the prototype of a class of geometric objects called *covariant vectors*. A vector $\boldsymbol{T} = T_m \boldsymbol{e}^m$ has covariant components T_i if they transform as $T_{n'} = A^m_{n'} T_m$, so that $\boldsymbol{T} = T_m \boldsymbol{e}^m = T_m A^m_{n'} \boldsymbol{e}^{n'} = T_{n'} \boldsymbol{e}^{n'}$, from Eq. (5.34). Any set of n quantities $\{T_i\}$ that transform like

$$T_{j'} = A^i_{j'} T_i = \frac{\partial x^i}{\partial x^{j'}} T_i \tag{5.38}$$

are the *covariant components of a vector*—they "co-vary" with the basis vectors[28] $\boldsymbol{e}_i$.

A set of $(n)^2$ objects $\{T_{ij}\}$ that transform like

$$T_{i'j'} = A^k_{i'} A^l_{j'} T_{kl} \tag{5.39}$$

are *covariant components of a second-rank tensor*. A second-rank tensor with covariant components, $\mathbf{T} \equiv T_{ij} \boldsymbol{e}^i \boldsymbol{e}^j$, is independent of basis (as can readily be shown). A set of $(n)^r$ objects $\{T_{k_1 \cdots k_r}\}$ that transform like $T_{k'_1 \cdots k'_r} = A^{m_1}_{k'_1} \cdots A^{m_r}_{k'_r} T_{m_1 \cdots m_r}$ are the covariant components of a tensor of rank r, $\mathbf{T} = T_{k_1 \cdots k_r} \boldsymbol{e}^{k_1} \cdots \boldsymbol{e}^{k_r}$.

We now define *mixed tensors*. A set of $(n)^3$ objects $\{T^i_{\ jk}\}$ that transform as

$$T^{i'}_{\ j'k'} = A^{i'}_p A^l_{j'} A^m_{k'} T^p_{\ lm} \tag{5.40}$$

are the components of a third-rank tensor with one contravariant and two covariant indices. Notationally, the upper and lower indices are set apart, $T^i_{\ jk}$. It's *good hygiene* in writing the components of mixed tensors not to put superscript indices aligned with subscript indices (as in T^i_{jk}); adopting this convention helps avoid mistakes.[29] The components of a mixed tensor of type (p, q) (having p contravariant indices and q covariant indices) are a set of $n^{(p+q)}$ objects $\{T^{i_1 \cdots i_p}_{\ \ \ \ \ \ \ j_1 \cdots j_q}\}$ that transform as

$$T^{k'_1 \cdots k'_p}_{\ \ \ \ \ \ \ \ m'_1 \cdots m'_q} = A^{k'_1}_{t_1} \cdots A^{k'_p}_{t_p} A^{s_1}_{m'_1} \cdots A^{s_q}_{m'_q} T^{t_1 \cdots t_p}_{\ \ \ \ \ \ \ s_1 \cdots s_q} . \tag{5.41}$$

The tensor of type (p, q) is $\mathbf{T} = T^{k_1 \cdots k_p}_{\ \ \ \ \ \ \ m_1 \cdots m_q} \boldsymbol{e}_{k_1} \cdots \boldsymbol{e}_{k_p} \boldsymbol{e}^{m_1} \cdots \boldsymbol{e}^{m_q}$. A *tensor* as a geometric object is independent of the basis vectors used to represent it: $\mathbf{T} = T_{ij} \boldsymbol{e}^i \boldsymbol{e}^j = T^i_{\ j} \boldsymbol{e}_i \boldsymbol{e}^j = T_i^{\ j} \boldsymbol{e}^i \boldsymbol{e}_j = T^{ij} \boldsymbol{e}_i \boldsymbol{e}_j$.

Is $\delta^i_{\ j}$ an element of a second-rank mixed tensor as the notation suggests? How does it transform? Using Eq. (5.41), $(\delta^i_{\ j})' \equiv A^{i'}_k A^m_{j'} \delta^k_{\ m} = A^{i'}_k A^k_{j'} = \delta^{i'}_{\ j'}$, from Eq. (5.27). The transformation of δ^i_j, $(\delta^i_j)'$, has the value of δ^i_j in the new frame. The Kronecker delta is a *constant* tensor, a tensor with elements that are numerically the same in every coordinate system. (The same is not true of δ_{ij}.[30]) Equation (5.10) defines the elements of the *mixed metric tensor*, $g^i_{\ j} = \boldsymbol{e}^i \cdot \boldsymbol{e}_j = \delta^i_{\ j}$.

5.1.5.4 *Inner product is a scalar*

We now show that the inner product defined by Eq. (5.2) is invariant. Using Eq. (5.20),

$$(\boldsymbol{T} \cdot \boldsymbol{U})' = T_{i'} U^{i'} = A^k_{i'} A^{i'}_j T_k U^j = \delta^k_{\ j} T_k U^j = T_j U^j = \boldsymbol{T} \cdot \boldsymbol{U} , \tag{5.42}$$

where we've used Eqs. (5.35), (5.38), and (5.27). *If we know the value of the inner product in one coordinate system, we know it in all coordinate systems.* Note that the metric tensor is lurking in Eq. (5.42) from lowering indices: $\boldsymbol{T} \cdot \boldsymbol{U} = g_{\alpha\beta} T^\alpha U^\beta = T_\beta U^\beta$.

---

[28] A useful mnemonic for the placement of indices is "co goes below."

[29] For example, by writing $g_{\nu\alpha} C^{\mu\alpha\lambda} = C^{\mu\ \lambda}_{\ \nu}$ we know where ν "comes from." Had we written $C^{\mu\lambda}_\nu$, where would ν "go" if later we decide to raise the index?

[30] The Kronecker symbol was defined in Eq. (5.10) as a mixed tensor. The Kronecker symbol as it's usually written, δ_{ij}, is in general tensor analysis obtained by lowering an index: $\delta_{ij} = g_{ik} \delta^k_{\ j} = g_{ij}$.

5.1.6 Tensor contraction and outer product

When a contravariant (upper) index is set equal to a covariant (lower) index and summed over, it reduces a tensor of type (p, q) to one of type $(p-1, q-1)$, i.e., *it lowers the tensor rank by two*, a process called *contraction*. Consider $T^i_{\ j} \equiv U^i V_j$, a second-rank tensor formed from the product of vector components U^i and V_j. If we set $j = i$ and sum over i, $T^i_{\ i} \equiv U^i V_i$, we form a scalar. The inverse process, of forming the components of higher-rank tensors from products of the components of lower-rank tensors, is called the *outer product*. The product of the components of a tensor of type (r, s) with the components of a tensor of type (p, q) form the components a new tensor of type $(r+p, s+q)$. For example, the quantities $T^{ij}_{\ \ k} = U^{ij} V_k$ are the components of a third-rank tensor. If we set $j = k$ and sum, we lower the rank by two to form a vector (first-rank tensor), $T^i \equiv U^{ik} V_k$. To prove that T^i is the component of a vector, we must show that it transforms like one,

$$\begin{aligned} T^{i'} &= U^{i'k'} V_{k'} = A^{i'}_l A^{k'}_m U^{lm} A^n_{k'} V_n = A^{i'}_l \left(A^{k'}_m A^n_{k'} \right) U^{lm} V_n = A^{i'}_l \delta^n_{\ m} U^{lm} V_n \\ &= A^{i'}_l U^{ln} V_n = A^{i'}_l T^l \ , \end{aligned}$$

where we've used Eqs. (5.36), (5.38), and (5.27). The *tensor* that results from components obtained through outer products is called the *tensor product*, $\mathbf{C} = \boldsymbol{A} \otimes \boldsymbol{B}$. If $\boldsymbol{A} = \alpha^i \boldsymbol{e}_i$ and $\boldsymbol{B} = \beta^j \boldsymbol{e}_j$, $\mathbf{C} = \alpha^i \beta^j \boldsymbol{e}_i \otimes \boldsymbol{e}_j \equiv C^{ij} \boldsymbol{e}_i \otimes \boldsymbol{e}_j$. The quantities $\boldsymbol{e}_i \otimes \boldsymbol{e}_j$ are a basis for type $(2, 0)$ tensors formed from the basis vectors $\boldsymbol{e}_i$ for type $(1, 0)$ tensors.

5.1.7 Quotient theorem

The direct test for whether a set of mathematical objects form tensor components is to verify that they transform appropriately. There is an indirect method for checking the tensorial character of a set of quantities known loosely as the *quotient theorem*, which says that in an equation $UV = T$, if V and T are known to be elements of a tensor, then U is also a tensor element. With the quotient theorem, we use known tensors to ascertain the tensor character of putative tensors.

Suppose $\{T_r\}$ is a set of quantities we wish to test for its tensor character. Let $\{X^r\}$ be the components of a contravariant vector. If the sum $T_r X^r$ is an invariant, then by the quotient theorem, the quantities T_r form the elements of a covariant vector. From the given invariance, we have $T_r X^r = T_{s'} X^{s'}$. We can use the known transformation properties of X^r, Eq. (5.35), to write $T_r X^r = T_{s'} A^{s'}_j X^j$, or equivalently $\left(T_j - T_{s'} A^{s'}_j \right) X^j = 0$. Because the $\{X^j\}$ are arbitrary, the terms in parentheses must vanish, establishing $T_j = A^{s'}_j T_{s'}$ as the elements of a covariant vector.

Let's take a more challenging example. Suppose we run into an equation,

$$T^{mn}_{\ \ \ kl} = U^m S^n_{\ kl} \ , \tag{5.43}$$

where it's known that $\mathbf{T}$ is a type $(2, 2)$ tensor and $\boldsymbol{U}$ is a vector. By the quotient theorem we're entitled to conclude that $\mathbf{S}$ is a tensor of type $(1, 2)$. To show this, introduce contravariant vector components $\{x^i\}$ and covariant vector components $\{y_r\}$. Multiply Eq. (5.43) by $y_m y_n x^k x^l$ and contract,

$$T^{mn}_{\ \ \ kl} y_m y_n x^k x^l = U^m S^n_{\ kl} y_m y_n x^k x^l = (U^m y_m) S^n_{\ kl} y_n x^k x^l \ . \tag{5.44}$$

We take this step so that the left side of Eq. (5.44) is a scalar (because we have contracted all indices); $U^m y_m$ is also a scalar. We have therefore established that $S^n_{\ kl} y_n x^k x^l$ is a scalar, and thus $S^{n'}_{\ k'l'} y_{n'} x^{k'} x^{l'} = S^n_{\ kl} y_n x^k x^l$. Now use the transformation properties of x^i and y_m, Eqs. (5.35) and (5.38), $S^{n'}_{\ k'l'} A^i_{n'} y_i A^{k'}_j x^j A^{l'}_m x^m = S^i_{\ jm} y_i x^j x^m$. Because x^i and y_j are arbitrary, $S^{n'}_{\ k'l'} A^i_{n'} A^{k'}_j A^{l'}_m = S^i_{\ jm}$, establishing $\mathbf{S}$ as a type $(1, 2)$ tensor.

5.1.8 Geometric interpretation of covariant vectors

We've now met the main players: scalars, contravariant and covariant vectors, and their generalizations as tensors. Contravariant vectors share the attributes of the displacement vector and should simply be called vectors. What then are covariant vectors? Their components transform like those of the gradient vector, which doesn't immediately convey a picture of what they are. For many students a course in relativity is their first introduction to covariant vectors, and one might wonder how important they are, given that one has arrived this far in a scientific education without encountering them.[31] Can we provide a geometric interpretation of covariant vectors?

As we now show, *covariant vectors represent families of parallel planes.*[32] Figure 5.5 shows a

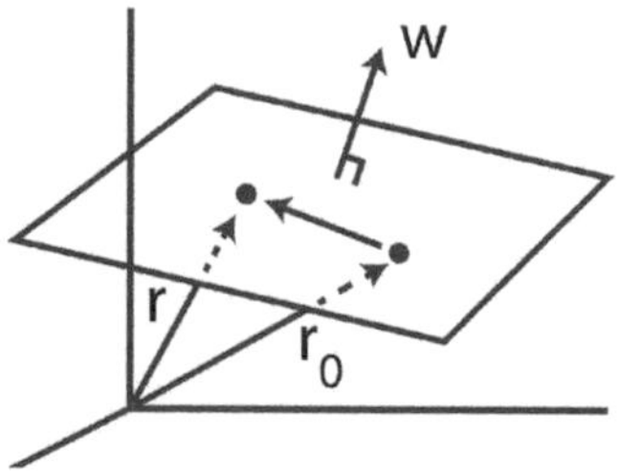

Figure 5.5 Plane in 3-space. The vector $\boldsymbol{r} - \boldsymbol{r}_0$ lies in the plane.

plane in 3-space. Locate a point on the plane having coordinates (x_0^1, x_0^2, x_0^3) with the fixed vector $\boldsymbol{r}_0$. Let $\boldsymbol{r}$ locate an arbitrary point *on the plane* with coordinates (x^1, x^2, x^3). Let there be a vector $\boldsymbol{w}$ perpendicular to the plane with components (w_1, w_2, w_3). The vector $\boldsymbol{r} - \boldsymbol{r}_0$ lies in the plane and thus $\boldsymbol{w} \cdot (\boldsymbol{r} - \boldsymbol{r}_0) = 0$. By the quotient theorem, $\boldsymbol{w}$ is a covariant vector. The coordinates $\{x^j\}$ of all points on the plane then satisfy the *equation of a plane* $w_i x^i = d$, where $d \equiv \boldsymbol{w} \cdot \boldsymbol{r}_0$ is a constant. The *intercept* p^i of the plane with the ith coordinate axis is found by setting all other coordinates $x^j = 0$ $(j \neq i)$, with the result that $p^i = d/w_i$ (see Fig. 5.6). For a plane parallel to

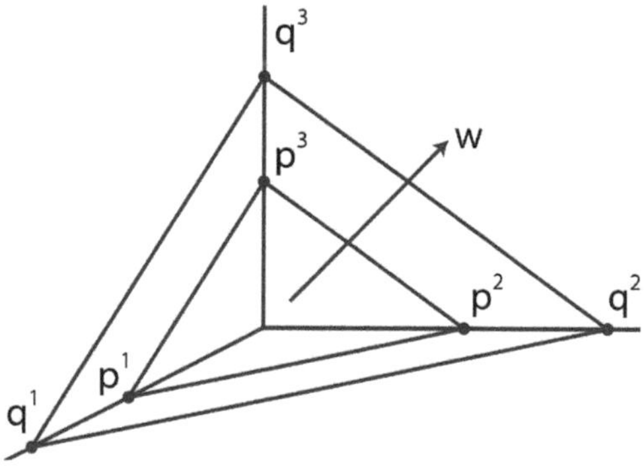

Figure 5.6 Covariant vectors $\boldsymbol{w}$ represent families of parallel planes.

the first, its coordinates satisfy $w_i x^i = d'$, where $d' \neq d$ is another constant. The intercepts of the parallel plane are given by $q^i \equiv d'/w_i$. Subtracting these equations, $w_i = (d' - d) / (q^i - p^i)$. The components w_i are therefore related to the intercepts made by a pair of parallel planes with the coordinate axes. The direction of $\boldsymbol{w}$ is perpendicular to the planes, and the magnitude is specified

[31] One reason covariant vectors are relatively unfamiliar is that the distinction between contravariant and covariant is unnecessary in orthogonal coordinate systems, and physics is most often done using orthogonal coordinate systems. We're marching towards GR, however, which touts itself as applying to *any* coordinate system.

[32] Planes are two-dimensional structures embedded in three-dimensional space. Planes in higher-dimensional spaces are called *hyperplanes*: $(n-1)$-dimensional structures embedded in n-dimensional spaces, with $n > 3$.

by the distance separating the planes in the family, with *magnitude inversely proportional to the interplanar separation.*[33]

The connection with gradients thus becomes apparent. The *level set* of a function is the locus of points such that $f(x^1, \cdots, x^n) = f_0$, where f_0 is a given constant value. As is well known, the gradient of a function is orthogonal to its level set.[34] Consider the change in a scalar field ϕ over a displacement $\mathrm{d}\boldsymbol{s} = \mathrm{d}u^i \boldsymbol{e}_i$, with

$$\mathrm{d}\phi = \frac{\partial \phi}{\partial u^i} \mathrm{d}u^i \,. \tag{5.45}$$

We can represent $\mathrm{d}\phi$ (a scalar) as an inner product between $\mathrm{d}\boldsymbol{s}$ and a new vector (the gradient) such that $\mathrm{d}\phi \equiv \boldsymbol{\nabla}\phi \cdot \mathrm{d}\boldsymbol{s}$. If $\mathrm{d}\boldsymbol{s}$ lies within the level set, $\mathrm{d}\phi = 0$, implying that $\boldsymbol{\nabla}\phi$ is orthogonal to the level set of ϕ. By the quotient theorem, $\boldsymbol{\nabla}\phi$ is a covariant vector which we can represent in the dual basis, $\boldsymbol{\nabla} \equiv \boldsymbol{e}^i \nabla_i$,

$$\mathrm{d}\phi = \boldsymbol{\nabla}\phi \cdot \mathrm{d}\boldsymbol{s} = (\nabla_i \phi)\boldsymbol{e}^i \cdot (\mathrm{d}u^j \boldsymbol{e}_j) = (\nabla_i \phi)\, \mathrm{d}u^j \delta^i{}_j = (\nabla_i \phi)\, \mathrm{d}u^i \,. \tag{5.46}$$

Comparing Eqs. (5.46) and (5.45), $\nabla_i \equiv \partial/\partial u^i$. Note the location of the indices: A derivative with respect to a contravariant component, u^i, is a covariant vector component, ∇_i. To show that ∇_i is the component of a covariant vector is simple; see Eq. (5.37), $\nabla_{i'} = A^j_{i'} \nabla_j$.

Example. The electric field $\boldsymbol{E}$ is a geometric object. It has a natural representation as a covariant vector $E_i = -\nabla_i \phi$ from its role as the gradient of the electrostatic potential $\phi(\boldsymbol{r})$. It's also naturally represented as a contravariant vector from its relation to the Newtonian equation of motion $E^i = (m/q)\mathrm{d}v^i/\mathrm{d}t$. The two quantities are related through the metric tensor, $E^i = g^{ij} E_j$. The distinction is necessary only in non-orthogonal coordinate systems.

Gradients provide a geometric interpretation of the dual basis vectors. Vectors normal to the level set of a function[35] $f(u, v, w)$ can be expressed in a basis of *normals to coordinate surfaces*,

$$\boldsymbol{e}^u \equiv \boldsymbol{\nabla} u \qquad \boldsymbol{e}^v \equiv \boldsymbol{\nabla} v \qquad \boldsymbol{e}^w \equiv \boldsymbol{\nabla} w \,. \tag{5.47}$$

A coordinate surface is the surface that results by holding one of the coordinates fixed.[36] A sphere, for example, results by holding the radial coordinate fixed and letting the coordinates θ, ϕ vary; the sphere is the coordinate surface associated with the radial coordinate.[37] Figure 5.7 illustrates the distinction between coordinate basis vectors $\boldsymbol{e}_\alpha$ (tangents to coordinate curves) and the dual basis vectors $\boldsymbol{e}^\beta$, orthogonal to coordinate surfaces. The vectors in Eq. (5.47) are dual to the basis vectors $\boldsymbol{e}_i$ in the sense of Eq. (5.10), $(u, v, w \equiv u^1, u^2, u^3)$

$$\boldsymbol{e}^i \cdot \boldsymbol{e}_j = \boldsymbol{\nabla} u^i \cdot \frac{\partial \boldsymbol{r}}{\partial u^j} = \frac{\partial u^i}{\partial x^k}\frac{\partial x^k}{\partial u^j} = \frac{\partial u^i}{\partial u^j} = \delta^i_j \,. \tag{5.48}$$

Example. Consider a coordinate system (u, v, w) defined by $x = u+v$, $y = u-v$, and $z = \alpha uv + w$, where α is a constant. These equations can be inverted, with

$$u = \frac{1}{2}(x+y) \qquad v = \frac{1}{2}(x-y) \qquad w = z - \frac{\alpha}{4}(x^2 - y^2) \,.$$

---

[33]Your inner mathematician would want to know that vectors defined from families of parallel planes can be added to other such vectors to produce new vectors of the same type. They can, as shown in the delightful book by Weinreich.[26]

[34]Anyone who's worked with topographic maps knows that a steeper terrain (gradient) is implied by contours of equal elevation spaced closer together.

[35]Defined with respect to a general (u, v, w) coordinate system.

[36]In an n-dimensional space with $n > 3$, $(n-1)$-dimensional coordinate surfaces—called *hypersurfaces*—result by holding one of the n coordinates fixed.

[37]For a sphere, the unit vector $\hat{\boldsymbol{r}}$ is both tangent to the radial coordinate curve and orthogonal to the radial coordinate surface: The distinction between the two types of vectors is unnecessary in orthogonal coordinate systems.

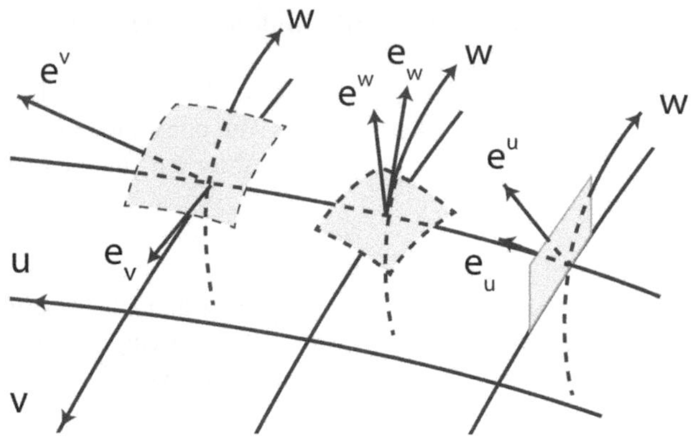

Figure 5.7 Vectors of the coordinate basis e_α are tangent to coordinate curves, vectors of the dual basis e^β are orthogonal to coordinate surfaces.

The coordinate surfaces for $u = u_0$ and $v = v_0$ are planes, while the surface for $w = w_0$ is a hyperbolic paraboloid. The position vector can be written

$$\boldsymbol{r} = (u+v)\hat{\boldsymbol{x}} + (u-v)\hat{\boldsymbol{y}} + (\alpha uv + w)\hat{\boldsymbol{z}} \,.$$

Using Eq. (5.9), we find the coordinate basis vectors

$$\boldsymbol{e}_u = \frac{\partial \boldsymbol{r}}{\partial u} = \hat{\boldsymbol{x}} + \hat{\boldsymbol{y}} + \alpha v \hat{\boldsymbol{z}} \qquad \boldsymbol{e}_v = \frac{\partial \boldsymbol{r}}{\partial v} = \hat{\boldsymbol{x}} - \hat{\boldsymbol{y}} + \alpha u \hat{\boldsymbol{z}} \qquad \boldsymbol{e}_w = \frac{\partial \boldsymbol{r}}{\partial w} = \hat{\boldsymbol{z}} \,.$$

It's easily shown that $\boldsymbol{e}_u \cdot \boldsymbol{e}_v = \alpha^2 uv$, $\boldsymbol{e}_u \cdot \boldsymbol{e}_w = \alpha v$, $\boldsymbol{e}_v \cdot \boldsymbol{e}_w = \alpha u$; this is not an orthogonal coordinate system. From Eq. (5.47),

$$\boldsymbol{e}^u = \boldsymbol{\nabla} u = \frac{1}{2}(\hat{\boldsymbol{x}} + \hat{\boldsymbol{y}}) \qquad \boldsymbol{e}^v = \boldsymbol{\nabla} v = \frac{1}{2}(\hat{\boldsymbol{x}} - \hat{\boldsymbol{y}}) \qquad \boldsymbol{e}^w = \boldsymbol{\nabla} w = \hat{\boldsymbol{z}} - \frac{\alpha}{2}(x\hat{\boldsymbol{x}} - y\hat{\boldsymbol{y}}) \,.$$

It can be verified that $\boldsymbol{e}^u \cdot \boldsymbol{e}_u = \boldsymbol{e}^v \cdot \boldsymbol{e}_v = \boldsymbol{e}^w \cdot \boldsymbol{e}_w = 1$ and $\boldsymbol{e}^u \cdot \boldsymbol{e}_v = \boldsymbol{e}^u \cdot \boldsymbol{e}_w = \boldsymbol{e}^v \cdot \boldsymbol{e}_w = 0$. Equation (5.10) is satisfied.

5.1.9 Connection with relativity

If a tensor equation is true in one reference frame, it's true in all reference frames. Suppose we have a relation between tensors, valid in one coordinate system, $A_{ij} = B_{ij}$. Write this equation as $D_{ij} = 0$, where $D_{ij} \equiv A_{ij} - B_{ij}$. If $D_{ij} = 0$ in one coordinate system, then $D_{i'j'} = 0$ in *any* coordinate system, because the tensor transformation equations are linear and homogeneous. Thus, $A_{i'j'} = B_{i'j'}$ in all coordinate systems. While the individual *components* A_{ij}, B_{ij} transform between frames, the *form* of the equation is the same in all coordinate systems.[38] For physical laws to be the same for all observers, they must be formulated in a covariant manner, which is why it's so important to be able to establish whether a given set of objects constitute a tensor.

Let's pause for a passage from Einstein's 1916 article on GR. Based on what we've covered in this chapter, you should be able to follow what he says:[9, p121]

[38] Tensor equations are called *covariant equations* because their form co-varies with transformations between coordinate systems.

> Let certain things ("tensors") be defined with respect to any system of coordinates by a number of functions of the coordinates, called the "components" of the tensor. There are then certain rules by which these components can be calculated for a new system of coordinates, if they are known for the original system of coordinates, and if the transformation connecting the two systems is known. The things hereafter called tensors are further characterized by the fact that the equations of transformation for their components are linear and homogeneous. Accordingly, all the components in the new system vanish, if they all vanish in the original system. If, therefore, a law of nature is expressed by equating all the components of a tensor to zero, it is generally covariant. By examining the laws of the formation of tensors, we acquire the means of formulating generally covariant laws.

We can write the LT in tensor notation as a coordinate transformation in MS:

$$x^{\mu'} = L^{\mu'}_{\nu} x^{\nu} \,. \tag{5.49}$$

Regardless of the details of the LT (whether simple, as in Eq. (3.17), or more complicated as in Eq. (3.24)), because the LT is linear, $L^{\mu'}_{\nu} = \partial x^{\mu'}/\partial x^{\nu}$, the same as Eq. (5.35). Thus, *we can use all the apparatus of tensor analysis in SR with the LT as the Jacobian matrix*, A^i_j, and indeed we *must* use tensor analysis in relativity to formulate covariant equations. The inverse of Eq. (5.49) is $x^{\mu} = L^{\mu}_{\nu'} x^{\nu'}$ where $L^{\mu}_{\nu'}$ is obtained from $L^{\mu'}_{\nu}$ by letting $\beta \to -\beta$. The analog of Eq. (5.27) is $L^{\mu}_{\nu'} L^{\nu'}_{\alpha} = \delta^{\mu}{}_{\alpha}$.

By definition the LT satisfies Eq. (4.12), $L^{\mathrm{T}}\eta L = \eta$, a matrix equation. In terms of tensor components (using $(L^T)^{\mu}_{\lambda} = L^{\lambda}_{\mu}$), Eq. (4.12) is equivalent to $\eta_{\mu\nu} = L^{\kappa}_{\mu} L^{\lambda}_{\nu} \eta_{\kappa\lambda}$. The defining requirement of a LT is none other than the transformation equation for the Lorentz metric! The Lorentz metric is the *same* in all IRFs: The principle of relativity requires the invariance of the spacetime interval $(\mathrm{d}s)^2$. The Lorentz metric is thus a constant tensor in MS. If x^{μ} transforms as in Eq. (5.49), the basis vectors transform inversely, showing that the LT is equivalently a change of basis vectors,

$$\boldsymbol{e}_{\alpha'} = L^{\beta}_{\alpha'} \boldsymbol{e}_{\beta} \,. \tag{5.50}$$

The time axis in an IRF is perpendicular to the spatial axes, so that $\eta_{0i} = \boldsymbol{e}_0 \cdot \boldsymbol{e}_i = 0$. It would not appear from Fig. 2.9 and similar figures that time is orthogonal to space in the transformed frame. Nevertheless, as we now show, in the transformed frame $\eta_{0'1'} = 0$. Using Eq. (5.50),

$$\begin{aligned} \eta_{1'0'} \equiv \boldsymbol{e}_{1'} \cdot \boldsymbol{e}_{0'} &= L^{\alpha}_{1'} \boldsymbol{e}_{\alpha} \cdot L^{\beta}_{0'} \boldsymbol{e}_{\beta} = L^{\alpha}_{1'} L^{\beta}_{0'} \eta_{\alpha\beta} \\ &= L^{0}_{1'} L^{0}_{0'} \eta_{00} + L^{1}_{1'} L^{1}_{0'} \eta_{11} + L^{2}_{1'} L^{2}_{0'} \eta_{22} + L^{3}_{1'} L^{3}_{0'} \eta_{33} \,, \end{aligned} \tag{5.51}$$

where we've used that $[\eta]$ is diagonal. Thus, $\eta_{1'0'} = (\beta\gamma)(\gamma)(-1) + (\gamma)(\beta\gamma)(1) = 0$.

5.2 TENSOR DENSITIES, INVARIANT VOLUME ELEMENT

We now bring onto the stage another member from our cast of mathematical players, *densities* (the final member of the "fab four" prototypes of physical quantities, in addition to scalars, covariant, and contravariant vectors). Consider the integral of a scalar field, $\int \phi(x) \mathrm{d}^n x$. Is the *integral* a scalar? Not in general. While $\phi'(\boldsymbol{r}') = \phi(\boldsymbol{r})$ under a coordinate transformation, we have to take into account the transformation of the volume element. Under the change of variables $x^i = x^i(x^{j'})$, the volume element of a multiple integral transforms as

$$\mathrm{d}^n x \equiv \mathrm{d}x^1 \cdots \mathrm{d}x^n = J \mathrm{d}x^{1'} \cdots \mathrm{d}x^{n'} \equiv J \mathrm{d}^n x' \tag{5.52}$$

(so that the integral transforms as $\int \phi(x^i)\mathrm{d}^n x = \int \phi(x^i(x^{j'}))J\mathrm{d}^n x' \equiv \int \phi'(x^{j'})J\mathrm{d}^n x'$), where "the Jacobian" J is the determinant of the Jacobian matrix $A^i_{j'}$,

$$J = \begin{vmatrix} \dfrac{\partial x^1}{\partial x^{1'}} & \cdots & \dfrac{\partial x^1}{\partial x^{n'}} \\ \vdots & & \vdots \\ \dfrac{\partial x^n}{\partial x^{1'}} & \cdots & \dfrac{\partial x^n}{\partial x^{n'}} \end{vmatrix} \equiv \left| \frac{\partial(x^1, \cdots, x^n)}{\partial(x^{1'}, \cdots, x^{n'})} \right| \equiv \left| \frac{\partial x^i}{\partial x^{j'}} \right| = \left| A^i_{j'} \right| . \tag{5.53}$$

In general we work with *oriented* volume elements, implying that we *don't take the absolute value of the Jacobian determinant.*[39]

Relative tensors of weight w have components that (by definition) transform according to the rules we have developed (such as Eq. (5.41)), with the additional requirement of the Jacobian raised to an integer power, w:[40]

$$T^{k'_1 \cdots k'_p}{}_{m'_1 \cdots m'_q} = J^w A^{k'_1}_{t_1} \cdots A^{k'_p}_{t_p} A^{s_1}_{m'_1} \cdots A^{s_q}_{m'_q} T^{t_1 \cdots t_p}{}_{s_1 \cdots s_q} . \tag{5.54}$$

Linear combinations of tensors *of the same weight* produce new tensors with that weight. Products of tensors of weights w_1 and w_2 produce tensors of weight $w_1 + w_2$. Contractions of relative tensors do not change the weight. *A tensor equation must be among tensors of the same weight.* We require that tensor equations valid in one coordinate system be valid in all others; this property would be lost in an equation among tensors of different weights. Relative tensors with $w = \pm 1$ occur frequently, what we'll call *tensor densities*. Tensors that transform with $w = 0$ are called *absolute tensors.*

The covariant metric tensor is an absolute tensor: From Eq. (5.39),

$$g_{i'j'} = A^l_{i'} A^m_{j'} g_{lm} . \tag{5.55}$$

The *determinant* of the metric tensor, however, is a relative scalar of weight[41] $w = 2$. Let g denote the determinant of the *covariant* metric tensor (a convention we adhere to). Applying the product rule for determinants to Eq. (5.55),

$$g' = J^2 g , \tag{5.56}$$

where we have used Eq. (5.53). *The sign of g is an absolute quantity, invariant under coordinate transformations.* Equation (5.56) then provides an alternate expression for the Jacobian, one that separates the contributions from the coordinate systems it connects: $J = \sqrt{g'/g}$. For positive definite metrics, $g > 0$; for the Lorentz metric, $g = -1$. Using Eq. (5.56), $\sqrt{|g'|} = J\sqrt{|g|}$ and thus $\sqrt{|g|}$ is a scalar density (transforms with $w = 1$).

Combining $J = \sqrt{g'/g}$ with Eq. (5.52), we have the *invariant volume element*

$$\sqrt{|g'|}\mathrm{d}y^1 \cdots \mathrm{d}y^n = \sqrt{|g|}\mathrm{d}x^1 \cdots \mathrm{d}x^n . \tag{5.57}$$

Note how Eq. (5.57) has a net weight of $w = 0$: $\sqrt{|g|}\mathrm{d}^n x$ is an absolute scalar. (Under $x \to y$, $\mathrm{d}^n y = J^{-1}\mathrm{d}^n x$ from Eq. (5.52).) Thus, the integral of a scalar field $\int \phi \mathrm{d}^n x$ is not invariant, but $\int \phi\sqrt{|g|}\mathrm{d}^n x$ is, something we make frequent use of in GR; in SR it's unnecessary because $|g| = 1$.

Substituting $J = \sqrt{g'/g}$ in Eq. (5.54), we find that

$$(|g'|)^{-w/2} T^{k'_1 \cdots k'_p}{}_{m'_1 \cdots m'_q} = A^{k'_1}_{t_1} \cdots A^{k'_p}_{t_p} A^{s_1}_{m'_1} \cdots A^{s_q}_{m'_q} \left((|g|)^{-w/2} T^{t_1 \cdots t_p}{}_{s_1 \cdots s_q} \right) .$$

---

[39] By not taking the absolute value of the determinant, we allow for the possibility of transformations with $J < 0$. Transformations for which $J < 0$ allow us to further classify tensors as pseudotensors, those that transform as tensors when $J > 0$, but transform with an additional change of sign when $J < 0$.

[40] Beware: Relative tensors are also defined with w replaced by $-w$. I have adopted a definition that leads to $w = +2$ for the determinant of the covariant elements of the metric tensor.

[41] The same is true of the determinant of any covariant second-rank absolute tensor.

For a tensor $\mathbf{T}$ of weight w, $(|g|)^{-w/2}\mathbf{T}$ transforms as an absolute tensor. Conversely, an absolute tensor $\mathbf{U}$ when multiplied by $(|g|)^{w/2}$ becomes a tensor of weight w. In particular, $\sqrt{|g|}\mathbf{U}$ is a tensor density.

A notational issue arises if $J = 1$. The Jacobian of proper LTs is unity, for example (Section 4.3). In that case densities "fly under the radar": Physical quantities that properly are tensor densities nominally transform as absolute tensors when $J = 1$. It's traditional in the theory of tensor analysis to indicate densities with a special notation, with Gothic letters: $\mathfrak{T}$ instead of $\mathbf{T}$. I will use this notation sparingly, but it can come in handy; without it, one has to keep calling attention to the fact that certain symbols represent tensor densities.

5.3 DERIVATIVES OF TENSORS AND THE FOUR-WAVEVECTOR

5.3.1 Derivatives of tensors

Is the derivative of a tensor a tensor? How would we answer such a question? I hope you're saying, "How does it transform?". Before delving into that question, we need to establish some notation.

In Section 5.1 we used the gradient of a scalar function to motivate the concept of covariant vector, $\boldsymbol{\nabla} = \boldsymbol{e}^i \nabla_i$. Because a geometric object is independent of basis, however, we *could* have declared it to be a contravariant vector, $\boldsymbol{\nabla} = \boldsymbol{e}_i \nabla^i$—the contravariant components of a vector can always be found from the covariant components by raising the index: $\nabla^j = g^{jk}\nabla_k$. For the gradient as a contravariant vector, the change in a scalar function $\mathrm{d}\phi = \boldsymbol{\nabla}\phi \cdot \mathrm{d}\boldsymbol{s}$ would require that we express $\mathrm{d}\boldsymbol{s} = \boldsymbol{e}^i \mathrm{d}x_i$ as a covariant vector, with the result that $\mathrm{d}\phi = \left(\nabla^i \phi\right) \mathrm{d}x_i$, in which case we would conclude that $\nabla^i = \partial/\partial x_i$ (note the placement of the indices). The quantity ∇^i, being the contravariant component of a vector, must transform as such,

$$\nabla^{i'} = \frac{\partial}{\partial x_{i'}} = \frac{\partial x_k}{\partial x_{i'}} \frac{\partial}{\partial x_k} = \left(\frac{\partial x_k}{\partial x_{i'}}\right) \nabla^k . \tag{5.58}$$

By Eq. (5.35), however, Eq. (5.58) should read $\nabla^{i'} = A^{i'}_k \nabla^k$. Comparing Eqs. (5.58) and (5.35), we conclude that $A^{i'}_k \equiv \partial x^{i'}/\partial x^k = \partial x_k / \partial x_{i'}$. Using Eq. (5.27), $A^k_{j'} \equiv \partial x^k / \partial x^{j'} = \partial x_{j'}/\partial x_k$. Note how the indices work here.

We now define the *four-gradient*, for which we switch to a fairly standard notation. Let ∂_μ denote the covariant four-vector of partial derivatives $\partial/\partial x^\mu$ (instead of ∇_μ which will be used in Chapter 14 for another purpose),[42]

$$\partial_\mu \equiv \frac{\partial}{\partial x^\mu} = \left(\frac{\partial}{\partial x^0}, \boldsymbol{\nabla}\right) = \left(\frac{\partial}{\partial (ct)}, \boldsymbol{\nabla}\right) = (\partial_0, \boldsymbol{\nabla}) .$$

Likewise, let ∂^μ denote the contravariant version. However, instead of $\partial^\mu = \partial/\partial x_\mu$ (which is correct), use the fact that it can be obtained by raising the index, $\partial^\mu = g^{\mu\nu}\partial_\nu$. Using the Lorentz metric,

$$\partial^\mu = \eta^{\mu\nu}\partial_\nu = \left(-\frac{\partial}{\partial x^0}, \boldsymbol{\nabla}\right) = \left(-\frac{\partial}{\partial (ct)}, \boldsymbol{\nabla}\right) = (-\partial_0, \boldsymbol{\nabla}) = (\partial^0, \boldsymbol{\nabla}) .$$

The only effective difference between ∂^μ and ∂_μ is in the time component, $\partial^0 = -\partial_0$. The inner product $\partial_\mu \partial^\mu$ generates the wave-equation operator,

$$\partial^\mu \partial_\mu = \partial_\mu \partial^\mu = -\frac{\partial^2}{\partial (x^0)^2} + \nabla^2 = -\frac{1}{c^2}\frac{\partial^2}{\partial t^2} + \nabla^2 . \tag{5.59}$$

[42] Another prevalent notation for the partial derivative is to write $\partial\phi/\partial x^\mu$ (what we're calling $\partial_\mu \phi$) as $\phi_{,\mu}$. This taxes everyone's eyesight. In this notation, $\partial A_\nu / \partial x^\mu \equiv A_{\nu,\mu}$, what we'll write as $\partial_\mu A_\nu$, which causes less eye strain.

As an inner product, *the wave-equation operator is invariant.* The wave equation is preserved under the LT (Appendix A); here we see that it can be written compactly in covariant form,[43] $\partial^\mu\partial_\mu$.

Getting back to the issue, consider the partial derivative of a tensor, $\partial T^\rho/\partial x^\beta \equiv F^\rho_\beta$. We have written F^ρ_β in tensor notation, but is it a tensor? How does it transform? By differentiating Eq. (5.35), we derive the transformation equation for the derivative of a tensor:

$$\frac{\partial T^{\lambda'}}{\partial x^{\alpha'}} = \frac{\partial x^\beta}{\partial x^{\alpha'}}\frac{\partial}{\partial x^\beta}\left(A^{\lambda'}_\rho T^\rho\right) = A^\beta_{\alpha'}A^{\lambda'}_\rho\frac{\partial T^\rho}{\partial x^\beta} + A^\beta_{\alpha'}\frac{\partial^2 x^{\lambda'}}{\partial x^\beta \partial x^\rho}T^\rho . \tag{5.60}$$

Equation (5.60) is not in the form of a homogeneous transformation that we require of tensors. *The derivative of a tensor is not in general a tensor* (at least, not for the partial derivative). For SR, however, with its flat geometry (MS) and *linear* coordinate transformations, the inhomogeneous term in Eq. (5.60) vanishes, $\partial_\beta A^{\lambda'}_\rho = 0$. *Within the confines of SR we can treat partial derivatives of tensors as tensors.* When we venture into GR, however, which is based on more general coordinate transformations, we'll have to face the matter of how to define the derivative of a tensor so that it transforms as a tensor, yet reduces to the partial derivative on a flat geometry (see Chapter 14).

5.3.2 The four-wavevector

Among solutions to the homogeneous wave equation $\partial_\mu\partial^\mu\phi(x^\alpha) = 0$ are monochromatic plane waves $\phi = \phi_0 e^{i(\boldsymbol{k}\cdot\boldsymbol{x}-\omega t)}$, which we can write in the relativistically suggestive form $\phi = \phi_0 e^{i(\boldsymbol{k}\cdot\boldsymbol{x}-(\omega/c)x^0)}$. The three-wavevector $\boldsymbol{k} = k\hat{\boldsymbol{n}}$ has magnitude $k = 2\pi/\lambda$, with the propagation direction represented by the unit vector $\hat{\boldsymbol{n}}$. To satisfy the wave equation, there must be a relation between ω and k, the *dispersion relation*, $\omega = ck$. We can write the phase factor in covariant form if we define the *four-wavevector*

$$k_\mu \equiv \left(-\frac{\omega}{c}, \boldsymbol{k}\right) = \frac{\omega}{c}\left(-1, \hat{\boldsymbol{n}}\right) . \tag{5.61}$$

In that way, $k_\mu x^\mu = -\omega t + \boldsymbol{k}\cdot\boldsymbol{x}$. Whenever we add a new four-vector to the pantheon of four-vectors (which right now consists of the prototype contravariant and covariant four-vectors, $\mathrm{d}x^\mu$ and ∂_μ), we must provide justification for why we're entitled to do so. In this case we have a scalar field ϕ and thus the phase $k_\mu x^\mu$ is a scalar. By the quotient theorem k_μ is a covariant vector. The spatial parts of k_μ meet our expectation of covariant vector (Section 5.1.8): $k = 2\pi/\lambda$ is proportional to the *density* of waves (number of wave crests per unit distance, $1/\lambda$), and $\boldsymbol{k}\cdot\boldsymbol{x}$ is fixed for a plane perpendicular to $\boldsymbol{k}$ (equation of a plane). The time component of k_μ is proportional to the density of waves *in time*, $k_0 = -(2\pi/c)f$ (with the frequency f the number of wave crests per unit time). The contravariant version $k^\mu = \eta^{\mu\nu}k_\nu = (\omega/c)(1, \hat{\boldsymbol{n}})$, and thus k_μ is a null vector, $k_\mu k^\mu = 0$.

We can now use the fact that because k^μ is a four-vector, it transforms as such under the LT, $k^\mu = L^\mu_{\nu'}k^{\nu'}$. For S and S' in standard configuration, let $k^{\nu'} = (\omega'/c)(1, \cos\theta', \sin', 0)$ in S' and $k^\mu = (\omega/c)(1, \cos\theta, \sin\theta, 0)$ in S. Under the LT,

$$(\omega/c)\begin{pmatrix}1\\ \cos\theta\\ \sin\theta\\ 0\end{pmatrix} = (\omega'/c)\begin{pmatrix}\gamma & \beta\gamma & 0 & 0\\ \beta\gamma & \gamma & 0 & 0\\ 0 & 0 & 1 & 0\\ 0 & 0 & 0 & 1\end{pmatrix}\begin{pmatrix}1\\ \cos\theta'\\ \sin\theta'\\ 0\end{pmatrix} . \tag{5.62}$$

The time component of Eq. (5.62) yields $\omega = \omega'\gamma\left(1 + \beta\cos\theta'\right)$, the relativistic Doppler effect, Eq. (3.41), while the spatial parts of Eq. (5.62) generate Eq. (3.37), the relativistic aberration formula. With one equation, the LT of the four-wavevector, we obtain the Doppler effect and aberration formulas, which were obtained in Section 3.4 by a more laborious procedure. What in one IRF is partitioned into frequency and direction of propagation has a different partition in another IRF.

[43] The wave equation operator, which can be considered the Laplacian operator in a four-dimensional space with Lorentz metric, is sometimes written $\Box^2$, which is termed the *d'Alembertian operator*. I prefer $\partial^\mu\partial_\mu$—which is easier to write.

5.4 INTERLUDE: DRAW A LINE HERE

The rest of this chapter contains more advanced material on tensors; it could be skipped on a first encounter—read it when you need it. What we've covered provides a foundation for the use of tensors in SR. For GR a deeper understanding of tensors must be developed. If you're not familiar with the dual vector space, now would be a good time to consult Appendix C. From here on, we step up the level of mathematical sophistication, what the subject requires us to do, and which we'll do again[44] in Chapter 13. What we develop in the rest of Chapter 5 is used in Chapters 13 and 14.

5.5 TENSORS AS MULTILINEAR MAPPINGS

So what are tensors, really? Our treatment has emphasized the transformation properties of tensors, that tensors consist of sets of quantities that transform according to certain rules. This approach leaves one with a lifeless view of tensors. In this section we give a definition that's not circular, as in a tensor is anything that acts like a tensor. We show that tensors are linear mappings between vectors, scalars, and other tensors. That tensors are operators can be seen from their definition as geometric objects, $\mathbf{T} = T^{ij}\boldsymbol{e}_i\boldsymbol{e}_j$, and that dyadic sums are operators (page 75).

5.5.1 Bilinearity defined

For vector spaces V_1, V_2, a function F that acts on elements of the set $V_1 \times V_2$ (see Appendix C) to produce a number, $F : V_1 \times V_2 \to \mathbb{R}$, is *bilinear* if it's linear in both arguments:

$$\left.\begin{aligned} F(a\boldsymbol{x}_1 + b\boldsymbol{x}_2, \boldsymbol{y}) &= aF(\boldsymbol{x}_1, \boldsymbol{y}) + bF(\boldsymbol{x}_2, \boldsymbol{y}) \\ F(\boldsymbol{x}, a\boldsymbol{y}_1 + b\boldsymbol{y}_2) &= aF(\boldsymbol{x}, \boldsymbol{y}_1) + bF(\boldsymbol{x}, \boldsymbol{y}_2) \end{aligned}\right\} \quad \text{bilinear function}$$

for $a, b \in \mathbb{R}$, $\boldsymbol{x}, \boldsymbol{x}_1, \boldsymbol{x}_2 \in V_1$, and $\boldsymbol{y}, \boldsymbol{y}_1, \boldsymbol{y}_2 \in V_2$.

5.5.2 Tensor product space

Let V be a vector space with basis $\{\boldsymbol{e}_i\}_{i=1}^n$, with V^* the space dual to V with basis $\{\boldsymbol{e}^j\}_{j=1}^n$, so that[45] $\boldsymbol{e}^j(\boldsymbol{e}_i) = \delta_i^j$. The set of all bilinear functions that act on $V_1 \times V_2$ *themselves form a vector space*, the *tensor-product space* denoted $V_1^* \otimes V_2^*$ (similar to V^*, the set of all functions $\boldsymbol{\omega} : V \to \mathbb{R}$). For dual vectors $\boldsymbol{\omega}_1 \in V_1^*$ and $\boldsymbol{\omega}_2 \in V_2^*$, define bilinear functions $\boldsymbol{\omega}_1 \otimes \boldsymbol{\omega}_2 \in V_1^* \otimes V_2^*$ (the *tensor product* of $\boldsymbol{\omega}_1$ and $\boldsymbol{\omega}_2$), $\boldsymbol{\omega}_1 \otimes \boldsymbol{\omega}_2 : V_1 \times V_2 \to \mathbb{R}$ such that, for $\boldsymbol{v}_1 \in V_1$ and $\boldsymbol{v}_2 \in V_2$,

$$\boldsymbol{\omega}_1 \otimes \boldsymbol{\omega}_2(\boldsymbol{v}_1, \boldsymbol{v}_2) \equiv \boldsymbol{\omega}_1(\boldsymbol{v}_1)\boldsymbol{\omega}_2(\boldsymbol{v}_2) \,. \tag{5.63}$$

Bilinear functions that act on $V_1 \times V_2$ can be represented as linear combinations of tensor products. For $F : V_1 \times V_2 \to \mathbb{R}$ a bilinear function, let the numbers $F_{\mu\nu} \equiv F(\boldsymbol{e}_\mu, \boldsymbol{e}_\nu)$ be the result of F acting on basis vectors of V_1, V_2. The function (actually, tensor) can be constructed from[46] $F = F_{\mu\nu}\boldsymbol{e}^\mu \otimes \boldsymbol{e}^\nu$. For $\boldsymbol{x} \in V_1$ and $\boldsymbol{y} \in V_2$,

$$\begin{aligned} F(\boldsymbol{x}, \boldsymbol{y}) &= F(x^{i_1}\boldsymbol{e}_{i_1}, y^{i_2}\boldsymbol{e}_{i_2}) = x^{i_1}y^{i_2}F(\boldsymbol{e}_{i_1}, \boldsymbol{e}_{i_2}) \equiv x^{i_1}y^{i_2}F_{i_1 i_2} = F_{i_1 i_2}\boldsymbol{e}^{i_1} \otimes \boldsymbol{e}^{i_2}(\boldsymbol{x}, \boldsymbol{y}) \\ &= F_{i_1 i_2}\boldsymbol{e}^{i_1}(\boldsymbol{x})\boldsymbol{e}^{i_2}(\boldsymbol{y}) \,, \end{aligned}$$

---

[44] An analogy from hiking is apt: The top of the mountain is never where you want it to be.

[45] We use the notation of Appendix C where Eq. (C.1) replaces Eq. (5.10).

[46] What's the difference between expressing a tensor as $\mathbf{F} = F_{\mu\nu}\boldsymbol{e}^\mu \otimes \boldsymbol{e}^\nu$ and what we had earlier in this chapter, $\mathbf{F} = F_{\mu\nu}\boldsymbol{e}^\mu\boldsymbol{e}^\nu$? Answer: Nothing, really, except that the tensor product notation enforces a sense of order, of where the basis vectors "come from." The order of the basis vectors matters in cases where $F_{\mu\nu} \neq F_{\nu\mu}$.

where we've used Eq. (5.63) and $\boldsymbol{e}^i(\boldsymbol{x}) = \boldsymbol{e}^i(x^j\boldsymbol{e}_j) = x^j\delta^i_j = x^i$. The terms $\boldsymbol{e}^{\mu_1} \otimes \boldsymbol{e}^{\mu_2}$ are linearly independent and form a basis for the space $V_1^* \otimes V_2^*$. The dimension of $V_1^* \otimes V_2^*$ is the product of the dimensions of V_1 and V_2.

The tensor product has the properties (for any scalar λ and any vectors $\boldsymbol{u}$, $\boldsymbol{v}$, $\boldsymbol{w}$):

$$\left.\begin{aligned} \boldsymbol{u} \otimes (\boldsymbol{v} \otimes \boldsymbol{w}) &= (\boldsymbol{u} \otimes \boldsymbol{v}) \otimes \boldsymbol{w} \\ \lambda (\boldsymbol{v} \otimes \boldsymbol{w}) &= (\lambda \boldsymbol{v}) \otimes \boldsymbol{w} = \boldsymbol{v} \otimes (\lambda \boldsymbol{w}) \\ (\boldsymbol{u} + \boldsymbol{v}) \otimes \boldsymbol{w} &= \boldsymbol{u} \otimes \boldsymbol{w} + \boldsymbol{v} \otimes \boldsymbol{w} \\ \boldsymbol{u} \otimes (\boldsymbol{v} + \boldsymbol{w}) &= \boldsymbol{u} \otimes \boldsymbol{v} + \boldsymbol{u} \otimes \boldsymbol{w} \,. \end{aligned}\right\} \quad \text{rules of the tensor product}$$

It's associative and distributive but *not commutative*, $\boldsymbol{u} \otimes \boldsymbol{v} \neq \boldsymbol{v} \otimes \boldsymbol{u}$ (the tensor product is based on *ordered* pairs). Consider $(\boldsymbol{v}_1 - \boldsymbol{v}_2) \otimes (\boldsymbol{w}_1 - \boldsymbol{w}_2) = \boldsymbol{v}_1 \otimes \boldsymbol{w}_1 - \boldsymbol{v}_2 \otimes \boldsymbol{w}_1 - \boldsymbol{v}_1 \otimes \boldsymbol{w}_2 + \boldsymbol{v}_2 \otimes \boldsymbol{w}_2$. This expression cannot be simplified further; each tensor product is distinct, with its own identity, an element of a larger-dimensional vector space.

5.5.3 Second-rank tensors as bilinear functions

Elements of the tensor-product space $V_1^* \otimes V_2^*$ are second-rank covariant tensors. Using Eqs. (5.34) and (5.39), $F_{\mu\nu}\boldsymbol{e}^\mu \otimes \boldsymbol{e}^\nu$ transforms as a geometric object:

$$\mathbf{F} = F_{\mu\nu}\boldsymbol{e}^\mu \otimes \boldsymbol{e}^\nu = F_{\mu\nu}A^\mu_{\mu'}\boldsymbol{e}^{\mu'} \otimes A^\nu_{\nu'}\boldsymbol{e}^{\nu'} = F_{\mu'\nu'}\boldsymbol{e}^{\mu'} \otimes \boldsymbol{e}^{\nu'} \,.$$

We have an intrinsic definition of tensor (akin to a vector being an element of a vector space): A second-rank covariant tensor $\mathbf{T}$ is an element of the tensor-product space $V_1^* \otimes V_2^*$, the space of bilinear functions $\mathbf{T} : V_1 \times V_2 \to \mathbb{R}$. The action of $\mathbf{T}$ on a pair of vectors can be symbolized in a coordinate-independent way as $\mathbf{T}(\boldsymbol{u}, \boldsymbol{v})$. In a basis, $\mathbf{T}(\boldsymbol{u}, \boldsymbol{v}) = T_{\mu\nu}\boldsymbol{e}^\mu \otimes \boldsymbol{e}^\nu \left(u^i\boldsymbol{e}_i, v^j\boldsymbol{e}_j\right) = T_{\mu\nu}u^iv^j\boldsymbol{e}^\mu(\boldsymbol{e}_i)\boldsymbol{e}^\nu(\boldsymbol{e}_j) = T_{ij}u^iv^j$.

Contravariant tensors are likewise elements of a tensor-product space. For $\boldsymbol{v}_1 \in V_1$ and $\boldsymbol{v}_2 \in V_2$, we can construct a bilinear function that acts on $V_1^* \times V_2^*$, $\boldsymbol{v}_1 \otimes \boldsymbol{v}_2 : V_1^* \times V_2^* \to \mathbb{R}$, such that the action of $\boldsymbol{v}_1 \otimes \boldsymbol{v}_2$ on an element $(\boldsymbol{\omega}_1, \boldsymbol{\omega}_2) \in V_1^* \times V_2^*$ has the value

$$\boldsymbol{v}_1 \otimes \boldsymbol{v}_2(\boldsymbol{\omega}_1, \boldsymbol{\omega}_2) = \boldsymbol{v}_1(\boldsymbol{\omega}_1)\boldsymbol{v}_2(\boldsymbol{\omega}_2) = \boldsymbol{\omega}_1(\boldsymbol{v}_1)\boldsymbol{\omega}_2(\boldsymbol{v}_2) \,,$$

where we've used Eq. (C.2). Bilinear functions that act on $V_1^* \times V_2^*$ form a vector space, $V_1 \otimes V_2$, the elements of which are second-rank contravariant tensors, $F^{\mu\nu}\boldsymbol{e}_\mu \otimes \boldsymbol{e}_\nu$. Mixed tensors have similar definitions. The space of all bilinear functions $\boldsymbol{v} \otimes \boldsymbol{\omega} : V^* \times V \to \mathbb{R}$ is $V \otimes V^*$, the elements of which are the tensors $F^\mu{}_\nu\boldsymbol{e}_\mu \otimes \boldsymbol{e}^\nu$. The space of all bilinear functions $\boldsymbol{\omega} \otimes \boldsymbol{v} : V \times V^* \to \mathbb{R}$ is $V^* \otimes V$, the elements of which are the tensors $F_\mu{}^\nu\boldsymbol{e}^\mu \otimes \boldsymbol{e}_\nu$.

5.5.4 Higher-rank tensors: Multilinear mappings

A map $T : V_1 \times \cdots \times V_r \to \mathbb{R}$ is *multilinear* if it's linear in each argument, $T(\boldsymbol{v}_1, \cdots, a\boldsymbol{v}_i + b\boldsymbol{v}'_i, \cdots, \boldsymbol{v}_r) = aT(\boldsymbol{v}_1, \cdots, \boldsymbol{v}_i, \cdots, \boldsymbol{v}_r) + bT(\boldsymbol{v}_1, \cdots, \boldsymbol{v}'_i, \cdots, \boldsymbol{v}_r)$, $1 \leq i \leq r$, where $\boldsymbol{v}_i \in V_i$ and $(a, b) \in \mathbb{R}$. Multilinear mappings that act on products of vector spaces, for $\boldsymbol{v} \in V$ and $\boldsymbol{\omega} \in V^*$,

$$\underbrace{\boldsymbol{v} \otimes \cdots \otimes \boldsymbol{v}}_{k} \otimes \underbrace{\boldsymbol{\omega} \otimes \cdots \otimes \boldsymbol{\omega}}_{l} \left(\underbrace{V^* \times \cdots \times V^*}_{k} \times \underbrace{V \times \cdots \times V}_{l} \right) \to \mathbb{R} \tag{5.64}$$

form a vector space denoted $\mathcal{T}^k_l(V)$, the elements of which are type (k, l) tensors, $\mathbf{T}^k_l : (V^*)^k \times (V)^l \to \mathbb{R}$; $\mathbf{T}^k_l$ thus operates on k dual vectors and l vectors and produces a number. A type $(0, 1)$ tensor is a dual vector (check it!). A type $(1, 0)$ tensor is an element of V^{**} (isomorphic to V,

Appendix C). A basis for $\mathcal{T}_l^k(V)$ can be constructed out of tensor products of the basis vectors for V and V^*, $e_{i_1} \otimes \cdots \otimes e_{i_k} \otimes e^{j_1} \otimes \cdots \otimes e^{j_l}$. The tensor is the linear combination $\mathbf{T}_l^k = T^{i_1 \cdots i_k}{}_{j_1 \cdots j_l} e_{i_1} \otimes \cdots \otimes e_{i_k} \otimes e^{j_1} \otimes \cdots \otimes e^{j_l}$, with $T^{m_1 \cdots m_k}{}_{n_1 \cdots n_l} = \mathbf{T}_l^k(e^{m_1}, \cdots, e^{m_k}, e_{n_1}, \cdots, e_{n_l})$. The tensor product of a tensor $\mathbf{A}$ of type (r, s) and a tensor $\mathbf{B}$ of type (k, l) is a new tensor, $\mathbf{A} \otimes \mathbf{B}$, of type $(r + k, s + l)$, an operator on $(V^*)^{r+k} \times (V)^{s+l}$ such that

$$\begin{aligned} &\mathbf{A} \otimes \mathbf{B}(\boldsymbol{\omega}^1, \cdots, \boldsymbol{\omega}^{r+k}, \boldsymbol{v}_1, \cdots, \boldsymbol{v}_{s+l}) \\ &\quad = \mathbf{A}(\boldsymbol{\omega}^1, \cdots, \boldsymbol{\omega}^r, \boldsymbol{v}_1, \cdots, \boldsymbol{v}_s)\mathbf{B}(\boldsymbol{\omega}^{r+1}, \cdots, \boldsymbol{\omega}^{r+k}, \boldsymbol{v}_{s+1}, \cdots, \boldsymbol{v}_{s+l}). \end{aligned}$$

Tensors are mappings between products of vector spaces and numbers. Another role of tensors, however, is that they can map tensors to tensors. A type $(1, 2)$ tensor, $\mathbf{T} = T^i{}_{jk} e_i \otimes e^j \otimes e^k$, when acting on $(\boldsymbol{\omega}, \boldsymbol{v}_1, \boldsymbol{v}_2)$ produces a number, $\mathbf{T} : V^* \times V \times V \to \mathbb{R}$. The action of $\mathbf{T}$ on a dual vector, however, $\mathbf{T}(\boldsymbol{\omega}, \cdot, \cdot) = T^i{}_{jk} e_i \otimes e^j \otimes e^k(\boldsymbol{\omega}) = (T^i{}_{jk} \boldsymbol{\omega}(e_i)) e^j \otimes e^k = (T^i{}_{jk} \omega_i) e^j \otimes e^k \equiv T_{jk} e^j \otimes e^k$, produces a covariant tensor. Thus we can equally well express the action of $\mathbf{T}$ as $\mathbf{T} : V^* \to V^* \otimes V^*$. The same $\mathbf{T}$ acting on vectors produces a vector, $\mathbf{T}(\cdot, \boldsymbol{v}_1, \boldsymbol{v}_2) = T^i{}_{jk} e_i \otimes e^j \otimes e^k(\boldsymbol{v}_1, \boldsymbol{v}_2) = \left(T^i{}_{jk} e^j(\boldsymbol{v}_1) e^k(\boldsymbol{v}_2)\right) e_i \equiv \left(T^i{}_{jk} a^j b^k\right) e_i \equiv T^i e_i$, which we can express as $\mathbf{T} : V \times V \to V$. Tensors wear many hats.

The mathematical style of this section may be unfamiliar, yet it's on par with the level of mathematics in graduate texts in GR. GR requires a level of mathematical maturity a notch higher than in other branches of physics. You should take away that tensors are multilinear "machines" that effect mappings between geometric objects (scalars, vectors, and other tensors). The components of a tensor $\mathbf{T}$ in a given basis are the values it produces acting on basis vectors, e.g., $T_{ijk \cdots l} = \mathbf{T}(e_i, e_j, e_k, \cdots, e_l)$.

5.6 METRIC TENSOR REVISITED

That tensors can wear different hats (mappings onto numbers, mapping between spaces) is exemplified by the metric tensor, a type $(0, 2)$ symmetric, bilinear function $\mathbf{g} : V \times V \to \mathbb{R}$, i.e., it takes a pair of vectors (in either order) and produces a number. That idea generalizes our original definition as the dot product between basis vectors, Eq. (5.3). We can, however, also let $\mathbf{g}$ act on a single vector $\mathbf{g}(\boldsymbol{v}, \cdot) = g_{\mu\nu} e^\mu \otimes e^\nu (v^i e_i) = g_{\mu\nu} v^i \delta_i^\mu e^\nu = g_{\mu\nu} v^\mu e^\nu \equiv v_\nu e^\nu$, and we see that $\mathbf{g} : V \to V^*$, thus providing a natural accounting for the lowering of an index. What about *raising* indices, can $\mathbf{g}$ act on a dual vector, $\mathbf{g}(\boldsymbol{\omega}, \cdot)$? That operation isn't defined. We know, however (Appendix C), that there's a one-to-one correspondence between v_ν and v^ν. *The mapping* $\mathbf{g} : V \to V^*$ *must be invertible*; there must exist a unique linear mapping $\mathbf{g}^{-1} : V^* \to V$ such that

$$\begin{aligned} \boldsymbol{v} = v^i e_i = \mathbf{g}^{-1}(\mathbf{g}(\boldsymbol{v})) = \mathbf{g}^{-1}(v_j e^j) = v_j \mathbf{g}^{-1}(e^j) = v_j \left[\left(\mathbf{g}^{-1}\right)^{kl} e_k \otimes e_l\right] e^j \\ = v_j \left(\mathbf{g}^{-1}\right)^{kl} e_k \delta_l^j = v_j \left(\mathbf{g}^{-1}\right)^{kj} e_k \equiv v^k e_k \, . \end{aligned}$$

Thus, $\left(\mathbf{g}^{-1}\right)^{kj} v_j \equiv g^{kj} v_j = v^k$. It's customary to omit the inverse symbol, where it's understood that the contravariant tensor ($\mathbf{g}$ with upper indices) is the inverse of $\mathbf{g}$ with lower indices, Eq. (5.18). The lowering of an index on g^{jk} produces: $g^j{}_k = g_{kl} g^{jl} = \left(g^{-1}\right)^{jl} g_{kl} = \delta_k^j$, which was defined in Eq. (5.10), $e^j \cdot e_k$.

In a sense we've come full circle. We started this chapter with the inner product of basis vectors, Eq. (5.3). We now know that $\mathbf{g}$ *is a bilinear invertible function*, what's called *nondegenerate*. We also know that the elements of any tensor are specified by its action on the basis vectors $g_{\alpha\beta} = \mathbf{g}(e_\alpha, e_\beta)$. The metric $\mathbf{g}$ also connects the elements of V and V^* in a natural way (raising and lowering indices).

A symmetric bilinear function $\mathbf{g}$ is nondegenerate (invertible) when:

- The matrix of components $[g_{ij}]$ has a nonvanishing determinant, $g \neq 0$;
- For every nonzero $\boldsymbol{v} \in V$, there exists a $\boldsymbol{w} \in V$ such that $\mathbf{g}(\boldsymbol{v}, \boldsymbol{w}) \neq 0$.

These are equivalent aspects of a transformation being invertible. A matrix A has an inverse if and only if $\det A \neq 0$. The second requirement follows from the fact that $\mathbf{g} : V \to V^*$ is invertible if and only if the *nullity* (dimension of the *null space*) of $\mathbf{g}$ is zero. In that case, for every nonzero $\boldsymbol{v} \in V$, $\mathbf{g}(\boldsymbol{v}) \neq 0$; $\mathbf{g}(\boldsymbol{v})$ is not the zero functional. Thus, for a vector $\boldsymbol{w} \in V$, $[\mathbf{g}(\boldsymbol{v})](\boldsymbol{w}) \equiv \mathbf{g}(\boldsymbol{v}, \boldsymbol{w}) \neq 0$. Symmetric bilinear functions $\mathbf{g}$ can be classified as follows. For nonzero $\boldsymbol{v} \in V$,

- $\mathbf{g}$ is *positive (negative) definite* if $\mathbf{g}(\boldsymbol{v}, \boldsymbol{v}) > 0$ $(\mathbf{g}(\boldsymbol{v}, \boldsymbol{v}) < 0)$;
- $\mathbf{g}$ is *definite* if it is either positive definite or negative definite;
- $\mathbf{g}$ is *positive (negative) semidefinite* if $\mathbf{g}(\boldsymbol{v}, \boldsymbol{v}) \geq 0$ $(\mathbf{g}(\boldsymbol{v}, \boldsymbol{v}) \leq 0)$;
- $\mathbf{g}$ is *semidefinite* if either positive semidefinite or negative semidefinite;
- $\mathbf{g}$ is *indefinite* if not definite.

Vectors $\boldsymbol{u}, \boldsymbol{v} \in V$ are called *g-orthogonal* if $\mathbf{g}(\boldsymbol{u}, \boldsymbol{v}) = 0$. A *null vector* of $\mathbf{g}$ is orthogonal to itself, $\mathbf{g}(\boldsymbol{v}, \boldsymbol{v}) = 0$. If $\mathbf{g}$ is definite, the only null vector is the zero vector. If $\mathbf{g}$ is indefinite, *it must have a nonzero null vector.*[47] A basis $\{\boldsymbol{e}_i\}_{i=1}^n$ is called *g-orthonormal* if $\mathbf{g}(\boldsymbol{e}_i, \boldsymbol{e}_j) = 0$ for $i \neq j$, and $\mathbf{g}(\boldsymbol{e}_i, \boldsymbol{e}_i)$ (no sum) has one of the values $+1$, -1, or 0. Let n_+, n_-, and n_0 denote the number of vectors $\boldsymbol{e}_j$ for which $\mathbf{g}(\boldsymbol{e}_j, \boldsymbol{e}_j)$ is respectively $+1$, -1, or 0, where $n_+ + n_- + n_0 = n$. Every symmetric bilinear function $\mathbf{g}$ on V has an orthonormal basis; moreover, the numbers n_+, n_-, and n_0 are the *same* for all square matrices obtained from $[g_{ij}]$ by a transformation SgS^T, where S is a non-singular matrix (Sylvester's "law of inertia").[48] The integer $s \equiv n_+ - n_-$ is called the *signature* of $\mathbf{g}$. For $\mathbf{g}$ nondegenerate, an orthonormal basis must have $n_0 = 0$ and thus its determinant $g = (-1)^{n_-}$.

5.7 SYMMETRY OPERATIONS ON TENSORS

A tensor whose components remain unchanged when two of its covariant or two of its contravariant arguments are transposed is said to be *symmetric in these two arguments*. For example, if

$$\mathbf{T}\left(\boldsymbol{\omega}^1, \cdots, \boldsymbol{\omega}^k, \boldsymbol{v}_1, \cdots, \boldsymbol{v}_i, \cdots, \boldsymbol{v}_j, \cdots, \boldsymbol{v}_l\right) = \mathbf{T}\left(\boldsymbol{\omega}^1, \cdots, \boldsymbol{\omega}^k, \boldsymbol{v}_1, \cdots, \boldsymbol{v}_j, \cdots, \boldsymbol{v}_i, \cdots, \boldsymbol{v}_l\right),$$

then $\mathbf{T}$ is *symmetric* in contravariant arguments i and j. A tensor whose values *change sign* when two of its covariant or two of its contravariant arguments are transposed is *antisymmetric* in these arguments. Transposing covariant *and* contravariant arguments makes no sense, as they are defined with respect to different basis sets; *symmetry and antisymmetry are with respect to covariant or contravariant arguments only. The symmetry or antisymmetry of a tensor is a geometric property,* independent of basis transformations. A tensor antisymmetric (symmetric) in *all* of its arguments is said to be *totally antisymmetric* (totally symmetric).[49]

We now introduce a notation for signifying symmetric and antisymmetric tensors; its use can save much writing in complicated expressions. First, a *permutation* π of the integers $(1, \cdots, n)$ is a one-to-one mapping of the set onto itself with the values $\pi(1), \cdots, \pi(n)$. (We use 0123 as the reference list in relativity.) The permutation $\pi : 1234 \to 4132$ has values $\pi(1) = 4$, $\pi(2) = 1$,

[47] *Proof*: Because $\mathbf{g}$ is not positive definite, there is a nonzero vector $\boldsymbol{v} \in V$ such that $\mathbf{g}(\boldsymbol{v}, \boldsymbol{v}) \leq 0$. Because $\mathbf{g}$ is not negative definite, there is a nonzero vector $\boldsymbol{w} \in V$ such that $\mathbf{g}(\boldsymbol{w}, \boldsymbol{w}) \geq 0$. Let $\boldsymbol{u} \equiv \alpha\boldsymbol{v} + (1 - \alpha)\boldsymbol{w}$ so that $\mathbf{g}(\boldsymbol{u}, \boldsymbol{u}) = \alpha^2\mathbf{g}(\boldsymbol{v}, \boldsymbol{v}) + 2\alpha(1 - \alpha)\mathbf{g}(\boldsymbol{v}, \boldsymbol{w}) + (1 - \alpha)^2\mathbf{g}(\boldsymbol{w}, \boldsymbol{w})$. For $\alpha = 0$, $\mathbf{g}(\boldsymbol{u}, \boldsymbol{u}) = \mathbf{g}(\boldsymbol{w}, \boldsymbol{w}) \geq 0$, and for $\alpha = 1$, $\mathbf{g}(\boldsymbol{u}, \boldsymbol{u}) = \mathbf{g}(\boldsymbol{v}, \boldsymbol{v}) \leq 0$. With an appeal to continuity, there must exist some value of α for which $\mathbf{g}(\boldsymbol{u}, \boldsymbol{u}) = 0$.

[48] For a real symmetric matrix A and for S an invertible matrix such that $D = SAS^T$ is diagonal, the number of negative elements in D is the same for all S. That's the "inertia" in Sylvester's law of inertia—the invariance of the numbers n_-, n_+ and n_0.[27, p360]

[49] The distinction is often lost between tensors that are symmetric or antisymmetric in various indices versus totally symmetric or antisymmetric tensors; it's often just assumed that an "antisymmetric tensor" is totally antisymmetric.

$\pi(3) = 3$, and $\pi(4) = 2$. There are $n!$ permutations of n integers. The *symmetric part of a tensor* (with respect to indices $i_1 \cdots i_n$) is defined with (parentheses around the indices)

$$T_{(i_1 \cdots i_n)} \equiv \frac{1}{n!} \sum_{\pi} T_{i_{\pi(1)} \cdots i_{\pi(n)}} ,$$

where the sum is over the $n!$ permutations of $(1 \cdots n)$. The *antisymmetric part* (with respect to $i_1 \cdots i_n$) is defined with [square brackets around the indices]

$$T_{[i_1 \cdots i_n]} \equiv \frac{1}{n!} \sum_{\pi} \delta_{\pi} T_{i_{\pi(1)} \cdots i_{\pi(n)}} ,$$

where δ_π, the *parity* (also called the *signum* (sign)) of the permutation is $+1$ for even permutations and -1 for odd permutations.[50] An even (odd) permutation is a permutation obtained through an even (odd) number of pairwise interchanges of the numbers, starting from the reference sequence $(1 \cdots n)$. The sequence 2413 is an odd permutation of 1234. A second-rank tensor can be decomposed into symmetric and antisymmetric parts, $T_{ij} = T_{(ij)} + T_{[ij]}$. The same is not true of higher-rank tensors. For example, $T_{ijk} \neq T_{(ijk)} + T_{[ijk]}$. The notation can apply to any groups of indices. For example, $T^{(ij)k}{}_{[lm]} \equiv \frac{1}{4}(T^{ijk}{}_{lm} + T^{jik}{}_{lm} - T^{ijk}{}_{ml} - T^{jik}{}_{ml})$ denotes a tensor symmetric in its first two contravariant indices and antisymmetric in its covariant indices.

5.8 LEVI-CIVITA TENSOR AND DETERMINANTS

Totally antisymmetric tensors and determinants play an important role in differential geometry. Determinants share with totally antisymmetric tensors the property of being antisymmetric under interchange of columns. In this section we introduce some concepts from the theory of determinants and allied expressions.

5.8.1 Generalized Kronecker delta

Generalized Kronecker deltas $\delta^{i_1 \cdots i_k}_{j_1 \cdots j_k}$ have k superscripts and k subscripts, with $k \leq n$, where each index ranges from 1 to n such that:

$$\delta^{i_1 \cdots i_k}_{j_1 \cdots j_k} \equiv \begin{cases} +1 & \text{if } i_1 \cdots i_k \text{ is an even permutation of } j_1 \cdots j_k, \\ -1 & \text{if } i_1 \cdots i_k \text{ is an odd permutation of } j_1 \cdots j_k, \\ 0 & \text{if } i_1 \cdots i_k \text{ is not a permutation of } j_1 \cdots j_k \text{ or if } i_1 \cdots i_k \text{ are not all distinct.} \end{cases}$$

For $n = 3$ and $k = 2$, $\delta^{11}_{ij} = \delta^{22}_{ij} = \delta^{ij}_{11} = \delta^{12}_{13} = 0$, but $\delta^{12}_{12} = \delta^{13}_{13} = \delta^{21}_{21} = 1$ and $\delta^{12}_{21} = \delta^{13}_{31} = \delta^{21}_{12} = -1$. These quantities can be used to produce antisymmetrized expressions. Any three-rowed determinant (for example) can be constructed from among the quantities $\begin{pmatrix} A_1 & \cdots & A_n \\ B_1 & \cdots & B_n \\ C_1 & \cdots & C_n \end{pmatrix}$:

$$\delta^{abc}_{ijk} A_a B_b C_c = A_i \left(B_j C_k - B_k C_j\right) + A_j \left(B_k C_i - B_i C_k\right) + A_k \left(B_i C_j - B_j C_i\right) = \begin{vmatrix} A_i & A_j & A_k \\ B_i & B_j & B_k \\ C_i & C_j & C_k \end{vmatrix} .$$

The quantity $\delta^{k_1 \cdots k_m}_{s_1 \cdots s_m}$ is a type (m, m) tensor.[51]

[50] The parity of a permutation is unique. While there are many ways of realizing a given permutation of the reference sequence through pairwise exchanges, all ways require either an even or an odd number of exchanges.[28, p47]

[51] *Proof*: Let $T^{s_1 \cdots s_m}$ be the components of a type $(m, 0)$ tensor. Then, $\delta^{k_1 \cdots k_m}_{s_1 \cdots s_m} T^{s_1 \cdots s_m}$ generates an alternating sum of $m!$ terms, $T^{k_1 k_2 \cdots k_m} - T^{k_2 k_1 \cdots k_m} + \cdots$, i.e., a linear combination of $(m, 0)$ tensors, and hence an $(m, 0)$ tensor. By the quotient theorem, $\delta^{k_1 \cdots k_m}_{s_1 \cdots s_m}$ is a type (m, m) absolute tensor.

5.8.2 Levi-Civita symbols

The *permutation symbols*, or the *totally antisymmetric symbols* or the *Levi-Civita symbols*, are a special case of the generalized Kronecker delta:

$$\varepsilon_{i_1\cdots i_n} \equiv \delta^{1\cdots n}_{i_1\cdots i_n} = \delta^{i_1\cdots i_n}_{1\cdots n} \equiv \varepsilon^{i_1\cdots i_n} = \begin{cases} +1 & \text{if } i_1\cdots i_n \text{ is an even permutation of } 1\cdots n, \\ -1 & \text{if } i_1\cdots i_n \text{ is an odd permutation of } 1\cdots n, \\ 0 & \text{if two or more indices are equal.} \end{cases}$$

Permutation symbols always have n indices (whereas $\delta^{j_1\cdots j_k}_{i_1\cdots i_k}$ is such that $k \leq n$). For $n = 3$, out of the $3^3 = 27$ possible values of ε_{ijk}, only $3! = 6$ are non-zero, with $\varepsilon_{123} = \varepsilon_{231} = \varepsilon_{312} = 1$ and $\varepsilon_{132} = \varepsilon_{213} = \varepsilon_{321} = -1$. There is only one independent element of ε_{ijk} or ε^{ijk}.

For an $n \times n$ matrix A with elements A^{α}_{β}, define a function of its matrix elements:

$$D_A(\beta_1, \cdots, \beta_n) \equiv \varepsilon_{\alpha_1\cdots\alpha_n} A^{\alpha_1}_{\beta_1} \cdots A^{\alpha_n}_{\beta_n} .$$

This expression implies an alternating sum of $n!$, n-tuple products of the matrix elements (there are $n!$ ways to choose $\alpha_1 \cdots \alpha_n$ so that they're all distinct). The terms within each n-tuple come from a different row of the matrix ($\alpha_1 \cdots \alpha_n$ are distinct) and, as we'll see, a different column. The function $D_A(\beta_1, \cdots, \beta_n)$ is antisymmetric in all column indices β.[52] Thus, $D_A(\beta_1, \cdots, \beta_n)$ vanishes if any two indices β_i and β_j are equal. The quantity $D_A(\beta_1, \cdots, \beta_n)$ has the properties of the determinant[53] of A, and in fact equals $\det A$ ($-\det A$) when $\beta_1 \cdots \beta_n$ is an even (odd) permutation of $1 \cdots n$. Thus,

$$\varepsilon_{\alpha_1\cdots\alpha_n} A^{\alpha_1}_{\beta_1} \cdots A^{\alpha_n}_{\beta_n} = \varepsilon_{\beta_1\cdots\beta_n} \det A . \tag{5.65}$$

Through an analogous argument it can be shown that

$$\varepsilon^{\alpha_1\cdots\alpha_n} A^{\beta_1}_{\alpha_1} \cdots A^{\beta_n}_{\alpha_n} = \varepsilon^{\beta_1\cdots\beta_n} \det A . \tag{5.66}$$

Equations (5.65) and (5.66) generalize the more-frequently encountered definitions of determinant: $\det A = \varepsilon_{i_1\cdots i_n} A^{i_1}_{1} \cdots A^{i_n}_{n} = \varepsilon^{i_1\cdots i_n} A^{1}_{i_1} \cdots A^{n}_{i_n}$.[30, p30] The product rule for determinants is easily derived with these results: For $n \times n$ matrices A, B, C such that $C = AB$,

$$\begin{aligned} \det A \det B &= \det A \varepsilon_{\beta_1\cdots\beta_n} B^{\beta_1}_{1} \cdots B^{\beta_n}_{n} = \varepsilon_{\alpha_1\cdots\alpha_n} A^{\alpha_1}_{\beta_1} \cdots A^{\alpha_n}_{\beta_n} B^{\beta_1}_{1} \cdots B^{\beta_n}_{n} \\ &= \varepsilon_{\alpha_1\cdots\alpha_n} \left(A^{\alpha_1}_{\beta_1} B^{\beta_1}_{1}\right) \cdots \left(A^{\alpha_n}_{\beta_n} B^{\beta_n}_{n}\right) \equiv \varepsilon_{\alpha_1\cdots\alpha_n} C^{\alpha_1}_{1} \cdots C^{\alpha_n}_{n} = \det C . \end{aligned}$$

The generalized Kronecker delta is the determinant of Kronecker deltas (upper indices are totally antisymmetric, which must be a permutation of the lower indices),

$$\delta^{\alpha_1\cdots\alpha_k}_{\beta_1\cdots\beta_k} = \varepsilon_{a_1\cdots a_k} \delta^{\alpha_{a_1}}_{\beta_1} \delta^{\alpha_{a_2}}_{\beta_2} \cdots \delta^{\alpha_{a_k}}_{\beta_k} = \begin{vmatrix} \delta^{\alpha_1}_{\beta_1} & \cdots & \delta^{\alpha_1}_{\beta_k} \\ \vdots & & \vdots \\ \delta^{\alpha_k}_{\beta_1} & \cdots & \delta^{\alpha_k}_{\beta_k} \end{vmatrix} = k! \delta^{[\alpha_1}_{\beta_1} \cdots \delta^{\alpha_k]}_{\beta_k} . \tag{5.67}$$

For example, $\delta^{i_1 i_2}_{j_1 j_2} = \delta^{i_1}_{j_1}\delta^{i_2}_{j_2} - \delta^{i_1}_{j_2}\delta^{i_2}_{j_1}$ is a generalized Kronecker delta.

---

[52] *Proof*: Interchange *indices* $\alpha_i \leftrightarrow \alpha_j$ among the matrix elements $A^{\alpha_i}_{\beta_k}$ in the n-tuple, then restore the *matrix elements* back to their previous positions with respect to the order of the indices in $\varepsilon_{\alpha_1\cdots\alpha_n}$, and now interchange $\alpha_i \leftrightarrow \alpha_j$ in $\varepsilon_{\alpha_1\cdots\alpha_n}$; $D_A(\beta_1, \cdots, \beta_n)$ is antisymmetric under $\beta_i \leftrightarrow \beta_j$.

[53] The determinant of an $n \times n$ matrix $[A^i_j]$ can be defined as a function of its columns $\det A = f(C_1, \cdots, C_n)$ with $C_1 \equiv A^i_1, \cdots, C_n \equiv A^i_n$. The function f is determined by the properties that 1) it's multilinear and antisymmetric in its arguments, and 2) produces the value 1 for the identity matrix.[29, p99]

Equations (5.65) and (5.66) hold for any $n \times n$ matrix. Applying Eq. (5.65) to the Jacobian matrix we have (see Eq. (5.53)) $\varepsilon_{\alpha_1\cdots\alpha_n} A^{\alpha_1}_{\beta'_1} \cdots A^{\alpha_n}_{\beta'_n} = \varepsilon_{\beta'_1\cdots\beta'_n} J$. The permutation symbol thus has the transformation property:

$$\varepsilon_{\beta'_1\cdots\beta'_n} = J^{-1} A^{\alpha_1}_{\beta'_1} \cdots A^{\alpha_n}_{\beta'_n} \varepsilon_{\alpha_1\cdots\alpha_n} \,. \tag{5.68}$$

Comparing with Eq. (5.54), $\varepsilon_{\alpha_1\cdots\alpha_n}$ *is a covariant tensor with weight* $w = -1$. From Eq. (5.66) it follows that $\varepsilon^{\gamma'_1\cdots\gamma'_n} = J A^{\gamma'_1}_{\beta_1} \cdots A^{\gamma'_n}_{\beta_n} \varepsilon^{\beta_1\cdots\beta_n}$; $\varepsilon^{i_1\cdots i_n}$ is therefore *a contravariant tensor of weight*[54] $w = +1$. Combining Eqs. (5.68) and (5.65), $(\varepsilon_{\beta_1\cdots\beta_n})' = J^{-1} A^{\alpha_1}_{\beta'_1} \cdots A^{\alpha_n}_{\beta'_n} \varepsilon_{\alpha_1\cdots\alpha_n} = J^{-1} J \varepsilon_{\beta'_1\cdots\beta'_n}$; thus $(\varepsilon_{\beta_1\cdots\beta_n})' = \varepsilon_{\beta'_1\cdots\beta'_n}$, the transformed permutation symbol has the value of $\varepsilon_{\beta_1\cdots\beta_n}$ in the new frame. *The permutation symbol* $\varepsilon_{\beta_1\cdots\beta_n}$ *is a constant tensor*; ditto for $\varepsilon^{\beta_1\cdots\beta_n}$.

By the rules established in Section 5.2, we obtain an absolute covariant tensor by multiplying $\varepsilon_{i_1\cdots i_n}$ with $\sqrt{|g|}$. We define the covariant *Levi-Civita tensor* as

$$\epsilon_{i_1\cdots i_n} \equiv \sqrt{|g|}\varepsilon_{i_1\cdots i_n} \,. \tag{5.69}$$

Note the change in notation from $\varepsilon_{i_1\cdots i_n}$ (permutation symbol, tensor density), to $\epsilon_{i_1\cdots i_n}$, absolute tensor. The contravariant Levi-Civita tensor follows from raising indices on $\epsilon_{i_1\cdots i_n}$. From Eq. (5.69),

$$\epsilon^{j_1\cdots j_n} \equiv g^{j_1 i_1} \cdots g^{j_n i_n} \epsilon_{i_1\cdots i_n} = \sqrt{|g|} g^{j_1 i_1} \cdots g^{j_n i_n} \varepsilon_{i_1\cdots i_n} = \sqrt{|g|} \left(\det g^{ij}\right) \varepsilon^{j_1\cdots j_n} \,,$$

where in the last equality we have the determinant of the *contravariant* metric tensor, multiplied by $\varepsilon^{j_1\cdots j_n}$. Using Eq. (5.18), $\det g^{ij} = g^{-1}$, the inverse of g (determinant of the covariant metric tensor). Writing $g = \text{sgn}(g)|g|$ (where $\text{sgn}(g)$ is the *sign* of g), we have for the contravariant tensor

$$\epsilon^{j_1\cdots j_n} = \frac{\text{sgn}(g)}{\sqrt{|g|}} \varepsilon^{j_1\cdots j_n} = (-1)^{n_-} \frac{1}{\sqrt{|g|}} \varepsilon^{j_1\cdots j_n} \,, \tag{5.70}$$

where we have used $\text{sgn}(g) = (-1)^{n_-}$ (the sign of g is invariant, Section 5.2, which in a g-orthornormal basis is $(-1)^{n_-}$, Section 5.6). In Minkowski space, $\epsilon^{0123} = -\epsilon_{0123}$.

A type (n, n) totally antisymmetric tensor (in n-dimensional space) can have only one independent element: all nonzero tensor elements are equal to plus or minus times the same quantity. The generalized Kronecker delta $\delta^{\alpha_1\cdots\alpha_n}_{\beta_1\cdots\beta_n}$ is therefore proportional to the outer product of permutation tensors $\delta^{\alpha_1\cdots\alpha_n}_{\beta_1\cdots\beta_n} = \lambda \epsilon^{\alpha_1\cdots\alpha_n} \epsilon_{\beta_1\cdots\beta_n}$, where λ is a scalar. The proportionality constant can be evaluated with any set of indices for which both sides of the formula are nonzero. From Eqs. (5.69) and (5.70), consistency requires that $\lambda = (-1)^{n_-}$. We then have the useful result, using Eq. (5.67),

$$\delta^{i_1\cdots i_n}_{j_1\cdots j_n} = (-1)^{n_-} \epsilon^{i_1\cdots i_n} \epsilon_{j_1\cdots j_n} = \varepsilon^{i_1\cdots i_n} \varepsilon_{j_1\cdots j_n} = \begin{vmatrix} \delta^{i_1}_{j_1} & \cdots & \delta^{i_1}_{j_n} \\ \vdots & & \vdots \\ \delta^{i_n}_{j_1} & \cdots & \delta^{i_n}_{j_n} \end{vmatrix} \,. \tag{5.71}$$

5.9 PSEUDOTENSORS

Consider a right-handed three-dimensional Cartesian coordinate system as in Fig. 5.8. Associated with the vectors $\boldsymbol{x}$, $\boldsymbol{y}$ is the third $\boldsymbol{z} = \boldsymbol{x} \times \boldsymbol{y}$. The vector cross product is antisymmetric in its operands ($\boldsymbol{A} \times \boldsymbol{B} = -\boldsymbol{B} \times \boldsymbol{A}$), and there is a definite (but conventional) role assigned to the vectors participating in the cross product, $\textbf{Vector}_1 \times \textbf{Vector}_2$. The direction of $\boldsymbol{A} \times \boldsymbol{B}$ is that of the thumb on a human's right hand as $\boldsymbol{A}$ is "crossed" into $\boldsymbol{B}$. The cross product is a different kind

[54] We could use a Gothic letter, such as $\mathfrak{E}$ or $\mathfrak{e}$, for the Levi-Civita symbols. We conform to more standard notation and use ε to denote the Levi-Civita symbol (a tensor density), and ϵ to denote the absolute Levi-Civita tensor.

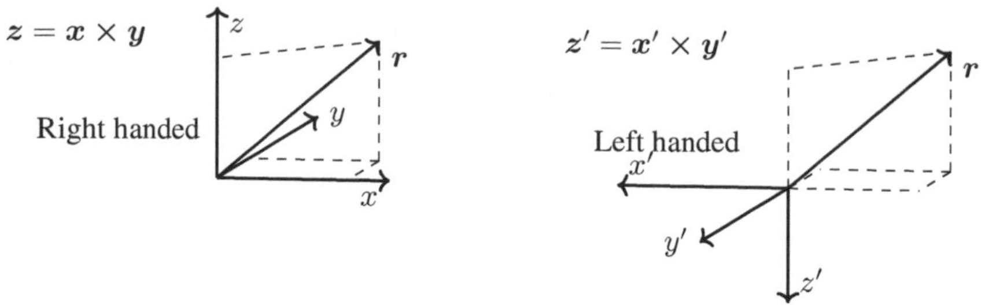

Figure 5.8 Right-handed and left-handed three-dimensional coordinate systems.

of vector, one derived from *two* vectors. An antisymmetric combination of two vectors is called a *bivector* (which is also an antisymmetric tensor, see Section 5.10.2).

An *inversion* of the coordinate axes in a right-handed system produces a *left-handed coordinate system*, with $\boldsymbol{z}' = \boldsymbol{x}' \times \boldsymbol{y}'$ given by the direction of the thumb on the left hand; Fig. 5.8. Under an inversion, the coordinates of the position vector $\boldsymbol{r}$ in the transformed system are the *negative* of their values in the original system: $(x', y', z') = (-x, -y, -z)$. Vectors with components that transform under inversion like the components of $\boldsymbol{r}$ are called *polar vectors*. Note that $\boldsymbol{r}$ is the *same vector* before and after the inversion (passive transformation), in keeping with the requirement that a vector is a geometric object that maintains its identity under a change of basis.

Are all vectors polar? Yes (on semantic grounds) and no because nonpolar vector-like quantities exist. Figure 5.9 shows the cross product $\boldsymbol{A} \times \boldsymbol{B}$ between polar vectors. Under inversion, the com-

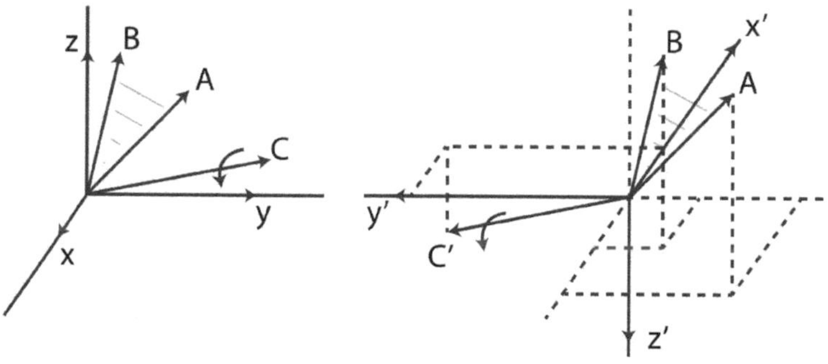

Figure 5.9 The vector cross product is a pseudovector.

ponents of $\boldsymbol{A}$ and $\boldsymbol{B}$ are negative relative to the inverted coordinate axes, but those of $\boldsymbol{C} = \boldsymbol{A} \times \boldsymbol{B}$ are *positive*. The cross product of polar vectors is therefore a fundamentally different kind of object: It's not a polar vector and *should not be called a vector.* Vectors with components that do not change sign under inversion are called *axial vectors* or *pseudovectors.*

Pseudotensors transform as tensors *when the Jacobian of the transformation is positive*, but transform with an additional change of sign for $J < 0$. For $J > 0$, pseudotensors have all the invariance properties expected of tensors.[55] We infer from its transformation property, Eq. (5.68), that the permutation symbol $\varepsilon_{\alpha_1 \cdots \alpha_n}$ is a pseudotensor *in spaces of an odd number of dimensions.* For an inversion of a three-dimensional Cartesian system, the Jacobian matrix has $(-1, -1, -1)$ on its diagonal (zeros otherwise), and hence $J = -1$. A third-rank tensor made from three covariant vectors $T_{ijk} = U_i V_j W_k$ (not a pseudotensor) *would* change sign under inversion, but not ε_{ijk}. The components of the cross product $\boldsymbol{C} \equiv \boldsymbol{F} \times \boldsymbol{G}$ (for $\boldsymbol{F}$ and $\boldsymbol{G}$ contravariant vectors) can be written

[55]The term pseudotensor is unfortunate, implying *false* tensor; perhaps "half-tensor" or "demi-tensor" would be better.

$C_i = \varepsilon_{ijk} F^j G^k$. It's readily shown using Eq. (5.68) that C_i transforms as $C_{i'} = J^{-1} A^j_{i'} C_j$. The cross product, which only lives in three dimensions—see below—is a pseudovector.

Polar vectors are related to the prototype vector, $\boldsymbol{r}$: velocity, $\boldsymbol{v} = \dot{\boldsymbol{r}}$; acceleration, $\boldsymbol{a} = \dot{\boldsymbol{v}} = \ddot{\boldsymbol{r}}$; force, $\boldsymbol{F} = m\boldsymbol{a}$; electric field, $\boldsymbol{E} = \boldsymbol{F}/q$; current density, $\boldsymbol{J} = \rho\boldsymbol{v}$. Axial vectors are associated with a cross product: torque, $\boldsymbol{\tau} = \boldsymbol{r} \times \boldsymbol{F}$; angular velocity, $\boldsymbol{v} = \boldsymbol{\omega} \times \boldsymbol{r}$; angular momentum, $\boldsymbol{L} = \boldsymbol{r} \times \boldsymbol{p}$; magnetic field, $\boldsymbol{F} = q\boldsymbol{v} \times \boldsymbol{B}$. Generally we have the rules for three-vectors:

- polar vector × polar vector = pseudovector
- pseudovector × pseudovector = pseudovector
- polar vector × pseudovector = polar vector.

Note that this classification relates to the behavior of vectors under inversion of *spatial* axes.

A *pseudoscalar* results from the inner product between a polar vector and an axial vector, such as between electric and magnetic field vectors, $\boldsymbol{E} \cdot \boldsymbol{B}$. *A pseudoscalar reverses sign under inversion.* For $\boldsymbol{A}$, $\boldsymbol{B}$, and $\boldsymbol{C}$ polar vectors, $\boldsymbol{A} \cdot (\boldsymbol{B} \times \boldsymbol{C})$ is a pseudoscalar.

Should pseudotensors concern us? Do the equations of physics depend on our choice of coordinate system? Pseudovectors depend on a *handedness convention*, which is arbitrary. The cross product associates a vector with a plane, and there are two sides to a plane. Does the universe care which hand we use? If not, then *a valid equation of physics cannot equate tensors with pseudotensors*. Faraday's law relates two *axial* vectors, $\boldsymbol{\nabla} \times \boldsymbol{E} = -\partial \boldsymbol{B}/\partial t$. Note that the curl of the curl does not change the vector character of what it operates on, as in the free-space wave equations: $\boldsymbol{\nabla} \times \boldsymbol{\nabla} \times \boldsymbol{E} = -(1/c^2)\partial^2 \boldsymbol{E}/\partial t^2$ and $\boldsymbol{\nabla} \times \boldsymbol{\nabla} \times \boldsymbol{B} = -(1/c^2)\partial^2 \boldsymbol{B}/\partial t^2$.

5.10 TOTALLY ANTISYMMETRIC TENSORS

Totally antisymmetric tensors are a significant part of tensor analysis. On the physics side, Maxwell's equations can be put in covariant form using antisymmetric tensors. On the math side, the generalization of the classic theorems of vector calculus to higher dimensions can be given systematic expression with antisymmetric tensors. We document their properties in this section.

5.10.1 Dual tensors

Antisymmetric tensors have the property that from them *new tensors can be defined known as dual tensors*.[56] The idea is based on a simple property: The contraction of an antisymmetric tensor $A_{\mu\nu}$ with a symmetric tensor $S^{\mu\nu}$ produces zero.[57] Consider the tensor comprised of the product of X^j and Y^k, the components of three-vectors. The contraction of $X^j Y^k$ with ϵ_{ijk} picks out the antisymmetric part,

$$\epsilon_{ijk} X^j Y^k = \tfrac{1}{2}\left[\epsilon_{ijk}\left(X^j Y^k - X^k Y^j\right) + \epsilon_{ijk}\left(X^j Y^k + X^k Y^j\right)\right]$$

because the contraction of ϵ_{ijk} with a symmetric tensor is zero. The elements of the antisymmetric tensor $T^{jk} \equiv X^j Y^k - X^k Y^j$ are therefore naturally associated with the elements of a three-vector,

$$P_i \equiv \tfrac{1}{2}\epsilon_{ijk} T^{jk} \,. \tag{5.72}$$

The vector $\boldsymbol{P}$ is called the *dual* of the tensor $\mathbf{T}$, denoted $*\mathbf{T} \equiv \boldsymbol{P}$. Equation (5.72) can be written $(*\mathbf{T})_i = \frac{1}{2}\epsilon_{ijk}T^{jk}$. (By raising and lowering indices, $(*\mathbf{T})^i = \frac{1}{2}\epsilon^i{}_{jk}T^{jk} = \frac{1}{2}\epsilon^{ijk}T_{jk}$.) Does it go

[56]The use of the word dual in dual tensor has nothing to do with the dual in dual space. Beware.

[57]*Proof*: $A_{\mu\nu}S^{\mu\nu} = -A_{\nu\mu}S^{\mu\nu} = -A_{\nu\mu}S^{\nu\mu} = -A_{\mu\nu}S^{\mu\nu}$, where in the last equality we have let $\mu \leftrightarrow \nu$.

the other way, can an antisymmetric tensor be associated with the components of a vector? We need a way to invert Eq. (5.72). Using Eq. (5.71),

$$\epsilon^{lmi}\epsilon_{ijk} = \delta^l_{\ j}\delta^m_{\ k} - \delta^l_{\ k}\delta^m_{\ j} \,. \tag{5.73}$$

Contracting Eq. (5.72) with ϵ^{lmi} and using Eq. (5.73), it's readily shown that $T^{lm} = \epsilon^{lmi}P_i$. One calls $\mathbf{T}$ the dual of $\boldsymbol{P}$, $\mathbf{T} = *\boldsymbol{P}$, with $(*\boldsymbol{P})^{ij} = \epsilon^{ijk}P_k$ (equivalently $(*\boldsymbol{P})_{ij} = \epsilon_{ijk}P^k$). It can be shown that $*(*\boldsymbol{P}) = \boldsymbol{P}$ and $*(*\mathbf{T}) = \mathbf{T}$. The dual-tensor pairs (not densities) for $n = 3$ and $n_- = 0$ are shown in Table 5.1 ($\epsilon_{ijk} = \sqrt{g}\varepsilon_{ijk}$).

Table 5.1 Tensor-dual tensor pairs for $n = 3$, $n^- = 0$

Tensor	Dual tensor
$\left[A^i\right] = \left(A^1, A^2, A^3\right)$	$\left[(*A)_{ij}\right] = \sqrt{g}\begin{pmatrix} 0 & A^3 & -A^2 \\ -A^3 & 0 & A^1 \\ A^2 & -A^1 & 0 \end{pmatrix}$
$\left[A^{ij}\right] = \begin{pmatrix} 0 & A^{12} & A^{13} \\ -A^{12} & 0 & A^{23} \\ -A^{13} & -A^{23} & 0 \end{pmatrix}$	$\left[(*A)_i\right] = \sqrt{g}\left(A^{23}, -A^{13}, A^{12}\right)$

In four dimensions, associated with a second-rank antisymmetric tensor $\mathbf{T}$ is its dual, $*\mathbf{T}$, another second-rank antisymmetric tensor with elements

$$(*\mathbf{T})_{mn} \equiv \tfrac{1}{2}\epsilon_{mnrt}T^{rt} \,. \tag{5.74}$$

By raising and lowering indices on Eq. (5.74),

$$(*\mathbf{T})^{rt} = \tfrac{1}{2}\epsilon^{rt}_{\ \ mn}T^{mn} = \tfrac{1}{2}\epsilon^{rtmn}T_{mn} \,. \tag{5.75}$$

We thus have a dual pair of antisymmetric second-rank tensors, the elements of which are given in Table 5.2. Each has six independent elements; each is a "repackaging" of the same information. In this case $*(*\mathbf{T}) = -\mathbf{T}$, which can be shown using the identity obtained from Eq. (5.71) with $n = 4$ and $n_- = 1$, $\epsilon_{ksmn}\epsilon^{rtmn} = -2\left(\delta^r_{\ k}\delta^t_{\ s} - \delta^r_{\ s}\delta^t_{\ k}\right)$.

Table 5.2 Tensor-dual tensor pairs for $n = 4$, $n^- = 1$

Tensor	Dual tensor
$\left[T_{\alpha\beta}\right] = \begin{pmatrix} 0 & T_{01} & T_{02} & T_{03} \\ -T_{01} & 0 & T_{12} & T_{13} \\ -T_{02} & -T_{12} & 0 & T_{23} \\ -T_{03} & -T_{13} & -T_{23} & 0 \end{pmatrix}$	$\left[(*T)^{\alpha\beta}\right] = \dfrac{1}{\sqrt{\lvert g\rvert}}\begin{pmatrix} 0 & -T_{23} & T_{13} & T_{12} \\ T_{23} & 0 & -T_{03} & T_{02} \\ -T_{13} & T_{03} & 0 & -T_{01} \\ -T_{12} & -T_{02} & T_{01} & 0 \end{pmatrix}$

For a rank-m totally antisymmetric tensor in n dimensions ($m \le n$) with elements $T^{r_1\cdots r_m}$, the elements of its dual, a rank $(n-m)$ tensor, are defined by

$$(*\mathbf{T})_{r_1\cdots r_{n-m}} \equiv \frac{1}{m!}\epsilon_{s_1\cdots s_m r_1\cdots r_{n-m}}T^{s_1\cdots s_m} \,. \tag{5.76}$$

By raising and lowering indices (and by renaming indices),

$$(*\mathbf{T})^{r_1\cdots r_m} = \frac{1}{(n-m)!}\epsilon^{s_1\cdots s_{n-m} r_1\cdots r_m}T_{s_1\cdots s_{n-m}} \,. \tag{5.77}$$

In general, $*(*\mathbf{T}) = (-1)^{m(n-m)+n_-}\mathbf{T}$. (Respect the order of the indices in the Levi-Civita tensor and use Eq. (5.71)). For $n = 4$ and $m = 1$ there's a dual relation between a four-vector and a third-rank tensor, $(*\mathbf{T})^{r_1} = \frac{1}{3!}\epsilon^{s_1 s_2 s_3 r_1} T_{s_1 s_2 s_3}$, $(*\mathbf{T})_{r_1 r_2 r_3} = \epsilon_{s_1 r_1 r_2 r_3} T^{s_1}$. An antisymmetric third-rank tensor in four dimensions has "four choose three" $\binom{4}{3} = 4$ independent components, just enough to associate with the components of a four-vector. A scalar (denoted $*T$) can be obtained by setting $m = n = 4$ in Eq. (5.76), $*T = \frac{1}{4!}\epsilon_{s_1 s_2 s_3 s_4} T^{s_1 s_2 s_3 s_4} = \frac{1}{4!}\epsilon^{s_1 s_2 s_3 s_4} T_{s_1 s_2 s_3 s_4}$. The "taxonomy" of antisymmetric tensors and their duals will become clearer when we introduce the Hodge star operator, below.

5.10.2 Exterior algebra: wedge products, k-multivectors, and k-forms

Recall (Section 5.5.2) that for a vector space V, the tensor product space $V \otimes V$ is the space of *all* bilinear functions (second-rank contravariant tensors) $F \in V \otimes V : V^* \times V^* \to \mathbb{R}$. The *subspace* of $V \otimes V$ obtained from the restriction to antisymmetric bilinear functions, denoted $V \wedge V$ or $\wedge^2 V$, is the space of *antisymmetric* second-rank contravariant tensors, the *wedge product* (or *exterior product*[58]) space. The elements of $\wedge^2 V$ are constructed from *wedge products* of vectors $\boldsymbol{v}_1, \boldsymbol{v}_2 \in V$:

$$\boldsymbol{v}_1 \wedge \boldsymbol{v}_2 \equiv \boldsymbol{v}_1 \otimes \boldsymbol{v}_2 - \boldsymbol{v}_2 \otimes \boldsymbol{v}_1$$

so that $\boldsymbol{v}_1 \wedge \boldsymbol{v}_2(\omega_1, \omega_2) = \boldsymbol{v}_1(\omega_1)\boldsymbol{v}_2(\omega_2) - \boldsymbol{v}_2(\omega_1)\boldsymbol{v}_1(\omega_2)$. The wedge product exploits the *non-commutativity* of tensor products, $\boldsymbol{v}_1 \otimes \boldsymbol{v}_2 \neq \boldsymbol{v}_2 \otimes \boldsymbol{v}_1$, and satisfies for $\boldsymbol{u}, \boldsymbol{v}, \boldsymbol{w} \in V$ and $\lambda \in \mathbb{R}$: (same as the rules for the tensor product, page 90, except with the imposition of antisymmetry)

$$\begin{aligned} \boldsymbol{u} \wedge \boldsymbol{v} &= -\boldsymbol{v} \wedge \boldsymbol{u} && \text{antisymmetry} \\ (\boldsymbol{u} \wedge \boldsymbol{v}) \wedge \boldsymbol{w} &= \boldsymbol{u} \wedge (\boldsymbol{v} \wedge \boldsymbol{w}) = \boldsymbol{u} \wedge \boldsymbol{v} \wedge \boldsymbol{w} && \text{associativity} \\ \left.\begin{aligned} (\lambda \boldsymbol{u}) \wedge \boldsymbol{v} &= \boldsymbol{u} \wedge (\lambda \boldsymbol{v}) = \lambda (\boldsymbol{u} \wedge \boldsymbol{v}) \\ (\boldsymbol{u} + \boldsymbol{v}) \wedge \boldsymbol{w} &= \boldsymbol{u} \wedge \boldsymbol{w} + \boldsymbol{v} \wedge \boldsymbol{w} \\ \boldsymbol{u} \wedge (\boldsymbol{v} + \boldsymbol{w}) &= \boldsymbol{u} \wedge \boldsymbol{v} + \boldsymbol{u} \wedge \boldsymbol{w}\,. \end{aligned}\right\} & && \text{bilinearity} \end{aligned}$$

Note that $\boldsymbol{v} \wedge \boldsymbol{v} = 0$. Elements of $\wedge^2 V$ are called *bivectors*.

Let $\{\boldsymbol{e}_i\}_{i=1}^n$ be a basis for V. For $\boldsymbol{u} = u^i \boldsymbol{e}_i$ and $\boldsymbol{v} = v^j \boldsymbol{e}_j$, $\boldsymbol{u} \wedge \boldsymbol{v} = (u^i \boldsymbol{e}_i) \wedge (v^j \boldsymbol{e}_j) = u^i v^j (\boldsymbol{e}_i \wedge \boldsymbol{e}_j)$. For each "diagonal" term $\boldsymbol{e}_i \wedge \boldsymbol{e}_i = 0$; moreover $\boldsymbol{e}_j \wedge \boldsymbol{e}_i = -\boldsymbol{e}_i \wedge \boldsymbol{e}_j$ for $i < j$, so that for the bivector $\mathbf{A}$

$$\mathbf{A} \equiv \boldsymbol{u} \wedge \boldsymbol{v} = \sum_{i<j} \left(u^i v^j - u^j v^i\right) \boldsymbol{e}_i \wedge \boldsymbol{e}_j \equiv \sum_{i<j} A^{ij} \boldsymbol{e}_i \wedge \boldsymbol{e}_j = \frac{1}{2} A^{ij} \boldsymbol{e}_i \wedge \boldsymbol{e}_j\,. \tag{5.78}$$

The typical element of $\wedge^2 V$ is a linear combination of wedge products $\boldsymbol{e}_i \wedge \boldsymbol{e}_j$, $1 \leq i < j \leq n$, *which form a basis*. The dimension of $\wedge^2 V$ is thus $\binom{n}{2} = \frac{1}{2}n(n-1)$. The components A^{ij} in Eq. (5.78) are antisymmetric, as are the basis vectors $\boldsymbol{e}_i \wedge \boldsymbol{e}_j$, hence the factor of $\frac{1}{2}$ in front of the *unrestricted* sum over i and j. The *components* of $\boldsymbol{u} \wedge \boldsymbol{v}$ (an element of $\wedge^2 V$, in the basis $\{\boldsymbol{e}_i \wedge \boldsymbol{e}_j\}$) are antisymmetric combinations of the components of $\boldsymbol{u}$ and $\boldsymbol{v}$ (elements of V, in the basis $\{\boldsymbol{e}_i\}$), $(\boldsymbol{u} \wedge \boldsymbol{v})^{ij} = \delta^{ij}_{lm} u^l v^m$. The virtue of wedge-product basis vectors is that they naturally carry information about *orientation*—see Fig. 5.10, a topic we discuss below.

The wedge product extends to more than two vectors. The kth-wedge product of vectors $\boldsymbol{v} \in V$, $\boldsymbol{v}_1 \wedge \cdots \wedge \boldsymbol{v}_k \equiv \delta^{i_1 \cdots i_k}_{1 \cdots k} \boldsymbol{v}_{i_1} \otimes \cdots \otimes \boldsymbol{v}_{i_k}$, termed a *k-multivector*,[59] is an element of the space $\wedge^k V$,

[58] The term *exterior* was introduced by the mathematician Hermann Grassmann in 1844. The wedge product of two bivectors containing a common vector $(\boldsymbol{a} \wedge \boldsymbol{b}) \wedge (\boldsymbol{a} \wedge \boldsymbol{c}) = 0$. The wedge product of bivectors is *nonzero* only if they have no vectors in common, or that the wedge product is "exterior" to (outside of) each other. The exterior product should not be confused with the outer product defined in Section 5.1.6.

[59] We've used the term k-multivector instead of what they are often called, *k-vectors*, to avoid confusion with 4-vectors, vectors in Minkowski space.

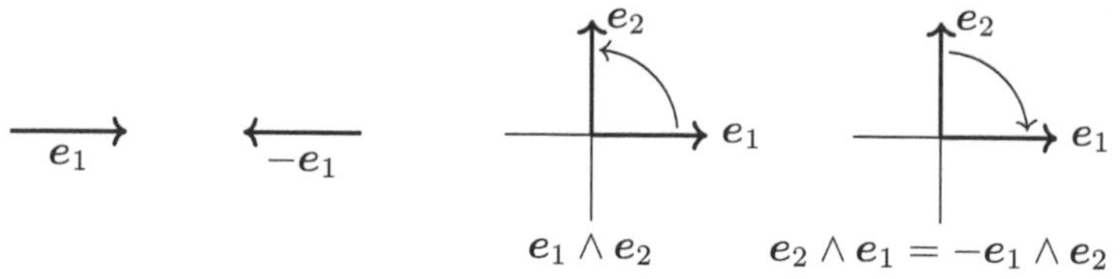

Figure 5.10 Oriented basis vectors, e_1, $-e_1$, $e_1 \wedge e_2$, $e_2 \wedge e_1$.

the space of all totally antisymmetric type $(k, 0)$ tensors.[60] For ease of notation, $\wedge^0 V \equiv \mathbb{R}$ and $\wedge^1 V \equiv V$. For fixed dimension n of the base space V, we can only have type $(k, 0)$ antisymmetric tensors defined on products of V for[61] $k \leq n$. The dimension of $\wedge^k V$ is $\binom{n}{k}$.

Example. The elements of $\wedge^3 V$ consist of totally antisymmetric combinations of products of three vectors called *trivectors*:

$$a \wedge b \wedge c \equiv a \otimes b \otimes c - a \otimes c \otimes b + c \otimes a \otimes b - c \otimes b \otimes a + b \otimes c \otimes a - b \otimes a \otimes c \, .$$

Using the basis for V, an element of $\wedge^3 V$ can be expressed as an unrestricted sum over indices

$$a \wedge b \wedge c = \frac{1}{3!} (a \wedge b \wedge c)^{ijk} \, e_i \wedge e_j \wedge e_k \, ,$$

where the components of the trivector are totally antisymmetric combinations of the components of the individual vectors (i.e., the determinant), $(a \wedge b \wedge c)^{ijk} = \delta^{ijk}_{lmn} a^l b^m c^n$. The wedge product between a vector and a bivector is a trivector: $(a \wedge b) \wedge c = a \wedge (b \wedge c) = a \wedge b \wedge c$.

The wedge product vanishes among linearly dependent vectors. If $u = \lambda v$, clearly $u \wedge v = 0$. A set of vectors $\{v_1, \cdots, v_k\}$ is linearly independent if (and only if) $v_1 \wedge \cdots \wedge v_k \neq 0$, i.e., if it's a nonzero element (tensor) of $\wedge^k V$.

Example. In $\mathbb{R}^3$ with basis vectors $\{e_1, e_2, e_3\}$, let $a = \frac{1}{2}(e_2 - e_3)$, $b = 2(e_1 - e_2)$, and $c = -2e_1 + 5e_2 - 3e_3$. The wedge product between a and b is $a \wedge b = -e_1 \wedge e_2 - e_2 \wedge e_3 - e_3 \wedge e_1$. The trivector $a \wedge b \wedge c = (3 + 2 - 5)\, e_1 \wedge e_2 \wedge e_3 = 0$. In this case $c = 6a - b$; c is not linearly independent of a and b.

Wedge-product spaces can equally well be constructed out of *dual vectors*, denoted $\wedge^k V^*$, the space of all antisymmetric type $(0, k)$ tensors, the elements of which are called *k-forms*. The Levi-Civita tensor $\epsilon_{\alpha\beta\gamma\delta}$ is a 4-form. All results for k-multivectors hold for k-forms by interchanging the roles of V and V^*. Dual vectors are called 1-forms because $\wedge^1 V^* \equiv V^*$.

5.10.3 Oriented area: Generalization of $a \times b$ to arbitrary dimension

We need to be able to describe *areas* in a covariant manner, i.e., using tensors. In three-dimensional space a pair of vectors (a, b) specify an area, that enclosed by the parallelogram they span (see Fig. 5.11), with magnitude found from the cross product $|a \times b|$. As we'll explain, *the cross product is valid only in three dimensions*. In this subsection we show that a generalization of $a \times b$ (no

[60] The kth-wedge product $v_1 \wedge \cdots \wedge v_k$ involves $v_{i_1} \otimes \cdots \otimes v_{i_k}$ and hence is a multilinear function on $\otimes^k V^*$, and thus is a type $(k, 0)$ tensor, restricted by $\delta^{i_1 \cdots i_k}_{1 \cdots k}$ to be totally antisymmetric in all arguments.

[61] For $k > n$, we run out of distinct indices so that antisymmetric tensors of order greater than n are zero.

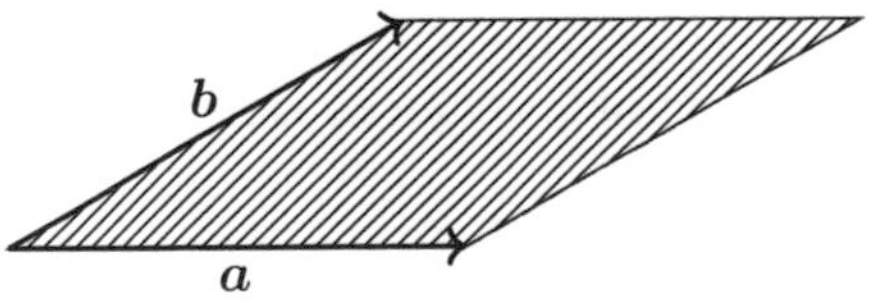

Figure 5.11 Area of the parallelogram spanned by $\boldsymbol{a}$ and $\boldsymbol{b}$ equal to $|\boldsymbol{a} \times \boldsymbol{b}|$.

absolute value), valid in any dimension, is the *oriented area* $\boldsymbol{a} \wedge \boldsymbol{b}$—a *signed* area—which is a second-rank antisymmetric tensor, and which in three dimensions is dual to a vector (Table 5.1). We're accustomed to the idea of surface areas as vectors for $n = 3$. For arbitrary n, the oriented area is (as we show) an element of a vector space of dimension $\frac{1}{2}n(n-1)$. The case of $n = 3$ is special, the non-trivial solution of $n = \frac{1}{2}n(n-1)$.

The oriented area spanned by vectors $(\boldsymbol{a}, \boldsymbol{b})$ is defined through the requirement that it be specified by an antisymmetric, bilinear function $A(\boldsymbol{a}, \boldsymbol{b})$, such that for $\lambda, \mu \in \mathbb{R}$:

$$A(\boldsymbol{a}, \boldsymbol{b}) = -A(\boldsymbol{b}, \boldsymbol{a}) \qquad \text{antisymmetry}$$

$$\left.\begin{aligned} A(\boldsymbol{a}, \lambda\boldsymbol{b} + \mu\boldsymbol{c}) &= \lambda A(\boldsymbol{a}, \boldsymbol{b}) + \mu A(\boldsymbol{a}, \boldsymbol{c}) \\ A(\lambda\boldsymbol{a} + \mu\boldsymbol{b}, \boldsymbol{c}) &= \lambda A(\boldsymbol{a}, \boldsymbol{c}) + \mu A(\boldsymbol{b}, \boldsymbol{c})\,. \end{aligned}\right\} \qquad \text{bilinearity}$$

We shouldn't be surprised if we find that $A(\boldsymbol{a}, \boldsymbol{b}) = \boldsymbol{a} \wedge \boldsymbol{b}$ because wedge products share the same properties (see page 99). Let's show first that the requirements make sense.

If $\boldsymbol{a}$ or $\boldsymbol{b}$ is scaled by a factor λ, we want the area they span to scale by the same factor (see Fig. 5.12), a requirement that argues for bilinearity. For *negative* λ, the orientation of the parallelogram

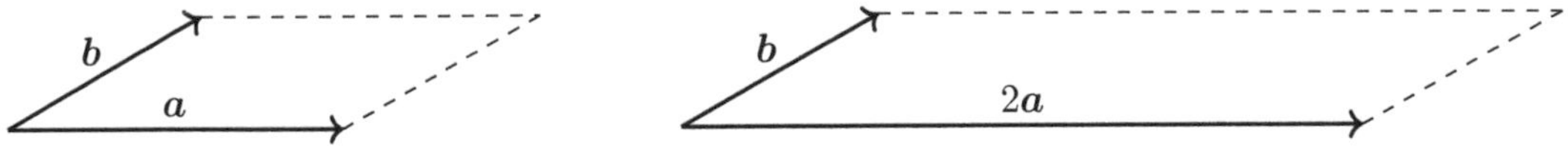

Figure 5.12 Area scales linearly with the size of the vector: $A(2\boldsymbol{a}, \boldsymbol{b}) = 2A(\boldsymbol{a}, \boldsymbol{b})$.

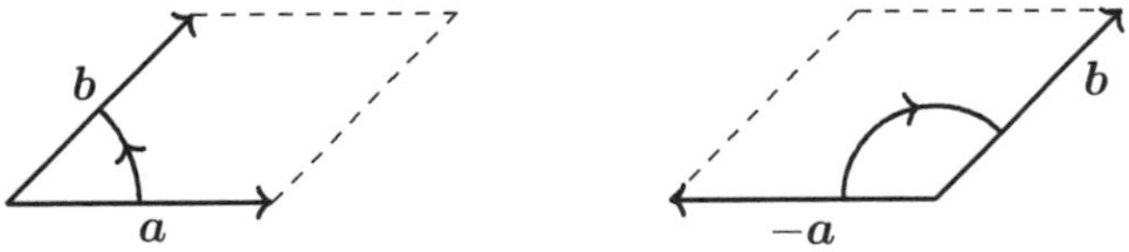

Figure 5.13 Oriented area: $A(-\boldsymbol{a}, \boldsymbol{b}) = -A(\boldsymbol{a}, \boldsymbol{b})$.

is reversed, and the oriented area changes sign (Fig. 5.13). If we didn't allow for a signed area $A(\boldsymbol{a}, \boldsymbol{b})$ wouldn't behave properly. From $A(\boldsymbol{a}, \boldsymbol{b}+\boldsymbol{c}) = A(\boldsymbol{a}, \boldsymbol{b}) + A(\boldsymbol{a}, \boldsymbol{c})$ set $\boldsymbol{c} = -\boldsymbol{b}$; if area was strictly a positive quantity we'd have the nonsensical result $A(\boldsymbol{a}, 0) = 2A(\boldsymbol{a}, \boldsymbol{b}) \neq 0$. With $A(\boldsymbol{a}, \boldsymbol{b})$ a bilinear function, with $\boldsymbol{c} = -\boldsymbol{b}$, $A(\boldsymbol{a}, 0) = A(\boldsymbol{a}, \boldsymbol{b}) - A(\boldsymbol{a}, \boldsymbol{b}) = 0$. We also want $A(\boldsymbol{a}, \boldsymbol{a}) = 0$: zero area enclosed by a single vector. By writing $\boldsymbol{a} = \boldsymbol{b} + \boldsymbol{c}$, $A(\boldsymbol{a}, \boldsymbol{a}) = 0$ implies $A(\boldsymbol{b}, \boldsymbol{c}) = -A(\boldsymbol{c}, \boldsymbol{b})$; bilinearity of $A(\boldsymbol{a}, \boldsymbol{b})$ implies antisymmetry.

The quantity $|\boldsymbol{a}||\boldsymbol{b}| \sin\theta$ (no absolute value sign) satisfies the requirements for an oriented area. Antisymmetry follows from the sign of the angle θ. Additivity follows from the property of parallelograms that the area spanned by its sides is invariant under a *shear transformation* where one of its sides is displaced in the direction parallel to the other side,[62] $A(\boldsymbol{a}, \boldsymbol{b}+\alpha\boldsymbol{a}) = A(\boldsymbol{a}, \boldsymbol{b})+\alpha A(\boldsymbol{a}, \boldsymbol{a}) = A(\boldsymbol{a}, \boldsymbol{b})$; see Fig. 5.14. A geometric proof of additivity is shown in Figure 5.15.

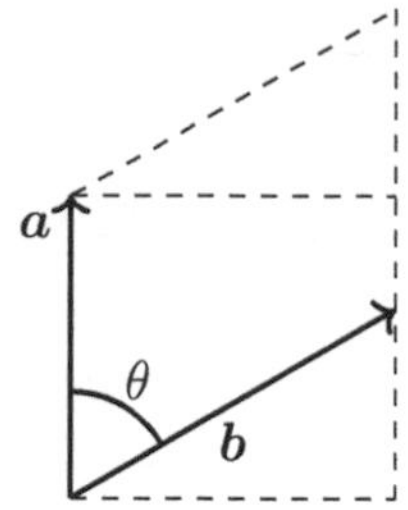

Figure 5.14 Oriented area is preserved under $\boldsymbol{b} \to \boldsymbol{b} + \alpha\boldsymbol{a}$: $A(\boldsymbol{a}, \boldsymbol{b}) = A(\boldsymbol{a}, \boldsymbol{b} + \alpha\boldsymbol{a})$.

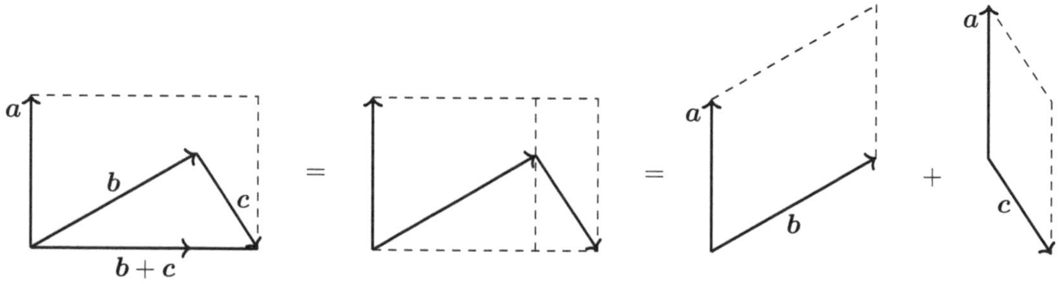

Figure 5.15 Oriented areas are additive: $A(\boldsymbol{a}, \boldsymbol{b} + \boldsymbol{c}) = A(\boldsymbol{a}, \boldsymbol{b}) + A(\boldsymbol{a}, \boldsymbol{c})$.

Let $(\mathbf{e}_1, \mathbf{e}_2)$ be a two-dimensional orthonormal basis with $\boldsymbol{a} = a^i\mathbf{e}_i$ and $\boldsymbol{b} = b^j\mathbf{e}_j$. The oriented area spanned by $\boldsymbol{a}$ and $\boldsymbol{b}$ is, using bilinearity and antisymmetry, given by the expression $A(\boldsymbol{a}, \boldsymbol{b}) = A(a^i\mathbf{e}_i, b^j\mathbf{e}_j) = a^ib^jA(\mathbf{e}_i, \mathbf{e}_j) = \left(a^1b^2 - a^2b^1\right)A(\mathbf{e}_1, \mathbf{e}_2) = \left|\begin{smallmatrix} a^1 & a^2 \\ b^1 & b^2 \end{smallmatrix}\right| A(\mathbf{e}_1, \mathbf{e}_2)$. Thus, the oriented area (spanned by $\boldsymbol{a}$ and $\boldsymbol{b}$) is a multiple (the determinant of the vector components) of that spanned by $\mathbf{e}_1$ and $\mathbf{e}_2$ (see Fig. 5.16), which we can take to be unity because the basis is orthonormal. This

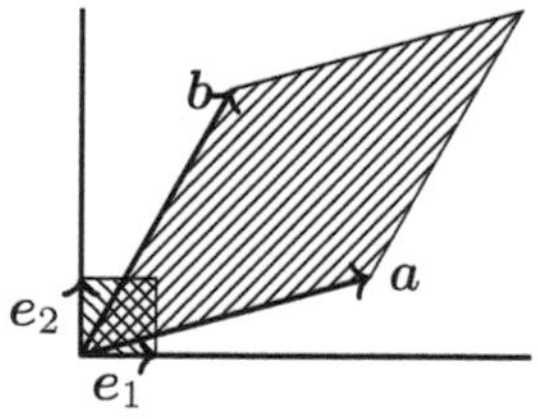

Figure 5.16 Oriented area spanned by $\boldsymbol{a}, \boldsymbol{b}$ is proportional to that spanned by $\mathbf{e}_1, \mathbf{e}_2$.

expression for $A(\boldsymbol{a}, \boldsymbol{b})$ is *almost* the formula for $\boldsymbol{a} \times \boldsymbol{b}$ (when computed in terms of the components of $\boldsymbol{a}$ and $\boldsymbol{b}$), *except* that we have not equated $A(\mathbf{e}_1, \mathbf{e}_2)$ with a unit vector perpendicular to the plane. Instead, $A(\mathbf{e}_1, \mathbf{e}_2)$ carries the *orientation* specified by $\mathbf{e}_1$ and $\mathbf{e}_2$.

[62] Invariance under shear transformations is analogous to the invariance of the value of a determinant of a matrix to adding multiples of one column of the matrix to another column. It's also a property of the wedge product: $\boldsymbol{a} \wedge (\boldsymbol{b} + \alpha\boldsymbol{a}) = \boldsymbol{a} \wedge \boldsymbol{b}$.

Now let $\{e_i\}_{i=1}^3$ be an orthonormal basis, with $\boldsymbol{a} = a^i \boldsymbol{e}_i$ and $\boldsymbol{b} = b^j \boldsymbol{e}_j$. We have

$$\begin{aligned} A(\boldsymbol{a},\boldsymbol{b}) &= a^i b^j A(\boldsymbol{e}_i,\boldsymbol{e}_j) = (a^1b^2 - a^2b^1)A(\boldsymbol{e}_1,\boldsymbol{e}_2) + (a^2b^3 - a^3b^2)A(\boldsymbol{e}_2,\boldsymbol{e}_3) \\ &\quad + (a^3b^1 - a^1b^3)A(\boldsymbol{e}_3,\boldsymbol{e}_1) \\ &= \begin{vmatrix} a^1 & a^2 \\ b^1 & b^2 \end{vmatrix} A(\boldsymbol{e}_1,\boldsymbol{e}_2) + \begin{vmatrix} a^2 & a^3 \\ b^2 & b^3 \end{vmatrix} A(\boldsymbol{e}_2,\boldsymbol{e}_3) + \begin{vmatrix} a^3 & a^1 \\ b^3 & b^1 \end{vmatrix} A(\boldsymbol{e}_3,\boldsymbol{e}_1) \,. \end{aligned} \tag{5.79}$$

Referring to Fig. 5.17, the area enclosed by the cross-hatched parallelogram is a *superposition* of the

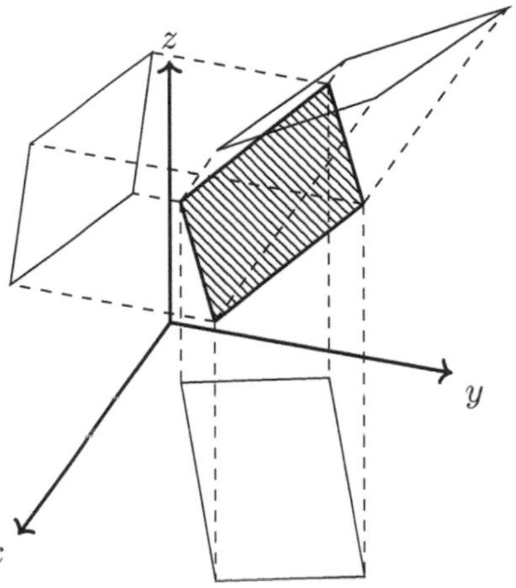

Figure 5.17 Projection of a parallelogram onto the three coordinate planes.

oriented areas obtained from projections onto each of the planes defined by pairs of coordinate axes. Each of the three projections is the oriented area obtained from the restrictions of the vectors $\boldsymbol{a}, \boldsymbol{b}$ to two dimensions. Projections are linear operations: parallelograms are projected onto parallelograms. Equation (5.79) indicates that *oriented areas add like vectors.* The three projected areas coincide with the Euclidean components of $\boldsymbol{a} \times \boldsymbol{b}$, except that we haven't equated with *spatial* unit vectors the oriented areas spanned by the basis vectors in the coordinate planes, $A(\boldsymbol{e}_1, \boldsymbol{e}_2)$, etc. Rather, the three areas $A(\boldsymbol{e}_i, \boldsymbol{e}_j)$ serve as a *basis* for a *space of oriented areas.* Equation (5.78) (three-dimensional space), has the same form as the expression for $\boldsymbol{a} \wedge \boldsymbol{b}$ in an n-dimensional space, Eq. (5.79).

We can now generalize to arbitrary dimension. Consider a parallelogram[63] spanned by $\boldsymbol{a}$ and $\boldsymbol{b}$ in an n-dimensional space with basis $\{\boldsymbol{e}_1, \cdots, \boldsymbol{e}_n\}$. There are $\binom{n}{2} = \frac{1}{2}n(n-1)$ planes spanned by $\{\boldsymbol{e}_i, \boldsymbol{e}_j\}$ with $1 \leq i < j \leq n$. Projections onto planes spanned by $\{\boldsymbol{e}_i, \boldsymbol{e}_j\}$ are obtained by omitting all components of $\boldsymbol{a}$ and $\boldsymbol{b}$ except those for $\boldsymbol{e}_i$ and $\boldsymbol{e}_j$. We may regard the $\binom{n}{2}$ projections as the components of a vector representing the oriented area *in a new vector space*, the space of oriented areas. Each of these components is necessary to specify the geometric orientation in n dimensions.[64] Only in three dimensions does the number of coordinate planes equal the dimension of the space, and thus only in three dimensions can we associate the oriented area with a vector $\boldsymbol{a} \times \boldsymbol{b}$ of the same dimension as the space of the vectors $\boldsymbol{a}$ and $\boldsymbol{b}$.

The oriented area spanned by n-dimensional vectors $\boldsymbol{x}, \boldsymbol{y} \in V$ is a bilinear antisymmetric function $A(\boldsymbol{x}, \boldsymbol{y})$, whose value is a vector in the space of oriented areas of dimension $\frac{1}{2}n(n-1)$. The wedge-product space $\wedge^2 V$, the space of antisymmetric vector products, is also of dimension $\frac{1}{2}n(n-1)$. Are the two the same? Vector spaces are *isomorphic* if (and only if) they have the same dimension. The space $\wedge^2 V$ represents *every possible* bilinear antisymmetric product of vectors, and

[63]The generalization of parallelograms to higher dimensions are known as *parallelotopes*.[31, p122]

[64]The orientation of a parallelogram in four dimensions is specified by six projections onto coordinate planes.

thus any antisymmetric, bilinear function $A(\boldsymbol{x}, \boldsymbol{y})$ is proportional to $\boldsymbol{x} \wedge \boldsymbol{y}$, implying that the spaces are the same.

Example. In $\mathbb{R}^2$, with $\boldsymbol{a} = a^i \boldsymbol{e}_i$ and $\boldsymbol{b} = b^j \boldsymbol{e}_j$, it's straightforward to show that $\boldsymbol{a} \wedge \boldsymbol{b} = \begin{vmatrix} a^1 & a^2 \\ b^1 & b^2 \end{vmatrix} \boldsymbol{e}_1 \wedge \boldsymbol{e}_2$. As shown on page 102, $A(\boldsymbol{a}, \boldsymbol{b}) = \begin{vmatrix} a^1 & a^2 \\ b^1 & b^2 \end{vmatrix} A(\boldsymbol{e}_1, \boldsymbol{e}_2)$.

5.10.4 Oriented volume: Generalization of $\boldsymbol{a} \cdot (\boldsymbol{b} \times \boldsymbol{c})$

Let vectors $\boldsymbol{a}, \boldsymbol{b}, \boldsymbol{c} \in \mathbb{R}^3$ span a parallelepiped (see Fig. 5.18), the volume of which we denote as

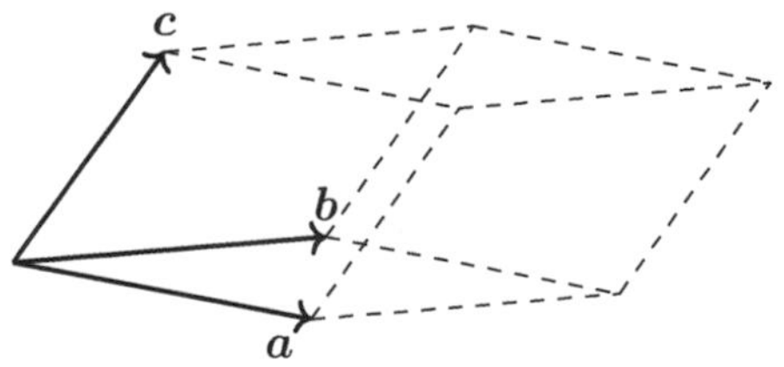

Figure 5.18 Parallelepiped formed from vectors $\boldsymbol{a}$, $\boldsymbol{b}$, $\boldsymbol{c}$.

Vol$(\boldsymbol{a}, \boldsymbol{b}, \boldsymbol{c})$. For vectors in three dimensions, Vol$(\boldsymbol{a}, \boldsymbol{b}, \boldsymbol{c}) = |\boldsymbol{a} \cdot (\boldsymbol{b} \times \boldsymbol{c})|$. Because the triple vector product is cyclically invariant, however $[\boldsymbol{a} \cdot (\boldsymbol{b} \times \boldsymbol{c}) = \boldsymbol{b} \cdot (\boldsymbol{c} \times \boldsymbol{a}) = \boldsymbol{c} \cdot (\boldsymbol{a} \times \boldsymbol{b})]$, $\boldsymbol{a} \cdot (\boldsymbol{b} \times \boldsymbol{c})$ (no absolute value) is totally antisymmetric, *similar* to $\boldsymbol{a} \wedge \boldsymbol{b} \wedge \boldsymbol{c}$. That would suggest (but not prove) the *oriented* (signed) volume spanned by $\boldsymbol{a}$, $\boldsymbol{b}$, $\boldsymbol{c}$ is found from the expression $\boldsymbol{a} \wedge \boldsymbol{b} \wedge \boldsymbol{c}$. We show that is indeed the case, in n-dimensional space *the oriented volume spanned by vectors* $\{\boldsymbol{v}_1, \cdots, \boldsymbol{v}_n\}$ *is given by the wedge product* $\boldsymbol{v}_1 \wedge \cdots \wedge \boldsymbol{v}_n$.

We're guided by two requirements, that for Vol$(\boldsymbol{v}_1, \cdots, \boldsymbol{v}_n)$:

- If a vector is scaled by a factor λ, the volume is scaled by the same factor, Vol$(\boldsymbol{v}_1, \cdots, \lambda\boldsymbol{v}_i, \cdots, \boldsymbol{v}_n) = \lambda$Vol$(\boldsymbol{v}_1, \cdots, \boldsymbol{v}_i, \cdots, \boldsymbol{v}_n)$, implying multilinearity.
- If $\boldsymbol{v}_i = \boldsymbol{v}_j$ for $j \neq i$, we require that Vol$(\boldsymbol{v}_1, \cdots, \boldsymbol{v}_i, \cdots, \boldsymbol{v}_j, \cdots, \boldsymbol{v}_n) = 0$, implying Vol$(\boldsymbol{v}_1, \cdots, \boldsymbol{v}_n)$ is a totally antisymmetric function—the same as $\boldsymbol{v}_1 \wedge \cdots \wedge \boldsymbol{v}_n$.

These requirements imply that parallelepipeds spanned by the vectors $\{\boldsymbol{v}_1, \boldsymbol{v}_2, \cdots, \boldsymbol{v}_n\}$ and $\{\boldsymbol{v}_1 + \lambda\boldsymbol{v}_2, \boldsymbol{v}_2, \cdots, \boldsymbol{v}_n\}$ have the same volume for any value of λ.

A theorem that we sketch the proof of, is that for vectors in an n-dimensional space, parallelepipeds spanned by $\{\boldsymbol{u}_1, \cdots, \boldsymbol{u}_n\}$ and $\{\boldsymbol{v}_1, \cdots, \boldsymbol{v}_n\}$ have equal volumes if the n-fold wedge products are equal up to a sign, $\boldsymbol{u}_1 \wedge \cdots \wedge \boldsymbol{u}_n = \pm\boldsymbol{v}_1 \wedge \cdots \wedge \boldsymbol{v}_n$. To show this, we transform the set of vectors $\{\boldsymbol{v}_i\}$ into the set $\{\boldsymbol{u}_j\}$ through a sequence of transformations, of two types. Either multiply $\boldsymbol{v}_j$ by a number λ, or add $\lambda\boldsymbol{v}_j$ to another vector $\boldsymbol{v}_k$ (shear transformation). The substitutions $\boldsymbol{u}_i \to \lambda\boldsymbol{u}_i$ and $\boldsymbol{u}_i \to \boldsymbol{u}_i + \lambda\boldsymbol{u}_k$ are such that $\boldsymbol{u}_1 \wedge \cdots \wedge \boldsymbol{u}_n$ changes in the same way as Vol$(\boldsymbol{u}_1, \cdots, \boldsymbol{u}_n)$. Under successive transformations $\boldsymbol{v}_i \to \lambda\boldsymbol{u}_i$ or $\boldsymbol{v}_j \to \boldsymbol{u}_j + \lambda\boldsymbol{u}_k$, $\boldsymbol{v}_1 \wedge \cdots \wedge \boldsymbol{v}_n$ and Vol$(\boldsymbol{v}_1, \cdots, \boldsymbol{v}_n)$ transform *in the same way* until $\{\boldsymbol{v}_i\}$ has become $\{\boldsymbol{u}_i\}$ (up to the ordering of vectors), with $\boldsymbol{u}_1 \wedge \cdots \wedge \boldsymbol{u}_n = \alpha\boldsymbol{v}_1 \wedge \cdots \wedge \boldsymbol{v}_n$ and Vol$(\boldsymbol{u}_1, \cdots, \boldsymbol{u}_n) = \alpha$Vol$(\boldsymbol{v}_1, \cdots, \boldsymbol{v}_n)$ where α is a number. The two oriented volumes are the same if $\alpha = \pm 1$.

We therefore have a means of comparing n-dimensional volumes:

$$\text{Vol}(\boldsymbol{v}_1, \cdots, \boldsymbol{v}_n) = \frac{\boldsymbol{v}_1 \wedge \cdots \wedge \boldsymbol{v}_n}{\boldsymbol{u}_1 \wedge \cdots \wedge \boldsymbol{u}_n} \text{Vol}(\boldsymbol{u}_1, \cdots, \boldsymbol{u}_n) , \tag{5.80}$$

but not (yet) of determining the volume. We're allowed to "divide tensors" in Eq. (5.80) because for V an n-dimensional space, the highest non-trivial wedge-product space is $\wedge^n V$, the dimension of which is $\binom{n}{n} = 1$. *Any* n-multivector $\boldsymbol{v}_1 \wedge \cdots \wedge \boldsymbol{v}_n$ can serve as a basis for $\wedge^n V$.

Let $\{\boldsymbol{e}_i\}_{i=1}^n$ be an orthonormal basis for V. The parallelepiped spanned by these vectors can be taken to have unit volume for an n-dimensional space. With the vectors $\{\boldsymbol{u}_i\}$ in Eq. (5.80) given by the basis vectors $\{\boldsymbol{e}_i\}$, fix the proportionality by setting $\text{Vol}(\boldsymbol{e}_1, \cdots, \boldsymbol{e}_n) = (1)\boldsymbol{e}_1 \wedge \cdots \wedge \boldsymbol{e}_n$. The oriented volume of an n-dimensional parallelepiped is then (relative to the unit volume)

$$\text{Vol}(\boldsymbol{v}_1, \cdots, \boldsymbol{v}_n) = \boldsymbol{v}_1 \wedge \cdots \wedge \boldsymbol{v}_n \, . \tag{5.81}$$

In what follows, we need the use of a theorem. Let $\mathcal{A}$ be a linear operator on the n-dimensional space V, $\mathcal{A}: V \to V$, and let V have the basis $\{\boldsymbol{e}_i\}_{i=1}^n$. Then,

$$(\mathcal{A}\boldsymbol{e}_1) \wedge \cdots \wedge (\mathcal{A}\boldsymbol{e}_n) = (\det \mathcal{A})\, \boldsymbol{e}_1 \wedge \cdots \wedge \boldsymbol{e}_n \, , \tag{5.82}$$

where $\det \mathcal{A}$ is the determinant of the matrix elements[65] of $\mathcal{A}$. As an immediate application, expressing the ith contravariant vector in the basis $\{\boldsymbol{e}_j\}$, $\boldsymbol{v}_i = v_i^j \boldsymbol{e}_j$, we have combining Eqs. (5.82) and (5.81),

$$\text{Vol}(\boldsymbol{v}_1, \cdots, \boldsymbol{v}_n) = \boldsymbol{v}_1 \wedge \cdots \wedge \boldsymbol{v}_n = \left(\det v_i^j\right) \boldsymbol{e}_1 \wedge \cdots \wedge \boldsymbol{e}_n \, . \tag{5.83}$$

The determinant of the vector components $\det v_j^i$ can be interpreted as a *volume change factor*.

Example. In Cartesian coordinates, let $\boldsymbol{a} = a\hat{\boldsymbol{x}}$, $\boldsymbol{b} = b\hat{\boldsymbol{y}}$, and $\boldsymbol{c} = c\hat{\boldsymbol{z}}$. Using Eq. (5.81), we have that $\text{Vol}(\boldsymbol{a}, \boldsymbol{b}, \boldsymbol{c}) = abc\, \hat{\boldsymbol{x}} \wedge \hat{\boldsymbol{y}} \wedge \hat{\boldsymbol{z}}$, the same as obtained from Eq. (5.83).

One could worry that the unit-volume tensor $\boldsymbol{e}_1 \wedge \cdots \wedge \boldsymbol{e}_n$ has been chosen arbitrarily and will change if we select another orthonormal basis. It turns out that $\boldsymbol{e}_1 \wedge \cdots \wedge \boldsymbol{e}_n$ is the same for any orthonormal basis, *up to a sign*. Let there be two orthonormal bases, $\{\boldsymbol{e}_i\}$ and $\{\boldsymbol{f}_j\}$. Then there exists an orthogonal transformation $\mathcal{R}$ that maps $\{\boldsymbol{e}_i\}$ into $\{\boldsymbol{f}_j\}$ such that[66] $\det \mathcal{R} = \pm 1$. We then have as a special case of Eq. (5.82)

$$\boldsymbol{f}_1 \wedge \cdots \wedge \boldsymbol{f}_n = \det \mathcal{R} \boldsymbol{e}_1 \wedge \cdots \wedge \boldsymbol{e}_n = \pm \boldsymbol{e}_1 \wedge \cdots \wedge \boldsymbol{e}_n \, . \tag{5.84}$$

We're entitled to refer to the volume given by Eq. (5.81) as *the* volume.

5.10.5 Ordered basis

The $\pm$ factor in Eq. (5.84) is an essential ambiguity that cannot be avoided. A nonzero element θ of $\wedge^n V$ specifies an *orientation* of V. *Ordered bases* fall into two classes: those in the orientation and those not. An ordered basis $\{\boldsymbol{e}_i\}_{i=1}^n$ is in the orientation specified by θ if $\boldsymbol{e}_1 \wedge \cdots \wedge \boldsymbol{e}_n = \alpha\theta$, where $\alpha > 0$. *The orientation is the collection of bases having the same sign as* θ. There are only

---

[65] *Proof*: Let $\mathcal{A}\boldsymbol{e}_r = A_r^j \boldsymbol{e}_j$, $r = 1, \cdots, n$, i.e., the action of $\mathcal{A}$ on $\boldsymbol{e}_r$ is to produce an element of V, which has its own expansion in the basis $\{\boldsymbol{e}_i\}$. From the multilinearity of the wedge product,

$$(\mathcal{A}\boldsymbol{e}_1) \wedge \cdots \wedge (\mathcal{A}\boldsymbol{e}_n) = A_1^{i_1} \cdots A_n^{i_n} \boldsymbol{e}_{i_1} \wedge \cdots \wedge \boldsymbol{e}_{i_n} = A_1^{i_1} \cdots A_n^{i_n} \varepsilon_{i_1 \cdots i_n} \boldsymbol{e}_1 \wedge \cdots \wedge \boldsymbol{e}_n = (\det \mathcal{A})\, \boldsymbol{e}_1 \wedge \cdots \wedge \boldsymbol{e}_n \, ,$$

where we've used $\boldsymbol{e}_{i_1} \wedge \cdots \wedge \boldsymbol{e}_{i_n} = \varepsilon_{i_1 \cdots i_n} \boldsymbol{e}_1 \wedge \cdots \wedge \boldsymbol{e}_n$ and Eq. (5.65). Your inner mathematician might worry that $\det \mathcal{A}$ is basis dependent. It turns out the value of a determinant is the same in all bases. If the abstract operator $\mathcal{A}$ is represented by a matrix A in one basis and by A' in another, then there exists an invertible matrix R such that $A' = RAR^{-1}$. Taking determinants, $\det A' = \det R \det A\, (\det R)^{-1} = \det A$. Why the concern with basis independence? As budding relativists, you seek what is the same in all reference frames. Accept no substitute.

[66] *Proof*: Let $\boldsymbol{f}_i = R_i^j \boldsymbol{e}_j$, where $\{\boldsymbol{f}_i\}$ and $\{\boldsymbol{e}_j\}$ are both orthonormal bases under the inner product $\mathbf{g}$. Then, $\mathbf{g}(\boldsymbol{f}_i, \boldsymbol{f}_j) = \delta_{ij} = R_i^k R_j^l \mathbf{g}(\boldsymbol{e}_k, \boldsymbol{e}_l) = R_i^k R_j^l \delta_{kl} = \sum_k R_i^k R_j^k = R_i^k (R_k^j)^T$. Thus, $\det \delta_{ij} = 1 = \left[\det R_j^i\right]^2$.

two orientations, $\alpha > 0$ or $\alpha < 0$; there is no third alternative.[67] Orientations with $\alpha > 0$ ($\alpha < 0$) are called right-handed (left-handed). For example, in $\mathbb{R}^3$ the bases $\{\boldsymbol{e}_x, \boldsymbol{e}_y, \boldsymbol{e}_z\}$ and $\{\boldsymbol{e}_y, \boldsymbol{e}_x, \boldsymbol{e}_z\}$ are in different orientations because $\boldsymbol{e}_x \wedge \boldsymbol{e}_y \wedge \boldsymbol{e}_z = -\boldsymbol{e}_y \wedge \boldsymbol{e}_x \wedge \boldsymbol{e}_z$.

For each n-dimensional space[68] V, elements of the highest wedge-product space $\wedge^n V$ specify oriented volumes $\boldsymbol{v}_1 \wedge \cdots \wedge \boldsymbol{v}_n$, what we'll refer to as n-volumes.

Table 5.3 Volume elements of n-dimensional oriented vector spaces.

n-volume	The old way	Covariant way
2-volume	$\boldsymbol{a} \times \boldsymbol{b}$	$\boldsymbol{a} \wedge \boldsymbol{b}$
3-volume	$\boldsymbol{a} \cdot (\boldsymbol{b} \times \boldsymbol{c})$	$\boldsymbol{a} \wedge \boldsymbol{b} \wedge \boldsymbol{c}$
4-volume	?	$\boldsymbol{a} \wedge \boldsymbol{b} \wedge \boldsymbol{c} \wedge \boldsymbol{d}$

5.10.6 Hodge star operator

For V an n-dimensional space, $\wedge^p V$ and $\wedge^{n-p} V$ have the same dimension—and are isomorphic—because of the symmetry of the binomial coefficients, $\binom{n}{p} = \binom{n}{n-p}$. The *Hodge star operator*, denoted simply as $*$, is a linear mapping $* : \wedge^p \to \wedge^{n-p}$ that implements the isomorphism.[69] This mapping accounts for the "taxonomy" among tensors and their duals noted in Sec 5.10.1.

To proceed, we must define an inner product for multivectors. For V a two-dimensional space, denote the inner product for $\wedge^2 V$ as $\langle \boldsymbol{v}_1 \wedge \boldsymbol{v}_2, \boldsymbol{v}_3 \wedge \boldsymbol{v}_4 \rangle$ to distinguish it from $\mathbf{g}(\boldsymbol{v}_1, \boldsymbol{v}_2)$. We require $\langle \cdot, \cdot \rangle$ to be symmetric and bilinear (Section 5.6). Let V have an orthonormal basis $\{\boldsymbol{e}_i\}$ (with respect to $\mathbf{g}$). For simplicity take $\langle \boldsymbol{e}_1 \wedge \boldsymbol{e}_2, \boldsymbol{e}_1 \wedge \boldsymbol{e}_2 \rangle = 1$. Then, for $\boldsymbol{u} = u^i \boldsymbol{e}_i$ and $\boldsymbol{v} = v^j \boldsymbol{e}_j$,

$$\langle \boldsymbol{u} \wedge \boldsymbol{v}, \boldsymbol{e}_1 \wedge \boldsymbol{e}_2 \rangle = \left(u^1 v^2 - u^2 v^1\right) \langle \boldsymbol{e}_1 \wedge \boldsymbol{e}_2, \boldsymbol{e}_1 \wedge \boldsymbol{e}_2 \rangle = \begin{vmatrix} u^1 & u^2 \\ v^1 & v^2 \end{vmatrix} = \begin{vmatrix} \mathbf{g}(\boldsymbol{u}, \boldsymbol{e}_1) & \mathbf{g}(\boldsymbol{u}, \boldsymbol{e}_2) \\ \mathbf{g}(\boldsymbol{v}, \boldsymbol{e}_1) & \mathbf{g}(\boldsymbol{v}, \boldsymbol{e}_2) \end{vmatrix} ,$$

where $\mathbf{g}(\boldsymbol{u}, \boldsymbol{e}_i) = u^i$. Generalizing, the inner product for $\wedge^k V$ is defined as[70]

$$\langle \boldsymbol{u}_1 \wedge \cdots \wedge \boldsymbol{u}_k, \boldsymbol{v}_1 \wedge \cdots \wedge \boldsymbol{v}_k \rangle \equiv \det \mathbf{g}(\boldsymbol{u}_i, \boldsymbol{v}_j) . \tag{5.85}$$

For each k, therefore,

$$\langle \boldsymbol{e}_1 \wedge \cdots \wedge \boldsymbol{e}_k, \boldsymbol{e}_1 \wedge \cdots \wedge \boldsymbol{e}_k \rangle = (-1)^{n_-^k} , \tag{5.86}$$

where n_-^k is the number of minus signs in the $\mathbf{g}$-orthonormal basis $\{\boldsymbol{e}_i\}_{i=1}^k$.

Example. Consider the bivector $\boldsymbol{A} = \frac{1}{2} A^{ij} \boldsymbol{e}_i \wedge \boldsymbol{e}_j$, Eq. (5.78). Using Eq. (5.85),

$$\begin{aligned} \langle \boldsymbol{e}_l \wedge \boldsymbol{e}_m, \boldsymbol{A} \rangle &= \tfrac{1}{2} A^{ij} \langle \boldsymbol{e}_l \wedge \boldsymbol{e}_m, \boldsymbol{e}_i \wedge \boldsymbol{e}_j \rangle = \tfrac{1}{2} A^{ij} \begin{vmatrix} g_{li} & g_{lj} \\ g_{mi} & g_{mj} \end{vmatrix} = \tfrac{1}{2} A^{ij} \left(g_{li} g_{mj} - g_{mi} g_{lj}\right) \\ &= \tfrac{1}{2} \left(A_{lm} - A_{ml}\right) = A_{lm} . \end{aligned}$$

The inner product returns A_{lm}, similar to $\boldsymbol{e}_i \cdot \boldsymbol{B} = \boldsymbol{e}_i \cdot B^j \boldsymbol{e}_j = B^j g_{ij} = B_i$.

For $\{\boldsymbol{e}_i\}_{i=1}^n$ an orthonormal basis on V, let $\boldsymbol{\omega} \equiv \boldsymbol{e}_1 \wedge \cdots \wedge \boldsymbol{e}_n$ be a basis for the "top space" $\wedge^n V$; $\boldsymbol{\omega}$ is unique up to a choice of sign,[71] which we take to be positive. The *idea* behind the star

[67] Because $\{\boldsymbol{e}_i\}$ is linearly independent—a basis.

[68] A proviso should be added: vector spaces V with inner product $\mathbf{g}(\boldsymbol{v}_i, \boldsymbol{v}_j)$.

[69] Two spaces related by the Hodge star operator are called Hodge duals. The dual in Hodge dual (those spaces related by $* : \wedge^k \to \wedge^{n-k}$) has nothing to do with the dual in dual spaces, the distinction between V and V^*.

[70] For $k = 1$ Eq. (5.85) reduces to the inner product between single vectors.

[71] Shown in Section 5.10.4.

operator $* : \wedge^k V \to \wedge^{n-k} V$ is as follows. For $\boldsymbol{\alpha} \equiv \boldsymbol{e}_1 \wedge \cdots \wedge \boldsymbol{e}_k$, we'd "like" $\boldsymbol{\alpha} \wedge (*\boldsymbol{\alpha}) = \boldsymbol{\omega}$, so that $*(\boldsymbol{e}_1 \wedge \cdots \wedge \boldsymbol{e}_k) = \boldsymbol{e}_{k+1} \wedge \cdots \wedge \boldsymbol{e}_n$—that's the gist of the idea.

The Hodge operator is defined with an extra requirement so that for two k-multivectors $\boldsymbol{\alpha} = \boldsymbol{v}_1 \wedge \cdots \wedge \boldsymbol{v}_k$ and $\boldsymbol{\beta} = \boldsymbol{u}_1 \wedge \cdots \wedge \boldsymbol{u}_k$, with $1 \le k \le n$,

$$\boldsymbol{\alpha} \wedge (*\boldsymbol{\beta}) \equiv \langle \boldsymbol{\alpha}, \boldsymbol{\beta} \rangle \boldsymbol{\omega} \; . \tag{5.87}$$

From Eq. (5.87), $\boldsymbol{\omega} \wedge (*\boldsymbol{\omega}) = \langle \boldsymbol{\omega}, \boldsymbol{\omega} \rangle \boldsymbol{\omega} = (-1)^{n_-} \boldsymbol{\omega}$, where we've used Eq. (5.86) and n_- is the number of minus signs in the $\mathbf{g}$-orthonormal basis for V. Thus, $*\boldsymbol{\omega} = (-1)^{n_-} \mathbb{1}$, where $\mathbb{1}$ denotes the basis vector of $\wedge^0 V$, the dual of $\wedge^n V$. Using Eq. (5.87), $\mathbb{1} \wedge (*\mathbb{1}) = g(\mathbb{1}, \mathbb{1}) \boldsymbol{\omega} = \boldsymbol{\omega}$, where $g(\mathbb{1}, \mathbb{1}) \equiv 1$. Thus, $*\mathbb{1} = \boldsymbol{\omega}$; the dual of $\wedge^0$ is $\wedge^n$. From Eq. (5.87): $\langle \boldsymbol{\alpha}, \boldsymbol{\beta} \rangle = (-1)^{n_-} * (\boldsymbol{\alpha} \wedge *\boldsymbol{\beta})$.

Let $\boldsymbol{e}_I \equiv \boldsymbol{e}_1 \wedge \cdots \wedge \boldsymbol{e}_k$ and $\boldsymbol{e}_J \equiv \boldsymbol{e}_{k+1} \wedge \cdots \wedge \boldsymbol{e}_n$ so that[72] $\boldsymbol{e}_I \wedge \boldsymbol{e}_J = \boldsymbol{\omega}$. Using Eq. (5.87), $\boldsymbol{e}_I \wedge (*\boldsymbol{e}_I) = \langle \boldsymbol{e}_I, \boldsymbol{e}_I \rangle \boldsymbol{\omega} = \langle \boldsymbol{e}_I, \boldsymbol{e}_I \rangle \boldsymbol{e}_I \wedge \boldsymbol{e}_J$ so that $*\boldsymbol{e}_I = \langle \boldsymbol{e}_I, \boldsymbol{e}_I \rangle \boldsymbol{e}_J$. The wedge product $\boldsymbol{e}_J \wedge (*\boldsymbol{e}_J) = \langle \boldsymbol{e}_J, \boldsymbol{e}_J \rangle \boldsymbol{e}_I \wedge \boldsymbol{e}_J$. It's straightforward to show that[73] $\boldsymbol{e}_I \wedge \boldsymbol{e}_J = (-1)^{k(n-k)} \boldsymbol{e}_J \wedge \boldsymbol{e}_I$, so that $*\boldsymbol{e}_J = (-1)^{k(n-k)} \langle \boldsymbol{e}_J, \boldsymbol{e}_J \rangle \boldsymbol{e}_I$. These results are summarized in Table 5.4.

Table 5.4 Hodge duals.

$*\mathbb{1} = \boldsymbol{\omega} \equiv \boldsymbol{e}_I \wedge \boldsymbol{e}_J$	$*\boldsymbol{\omega} = (-1)^{n_-} \mathbb{1}$
$*\boldsymbol{e}_I = \langle \boldsymbol{e}_I, \boldsymbol{e}_I \rangle \boldsymbol{e}_J$	$*\boldsymbol{e}_J = (-1)^{k(n-k)} \langle \boldsymbol{e}_J, \boldsymbol{e}_J \rangle \boldsymbol{e}_I$

By combining these results, $*(*\boldsymbol{e}_I) = (-1)^{k(n-k)} \langle \boldsymbol{e}_I, \boldsymbol{e}_I \rangle \langle \boldsymbol{e}_J, \boldsymbol{e}_J \rangle \boldsymbol{e}_I$. But, for an orthonormal basis, $\langle \boldsymbol{e}_I, \boldsymbol{e}_I \rangle \langle \boldsymbol{e}_J, \boldsymbol{e}_J \rangle = \det \mathbf{g}(\boldsymbol{e}_i, \boldsymbol{e}_i) \det \mathbf{g}(\boldsymbol{e}_j, \boldsymbol{e}_j) = \langle \boldsymbol{\omega}, \boldsymbol{\omega} \rangle = (-1)^{n_-}$. For k-multivectors, therefore, the star operator composed with itself satisfies

$$* \circ * = (-1)^{n_- + k(n-k)} \; . \tag{5.88}$$

We found the same relation in our discussion of dual tensors, near Eq. (5.77).

For $n = 3$, we have the wedge-product spaces $\wedge^0, \wedge^1, \wedge^2, \wedge^3$, all having an odd number of dimensions, $1, 3, 3, 1$. There is therefore a dual relation between $\wedge^0$ and $\wedge^3$ and between $\wedge^1$ and $\wedge^2$. From Eq. (5.88) with $n_- = 0$, $* \circ * = 1$ for all k (check it!). In this case, $*\boldsymbol{e}_1 = \boldsymbol{e}_2 \wedge \boldsymbol{e}_3$, $*\boldsymbol{e}_2 = \boldsymbol{e}_3 \wedge \boldsymbol{e}_1$ and $*\boldsymbol{e}_3 = \boldsymbol{e}_1 \wedge \boldsymbol{e}_2$. There is thus a close resemblance between the cross product and the wedge product. They are in fact Hodge duals, $* (\boldsymbol{a} \times \boldsymbol{b}) = \boldsymbol{a} \wedge \boldsymbol{b}$. The space $\wedge^0$ is a space of scalars but its dual $\wedge^3$ is a space of pseudoscalars (see Section 5.9). This can be seen from the relation among basis vectors, $*\mathbb{1} \to \boldsymbol{e}_1 \wedge \boldsymbol{e}_2 \wedge \boldsymbol{e}_3$. Under an inversion $\mathbb{1}$ is invariant (scalar) but $\boldsymbol{\omega}$ changes sign (pseudoscalar). Likewise, $\wedge^1$ is a space of vectors and $\wedge^2$ is a space of pseudovectors, as can be seen from the basis vectors $*\boldsymbol{e}_1 \to \boldsymbol{e}_2 \wedge \boldsymbol{e}_3$: Under inversion $\boldsymbol{e}_1$ changes sign (vector) but $\boldsymbol{e}_2 \wedge \boldsymbol{e}_3$ does not (pseudovector).

For Minkowski space, we have the wedge-product spaces $\wedge^0, \wedge^1, \wedge^2, \wedge^3, \wedge^4$ with dimensions $1, 4, 6, 4, 1$. There is an isomorphism between vectors and totally antisymmetric third-rank tensors $(*\wedge^1 \to \wedge^3)$, and a relation between second-rank antisymmetric tensors, $*\wedge^2 \to \wedge^2$. It should be noted that the isomorphism $*\wedge^2 \to \wedge^2$ does not imply the *identity* of the two second-rank antisymmetric tensors, only that they are isomorphic.

---

[72]The notation $I = (i_1, \cdots, i_k)$ is a *multi-index*, a strictly-increasing string of integers $1 \le i_1 < i_2 < \cdots < i_k \le n$.

[73]Consider that $\boldsymbol{u} \wedge \boldsymbol{v}_1 \wedge \cdots \wedge \boldsymbol{v}_k = (-1)^k \boldsymbol{v}_1 \wedge \cdots \wedge \boldsymbol{v}_k \wedge \boldsymbol{u}$, i.e., "pulling" a vector through a k-multivector entails k interchanges and hence k minus signs. For $\boldsymbol{a} \in \wedge^p V$ and $\boldsymbol{b} \in \wedge^q V$, $\boldsymbol{a} \wedge \boldsymbol{b} \in \wedge^{p+q} V$, unless $p + q > n$ in which case it is zero. The wedge product satisfies the antisymmetry condition $\boldsymbol{a} \wedge \boldsymbol{b} = (-1)^{pq} \boldsymbol{b} \wedge \boldsymbol{a}$.

5.11 INTEGRATION ON MINKOWSKI SPACE

5.11.1 Differential volume elements

One dimensional

The "volume element" along a curve is the differential displacement vector $\mathrm{d}\boldsymbol{x}$ with components $\mathrm{d}\Sigma^\alpha = \mathrm{d}x^\alpha$ (in a notation we'll use for the higher-dimensional volume elements). The direction in which the curve is traversed specifies the orientation. If the orientation is reversed, the signs of $\mathrm{d}\Sigma^\alpha$ are reversed.

Two dimensional

A pair of infinitesimal contravariant vectors $\mathrm{d}\boldsymbol{x}_{(1)}$ and $\mathrm{d}\boldsymbol{x}_{(2)}$ specify an oriented area $\mathrm{d}\boldsymbol{x}_{(1)} \wedge \mathrm{d}\boldsymbol{x}_{(2)}$, the components of which are $\mathrm{d}\Sigma^{\alpha\beta} \equiv \left(\mathrm{d}\boldsymbol{x}_{(1)} \wedge \mathrm{d}\boldsymbol{x}_{(2)}\right)^{\alpha\beta} = \delta^{\alpha\beta}_{\mu\nu}\mathrm{d}x^\mu_{(1)}\mathrm{d}x^\nu_{(2)}$. The vectors $\mathrm{d}\boldsymbol{x}_{(i)}$ have components $\mathrm{d}x^\mu_{(i)}$ ($\mu = 0, 1, 2, 3$), relative to a coordinate system in MS. The order of the vectors is important and specifies the orientation of $\mathrm{d}\boldsymbol{x}_{(1)} \wedge \mathrm{d}\boldsymbol{x}_{(2)}$.

Three dimensional

Three vectors $\mathrm{d}\boldsymbol{x}_{(1)}$, $\mathrm{d}\boldsymbol{x}_{(2)}$, and $\mathrm{d}\boldsymbol{x}_{(3)}$ specify an oriented volume $\mathrm{d}\boldsymbol{x}_{(1)} \wedge \mathrm{d}\boldsymbol{x}_{(2)} \wedge \mathrm{d}\boldsymbol{x}_{(3)}$, with $\mathrm{d}\Sigma^{\alpha\beta\gamma} \equiv \left(\mathrm{d}\boldsymbol{x}_{(1)} \wedge \mathrm{d}\boldsymbol{x}_{(2)} \wedge \mathrm{d}\boldsymbol{x}_{(3)}\right)^{\alpha\beta\gamma} = \delta^{\alpha\beta\gamma}_{\mu\nu\sigma}\mathrm{d}x^\mu_{(1)}\mathrm{d}x^\nu_{(2)}\mathrm{d}x^\sigma_{(3)}$. The order in which the vectors are written determines the orientation. The quantity $\mathrm{d}\Sigma^{\alpha\beta\gamma}$, a third-rank antisymmetric tensor, has as its dual a four-vector with components $(\mathrm{d}\Sigma^*)_\lambda \equiv \frac{1}{3!}\epsilon_{\alpha\beta\gamma\lambda}\mathrm{d}\Sigma^{\alpha\beta\gamma}$ (from Eq. (5.77)).[74] Consider an arbitrary linear combination the vectors $\mathrm{d}\boldsymbol{x}_{(i)}$, $\boldsymbol{A} \equiv \sum_{i=1}^3 a_{(i)}\mathrm{d}\boldsymbol{x}_{(i)}$; $\mathrm{d}\Sigma^*$ is orthogonal[75] to $\boldsymbol{A}$. Thus, $\mathrm{d}\Sigma^*$ is normal to every vector lying in the surface having oriented volume $\mathrm{d}\Sigma$.

Four dimensional

In four dimensions, four vectors $\mathrm{d}\boldsymbol{x}_{(1)}$, $\mathrm{d}\boldsymbol{x}_{(2)}$, $\mathrm{d}\boldsymbol{x}_{(3)}$, $\mathrm{d}\boldsymbol{x}_{(4)}$ determine an oriented volume $\mathrm{d}\boldsymbol{x}_{(1)} \wedge \mathrm{d}\boldsymbol{x}_{(2)} \wedge \mathrm{d}\boldsymbol{x}_{(3)} \wedge \mathrm{d}\boldsymbol{x}_{(4)}$, the components of which are

$$\mathrm{d}\Sigma^{\alpha\beta\gamma\delta} \equiv \left(\mathrm{d}\boldsymbol{x}_{(1)} \wedge \mathrm{d}\boldsymbol{x}_{(2)} \wedge \mathrm{d}\boldsymbol{x}_{(3)} \wedge \mathrm{d}\boldsymbol{x}_{(4)}\right)^{\alpha\beta\gamma\delta} = \delta^{\alpha\beta\gamma\delta}_{\mu\nu\sigma\tau}\mathrm{d}x^\mu_{(1)}\mathrm{d}x^\nu_{(2)}\mathrm{d}x^\sigma_{(3)}\mathrm{d}x^\tau_{(4)} \,.$$

The order in which the vectors are written determines the orientation. The orientation is usually chosen so that $\mathrm{d}\Sigma^{0123} > 0$. We can associate a scalar with $\mathrm{d}\Sigma^{\alpha\beta\gamma\delta}$, its dual $\mathrm{d}\Sigma^* \equiv \frac{1}{4!}\epsilon_{\alpha\beta\gamma\delta}\mathrm{d}\Sigma^{\alpha\beta\gamma\delta}$.

5.11.2 Stokes's theorem

Stokes's theorem relates the integral of the *gradient* of a tensor over the volume of an n-dimensional space V_n to an integral of the tensor evaluated at the surface S_{n-1} that bounds[76] V_n, which we indicate schematically as $\int_{V_n} \nabla(\mathbf{T})\mathrm{d}V^n = \int_{S_{n-1}} \mathbf{T}\mathrm{d}S^{(n-1)}$. To apply the theorem, *the orientations of S_{n-1} and V_n must agree.* Let $\mathrm{d}\boldsymbol{x}_{(1)} \wedge \cdots \wedge \mathrm{d}\boldsymbol{x}_{(n)}$ be an oriented volume element of V_n and $\mathrm{d}\boldsymbol{y}_{(1)} \wedge \cdots \wedge \mathrm{d}\boldsymbol{y}_{(n-1)}$ an oriented volume element of S_{n-1}. Let $\mathrm{d}\boldsymbol{y}_{(0)}$ be a vector in V_n that's orthogonal to every vector in the surface S_{n-1} and that's *outwardly pointing*. The relation between

[74] We're writing the dual of $\mathrm{d}\Sigma^{\alpha\beta\gamma}$ as $(\mathrm{d}\Sigma^*)_\lambda$ instead of $(*\mathrm{d}\Sigma)_\lambda$ to avoid confusion with another use of the symbol $*\mathrm{d}$, the exterior derivative—Section 13.7.

[75] *Proof*: Consider the inner product $A^\lambda\,(\mathrm{d}\Sigma^*)_\lambda = \sum_{i=1}^3 a_{(i)}\mathrm{d}x^\lambda_{(i)}\,(\mathrm{d}\Sigma^*)_\lambda$. There is, for each value of i, a term of the form $\mathrm{d}x^\lambda_{(i)}\epsilon_{\alpha\beta\gamma\lambda}\mathrm{d}x^\alpha_{(1)}\mathrm{d}x^\beta_{(2)}\mathrm{d}x^\gamma_{(3)}$. For each i, there are terms of the form $\mathrm{d}x^\lambda_{(i)}\mathrm{d}x^\mu_{(i)}$ (μ equals one of the indices (α, β, γ)), symmetric in λ and μ, which vanish when contracted with $\epsilon_{\alpha\beta\gamma\lambda}$.

[76] Stokes's theorem is the fundamental theorem of calculus in higher dimensions: $\int_a^b (\mathrm{d}f/\mathrm{d}x)\mathrm{d}x = f(b) - f(a)$.

vectors lying in the surface S_{n-1} and those belonging to the space V_n is illustrated schematically in Fig. 5.19 (schematic because the figure shows only part of the surface: S_{n-1} is a closed surface).

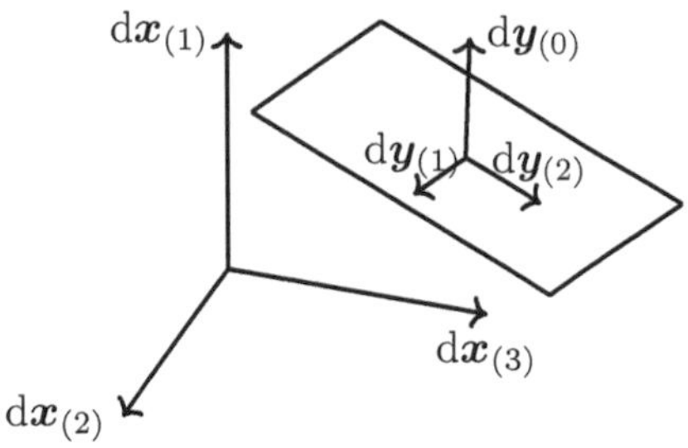

Figure 5.19 Vectors belonging to the surface $(\mathrm{d}\boldsymbol{y}_{(1)}, \mathrm{d}\boldsymbol{y}_{(2)})$ and the volume $(\mathrm{d}\boldsymbol{x}_{(i)}, \mathrm{d}\boldsymbol{y}_{(0)})$.

The orientation of $\mathrm{d}\boldsymbol{y}_{(0)} \wedge \mathrm{d}\boldsymbol{y}_{(1)} \wedge \cdots \wedge \mathrm{d}\boldsymbol{y}_{(n-1)}$ must match that of $\mathrm{d}\boldsymbol{x}_{(1)} \wedge \cdots \wedge \mathrm{d}\boldsymbol{x}_{(n)}$.

We state without proof Stokes's theorem, that for a tensor **F** defined on V_n,

$$\begin{aligned}
\int_{V_2} \partial_\gamma F^\alpha{}_\beta \mathrm{d}\Sigma^{\gamma\beta} &= \oint_{S_1} F^\alpha{}_\beta \mathrm{d}\Sigma^\beta && (n=2)\\
\int_{V_3} \partial_\sigma F^\alpha{}_{\beta\gamma} \mathrm{d}\Sigma^{\sigma\beta\gamma} &= \oint_{S_2} F^\alpha{}_{\beta\gamma} \mathrm{d}\Sigma^{\beta\gamma} && (n=3)\\
\int_{V_4} \partial_\tau F^\alpha{}_{\beta\gamma\sigma} \mathrm{d}\Sigma^{\tau\beta\gamma\sigma} &= \oint_{S_3} F^\alpha{}_{\beta\gamma\sigma} \mathrm{d}\Sigma^{\beta\gamma\sigma}\,, && (n=4) \qquad (5.89)
\end{aligned}$$

where α can be absent or present or could be replaced by two or more free indices.

Let's show how this works for $n = 2$, where S_1 is chosen parallel to the coordinate axes. Let V_2 be a square where S_1 is traversed in the clockwise direction (see Fig. 5.20). The vector $\mathrm{d}\boldsymbol{y}_0$ must be

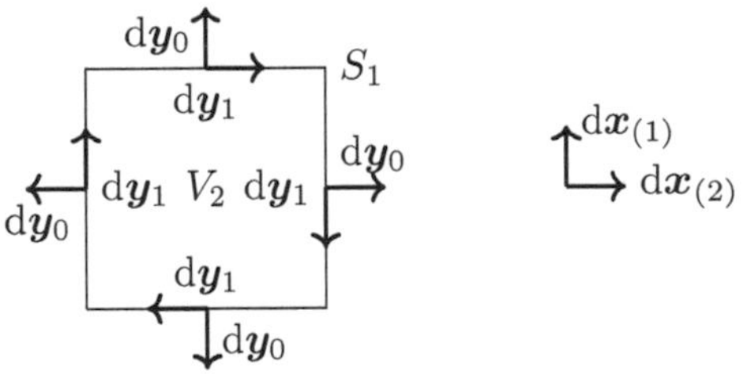

Figure 5.20 Orientation of $\mathrm{d}\boldsymbol{y}_0$ and $\mathrm{d}\boldsymbol{y}_1$ on S_1 everywhere matches that of $\mathrm{d}\boldsymbol{x}_{(1)}$ and $\mathrm{d}\boldsymbol{x}_{(2)}$.

outwardly pointing at all points of S_1. Let the direction of S_1 fix the orientation of V_2: $\mathrm{d}\boldsymbol{x}_{(1)} \wedge \mathrm{d}\boldsymbol{x}_{(2)}$ must have the same orientation as $\mathrm{d}\boldsymbol{y}_0 \wedge \mathrm{d}\boldsymbol{y}_1$. Let $\mathrm{d}\boldsymbol{x}_{(1)} = (0, 0, \mathrm{d}y, 0)$ and $\mathrm{d}\boldsymbol{x}_{(2)} = (0, \mathrm{d}x, 0, 0)$. Thus, $\mathrm{d}\Sigma^{\gamma\beta} = \delta^{\gamma\beta}_{\mu\nu} dx^\mu_{(1)} \mathrm{d}x^\nu_{(2)} = \delta^{\gamma\beta}_{21} \mathrm{d}y\mathrm{d}x$. Applying Eq. (5.89) to $F^\alpha{}_\beta$ defined on V_2,

$$\begin{aligned}
\int_{V_2} \partial_\gamma F^\alpha{}_\beta \mathrm{d}\Sigma^{\gamma\beta} &= \int_{V_2} \partial_\gamma F^\alpha{}_\beta \delta^{\gamma\beta}_{21} \mathrm{d}y\mathrm{d}x = \int_{V_2} (\partial_2 F^\alpha{}_1 - \partial_1 F^\alpha{}_2)\, \mathrm{d}y\mathrm{d}x\\
&= \int_{\text{top}} F^\alpha{}_1 \mathrm{d}x - \int_{\text{bottom}} F^\alpha{}_1 \mathrm{d}x - \int_{\text{right}} F^\alpha{}_2 \mathrm{d}y + \int_{\text{left}} F^\alpha{}_2 \mathrm{d}y \equiv \oint_{S_1} F^\alpha{}_\mu \mathrm{d}x^\mu\,.
\end{aligned}$$

Gauss's theorem

Gauss's theorem relates the integral over V_n of the *divergence* of a tensor to an integral of the tensor evaluated at the surface S_{n-1}. For $n = 4$ it states that:

$$\int_{V_4} \partial_\beta F^{\alpha\beta} \mathrm{d}\Sigma^* = \oint_{S_3} F^{\alpha\beta} \mathrm{d}\Sigma^*_\beta \, , \tag{5.90}$$

where $\mathrm{d}\Sigma^*$ is the dual of $\mathrm{d}\Sigma^{\alpha\beta\gamma\delta}$ and $\mathrm{d}\Sigma^*_\beta$ is the dual of $\mathrm{d}\Sigma^{\rho\gamma\sigma}$. The free index α gets a "free ride"; it can be absent, present, or replaced by two or more free indices.

5.12 THE GHOST OF TENSORS YET TO COME

The title of this chapter is tensors on flat spaces, yet no explicit use was made of the geometry being flat, and in fact we won't officially define flatness until Chapter 14 (vanishing of the Riemann tensor). Where the property of flatness lurks is in the implicit assumption that spacetime can be covered by a single coordinate system—sufficient and appropriate for SR where inertial observers see force-free motion as straight lines in spacetime. GR is based on the recognition that IRFs are an idealization having only local, approximate validity. In Chapter 13 we consider tensors on more general spaces known as manifolds. In SR we can assume that partial derivatives of tensors are tensors because of the strictly linear coordinate transformations between IRFs (Section 5.3). Derivatives of tensors on manifolds are treated in Chapters 13 and 14.

SUMMARY

This chapter comprised our first look at tensors, a topic of considerable importance to relativity.

- Minkowski space is a four-dimensional vector space having one timelike $\boldsymbol{e}_0$ and three spacelike basis vectors $\boldsymbol{e}_i$, $i = 1, 2, 3$, where basis vectors are labeled with subscripts. Points in MS (events) are located relative to an origin-event with position four-vectors, $\boldsymbol{r} = x^\mu \boldsymbol{e}_\mu$, where coordinates (x^0, x^1, x^2, x^3) are labeled with superscripts. We adhere to the Einstein summation convention that repeated raised and lowered indices imply a summation.

- Coordinate basis vectors $\{\boldsymbol{e}_\alpha\}$ are locally tangent to coordinate curves. The position four-vector $x^\mu \boldsymbol{e}_\mu$ has the geometric character of a line element; it has contravariant components, where contravariant and covariant refer to behavior under coordinate transformations. Associated with the coordinate basis is another set of vectors $\{\boldsymbol{e}^\beta\}$, the dual basis (labeled with superscripts) that are everywhere orthogonal to the vectors $\{\boldsymbol{e}_\alpha\}$, such that $\boldsymbol{e}^\beta \cdot \boldsymbol{e}_\alpha = \delta^\beta_{\ \alpha}$. A vector expressed in the dual basis $\boldsymbol{B} = B_\beta \boldsymbol{e}^\beta$ (with components labeled by subscripts) is called a covariant vector. Covariant vectors are orthogonal to coordinate surfaces and have the geometric character of a wave vector, where consecutive planes of equal phase become closer as the magnitude of the wave vector increases. The distinction between covariant and contravariant vectors is unnecessary in orthogonal coordinate systems.

- The components of the metric tensor are the inner products between basis vectors, either $g_{\alpha\beta} = \boldsymbol{e}_\alpha \cdot \boldsymbol{e}_\beta$ or $g^{\alpha\beta} = \boldsymbol{e}^\alpha \cdot \boldsymbol{e}^\beta$. The mixed metric tensor $g^\alpha_{\ \beta} = \boldsymbol{e}^\alpha \cdot \boldsymbol{e}_\beta = \delta^\alpha_{\ \beta}$. The metric tensor connects the contravariant and covariant components of vectors (raise and lower indices), $A^\alpha = g^{\alpha\beta} A_\beta$ and $A_\mu = g_{\mu\nu} A^\nu$. The inner product between vectors is expressed using the metric tensor, $\boldsymbol{A} \cdot \boldsymbol{B} = g_{\alpha\beta} A^\alpha B^\beta = A^\alpha B_\alpha$.

- Transformations between coordinate systems are smooth, invertible functions, $x^{j'} = x^{j'}(x^i)$ and $x^i = x^i(x^{j'})$, where we use the Schouten index convention. A transformation is invertible when the Jacobian determinant of the transformation does not vanish identically.

- The importance of tensors to relativity is that if a tensor equation is valid in one reference frame, it's valid in any reference frame. It's important therefore to be able to establish whether a given set of mathematical objects constitute a tensor. The quotient theorem provides an indirect test for tensor character.

- The Lorentz metric is invariant under the LT, $\boldsymbol{e}_{\mu'} \cdot \boldsymbol{e}_{\nu'} = \boldsymbol{e}_\mu \cdot \boldsymbol{e}_\nu = \eta_{\mu\nu}$. It's a constant tensor in MS. Other constant tensors are the Kronecker delta δ^i_j and the Levi-Civita symbol, $\varepsilon_{i_1 \cdots i_n}$.

- Spacetime gradients come in two varieties, covariant $\partial_\mu \equiv \partial/\partial x^\mu$, and contravariant, $\partial^\mu \equiv \partial/\partial x_\mu$. The are related by raising an index, $\partial^\mu = g^{\mu\nu}\partial_\nu$. The construct $\partial_\mu \partial^\mu = \partial^\mu \partial_\mu$ generates the wave equation operator and is a Lorentz invariant.

- Relative tensors transform with the Jacobian raised to an integer power w, J^w. Tensors that transform with $w = \pm 1$ are called tensor densities. Tensors that transform with $w = 0$ are called absolute tensors. The quantity $\sqrt{|g|}\mathrm{d}^n x$ is an absolute scalar, where g is the determinant of the covariant components of the metric tensor. The sign of g is an absolute quantity.

- Antisymmetric tensors have associated with them new tensors called dual tensors.

EXERCISES

5.1 Show that $(a_{rst} + a_{str} + a_{srt})\, x^r x^s x^t = 3a_{rst}x^r x^s x^t$.

5.2 Show that a symmetric second-rank tensor in n dimensions has $n(n+1)/2$ independent components.

5.3 What is the metric tensor for cylindrical coordinates, (ρ, ϕ, z)? Hint: The line element is $\mathrm{d}\boldsymbol{s} = \mathrm{d}\rho\hat{\boldsymbol{\rho}} + \rho\mathrm{d}\phi\hat{\boldsymbol{\phi}} + \mathrm{d}z\hat{\boldsymbol{z}}$.

5.4 Simplify the expression $g_{ij}A^iB^j - A^kB_k$.

5.5 Show in polar coordinates that if $\mathrm{d}x^i = \begin{pmatrix} \mathrm{d}r \\ \mathrm{d}\theta \end{pmatrix}$ then $\mathrm{d}x_i = \begin{pmatrix} \mathrm{d}r & r^2\mathrm{d}\theta \end{pmatrix}$.

5.6 Of the three four-vectors, $\boldsymbol{A} = 4\boldsymbol{e}_0 + 3\boldsymbol{e}_1 + 2\boldsymbol{e}_2 + \boldsymbol{e}_3$, $\boldsymbol{B} = 5\boldsymbol{e}_0 + 4\boldsymbol{e}_1 + 3\boldsymbol{e}_2$, and $\boldsymbol{C} = \boldsymbol{e}_0 + 2\boldsymbol{e}_1 + 3\boldsymbol{e}_2 + 4\boldsymbol{e}_3$, which is timelike, which is lightlike, and which is spacelike?

5.7 Show that for $x^\mu = (ct, \boldsymbol{r})$, then $x_\mu = (-ct, \boldsymbol{r})$.

5.8 Show that the metric tensor g_{ij} transforms as a second-order covariant tensor.

5.9 If $[g_{ij}] = \begin{pmatrix} A & 1 \\ 1 & 0 \end{pmatrix}$, what is $[g^{ij}]$?

5.10 Show that the metric tensor $\mathbf{g} = g_{ij}\boldsymbol{e}^i\boldsymbol{e}^j$ is actually the identity $\mathbf{I}$. Hint: Use Eq. (5.21).

5.11 Show that if δ_{ij} has the usual values for the Kronecker delta in one frame, it does not have those values in another frame. Show that $(\delta_{ij})' = \sum_a A^a_i A^a_j$.

5.12 In an n-dimensional space, evaluate $\delta^k_i \delta^i_k$. (Answer: n. Why?)

5.13 Verify that $L^\kappa_{\bar{0}}\eta_{\kappa\lambda}L^\lambda_{\bar{0}} = -1$.

5.14 If $\gamma = 2$ and $x^\mu = \begin{pmatrix} 1 & 1 \end{pmatrix}$, show that $x^{0'} = x^{1'} = 2 - \sqrt{3}$. Show that the transformed basis vectors are $\boldsymbol{e}_{0'} = 2\boldsymbol{e}_0 + \sqrt{3}\boldsymbol{e}_1$ and $\boldsymbol{e}_{1'} = \sqrt{3}\boldsymbol{e}_0 + 2\boldsymbol{e}_1$. Draw a spacetime diagram showing all quantities. Show explicitly that $x^{i'}\boldsymbol{e}_{i'} = x^j\boldsymbol{e}_j$.

5.15 Verify using Eq. (5.72) that for three-dimensional Euclidean space, $P_1 = T^{23}$, $P_2 = -T^{13}$, and $P_3 = T^{12}$.

5.16 Show that in three-dimensional Euclidean space, $*(*P)^i = P^i$. You'll need $\epsilon^{ijk}\epsilon_{jkm} = 2\delta^i_m$ (from Eq. (5.71)). Show that $*(*T_{lm}) = T_{lm}$. Use Eq. (5.73).

5.17 Show that in MS for a second-rank antisymmetric tensor, $*(*T)^{mn} = -T^{mn}$. Use Eqs. (5.74) and (5.75) and the identity $\epsilon_{ksmn}\epsilon^{rtmn} = -2\left(\delta^r{}_k\delta^t{}_s - \delta^r{}_s\delta^t{}_k\right)$.

5.18 Show that $\epsilon_{i_1\cdots i_n}\epsilon^{i_1\cdots i_n} = (-1)^{n-}n!$.

5.19 Show that if the quantities $\{A_{i_1\cdots i_k}\}$ are totally antisymmetric in their indices, then $\delta^{i_1\cdots i_k}_{j_1\cdots j_k}A_{i_1\cdots i_k} = k!A_{j_1\cdots j_k}$.

5.20 Show for the generalized Kronecker delta in an n-dimensional space that

$$\delta^{i_1\cdots i_r i_{r+1}\cdots i_k}_{j_1\cdots j_r i_{r+1}\cdots i_k} = \frac{(n-r)!}{(n-k)!}\delta^{i_1\cdots i_r}_{j_1\cdots j_r}.$$

Hint: The indices $i_{r+1}\cdots i_k$ must all be distinct, and they must also be different from any of the indices $i_1\cdots i_r$. Consider the binomial coefficient $\binom{n-r}{k-r}$, the number of ways to choose $(k-r)$ indices from $(n-r)$ where order is immaterial. There is then a sum over $(k-r)!$ permutations of the indices $i_{r+1}\cdots i_k$.

5.21 Let $\mathbf{F}$ be a second-rank antisymmetric covariant tensor with respect to the basis $\mathbf{e}^\mu$, i.e., $F_{\mu\nu} = -F_{\nu\mu}$. Show that the contravariant components $F^{\mu\nu}$ are also antisymmetric. Show that $\mathbf{F}$ is antisymmetric in any coordinate system, i.e., show $F^{\mu'\nu'} = -F^{\nu'\mu'}$ and $F_{\mu'\nu'} = -F_{\nu'\mu'}$. Antisymmetry is thus an intrinsic (coordinate-independent) property.

5.22 Show from the fact that antisymmetry is coordinate independent for a second-rank tensor, a symmetric tensor $F_{\mu\nu} = F_{\nu\mu}$ is symmetric in any coordinate system.

5.23 Show that in MS,

$$\begin{aligned}\epsilon^{s_1r_1r_2r_3}\epsilon_{s_1a_1a_2a_3} = &-\left[\delta^{r_1}_{a_1}\left(\delta^{r_2}_{a_2}\delta^{r_3}_{a_3} - \delta^{r_2}_{a_3}\delta^{r_3}_{a_2}\right) + \delta^{r_1}_{a_2}\left(\delta^{r_2}_{a_3}\delta^{r_3}_{a_1} - \delta^{r_2}_{a_1}\delta^{r_3}_{a_3}\right)\right.\\ &\left.+\,\delta^{r_1}_{a_3}\left(\delta^{r_2}_{a_1}\delta^{r_3}_{a_2} - \delta^{r_2}_{a_2}\delta^{r_3}_{a_1}\right)\right].\end{aligned}$$

Hint: The indices take on four values. How do we know that? Well, you are told this is MS (four dimensions) *and* you noticed that for an n-dimensional space the Levi-Civita symbol $\varepsilon_{i_1\cdots i_n}$ has n indices, each of which takes n values (as opposed to the generalized Kronecker symbol which has $k \le n$ symbols, each of which takes n values).

5.24 Show that a tensor of type $(1,1)$ maps a vector onto a vector, $\mathbf{T}^1_1(\boldsymbol{v}) : V \to V$.

5.25 Consider a three-dimensional spacetime, with one timelike unit vector $\mathbf{e}_0$ and two spacelike unit vectors, $\mathbf{e}_1$ and $\mathbf{e}_2$. The metric relation for this space is $(\mathrm{d}s)^2 = -(\mathrm{d}x_0)^2 + (\mathrm{d}x_1)^2 + (\mathrm{d}x_2)^2$. Starting from the volume element $\boldsymbol{\omega} = \mathbf{e}_0 \wedge \mathbf{e}_1 \wedge \mathbf{e}_2$, show that the Hodge duals are given by $*\mathbf{e}_0 = -\mathbf{e}_1 \wedge \mathbf{e}_2$, $*\mathbf{e}_1 = \mathbf{e}_2 \wedge \mathbf{e}_0$, and $*\mathbf{e}_2 = \mathbf{e}_0 \wedge \mathbf{e}_1$. Hint: For sets of wedge products $\boldsymbol{e}_I \wedge \boldsymbol{e}_J = \boldsymbol{\omega}$, $*\boldsymbol{e}_I = \langle \boldsymbol{e}_I, \boldsymbol{e}_I\rangle \boldsymbol{e}_J$ (Section 5.10.6).

CHAPTER 6

Lorentz transformation, II

Now that we've had an exposure to tensors, we delve deeper into the properties of Lorentz transformations, freely making use of the concepts and notation developed in Chapter 5.

6.1 DECOMPOSITION INTO ROTATIONS AND BOOSTS

We show that *Lorentz transformations can be uniquely decomposed into the product of a boost and a rotation*. The proof begins at Eq. (6.17). First we consider rotations and boosts separately.

6.1.1 Rotations

Rotations about *spatial* axes can be represented as spacetime transformations using 4×4 real matrices with the $(0,0)$ element equal to unity. Such operations are passive transformations involving the mapping of coordinate axes (see Fig. 6.1). The 4×4 matrices for *counterclockwise* rotations

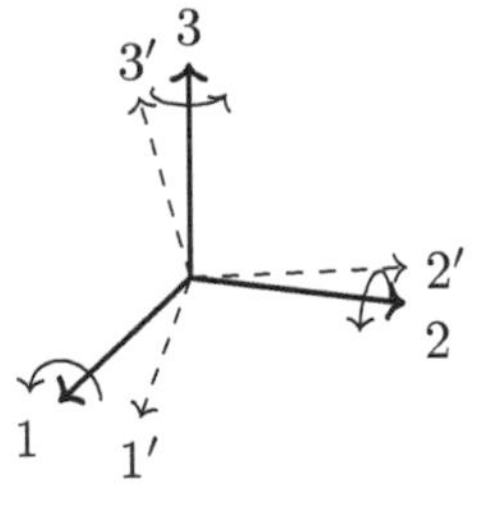

Figure 6.1 Rotation of 1-2-3 coordinate system into $1'$-$2'$-$3'$ system.

about each of the $(1,2,3)$-axes through angle θ are as follows:

$$R_1(\theta) = \begin{pmatrix} 1 & 0 & 0 & 0 \\ 0 & 1 & 0 & 0 \\ 0 & 0 & \cos\theta & \sin\theta \\ 0 & 0 & -\sin\theta & \cos\theta \end{pmatrix} = I + \theta \begin{pmatrix} 0 & 0 & 0 & 0 \\ 0 & 0 & 0 & 0 \\ 0 & 0 & 0 & 1 \\ 0 & 0 & -1 & 0 \end{pmatrix} + O(\theta^2) \equiv I + \theta J_1 + O(\theta^2)$$

$$R_2(\theta) = \begin{pmatrix} 1 & 0 & 0 & 0 \\ 0 & \cos\theta & 0 & -\sin\theta \\ 0 & 0 & 1 & 0 \\ 0 & \sin\theta & 0 & \cos\theta \end{pmatrix} = I + \theta \begin{pmatrix} 0 & 0 & 0 & 0 \\ 0 & 0 & 0 & -1 \\ 0 & 0 & 0 & 0 \\ 0 & 1 & 0 & 0 \end{pmatrix} + O(\theta^2) \equiv I + \theta J_2 + O(\theta^2)$$

$$R_3(\theta) = \begin{pmatrix} 1 & 0 & 0 & 0 \\ 0 & \cos\theta & \sin\theta & 0 \\ 0 & -\sin\theta & \cos\theta & 0 \\ 0 & 0 & 0 & 1 \end{pmatrix} = I + \theta \begin{pmatrix} 0 & 0 & 0 & 0 \\ 0 & 0 & 1 & 0 \\ 0 & -1 & 0 & 0 \\ 0 & 0 & 0 & 0 \end{pmatrix} + O(\theta^2) \equiv I + \theta J_3 + O(\theta^2) \, . \quad (6.1)$$

We've written $R_i(\theta)$ as the sum of the identity matrix I and a matrix θJ_i containing the terms that are first order in *small values of* θ. Is a rotation a LT? A LT must satisfy Eq. (4.12), $L^T \eta L = \eta$, and $(L^0_0)^2 \geq 1$ (Exercise 4.1). These requirements are met by the matrices $R_i(\theta)$ in Eq. (6.1). *Rotations are valid LTs.*

The matrices J_i in Eq. (6.1) are known as the *infinitesimal rotation generators*,

$$J_1 = \begin{pmatrix} 0 & 0 & 0 & 0 \\ 0 & 0 & 0 & 0 \\ 0 & 0 & 0 & 1 \\ 0 & 0 & -1 & 0 \end{pmatrix} \quad J_2 = \begin{pmatrix} 0 & 0 & 0 & 0 \\ 0 & 0 & 0 & -1 \\ 0 & 0 & 0 & 0 \\ 0 & 1 & 0 & 0 \end{pmatrix} \quad J_3 = \begin{pmatrix} 0 & 0 & 0 & 0 \\ 0 & 0 & 1 & 0 \\ 0 & -1 & 0 & 0 \\ 0 & 0 & 0 & 0 \end{pmatrix} . \quad (6.2)$$

They satisfy the commutation relations $[J_i, J_j] = -\varepsilon_{ijk} J_k$. The operator for *finite-angle* rotations about an arbitrary axis can be developed through repeated applications of infinitesimal rotations, as we now show. The matrices J_i are antisymmetric.

Rotations do not commute in general: $R_i R_j \neq R_j R_i$. They *do* commute, however, for infinitesimal rotation angles. By writing $R_i \approx I + \mathrm{d}\theta^i J_i$ (no sum),

$$\begin{aligned} R_i R_j &\approx \left(I + \mathrm{d}\theta^i J_i\right)\left(I + \mathrm{d}\theta^j J_j\right) = I^2 + \mathrm{d}\theta^i J_i I + \mathrm{d}\theta^j I J_j + \mathrm{d}\theta^i \mathrm{d}\theta^j J_i \circ J_j \\ &\approx I + \mathrm{d}\theta^i J_i + \mathrm{d}\theta^j J_j \approx R_j R_i \, . \end{aligned}$$

The non-commuting part begins at second order in infinitesimal quantities. To first order in small angles, $R_i R_j = R_j R_i$. An infinitesimal rotation can be built up as (where we keep terms to first order in small quantities)

$$\begin{aligned} R(\mathrm{d}\theta) \equiv& R_1(\mathrm{d}\theta_1) R_2(\mathrm{d}\theta_2) R_3(\mathrm{d}\theta_3) = I + \mathrm{d}\theta^1 J_1 + \mathrm{d}\theta^2 J_2 + \mathrm{d}\theta^3 J_3 \\ =& I + \begin{pmatrix} 0 & 0 & 0 & 0 \\ 0 & 0 & \mathrm{d}\theta^3 & -\mathrm{d}\theta^2 \\ 0 & -\mathrm{d}\theta^3 & 0 & \mathrm{d}\theta^1 \\ 0 & \mathrm{d}\theta^2 & -\mathrm{d}\theta^1 & 0 \end{pmatrix} \equiv I + \epsilon \, , \end{aligned} \quad (6.3)$$

where ϵ denotes the matrix of infinitesimal angles in Eq. (6.3). Rotation matrices satisfy $R^T R = I$, Eq. (4.5), and thus for infinitesimal rotations $(I + \epsilon^T)(I + \epsilon) = I$, which will be satisfied to first order if ϵ is antisymmetric, $\epsilon^T = -\epsilon$ (which we see explicitly in Eq. (6.3)).

Applying the infinitesimal rotation operator $R(\mathrm{d}\theta) = I + \epsilon$ to the position vector $\boldsymbol{r}$, $\boldsymbol{r}' = R(\mathrm{d}\theta)\boldsymbol{r} = (I + \epsilon)\boldsymbol{r}$, or $\mathrm{d}\boldsymbol{r} \equiv \boldsymbol{r}' - \boldsymbol{r} = \epsilon \boldsymbol{r}$. In terms of components,

$$\begin{pmatrix} 0 \\ \mathrm{d}x \\ \mathrm{d}y \\ \mathrm{d}z \end{pmatrix} = \begin{pmatrix} 0 & 0 & 0 & 0 \\ 0 & 0 & \mathrm{d}\theta^3 & -\mathrm{d}\theta^2 \\ 0 & -\mathrm{d}\theta^3 & 0 & \mathrm{d}\theta^1 \\ 0 & \mathrm{d}\theta^2 & -\mathrm{d}\theta^1 & 0 \end{pmatrix} \begin{pmatrix} 0 \\ x \\ y \\ z \end{pmatrix} = \begin{pmatrix} 0 \\ y\mathrm{d}\theta^3 - z\mathrm{d}\theta^2 \\ z\mathrm{d}\theta^1 - x\mathrm{d}\theta^3 \\ x\mathrm{d}\theta^2 - y\mathrm{d}\theta^1 \end{pmatrix} . \quad (6.4)$$

Equation (6.4) is equivalent to $\mathrm{d}\boldsymbol{r} = \boldsymbol{r} \times \mathbf{d}\boldsymbol{\theta}$, where we've made a *vector* out of the infinitesimal rotation angles, $\mathbf{d}\boldsymbol{\theta} = \mathrm{d}\theta^1 \mathbf{e}_1 + \mathrm{d}\theta^2 \mathbf{e}_2 + \mathrm{d}\theta^3 \mathbf{e}_3$ (permissible *only* for infinitesimal angles).[1] The matrices R_i in Eq. (6.1) describe counterclockwise, passive transformations of the coordinate axes. When applied to the vector $\boldsymbol{r}$, however, Eq. (6.4) describes an active, *clockwise* rotation of $\boldsymbol{r}$ through the infinitesimal angle $\mathrm{d}\theta$.

[1] Technically $\mathbf{d}\boldsymbol{\theta}$ is a pseudovector; see the rules on page 97.

For given angles $\mathrm{d}\theta^1$, $\mathrm{d}\theta^2$, $\mathrm{d}\theta^3$, $\mathbf{d\theta}$ defines a direction in space, the *axis of rotation*. Instead of specifying $\mathbf{d\theta}$, however, it's easier to first specify the axis of rotation with a unit vector $\hat{\boldsymbol{n}}$, and define the rotation about $\hat{\boldsymbol{n}}$. Let $\mathbf{d\theta} = \hat{\boldsymbol{n}}\mathrm{d}\phi$, which defines the angle of rotation $\mathrm{d}\phi$ about $\hat{\boldsymbol{n}}$. Clearly, $\mathrm{d}\phi = |\mathbf{d\theta}|$. Let n^1, n^2, n^3 be the components of $\hat{\boldsymbol{n}}$ with respect to the coordinate system, $\hat{\boldsymbol{n}} = n^i\hat{\boldsymbol{x}}_i$ (sum implied), with $\mathrm{d}\theta^i = n^i\mathrm{d}\phi$. Define the generator for infinitesimal rotations about $\hat{\boldsymbol{n}}$,

$$J(\hat{\boldsymbol{n}}) \equiv n^i J_i = \begin{pmatrix} 0 & 0 & 0 & 0 \\ 0 & 0 & n^3 & -n^2 \\ 0 & -n^3 & 0 & n^1 \\ 0 & n^2 & -n^1 & 0 \end{pmatrix} . \tag{6.5}$$

The operator for an infinitesimal rotation $\mathrm{d}\phi$ about $\hat{\boldsymbol{n}}$ is then

$$R(\hat{\boldsymbol{n}}, \mathrm{d}\phi) = I + \mathrm{d}\phi J(\hat{\boldsymbol{n}}) . \tag{6.6}$$

Rotations through a *finite* angle ϕ about the fixed axis $\hat{\boldsymbol{n}}$ can be realized from a succession of infinitesimal rotations about the same axis. Define the rotation operator as the limit of N rotations through angle ϕ/N as $N \to \infty$. Using Eq. (6.6):

$$R(\hat{\boldsymbol{n}}\phi) \equiv \lim_{N\to\infty} \left(1 + \frac{\phi}{N} J(\hat{\boldsymbol{n}})\right)^N = \mathrm{e}^{\phi J(\hat{\boldsymbol{n}})} , \tag{6.7}$$

where we've used the Euler definition of the exponential function. *The operator for a rotation ϕ about a fixed axis $\hat{\boldsymbol{n}}$ is the exponential of the generator.* The exponential of an operator is defined from the power series,

$$\exp(\phi J(\hat{\boldsymbol{n}})) = \sum_{m=0}^{\infty} \frac{\phi^m}{m!} J(\hat{\boldsymbol{n}})^m , \tag{6.8}$$

and hence we need all powers of $J(\hat{\boldsymbol{n}})$ for Eq. (6.8) to serve as a practical expression for rotations. Fortunately, Eq. (6.8) can be summed because the powers of $J(\hat{\boldsymbol{n}})$ close among themselves in a simple way. By the Cayley-Hamilton theorem, a square matrix satisfies its own characteristic equation, which from Eq. (6.5) is $\lambda^3 = -\lambda$. Therefore, $J(\hat{\boldsymbol{n}})^3 = -J(\hat{\boldsymbol{n}})$. For $n = 1, 2, \cdots$,

$$J(\hat{\boldsymbol{n}})^{2n} = (-1)^{n+1} J(\hat{\boldsymbol{n}})^2 \qquad J(\hat{\boldsymbol{n}})^{2n+1} = (-1)^n J(\hat{\boldsymbol{n}}) . \tag{6.9}$$

Combining Eq. (6.9) with Eq. (6.8) and separating even and odd powers,

$$\begin{aligned} R(\hat{\boldsymbol{n}}\phi) =& \mathrm{e}^{\phi J(\hat{\boldsymbol{n}})} = \sum_{m=0}^{\infty} \frac{\phi^m}{m!} J(\hat{\boldsymbol{n}})^m = I + \sum_{n=0}^{\infty} \frac{\phi^{2n+1}}{(2n+1)!} J(\hat{\boldsymbol{n}})^{2n+1} + \sum_{n=1}^{\infty} \frac{\phi^{2n}}{(2n)!} J(\hat{\boldsymbol{n}})^{2n} \\ =& I + J(\hat{\boldsymbol{n}}) \sum_{n=0}^{\infty} (-1)^n \frac{\phi^{2n+1}}{(2n+1)!} + J(\hat{\boldsymbol{n}})^2 \sum_{n=1}^{\infty} (-1)^{n+1} \frac{\phi^{2n}}{(2n)!} \\ =& I + J(\hat{\boldsymbol{n}}) \sin\phi + J(\hat{\boldsymbol{n}})^2 (1 - \cos\phi) . \end{aligned} \tag{6.10}$$

A rotation is thus characterized by three parameters: the angle ϕ and the unit vector $\hat{\boldsymbol{n}}$. Because $\sum_{i=1}^{3} (n^i)^2 = 1$, there are only two independent components of $\hat{\boldsymbol{n}}$.

From Eq. (6.10), $R(\hat{\boldsymbol{n}}\phi)$ is the 4×4 matrix

$$R(\hat{\boldsymbol{n}}, \phi) = \left(\begin{array}{c|ccc} 1 & 0 & 0 & 0 \\ \hline 0 & & & \\ 0 & & R^i_{\ j} & \\ 0 & & & \end{array}\right) , \tag{6.11}$$

where $R^i_j = \delta^i_j \cos\phi + (1-\cos\phi)n^i n^j + \sin\phi \epsilon^i{}_{jk} n^k$. Written out as a matrix,

$$[R^i_j] = \begin{pmatrix} \cos\phi + (n^1)^2(1-\cos\phi) & n^3\sin\phi + n^1n^2(1-\cos\phi) & -n^2\sin\phi + n^1n^3(1-\cos\phi) \\ -n^3\sin\phi + n^1n^2(1-\cos\phi) & \cos\phi + (n^2)^2(1-\cos\phi) & n^1\sin\phi + n^2n^3(1-\cos\phi) \\ n^2\sin\phi + n^1n^3(1-\cos\phi) & -n^1\sin\phi + n^3n^2(1-\cos\phi) & \cos\phi + (n^3)^2(1-\cos\phi) \end{pmatrix}.$$

For finite rotations, the rotation operator has no particular symmetry; it is, however, orthogonal $R(-\phi) = R^T(\phi)$ (show this).

6.1.2 Boosts

The LT for arbitrary boosts[2] was given in Eq. (3.24). We now obtain the same result through a succession of infinitesimal boosts. Using the hyperbolic form of the LT, Eq. (4.6), where the parameter θ is related to the velocity by $\tanh\theta = -\beta$, the 4×4 matrices for boosts in each of the three directions are

$$L_x(\theta) = \begin{pmatrix} \cosh\theta & \sinh\theta & 0 & 0 \\ \sinh\theta & \cosh\theta & 0 & 0 \\ 0 & 0 & 1 & 0 \\ 0 & 0 & 0 & 1 \end{pmatrix} = I + \theta \begin{pmatrix} 0 & 1 & 0 & 0 \\ 1 & 0 & 0 & 0 \\ 0 & 0 & 0 & 0 \\ 0 & 0 & 0 & 0 \end{pmatrix} + O(\theta^2) \equiv I + \theta K_1 + O(\theta^2)$$

$$L_y(\theta) = \begin{pmatrix} \cosh\theta & 0 & \sinh\theta & 0 \\ 0 & 1 & 0 & 0 \\ \sinh\theta & 0 & \cosh\theta & 0 \\ 0 & 0 & 0 & 1 \end{pmatrix} = I + \theta \begin{pmatrix} 0 & 0 & 1 & 0 \\ 0 & 0 & 0 & 0 \\ 1 & 0 & 0 & 0 \\ 0 & 0 & 0 & 0 \end{pmatrix} + O(\theta^2) \equiv I + \theta K_2 + O(\theta^2)$$

$$L_z(\theta) = \begin{pmatrix} \cosh\theta & 0 & 0 & \sinh\theta \\ 0 & 1 & 0 & 0 \\ 0 & 0 & 1 & 0 \\ \sinh\theta & 0 & 0 & \cosh\theta \end{pmatrix} = I + \theta \begin{pmatrix} 0 & 0 & 0 & 1 \\ 0 & 0 & 0 & 0 \\ 0 & 0 & 0 & 0 \\ 1 & 0 & 0 & 0 \end{pmatrix} + O(\theta^2) \equiv I + \theta K_3 + O(\theta^2).$$

In analogy with Eq. (6.2) we define the *boost generators*,

$$K_1 = \begin{pmatrix} 0 & 1 & 0 & 0 \\ 1 & 0 & 0 & 0 \\ 0 & 0 & 0 & 0 \\ 0 & 0 & 0 & 0 \end{pmatrix} \quad K_2 = \begin{pmatrix} 0 & 0 & 1 & 0 \\ 0 & 0 & 0 & 0 \\ 1 & 0 & 0 & 0 \\ 0 & 0 & 0 & 0 \end{pmatrix} \quad K_3 = \begin{pmatrix} 0 & 0 & 0 & 1 \\ 0 & 0 & 0 & 0 \\ 0 & 0 & 0 & 0 \\ 1 & 0 & 0 & 0 \end{pmatrix}. \tag{6.12}$$

The matrices K_i have the commutation relation $[K_i, K_j] = \varepsilon_{ijk} J_k$. The boost generators and the rotation generators have the commutation property $[J_i, K_j] = -\varepsilon_{ijk} K_k$. The matrices K_i are symmetric.

Infinitesimal boosts in different directions commute,

$$L_i L_j = (I + \mathrm{d}\theta^i K_i)(I + \mathrm{d}\theta^j K_j) \approx I + \mathrm{d}\theta^i K_i + \mathrm{d}\theta^j K_j \approx L_j L_i\,,$$

where $\mathrm{d}\theta^i = -\mathrm{d}\beta^i$. Finite-velocity boosts, however, do not commute for different directions, as evidenced by the non-associative addition of non-colinear velocities, Eq. (3.31). A general infinitesimal boost can be built up as

$$\begin{aligned} L \equiv L_1 L_2 L_3 &= I + \mathrm{d}\theta^1 K_1 + \mathrm{d}\theta^2 K_2 + \mathrm{d}\theta^3 K_3 \\ &= I + \begin{pmatrix} 0 & \mathrm{d}\theta^1 & \mathrm{d}\theta^2 & \mathrm{d}\theta^3 \\ \mathrm{d}\theta^1 & 0 & 0 & 0 \\ \mathrm{d}\theta^2 & 0 & 0 & 0 \\ \mathrm{d}\theta^3 & 0 & 0 & 0 \end{pmatrix}. \end{aligned}$$

[2]Refer to Fig. 1.1, with $\boldsymbol{R} = \beta ct$.

Specify a velocity direction with the unit vector $\hat{\boldsymbol{\beta}}$, where $\mathbf{d}\boldsymbol{\theta} = \hat{\boldsymbol{\beta}}\mathrm{d}\theta$. Let $\hat{\boldsymbol{\beta}}$ have components relative to the coordinate axes, b^1, b^2, b^3, where $\mathrm{d}\theta^i = b^i\mathrm{d}\theta$. Define $K(\hat{\boldsymbol{\beta}})$ as the generator of infinitesimal boosts in the direction $\hat{\boldsymbol{\beta}}$,

$$K(\hat{\boldsymbol{\beta}}) \equiv b^i K_i = \begin{pmatrix} 0 & b^1 & b^2 & b^3 \\ b^1 & 0 & 0 & 0 \\ b^2 & 0 & 0 & 0 \\ b^3 & 0 & 0 & 0 \end{pmatrix} . \tag{6.13}$$

The operator for an infinitesimal boost $\mathrm{d}\theta$ in direction $\hat{\boldsymbol{\beta}}$ is then (analogous to Eq. (6.6))

$$L(\hat{\boldsymbol{\beta}}, \mathrm{d}\theta) = I + \mathrm{d}\theta K(\hat{\boldsymbol{\beta}}) . \tag{6.14}$$

We can "integrate" Eq. (6.14) to obtain a finite boost in the fixed direction $\hat{\boldsymbol{\beta}}$:

$$L(\boldsymbol{\beta}) \equiv \lim_{N\to\infty} \left(1 + \frac{\theta}{N} K(\hat{\boldsymbol{\beta}})\right)^N = \mathrm{e}^{\theta K(\hat{\boldsymbol{\beta}})} . \tag{6.15}$$

The operator for a boost $\boldsymbol{\beta}$ is the exponential of the generator, analogous to Eq. (6.7). Using Eq. (6.13) it's simple to show that $K(\hat{\boldsymbol{\beta}})^{2n} = K(\hat{\boldsymbol{\beta}})^2$ and $K(\hat{\boldsymbol{\beta}})^{2n+1} = K(\hat{\boldsymbol{\beta}})$. We can therefore sum the infinite series implicit in Eq. (6.15),

$$\begin{aligned} L(\boldsymbol{\beta}) &= I + K(\hat{\boldsymbol{\beta}}) \sum_{n=0}^{\infty} \frac{\theta^{2n+1}}{(2n+1)!} + K(\hat{\boldsymbol{\beta}})^2 \sum_{n=1}^{\infty} \frac{\theta^{2n}}{(2n)!} \\ &= I + K(\hat{\boldsymbol{\beta}}) \sinh\theta + (\cosh\theta - 1) K(\hat{\boldsymbol{\beta}})^2 = I - \beta\gamma K(\hat{\boldsymbol{\beta}}) + (\gamma - 1) K(\hat{\boldsymbol{\beta}})^2 , \end{aligned}$$

where $\sinh\theta = -\beta\gamma$ and $\cosh\theta = \gamma$. A boost is thus determined by three parameters, $\boldsymbol{\beta} = \beta\hat{\boldsymbol{\beta}}$, where $|\hat{\boldsymbol{\beta}}| = 1$ has only two independent components. As a matrix $L(\boldsymbol{\beta})$ is

$$L(\boldsymbol{\beta}) = \left(\begin{array}{c|ccc} \gamma & -\gamma\beta^1 & -\gamma\beta^2 & -\gamma\beta^3 \\ \hline -\gamma\beta^1 & & & \\ -\gamma\beta^2 & & L^i_j & \\ -\gamma\beta^3 & & & \end{array} \right) , \tag{6.16}$$

where $L^i_j = \delta^i_j + \frac{\gamma^2}{1+\gamma}\beta^i\beta^j$ is the space-space part

$$[L^i_j] = \begin{pmatrix} 1 + \alpha(\beta^1)^2 & \alpha\beta^1\beta^2 & \alpha\beta^1\beta^3 \\ \alpha\beta^2\beta^1 & 1 + \alpha(\beta^2)^2 & \alpha\beta^2\beta^3 \\ \alpha\beta^3\beta^1 & \alpha\beta^3\beta^2 & 1 + \alpha(\beta^3)^2 \end{pmatrix} ,$$

where $\alpha \equiv \gamma^2/(1+\gamma)$. Equation (6.16) is identical to Eq. (3.24). The matrix $[L^i_j]$ is symmetric.

6.1.3 Decomposition into a rotation and a boost

We now show that an arbitrary LT, denoted Λ, can be decomposed into a boost and a rotation,

$$\Lambda = L(\boldsymbol{\beta}) R(\hat{\boldsymbol{n}}\phi) . \tag{6.17}$$

Note what this is *not* saying: Equation (6.17) is not informing us that a rotation followed by a boost constitutes a LT, what could be symbolized as $LR \to \Lambda$. *We already know that* because of the group property satisfied by LTs (Section 4.3). The theorem says something much stronger, that a *given* Λ can be decomposed into the product of a boost and a rotation, $\Lambda \to LR$. Because Λ is a LT it satisfies Eq. (4.12), $\Lambda^T\eta\Lambda = \eta$. There are 16 elements of Λ, yet because η is symmetric only 10 equations are implied by Eq. (4.12). *The most general LT must contain six parameters.* These are the three parameters that specify the boost $\boldsymbol{\beta}$, and the three that specify the rotation, $\phi\hat{\boldsymbol{n}}$.

Multiply the matrices representing rotations and boosts, Eq. (6.11) and Eq. (6.16):

$$\Lambda = \left(\begin{array}{c|ccc} \Lambda^0_0 & \Lambda^0_1 & \Lambda^0_2 & \Lambda^0_3 \\ \hline \Lambda^1_0 & & & \\ \Lambda^2_0 & & \Lambda^i_j & \\ \Lambda^3_0 & & & \end{array}\right) = \left(\begin{array}{c|ccc} \gamma & -\gamma\beta^1 & -\gamma\beta^2 & -\gamma\beta^3 \\ \hline -\gamma\beta^1 & & & \\ -\gamma\beta^2 & & L^i_k & \\ -\gamma\beta^3 & & & \end{array}\right) \left(\begin{array}{c|ccc} 1 & 0 & 0 & 0 \\ \hline 0 & & & \\ 0 & & R^k_j & \\ 0 & & & \end{array}\right). \tag{6.18}$$

We find for the time components of Eq. (6.18),

$$\gamma = \Lambda^0_0 \qquad -\gamma\beta^i = \Lambda^i_0 \qquad -\gamma\beta_j R^j_i = \Lambda^0_i\,. \tag{6.19}$$

The three boost parameters are thus determined by the first column of Λ, Λ^μ_0:

$$\beta^i = -\Lambda^i_0/\Lambda^0_0\,. \tag{6.20}$$

The complete boost matrix $L(\boldsymbol{\beta})$ is specified by the first column of Λ. Using the results of Eq. (6.19) in Eq. (6.16),

$$L = \left(\begin{array}{c|ccc} \Lambda^0_0 & \Lambda^1_0 & \Lambda^2_0 & \Lambda^3_0 \\ \hline \Lambda^1_0 & & & \\ \Lambda^2_0 & & \delta^i_j + \dfrac{\Lambda^i_0\Lambda^j_0}{1+\Lambda^0_0} & \\ \Lambda^3_0 & & & \end{array}\right). \tag{6.21}$$

Note that L is symmetric for any $\boldsymbol{\beta}$. From the $(0,0)$ component of $\eta_{\alpha\beta} = \Lambda^\gamma_\alpha\Lambda^\rho_\beta\eta_{\gamma\rho}$,

$$(\Lambda^0_0)^2 - \sum_i \left(\Lambda^i_0\right)^2 = 1\,, \tag{6.22}$$

from which we infer that $\left(\Lambda^0_0\right)^2 \geq 1$, while the $(0,i)$ components imply that

$$\Lambda^0_0\Lambda^0_i - \sum_j \Lambda^j_0\Lambda^j_i = 0\,. \tag{6.23}$$

Combining Eq. (6.20) with Eq. (6.22), we have (because $\left(\Lambda^0_0\right)^2 \geq 1$)

$$\beta^2 \equiv \sum_{i=1}^3 (\beta^i)^2 = 1 - \frac{1}{\left(\Lambda^0_0\right)^2} \leq 1\,.$$

The boost parameter defined by Eq. (6.20) is therefore physically admissible.

Now, invert Eq. (6.17)

$$R = L^{-1}(\boldsymbol{\beta})\Lambda = L(-\boldsymbol{\beta})\Lambda \tag{6.24}$$

and show that the matrix R so obtained represents a rotation. Because L is completely determined in terms of the elements of Λ, using Eq. (6.21), Eq. (6.24) is equivalent to

$$R = \left(\begin{array}{c|ccc} \Lambda^0_0 & -\Lambda^1_0 & -\Lambda^2_0 & -\Lambda^3_0 \\ \hline -\Lambda^1_0 & & & \\ -\Lambda^2_0 & & \delta^i_j + \dfrac{\Lambda^i_0\Lambda^j_0}{1+\Lambda^0_0} & \\ -\Lambda^3_0 & & & \end{array}\right) \left(\begin{array}{c|ccc} \Lambda^0_0 & \Lambda^0_1 & \Lambda^0_2 & \Lambda^0_3 \\ \hline \Lambda^1_0 & & & \\ \Lambda^2_0 & & \Lambda^i_j & \\ \Lambda^3_0 & & & \end{array}\right). \tag{6.25}$$

Multiplying these matrices, we find from the $(0,0)$ element of Eq. (6.25),

$$R^0_0 = (\Lambda^0_0)^2 - \sum_{j=1}^{3} \left(\Lambda^j_0\right)^2 = 1 ,$$

where the second equality follows from Eq. (6.22). Likewise, from Eq. (6.25),

$$R^0_i = \Lambda^0_0 \Lambda^0_i - \sum_{j=1}^{3} \Lambda^j_0 \Lambda^j_i = 0 ,$$

which follows from Eq. (6.23). It can be shown that $R^i_0 = 0$, but it takes a few more steps. The space-space part of Eq. (6.25) is given by

$$\begin{aligned} R^j_k &= -\Lambda^j_0 \Lambda^0_k + \sum_{l=1}^{3} \left(\delta^j_l + \frac{\Lambda^j_0 \Lambda^l_0}{1+\Lambda^0_0} \right) \Lambda^l_k = -\Lambda^j_0 \Lambda^0_k + \Lambda^j_k + \frac{\Lambda^j_0}{1+\Lambda^0_0} \sum_{l=1}^{3} \Lambda^l_0 \Lambda^l_k \\ &= \Lambda^j_k - \frac{\Lambda^j_0 \Lambda^0_k}{1+\Lambda^0_0} , \end{aligned} \tag{6.26}$$

where we've used Eq. (6.23) between the first and second lines. The rotation matrix is thus completely determined by the elements of Λ,

$$R = \left(\begin{array}{c|ccc} 1 & 0 & 0 & 0 \\ \hline 0 & & & \\ 0 & & \Lambda^j_k - \dfrac{\Lambda^j_0 \Lambda^0_k}{1+\Lambda^0_0} & \\ 0 & & & \end{array} \right) . \tag{6.27}$$

It can be shown from Eq. (6.27), using the space-space part of $\eta_{\alpha\beta} = \Lambda^\gamma_\alpha \Lambda^\rho_\beta \eta_{\gamma\rho}$, that $R^T R = I$ and thus Eq. (6.27) satisfies the requirement for a rotation matrix, Eq. (4.5).

We have to show uniqueness. Assume there are two decompositions of the same Λ, $\Lambda = L(\boldsymbol{\beta})R$ and $\Lambda = L(\boldsymbol{\beta}')R'$. If this were true, it would imply that

$$I = L^{-1}(\boldsymbol{\beta})\Lambda R^{-1} = L^{-1}(\boldsymbol{\beta})L(\boldsymbol{\beta}')R'R^{-1} = L(-\boldsymbol{\beta})L(\boldsymbol{\beta}')R'R^{-1} . \tag{6.28}$$

The $(0,0)$ component of Eq. (6.28) is equivalent to $1 = L(-\boldsymbol{\beta})^0_\alpha L(\boldsymbol{\beta}')^\alpha_0$, which in turn is equivalent to $\gamma\gamma'(1 - \boldsymbol{\beta}\cdot\boldsymbol{\beta}') = 1$, or

$$1 - \boldsymbol{\beta}\cdot\boldsymbol{\beta}' = \sqrt{(1-\beta^2)(1-\beta'^2)} . \tag{6.29}$$

The only solution to Eq. (6.29) is $\boldsymbol{\beta}' = \boldsymbol{\beta}$, implying from Eq. (6.28) that $R' = R$.

6.2 INFINITESIMAL LORENTZ TRANSFORMATION

We've considered separately infinitesimal rotations and boosts; let's combine them to characterize the most general *infinitesimal LT*. Start by writing

$$\Lambda^\mu_\nu = \delta^\mu_\nu + \lambda^\mu_\nu , \tag{6.30}$$

where $[\lambda^\mu_\nu]$ is a matrix containing infinitesimal parameters. For any LT $\eta_{\alpha\beta} = \Lambda^\mu_\alpha \Lambda^\nu_\beta \eta_{\mu\nu}$, which, using Eq. (6.30), implies to first order that $\eta_{\alpha\nu}\lambda^\nu_\beta + \eta_{\beta\mu}\lambda^\mu_\alpha = 0$. The matrix $[\lambda^\mu_\nu]$ must be such that $\eta_{\alpha\nu}\lambda^\nu_\beta$ is antisymmetric in (α,β). Thus, $\lambda^\alpha_\alpha = 0$ (no sum), $\lambda^i_0 = \lambda^0_i$, and $\lambda^i_j = -\lambda^j_i$.

The matrix $[\lambda^\mu_\nu]$ is therefore a 4×4 matrix with zeros on the diagonal having six independent elements—just enough for the six parameters of a LT. The combined set of rotation and boost generators, Eqs. (6.2) and (6.12), is a *natural six-dimensional basis for matrices of this type*. Thus, for infinitesimal parameters β^i and θ^i,

$$[\lambda^\mu_\nu] = \sum_{i=1}^{3} \left(\theta^i \, [J_i]^\mu_\nu - \beta^i \, [K_i]^\mu_\nu\right) = \begin{pmatrix} 0 & -\beta^1 & -\beta^2 & -\beta^3 \\ -\beta^1 & 0 & \theta^3 & -\theta^2 \\ -\beta^2 & -\theta^3 & 0 & \theta^1 \\ -\beta^3 & \theta^2 & -\theta^1 & 0 \end{pmatrix} . \tag{6.31}$$

Sometimes it's convenient to "package" the parameters of an infinitesimal LT into a quantity that's antisymmetric. Define $\omega_{\alpha\beta} \equiv \eta_{\alpha\nu}\lambda^\nu_\beta$,

$$[\omega_{\alpha\beta}] = \begin{pmatrix} 0 & \beta^1 & \beta^2 & \beta^3 \\ -\beta^1 & 0 & \theta^3 & -\theta^2 \\ -\beta^2 & -\theta^3 & 0 & \theta^1 \\ -\beta^3 & \theta^2 & -\theta^1 & 0 \end{pmatrix} . \tag{6.32}$$

Elementary boosts act as rotations where one of the axes affected is the time axis (Chapter 4). Spatial rotations about coordinate axes affect the other two axes. For either elementary boosts or rotations, two axes are involved. The quantities $\omega_{\alpha\beta}$ specify the parameter involved with the transformation affecting the α and β-axes.

6.3 SPINOR REPRESENTATION OF LORENTZ TRANSFORMATIONS

It hardly needs to be said that the LT has the form of a 4×4 real matrix. It turns out that LTs can also be obtained from 2×2 complex matrices with unit determinant. If that elicits a "so-what" response, stay tuned: Experience shows that the more ways you have of looking at something, the better.[3]

To start, we need the *Pauli spin matrices* σ_i $(i = 1, 2, 3)$:

$$\sigma_1 \equiv \begin{pmatrix} 0 & 1 \\ 1 & 0 \end{pmatrix} \qquad \sigma_2 \equiv \begin{pmatrix} 0 & -\mathrm{i} \\ \mathrm{i} & 0 \end{pmatrix} \qquad \sigma_3 \equiv \begin{pmatrix} 1 & 0 \\ 0 & -1 \end{pmatrix} . \tag{6.33}$$

If we add to the collection of Pauli matrices the 2×2 identity matrix $\sigma_0 \equiv \left(\begin{smallmatrix} 1 & 0 \\ 0 & 1 \end{smallmatrix}\right)$, the set $\{\sigma_0, \sigma_1, \sigma_2, \sigma_3\} \equiv \{\sigma_\mu\}$ is a basis for complex 2×2 matrices (Exercise 6.4). The Pauli matrices are Hermitian; they're also unitary because $(\sigma_i)^2 = \sigma_0$. They have the properties:

$$\{\sigma_l, \sigma_m\} = 2\delta_{lm}\sigma_0 \qquad [\sigma_l, \sigma_m] = 2\mathrm{i}\varepsilon_{lm}{}^{n}\sigma_n \qquad \sigma_j\sigma_k = \sigma_0\delta_{jk} + \mathrm{i}\varepsilon_{jk}{}^{l}\sigma_l , \tag{6.34}$$

where $\{A, B\} \equiv AB + BA$ is the *anticommutator*, in addition to the commutator, $[A, B] \equiv AB - BA$. They have the *trace properties*,

$$\mathrm{Tr}\,\sigma_i = 0 \qquad \mathrm{Tr}\,\sigma_i\sigma_j = 2\delta_{ij} \qquad \mathrm{Tr}\,\sigma_j\sigma_k\sigma_l = 2\mathrm{i}\varepsilon_{jkl} , \tag{6.35}$$

where $\mathrm{Tr}\,A \equiv \sum_i A_{ii}$ indicates the sum of diagonal elements.

Spacetime coordinates x^μ can be "encoded" in a 2×2 Hermitian matrix, X, using the basis $\{\sigma_\mu\}$: (for real coefficients, $\{\sigma_\mu\}$ is a basis for 2×2 Hermitian matrices)

$$X \equiv x^\mu\sigma_\mu = \begin{pmatrix} x^0 + x^3 & x^1 - \mathrm{i}x^2 \\ x^1 + \mathrm{i}x^2 & x^0 - x^3 \end{pmatrix} . \tag{6.36}$$

[3] As noted by Richard Feynman: "It always seems odd to me that the fundamental laws of physics, when discovered, can appear in so many different forms that are not apparently identical at first, but, with a little mathematical fiddling you can show the relationship. ... There is always another way to say the same thing that doesn't look at all like the way you said it before. I don't know what the reason for this is. I think it is somehow a representation of the simplicity of nature. ... Perhaps a thing is simple if you can describe it fully in several different ways without immediately knowing that you are describing the same thing."[32]

The coordinates x^μ can be recovered from X with the relation

$$x^\mu = \tfrac{1}{2}\operatorname{Tr}\sigma_\mu X \ . \tag{6.37}$$

Equation (6.37) can be verified directly from Eq. (6.36), or using $\frac{1}{2}\operatorname{Tr}\sigma_\mu x^\nu \sigma_\nu = \frac{1}{2}x^\nu \operatorname{Tr}\sigma_\mu\sigma_\nu = \frac{1}{2}(2\delta_{\mu\nu})x^\nu = x^\mu$. *There is a one-to-one correspondence between the set of all* 2×2 *Hermitian matrices (call it* $\mathcal{H}_2$*) and* $\mathbb{R}^4$: Every matrix $X \in \mathcal{H}_2$ is associated with a point $x^\mu \in \mathbb{R}^4$ through Eq. (6.37), and the same point in $\mathbb{R}^4$ implies the same matrix X through Eq. (6.36).

Now define the "transform" of X (which effects a LT, as we'll show),

$$X' \equiv MXM^\dagger \ , \tag{6.38}$$

where M is a 2×2 complex matrix with unit determinant (*unimodular*), $\det M = 1$. Like X, X' is Hermitian.[4] The set of all complex 2×2 unimodular matrices M comprises a group known as[5] $\mathrm{SL}(2,\mathbb{C})$. The coordinates x'^μ associated with X' can be found from Eq. (6.37): $x'^\mu \equiv \frac{1}{2}\operatorname{Tr}\sigma_\mu X'$. The quantities x'^μ are real from the hermiticity[6] of X' and σ_μ. Equation (6.38) implies[7] $\det X' = \det X$ The determinant of X is a Lorentz invariant: $\det X = (x^0)^2 - (x^1)^2 - (x^2)^2 - (x^3)^2 = -x_\mu x^\mu$. *The construct* $X' = MXM^\dagger$ *thus effects a LT*, a linear mapping between spacetime coordinates that preserves the spacetime interval.[8] Combining Eqs. (6.38), (6.36), and (6.37),

$$x'^\mu = \tfrac{1}{2}\operatorname{Tr}\sigma_\mu X' = \tfrac{1}{2}\operatorname{Tr}\sigma_\mu MXM^\dagger = \tfrac{1}{2}\operatorname{Tr}\sigma_\mu M x^\nu \sigma_\nu M^\dagger = \tfrac{1}{2}\left(\operatorname{Tr}\sigma_\mu M\sigma_\nu M^\dagger\right)x^\nu \ .$$

Remarkably, there's an association between matrices $M \in \mathrm{SL}(2,\mathbb{C})$ and LTs:

$$\Lambda(M)^\mu_{\ \nu} \equiv \tfrac{1}{2}\operatorname{Tr}\sigma_\mu M\sigma_\nu M^\dagger \ . \tag{6.39}$$

How is this possible? A complex 2×2 matrix is associated with eight real numbers, but because we require $\det M = 1 + 0\mathrm{i}$ there are only *six* independent parameters, just enough to encode the six parameters of a LT. Equation (6.38) defines, for each $M \in \mathrm{SL}(2,\mathbb{C})$, a mapping from the set of 2×2 Hermitian matrices onto itself, $\mathcal{H}_2 \to \mathcal{H}_2$. It's an *active* LT: Under $X \to X'$ induced by M, we have the coordinate transformation $x^\mu \to x'^\mu$, where x^μ and x'^μ are referenced to the same basis. Proper, orthochronous LTs have $\det\left[\Lambda^\mu_{\ \nu}\right] = 1$ with $\Lambda^0_{\ 0} \geq 1$ (Section 4.3). The LT defined by Eq. (6.39) has those properties (Exercise 6.14). The group property of LTs is preserved by Eq. (6.38): If $X' = MXM^\dagger$ and $X'' = NX'N^\dagger$, with $N, M \in \mathrm{SL}(2,\mathbb{C})$, then $X'' = NMXM^\dagger N^\dagger = (NM)X(NM)^\dagger$, where $NM \in \mathrm{SL}(2,\mathbb{C})$. The LTs induced by Eq. (6.38) therefore have the property $\Lambda(MN) = \Lambda(M)\Lambda(N)$. A correspondence thus exists between $\mathrm{SL}(2,\mathbb{C})$ and the set of all LTs, the Lorentz group, call it $\mathcal{L}$. *Every LT can be seen as arising from an element of* $SL(2,\mathbb{C})$. The correspondence is not one-to-one, however: For each $M \in \mathrm{SL}(2,\mathbb{C})$, $-M$ induces the same LT.[9]

Any matrix $M \in \mathrm{SL}(2,\mathbb{C})$ generates a LT through Eq. (6.39).[10] We'd like to be able to choose these matrices in a systematic way so that the LTs they generate are in a form we recognize. Let's

---

[4] $(X')^\dagger = (MXM^\dagger)^\dagger = (M^\dagger)^\dagger X^\dagger M^\dagger = MXM^\dagger = X'$.

[5] The product of matrices $M_1, M_2 \in \mathrm{SL}(2,\mathbb{C})$ is an element of the group and the inverse of any matrix in $\mathrm{SL}(2,\mathbb{C})$ is an element of the group: $\det M^{-1} = (\det M)^{-1} = 1$.

[6] $(x'^\mu)^* = \frac{1}{2}\operatorname{Tr}\sigma_\mu^* X'^* = \frac{1}{2}\sum_{ij}(\sigma_\mu)^*_{ij}X'^*_{ji} = \frac{1}{2}\sum_{ji}(\sigma_\mu)_{ji}X'_{ij} = \frac{1}{2}\operatorname{Tr}\sigma_\mu X' = x'^\mu$.

[7] Because $\det M = 1$: $\det X' = \det MXM^\dagger = \det M \det X \det M^\dagger = |\det M|^2 \det X = \det X$.

[8] Linearity: $M\,(X + \alpha Y)\,M^\dagger = MXM^\dagger + \alpha MYM^\dagger$.

[9] A property-preserving mapping between algebraic structures (such as between groups) is known as a *homomorphism*.

[10] Now would be a good time to interject the distinction between a group and its *representations*. Groups of transformations are *abstract*, whereas representations are explicit instantiations of the group elements, often in the form of matrices. The term "$\mathrm{SL}(2,\mathbb{C})$ representation" is loose (but common) jargon. Elements $M \in \mathrm{SL}(2,\mathbb{C})$ do not *represent* LTs in the sense of group representations. Yet, there's a LT for every $M \in \mathrm{SL}(2,\mathbb{C})$. A careful statement is: The Lorentz group has representations given by Eq. (6.38) for every $M \in \mathrm{SL}(2,\mathbb{C})$.

guess, based on Eqs. (6.7) and (6.15), that the matrices M can be expressed as exponentials of appropriate generators,

$$M = \exp\left(\sum_{i=1}^{6} f_i G_i\right) . \tag{6.40}$$

We know that matrices $M \in \mathrm{SL}(2,\mathbb{C})$ are associated with six independent real numbers; let these be represented by the quantities $\{f_i\}_{i=1}^{6}$ in Eq. (6.40). The strategy is to let the parameters f_i be infinitesimals and relate the G_i to the boost and rotation generators K_i and J_i.

For an infinitesimal LT, we have using Eqs. (6.38), (6.36), (6.40) to first order, and (6.30):

$$X' = MXM^\dagger = \left(I + \sum_i f_i G_i\right)(x^\nu \sigma_\nu)\left(I + \sum_i f_i G_i^\dagger\right) = x^\nu \left[\sigma_\nu + \sum_i f_i \left(\sigma_\nu G_i^\dagger + G_i \sigma_\nu\right)\right]$$
$$= x'^\mu \sigma_\mu = (L^\mu_\nu x^\nu)\sigma_\mu = (\delta^\mu_\nu + \lambda^\mu_\nu)\, x^\nu \sigma_\mu = x^\mu \sigma_\mu + x^\nu \lambda^\mu_\nu \sigma_\mu ,$$

implying that to first order in small quantities,

$$\lambda^\mu_\nu \sigma_\mu = \sum_{i=1}^{6} f_i \left(\sigma_\nu G_i^\dagger + G_i \sigma_\nu\right) . \tag{6.41}$$

Equation (6.41) connects the generators G_i with the generators of the LT; see Eq. (6.31).

Let G_1 be associated with an infinitesimal boost in the x^1-direction. With $\lambda^\mu_\nu = \beta^1 \left[K_1\right]^\mu_\nu$ from Eq. (6.31),[11] we have from Eq. (6.41)

$$\beta^1 \left[K_1\right]^\mu_\nu \sigma_\mu = f_1 \left(\sigma_\nu G_1^\dagger + G_1 \sigma_\nu\right)$$

which is satisfied by $f_1 = \beta^1$ and $G_1 = \frac{1}{2}\sigma_1$. The same holds for the other directions: $f_i = \beta^i$ and $G_i = \frac{1}{2}\sigma_i$, $i = 1, 2, 3$. The form of M that generates boosts is (with $\boldsymbol{b} = \boldsymbol{\beta}/\beta$ and $\boldsymbol{\sigma} \equiv (\sigma_1, \sigma_2, \sigma_3)$) the Hermitian matrix

$$\begin{aligned} M_B(\boldsymbol{\beta}) &= \mathrm{e}^{(\boldsymbol{b}\cdot\boldsymbol{\sigma})\theta/2} = I \cosh\theta/2 + (\boldsymbol{b}\cdot\boldsymbol{\sigma})\sinh\theta/2 \\ &= \begin{pmatrix} \cosh\theta/2 + b^3 \sinh\theta/2 & (b^1 - \mathrm{i}b^2)\sinh\theta/2 \\ (b^1 + \mathrm{i}b^2)\sinh\theta/2 & \cosh\theta/2 - b^3\sinh\theta/2 \end{pmatrix} , \end{aligned} \tag{6.42}$$

where $(\boldsymbol{b}\cdot\boldsymbol{\sigma})^{2n} = I$ and $(\boldsymbol{b}\cdot\boldsymbol{\sigma})^{2n+1} = \boldsymbol{b}\cdot\boldsymbol{\sigma}$. Combining Eqs. (6.42) and (6.39), it can be shown that

$$\Lambda(\boldsymbol{\beta}) = \left(\begin{array}{c|ccc} \cosh\theta & b^1\sinh\theta & b^2\sinh\theta & b^3\sinh\theta \\ \hline b^1\sinh\theta & & & \\ b^2\sinh\theta & & \delta^i_j + b^i b^j(\cosh\theta - 1) & \\ b^3\sinh\theta & & & \end{array}\right) , \tag{6.43}$$

which is identical to Eq. (6.16) with $\tanh\theta = -\beta$. Contrast Eq. (6.42) with Eq. (6.15), both involve exponentials of a matrix: $L = \mathrm{e}^{\theta K}$ involves the 4×4 boost generator K, whereas M involves the 2×2 Pauli matrices with *half* the boost parameter θ.

Example. Let

$$M(\theta) = \begin{pmatrix} \cosh\theta/2 & \sinh\theta/2 \\ \sinh\theta/2 & \cosh\theta/2 \end{pmatrix} .$$

[11] There's a slippery minus sign lurking here. Equation (6.31) specifies a *passive* infinitesimal LT, whereas Eq. (6.38) refers to an *active* LT, with the two related by $L^{\text{passive}}(\beta) = L^{\text{active}}(-\beta)$. That's why we chose $\lambda^\mu_\nu = +\beta^1 \left[K_1\right]^\mu_\nu$.

Note that $\det M = 1$. Using the result of Exercise 6.14, the LT induced by $M(\theta)$ is

$$\begin{pmatrix} x'^0 \\ x'^1 \\ x'^2 \\ x'^3 \end{pmatrix} = \begin{pmatrix} \cosh\theta & \sinh\theta & 0 & 0 \\ \sinh\theta & \cosh\theta & 0 & 0 \\ 0 & 0 & 1 & 0 \\ 0 & 0 & 0 & 1 \end{pmatrix} \begin{pmatrix} x^0 \\ x^1 \\ x^2 \\ x^3 \end{pmatrix},$$

a boost in the x^1-direction.

Let G_4 be associated with an infinitesimal rotation about the x^1-axis. With $\lambda^\mu_\nu = \theta^1 \left[J_1\right]^\mu_\nu$ from Eq. (6.31), we have from Eq. (6.41)

$$\theta^1 \left[J_1\right]^\mu_\nu \sigma_\mu = f_4(\sigma_\nu G_4^\dagger + G_4 \sigma_\nu)$$

which is satisfied by $f_4 = \theta^1$ and $G_4 = \frac{\mathrm{i}}{2}\sigma_1$. The same holds for the other directions: $f_{j+3} = \theta^j$ and $G_{j+3} = \frac{\mathrm{i}}{2}\sigma_j$, $j = 1, 2, 3$. The form of M that generates rotations is the unitary matrix

$$\begin{aligned} M_R(\hat{\boldsymbol{n}}\phi) &= \mathrm{e}^{\mathrm{i}(\hat{\boldsymbol{n}}\cdot\boldsymbol{\sigma})\phi/2} = I\cos\phi/2 + \mathrm{i}(\hat{\boldsymbol{n}}\cdot\boldsymbol{\sigma})\sin\phi/2 \\ &= \begin{pmatrix} \cos\phi/2 + \mathrm{i}n^3\sin\phi/2 & (\mathrm{i}n^1 + n^2)\sin\phi/2 \\ (\mathrm{i}n^1 - n^2)\sin\phi/2 & \cos\phi/2 - \mathrm{i}n^3\sin\phi/2 \end{pmatrix} \end{aligned} \tag{6.44}$$

where $\hat{\boldsymbol{n}}\cdot\boldsymbol{\sigma} = n^i\sigma_i$, with n^i the components of the rotation axis $\hat{\boldsymbol{n}}$. Combining Eq. (6.44) with Eq. (6.39), we find, identical to Eq. (6.11),

$$\Lambda(\hat{\boldsymbol{n}}\phi) = \left(\begin{array}{c|ccc} 1 & 0 & 0 & 0 \\ \hline 0 & & & \\ 0 & & R^i_j & \\ 0 & & & \end{array}\right).$$

With $M = HU$ expressed as the product of a Hermitian matrix H and a unitary matrix U (each having unit determinant), we recover the decomposition of LTs as the product of a rotation and a boost, $\Lambda(HU) = \Lambda(H)\Lambda(U) = LR$.

6.4 THOMAS-WIGNER ROTATION

Let an accelerated particle[12] have velocity $\boldsymbol{u}$ at time t in IRF K. We show in Chapter 7 that there is always an IRF frame K' in which the particle is *instantaneously at rest*. At $t + \mathrm{d}t$ the particle has velocity $\boldsymbol{u}+\mathrm{d}\boldsymbol{u}$ in K and appears instantaneously at rest in another frame, K''. Let K' be connected to K by a boost $L(\boldsymbol{u})$ and K'' connected to K by a boost $L(\boldsymbol{u}+\mathrm{d}\boldsymbol{u})$. The components of a four-vector A in K' and K'' are related to those in K by $A' = L(\boldsymbol{u})A$ and $A'' = L(\boldsymbol{u}+\mathrm{d}\boldsymbol{u})A$. The components of A'' are thus related to those of A' by $A'' = L(\boldsymbol{u}+\mathrm{d}\boldsymbol{u})L(-\boldsymbol{u})A'$. Is the compound transformation $\Lambda_T \equiv L(\boldsymbol{u}+\mathrm{d}\boldsymbol{u})L(-\boldsymbol{u})$ equivalent to a boost with velocity $\mathrm{d}\boldsymbol{u}$? Surprisingly, the answer is *No*. For $|\mathrm{d}\boldsymbol{u}| \ll |\boldsymbol{u}|$, Λ_T is (as we show) a LT with a new boost velocity $\mathrm{d}\boldsymbol{w}$ *and* a rotation $\hat{\boldsymbol{n}}\mathrm{d}\phi$, where the rotation is about the direction $\boldsymbol{u}\times\mathrm{d}\boldsymbol{u}$ with angle $\mathrm{d}\phi = \left[\gamma^2/((1+\gamma)c^2)\right]|\boldsymbol{u}\times\mathrm{d}\boldsymbol{u}|$, and where $\mathrm{d}\boldsymbol{w} = \gamma^2\mathrm{d}\boldsymbol{u}_\parallel + \gamma\mathrm{d}\boldsymbol{u}_\perp$, with $\mathrm{d}\boldsymbol{u} = \mathrm{d}\boldsymbol{u}_\parallel + \mathrm{d}\boldsymbol{u}_\perp$ decomposed into vectors parallel and perpendicular to $\boldsymbol{u}$ (see Fig. 6.2). The infinitesimal velocity $\mathrm{d}\boldsymbol{w}$ is, in the notation of Eq. (3.30), the velocity such that $\boldsymbol{u}\oplus\mathrm{d}\boldsymbol{w} = \boldsymbol{u}+\mathrm{d}\boldsymbol{u}$ to first order (Exercise 6.10). The composition of non-colinear boosts thus results in a LT that is the product of a boost and a rotation, the *Thomas-Wigner rotation.*[13] The Thomas-Wigner rotation is one of the more subtle effects of physics; it's a manifestation of the non-associativity of the addition of non-colinear velocities.

---

[12]Accelerated motion is treated in Chapter 12. In that chapter we'll make use of results obtained in this section.
[13]Discovered by L.H. Thomas in 1926[33] and later derived by Wigner.[34]

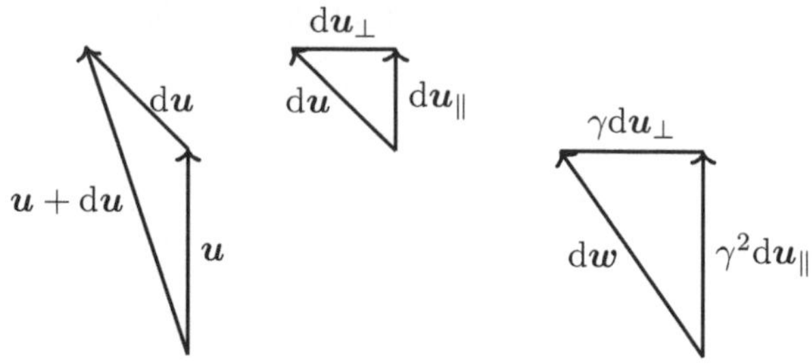

Figure 6.2 A particle has velocity $\boldsymbol{u}$ at time t and $\boldsymbol{u}+\mathrm{d}\boldsymbol{u}$ at time $t+\mathrm{d}t$. The velocity change $\mathrm{d}\boldsymbol{u}$ can be decomposed into vectors parallel and perpendicular to $\boldsymbol{u}$: $\mathrm{d}\boldsymbol{u} = \mathrm{d}\boldsymbol{u}_\parallel + \mathrm{d}\boldsymbol{u}_\perp$. The boost $\mathrm{d}\boldsymbol{w} = \gamma^2\mathrm{d}\boldsymbol{u}_\parallel + \gamma\mathrm{d}\boldsymbol{u}_\perp$ is such that $\boldsymbol{u} \oplus \mathrm{d}\boldsymbol{w} = \boldsymbol{u} + \mathrm{d}\boldsymbol{u}$ to first order.

Our goal is to find $\Lambda_T \equiv L(\boldsymbol{u}+\mathrm{d}\boldsymbol{u})L(-\boldsymbol{u})$ for $\mathrm{d}\boldsymbol{u} \ll |\boldsymbol{u}|$, which, if we were to work with LTs in the form of 4×4 matrices would be a task of considerable complexity. Exhibiting the Thomas rotation simplifies if we use the $\mathrm{SL}(2,\mathbb{C})$ representation of LTs (Section 6.3). Any conclusions we draw from $M_T \equiv M(\boldsymbol{u}+\mathrm{d}\boldsymbol{u})M(-\boldsymbol{u})$ will apply to Λ_T because Eq. (6.39) is a homomorphism between $\mathrm{SL}(2,\mathbb{C})$ and the Lorentz group $L_+^\uparrow$ (Section 4.3).

The boost matrix M_B in Eq. (6.42) can be written in the form (Exercise 6.8), converting from $\theta = -\tanh\beta$ to the Lorentz factor γ:

$$M_B(\boldsymbol{u}) = \sqrt{\frac{1+\gamma}{2}}I - \frac{\gamma}{c\sqrt{2(1+\gamma)}}\boldsymbol{u}\cdot\boldsymbol{\sigma}\,. \tag{6.45}$$

Let's write, to first order, $M_B(\boldsymbol{u}+\mathrm{d}\boldsymbol{u}) \equiv M_B(\boldsymbol{u}) + \mathrm{d}M$, with $\mathrm{d}M$ (a matrix) the differential of Eq. (6.45). With this definition, M_T can be written

$$M_T = (M(\boldsymbol{u}) + \mathrm{d}M)\circ M(-\boldsymbol{u}) = I + \mathrm{d}M \circ M(-\boldsymbol{u})\,. \tag{6.46}$$

The differential $\mathrm{d}M$ is, from Eq. (6.45),

$$\mathrm{d}M = \frac{\mathrm{d}\gamma}{2\sqrt{2(1+\gamma)}}I - \frac{(2+\gamma)\mathrm{d}\gamma}{c(2(1+\gamma))^{3/2}}\boldsymbol{u}\cdot\boldsymbol{\sigma} - \frac{\gamma}{c\sqrt{2(1+\gamma)}}\mathrm{d}\boldsymbol{u}\cdot\boldsymbol{\sigma}\,, \tag{6.47}$$

with $\mathrm{d}\gamma = (\gamma^3/c^2)\boldsymbol{u}\cdot\mathrm{d}\boldsymbol{u}$. Using Eq. (6.46), it can be shown that by combining Eqs. (6.47) and (6.45) and making use of the result of Exercise 6.5:

$$M_T = I - \frac{\gamma}{2c}\mathrm{d}\boldsymbol{u}\cdot\boldsymbol{\sigma} - \frac{\gamma^3}{2c^3(1+\gamma)}(\boldsymbol{u}\cdot\mathrm{d}\boldsymbol{u})(\boldsymbol{u}\cdot\boldsymbol{\sigma}) + \mathrm{i}\frac{\gamma^2}{2c^2(1+\gamma)}(\boldsymbol{u}\times\mathrm{d}\boldsymbol{u})\cdot\boldsymbol{\sigma}\,. \tag{6.48}$$

The following relation can be derived (Exercise 6.9),

$$\gamma\mathrm{d}\boldsymbol{u}\cdot\boldsymbol{\sigma} + \frac{\gamma^3}{c^2(1+\gamma)}(\boldsymbol{u}\cdot\mathrm{d}\boldsymbol{u})(\boldsymbol{u}\cdot\boldsymbol{\sigma}) = [\gamma\mathrm{d}\boldsymbol{u}_\perp + \gamma^2\mathrm{d}\boldsymbol{u}_\parallel]\cdot\boldsymbol{\sigma} \equiv \mathrm{d}\boldsymbol{w}\cdot\boldsymbol{\sigma}\,, \tag{6.49}$$

where $\mathrm{d}\boldsymbol{u} = \mathrm{d}\boldsymbol{u}_\perp + \mathrm{d}\boldsymbol{u}_\parallel$ is resolved into vectors perpendicular and parallel to $\boldsymbol{u}$. Combining Eqs. (6.48) and (6.49), the matrix M_T is, to first order in small quantities,

$$M_T = I + \frac{1}{2}\left(-\frac{\mathrm{d}\boldsymbol{w}}{c} + \mathrm{i}\frac{\gamma^2}{c^2(1+\gamma)}\boldsymbol{u}\times\mathrm{d}\boldsymbol{u}\right)\cdot\boldsymbol{\sigma}\,. \tag{6.50}$$

Now examine the form of the boost and rotation matrices, Eqs. (6.42) and Eq. (6.44), for infinitesimal parameters: $M_B(\mathrm{d}\boldsymbol{\beta}) \approx I - \frac{1}{2}(\mathrm{d}\boldsymbol{\beta}\cdot\boldsymbol{\sigma})$ and $M_R(\hat{\boldsymbol{n}}\mathrm{d}\phi) \approx I + \frac{\mathrm{i}}{2}\mathrm{d}\phi(\hat{\boldsymbol{n}}\cdot\boldsymbol{\sigma})$ (for small arguments). Thus, M_T factorizes (correct to first order), $M_T \approx M_B(\mathrm{d}\boldsymbol{w})M_R(\hat{\boldsymbol{n}}\mathrm{d}\phi)$, corresponding to an infinitesimal velocity boost $\mathrm{d}\boldsymbol{w} = \gamma \mathrm{d}\boldsymbol{u}_\perp + \gamma^2 \mathrm{d}\boldsymbol{u}_\parallel$ and an infinitesimal rotation $\hat{\boldsymbol{n}}\mathrm{d}\phi = \gamma^2(\boldsymbol{u}\times \mathrm{d}\boldsymbol{u})/(c^2(1+\gamma))$, so that the rotation is about the direction of $\boldsymbol{u}\times\mathrm{d}\boldsymbol{u}$ with magnitude (Thomas rotation)

$$\mathrm{d}\phi = \frac{\gamma^2}{c^2(1+\gamma)} |\boldsymbol{u}\times \mathrm{d}\boldsymbol{u}| \ , \tag{6.51}$$

The *rate* of rotation is

$$\frac{\mathrm{d}\phi}{\mathrm{d}t} = \frac{\gamma^2}{c^2(1+\gamma)} |\boldsymbol{u}\times \boldsymbol{a}| \ , \tag{6.52}$$

where $\boldsymbol{a}$ is the acceleration. Thomas rotation is purely kinematical: It occurs regardless of the cause of acceleration as long as $\boldsymbol{u}\times\boldsymbol{a} \neq 0$.

Using the correspondence $M \to \Lambda$ we have, correct to first order, that

$$\Lambda_T \equiv L(\boldsymbol{u}+\mathrm{d}\boldsymbol{u})L(-\boldsymbol{u}) = L(\mathrm{d}\boldsymbol{w})R(\hat{\boldsymbol{n}}\mathrm{d}\phi) = R(\hat{\boldsymbol{n}}\mathrm{d}\phi)L(\mathrm{d}\boldsymbol{w}) \ . \tag{6.53}$$

Equation (6.53) implies

$$L(\mathrm{d}\boldsymbol{w})L(\boldsymbol{u}) = R(-\hat{\boldsymbol{n}}\mathrm{d}\phi)L(\boldsymbol{u}+\mathrm{d}\boldsymbol{u}) \ . \tag{6.54}$$

The two boosts on the left of Eq. (6.54) affect the relativistic addition of velocities, $\boldsymbol{u}\oplus\mathrm{d}\boldsymbol{w} = \boldsymbol{u}+\mathrm{d}\boldsymbol{u}$. This compound LT, call it $\Lambda \equiv L(\mathrm{d}\boldsymbol{w})L(\boldsymbol{u})$, can be decomposed into a boost followed by a rotation, what we have in Eq. (6.54). *A boost followed by a boost is not a boost for non-colinear velocities*. Boost transformations are represented by symmetric matrices. The product of symmetric matrices, however, is not necessarily symmetric so that the product of non-colinear boosts is not itself a boost. Indeed, the antisymmetric part of the product of symmetric matrices is related to the commutator between the matrices (Exercise 6.13), and LTs in different directions do not commute.

SUMMARY

- Rotations are LTs. A rotation through an infinitesimal angle $\mathrm{d}\phi$ about the direction $\hat{\boldsymbol{n}}$ is described by the operator $R(\hat{\boldsymbol{n}},\mathrm{d}\phi) = I + \mathrm{d}\phi J(\hat{\boldsymbol{n}})$, where $J(\hat{\boldsymbol{n}})$ is the infinitesimal rotation generator. The operator for finite rotations about $\hat{\boldsymbol{n}}$ is $R(\hat{\boldsymbol{n}}\phi) = \mathrm{e}^{\phi J(\hat{\boldsymbol{n}})}$. Rotations are specified by three parameters: ϕ and $\hat{\boldsymbol{n}}$. There are only two independent components to $\hat{\boldsymbol{n}}$ because of the constraint $\sum_{i=1}^{3}(n^i)^2 = 1$. The rotation matrix has no particular symmetry, but is orthogonal $R^T(\phi) = R^{-1}(\phi) = R(-\phi)$.

- The LT for an infinitesimal boost is $L(\hat{\boldsymbol{\beta}},\mathrm{d}\theta) = I + \mathrm{d}\theta K(\hat{\boldsymbol{\beta}})$ where $\hat{\boldsymbol{\beta}}$ is a unit vector in the direction of the relative velocity, $\mathrm{d}\theta$ is the infinitesimal change in the parameter θ, with $\tanh\theta = -\beta$, and where $K(\hat{\boldsymbol{\beta}})$ is the boost generator. The operator for a finite boost is $L(\boldsymbol{\beta}) = \mathrm{e}^{\theta K(\hat{\boldsymbol{\beta}})}$. Boosts are specified by the three components of $\boldsymbol{\beta}$. The boost matrix is symmetric.

- An arbitrary LT, Λ, can be uniquely decomposed into the product of a rotation and a boost, $\Lambda = L(\boldsymbol{\beta})R(\hat{\boldsymbol{n}}\phi)$. Given the elements of Λ, the elements of $L(\boldsymbol{\beta})$ and the elements of $R(\hat{\boldsymbol{n}}\phi)$ are uniquely determined.

- The composition of non-colinear boosts is not a boost, but rather is a boost followed by a rotation, $L(\boldsymbol{u}+\mathrm{d}\boldsymbol{u})L(-\boldsymbol{u}) = L(\mathrm{d}\boldsymbol{w})R(\hat{\boldsymbol{n}}\mathrm{d}\phi)$, where $\mathrm{d}\boldsymbol{w} = \gamma^2\mathrm{d}\boldsymbol{u}_\parallel + \gamma\mathrm{d}\boldsymbol{u}_\perp$, with $\mathrm{d}\boldsymbol{u} = \mathrm{d}\boldsymbol{u}_\parallel + \mathrm{d}\boldsymbol{u}_\perp$ decomposed into vectors parallel and perpendicular to $\boldsymbol{u}$. The infinitesimal velocity $\mathrm{d}\boldsymbol{w}$ is such that the relativistic addition of velocities $\boldsymbol{u}\oplus\mathrm{d}\boldsymbol{w} = \boldsymbol{u}+\mathrm{d}\boldsymbol{u}$. The rotation is about the direction $\boldsymbol{u}\times\mathrm{d}\boldsymbol{u}$, through the angle $\mathrm{d}\phi = \gamma^2|\boldsymbol{u}\times\mathrm{d}\boldsymbol{u}|/(c^2(1+\gamma))$. If $\mathrm{d}\boldsymbol{u}$ is parallel to $\boldsymbol{u}$, $\mathrm{d}\phi = 0$ and $\mathrm{d}\boldsymbol{w} = \gamma^2\mathrm{d}\boldsymbol{u}$.

EXERCISES

6.1 Show that the characteristic polynomial of the rotation generator matrix Eq. (6.5) is $\lambda^3 + \lambda = 0$. The components of the unit vector $\hat{\boldsymbol{n}}$ are constrained so that $\sum_{i=1}^{3}(n^i)^2 = 1$.

6.2 Show that Eq. (6.10) for a rotation of 90° about the x-axis reduces to $R_x(\pi/2)$ in Eq. (6.1).

6.3 Verify that Eq. (6.21) is a symmetric matrix. Does the matrix R in Eq. (6.27) have any particular symmetry?

6.4 a. Show that the Pauli matrices σ_i, Eq. (6.33), together with the 2×2 unit matrix, σ_0, form a basis set for 2×2 complex matrices. That is, show that any possible complex 2×2 matrix can be expressed as a linear combination of the basis matrices.

b. Show that the set $\{\sigma_0, \sigma_1, \sigma_2, \sigma_3\}$ is a basis for 2×2 Hermitian matrices when the expansion coefficients are real numbers.

6.5 Show for arbitrary three-vectors $\boldsymbol{A}$ and $\boldsymbol{B}$ that $(\boldsymbol{\sigma}\cdot\boldsymbol{A})(\boldsymbol{\sigma}\cdot\boldsymbol{B}) = (\boldsymbol{A}\cdot\boldsymbol{B})I + \mathrm{i}\boldsymbol{\sigma}\cdot(\boldsymbol{A}\times\boldsymbol{B})$, where $\boldsymbol{\sigma} \equiv (\sigma_1, \sigma_2, \sigma_3)$ is a "vector" of Pauli matrices. Use the properties of the Pauli matrices in Eq. (6.34). Note that the final result is not symmetric in $\boldsymbol{A}$ and $\boldsymbol{B}$.

6.6 Show using Eqs. (6.38) and (6.42) that $\Lambda(M)^0_0 = \cosh\theta$.

6.7 Show for $n = 1, 2, \cdots$ that $(\boldsymbol{b}\cdot\boldsymbol{\sigma})^{2n} = I$ where $\boldsymbol{b}$ is a unit vector.

6.8 Show that M_B in Eq. (6.42) can be written

$$M_B(\boldsymbol{u}) = \sqrt{\frac{1+\gamma}{2}}I - \frac{\gamma}{c\sqrt{2(1+\gamma)}}\boldsymbol{u}\cdot\boldsymbol{\sigma}\,.$$

The quantities b^i are given by $(\boldsymbol{\beta})^i/\beta = u^i/(c\beta)$.

6.9 Derive Eq. (6.49). Decompose $\mathrm{d}\boldsymbol{u}$ into vectors parallel and perpendicular to $\boldsymbol{u}$: $\mathrm{d}\boldsymbol{u} = \mathrm{d}\boldsymbol{u}_{\|} + \mathrm{d}\boldsymbol{u}_{\perp}$. Hint: $(\boldsymbol{u}\cdot\mathrm{d}\boldsymbol{u}_{\|})\,\boldsymbol{u} = u^2\mathrm{d}\boldsymbol{u}_{\|}$.

6.10 Show that $\mathrm{d}\boldsymbol{w} \equiv \gamma^2\mathrm{d}\boldsymbol{u}_{\|} + \gamma\mathrm{d}\boldsymbol{u}_{\perp}$ is such that $\boldsymbol{u}\oplus\mathrm{d}\boldsymbol{w} = \boldsymbol{u} + d\boldsymbol{u}$ where the direct sum is defined in Eq. (3.30).

6.11 Show that the generators of 2×2 unimodular matrices found in Section 6.3 satisfy the same commutation relations as the generators of boost and rotation LTs.

6.12 Commutators among the boost generators K_i do not close among themselves (whereas they do among the rotation generators J_i). Define new generators

$$L_i \equiv \tfrac{1}{2}(J_i + \mathrm{i}K_i) \qquad R_i \equiv \tfrac{1}{2}(J_i - \mathrm{i}K_i)\,.$$

Show that

$$[L_i, L_j] = -\varepsilon_{ijk}L_k \qquad [R_i, R_j] = -\varepsilon_{ijk}R_k \qquad [L_i, R_j] = 0\,.$$

6.13 Let M and N be symmetric matrices, $M_{ij} = M_{ji}$ and $N_{ij} = N_{ji}$. You're going to show that the product of symmetric matrices is not necessarily symmetric, where the antisymmetric part of the product of symmetric matrices is related to the commutator between the matrices.

a. Show that $(MN)_{ij} = (NM)_{ji}$.

b. Define, for any matrix D, its symmetric and antisymmetric parts $D_{ij} = \frac{1}{2}(D_{ij} + D_{ji}) + \frac{1}{2}(D_{ij} - D_{ji}) \equiv D^S_{ij} + D^A_{ij}$. Define the commutator of two matrices as $[M, N]_{ij} \equiv (MN)_{ij} - (NM)_{ij}$. Show that the commutator of two *symmetric* matrices is given by

$$[M, N]_{ij} = 2(MN)^A_{ij} .$$

The composition of two boosts is the product of two symmetric matrices. You've just shown that the product of two symmetric matrices (MN) does not have to be symmetric, and the antisymmetric part of (MN) is related to the commutator of M and N.

6.14 The terms Λ^μ_ν in Eq. (6.39) effect a linear transformation of coordinates preserving the spacetime interval, and hence is a LT for any 2×2 complex unimodular matrix M. The determinant of proper LTs is $+1$ (Chapter 4). Can we calculate the determinant of the matrix $[\Lambda^\mu_\nu]$ with elements specified by Eq. (6.39)? The direct approach would be to use Eq. (5.65), $\varepsilon_{\mu_0\mu_1\mu_2\mu_3}\Lambda^{\mu_0}_0\Lambda^{\mu_1}_1\Lambda^{\mu_2}_2\Lambda^{\mu_3}_3$, which would be quite difficult. A simpler approach (although not simple) is to show that the matrix $[\Lambda^\mu_\nu]$ can be factored as the product of two matrices. Let M have elements $M = \begin{pmatrix} a & b \\ c & d \end{pmatrix}$.

a. Work out directly the eight matrices specified by $\sigma_\mu M$ and $\sigma_\nu M^\dagger$.

b. Show using Eq. (6.39) that

$$\begin{pmatrix} \Lambda^0_0 & \Lambda^0_1 & \Lambda^0_2 & \Lambda^0_3 \\ \Lambda^1_0 & \Lambda^1_1 & \Lambda^1_2 & \Lambda^1_3 \\ \Lambda^2_0 & \Lambda^2_1 & \Lambda^2_2 & \Lambda^2_3 \\ \Lambda^3_0 & \Lambda^3_1 & \Lambda^3_2 & \Lambda^3_3 \end{pmatrix} = \frac{1}{2}\begin{pmatrix} a & b & c & d \\ c & d & a & b \\ -\mathrm{i}c & -\mathrm{i}d & \mathrm{i}a & \mathrm{i}b \\ a & b & -c & -d \end{pmatrix}\begin{pmatrix} a^* & b^* & -\mathrm{i}b^* & a^* \\ b^* & a^* & \mathrm{i}a^* & -b^* \\ c^* & d^* & -\mathrm{i}d^* & c^* \\ d^* & c^* & \mathrm{i}c^* & -d^* \end{pmatrix} .$$

c. Write this as $\Lambda = \frac{1}{2}AB$. Show that $\det A = -4\mathrm{i}\,(\det M)^2$ and $\det B = 4\mathrm{i}\,(\det M^*)^2$. Conclude that $\det \Lambda = |\det M|^4$. Only when we specify $\det M = 1$ do we have a proper LT, what we inferred from Eq. (6.38).

d. For orthochronous LTs, $\Lambda^0_0 \geq 1$ (Exercise 4.1). Show this condition is satisfied by Λ^0_0 as specified by Eq. (6.39) if $\det M = 1$. This problem is tantamount to showing that because $\det M = 1$, the elements of M satisfy the inequality $|a|^2 + |b|^2 + |c|^2 + |d|^2 \geq 2$. Hint: For positive real quantities x and y, $x + y \geq 2\sqrt{xy}$, with equality holding if $y = x$ (show this). Write down the sum of the magnitude squared of four complex numbers (a, b, c, d): $|a|^2 + |b|^2 + |c|^2 + |d|^2$. Use the inequality to show that $|a|^2 + |b|^2 + |c|^2 + |d|^2 \geq 2\,(|ad| + |bc|)$. Now invoke $\det M = 1$, $ad - bc = 1$. Prove the inequality $|1 + z| \geq 1 - |z|$ for any complex number z.

6.15 Show that the matrix

$$M = \begin{pmatrix} \mathrm{e}^{\theta/2} & 0 \\ 0 & \mathrm{e}^{-\theta/2} \end{pmatrix}$$

generates a boost in the x^3-direction. It may be convenient to use the factorization derived in Exercise 6.14.

6.16 Minkowski space can have bases consisting of four null (lightlike) vectors, a *null basis*. Just as with any basis, there's not a unique null basis. As an example, define the four vectors (where $\{\boldsymbol{e}_\mu\}^3_{\mu=0}$ is the usual orthogonal basis that generates the Lorentz metric $\eta = \mathrm{diag}(-1, 1, 1, 1)$)

$$\hat{\boldsymbol{e}}_0 = \boldsymbol{e}_0 + \boldsymbol{e}_1 \qquad \hat{\boldsymbol{e}}_1 = \boldsymbol{e}_0 - \boldsymbol{e}_1 \qquad \hat{\boldsymbol{e}}_2 = \boldsymbol{e}_0 + \boldsymbol{e}_2 \qquad \hat{\boldsymbol{e}}_3 = \boldsymbol{e}_0 + \boldsymbol{e}_3 .$$

a. Show that the vectors $\hat{\mathbf{e}}_\mu$ are each null, i.e., $\hat{\mathbf{e}}_\mu \cdot \hat{\mathbf{e}}_\mu = 0$ (no sum).

b. Are the vectors $\hat{\mathbf{e}}_\mu$ linearly independent? If so, we can speak of a null basis, along with the conclusion that *a basis of null vectors spans Minkowski space*. Hint: Work out the wedge product $\hat{\mathbf{e}}_0 \wedge \hat{\mathbf{e}}_1 \wedge \hat{\mathbf{e}}_2 \wedge \hat{\mathbf{e}}_3$. Equivalently, consider the basis transformation in matrix form:

$$\begin{pmatrix} \hat{\mathbf{e}}_0 \\ \hat{\mathbf{e}}_1 \\ \hat{\mathbf{e}}_2 \\ \hat{\mathbf{e}}_3 \end{pmatrix} = \begin{pmatrix} 1 & 1 & 0 & 0 \\ 1 & -1 & 0 & 0 \\ 1 & 0 & 1 & 0 \\ 1 & 0 & 0 & 1 \end{pmatrix} \begin{pmatrix} \mathbf{e}_0 \\ \mathbf{e}_1 \\ \mathbf{e}_2 \\ \mathbf{e}_3 \end{pmatrix} .$$

Evaluate the determinant of this matrix.

c. Show that the metric tensor $\hat{g}_{\alpha\beta}$ defined by these basis vectors has the form

$$[\hat{g}]_{\alpha\beta} \equiv \hat{\mathbf{e}}_\alpha \cdot \hat{\mathbf{e}}_\beta = -\begin{pmatrix} 0 & 2 & 1 & 1 \\ 2 & 0 & 1 & 1 \\ 1 & 1 & 0 & 1 \\ 1 & 1 & 1 & 0 \end{pmatrix} .$$

In Section 5.6 we mentioned Sylvester's law of inertia, that the signature of the metric tensor is an invariant. Show that the metric tensor $\hat{g}_{\alpha\beta}$ obeys Sylvester's theorem. (Hint: Find the eigenvalues.) The null basis is clearly not an orthogonal basis. We showed in Section 4.4 that null vectors can be orthogonal if and only if they're proportional to each other, i.e., not linearly independent. We cannot have an orthogonal, null basis.

6.17 Show that the matrices in Eq. (6.42) and Eq. (6.44) have unit determinant. Show that the matrix in Eq. (6.44) is unitary.

6.18 Show that Eq. (6.54) follows from Eq. (6.53).

6.19 a. Show to first order in small quantities that

$$M_B(\boldsymbol{u} + \mathrm{d}\boldsymbol{u}) M_R(\hat{\boldsymbol{n}}\mathrm{d}\phi) M_B(-\boldsymbol{u}) = I + \mathrm{i}\frac{\gamma^2}{2c^2}(\boldsymbol{u} \times \mathrm{d}\boldsymbol{u}) \cdot \boldsymbol{\sigma} - \frac{\gamma^2}{2c}\mathrm{d}\boldsymbol{u} \cdot \boldsymbol{\sigma} , \qquad \text{(P6.1)}$$

where $\hat{\boldsymbol{n}}\mathrm{d}\phi = \gamma^2(\boldsymbol{u} \times \mathrm{d}\boldsymbol{u})/(c^2(1+\gamma))$. Make liberal use of the result of Exercise 6.5.

b. Using the correspondence $M \to \Lambda$, including the correspondence of the generators $\frac{1}{2}\boldsymbol{\sigma} \to \boldsymbol{K}$ and $\frac{\mathrm{i}}{2}\boldsymbol{\sigma} \to \boldsymbol{J}$, where $\boldsymbol{K}$ and $\boldsymbol{J}$ are "vectors" of the boost and rotation generators, show that Eq. (P6.1) is equivalent to

$$L(\boldsymbol{u} + \mathrm{d}\boldsymbol{u}) R(\hat{\boldsymbol{n}}\mathrm{d}\phi) L(-\boldsymbol{u}) = I + \frac{\gamma^2}{c^2}(\boldsymbol{u} \times \mathrm{d}\boldsymbol{u}) \cdot \boldsymbol{J} - \frac{\gamma^2}{c}\boldsymbol{K} \cdot \mathrm{d}\boldsymbol{u} . \qquad \text{(P6.2)}$$

We'll use this formula in Chapter 12.

CHAPTER 7

Particle dynamics

We take up the *dynamical* description of motion involving *four-momentum*, the generalization of Newton's quantity of motion that brings together energy and momentum.

7.1 PROPER TIME, FOUR-VELOCITY, AND FOUR-ACCELERATION

Proper time

Time is not a parameter provided by the universe, something external to one's frame of reference, rather it's *local* to each reference frame, and the most local time is the *proper time*, what's measured *on the wordline of a clock*, the time between events in the frame of the clock that are without spatial separation. Proper time can be defined on accelerated worldlines if we work with infinitesimal quantities—over short enough time intervals accelerated motion can be approximated as having uniform speed, where SR holds sway. Figure 7.1 shows an accelerated worldline. One can always

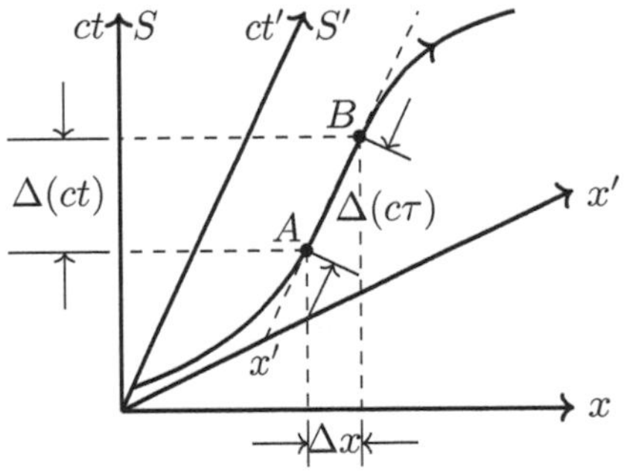

Figure 7.1 Instantaneous rest frame S', in which events A and B are co-local.

find an IRF, an *instantaneous rest frame*, that briefly serves as a proxy for the rest frame of a clock, in which events appear at the same location.[1] In Fig. 7.1, events A and B occur at the same location in S' for the time interval $\Delta(c\tau)$. From the invariance of the spacetime separation between IRFs (here S and S'), we have referring to Fig. 7.1,

$$-(c\Delta\tau)^2 = -(c\Delta t)^2 + (\Delta r)^2 = \Delta x_\mu \Delta x^\mu \ , \tag{7.1}$$

[1]We can find an instantaneous rest frame because the tangent to the worldline of a material particle, the four-velocity, is a timelike vector (see below). The worldline shown in Fig. 7.1 is notional and not accurately drawn.

where $\Delta r \equiv \sqrt{(\Delta x)^2 + (\Delta y)^2 + (\Delta z)^2}$. Equation (7.1) implies

$$(\Delta\tau)^2 = (\Delta t)^2 \left[1 - \frac{1}{c^2}\left(\frac{\Delta r}{\Delta t}\right)^2\right] = -\frac{1}{c^2}\Delta x_\mu \Delta x^\mu \,. \tag{7.2}$$

Letting all quantities become infinitesimal, we have a differential form of time dilation

$$\mathrm{d}\tau = \mathrm{d}t\sqrt{1-\beta^2(t)} = \frac{1}{c}\sqrt{-\mathrm{d}x_\mu \mathrm{d}x^\mu} \,, \tag{7.3}$$

where $\beta(t) \equiv u(t)/c$, with $u(t) = \mathrm{d}r/\mathrm{d}t$ the instantaneous velocity as seen in S. If $u(t)$ is known for $t_i < t < t_j$, the elapsed proper time can be found by integrating Eq. (7.3):

$$\tau_j - \tau_i = \int_{t_i}^{t_j} \mathrm{d}t' \sqrt{1-\beta^2(t')} \,. \tag{7.4}$$

The left portion of Fig. 7.2 shows the proper time in the frame of an accelerated clock obtained

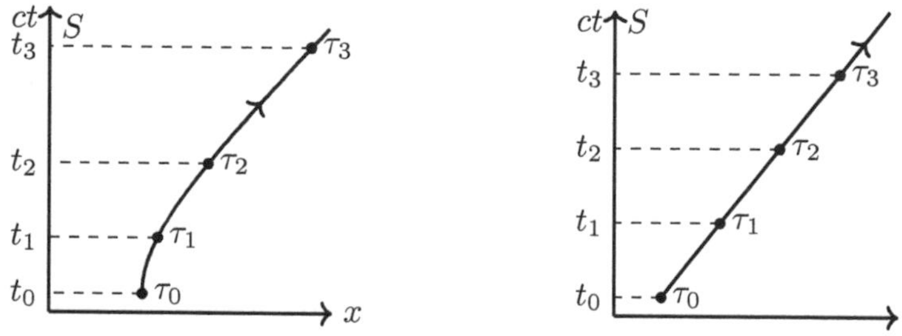

Figure 7.2 Proper time along an accelerated worldline (left) and a straight worldline (right).

by integrating Eq. (7.4), and the associated times in S. The differences between the proper times shown in Fig. 7.2 represent equal "ticks" of the clock in its rest frame, as it's seen to accelerate in S: $\tau_3 - \tau_2 = \tau_2 - \tau_1 = \tau_1 - \tau_0$. The right part of Fig. 7.2 shows an unaccelerated worldline for which there's a linear relation between proper time and coordinate time, $t = \gamma\tau$; for accelerated worldlines the connection between t and τ is nonlinear.

In Newtonian mechanics, particle trajectories $x^i(t)$ are parameterized by absolute time, which in pre-relativistic physics is the same for all observers. In relativity, time is a coordinate, implying that spacetime trajectories (worldlines) be given as parameterized curves, $x^\mu = x^\mu(\lambda)$, where λ is an observer-independent parameter.[2] The proper time $c\mathrm{d}\tau = \sqrt{-\mathrm{d}x_\mu \mathrm{d}x^\mu}$ is a Lorentz-invariant measure of the *arc length* of worldlines. All observers agree on the proper time along *a given worldline*. However, like any line integral, the path length (proper time) *depends on the curve*, which has an interesting twist in the non-Euclidean geometry of spacetime. Figure 7.3 shows two paths (I and II) between the same endpoints A and B. Using Eq. (7.4), we have the inequality

$$\Delta\tau_I = \int_{A,I}^{B} \mathrm{d}t > \int_{A,II}^{B} \mathrm{d}t\sqrt{1-\beta^2(t)} = \Delta\tau_{II} \,, \tag{7.5}$$

because $\beta \neq 0$ along path II. What appears to be the *shortest* path on a spacetime diagram is actually the path with the *longest* elapsed proper time! *Among all worldlines connecting timelike-separated events, the straight worldline has the longest elapsed proper time.* Basically this is the twin paradox, which we analyze in a later chapter. In Euclidean space, a straight line is the shortest distance between two points; in MS a straight line is the *longest* distance between timelike-separated points. Expectations of distance between points based on our experience with space-only geometry lead to erroneous conclusions in the geometry of spacetime.

[2]The worldlines of free particles have a simple parameterization in terms of the proper time: $t = \gamma\tau$ and $x = \beta c\gamma\tau$.

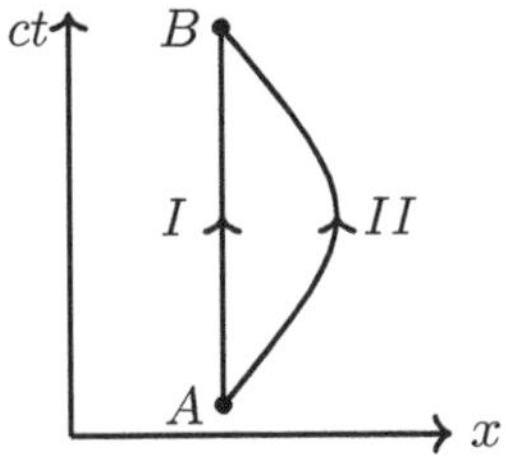

Figure 7.3 The straightest path in Minkowski space is the longest path.

Four-velocity

The *four-velocity* is the rate of change of the coordinate differentials dx^α on the worldline with respect to the proper time

$$U^\alpha \equiv \frac{\mathrm{d}}{\mathrm{d}\tau} x^\alpha(\tau) \,. \tag{7.6}$$

We're entitled to call U^α a four-vector because we have differentiated the prototype four-vector x^μ with respect to an invariant quantity,[3] τ. Using Eq. (7.3), we have the useful result

$$\frac{\mathrm{d}}{\mathrm{d}\tau} = \frac{\mathrm{d}t}{\mathrm{d}\tau}\frac{\mathrm{d}}{\mathrm{d}t} = \gamma \frac{\mathrm{d}}{\mathrm{d}t} \,. \tag{7.7}$$

Combining Eq. (7.7) with Eq. (7.6),

$$U^\alpha = \frac{\mathrm{d}x^\alpha}{\mathrm{d}\tau} = \gamma \frac{\mathrm{d}x^\alpha}{\mathrm{d}t} = \gamma \left(c, \boldsymbol{u}\right) \,, \tag{7.8}$$

where $\boldsymbol{u} \equiv \mathrm{d}\boldsymbol{r}/\mathrm{d}t$ is the three-velocity. Note that $U^\mu \to (c, 0)$ as $\boldsymbol{u} \to 0$. From Eq. (7.8), $U_\alpha = \eta_{\alpha\beta} U^\beta = \gamma \left(-c, \boldsymbol{u}\right)$. The norm of the four-velocity (an invariant) is $U_\alpha U^\alpha = -c^2$. *The four-velocity is a timelike vector*; if we worked in units where $c = 1$ it would be a timelike unit vector. The four-velocity is tangent to the worldline at every point. On an accelerated worldline the direction of U changes, but not its magnitude—see Fig. 7.4.

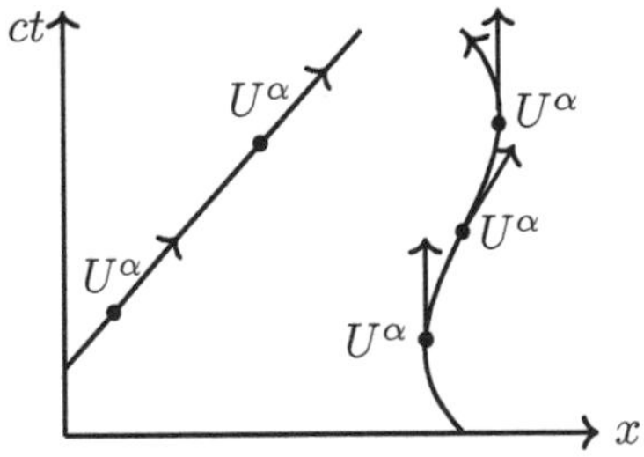

Figure 7.4 Four-velocity vector is tangent to the worldline at every point.

Four-acceleration

The *four-acceleration* is similarly defined,

$$A^\alpha \equiv \frac{\mathrm{d}}{\mathrm{d}\tau} U^\alpha = \gamma \frac{\mathrm{d}}{\mathrm{d}t} \left(\gamma \left(c, \boldsymbol{u}\right)\right) \,,$$

[3]Differentiating the LT $x^\mu = L^\mu_{\nu'} x^{\nu'}$ with respect to the invariant parameter τ produces $U^\mu = L^\mu_{\nu'} U^{\nu'}$, the transformation formula for the four-velocity. The four-velocity transforms like a contravariant four-vector.

which is a four-vector for the same reason that U^α is. The components of $\mathcal{A}^\alpha$ are $\mathcal{A}^0 = \gamma c\dot{\gamma}$, $\mathcal{A}^i = \gamma^2 a^i + \gamma u^i \dot{\gamma}$, where $\boldsymbol{a}$ is the three-vector acceleration $\boldsymbol{a} \equiv \mathrm{d}\boldsymbol{u}/\mathrm{d}t$ and $\dot{\gamma} = \gamma^3(\boldsymbol{u}\cdot\boldsymbol{a})/c^2$. The four-acceleration is thus

$$\mathcal{A}^\mu = \gamma^2\left(\gamma^2\frac{\boldsymbol{u}\cdot\boldsymbol{a}}{c}, \boldsymbol{a} + \gamma^2\boldsymbol{u}\frac{\boldsymbol{u}\cdot\boldsymbol{a}}{c^2}\right) . \tag{7.9}$$

Note that $\mathcal{A}^\mu \to (0, \boldsymbol{a})$ as $\boldsymbol{u} \to 0$. The four-acceleration is spacelike. Differentiate $U_\beta U^\beta = -c^2$: $U_\beta \mathcal{A}^\beta + \mathcal{A}_\beta U^\beta = 2U_\beta \mathcal{A}^\beta = 0$. Thus, $\mathcal{A}^\alpha$ *is orthogonal to* U^β (in any IRF), and a vector orthogonal to a timelike vector is spacelike (Section 4.4). One can show explicitly that $\mathcal{A}_\alpha\mathcal{A}^\alpha > 0$.

7.2 THE ENERGY-MOMENTUM FOUR-VECTOR

We discussed in Section 1.2 how the non-invariance of the wave equation under the GT implies, if mechanics and electrodynamics are to be subject to the principle of relativity, that Newton's second law must be modified to be relativistically correct. One should recognize that any truly *new* equation in physics cannot be derived from something more fundamental, it can only be motivated and made plausible, and that ultimately its validity can only be established through testing against experiment.

With that said, what guidelines do we have in producing a relativistically correct mechanics? First, the equation of motion should have the same form in every IRF, which can be achieved through the use of four-vectors, or more generally tensors. Second, for a free particle it must reduce to $\mathcal{A}^\mu = 0$. If $\mathcal{A}^\mu$ vanishes in one frame it vanishes in all IRFs (because it's a four-vector). Third, the equation of motion should reduce to $\boldsymbol{F} = m\boldsymbol{a}$ in any frame where the speed of the particle is much less than the speed of light. Of course, $\boldsymbol{F} = m\boldsymbol{a}$ is a relation among three-vectors, not four-vectors. These considerations suggest $K^\mu \equiv m\mathcal{A}^\mu$ as a covariant equation of motion for a *suitably defined four-force*, K^μ (the *Minkowski force*); K^μ is a four-vector because m is an invariant and $\mathcal{A}^\mu$ is a four-vector. The *four-momentum* defined as $P^\mu \equiv mU^\mu = m\gamma(c, \boldsymbol{u})$ is a four-vector for the same reason; P^μ is timelike, $P^\alpha P_\alpha = m^2 U^\alpha U_\alpha = -m^2c^2$. Putting together these definitions,

$$\frac{\mathrm{d}}{\mathrm{d}\tau}P^\mu = K^\mu , \tag{7.10}$$

would be a suitable generalization of Newton's second law. Too bad we don't know what K^μ is.

Ready for the new physics? The quantities U^μ and $\mathcal{A}^\mu$ are *kinematic* descriptions of worldlines; where's the dynamics? We're going to *redefine* the three-momentum as

$$\boldsymbol{p} = m\gamma\boldsymbol{u} = \frac{m\boldsymbol{u}}{\sqrt{1-\beta^2}} . \tag{7.11}$$

The transition to relativistic mechanics comes from redefining the three-momentum from $\boldsymbol{p} = m\boldsymbol{u}$ to $\boldsymbol{p} = m\gamma\boldsymbol{u}$; Newton was understandably wrong with his *quantity of motion*. Defining $\boldsymbol{p} = m\gamma\boldsymbol{u}$ is consistent with the definition $P^\mu = mU^\mu$ so that $P^\mu = (m\gamma c, \boldsymbol{p})$. We require that the three-vector *form* of Newton's second law still holds

$$\frac{\mathrm{d}\boldsymbol{p}}{\mathrm{d}t} = \boldsymbol{F} , \tag{7.12}$$

with $\boldsymbol{p}$ given by Eq. (7.11). The spatial part of K^μ must then be $\gamma\boldsymbol{F}$, i.e., $K^\mu = (K^0, \gamma\boldsymbol{F})$, so that the spatial part of Eq. (7.10) is consistent with Eq. (7.12). The time component K^0 can be found from $K^\mu U_\mu = m\mathcal{A}^\mu U_\mu = 0$, implying that $K^0 = \gamma\boldsymbol{F}\cdot\boldsymbol{u}/c$. Thus, K^0c has the dimension of *power* delivered by the force.[4] The four-force is now determined: $K^\mu = \gamma(\boldsymbol{F}\cdot\boldsymbol{u}/c, \boldsymbol{F})$; K^μ is also called the *power-force four-vector*.

[4] While K^0c has the interpretation of the power delivered by the force, K^0 has the dimension of force. All components of a four-vector must have the same physical dimension.

We're lacking a physical interpretation of $P^0 = m\gamma c$. From the time component of Eq. (7.10),

$$\frac{\mathrm{d}P^0}{\mathrm{d}\tau} = \gamma\frac{\mathrm{d}P^0}{\mathrm{d}t} = K^0 = \frac{\gamma}{c}\boldsymbol{F}\cdot\boldsymbol{u} \implies \frac{\mathrm{d}}{\mathrm{d}t}\left(P^0 c\right) = \boldsymbol{F}\cdot\boldsymbol{u}\,. \tag{7.13}$$

Because $\boldsymbol{F}\cdot\boldsymbol{u}$ is the rate of work done, $P^0 c$ can be interpreted as the *energy*

$$E = P^0 c = m\gamma c^2\,. \tag{7.14}$$

Equation (7.14) is of course Einstein's famous equation showing the equivalence of mass and energy. A constant could seemingly be added to Eq. (7.14) and satisfy Eq. (7.13), a constant having the dimension of energy. Such a constant, however, must be zero;[5] $E = m\gamma c^2$ already captures a constant energy associated with a particle, mc^2.

The kinetic energy of a particle, T, is the work done on it from rest,

$$\begin{aligned} T &= \int \boldsymbol{F}\cdot\mathrm{d}\boldsymbol{r} = \int \frac{\mathrm{d}\boldsymbol{p}}{\mathrm{d}t}\cdot\mathrm{d}\boldsymbol{r} = \int \frac{\mathrm{d}\boldsymbol{p}}{\mathrm{d}t}\cdot\boldsymbol{u}\mathrm{d}t = \int \mathrm{d}\boldsymbol{p}\cdot\boldsymbol{u} = \int \mathrm{d}(m\gamma\boldsymbol{u})\cdot\boldsymbol{u} \\ &= m\int(\gamma\mathrm{d}\boldsymbol{u} + \boldsymbol{u}\mathrm{d}\gamma)\cdot\boldsymbol{u} = m\int\left(\gamma\mathrm{d}\boldsymbol{u} + \boldsymbol{u}\gamma^3\frac{\boldsymbol{u}\cdot\mathrm{d}\boldsymbol{u}}{c^2}\right)\cdot\boldsymbol{u} = m\int\gamma\left(1+\frac{u^2}{c^2}\gamma^2\right)\boldsymbol{u}\cdot\mathrm{d}\boldsymbol{u} \\ &= m\int\gamma^3\tfrac{1}{2}\mathrm{d}(u^2) = \frac{m}{2}\int\frac{\mathrm{d}(u^2)}{(1-u^2/c^2)^{3/2}} = \gamma mc^2 + \alpha = E + \alpha\,, \end{aligned}$$

where we've used the trick $\boldsymbol{u}\cdot\mathrm{d}\boldsymbol{u} = \frac{1}{2}\mathrm{d}(\boldsymbol{u}\cdot\boldsymbol{u}) = \frac{1}{2}\mathrm{d}(u^2)$ and α is an integration constant. Because $T = 0$ if $\gamma = 1$, $\alpha = -mc^2$. We thus have the expression for the kinetic energy,

$$T = (\gamma - 1)mc^2 \implies E = T + mc^2\,. \tag{7.15}$$

Example. At CERN electrons can be accelerated to energies exceeding 100 GeV. What is the speed of an electron of 100 GeV total energy? Use $E = \gamma mc^2$ to find γ,

$$\gamma = \frac{E}{mc^2} = \frac{10^2\ \text{GeV}}{0.511\times10^{-3}\ \text{GeV}} = 1.96\times10^5\,.$$

Take apart the Lorentz factor to find the speed:

$$\beta = \sqrt{1-\frac{1}{\gamma^2}} \approx 1 - \frac{1}{2\gamma^2} = 1 - 1.3\times10^{-11}\,.$$

The electron speed differs from the speed of light by one part in 10^{11}.

The four-momentum is now specified in terms of dynamical quantities:

$$P^\mu = (E/c, \boldsymbol{p})\ , \tag{7.16}$$

where $\boldsymbol{p} = m\gamma\boldsymbol{u}$ and $E = m\gamma c^2$. *Energy and momentum are components of a four-vector* (the "momenergy"), and transform together under a LT, $P^{\mu'} = L^{\mu'}_{\nu}P^\nu$. For this reason, the four-momentum is called the *energy-momentum four-vector.* From Eq. (7.16), $P^\mu P_\mu = -m^2c^2 = -(E/c)^2 + p^2$ or[6]

$$E = \sqrt{(pc)^2 + (mc^2)^2}\,. \tag{7.17}$$

[5] At lowest order in β, the spatial parts of the LT for P^μ ($P^{\mu'} = L^{\mu'}_{\nu}P^\nu$) imply the three-vector relation $\boldsymbol{p}' = \boldsymbol{p} - \boldsymbol{\beta}P^0$. On the other hand, the GT of the momentum is $\boldsymbol{p}' = \boldsymbol{p} - mc\boldsymbol{\beta}$ for all β (multiply the Galilean velocity formula by m). We require therefore that in the limit of small speeds, $\lim_{\beta\to0}P^0 = mc$. The LT of P^μ would not have the correct non-relativistic limit if we took $E = P^0c + \alpha$, where α is a nonzero constant.

[6] When we solve $P^\mu P_\mu = -E^2/c^2 + p^2 = -m^2c^2$ for E, we obtain $E = \pm\sqrt{p^2c^2 + m^2c^4}$. One would normally exclude the negative sign as unphysical. In relativistic quantum mechanics, the complete set of solutions unavoidably contains negative energies. In 1928 Dirac postulated *antiparticles* based on this "doubling" of the solutions to the equations of relativistic quantum mechanics. The antiparticle to the electron, the positron, was discovered in 1932.

We can set $m = 0$ in Eq. (7.17) and extend four-momentum to photons.[7] With $p = E/c$ in Eq. (7.16), the four-momentum of a photon $Q^\mu \equiv (E/c)(1, \hat{\boldsymbol{n}})$ is a null vector, with $\hat{\boldsymbol{n}}$ a unit vector specifying the direction of propagation. We see that Q^μ is proportional to the four-wavevector, $k^\mu = (\omega/c)(1, \hat{\boldsymbol{n}})$, Eq. (5.61). A proportionality factor is naturally suggested by the Planck relation $E = \hbar\omega$ for the energy of a photon, and thus

$$Q^\mu = \frac{\hbar\omega}{c}(1, \hat{\boldsymbol{n}}) = \hbar k^\mu . \tag{7.18}$$

The identity $Q^\mu = \hbar k^\mu$ for photons is "one half" of wave-particle duality, with the three-momentum of photons (particles) given in terms of a wave property $\boldsymbol{p} = \hbar\boldsymbol{k}$. Does the converse relation hold for massive particles ("matter waves") with wave properties ascribed to particles, $k^\mu = P^\mu/\hbar$? The extension of $\boldsymbol{p} = \hbar\boldsymbol{k}$ to particles of nonzero mass, $\boldsymbol{k} = \boldsymbol{p}/\hbar = m\gamma\boldsymbol{v}/\hbar$, confirmed in the Davisson-Germer experiment, contributed decisively to the development of quantum mechanics. Wave-particle duality is anticipated by SR.

7.3 ACTION PRINCIPLE FOR PARTICLES

The methods of analytical mechanics apply in SR with suitable modifications. By Hamilton's principle, Eq. (D.17), of all the paths $x(t)$ a particle *might* take between fixed endpoints, the actual path is the one that *extremizes* the action integral

$$S[x] \equiv \int_{t_0}^{t_1} L(x, \dot{x})\mathrm{d}t , \tag{7.19}$$

where the integrand L is the Lagrangian function. In pre-relativistic mechanics L is the difference between the kinetic and potential energies, $L(x, \dot{x}) = T(\dot{x}) - V(x)$; we'll show how that's generalized in SR. Analytical mechanics specifies the canonical momentum, Eq. (D.25), and the Hamiltonian function H, Eq. (D.26):

$$p \equiv \frac{\partial L}{\partial \dot{x}} \qquad H \equiv p\dot{x} - L ; \tag{7.20}$$

definitions that are retained in SR. The Hamiltonian $H = H(p, x)$ is a function of p and x and is a constant of the motion, Eq. (D.32), a constant we can take to be the energy. The extremal path is such that first-order variations $\delta x(t)$ around it lead to second-order variations of the action integral, a condition signified by writing $\delta S = 0$,

$$\delta S = \int \frac{\delta L}{\delta x(t)}\delta x(t)\mathrm{d}t = \int \left[\frac{\partial L}{\partial x} - \frac{\mathrm{d}}{\mathrm{d}t}\left(\frac{\partial L}{\partial \dot{x}}\right)\right]\delta x(t)\mathrm{d}t = 0 , \tag{7.21}$$

where we've used Eq. (D.10). By requiring the integrand in Eq. (7.21) to vanish, we obtain the Euler-Lagrange equation, the differential equation that describes the extremal path

$$\frac{\partial L}{\partial x} = \frac{\mathrm{d}}{\mathrm{d}t}\left(\frac{\partial L}{\partial \dot{x}}\right) . \tag{7.22}$$

The Lagrangian formalism is a trusty machine for generating equations of motion.

To what extent do these ideas work in SR? Consider a free particle in one dimension. It would seem natural to take the Lagrangian to be the relativistic kinetic energy, Eq. (7.15):

$$L \stackrel{?}{=} mc^2(\gamma - 1) = mc^2\left[\frac{1}{\sqrt{1 - \dot{x}^2/c^2}} - 1\right] . \qquad \text{(wrong!)} \tag{7.23}$$

[7]The particle nature of photons is established in Compton scattering experiments.

What is the canonical momentum obtained from Eq. (7.23)? Using Eq. (7.20),

$$p = \frac{\partial L}{\partial \dot{x}} = mc^2 \frac{\partial \gamma}{\partial \dot{x}} = m\gamma^3 \dot{x} \,. \qquad \text{(wrong!)} \qquad (7.24)$$

Equation (7.24) is not the same as Eq. (7.11); it's not right. What about the Hamiltonian? Using Eqs. (7.20), (7.23), and (7.24), $H = p^2/(m\gamma^3) - mc^2(\gamma - 1)$, which is not the same as Eq. (7.17). Something is wrong: *The Lagrangian of a free particle is not the kinetic energy*! *We can't generalize from nonrelativistic to relativistic physics*. It usually doesn't work that way.

At this point, let's *guess* at the Lagrangian for a free particle:[8]

$$L = -mc^2\sqrt{1 - v^2/c^2} \,. \qquad (7.25)$$

It's a simple matter to show that Eq. (7.25) correctly reproduces the relativistic momentum and the relativistic Hamiltonian:

$$p = \frac{\partial L}{\partial \dot{x}} = m\gamma\dot{x} \qquad H = p\dot{x} - L = c\sqrt{p^2 + m^2c^2} = cP^0 \,.$$

The methods of analytical mechanics generate the components of the energy-momentum four-vector using Eq. (7.25) as the Lagrangian. Guessing, however, is not satisfactory; we'd like to know how to derive Eq. (7.25).

What do we want from a relativistic action principle? The action, a scalar, should not depend on our choice of coordinates; it should therefore be expressed in terms of Lorentz invariants. The simplest scalar along worldlines is the proper time τ, the arc length of the worldline. Take as the action of a free particle

$$S = -\alpha \int_A^B \mathrm{d}\tau \,, \qquad (7.26)$$

where the minus sign gets the nonrelativistic limit right and α is a constant, characteristic of a particle. (We show that $\alpha = mc^2$, where we stick to the convention that action has the dimension of energy $\times$ time.) We can put Eq. (7.26) into the same form as Eq. (7.19) by writing

$$S = -\alpha \int \mathrm{d}\tau = -\alpha \int \frac{\mathrm{d}\tau}{\mathrm{d}t}\mathrm{d}t \equiv \int L\mathrm{d}t \,.$$

The Lagrangian is then

$$L = -\alpha \frac{\mathrm{d}\tau}{\mathrm{d}t} = -\alpha\sqrt{1 - \beta^2} = -mc^2\sqrt{1 - \beta^2} \,, \qquad (7.27)$$

the same as Eq. (7.25), where we've used Eq. (7.2). The constant α can be obtained from the form of L for slow speeds: $L = -\alpha + \frac{1}{2}\alpha\beta^2 + O(\beta^4)$. Taking $\alpha = mc^2$ establishes contact with the pre-relativistic Lagrangian, $L = -mc^2 + \frac{1}{2}mu^2$; a constant added to the Lagrangian does not affect the equations of motion.

The Lagrangian of a free particle is not the kinetic energy. In units of mc^2, $T = \gamma - 1$ while $L = -\gamma^{-1}$. As $\beta \to 0$, $\gamma - 1 = \frac{1}{2}\beta^2 + \frac{3}{8}\beta^4 + O(\beta^6)$ and $-\gamma^{-1} = -1 + \frac{1}{2}\beta^2 + \frac{1}{8}\beta^4 + O(\beta^6)$. The canonical momentum is basically $\partial L/\partial\beta$. We see that $\partial T/\partial\beta$ matches $\partial L/\partial\beta$ as $\beta \to 0$, and hence we may replace the Lagrangian with the nonrelativistic kinetic energy for sufficiently slow speeds. The two functions are completely different for large β, however: T diverges as $\beta \to 1$, while $L \to 0$ as $\beta \to 1$. In fact, L must vanish as $\beta \to 1$: The arc length of a lightlike curve is zero.

[8] Students sometimes leave off the prefactor of $-mc^2$ from the relativistic Lagrangian, Eq. (7.25). That term, however, is needed to make the identification of the canonical momentum with the spatial parts of P^μ and the identification of the Hamiltonian with the energy, as the time component cP^0. We need that factor!

For forces derivable from a potential energy function, $F_i = -\partial V/\partial x^i$, the relativistic equations of motion are generated by the Lagrangian[9]

$$L(x, \dot{x}) = -mc^2\sqrt{1-\beta^2} - V(x) \,. \tag{7.28}$$

From Eq. (7.28),

$$\frac{\partial L}{\partial \dot{x}^i} = m\gamma \dot{x}_i \qquad \frac{\partial L}{\partial x^i} = -\frac{\partial V}{\partial x^i} \,,$$

which when combined with Eq. (7.22) produces the equation of motion

$$\frac{\mathrm{d}}{\mathrm{d}t}\left(m\gamma\dot{x}_i\right) = -\frac{\partial V}{\partial x^i} \,, \tag{7.29}$$

the same as Eq. (7.12).

For a particle of charge q (see Eq. (D.41))

$$L(x, \dot{x}) = -mc^2\sqrt{1-\beta^2} - q\phi + q\boldsymbol{u}\cdot\boldsymbol{A} \,, \tag{7.30}$$

where ϕ is the scalar potential and $\boldsymbol{A}$ is the vector potential. From Eq. (7.30),

$$\frac{\partial L}{\partial \dot{x}^i} = m\gamma\dot{x}_i + qA_i \qquad \frac{\partial L}{\partial x^i} = -q\frac{\partial \phi}{\partial x^i} + q\boldsymbol{u}\cdot\frac{\partial \boldsymbol{A}}{\partial x^i} \,,$$

which when combined with Eq. (7.22) leads to the equation of motion

$$\frac{\mathrm{d}}{\mathrm{d}t}\left(m\gamma\dot{x}_i\right) = -q\frac{\partial\phi}{\partial x^i} - q\frac{\mathrm{d}A_i}{\mathrm{d}t} + q\boldsymbol{u}\cdot\frac{\partial\boldsymbol{A}}{\partial x^i} = q\left[\boldsymbol{E} + \boldsymbol{u}\times\boldsymbol{B}\right]_i \,. \tag{7.31}$$

The equivalence between Eq. (7.31) and the Lorentz force is shown in Appendix D.

7.4 KEPLER PROBLEM IN SPECIAL RELATIVITY

We analyze the Kepler problem in SR (we revisit this problem in GR). Consider a particle of mass m in the Newtonian gravitational potential produced by mass M, $V(r) = -k/r$, where $k \equiv GMm$. The motion is planar, just as in Newtonian mechanics; therefore, work in polar coordinates (r, θ) so that the velocity squared is $v^2 = \dot{r}^2 + r^2\dot{\theta}^2$. The Lagrangian, Eq. (7.28), is in this case

$$L(r, \dot{r}, \dot{\theta}) = -mc^2\sqrt{1 - \frac{1}{c^2}\left(\dot{r}^2 + r^2\dot{\theta}^2\right)} + \frac{k}{r} = -mc^2\gamma^{-1} + \frac{k}{r} \,. \tag{7.32}$$

The canonical momenta are therefore

$$p_r \equiv \frac{\partial L}{\partial \dot{r}} = m\gamma\dot{r} \qquad p_\theta \equiv \frac{\partial L}{\partial \dot{\theta}} = m\gamma r^2\dot{\theta} \,. \tag{7.33}$$

The angular momentum p_θ is constant because $\partial L/\partial\theta = 0$ (and hence the motion is planar). The Hamiltonian, another constant of the motion, is from Eq. (D.26):

$$H \equiv p_r\dot{r} + p_\theta\dot{\theta} + mc^2\gamma^{-1} - \frac{k}{r} = c\sqrt{m^2c^2 + p_r^2 + \frac{p_\theta^2}{r^2}} - \frac{k}{r} \,. \tag{7.34}$$

We seek the *orbit equation* $r(\theta)$. It turns out to be easier to work with the variable $u \equiv r^{-1}$ and then convert to $r = u^{-1}$ when required. Thus,

$$\frac{\mathrm{d}u}{\mathrm{d}\theta} = -\frac{1}{r^2}\frac{\mathrm{d}r}{\mathrm{d}\theta} = -\frac{1}{r^2}\frac{\dot{r}}{\dot{\theta}} = -\frac{p_r}{p_\theta} \,, \tag{7.35}$$

[9]Note that we've written force as a covariant vector because it's derived from the gradient of the potential energy function.

where we've used Eq. (7.33). An expression for p_r can be obtained from Eq. (7.34),

$$p_r^2 = \frac{1}{c^2}\left(H + \frac{k}{r}\right)^2 - \frac{p_\theta^2}{r^2} - m^2c^2 \,. \tag{7.36}$$

By squaring Eq. (7.35), using Eq. (7.36), and substituting $r^{-1} = u$ we have

$$\left(\frac{\mathrm{d}u}{\mathrm{d}\theta}\right)^2 = \frac{1}{p_\theta^2}p_r^2 = \frac{1}{p_\theta^2}\left[\frac{1}{c^2}(H + ku)^2 - p_\theta^2 u^2 - m^2c^2\right] \,. \tag{7.37}$$

Now differentiate Eq. (7.37) with respect to θ. We find

$$\frac{\mathrm{d}^2u}{\mathrm{d}\theta^2} + \alpha^2 u = \frac{kH}{p_\theta^2 c^2} \,, \tag{7.38}$$

where $\alpha^2 \equiv 1 - k^2/(p_\theta c)^2$ is a constant. The solution of Eq. (7.38) can be written

$$u(\theta) = \frac{1}{p}\left(1 + \epsilon \cos\alpha\left(\theta - \theta_0\right)\right) \,, \tag{7.39}$$

where $p \equiv \left(p_\theta^2 c^2 - k^2\right)/(kH)$ is a characteristic length, θ_0 is a constant we can take as zero, and ϵ, the *eccentricity*, is a parameter of the orbit obtained from $\epsilon = (r_{\text{max}} - r_{\text{min}})/(r_{\text{max}} + r_{\text{min}})$. The orbit equation is, from Eq. (7.39) with $\theta_0 = 0$,

$$r(\theta) = \frac{p}{1 + \epsilon\cos\alpha\theta} \,. \tag{7.40}$$

The orbit described by Eq. (7.40) differs from the classic Kepler orbit in one significant way: $\alpha \neq 1$. Relativistic Kepler motion is an ellipse with an advancing *line of the apsides* (see Fig. 7.5). The *apsis* (plural *apsides*) is a point on the orbit of either closest approach to the center of force

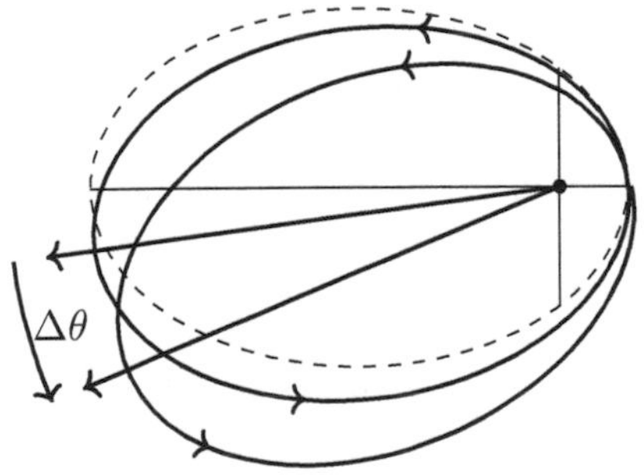

Figure 7.5 Precessing orbit, from Eq. (7.40) with $\epsilon = 0.75$ and $\alpha = 0.96$. Line of the apsides advances by $\Delta\theta$ per revolution. Classical Kepler orbit shown as a dashed curve.

(the *periapsis*) or the point furthest away (the *apoapsis*), with the line between them the line of the apsides. Successive points of apoapsis occur at angles $\theta = (2n+1)\pi/\alpha$, where n is an integer (show this). After one revolution, the line of the apsides advances through the angle

$$\Delta\theta = \frac{2\pi}{\alpha} \pmod{2\pi} = \frac{2\pi}{\sqrt{1 - k^2/(p_\theta c)^2}} \pmod{2\pi} \,. \tag{7.41}$$

To calculate $\Delta\theta$ requires that we know the angular momentum p_θ. For the classic Kepler problem ($\alpha = 1$) it's straightforward to show that $p_\theta^2 = kmp$. For the relativistic problem, however, determining p_θ entails a fair amount of algebra. At the turning points of the motion $p_r \propto \dot{r} = 0$, and thus from the constancy of H, Eq. (7.34),

$$c\sqrt{m^2c^2 + \frac{p_\theta^2}{r_{\text{min}}^2}} - \frac{k}{r_{\text{min}}} = c\sqrt{m^2c^2 + \frac{p_\theta^2}{r_{\text{max}}^2}} - \frac{k}{r_{\text{max}}} \,. \tag{7.42}$$

Equation (7.42) can be solved for p_θ, with the result[10]

$$p_\theta^2 = GMm^2p\left[\sqrt{1 + \frac{G^2M^2}{4a^2c^4}} + \frac{GM}{2ac^2}\frac{1+\epsilon^2}{1-\epsilon^2}\right] \,. \tag{7.43}$$

The nonrelativistic result is found by letting $c \to \infty$ in Eq. (7.43). Using the parameters of the orbit of Mercury and the mass of the sun, $GM/(ac^2) \approx 10^{-10}$; the nonrelativistic value for p_θ is thus an excellent approximation in this case. Using the nonrelativistic expression for p_θ in Eq. (7.41), $k^2/(p_\theta^2c^2) = GM/[ac^2(1-\epsilon^2)]$ which is also small. We may therefore approximate Eq. (7.41) with its Taylor expansion,

$$\Delta\theta \approx \frac{\pi GM}{ac^2(1-\epsilon^2)} \,. \tag{7.44}$$

The orbit of Mercury is known to exhibit an advancing perihelion,[11] the famous 43″ per century (an arcsecond is 4.848×10^{-6} radians). As we'll see in Chapter 17, GR also predicts an elliptical orbit featuring an advancing periapsis. For the orbit of Mercury the prediction of GR agrees with the observed value of 43″ per century. The formula for the precession angle obtained in GR is the same as Eq. (7.44) except for a factor of six; the precession angle predicted by SR is a *sixth* that predicted by GR—see Eq. (17.58).

7.5 COVARIANT EULER-LAGRANGE EQUATION

Equation (7.29) is the relativistic generalization of Newton's second law. It's not "fully" relativistic, however, in that it's a relation among three-vectors, not four-vectors. The interaction potential $V(r)$ (in the Kepler problem, for example) is a function only of the instantaneous spatial location of the particle. Neither the equations of motion, nor the Lagrangian have been expressed in *covariant form*. In this section we obtain the covariant Lagrangian of a free particle.

In pre-relativistic mechanics the action integral is stationary around a path in *configuration space* where the spatial location of a particle is specified by generalized coordinates, $\{q^i\}_{i=1}^n$. The points of the paths (actual and varied) are labeled by *absolute* time. In the path functional $S[\text{path}] = \int L(q^i(t), \dot{q}^i(t))\mathrm{d}t$, time is integrated out, leaving the action as a function solely of the path.

In the theory of relativity *time is another coordinate*, "configuration space" is *spacetime* and paths are worldlines. We therefore need a parameter other than time to label the points of worldlines (actual and varied). The proper time would meet this need except that τ *is path dependent* (Section 7.1). We assume that a parameter exists, call it λ, such that the coordinates of all worldlines can be parameterized $x^\mu(\lambda)$. At the end of the calculation we set $\lambda = \tau$.

A straight line in MS (inertial motion) has the *longest* elapsed proper time between the events that it connects (Section 7.1). In GR, the worldline of a free particle is a geodesic, the straightest possible curve on a curved geometry (Chapter 14). Extremal proper time is a property that distinguishes the worldlines of free particles. Using the spacetime metric, $-(c\mathrm{d}\tau)^2 = (\mathrm{d}s)^2 = g_{\mu\nu}\mathrm{d}x^\mu\mathrm{d}x^\nu$, implying $c\mathrm{d}\tau = \sqrt{-g_{\mu\nu}\mathrm{d}x^\mu\mathrm{d}x^\nu}$. The covariant generalization of Eq. (7.26) is[12]

$$S = -mc^2\int \mathrm{d}\tau = -mc\int\sqrt{-g_{\mu\nu}\frac{\mathrm{d}x^\mu}{\mathrm{d}\lambda}\frac{\mathrm{d}x^\nu}{\mathrm{d}\lambda}}\mathrm{d}\lambda \,. \tag{7.45}$$

[10] The quantity a is the *semi-major axis* of the orbit, $a = \frac{1}{2}(r_{\text{max}} + r_{\text{min}})$.

[11] Perihelion is the point of closest approach in the orbit of a planet around the sun.

[12] We "borrow" a factor of c from $\alpha = mc^2$ to include with $cd\tau$ in Eq. (7.45).

The covariant generalization of Eq. (7.27) is then:

$$L = -mc^2 \frac{\mathrm{d}\tau}{\mathrm{d}\lambda} = -mc\sqrt{-g_{\mu\nu}\frac{\mathrm{d}x^\mu}{\mathrm{d}\lambda}\frac{\mathrm{d}x^\nu}{\mathrm{d}\lambda}} \equiv -mc\sqrt{-g_{\mu\nu}U^\mu U^\nu} \,, \tag{7.46}$$

where the four-velocity $U^\alpha = \mathrm{d}x^\alpha/\mathrm{d}\lambda$ is tangent to the worldline. When in Chapter 8 we allow a coupling to the electromagnetic field, L will also contain the coordinates x^μ. Let's do the variational calculus with $L = L(x^\mu, U^\mu)$. In that case,

$$\delta S = -mc\int\left[\frac{\partial L}{\partial x^\mu}\delta x^\mu + \frac{\partial L}{\partial U^\mu}\delta U^\mu\right]\mathrm{d}\lambda = -mc\int\left[\frac{\partial L}{\partial x^\mu}\delta x^\mu + \frac{\partial L}{\partial U^\mu}\frac{\mathrm{d}}{\mathrm{d}\lambda}\delta x^\mu\right]\mathrm{d}\lambda \,, \tag{7.47}$$

where $\delta U^\mu = (\mathrm{d}/\mathrm{d}\lambda)\delta x^\mu(\lambda)$, the variation in the derivative is the derivative of the variation, Eq. (D.7). The second term in Eq. (7.47) can then be integrated by parts with the integrated term vanishing because the endpoints of the path are held fixed. Thus,

$$\delta S = -mc\int\left[\frac{\partial L}{\partial x^\mu} - \frac{\mathrm{d}}{\mathrm{d}\lambda}\left(\frac{\partial L}{\partial U^\mu}\right)\right]\delta x^\mu \mathrm{d}\lambda \,.$$

To have $\delta S = 0$ for arbitrary (and independent) variations δx^μ, the integrand must vanish, leaving us with the Euler-Lagrange equation in covariant form

$$\frac{\mathrm{d}}{\mathrm{d}\tau}\left(\frac{\partial L}{\partial U^\mu}\right) = \frac{\partial L}{\partial x^\mu} \,, \qquad \mu = 0, 1, 2, 3 \tag{7.48}$$

where we have identified $\lambda = \tau$ for the extremal worldline.

There's another way to define the free-particle Lagrangian, basically the *square* of Eq. (7.46). Substitute $(L)^2$ in Eq. (7.48),

$$\begin{aligned}\frac{\mathrm{d}}{\mathrm{d}\lambda}\left(\frac{\partial L^2}{\partial U^\mu}\right) - \frac{\partial L^2}{\partial x^\mu} &= \frac{\mathrm{d}}{\mathrm{d}\lambda}\left(2L\frac{\partial L}{\partial U^\mu}\right) - 2L\frac{\partial L}{\partial x^\mu} = 2L\left[\frac{\mathrm{d}}{\mathrm{d}\lambda}\left(\frac{\partial L}{\partial U^\mu}\right) - \frac{\partial L}{\partial x^\mu}\right] + 2\frac{\partial L}{\partial U^\mu}\frac{\mathrm{d}L}{\mathrm{d}\lambda} \\ &= -\,2mc^2\frac{\partial L}{\partial U^\mu}\frac{\mathrm{d}^2\tau}{\mathrm{d}\lambda^2} = 0 \,,\end{aligned}$$

where the terms in square brackets vanish by Eq. (7.48), and in the second line we have used $L = -mc^2\mathrm{d}\tau/\mathrm{d}\lambda$, Eq. (7.46). For the extremal path τ is at most a linear function of λ (shown in Chapter 14, the *affine parameterization*) and thus $\mathrm{d}^2\tau/\mathrm{d}\lambda^2 = 0$. Thus, we can equally well take for the free-particle Lagrangian

$$L = \frac{1}{2}mg_{\mu\nu}\frac{\mathrm{d}x^\mu}{\mathrm{d}\lambda}\frac{\mathrm{d}x^\nu}{\mathrm{d}\lambda} = \frac{1}{2}mg_{\mu\nu}U^\mu U^\nu \tag{7.49}$$

instead of (7.46). Equation (7.49) leads to the same equation of motion as (7.46), even in GR, and is more convenient to use in certain calculations.

Let's calculate the canonical momentum using the two Lagrangians. Using Eq. (7.46),

$$\begin{aligned}\frac{\partial L}{\partial U^\mu} &= -\,mc\frac{\partial}{\partial U^\mu}\sqrt{-g_{\rho\sigma}U^\rho U^\sigma} = \frac{-mc}{2\sqrt{-g_{\kappa\lambda}U^\kappa U^\lambda}}\left(-g_{\rho\sigma}U^\rho\delta^\sigma{}_\mu - g_{\rho\sigma}\delta^\rho{}_\mu U^\sigma\right) \\ &= \frac{-mc}{2\sqrt{-g_{\kappa\lambda}U^\kappa U^\lambda}}\left(-g_{\rho\mu}U^\rho - g_{\mu\sigma}U^\sigma\right) = \frac{mcg_{\mu\sigma}U^\sigma}{\sqrt{-g_{\kappa\lambda}U^\kappa U^\lambda}} = mU_\mu \,,\end{aligned}$$

where we have used $\partial U^\alpha/\partial U^\beta = \delta^\alpha_\beta$ and we have lowered the index on U^σ. Because this equation applies for the actual worldline, set $U^\beta U_\beta = -c^2$. If we use Eq. (7.49), $\partial L/\partial U^\mu$ leads us directly to mU_μ. For either form of the free-particle Lagrangian, Eq. (7.46) or Eq. (7.49), the Euler-Lagrange equation yields $\mathrm{d}P_\mu/\mathrm{d}\tau = 0$ or $\mathcal{A}_\mu = 0$, the law of inertia.

7.6 PARTICLE CONSERVATION LAWS

In pre-relativistic mechanics, conservation of linear momentum, angular momentum, and energy follow directly from Newton's second law. Another way to establish conservation laws, one that we can use in the theory of relativity, involves coordinate transformations that leave the action integral invariant. The path taken by a particle is the one that extremizes the action (Hamilton's principle). The *value* of the action integral is invariant under coordinate transformations (Appendix D). As we'll show, *invariance of the action implies the existence of conserved quantities along worldlines*. Conservation of four-momentum is implied by invariance under spacetime translations (homogeneity of spacetime) and conservation of angular momentum is implied by the invariance of the action under LTs.

Invariance, conservation—is there a difference? An invariant does not change between reference frames; a conserved quantity doesn't change during *processes*. Momentum may be conserved in a given reference frame, but is not an invariant (it transforms as a four-vector). In Chapter 9 we show that conservation laws associated with *fields* (as opposed to particles) can also be derived from invariance of the action.

Consider a nonrelativistic free particle in one dimension with action integral

$$S = \int_{t_i}^{t_f} \tfrac{1}{2} m\dot{x}^2 \mathrm{d}t \,. \tag{7.50}$$

The worldline of stationary action is obtained from Lagrange's equation, which in this case ($\ddot{x} = 0$) is a straight line $x(t) = \alpha + \beta t$, where α and β are integration constants, to be fit to the endpoints:

$$\alpha = \frac{x_i t_f - x_f t_i}{t_f - t_i} \equiv \frac{x_i t_f - x_f t_i}{\Delta t} \qquad \beta = \frac{x_f - x_i}{t_f - t_i} \equiv \frac{\Delta x}{\Delta t} \,.$$

Substituting $\dot{x} = \beta$ back into Eq. (7.50), we obtain the action function,

$$S = \frac{1}{2} m\beta^2 \Delta t = \frac{1}{2} m \frac{(\Delta x)^2}{\Delta t} = \frac{m}{2} \frac{(x_f - x_i)^2}{t_f - t_i} \,. \tag{7.51}$$

What transformations leave S invariant? We can shift the origin of the coordinates, $x \to x' = x + a$, and we can shift the origin of time, $t \to t' = t + b$; both transformations leave S unchanged. We show in Eq. (7.60) that the first symmetry implies conservation of momentum and the second implies conservation of energy.

Is the nonrelativistic action invariant under the GT? Under $x \to x' = x - vt$, $t \to t' = t$, we have from Eq. (7.51)

$$S \to S' = \frac{m}{2} \frac{(\Delta x')^2}{\Delta t'} = \frac{m}{2} \frac{(\Delta x - v\Delta t)^2}{\Delta t} = S - mv\Delta x + \frac{m}{2} v^2 \Delta t \,.$$

The nonrelativistic action is therefore *not* invariant under the GT ($S' \neq S$); *there are no conserved quantities implied by the GT*. The relativistic action, however, is a Lorentz invariant (by construction). For a free particle, from Eq. (7.26)

$$S = -mc^2 \int \mathrm{d}\tau = -mc^2 \Delta\tau = -mc^2 \sqrt{(\Delta t)^2 - (\Delta x)^2/c^2} \,. \tag{7.52}$$

The action function in Eq. (7.52) is invariant under a shift of the spacetime origin, which (as we have noted) implies conservation of energy and momentum.

As another example, take a nonrelativistic particle in a uniform gravitational field,

$$S = \int_{t_i}^{t_f} \left(\frac{m}{2} \dot{x}^2 - mgx \right) \mathrm{d}t \,. \tag{7.53}$$

Lagrange's equation gives $\ddot{x} = -g$, and hence $x(t) = \alpha + \beta t - \frac{1}{2}gt^2$. Solving for α and β in terms of the endpoints of the motion,

$$\alpha = \frac{t_f x_i - t_i x_f}{\Delta t} - \frac{1}{2} g t_i t_f \qquad \beta = \frac{\Delta x}{\Delta t} + \frac{1}{2} g \left(t_f + t_i\right) . \tag{7.54}$$

Substituting $x(t)$ back into Eq. (7.53), and using Eq. (7.54), we obtain

$$\frac{1}{m} S = \frac{1}{2} \frac{(\Delta x)^2}{\Delta t} - g \frac{\Delta t}{2} \left(x_f + x_i\right) - \frac{g^2}{4!} (\Delta t)^3 . \tag{7.55}$$

What are the symmetries of S now? Clearly, S is not invariant under a shift in the origin. Under $x \to x + a$ in Eq. (7.55), $S \to S - mga\Delta t$. Because S is not invariant under this transformation, momentum is not conserved along the worldline, which we should not expect for a particle in a gravitational field. Equation (7.55) however is invariant under time translations, $t \to t + b$, implying conservation of energy.

Let's now consider what's implied by invariance of the relativistic action integral under an infinitesimal change in spacetime coordinates, $x^\mu \to x^\mu + \delta x^\mu$. We require

$$\delta S = \int_{\lambda_1}^{\lambda_2} \left(\frac{\partial L}{\partial x^\mu} \delta x^\mu + \frac{\partial L}{\partial \dot{x}^\mu} \delta(\dot{x}^\mu) \right) \mathrm{d}\lambda = 0 . \tag{7.56}$$

Following the same steps as in Eq. (7.47), Eq. (7.56) is equivalent to

$$\delta S = \left. \frac{\partial L}{\partial (\dot{x}^\mu)} \delta x^\mu \right|_{\lambda_1}^{\lambda_2} + \int_{\lambda_1}^{\lambda_2} \left[\frac{\partial L}{\partial x^\mu} - \frac{\mathrm{d}}{\mathrm{d}\lambda} \left(\frac{\partial L}{\partial \dot{x}^\mu} \right) \right] \delta x^\mu \mathrm{d}\lambda = 0 . \tag{7.57}$$

In the usual variational problem (Appendix D), in order to discover the extremal path, we vary the path such that the variation vanishes at the endpoints. In that case, we would set the integrated part in Eq. (7.57) to zero, and we would require the vanishing of the integrand in Eq. (7.57) to derive Eq. (7.48). Here we are doing something different. We are changing variables *on* the extremal path and are demanding that the action be invariant. In that case, Euler's equations are satisfied and the integral in Eq. (7.57) vanishes, leaving us with the *conservation law* (set $\lambda = \tau$ and $U^\mu = \mathrm{d}x^\mu / \mathrm{d}\tau$)

$$\left. \frac{\partial L}{\partial U^\mu} \delta x^\mu \right|_{\tau_1}^{\tau_2} = 0 . \tag{7.58}$$

With $P_\mu = \partial L / \partial U^\mu$, Eq. (7.58) is equivalent to

$$\left. P_\mu \delta x^\mu \right|_{\tau_1}^{\tau_2} = 0 . \tag{7.59}$$

Invariance of the action under $x^\mu \to x^\mu + \delta x^\mu$ implies the existence of a quantity $P_\mu \delta x^\mu$ that is the same at the endpoints of the worldline, i.e., is conserved.

For δx^μ derived from constant shift in coordinates, $\delta x^\mu = (c\delta t, \delta \boldsymbol{r})$, Eq. (7.59) becomes

$$\left. P_\mu \delta x^\mu \right|_{\tau_1}^{\tau_2} = \left. \left(-E\delta t + \boldsymbol{p} \cdot \delta \boldsymbol{r}\right) \right|_{\tau_1}^{\tau_2} = 0 , \tag{7.60}$$

implying (as advertised) that *energy and momentum are conserved along the worldline if the action is invariant under a translation of spacetime coordinates*, i.e., if the action is independent of the location of the origin of the spacetime coordinate system.

What's implied by invariance under infinitesimal LTs, $\delta x^\mu = \lambda^\mu_\nu x^\nu$ (Section 6.2)? From Eq. (7.59), $P_\mu \delta x^\mu = P_\mu \lambda^\mu_\nu x^\nu = P^\alpha \eta_{\alpha\mu} \lambda^\mu_\nu x^\nu$. Because $\eta_{\alpha\mu} \lambda^\mu_\nu$ is antisymmetric in (α, ν) (Section 6.2), Eq. (7.59) picks out the antisymmetric part of the tensor $P^\alpha x^\nu$ and is equivalent to

$$\left. \frac{1}{2} \left(P^\mu x^\nu - P^\nu x^\mu\right) \eta_{\mu\rho} \lambda^\rho_\nu \right|_{\tau_1}^{\tau_2} = 0 .$$

Define the *angular momentum tensor*

$$M^{\mu\nu} \equiv x^\mu P^\nu - x^\nu P^\mu \,. \tag{7.61}$$

Invariance of the action under infinitesimal LTs implies that $M^{\mu\nu}$ *is conserved along worldlines.* An antisymmetric second-rank tensor in four dimensions has six independent components. Three are associated with the orbital angular momentum pseudovector, $L_i \equiv \frac{1}{2}\varepsilon_{ijk}M^{jk}$, the dual of the spatial parts of $M^{\mu\nu}$. The time components of Eq. (7.61) form a vector $M^{0i} = x^0 P^i - x^i P^0 = c\left(tP^i - x^i E/c^2\right)$ which implies the constancy of $t\boldsymbol{p} - \boldsymbol{r}E/c^2$. Because E and $\boldsymbol{p}$ are conserved, Eq. (7.59), a free particle moves with the constant velocity

$$\frac{\mathrm{d}\boldsymbol{r}}{\mathrm{d}t} = \frac{c^2\boldsymbol{p}}{E} \,. \tag{7.62}$$

Equation (7.62) is the same as Eq. (3.19). With $E = \gamma mc^2$, Eq. (7.62) implies $\boldsymbol{p} = m\gamma\boldsymbol{v}$.

More than one free particle

A system of n particles is described by n worldlines $x_i^\mu(\tau_i)$ $(i = 1, \ldots, n)$, each characterized by its own proper time τ_i. We must introduce a separate proper time for each particle; proper time is recorded in an instantaneous rest frame and we cannot find a frame instantaneously at rest with respect to more than one particle. Each worldline is parameterized separately,

$$c\mathrm{d}\tau_i = \sqrt{-g_{\mu\nu}\mathrm{d}x_i^\mu \mathrm{d}x_i^\nu} = \sqrt{-g_{\mu\nu}\frac{\mathrm{d}x_i^\mu}{\mathrm{d}\lambda_i}\frac{\mathrm{d}x_i^\nu}{\mathrm{d}\lambda_i}}\mathrm{d}\lambda_i = \sqrt{-g_{\mu\nu}U_i^\mu U_i^\nu}\mathrm{d}\lambda_i \,.$$

The action integral for non-interacting particles is the generalization of Eq. (7.45),

$$S = -\sum_i m_i c \int \mathrm{d}\lambda_i \sqrt{-g_{\mu\nu}U_i^\mu U_i^\nu} \,. \tag{7.63}$$

Hamilton's principle applied to the variation of the worldlines $x_i^\mu(\lambda_i)$ leads to a separate Euler-Lagrange equation for each particle. Without providing details, invariance of S under spacetime translations implies conservation of the total momentum $\sum_i \boldsymbol{p_i}$ and the total energy, $\sum_i E_i$. Invariance under infinitesimal LTs leads to conservation of the total angular momentum tensor, $\sum_i M_i^{\mu\nu}$, the spatial components of which imply conservation of the total angular momentum, and the time components imply that $\sum_i \left(t\boldsymbol{p}_i - E_i\boldsymbol{r}_i/c^2\right) = \text{constant}$. Because the total energy is a constant,

$$\frac{\sum_i \boldsymbol{r}_i E_i}{\sum_i E_i} = \frac{c^2 t \sum_i \boldsymbol{p}_i}{\sum_i E_i} + \text{constant} \,.$$

The *center of energy* $\boldsymbol{R} \equiv \sum_i \boldsymbol{r}_i E_i / \sum_i E_i$ moves with constant velocity,

$$\frac{\mathrm{d}\boldsymbol{R}}{\mathrm{d}t} = \frac{c^2 \sum_i \boldsymbol{p}_i}{\sum_i E_i} = \frac{\sum_i m_i \gamma_i \boldsymbol{v}_i}{\sum_i m_i \gamma_i} = \text{constant} \,. \tag{7.64}$$

Equation (7.64) is the analog of the center-of-mass theorem in pre-relativistic mechanics. For non-relativistic speeds $\gamma_i \to 1$ and $\boldsymbol{R} \to \sum_i m_i \boldsymbol{r}_i / \sum_i m_i$; the expression for $\boldsymbol{R}$ reduces to that of the Newtonian center of mass. Note that $\boldsymbol{R}$ is not the spatial part of a four-vector. The location of the center of energy is reference-frame dependent.

7.7 ENERGY-MOMENTUM CONSERVATION

Processes involving relativistic particles are studied in nuclear and high-energy physics, processes in which particles initially far enough apart to be considered free, come together and interact over small spacetime regions, and then move apart again as free particles. In such processes, the total four-momentum of the particles is conserved:

$$\sum_{i=1}^{N} P_i^\nu = \sum_{j=1}^{N'} P_j'^\nu \,, \tag{7.65}$$

where we allow for the possibility that the number of particles after the interaction N' is not the same as before, N. Equation (7.65) has been verified in countless experiments.

The analysis of Eq. (7.65) is often easiest in a particular frame of reference. We know that through the use of Lorentz invariants, the results obtained are independent of reference frame. *Evaluating* such invariants, however, is often easiest in a particular frame—why not use that one? In this section we review selected processes between relativistic particles and the reference frames used to analyze them.

7.7.1 Center of momentum frame

The *center of momentum* (CM) frame is the IRF in which the spatial part of the total four-momentum is zero, $\sum_i \boldsymbol{p}_i = 0$. The analysis of many processes simplifies in the CM frame. *It's always possible to find such a frame* because of the timelike character of P^μ for material particles. The total four-momentum in the CM frame is thus $P^\mu_{CM} = (E/c, 0)$, where $E = \sum_i E_i = \sum_i m_i \gamma_i c^2$. The "rest mass" in the CM frame, $M_{CM} = E/c^2 = \sum_i m_i \gamma_i$, is not the sum of the masses of the individual particles because of their velocities $\boldsymbol{u}_i$ relative to the CM frame. The center of energy is at rest in the CM frame.

7.7.2 Spontaneous decay

Almost all particles will decay into others if they can while not violating conservation laws—energy-momentum of course—but others like charge conservation. A particle of mass M can spontaneously decay into particles with masses m_1, m_2 and momenta $\boldsymbol{p}_1$, $\boldsymbol{p}_2$. Working in the CM frame, four-momentum conservation implies $Mc^2 = E_1 + E_2$ and $0 = \boldsymbol{p}_1 + \boldsymbol{p}_2$. Using Eq. (7.15), $Mc^2 = m_1c^2 + m_2c^2 + T_1 + T_2$. Because $T_1 + T_2 > 0$, spontaneous decay is possible only if $M > m_1 + m_2$. For charged pion decay $\pi^+ \to \mu^+ + \nu_\mu$, $\pi^- \to \mu^- + \overline{\nu}_\mu$, $M_{\pi^\pm} = 139.57\ \text{MeV}/c^2$, $m_{\mu^\pm} = 105.66\ \text{MeV}/c^2$, and $m_{\nu_\mu} < 0.17\ \text{MeV}/c^2$; the inequality $M > m_1 + m_2$ is easily satisfied.

7.7.3 Pair production, pair annihilation

The positively charged electron e^+ (*positron*) was discovered in 1932. Soon afterwards it was discovered that a photon could undergo *pair production*, where a photon is annihilated and an electron and a positron are created together, $\gamma \to e^+ + e^-$. One would think that a photon of energy $E = 2mc^2$ could convert its energy into an electron-positron pair (m is the electron mass). As we now show, pair production cannot occur spontaneously in free space; another particle is necessary.

Let $Q^\mu = (E/c)(1, \hat{\boldsymbol{n}})$ be the photon four-momentum, Eq. (7.18), the conservation of which in $\gamma \to e^+ + e^-$ would require $Q^\mu = P^\mu = ((E_+ + E_-)/c, \boldsymbol{p}_+ + \boldsymbol{p}_-)$. This equation cannot be satisfied (try it!). Before the reaction, the four-momentum is lightlike ($Q^\mu Q_\mu = 0$), and afterwards it is timelike ($P^\mu P_\mu < 0$). We cannot get into the CM frame of a photon (without killing it).

The problem can be circumvented by allowing another massive body to participate: The sum of a lightlike vector and a timelike vector is timelike (Exercise 4.6c.). Consider $\gamma + X \to X' + e^+ + e^-$, where X denotes another body of mass M. In the rest frame of X the total four-momentum $P^\mu = \left(E + Mc^2, E\hat{\boldsymbol{n}}\right)/c$ and thus

$$P^\mu P_\mu = -\frac{1}{c^2}\left(E + Mc^2\right)^2 + (E/c)^2 \ .$$

By four-momentum conservation $P'^\mu = P^\mu$, and by Lorentz invariance $P^\mu P_\mu = P'^\mu P'_\mu$. We are free to evaluate the invariant in an arbitrary inertial frame. In the CM frame $P'^\mu = ((E_{X'} + E_+ + E_-)/c, 0)$ and thus

$$P'^\mu P'_\mu = -\frac{1}{c^2}\left(E_{X'} + E_+ + E_-\right)^2 \ .$$

The *threshold energy* is the photon energy E such that the particles in the final state have no kinetic energy. At threshold, $\left(E + Mc^2\right)^2 - E^2 = \left(Mc^2 + 2mc^2\right)^2$ or

$$E = 2mc^2\left(1 + \frac{m}{M}\right) \ .$$

For $m/M \ll 1$ (such as in the vicinity of a nucleus), the threshold energy $E \gtrsim 2mc^2$. If the process occurs near another electron, the threshold energy $E = 4mc^2$.

The inverse process is *pair annihilation*, $e^+ + e^- \to \gamma + \gamma$. (The process $e^+ + e^- \to \gamma$ does not conserve energy-momentum.) Because *two* photons are created, we *can* get into the zero momentum frame associated with the final state. Let $\boldsymbol{p}_+$ and $\boldsymbol{p}_-$ be the momenta of the positron and electron, and let $\boldsymbol{q}_1$, $\boldsymbol{q}_2$ be the momenta of the photons. In the CM frame of the initial state one has $\boldsymbol{p}_- = -\boldsymbol{p}_+$. Momentum conservation requires that $\boldsymbol{q}_2 = -\boldsymbol{q}_1$, i.e., the photons are emitted in opposite directions with equal momenta and hence with equal energies. From energy conservation, the energy of each photon is $E_\gamma = \sqrt{(pc)^2 + m^2c^4}$. The smallest possible photon energy is obtained if $p = 0$, or $E_\gamma = mc^2$. The special case of a lightlike four-momentum vector with zero spatial part allows us to equate it to a timelike vector (P^μ for e^+, e^-).

7.7.4 Compton scattering: $\gamma + e \to \gamma' + e'$

Compton scattering is the inelastic scattering of a photon by a free particle. The effect gives experimental confirmation that the momentum of electromagnetic radiation is carried by *particles* called photons. In this process, a photon of energy E scatters from a particle (usually an electron) into a direction at an angle θ from the incident direction (see Fig. 7.6). Because the electron re-

Figure 7.6 Compton scattering of a photon from an electron.

coils, the scattered photon has an energy $E' < E$. Let the four-momentum of the photon (electron) before and after the collision be Q^μ and Q'^μ (P^μ and P'^μ). From four-momentum conservation $P^\mu + Q^\mu = P'^\mu + Q'^\mu$. We can eliminate P'^μ (four-momentum of recoil electron):

$$\begin{aligned} P'^\mu P'_\mu &= (P^\mu + Q^\mu - Q'^\mu)(P_\mu + Q_\mu - Q'_\mu) = P^\mu P_\mu + 2P^\mu(Q_\mu - Q'_\mu) + (Q^\mu - Q'^\mu)(Q_\mu - Q'_\mu) \\ &= P^\mu P_\mu + 2P^\mu(Q_\mu - Q'_\mu) + Q^\mu Q_\mu - 2Q^\mu Q'_\mu + Q'^\mu Q'_\mu \end{aligned}$$

This equation simplifies considerably because of the invariant magnitude of four-vectors: $P'^\mu P'_\mu = P^\mu P_\mu = -m^2c^2$ and $Q^\mu Q_\mu = Q'^\mu Q'_\mu = 0$ (null vector). Energy-momentum conservation therefore reduces to the equation

$$P^\mu(Q_\mu - Q'_\mu) = Q^\mu Q'_\mu \ . \tag{7.66}$$

In the rest frame of the electron $P^\mu = (mc, 0, 0, 0)$, $Q^\mu = (E/c)(1, 1, 0, 0)$, and $Q'^\mu = (E'/c)(1, \cos\theta, \sin\theta, 0)$. The inner products are thus $Q^\mu Q'_\mu = (EE'/c^2)(-1 + \cos\theta)$ and $P^\mu(Q_\mu - Q'_\mu) = m(E' - E)$. Substituting these results in Eq. (7.66), we find

$$\frac{1}{E'} - \frac{1}{E} = \frac{1}{mc^2}(1 - \cos\theta) \; .$$

Using the Planck relation $E = hc/\lambda$, we have the (experimentally verified) Compton scattering formula as it's usually written:

$$\lambda' - \lambda = \frac{h}{mc}(1 - \cos\theta) \equiv \lambda_C\,(1 - \cos\theta) \; ,$$

where $\lambda_C = h/(mc)$ is the *Compton wavelength*, the wavelength of a photon such that its energy is equal to mc^2. For electrons $\lambda_C \approx 2.4 \times 10^{-12}$ m. Note that nowhere in these formulas does the charge of the particle appear. Photons carry no charge!

SUMMARY

- The proper time $\mathrm{d}\tau$ is the time measured by a clock along its worldline. In an IRF, $\mathrm{d}\tau = \sqrt{-(\mathrm{d}s)^2}/c = \mathrm{d}t\sqrt{1-\beta^2}$, where $\beta = \boldsymbol{u}/c$ is the instantaneous velocity as seen from that frame. The proper time is an invariant quantity $c\mathrm{d}\tau = \sqrt{-\mathrm{d}x_\mu \mathrm{d}x^\mu}$ and provides an absolute (observer-independent) means to parameterize the worldline, $x^\mu = x^\mu(\tau)$.

- The elapsed proper time between events is path-dependent. Straight worldlines between events have the longest proper time; this is the basis of the so-called twin paradox.

- The four-velocity is defined as

$$U^\alpha = \frac{\mathrm{d}x^\alpha}{\mathrm{d}\tau} = \gamma\frac{\mathrm{d}}{\mathrm{d}t}(ct, \boldsymbol{x}) = \gamma(c, \boldsymbol{u}) \; .$$

  The four-velocity is a timelike vector ($U^\alpha U_\alpha = -c^2$), tangent to the worldline with an invariant magnitude.

- The four-momentum $P^\mu = mU^\mu = (E/c, \boldsymbol{p})$ where $E = m\gamma c^2$ and $\boldsymbol{p} = m\gamma\boldsymbol{u}$. The invariant magnitude of P^μ is given by $P^\mu P_\mu = -mc^2$. The connection between energy and momentum is $E = c\sqrt{p^2 + m^2c^2}$.

- The four-acceleration $\mathcal{A}^\mu = \mathrm{d}U^\mu/\mathrm{d}\tau$ is a spacelike vector orthogonal to U^μ, $\mathcal{A}_\mu U^\mu = 0$.

- The Lagrangian for a particle is $L = -mc^2\sqrt{1-\beta^2} - V(x)$. For a charged particle $L = -mc^2\sqrt{1-\beta^2} - q\phi + q\boldsymbol{u}\cdot\boldsymbol{A}$. The free-particle Lagrangian is the integrand of the action integral, $L = -mc^2(\mathrm{d}\tau/\mathrm{d}t)$.

- The covariant free-particle Lagrangian can be written in two ways, $L = -mc\sqrt{-g_{\mu\nu}U^\mu U^\nu}$ and $L = \frac{1}{2}mg_{\mu\nu}U^\mu U^\nu$. Both lead to the same equations of motion. The first form emphasizes the geometric feature that the elapsed proper time for a straight line in spacetime is the longest, the second leads to an extremum in the coordinate-invariant generalization of the kinetic energy.

- The Euler-Lagrange equations in covariant form are

$$\frac{\mathrm{d}}{\mathrm{d}\tau}\left(\frac{\partial L}{\partial U^\mu}\right) = \frac{\partial L}{\partial x^\mu} \cdot \qquad \mu = 0, 1, 2, 3$$

- The canonical momentum is $P_\mu = \partial L/\partial U^\mu$.
- The angular momentum tensor is $M^{\mu\nu} = x^\mu P^\nu - x^\nu P^\mu$. The angular momentum pseudovector $\boldsymbol{L}$ is the dual associated with the spatial parts of $M^{\mu\nu}$, $L_i = \frac{1}{2}\varepsilon_{ijk}M^{jk}$. The time components M^{0i} form a three-vector associated with the relativistic version of the center of mass theorem in mechanics.
- Invariance of the action integral under coordinate transformations is a recipe that allows us to identify conserved quantities along the worldline of a particle. Invariance under a shift in the origin of spacetime coordinates implies energy-momentum conservation, and invariance under infinitesimal LTs implies conservation of the angular momentum tensor.

EXERCISES

7.1 Show explicitly using (7.6) and (7.9) that the four-acceleration and the four-velocity are orthogonal, $U_\beta \mathcal{A}^\beta = 0$.

7.2 Derive the form for the relativistic Hamiltonian, $H = c\sqrt{p^2 + m^2c^2}$, starting with Eq. (7.25), the relativistic Lagrangian, and the definition of the Hamiltonian in Eq. (7.20). The Hamiltonian should be a function of position and momentum, $H = H(x, p)$.

7.3 Show from Eq. (7.52) that the relativistic action for a free particle becomes, an expansion for $\Delta x \ll c\Delta t$,

$$S_{\text{rel}} = -mc^2\Delta t + \frac{1}{2}m\frac{(\Delta x)^2}{\Delta t} + \cdots .$$

The nonrelativistic action, Eq. (7.51), thus "lives" inside the relativistic action.

7.4 Derive Eq. (7.55), the nonrelativistic action for a particle in a gravitational field.

7.5 Using Eq. (7.9), show explicitly that the four-acceleration is spacelike, $A_\alpha A^\alpha > 0$. Show either (or both) of these results:

$$A^\alpha A_\alpha = \gamma^6 \left[a^2 - \frac{1}{c^2}\left(\boldsymbol{u} \times \boldsymbol{a}\right)^2\right] = \gamma^4 \left(a^2 + \frac{\gamma^2}{c^2}(\boldsymbol{u} \cdot \boldsymbol{a})^2\right) .$$

Why are the terms in square brackets positive?

7.6 Show that a covariant definition of the energy (valid in any reference frame) is $E = -P^\mu U_\mu$. (This involves an inner product between four-vectors, an invariant quantity. Is there a particular reference frame that simplifies the expression?)

7.7 Show that an electron in vacuum cannot emit a photon, as in $e^- \to e^- + \gamma$.

7.8 Show explicitly that the process $\gamma \to e^+ + e^-$ cannot conserve energy and momentum in vacuum.

7.9 Let the four-acceleration in Eq. (7.9) be denoted $\mathcal{A}^\mu = (\mathcal{A}^0, \boldsymbol{\mathcal{A}})$. Show that $\mathcal{A}^0 = \boldsymbol{u} \cdot \boldsymbol{\mathcal{A}}/c$. Because the norm $\mathcal{A}_\mu \mathcal{A}^\mu$ is invariant, the four components of $\mathcal{A}^\mu$ cannot be independent.

7.10 Derive the Hamiltonian for the relativistic Kepler problem, Eq. (7.34).

7.11 Show that for a free particle the angular momentum tensor is constant, i.e., $\mathrm{d}M^{\mu\nu}/\mathrm{d}\tau = 0$, as we would expect from Eq. (7.61).

7.12 Show that $U_0\mathcal{A}_i - U_i\mathcal{A}_0 = -c\gamma^3 a_i$, where a_i is the three-acceleration. Use Eq. (7.9).

CHAPTER 8

Covariant electrodynamics

THE material in this chapter is often referred to as "relativistic electrodynamics." Such a term, however, is redundant: Electrodynamics is consistent with SR and has been since its inception; there never was a nonrelativistic predecessor. What's meant by relativistic electrodynamics is the *covariant theory* of the electromagnetic field. It's useful to see the covariant formulation of electromagnetism (essential, actually): constructs encountered here have analogs in GR, which is also a relativistic field theory.

8.1 ELECTROMAGNETISM IN SPACE AND TIME

The electromagnetic field is described by four interrelated equations known collectively as Maxwell's equations:

$$\begin{aligned} \nabla \cdot \boldsymbol{B} &= 0 \qquad & \nabla \times \boldsymbol{E} + \frac{\partial \boldsymbol{B}}{\partial t} &= 0 \\ \nabla \cdot \boldsymbol{D} &= \rho \qquad & \nabla \times \boldsymbol{H} - \frac{\partial \boldsymbol{D}}{\partial t} &= \boldsymbol{J} , \end{aligned} \tag{8.1}$$

where the charge density function $\rho(\boldsymbol{r})$ and the current density vector $\boldsymbol{J}(\boldsymbol{r})$ are the *sources* of the fields. Writing Maxwell's equations in their customary form—with three-vectors—carries with it the implication that space is decoupled from time. The covariant version of Maxwell's equations is developed in Section 8.5. Note that there are two sets of field vectors: $(\boldsymbol{E}, \boldsymbol{B})$, which are prescribed *independently of sources*, and $(\boldsymbol{D}, \boldsymbol{H})$ which are determined by $(\rho, \boldsymbol{J})$. We'll refer to the equations involving $(\boldsymbol{E}, \boldsymbol{B})$ as the *homogeneous Maxwell equations*, and those for $(\boldsymbol{D}, \boldsymbol{H})$ the *inhomogeneous Maxwell equations*. Maxwell's equations as written in Eq. (8.1) are explicitly free of parameters pertaining to a medium, *including free space*. In free space $\boldsymbol{D} = \epsilon_0 \boldsymbol{E}$ and $\boldsymbol{B} = \mu_0 \boldsymbol{H}$, where $\epsilon_0 = 10^7/(4\pi c^2)$ and $\mu_0 = 4\pi/10^7$ in SI units. By definition, $\epsilon_0 \mu_0 = c^{-2}$. $\boldsymbol{E}$ and $\boldsymbol{H}$ can arguably be considered the fundamental fields.[1]

[1] $\boldsymbol{H}$ is often called *the* magnetic field with $\boldsymbol{B}$ referred to as the magnetic induction. Edward Purcell remarked:[35, p392] "This seems clumsy and pedantic. If you go into the laboratory and ask a physicist what causes the pion trajectories in his bubble chamber to curve, he'll probably answer 'magnetic field,' not 'magnetic induction'." Purcell further noted: "It is only the names that give trouble, not the symbols." $\boldsymbol{B}$ has the fundamental property that it always has zero divergence, $\nabla \cdot \boldsymbol{B} = 0$. Moreover, the Lorentz force $q\boldsymbol{v} \times \boldsymbol{B}$ involves $\boldsymbol{B}$. $\boldsymbol{H}$ however responds to currents, independent of the medium, and currents are subject to experimental control; from that point of view $\boldsymbol{H}$ can be seen as fundamental. The two are related by $\boldsymbol{B} = \mu_0(\boldsymbol{H} + \boldsymbol{M})$ where $\boldsymbol{M}$ is the magnetization of the medium. For the two electric fields, $\boldsymbol{D}$ couples to free charges (independent of the medium), while $\boldsymbol{E}$ is controlled by potential gradients. In this case, potentials are easier to control experimentally than the placement of charge; $\boldsymbol{E}$ can be seen as fundamental. The two are related by $\boldsymbol{D} = \epsilon_0 \boldsymbol{E} + \boldsymbol{P}$, where $\boldsymbol{P}$ is the polarization of the medium. For the purposes of classifying fields, it should be recognized that $\boldsymbol{E}$ and $\boldsymbol{H}$ are vectors, whereas $\boldsymbol{D}$ and $\boldsymbol{B}$ are vector densities. From that point of view $(\boldsymbol{E}, \boldsymbol{H})$ should be treated on equal footing. Add to that the criterion of which fields are subject to experimental control and $\boldsymbol{E}$ and $\boldsymbol{H}$ are often taken as the fundamental fields.

Table 8.1 Tensor character of electromagnetic fields (↔ indicates isomorphism)

$\boldsymbol{E}$ = covariant vector	$\boldsymbol{B}$ = covariant bivector ↔ contravariant vector density
$\boldsymbol{D}$ = contravariant vector density	$\boldsymbol{H}$ =contravariant bivector density ↔ covariant vector

The pairs $(\boldsymbol{E},\boldsymbol{H})$ and $(\boldsymbol{D},\boldsymbol{B})$ have different tensor characters,[2] summarized in Table 8.1. To see this, introduce scalar and vector potential functions ϕ and $\boldsymbol{A}$ such that

$$\boldsymbol{B} = \boldsymbol{\nabla}\times\boldsymbol{A} \qquad \boldsymbol{E} = -\boldsymbol{\nabla}\phi - \frac{\partial \boldsymbol{A}}{\partial t}\,. \tag{8.2}$$

In that way the homogeneous Maxwell equations are satisfied identically. From Eq. (8.2) we see that the vector $\boldsymbol{E}$ in Faraday's law (the only place where $\boldsymbol{E}$ "officially" appears in Eq. (8.1)) occurs naturally in the theory as a *covariant vector* (related to a gradient, the prototype covariant vector). The quantity $\boldsymbol{A}$ in Eq. (8.2) must then be considered a covariant vector. So far so good: $\boldsymbol{E}$ and $\boldsymbol{A}$ are covariant vectors. The $\boldsymbol{B}$-field, however, (from $\boldsymbol{B} = \boldsymbol{\nabla}\times\boldsymbol{A}$) has the structure of a *covariant bivector*, $\boldsymbol{B} \leftrightarrow \frac{1}{2}B_{ij}\boldsymbol{e}^i\wedge\boldsymbol{e}^j$ (Section 5.10.2). Ordinarily such an observation would find no place in applications of electromagnetic theory; in our march to GR, however, we need to understand the fundamental transformation properties of the electromagnetic fields. This aspect of the $\boldsymbol{B}$-field, formally present in the non-covariant theory as an isomorphism with an antisymmetric tensor B_{ij} in three dimensions, presages what we'll find in the covariant formulation, that the electromagnetic field comprises a second-rank antisymmetric tensor in four dimensions. Such a tensor has six independent elements, just enough for the components of $\boldsymbol{E}$ and $\boldsymbol{B}$. In three dimensions, B_{jk} is isomorphic to a contravariant vector *density* $\mathfrak{B}^i = \frac{1}{2}\varepsilon^{ijk}B_{jk}$ (its dual, Section 5.10.1).[3] That $\boldsymbol{B}$ is a contravariant quantity can be seen from the integral relation $\oint_S \boldsymbol{B}\cdot\mathrm{d}\boldsymbol{a} = 0$: the surface area vector $\mathrm{d}\boldsymbol{a}$, being normal to S is a covariant vector, implying that $\boldsymbol{B}$ is a contravariant vector per area, or vector density. Likewise, $\boldsymbol{D}$ is a contravariant vector density: $\oint_S \boldsymbol{D}\cdot\mathrm{d}\boldsymbol{a} = Q$ (charge Q is a Lorentz scalar, Section 8.3). From the Ampère-Maxwell law we infer that $\boldsymbol{H}$ has the structure of a *contravariant bivector density*, $\mathfrak{H} = \frac{1}{2}\mathfrak{H}^{ij}\boldsymbol{e}_i\wedge\boldsymbol{e}_j$. The dual of $\mathfrak{H}^{ij}$ is a covariant vector, $H_i = \frac{1}{2}\varepsilon_{ijk}\mathfrak{H}^{jk}$, which can be inferred from $\oint \boldsymbol{H}\cdot\mathrm{d}\boldsymbol{l} = I$; a covariant vector *is* a vector per length. Note that $\boldsymbol{E}\cdot\boldsymbol{D}$ and $\boldsymbol{H}\cdot\boldsymbol{B}$ both generate scalar densities; $\frac{1}{2}(\boldsymbol{E}\cdot\boldsymbol{D}+\boldsymbol{H}\cdot\boldsymbol{B})$ is the energy density of the electromagnetic field.[4] That $(\boldsymbol{E},\boldsymbol{H})$ and $(\boldsymbol{D},\boldsymbol{B})$ have different tensor properties implies that constitutive relations *must be in the form of tensor equations*: $\mathfrak{D}^i = \epsilon^{ij}E_j$ and $\mathfrak{B}^k = \mu^{kl}H_l$, where ϵ^{ij} and μ^{kl} denote second-rank tensor densities, the permittivity and permeability tensors.[5]

[2]Maxwell [36, p11–13] distinguished two *types* of vectors, *forces* and *fluxes*, corresponding to our distinction between vectors $(\boldsymbol{E},\boldsymbol{H})$ and vector densities $(\boldsymbol{D},\boldsymbol{B})$. We do a disservice in teaching electromagnetism when we introduce $\boldsymbol{D}$ and $\boldsymbol{H}$ as "auxiliary"; they're an integral part of the framework of Maxwell's equations.

[3]The notation for vector densities is explained in Section 5.2, and ε^{ijk} transforms with $w = +1$, Section 5.8.2.

[4]Shown in any standard text on electromagnetism.

[5]Constitutive relations apply between the Fourier components of the fields. One typically works in the frequency domain and Fourier transforms the time coordinate. Such a linear operation does not change the tensor character of the fields. For free space, when we write $\boldsymbol{D} = \epsilon_0\boldsymbol{E}$ two things are happening: The covariant components E_j are being mapped into the contravariant components D^i, and we're converting a vector into a vector density. It might be thought for isotropic free space that the permittivity tensor ϵ^{ij} is the number ϵ_0 multiplied by a "unit tensor," δ^{ij}. The quantity δ^{ij} can be obtained by raising an index on the Kronecker delta δ^i_j, with $\delta^{ik} = g^{kj}\delta^i_j = g^{ki}$. The free-space fields are related by $D^i = \epsilon_0 g^{ik}E_k = \epsilon_0 E^i$. Thus, the free-space replacement $\boldsymbol{D}\to\epsilon_0\boldsymbol{E}$ is actually short hand for quite a bit: convert $\boldsymbol{E}$ (naturally a covariant vector, with units V/m) into its contravariant form, then multiply by ϵ_0 which brings to the party the right units to convert $\boldsymbol{E}$ into a vector density, $\boldsymbol{D}$ (C/m^2). Similar remarks apply for the free-space replacement $\boldsymbol{B}\to\mu_0\boldsymbol{H}$.

From here on we assume free-space fields ($\boldsymbol{D} = \epsilon_0 \boldsymbol{E}$ and $\boldsymbol{B} = \mu_0 \boldsymbol{H}$). Combining Eq. (8.2) with the inhomogeneous Maxwell equations, we obtain[6]

$$\nabla^2 \phi + \frac{\partial}{\partial t}(\boldsymbol{\nabla} \cdot \boldsymbol{A}) = -\frac{\rho}{\epsilon_0} \quad \nabla^2 \boldsymbol{A} - \frac{1}{c^2}\frac{\partial^2 \boldsymbol{A}}{\partial t^2} - \boldsymbol{\nabla}\left(\boldsymbol{\nabla} \cdot \boldsymbol{A} + \frac{1}{c^2}\frac{\partial \phi}{\partial t}\right) = -\mu_0 \boldsymbol{J} \,. \tag{8.3}$$

The differential equations for ϕ and $\boldsymbol{A}$ are thus coupled in a nontrivial manner, and one wonders whether introducing potential functions was a good idea. There is a way, however, that these equations simplify: Suppose that

$$\boldsymbol{\nabla} \cdot \boldsymbol{A} + \frac{1}{c^2}\frac{\partial \phi}{\partial t} = 0 \,, \tag{8.4}$$

the *Lorenz condition*. If Eq. (8.4) holds, the equations in Eq. (8.3) separate,

$$\nabla^2 \phi - \frac{1}{c^2}\frac{\partial^2 \phi}{\partial t^2} = -\frac{\rho}{\epsilon_0} \qquad \nabla^2 \boldsymbol{A} - \frac{1}{c^2}\frac{\partial^2 \boldsymbol{A}}{\partial t^2} = -\mu_0 \boldsymbol{J} \,. \tag{8.5}$$

While ϕ and $\boldsymbol{A}$ appear to be uncoupled in Eq. (8.5), they remain coupled through the Lorenz condition. The *sources* are also coupled through the *continuity equation* (see below).

The Lorenz condition can always be satisfied through a *gauge transformation*. The potentials $(\phi, \boldsymbol{A})$ are not unique in their determination of $(\boldsymbol{E}, \boldsymbol{B})$: The transformation

$$\boldsymbol{A} \to \boldsymbol{A}' = \boldsymbol{A} + \boldsymbol{\nabla}\chi \qquad \phi \to \phi' = \phi - \frac{\partial \chi}{\partial t} \,, \tag{8.6}$$

where $\chi(\boldsymbol{r}, t)$ is an arbitrary scalar function, leaves $\boldsymbol{E}$ and $\boldsymbol{B}$ unchanged; $\boldsymbol{E}$ and $\boldsymbol{B}$ are said to be *gauge invariant*. Suppose ϕ and $\boldsymbol{A}$ are such that $\boldsymbol{\nabla} \cdot \boldsymbol{A} + \partial(\phi/c^2)/\partial t = f(\boldsymbol{r}, t) \neq 0$. Under the transformation Eq. (8.6), the new potentials would satisfy

$$\boldsymbol{\nabla} \cdot \boldsymbol{A}' + \frac{1}{c^2}\frac{\partial \phi'}{\partial t} - \left(\nabla^2 \chi - \frac{1}{c^2}\frac{\partial^2 \chi}{\partial t^2}\right) = f(\boldsymbol{r}, t) \,.$$

By choosing χ to satisfy the inhomogeneous wave equation,

$$\nabla^2 \chi - \frac{1}{c^2}\frac{\partial^2 \chi}{\partial t^2} = -f(\boldsymbol{r}, t) \,,$$

ϕ' and $\boldsymbol{A}'$ now satisfy the Lorenz condition. Using potentials that satisfy Eq. (8.4) is said to be working in the *Lorenz gauge*, which we assume to be the case in what follows.

Balance equations and charge conservation

Conserved quantities satisfy continuity equations, and because conservation laws play an important role in SR and GR, let's take a moment and derive the continuity equation. *Balance equations* (for any quantity ψ) have the form

$$\frac{\mathrm{d}}{\mathrm{d}t}\int_V \rho_\psi \mathrm{d}V = -\oint_S \boldsymbol{J}_\psi \cdot \mathrm{d}\boldsymbol{S} + \int_V \sigma_\psi \mathrm{d}V \,. \tag{8.7}$$

Here V is a volume bounded by surface S having outward-pointing surface element $\mathrm{d}\boldsymbol{S}$, ρ_ψ is the density of ψ ($[\psi]$ m^{-3}), $\boldsymbol{J}_\psi$ is the *flux* of ψ through S ($[\psi]$ m^{-2} s^{-1}), and σ_ψ is a *source term* representing the rate per volume at which ψ is created or destroyed in V, ($[\psi]$ m^{-3} s^{-1}). Equation (8.7) specifies that the rate of change of the amount of ψ in V is accounted for by flows through S,

[6]The vector $\boldsymbol{A}$ appearing in Eq. (8.3) should be taken as its contravariant version: $\boldsymbol{A}$ divided by area must have the same tensor character as $\mu_0 \boldsymbol{J}$.

and by the net rate of creation of ψ in V by means other than flow; *there are no other possibilities*,[7] which is why it's called a continuity equation. When the transport of ψ occurs because of flow[8] (convection), $\boldsymbol{J}_\psi = \rho_\psi \boldsymbol{v}$. If $\oint \boldsymbol{J}_\psi \cdot \mathrm{d}\boldsymbol{S}$ is positive (negative), the amount of ψ in V decreases (increases) in time. By applying the divergence theorem to Eq. (8.7) (for fixed V), we arrive at the *local* form of a balance equation:

$$\frac{\partial \rho_\psi}{\partial t} + \boldsymbol{\nabla} \cdot \boldsymbol{J}_\psi = \sigma_\psi \, . \tag{8.8}$$

If $\sigma_\psi = 0$, ψ is *conserved*; in that case the only way ψ in V can change is to flow through S.

Charge is conserved and thus it obeys the continuity equation (in non-covariant form)

$$\frac{\partial \rho}{\partial t} + \boldsymbol{\nabla} \cdot \boldsymbol{J} = 0 \, . \tag{8.9}$$

Equation (8.9) is secretly buried in Maxwell's equations: Take the divergence of the Ampère-Maxwell law in Eq. (8.1) and make use of Gauss's law.

8.2 SOURCES IN SPACETIME: THE FOUR-CURRENT

How does *density* transform between reference frames? The topic was covered in Section 5.2, where we noted that because the Jacobian of the LT is unity, LTs do not highlight the density feature of physical quantities. Density is the amount of a substance, "stuff," per volume. While the amount of stuff is invariant, volume is not. We know from Eq. (5.57) that $\sqrt{|g|}\mathrm{d}^4x$ is invariant. In SR $|g| = 1$; 4-volume is thus invariant under the LT, $\mathrm{d}x^{0'}\mathrm{d}x^{1'}\mathrm{d}x^{2'}\mathrm{d}x^{3'} = \mathrm{d}x^0\mathrm{d}x^1\mathrm{d}x^2\mathrm{d}x^3$. Let the primed variables denote the rest frame. By time dilation, $\mathrm{d}x^0 = \gamma \mathrm{d}x^{0'}$, and thus charge density transforms under the LT as $\rho = \gamma\rho_0$, where ρ_0 is the *proper density*.

The *current four-vector density* or simply *four-current* is defined as

$$J^\mu \equiv (\rho c, \rho \boldsymbol{u}) = (\rho c, \boldsymbol{J}) \, . \tag{8.10}$$

As with the introduction of any four-vector, we must show that it belongs to the club, and here there's a quick way and a sophisticated way. The quick way is to show that it transforms as expected under the LT. In the rest frame, $J^{\nu'} = (\rho_0 c, 0)$. In another IRF connected by a boost, we have using Eq. (6.16), $J^\mu = L^\mu_{\nu'} J^{\nu'} = (\gamma\rho_0 c, \boldsymbol{\beta}\gamma\rho_0 c) = (\rho c, \rho\boldsymbol{u})$, where $\rho = \gamma\rho_0$. Note that $J^\mu = \rho_0(\gamma c, \gamma \boldsymbol{u}) = \rho_0 U^\mu$. The four-current is timelike, $J^\alpha J_\alpha = -\rho_0^2 c^2$ (from $U^\alpha U_\alpha = -c^2$), and is tangent to worldlines of charged particles. The continuity equation, (8.9), can then be written in covariant form,[9]

$$0 = \frac{\partial \rho}{\partial t} + \boldsymbol{\nabla}\cdot\boldsymbol{J} = \frac{\partial \rho c}{\partial (ct)} + \frac{\partial J^i}{\partial x^i} = \frac{\partial J^0}{\partial x^0} + \frac{\partial J^i}{\partial x^i} = \frac{\partial J^\mu}{\partial x^\mu} = \partial_\mu J^\mu \, . \tag{8.11}$$

Conservation is expressed covariantly as the vanishing *four-divergence* of J^μ.

The fancy way is to show that J^μ is related to the four-velocity in a Lorentz invariant manner. Let $\boldsymbol{z}(t)$ be the worldline of a particle in a particular reference frame. We can ascribe "density" to a *particle* through the use of the Dirac delta function,

$$\rho(\boldsymbol{r}, t) = q\delta^3(\boldsymbol{r} - \boldsymbol{z}(t)) \qquad \boldsymbol{J}(\boldsymbol{r}, t) = q\frac{\mathrm{d}\boldsymbol{z}}{\mathrm{d}t}\delta^3(\boldsymbol{r} - \boldsymbol{z}(t)) \, ,$$

[7] An example of a source term occurs in semiconductor physics, where additional charge carriers are generated by the interaction of electromagnetic radiation with the semiconductor.

[8] The Poynting vector S, the flux of electromagnetic energy, is given by $S = uc$ with u the density of electromagnetic energy. Not all currents are convective, however. There can be diffusive transport, such as in Ohm's law, $J = \sigma E$.

[9] Implying that if charge is conserved in one frame, it's conserved in all frames connected to the first by a LT.

where $\delta^3(x)$ is such that $\int d^3x f(\boldsymbol{x})\delta^3(\boldsymbol{x}-\boldsymbol{y}) = f(\boldsymbol{y})$. These expressions can be combined into a four-vector through the device of introducing the integral over time of another delta function and then changing variables to the proper time,

$$\rho(x^\mu) = q\int_{-\infty}^{\infty} dz^0 \delta(x^0 - z^0)\delta^3(\boldsymbol{r}-\boldsymbol{z}) = q\int_{-\infty}^{\infty} d\tau \frac{dz^0}{d\tau}\delta^4(x - z(\tau))$$
$$\boldsymbol{J}(x^\mu) = q\int_{-\infty}^{\infty} dz^0 \frac{d\boldsymbol{z}}{dt}\delta(x^0 - z^0)\delta^3(\boldsymbol{r}-\boldsymbol{z}) = qc\int_{-\infty}^{\infty} d\tau \frac{d\boldsymbol{z}}{d\tau}\delta^4(x - z(\tau)) ,$$

where $\delta^4(x-y) \equiv \delta(x^0-y^0)\delta(x^1-y^1)\delta(x^2-y^2)\delta(x^3-y^3)$ is the four-dimensional delta function and $z^\mu(\tau) = (z^0, \boldsymbol{z})$ is parameterized by the proper time. We have let $dz^0 \to (dz^0/d\tau)d\tau$ in the first integral and $dz^0(d\boldsymbol{z}/dt) = cd\boldsymbol{z} \to c(d\boldsymbol{z}/d\tau)d\tau$ in the second. Multiplying the expression for ρ by c, we have the four-current J^μ as an integral over the four-velocity $U^\mu = dz^\mu/d\tau$,

$$J^\mu(x^\alpha) = (\rho c, \boldsymbol{J}) = qc\int_{-\infty}^{\infty} d\tau U^\mu(\tau)\delta^4(x - z(\tau)) . \tag{8.12}$$

Before accepting Eq. (8.12) as a four-vector we should ascertain that the delta function is Lorentz invariant. The two requirements on $\delta^4(x-y)$ are that it's nonzero only if $x^\mu - y^\mu = 0$ and $\int d^4x\delta^4(x)f(x) = f(0)$. The first property is independent of reference frame: From $x^{\mu'} - y^{\mu'} = L^{\mu'}_{\nu}(x^\nu - y^\nu)$, if $x^\mu = y^\mu$ in one frame, the same is true in all frames. Thus, we can write $\delta^4(x'-y') = S\delta^4(x-y)$, where S is a constant. (Recall that $\delta(ax) = \delta(x)/|a|$; while $y' \to x'$ as $y \to x$, there could be scale factors.) It turns out that $S = 1$ for a LT. Writing $x' = Lx$ for $x^{\mu'} = L^{\mu'}_{\nu}x^\nu$,

$$f(y') = \int d^4x'\delta^4(x'-y')f(x') = \int d^4x J^{-1}S\delta^4(x-y)f(Lx) = J^{-1}Sf(y') ,$$

where J is the Jacobian, Eq. (5.52). Thus, $S = J$ and $J = 1$ for a LT. The Dirac delta function is therefore a Lorentz invariant.[10]

8.3 CONSERVATION IN SPACETIME: SPACELIKE HYPERSURFACES

A *hypersurface* Σ is an $(n-1)$-dimensional surface embedded in an n-dimensional space. For example, setting $x^0 = 0$ in MS specifies a hypersurface spanned by three spacelike vectors. At any point of Σ there's a vector n_α orthogonal to all vectors lying in Σ at that point.[11] Hypersurfaces are classified as spacelike, null, or timelike according to whether n_α is timelike, null, or spacelike (see Fig. 8.1). Spacelike hypersurfaces (SHs) have timelike normals, and timelike hypersurfaces have spacelike normals. For Σ not null, n_μ can be normalized so that $n_\mu n^\mu = \epsilon$, where $\epsilon = -1\ (+1)$ for Σ spacelike (timelike). SHs play an important role in the theory of relativity.

Let $\{y^i\}$ be coordinates intrinsic to hypersurface Σ. Displacements on Σ are described by $(ds)^2_\Sigma \equiv h_{ij}dy^i dy^j$, where h_{ij} is the *induced metric* on Σ, the restriction of $g_{\mu\nu}$ (the metric of the embedding space) to vectors tangent to Σ. We show in Section 13.6 that if $g_{\mu\nu}$ is positive definite, so is the induced metric on Σ. If, however, $g_{\mu\nu}$ is indefinite then the induced metric is indefinite on timelike hypersurfaces, but positive definite on SHs. On a three-dimensional hypersurface the invariant volume element is $d\Sigma \equiv \sqrt{|h|}d^3y$ where $h \equiv \det h_{ij}$. The *oriented* volume element can be

[10] The Dirac function $\delta^4(x)$ is *not* invariant under arbitrary coordinate transformations because d^4x is not invariant. However, $\sqrt{|g|}d^4x$ *is* an invariant, implying that $\delta^4(x)/\sqrt{|g|}$ is invariant under general coordinate transformations.

[11] The normal vector belongs to the n-dimensional embedding space. If for example the hypersurface is specified by $x^1 = 0$ in an n-dimensional space, the coordinates of a point in the hypersurface are $x^1 = 0, x^2, \cdots, x^n$. The normal vector n_α is such that $n_\alpha x^\alpha = 0$, where $n_1 \neq 0$. As discussed in Section 5.11 for the elements of the oriented 3-volume $d\Sigma^{\alpha\beta\gamma}$, its dual $(d\Sigma^*)_\lambda$ is a vector orthogonal to the hypersurface.

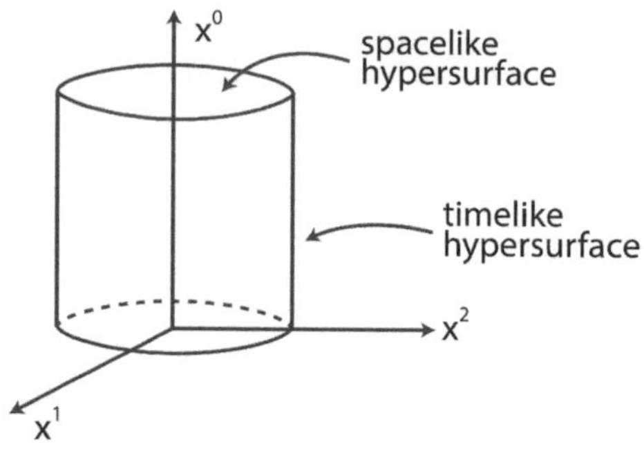

Figure 8.1 Hypersurfaces are spacelike or timelike if its normal is timelike or spacelike

written $d\Sigma_\mu \equiv n_\mu d\Sigma$. Note that $d\Sigma_\mu$ as a vector belongs to the embedding space. To apply Gauss's theorem, we need the orientations of Σ and the embedding space to agree (Section 5.11). We specify the orientation so that $n_\mu X^\mu > 0$ for a vector X^μ that points out[12] of Σ.

Combining the continuity equation, (8.11), with Gauss's theorem, Eq. (5.90), we have that *the net flux in spacetime is zero,*

$$\oint_S J^\mu d\Sigma_\mu = 0 \,, \tag{8.13}$$

where S denotes a closed hypersurface. *Nothing is lost or gained along the worldlines of a conserved quantity.* Consider a region of space large enough so that $J^\mu \to 0$ on timelike hypersurfaces (see Fig. 8.2). A physical four-current differs from zero only in a finite region of space; thus we

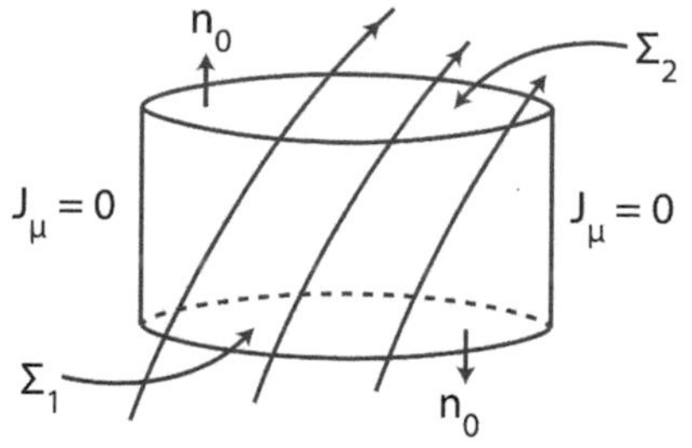

Figure 8.2 The net flux in spacetime is zero for conserved quantities

can choose spatial boundaries (timelike hypersurfaces) so that on them $J^\mu = 0$. A finite region of space, in spacetime is a tube of infinite length in the time direction, the worldtube. Because Eq. (8.13) requires a *closed* hypersurface, we "cap off" the worldtube by SHs, Σ_1 and Σ_2. Zero net flow then requires that $\int_{\Sigma_1} J^\mu d\Sigma_\mu + \int_{\Sigma_2} J^\mu d\Sigma_\mu = 0$. Taking the outwardly pointing normals, $d\Sigma_\mu^{(1)} = -\delta^0{}_\mu d^3y$ and $d\Sigma_\mu^{(2)} = \delta^0{}_\mu d^3y$, we have from Eq. (8.13)

$$\oint_S J^\beta d\Sigma_\beta = Q(\Sigma_2) - Q(\Sigma_1) = 0 \,, \tag{8.14}$$

where $Q(\Sigma) \equiv \int_\Sigma J^0 d^3y$. Equation (8.14) indicates that the total charge in a SH is *independent* of SH, i.e., conserved.

[12] If for a vector X^μ that points out of the volume we require $n_\mu X^\mu > 0$, then $n_0 > 0$ for a future pointing vector (take $X^\mu = (X^0, 0, 0, 0)$), with $X^0 > 0$. For a metric where $n^\alpha n_\alpha < 0$, n^α is *inwardly* pointing. See R.M. Wald[37, p434] or Hawking and Ellis.[38, p50]

A continuity equation implies the existence of a fixed quantity contained in a SH. In the rest frame take the outwardly pointing normal (to Σ) to be $n_\mu = (1,0,0,0)$. In another IRF, $n_{\mu'} = L^\nu_{\mu'} n_\nu = L^0_{\mu'} = \gamma(1, \boldsymbol{\beta})$ (Note that $n^{\mu'} n_{\mu'} = -1$; the spacelike character of Σ is preserved under LTs.) If in one frame $Q = \int J^0 \mathrm{d}^3 y$, then in any other frame (related by a LT) $Q = \int J^\mu \mathrm{d}\Sigma_\mu$ because of the Lorentz invariance of $J^\mu \mathrm{d}\Sigma_\mu$. Conserved quantities can then be written in covariant form, $Q = \int_\Sigma J^\mu \mathrm{d}\Sigma_\mu$ for SHs Σ. There are two interrelated ideas here: The invariance and the conservation of charge: Q is conserved because it's independent of the SH used to evaluate it (a SH is a slice of "now"), and Q is a Lorentz invariant because the LT maps SHs onto SHs.

8.4 THE FOUR-POTENTIAL

We can group the potentials ϕ and $\boldsymbol{A}$ into a single quantity, the *four-potential*[13]

$$A^\mu \equiv (\phi/c, \boldsymbol{A}) \tag{8.15}$$

(A^μ should not be confused with the four-acceleration $\mathcal{A}^\mu$). The wave equations for ϕ and $\boldsymbol{A}$ in Eq. (8.5) are the time and space components of a single wave equation for A^μ. Using Eqs. (8.15), (8.10), and (5.59), the four equations implied by Eq. (8.5) are equivalent to an inhomogeneous wave equation for the components of the four-potential,

$$\partial_\mu \partial^\mu A^\alpha = -\mu_0 J^\alpha \, . \tag{8.16}$$

Because $\partial_\mu \partial^\mu$ is Lorentz invariant, we have from Eq. (8.16) that A^α transforms the same as J^α and hence is a four-vector. The Lorenz condition in covariant form is,

$$0 = \frac{1}{c^2}\frac{\partial \phi}{\partial t} + \boldsymbol{\nabla}\cdot\boldsymbol{A} = \frac{\partial(\phi/c)}{\partial(ct)} + \boldsymbol{\nabla}\cdot\boldsymbol{A} = \frac{\partial A^0}{\partial x^0} + \frac{\partial A^i}{\partial x^i} = \partial_\mu A^\mu \, . \tag{8.17}$$

Note that the covariant four-potential $A_\mu = (-\phi/c, \boldsymbol{A})$.

8.5 MAXWELL EQUATIONS IN COVARIANT FORM: FIELD TENSOR

We define the *electromagnetic field tensor* (antisymmetric, covariant tensor):[14]

$$F_{\mu\nu} \equiv \partial_\mu A_\nu - \partial_\nu A_\mu \equiv (\partial \wedge A)_{\mu\nu} \, . \tag{8.18}$$

Derivatives of tensors transform as tensors under the restricted, linear coordinate transformations of SR (Section 5.3), but not under the more general transformations used in GR. It turns out that the *antisymmetric* combination of derivatives in Eq. (8.18) transforms as a tensor even on curved manifolds.[15] The field tensor is the generalization of the curl to four dimensions applied to the four-potential: It's a four-dimensional covariant bivector. Instead of $\boldsymbol{B} = \boldsymbol{\nabla} \times \boldsymbol{A}$, $F_{\mu\nu} = (\partial \wedge A)_{\mu\nu}$ covariantly "encodes" both of the relations in Eq. (8.2). An antisymmetric second-rank tensor in n dimensions has $n(n-1)/2$ independent elements, just enough (for $n = 4$) for the six components of $\boldsymbol{E}$ and $\boldsymbol{B}$!

We can evaluate $\{F_{\mu\nu}\}$ by combining Eqs. (8.18), (8.15), and (8.2):

$$[F_{\mu\nu}] = \begin{pmatrix} 0 & -E_1/c & -E_2/c & -E_3/c \\ E_1/c & 0 & B^3 & -B^2 \\ E_2/c & -B^3 & 0 & B^1 \\ E_3/c & B^2 & -B^1 & 0 \end{pmatrix} . \tag{8.19}$$

[13] The dimension of ϕ/c (Volt · s/m) is the same as that for $\boldsymbol{A}$, as can seen from the Lorenz condition Eq. (8.4).

[14] The field tensor is sometimes called the Faraday tensor.

[15] Thus, $F_{\mu\nu}$ is a bona fide tensor even in curved spaces.

To derive the top row of $[F_{\mu\nu}]$, $F_{0i} \equiv \partial_0 A_i - \partial_i A_0 = \partial A_i/\partial(ct) + \partial_i(\phi/c) = -E_i/c$. The diagonal elements $F_{\alpha\alpha}$ are identically zero (antisymmetry). The other independent elements of Eq. (8.19) follow from Eq. (8.18) using $B^i = \frac{1}{2}\varepsilon^{ijk}F_{jk}$.

The field tensor is gauge invariant. The four-vector version of the gauge transformation is[16]

$$A'_\alpha = A_\alpha + \partial_\alpha \chi \,. \tag{8.20}$$

Combining Eq. (8.20) with Eq. (8.18),

$$F'_{\mu\nu} = \partial_\mu A'_\nu - \partial_\nu A'_\mu = \partial_\mu (A_\nu + \partial_\nu \chi) - \partial_\nu (A_\mu + \partial_\mu \chi) = \partial_\mu A_\nu - \partial_\nu A_\mu + \partial_\mu \partial_\nu \chi - \partial_\nu \partial_\mu \chi = F_{\mu\nu} \,,$$

where we assume that $\partial_\mu \partial_\nu \chi = \partial_\nu \partial_\mu \chi$. From Eq. (8.20) it can be seen that A^μ satisfies the Lorenz condition, Eq. (8.17), when the gauge function χ satisfies $\partial_\alpha \partial^\alpha \chi = -\partial_\beta A^\beta$.

The homogeneous Maxwell equations can be expressed as a covariant equation involving $F_{\mu\nu}$:

$$\varepsilon^{\mu\nu\lambda\sigma}\partial_\nu F_{\lambda\sigma} = 0 \,. \qquad (\mu = 0,1,2,3) \tag{8.21}$$

For example, take $\mu = 0$:

$$\begin{aligned}
\varepsilon^{0\nu\lambda\sigma}\partial_\nu F_{\lambda\sigma} =& \varepsilon^{01\lambda\sigma}\partial_1 F_{\lambda\sigma} + \varepsilon^{02\lambda\sigma}\partial_2 F_{\lambda\sigma} + \varepsilon^{03\lambda\sigma}\partial_3 F_{\lambda\sigma} \\
=& \varepsilon^{0123}\partial_1 F_{23} + \varepsilon^{0132}\partial_1 F_{32} + \varepsilon^{0213}\partial_2 F_{13} + \varepsilon^{0231}\partial_2 F_{31} + \varepsilon^{0312}\partial_3 F_{12} + \varepsilon^{0321}\partial_3 F_{21} \\
=& 2\varepsilon^{0123}\partial_1 F_{23} + 2\varepsilon^{0231}\partial_2 F_{31} + 2\varepsilon^{0312}\partial_3 F_{12} \\
=& 2\varepsilon^{0123}\left(\partial_1 F_{23} + \partial_2 F_{31} + \partial_3 F_{12}\right) = 2\partial_i B^i = 2\boldsymbol{\nabla} \cdot \boldsymbol{B} = 0 \,,
\end{aligned}$$

where we have used $\varepsilon^{0123} = 1$ and the antisymmetry of $\varepsilon^{\mu\nu\lambda\sigma}$ and $F_{\lambda\sigma}$. For μ a spatial index, Eq. (8.21) generates the components of Faraday's law. In the notation of Section 5.7, the homogeneous Maxwell equations can be written $\partial_{[\alpha}F_{\beta\gamma]} = 0$.

A simpler way to write Eq. (8.21) is to use the *dual* of the field tensor (see Eq. (5.75)),

$$\epsilon^{\mu\nu\lambda\sigma}\partial_\nu F_{\lambda\sigma} \equiv 2\partial_\nu \left(*F\right)^{\mu\nu} = 0 \,. \qquad (\mu = 0,1,2,3) \tag{8.22}$$

Combining Eq. (8.19) with Eq. (5.75), we find that

$$[*F^{\mu\nu}] = \begin{pmatrix} 0 & B^1 & B^2 & B^3 \\ -B^1 & 0 & -E_3/c & E_2/c \\ -B^2 & E_3/c & 0 & -E_1/c \\ -B^3 & -E_2/c & E_1/c & 0 \end{pmatrix} . \tag{8.23}$$

The homogeneous Maxwell equations are the vanishing four-divergence of the dual field tensor, $\partial_\nu \left(*F\right)^{\mu\nu} = 0$. For $\mu = 0$, $\partial_\nu \left(*F^{0\nu}\right) = \partial_i \left(*F\right)^{0i} = \partial_i B^i = 0$.

The inhomogeneous Maxwell equations can be expressed in covariant form with the introduction of another antisymmetric tensor, (actually a tensor density, but we won't use the notation $\mathfrak{G}^{\mu\nu}$)

$$[G^{\mu\nu}] \equiv \begin{pmatrix} 0 & D^1 c & D^2 c & D^3 c \\ -D^1 c & 0 & H_3 & -H_2 \\ -D^2 c & -H_3 & 0 & H_1 \\ -D^3 c & H_2 & -H_1 & 0 \end{pmatrix} . \tag{8.24}$$

It can readily be verified that the inhomogeneous Maxwell equations are equivalent to

$$\partial_\nu G^{\mu\nu} = J^\mu \,. \qquad (\mu = 0,1,2,3) \tag{8.25}$$

[16]We put the primes here *not* on the index, but on the function. In the index convention of Section 5.1 we put the prime on an index to indicate a new coordinate system. Here we are transforming a function; hence the primes on the function.

As a check, set $\mu = 0$ in Eq. (8.25): $\partial_\nu G^{0\nu} = c\partial_i D^i = J^0 = \rho c$, Gauss's law. Can $G^{\mu\nu}$ be derived from $F_{\mu\nu}$? No, in the same way that the inhomogeneous Maxwell equations cannot be derived from the homogeneous. In free space, however, the replacements $\boldsymbol{D} \to \epsilon_0 \boldsymbol{E}$ and $\boldsymbol{H} \to \mu_0^{-1}\boldsymbol{B}$ are equivalent to the replacement $G^{\mu\nu} \to \mu_0^{-1}F^{\mu\nu}$, where $F^{\mu\nu} = \eta^{\mu\alpha}\eta^{\nu\beta}F_{\alpha\beta} = \partial^\mu A^\nu - \partial^\nu A^\mu$ is the contravariant version of $F_{\mu\nu}$:

$$[F^{\mu\nu}] = \begin{pmatrix} 0 & E^1/c & E^2/c & E^3/c \\ -E^1/c & 0 & B_3 & -B_2 \\ -E_2/c & -B_3 & 0 & B_1 \\ -E_3/c & B_2 & -B_1 & 0 \end{pmatrix} . \tag{8.26}$$

From Eq. (8.26), $E^i = cF^{0i}$ and $B_i = \frac{1}{2}\varepsilon_{ijk}F^{jk}$. The inhomogeneous Maxwell equations for fields in free space can then be expressed covariantly as

$$\partial_\nu F^{\mu\nu} = \mu_0 J^\mu . \qquad (\mu = 0, 1, 2, 3) \text{ (free space)} \tag{8.27}$$

Thus, Maxwell's equations can be expressed covariantly in terms of the divergence of four-dimensional second-rank antisymmetric tensors, $*F^{\mu\nu}$ and $G^{\mu\nu}$, Eqs. (8.22) and (8.25), or, for fields in free space, Eq. (8.27). The covariant form of Maxwell's equations was derived by Minkowski in 1907, a development that's sometimes criticized as not providing anything "new." We can hardly expect the covariant formulation of a theory that was already consistent with SR to predict new physical effects. Minkowski's contribution demonstrates that physical fields ($\boldsymbol{E}$ and $\boldsymbol{B}$) can be described as *tensor fields* on a four-dimensional spacetime manifold, which is a huge achievement. Can Newton's law of gravity be so expressed? That is in fact the goal of GR, and we'll be guided by what we learn in this chapter. We show in Eq. (8.43) that $\mathrm{d}P^\mu/\mathrm{d}\tau = qF^{\mu\nu}U_\nu$, that the Newton-Maxwell equation of motion is relativistically correct once the mechanical side of the equation is fixed up by packaging $\boldsymbol{p} = m\gamma\boldsymbol{v}$ and $E = m\gamma c^2$ into a four-vector.

8.6 LORENTZ TRANSFORMATION OF **E** AND **B** FIELDS

A side benefit of packaging the components of $\boldsymbol{E}$ and $\boldsymbol{B}$ into a tensor is that we immediately know the LT of $\boldsymbol{E}$ and $\boldsymbol{B}$ from the transformation property of tensors. Einstein used a different approach in his 1905 paper: He derived the transformation equations for $\boldsymbol{E}$ and $\boldsymbol{B}$ by applying the LT to the *coordinates* of the two curl equations in free space and demanding that the *form* of the equations be invariant (the principle of covariance). Einstein had not yet adopted the four-dimensional geometric view of spacetime. The elements of $F^{\mu\nu}$ transform under the LT as

$$F^{\mu'\nu'} = L^{\mu'}_{\alpha} L^{\nu'}_{\beta} F^{\alpha\beta} . \tag{8.28}$$

We cannot directly transform $\boldsymbol{E}$ and $\boldsymbol{B}$ under the LT because they are not four-vectors. As we'll see, the components of $\boldsymbol{E}$ and $\boldsymbol{B}$ transform among themselves, and thus we need a six-dimensional animal to express their transformation properties—but that's just what we have, an antisymmetric second-rank tensor in four dimensions.

Let's work through some examples using the LT for frames in standard configuration. From Eq. (8.28), with $\mu = 0$ and $\nu = 1$

$$\begin{aligned} F^{0'1'} &= \frac{E^{1'}}{c} = L^{0'}_{\alpha} L^{1'}_{\beta} F^{\alpha\beta} = L^{0'}_{\alpha}\left(L^{1'}_{0}F^{\alpha 0} + L^{1'}_{1}F^{\alpha 1}\right) = L^{0'}_{\alpha}\left(-\beta\gamma F^{\alpha 0} + \gamma F^{\alpha 1}\right) \\ &= -\beta\gamma L^{0'}_{1}F^{10} + \gamma L^{0'}_{0}F^{01} = (\beta\gamma)^2 F^{10} + \gamma^2 F^{01} = \gamma^2(1-\beta^2)\frac{E^1}{c} = \frac{E^1}{c} . \end{aligned}$$

The *longitudinal* component of $\boldsymbol{E}$ is invariant, $E^{1'} = E^1$; the same is true of the longitudinal component of $\boldsymbol{B}$. As another example, take $\mu = 0$ and $\nu = 2$

$$F^{0'2'} = \frac{E^{2'}}{c} = L^{0'}_{\alpha} L^{2'}_{\beta} F^{\alpha\beta} = L^{0'}_{\alpha} F^{\alpha 2} = L^{0'}_{0} F^{02} + L^{0'}_{1} F^{12} = \gamma F^{02} - \beta\gamma F^{12} = \gamma\frac{E^2}{c} - \beta\gamma B_3 ,$$

and hence $E^{2'} = \gamma\left(E^2 - vB_3\right)$. The *transverse* components of the fields thus transform among themselves. Note that the components of the field vectors transform differently under the LT than do components of the position vector. We find that for frames in standard configuration:

$$\begin{aligned} E^{1'} &= E^1 \quad E^{2'} = \gamma\left(E^2 - vB_3\right) \qquad E^{3'} = \gamma\left(E^3 + vB_2\right) \\ B_{1'} &= B_1 \quad B_{2'} = \gamma\left(B_2 + \frac{v}{c^2}E^3\right) \quad B_{3'} = \gamma\left(B_3 - \frac{v}{c^2}E^2\right) . \end{aligned} \tag{8.29}$$

We can put Eq. (8.29) in vector form. Decompose $\boldsymbol{E}$ and $\boldsymbol{B}$ into vectors parallel and perpendicular to $\boldsymbol{v}$, $\boldsymbol{E} = \boldsymbol{E}_{\parallel} + \boldsymbol{E}_{\perp}$ and $\boldsymbol{B} = \boldsymbol{B}_{\parallel} + \boldsymbol{B}_{\perp}$. From Eq. (8.29),

$$\begin{aligned} \boldsymbol{E}'_{\parallel} &= \boldsymbol{E}_{\parallel} \qquad \boldsymbol{E}'_{\perp} = \gamma\left(\boldsymbol{E}_{\perp} + \boldsymbol{v}\times\boldsymbol{B}\right) \\ \boldsymbol{B}'_{\parallel} &= \boldsymbol{B}_{\parallel} \qquad \boldsymbol{B}'_{\perp} = \gamma\left(\boldsymbol{B}_{\perp} - \frac{1}{c^2}\boldsymbol{v}\times\boldsymbol{E}\right) . \end{aligned} \tag{8.30}$$

The projection of $\boldsymbol{E}$ onto $\boldsymbol{v}$ is $\boldsymbol{E}_{\parallel} = (\boldsymbol{v}\cdot\boldsymbol{E})\boldsymbol{v}/v^2$, and $\boldsymbol{E}_{\perp} = \boldsymbol{E} - \boldsymbol{E}_{\parallel}$, allowing us to write

$$\begin{aligned} \boldsymbol{E}' &= \gamma\left(\boldsymbol{E} + \boldsymbol{v}\times\boldsymbol{B}\right) - (\gamma - 1)\left(\boldsymbol{v}\cdot\boldsymbol{E}\right)\frac{\boldsymbol{v}}{v^2} \\ \boldsymbol{B}' &= \gamma\left(\boldsymbol{B} - \frac{1}{c^2}\boldsymbol{v}\times\boldsymbol{E}\right) - (\gamma - 1)\left(\boldsymbol{v}\cdot\boldsymbol{B}\right)\frac{\boldsymbol{v}}{v^2} . \end{aligned} \tag{8.31}$$

Finally, using the "*BAC-CAB*" rule, an equivalent way of writing Eq. (8.31) is

$$\begin{aligned} \boldsymbol{E}' &= \boldsymbol{E} + \gamma(\boldsymbol{v}\times\boldsymbol{B}) - \frac{\gamma^2}{c^2(1+\gamma)}\boldsymbol{v}\times(\boldsymbol{v}\times\boldsymbol{E}) \\ \boldsymbol{B}' &= \boldsymbol{B} - \frac{\gamma}{c^2}(\boldsymbol{v}\times\boldsymbol{E}) - \frac{\gamma^2}{c^2(1+\gamma)}\boldsymbol{v}\times(\boldsymbol{v}\times\boldsymbol{B}) . \end{aligned} \tag{8.32}$$

To lowest order in $\boldsymbol{v}$, $\boldsymbol{E}' \approx \boldsymbol{E} + \boldsymbol{v}\times\boldsymbol{B}$ and $\boldsymbol{B}' \approx \boldsymbol{B} - (\boldsymbol{v}\times\boldsymbol{E})/c^2$.

Statements such as "the field is purely electric" (or magnetic) lack intrinsic meaning. While the magnetic field may be zero in a particular frame, it will be nonzero in another frame, as we see from Eq. (8.32). In general one has electric and magnetic fields; *the two fields do not separately exist,* they're aspects of a *single entity*, which all observers agree is a tensor quantity, $F^{\mu\nu}$.

8.7 LORENTZ FORCE AS A RELATIVISTIC EFFECT

The "relativism" between electric and magnetic fields can be illustrated by showing that the force experienced by a charge in motion relative to a current-carrying wire arises from an *electric* field in the rest frame of the charge. A current-carrying wire is electrically neutral (a charge *at rest* relative to such a wire is neither attracted nor repelled). We can conceive a current-carrying wire (in the lab frame) to be a line density λ (charge per length) of immobile positive charges, together with a line density of equal and opposite sign moving with a constant speed v to the left (the drift speed), producing a current $I = \lambda v$ directed to the right (see Fig. 8.3). Consider a positive charge q that in the lab frame moves at speed u to the right. In the rest frame of q (call it the q-frame), the positive charges of the wire move to the left with speed $v_+ = u$, but the electrons in the wire move to the left with a different speed v_-, which from the velocity addition theorem, Eq. (3.16), is

$$v_- = \frac{u+v}{1+uv/c^2} .$$

In the rest frame of the *electrons* in the wire, the charge contained in length l_0 is $\lambda_0 l_0 = \lambda l$, where $l = l_0\sqrt{1-\beta_v^2}$ is the contracted length in the lab frame ($\beta_v \equiv v/c$) and λ_0 is the line density in

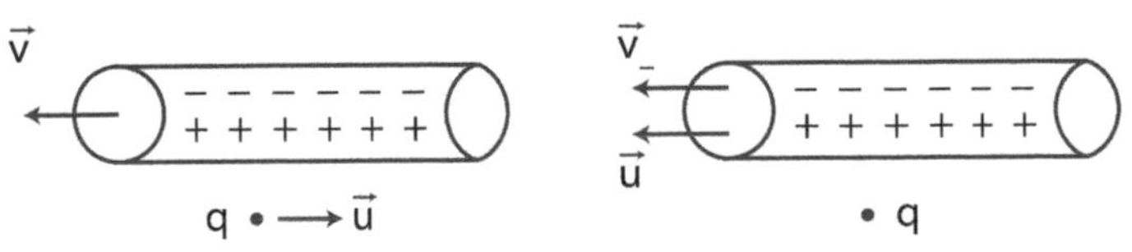

Figure 8.3 Left: Rest frame of wire. Right: Rest frame of charge

the electron frame. Thus, $\lambda_0 = \lambda\sqrt{1-\beta_v^2} \equiv \lambda/\gamma_v$ (how densities transform, Section 8.2). There are *three* frames of reference here: the lab frame, the rest frame of electrons in the wire, and the rest frame of q. The line density of electrons as observed in the q-frame is thus $\widetilde{\lambda}_- = \lambda_0\gamma_-$ where $\gamma_- \equiv (1-\beta_-^2)^{-1/2}$ with $\beta_- \equiv v_-/c$. Thus,

$$\widetilde{\lambda}_- = \gamma_-\lambda_0 = \frac{\gamma_-}{\gamma_v}\lambda = \gamma_u(1+uv/c^2)\lambda\,,$$

where we've used Eq. (3.14), $\gamma_- = \gamma_v\gamma_u(1+uv/c^2)$, with $\gamma_u \equiv 1/\sqrt{1-\beta_u^2}$ and $\beta_u \equiv u/c$. The line density of positive charges in the q-frame is $\widetilde{\lambda}_+ = \gamma_u\lambda$. The *net* line density in the q-frame is thus

$$\widetilde{\lambda} \equiv \widetilde{\lambda}_- - \widetilde{\lambda}_+ = \gamma_u\frac{uv}{c^2}\lambda\,.$$

The wire, which in the lab frame is electrically neutral, appears *negatively charged* in the q-frame because of the difference in transformations between the densities of positive and negative charges in the wire. As $u \to 0$, $\widetilde{\lambda} \to 0$.

In its frame, therefore, q experiences a force from the *electric field* at a distance r from a negatively charged wire, (the coordinate r is unaffected by the LT)

$$q\boldsymbol{E} = \frac{q\widetilde{\lambda}}{2\pi\epsilon_0 r}(-\hat{\boldsymbol{r}}) = \gamma_u\frac{quv\lambda}{2\pi\epsilon_0c^2r}(-\hat{\boldsymbol{r}}) = \gamma_u\frac{\mu_0Iqu}{2\pi r}(-\hat{\boldsymbol{r}}) = \gamma_u\frac{\mu_0Iqu}{2\pi r}\hat{\boldsymbol{z}}\times\hat{\boldsymbol{\phi}} \equiv \gamma_uq\boldsymbol{u}\times\boldsymbol{B}\,,$$

where $\hat{\boldsymbol{r}}$ points from the wire to q, $\boldsymbol{u} = u\hat{\boldsymbol{z}}$ ($\hat{\boldsymbol{z}}$ points along the wire) and $\boldsymbol{B} = \mu_0I\hat{\boldsymbol{\phi}}/(2\pi r)$ (the $\boldsymbol{B}$-field of a long wire). The spatial part of the four-force K^μ is $\gamma\boldsymbol{F}$ (Section 7.2). The three-force in the lab frame is therefore $q\boldsymbol{u}\times\boldsymbol{B}$. *The Lorentz force is a consequence of frame transformations.* The magnetic force is a *second-order* effect in powers of (speed/c)—the speed of the particle and the speed of the current; that it's an appreciable force at all in the lab frame is a testament to the strength of the Coulomb force.

8.8 INVARIANTS OF THE ELECTROMAGNETIC FIELD

The transformation properties of the electromagnetic field follow from the transformation properties of the field tensor, Eq. (8.28). A *scalar* derived from the field tensor would therefore imply a Lorentz-invariant relation among the field components.

There are two independent invariants that can be built this way, call them I_1 and I_2; any other invariants are functions of I_1 and I_2. The first is the scalar density

$$I_1 \equiv \frac{1}{2}G^{\mu\nu}F_{\mu\nu} = -E_iD^i + H_iB^i = -\boldsymbol{E}\cdot\boldsymbol{D} + \boldsymbol{H}\cdot\boldsymbol{B}\,, \tag{8.33}$$

where we've used Eqs. (8.24) and (8.19). For free-space fields, Eq. (8.33) is equivalent to $I_1 = F^{\mu\nu}F_{\mu\nu} = |B|^2 - |E|^2/c^2$. If for example $I_1 = 0$ in one reference frame, e.g., $|B| = |E|/c$, then that condition holds in any reference frame. The second invariant is the pseudoscalar density

$$I_2 \equiv -\frac{c}{4}(*F)^{\mu\nu}F_{\mu\nu} = E_iB^i = \boldsymbol{E}\cdot\boldsymbol{B}\,, \tag{8.34}$$

that is, $\boldsymbol{E}\cdot\boldsymbol{B}$ is the same in all IRFs.

Thesse invariants allow us to make qualitative statements about the fields that are valid in any IRF. If $I_2 = 0$ in one frame, i.e., $\boldsymbol{E}$ and $\boldsymbol{B}$ are orthogonal, they are orthogonal in all frames. Suppose in one frame that $I_1 > 0$, i.e., $|B| > |E|/c$. Then it would be possible to find a frame in which $\boldsymbol{E}$ vanishes, but not $\boldsymbol{B}$. Likewise, if $I_1 < 0$, it would be possible to find a frame in which $\boldsymbol{B} = 0$ but not $\boldsymbol{E}$. If $I_2 = 0$ and $I_1 = 0$, such as occur in electromagnetic plane waves, then $|E| = c|B|$ and $\boldsymbol{E}\cdot\boldsymbol{B} = 0$ in all IRFs.

8.9 ACTION PRINCIPLE FOR CHARGED PARTICLES

8.9.1 Covariant equation of motion for a charged particle

The non-covariant Lagrangian for a charged particle is $L = -mc^2\sqrt{1-\beta^2} + q(-\phi + \boldsymbol{u}\cdot\boldsymbol{A})$, Eq. (7.30). Noting that $q(-\phi + \boldsymbol{u}\cdot\boldsymbol{A})$ is basically $A_\mu J^\mu = \rho(-\phi + \boldsymbol{u}\cdot\boldsymbol{A})$, we are led to consider as a generalization of Eq. (7.45):

$$S = -mc\int\sqrt{-g_{\mu\nu}\mathrm{d}x^\mu\mathrm{d}x^\nu} + \int A_\mu J^\mu \mathrm{d}^3x\mathrm{d}t\,. \tag{8.35}$$

The new term in Eq. (8.35) involves the Lorentz invariants $\mathrm{d}^3x\mathrm{d}t$ and $A_\mu J^\mu$.[17] Because it involves the spacetime volume element, it's appropriate for describing the coupling of the current *density* to the field. We want the action for a *particle* coupled to the electromagnetic field. Using Eq. (8.12) for J^μ (the "density" of a particle)

$$\int A_\mu J^\mu \mathrm{d}^3x\mathrm{d}t = q\int A_\mu \int \mathrm{d}\tau U^\mu \delta^4(x - z(\tau))\mathrm{d}^4x = q\int \mathrm{d}\tau A_\mu U^\mu = q\int A_\mu \mathrm{d}x^\mu\,, \tag{8.36}$$

where $\mathrm{d}^3x\mathrm{d}t = \mathrm{d}^4x/c$. Combining Eqs. (8.36) and (8.35), we have the action integral of a point charge coupled to the field

$$S = \int\left[-mc\sqrt{-g_{\mu\nu}\frac{\mathrm{d}x^\mu}{\mathrm{d}\lambda}\frac{\mathrm{d}x^\nu}{\mathrm{d}\lambda}} + q\frac{\mathrm{d}x^\mu}{\mathrm{d}\lambda}A_\mu(x)\right]\mathrm{d}\lambda = \int\left[-mc\sqrt{-g_{\mu\nu}U^\mu U^\nu} + qU^\mu A_\mu(x)\right]\mathrm{d}\lambda\,, \tag{8.37}$$

and hence the generalization of Eq. (7.46),

$$L = -mc\sqrt{-g_{\mu\nu}U^\mu U^\nu} + qA_\mu U^\mu\,. \tag{8.38}$$

What equation of motion is implied by Hamilton's principle? Using Eq. (8.38),

$$\frac{\partial L}{\partial U^\mu} = mU_\mu + qA_\mu \qquad \frac{\partial L}{\partial x^\mu} = qU^\alpha\frac{\partial A_\alpha}{\partial x^\mu}\,. \tag{8.39}$$

The canonical momentum $\partial L/\partial U^\mu$ is the "kinetic" momentum mU_μ associated with an uncharged mass, together with the additional term qA_μ, a momentum associated with the coupling to the field. From the covariant Euler-Lagrange equation, Eq. (7.48),

$$\frac{\mathrm{d}}{\mathrm{d}\tau}(mU_\mu + qA_\mu) = qU^\alpha\frac{\partial A_\alpha}{\partial x^\mu}\,. \tag{8.40}$$

Equation (8.40) can be cast into a more transparent form using the chain rule

$$\frac{\mathrm{d}A_\mu}{\mathrm{d}\tau} = \frac{\partial A_\mu}{\partial x^\nu}\frac{\mathrm{d}x^\nu}{\mathrm{d}\tau} = U^\nu\frac{\partial A_\mu}{\partial x^\nu}\,. \tag{8.41}$$

---

[17] Equation (8.35) is valid only in SR. It should be written with the invariant volume element $\sqrt{|g|}\mathrm{d}^4x$ (Section 5.2). In SR $g = -1$. We use the more general formulation in GR.

Combining Eqs. (8.40) and (8.41),

$$m\frac{\mathrm{d}U_\mu}{\mathrm{d}\tau} = q\left(\frac{\partial A_\nu}{\partial x^\mu} - \frac{\partial A_\mu}{\partial x^\nu}\right)U^\nu = qF_{\mu\nu}U^\nu\ . \tag{8.42}$$

Equation (8.42) is the covariant equation of motion for a point charge. Equivalently, Eq. (8.42) can be written

$$\frac{\mathrm{d}P^\mu}{\mathrm{d}\tau} = qF^{\mu\nu}U_\nu\ . \tag{8.43}$$

The quantity $qF^{\mu\nu}U_\nu$ is the four-force experienced by a charged particle (see Eq. (7.10)).

The equation of motion Eq. (8.43) is gauge invariant because $F^{\mu\nu}$ is, yet the Lagrangian Eq. (8.38) is not. Huh? Two Lagrangians are *equivalent*—generate the same equations of motion—if they differ by the total time derivative of a function. That function is none other than the gauge function. Under $A_\mu \to A_\mu + \partial_\mu\chi$, $U^\mu A_\mu \to U^\mu A_\mu + \mathrm{d}x^\mu(\partial_\mu\chi)/\mathrm{d}\tau = U^\mu A_\mu + \mathrm{d}\chi/\mathrm{d}\tau$. For the action integral Eq. (8.35), the interaction term transforms as $J^\mu A_\mu \to J^\mu A_\mu + J^\mu\partial_\mu\chi = J^\mu A_\mu + \partial_\mu\left(J^\mu\chi\right) - \chi\partial_\mu J^\mu$. The last term vanishes by the continuity equation, (8.11). Under the gauge transformation, $J^\mu A_\mu$ generates the four-divergence $\partial_\mu\left(J^\mu\chi\right)$. By Gauss's theorem, $\int_{V_4}\mathrm{d}^4x\partial_\mu\left(J^\mu\chi\right) = \int_{S_3}\left(J^\mu\chi\right)\mathrm{d}\Sigma_\mu$. For the equation of motion derived from Hamilton's principle to be gauge invariant, the end points of the varied worldlines *must* be held fixed on the hypersurface enclosing the volume of integration.

8.9.2 Lagrangian density for the electromagnetic field

Equations (8.35) or (8.37) can be written schematically as $S = S_M + S_{FM}$, where S_M, the "matter" action, describes the motion of a free particle and S_{FM}, the "field-matter" action, contains the interaction between $A^\mu(x)$ and a charge or current. Here we add a third term, S_F, an action for the field, so that for a system consisting of a particle, the electromagnetic field, and the field-particle interaction, $S = S_M + S_F + S_{FM}$. When S is varied with respect to the worldlines x^μ, *treating* A^μ *as given*, Hamilton's principle leads to the particle equation of motion, Eq. (8.43). As we now show, when S is varied with respect to the generalized "coordinates" of the field, A^μ, treating the worldlines $x^\mu(\tau)$ (and hence the four-current J^μ) as given, Hamilton's principle produces the "equation of motion" for the field, Eq. (8.27), the Maxwell equation with sources.

To build the action for the field, we require that S_F be: 1) Lorentz invariant; 2) gauge invariant (so that it leads to the correct field strengths $\boldsymbol{E}$ and $\boldsymbol{B}$); and 3) at most quadratic in the fields so that it leads to equations of motion linear in the fields (the electromagnetic fields obey superposition). These requirements suffice to determine S_F.[18] The first requirement can be met by constructing S_F out of the invariants for the field, I_1 and I_2, Eqs. (8.33) and (8.34). We rule out I_2, however, because it's a pseudoscalar. We rule out nonlinear combinations such as $(I_1)^2$ and $(I_2)^2$ as they do not lead to linear equations of motion. We also rule out $A_\mu A^\mu$ as it's not gauge invariant. The action integral meeting these requirements is

$$S_F = \alpha\int F_{\mu\nu}F^{\mu\nu}\mathrm{d}^3x\mathrm{d}t\ , \tag{8.44}$$

where α is a constant to be determined (we show that $\alpha = -1/(4\mu_0)$).

Adding Eq. (8.44) to Eq. (8.35), we have the action integral for the combined system of matter, matter-field interactions, and the field:

$$S = S_M + \int\mathrm{d}^3x\mathrm{d}t\left[A_\mu J^\mu + \alpha F^{\mu\nu}F_{\mu\nu}\right]\ . \tag{8.45}$$

[18] S_F is determined up to the divergence of a four-vector which could depend on the fields. The surface integral that follows from Gauss's theorem does not contribute to the equations of motion because the variations vanish at the surface of the volume of integration.

The integrand in Eq. (8.45) is a *Lagrangian density* (Appendix D),

$$\mathscr{L} \equiv A_\mu J^\mu + \alpha F^{\mu\nu} F_{\mu\nu} \,. \tag{8.46}$$

A Lagrangian density is the Lagrangian for *fields*, where the generalized "coordinates" are the field degrees of freedom $A_\mu(x)$. The Lagrange equation for fields is Eq. (D.65),

$$\partial_\mu \left(\frac{\partial \mathscr{L}}{\partial(\partial_\mu A_\nu)} \right) = \frac{\partial \mathscr{L}}{\partial A_\nu} \,. \qquad (\nu = 0,1,2,3) \tag{8.47}$$

To take the derivative in Eq. (8.47), we need to express $F^{\alpha\beta}F_{\alpha\beta}$ in terms of A_ν:

$$F^{\alpha\beta}F_{\alpha\beta} = g^{\alpha\lambda}g^{\beta\rho}F_{\lambda\rho}F_{\alpha\beta} = g^{\alpha\lambda}g^{\beta\rho}\left(\partial_\lambda A_\rho - \partial_\rho A_\lambda\right)\left(\partial_\alpha A_\beta - \partial_\beta A_\alpha\right) \,.$$

Now we can take the derivative,

$$\frac{\partial}{\partial(\partial_\mu A_\nu)} F^{\alpha\beta}F_{\alpha\beta} = g^{\alpha\lambda}g^{\beta\rho}\left[F_{\lambda\rho}\left(\delta^\mu{}_\alpha \delta^\nu{}_\beta - \delta^\mu{}_\beta \delta^\nu{}_\alpha\right) + F_{\alpha\beta}\left(\delta^\mu{}_\lambda \delta^\nu{}_\rho - \delta^\mu{}_\rho \delta^\nu{}_\lambda\right)\right] = -4F^{\nu\mu} \,.$$

The derivative on the right of Eq. (8.47) is, using Eq. (8.46), $\partial\mathscr{L}/\partial A_\nu = J^\nu$. Equations (8.46) and (8.47) imply that $-4\alpha\partial_\mu F^{\nu\mu} = J^\nu$. By setting $-4\alpha = 1/\mu_0$ we reproduce Eq. (8.27), $\partial_\mu F^{\nu\mu} = \mu_0 J^\nu$. For future reference,

$$\frac{\partial \mathscr{L}_\text{F}}{\partial(\partial_\mu A_\nu)} = \frac{1}{\mu_0} F^{\nu\mu} \,, \tag{8.48}$$

where $\mathscr{L}_\text{F} \equiv -F^{\alpha\beta}F_{\alpha\beta}/(4\mu_0)$. Note the order of the indices in Eq. (8.48).

8.9.3 More than one charged particle

What should the action integral be for a system of charged particles, $\{q_i\}_{i=1}^n$? We know that n particles are described by n worldlines $\{z_i^\mu(\tau_i)\}_{i=1}^n$, each with its own proper time τ_i (Section 7.6). A straightforward generalization of Eq. (8.45) would be

$$S = -\sum_i m_i c \int \text{d}\lambda_i \sqrt{-g_{\mu\nu}U_i^\mu U_i^\nu} + \int \text{d}^3x\text{d}t \left(\sum_i A_\mu J_i^\mu - \frac{1}{4\mu_0} F^{\mu\nu}F_{\mu\nu} \right) \,, \tag{8.49}$$

where $J_i^\mu(x) = q_i c \int d\lambda_i U_i^\mu(\lambda_i)\delta^4(x - z_i(\lambda_i))$, Eq. (8.12). The first term in Eq. (8.49) represents the inertial properties of the particles, Eq. (7.63), the second term captures the coupling of charges to the field, while the third is the action integral for the free electromagnetic field. Varying S in Eq. (8.49) with respect to the fields A_μ, *for fixed worldlines*, Hamilton's principle leads (as in Eq. (8.47)) to the Maxwell equations

$$\partial_\nu F^{\mu\nu} = \mu_0 \sum_i J_i^\mu \,. \tag{8.50}$$

Varying S in Eq. (8.49) with respect to the worldlines for prescribed fields A_μ, we obtain an equation of motion à la Eq. (8.43) for each particle

$$\frac{\text{d}}{\text{d}\tau} P_i^\mu = q_i F^{\mu\nu} U_{i\nu} \,. \qquad (i = 1, \cdots, n) \tag{8.51}$$

The total current generates the field, Eq. (8.50), and particles respond to the field, Eq. (8.51). All is good, *right*? Well, since you asked, these equations are *incomplete*. Accelerated charges radiate energy (and the momentum that accompanies it); there is thus a loss of mechanical energy-momentum (*radiation damping*) not accounted for in Eq. (8.51). (Energy-momentum of fields is treated in Chapter 9.) Maxwell's equations are time-reversal invariant (Exercise 8.5), yet radiation is an *irreversible* transfer of energy between particles and fields. Irreversibility is a tough one to get right. Closely related is that the particle-field coupling in Eq. (8.49) is between charges and the field created by *other* particles. There's no prescription for a coupling between a charge and the field *it creates*, the *self-field*, yet it's the self-field that produces radiation damping. We're going to have to leave this issue here; length restrictions preclude us from venturing further down this fascinating path.

8.10 GAUGE INVARIANCE AND CHARGE CONSERVATION

What happens if we drop the requirement of gauge invariance? As we show, new physical possibilities present themselves, namely that of a *photon mass*. Non-gauge-invariance also *forces us* to adopt the Lorenz gauge to preserve charge conservation.

Let's add the non-gauge-invariant term $A_\mu A^\mu$ to the Lagrangian density (the *Proca Lagrangian*):

$$\mathscr{L} = -\frac{1}{4\mu_0} F^{\mu\nu} F_{\mu\nu} - \frac{1}{2\mu_0 \Lambda^2} A_\mu A^\mu + A_\mu J^\mu , \tag{8.52}$$

where Λ is an unknown parameter having the dimension of length.[19] The length Λ can be associated with an equivalent mass through the formula for the Compton wavelength, $\Lambda \equiv h/(m_\gamma c)$, where m_γ is the implied photon mass. Combining Eqs. (8.52) and (8.47), we obtain the field equation

$$\partial_\mu F^{\nu\mu} = \mu_0 J^\nu - (m_\gamma c/h)^2 A^\nu . \tag{8.53}$$

From Eq. (8.53), we obtain modifications to the inhomogeneous Maxwell equations:

$$\boldsymbol{\nabla} \cdot \boldsymbol{E} = \frac{\rho}{\epsilon_0} - (m_\gamma c/h)^2 \phi \qquad \boldsymbol{\nabla} \times \boldsymbol{B} = \frac{1}{c^2}\frac{\partial \boldsymbol{E}}{\partial t} + \mu_0 \boldsymbol{J} - (m_\gamma c/h)^2 \boldsymbol{A} . \tag{8.54}$$

The field vectors $(\boldsymbol{E}, \boldsymbol{B})$ therefore couple to the *potentials* $(\phi, \boldsymbol{A})$ as well as to the sources $(\rho, \boldsymbol{J})$; in this theory the potentials have an independent existence and are not merely artifacts that help us calculate the fields.[20]

By taking the divergence of the curl equation in Eq. (8.54),

$$\boldsymbol{\nabla} \cdot (\boldsymbol{\nabla} \times \boldsymbol{B}) = 0 = \mu_0 \boldsymbol{\nabla} \cdot \boldsymbol{J} + \frac{1}{c^2}\frac{\partial}{\partial t} \boldsymbol{\nabla} \cdot \boldsymbol{E} - (m_\gamma c/h)^2 \boldsymbol{\nabla} \cdot \boldsymbol{A} = \mu_0 \partial_\mu J^\mu - (m_\gamma c/h)^2 \partial_\mu A^\mu ,$$

where we've used Eq. (8.54) for $\boldsymbol{\nabla} \cdot \boldsymbol{E}$. *The Lorenz condition $\partial_\mu A^\mu = 0$ is a requirement for charge conservation*, $\partial_\mu J^\mu = 0$ (whereas for $m_\gamma = 0$ they are *separate* conditions). The Proca theory is not gauge invariant; it *requires* the Lorenz gauge to preserve charge conservation.

We can seek solutions to Eq. (8.54) by invoking the electromagnetic potentials, as in Eq. (8.2).[21] For the case of electrostatics, the Poisson equation becomes

$$\nabla^2 \phi - (m_\gamma c/h)^2 \phi = -\frac{\rho}{\epsilon_0} . \tag{8.55}$$

For a point charge Q, the solution of Eq. (8.55) is $\phi(r) = [Q/(4\pi\epsilon_0 r)] \exp(-m_\gamma c r/h)$, leading to a modification of the electrostatic field

$$\boldsymbol{E} = \frac{Q}{4\pi\epsilon_0}\left(\frac{1}{r^2} + \frac{m_\gamma c}{hr}\right) \exp(-m_\gamma c r/h) \hat{\boldsymbol{r}} . \tag{8.56}$$

---

[19] The term $A^\mu A_\mu/(\mu_0 \Lambda^2)$ must have the same dimension as $F^{\mu\nu} F_{\mu\nu}/\mu_0$. From Eq. (8.18), $F^{\mu\nu} F_{\mu\nu}$ has the same dimension as $A^\mu A_\mu$ divided by length squared. The quantity Λ must therefore have the dimension of length.

[20] In quantum mechanics the vector potential couples to the phase of an electron's wave function, leading to measurable effects such as the Aharonov-Bohm effect, even when the magnetic field is zero. This illustrates the physical reality of the potentials, which are not just calculational artifacts.

[21] That is to say, the homogeneous Maxwell equations are unaffected by the new term in the Lagrangian.

More generally, combining Eqs. (8.18) and (8.53) and using the Lorenz condition, we obtain the wave equation for the four-potential (compare with Eq. (8.16))[22]

$$\left(\partial_\mu \partial^\mu - (m_\gamma c/h)^2\right) A^\nu = -\mu_0 J^\nu . \tag{8.57}$$

Equation (8.57) supports plane-wave solutions, $A^\mu = A_0^\mu \exp(ik_\alpha x^\alpha)$, with the dispersion relation $\omega^2 = c^2k^2 + \left(m_\gamma c^2/h\right)^2$.

Does the photon have mass? The weak bosons $W^\pm$ and Z are massive (mass of $W^\pm$ = 80.4 GeV$/c^2$; mass of Z = 91.2 GeV$/c^2$), and thus a photon mass is not altogether out of the question. The range of the interaction, however, approximately the Compton wavelength associated with the mass of the mediating boson, suggests a small photon mass. Attempts to measure m_γ must search either for discrepancies to a r^{-2} distance dependence of the electric field, Eq. (8.56), or for a frequency-dependent speed of light. The group velocity $v_g = \partial\omega/\partial k = c(1 - \left(m_\gamma c^2/h\omega\right)^2)^{1/2}$, and thus for $h\omega \gg m_\gamma c^2$, $v_g \approx c\left(1 - \frac{1}{2}\left(m_\gamma c^2/h\omega\right)^2\right)$. For two wave packets with frequencies $\omega_1 > \omega_2 \gg (m_\gamma c^2)/h$, the difference in propagation speed is

$$\frac{\Delta v}{c} \approx \frac{1}{2}\left(m_\gamma c^2/h\right)^2 \left(\frac{1}{\omega_2^2} - \frac{1}{\omega_1^2}\right) \approx \frac{m_\gamma^2 c^2}{8\pi^2 h^2}\left(\lambda_2^2 - \lambda_1^2\right) .$$

For waves of different frequencies that travel the same distance L there is a difference in arrival times $\Delta t \approx L(\Delta v)/c^2$ which provides an estimate for m_γ. A study in 1999 of the difference in arrival times from gamma-ray bursts from supernovae placed an upper limit to the photon mass at $m_\gamma < 2.4 \times 10^{-11}$ eV$/c^2$.[39]

Gauge invariance, charge conservation, and a possible mass of the photon are thus interrelated concepts. As discussed in Section 3.1.7, Einstein's second postulate for SR (the universality of the speed of light) is equivalent to the photon having zero mass and experimentally it has been found that the universal limiting speed predicted by SR is equal to c with high precision.

SUMMARY

This chapter has treated electrodynamics from the covariant perspective.

- The current four-vector density (four-current) $J^\alpha = (\rho c, \rho \boldsymbol{u})$. The electromagnetic potentials form a four-vector, the four-potential $A^\mu = (\phi/c, \boldsymbol{A})$. In covariant form charge conservation and the Lorenz condition are written $\partial_\mu J^\mu = 0$ and $\partial_\mu A^\mu = 0$. The four-potential satisfies an inhomogeneous wave equation $\partial_\mu \partial^\mu A^\alpha = -\mu_0 J^\alpha$ in the Lorenz gauge.
- A continuity equation $\partial_\mu J^\mu = 0$ implies the existence of a conserved quantity $Q = \int_\Sigma J^\mu d\Sigma_\mu$ that's independent of the SH Σ.
- The tensor $F^{\mu\nu}$ and its dual $(*F)^{\mu\nu}$ allow Maxwell's equation to be written in covariant form, $\partial_\beta F^{\alpha\beta} = \mu_0 J^\alpha$ and $\partial_\beta (*F)^{\alpha\beta} = 0$, where $F^{\mu\nu} = \partial^\mu A^\nu - \partial^\nu A^\mu$. The components of the electric (magnetic) field are related to the elements of the field tensor through $E^i = cF^{0i}$ $(B_i = \frac{1}{2}\varepsilon_{ijk}F^{jk})$.
- The transformation equations for $\boldsymbol{E}$ and $\boldsymbol{B}$ under the LT are obtained from the transformation of the field tensor, $F^{\mu'\nu'} = L^{\mu'}_{\alpha} L^{\nu'}_{\beta} F^{\alpha\beta}$.

[22]Equation (8.57) should be contrasted with the *Klein-Gordon equation* (KG), the relativistic quantum equation in which the Compton wavelength naturally appears. The Hamiltonian for free particles is $H = cP^0 = \sqrt{(pc)^2 + (mc^2)^2}$. Instead of trying to solve the Schrödinger equation in the form $H\psi = E\psi$, reinterpret the equation as $H^2\psi = E^2\psi$. Making the usual correspondences, $E \to i\hbar\partial/\partial t$ and $\boldsymbol{p} \to -i\hbar\boldsymbol{\nabla}$, the KG equation is $\left(\partial_\mu\partial^\mu - (mc/\hbar)^2\right)\psi = 0$. The KG equation applies for a scalar field, whereas the Proca equation is for a vector field, A^μ; it also involves the sources J^μ.

- The quantities $I_1 = \frac{1}{2}F^{\mu\nu}F_{\mu\nu} = |B|^2 - |E|^2/c^2$ and $I_2 = (c/4)(*F)^{\mu\nu}F_{\mu\nu} = \boldsymbol{E}\cdot\boldsymbol{B}$ are relativistic invariants of the electromagnetic field.
- The covariant Lagrangian of a charged particle (in prescribed electromagnetic potentials A_μ) is $L = -mc\sqrt{-g_{\mu\nu}U^\mu U^\nu} + qA_\mu U^\mu$. The covariant equation of motion for a charged particle is $\mathrm{d}P^\mu/\mathrm{d}\tau = qF^{\mu\nu}U_\nu$.
- The Lagrangian density for the electromagnetic field $\mathscr{L} = -F^{\mu\nu}F_{\mu\nu}/(4\mu_0) + A_\mu J^\mu$.

EXERCISES

8.1 Show that the four-current as given by Eq. (8.12) satisfies the continuity equation, $\partial_\mu J^\mu = 0$. Use the identity

$$\frac{\mathrm{d}}{\mathrm{d}\tau}\delta\left(x - z(\tau)\right) = \frac{\mathrm{d}z^\alpha}{\mathrm{d}\tau}\frac{\partial}{\partial z^\alpha}\delta(x-z) = -U^\alpha\frac{\partial}{\partial x^\alpha}\delta(x-z)\,.$$

8.2 Show that $F^{lm} = \varepsilon^{ilm}B_i$. Use the relation $B_i = \frac{1}{2}\varepsilon_{ijk}F^{jk}$ and Eq. (5.73).

8.3 Show that the spatial components of Eq. (8.21) reproduce Faraday's law.

8.4 In a material medium there are bound charges and free charges, bound currents and free currents. The total charge density $\rho = \rho_f + \rho_b$ and the total current density $\boldsymbol{J} = \boldsymbol{J}_f + \boldsymbol{J}_b$, where $\rho_b = -\boldsymbol{\nabla}\cdot\boldsymbol{P}$ and $\boldsymbol{J}_b = \partial\boldsymbol{P}/\partial t + \boldsymbol{\nabla}\times\boldsymbol{M}$, with $\boldsymbol{P}$ and $\boldsymbol{M}$ the response of the medium to the applied fields, $\boldsymbol{E}$ and $\boldsymbol{H}$.

a. Can the three-vector relations $\boldsymbol{D} \equiv \epsilon_0\boldsymbol{E} + \boldsymbol{P}$ and $\boldsymbol{H} \equiv \boldsymbol{B}/\mu_0 - \boldsymbol{M}$ be packaged inside tensor equations? There are three pairs of three-vectors ($\boldsymbol{D}$, $\boldsymbol{H}$), ($\boldsymbol{E}$, $\boldsymbol{B}$), and ($\boldsymbol{P}$, $\boldsymbol{M}$), each pair with six quantities. This suggests the use of antisymmetric second-rank four-dimensional tensors to encode relations between pairs of three-vectors. Show that

$$G^{\mu\nu} = \frac{1}{\mu_0}F^{\mu\nu} - M^{\mu\nu}\,,$$

where $G^{\mu\nu}$ and $F^{\mu\nu}$ are given by Eqs. (8.24) and (8.26), and

$$M^{\mu\nu} \equiv \begin{pmatrix} 0 & -cP^1 & -cP^2 & -cP^3 \\ cP^1 & 0 & M_3 & -M_2 \\ cP^2 & -M_3 & 0 & M_1 \\ cP^3 & M_2 & -M_1 & 0 \end{pmatrix}.$$

b. Define the bound current four-vector density $J_b^\mu = (\rho_b c, \boldsymbol{J}_b)$. Show that $\partial_\nu M^{\mu\nu} = J_b^\mu$.

c. Show that $\partial_\mu J_b^\mu = 0$. Hint: This follows trivially from the antisymmetry of $M^{\mu\nu}$.

8.5 Maxwell's equations are invariant under the LT, Section 8.5. You're going to show that they're also invariant under the parity and time reversal operations P and T (Section 4.3). In Section 5.9 we introduced the distinction between polar and axial vectors, with polar vectors such as the electric field changing sign under P (what we can denote as $P(\boldsymbol{E}) = -\boldsymbol{E}$), and axial vectors like the magnetic field invariant under P ($P(\boldsymbol{B}) = +\boldsymbol{B}$). We can use a similar notation for the signature under time reversal.

a. Show that the entries in the table are correct, then verify that Maxwell's equations in Eq. (8.1) are separately invariant under P and T.

	$\boldsymbol{E}$	$\boldsymbol{D}$	$\boldsymbol{B}$	$\boldsymbol{H}$	ρ	$\boldsymbol{J}$	$\boldsymbol{\nabla}$	∂_t
P	−	−	+	+	+	−	−	+
T	+	+	−	−	+	−	+	−
PT	−	−	−	−	+	+	−	−

b. Suppose magnetic monopoles were found to exist with $\boldsymbol{\nabla} \cdot \boldsymbol{H} = \rho_m$. How would the monopole density ρ_m behave under P and T?

c. How do $\boldsymbol{A}$ and ϕ behave under P and T? Complete the following table:

	$\boldsymbol{A}$	ϕ
P		
T		

d. How does the four-current J^μ transform under PT, an inversion of both space and time axes? Ditto for the four-potential A^μ.

e. How does the four-gradient ∂_μ transform under PT? Show that the wave-equation operator $\partial^\mu \partial_\mu$ is invariant under PT.

8.6 Starting from Eq. (8.32), show that $\boldsymbol{E}' \cdot \boldsymbol{B}' = \boldsymbol{E} \cdot \boldsymbol{B}$.
Note that $(\boldsymbol{A} \times \boldsymbol{B}) \cdot (\boldsymbol{C} \times \boldsymbol{D}) = (\boldsymbol{A} \cdot \boldsymbol{C})(\boldsymbol{B} \cdot \boldsymbol{D}) - (\boldsymbol{A} \cdot \boldsymbol{D})(\boldsymbol{B} \cdot \boldsymbol{C})$.

8.7 What are the dimensions of the four-potential? Hint: $L = qA_\mu U^\mu$.

8.8 The Minkowski force was defined in Section 7.2 as $K^\mu = dP^\mu/d\tau$, where it was shown that K^μ must have the form $K^\mu = (\gamma \boldsymbol{F} \cdot \boldsymbol{u}/c, \gamma \boldsymbol{F})$ where $\boldsymbol{F}$ is the three-force. Equation (8.43) gives us an explicit expression for the Minkowski force, $K^\mu = qF^{\mu\nu}U_\nu$. Show that $qU_\nu F^{\mu\nu}$ meets the requirements of the Minkowski force. You should find $F^{0\nu}U_\nu = \dfrac{\gamma}{c}\boldsymbol{E} \cdot \boldsymbol{v}$ and $F^{i\nu}U_\nu = \gamma(\boldsymbol{E} + \boldsymbol{v} \times \boldsymbol{B})^i$. Use the result of Exercise 8.2.

8.9 Show that the time component of Eq. (8.43) reduces to $mc^2\dot{\gamma} = q\boldsymbol{u} \cdot \boldsymbol{E}$.

8.10 Work out the time component of Eq. (8.41). Interpret this equation.

CHAPTER 9

Energy-momentum of fields

CONSERVATION laws for particles (energy, momentum, angular momentum) follow from invariance of the action integral to transformations of spacetime coordinates (Section 7.6). In this chapter we show how conservation laws for *fields* follow from the same type of reasoning. To do so, we establish *Noether's theorem*, one of the central tools of field theory.[1]

9.1 SYMMETRIES AND CONSERVATION LAWS

The action integral associated with a set of physical fields $\{\phi_\alpha(x)\}$ is an integral of the Lagrangian density over spacetime (Appendix D), $S = \int \mathscr{L}(\phi_\alpha, \partial_\mu \phi_\alpha)\mathrm{d}^4x$. We're going to examine the behavior of S under infinitesimal coordinate transformations, which we'll write in the form

$$x^\mu \to \overline{x}^\mu = x^\mu + \epsilon^i \lambda_i^\mu(x) , \tag{9.1}$$

where the ϵ^i are infinitesimal parameters and the quantities $\lambda_i^\mu(x)$ are known functions. The range of the index i is at this point unspecified; the ϵ^i need not be part of a four-vector. The form of Eq. (9.1) is sufficiently general that it allows us to consider within the same formalism infinitesimal translations, $\lambda_\nu^\mu(x) = \delta_\nu^\mu$, and infinitesimal LTs, where the functions $\lambda_i^\mu(x)$ have more structure (Section 6.2). Invariance of S under Eq. (9.1) is called a *continuous symmetry* because the ϵ^i can vary continuously, as opposed to *discrete symmetries* such as parity and time reversal (Section 4.3). We allow for the possibility of infinitesimal variations in the fields under Eq. (9.1),

$$\phi_\alpha(x^\mu) \to \overline{\phi}_\alpha(\overline{x}^\mu) = \phi_\alpha(x^\mu) + \epsilon^i \Omega_{i\alpha}(x) , \tag{9.2}$$

where the $\Omega_{i\alpha}(x)$ are known functions. Noether's theorem is that *for every continuous symmetry of S there exists a conserved quantity*. The reader uninterested in the proof should skip to Eq. (9.10).

Invariance of the action integral under Eq. (9.1) requires that

$$\int_{V'} \overline{\mathscr{L}}(\overline{x})\mathrm{d}^4\overline{x} = \int_V \mathscr{L}(x)\mathrm{d}^4x , \tag{9.3}$$

where V is a spacetime volume, V' is obtained from V under Eq. (9.1), and $\overline{\mathscr{L}}$ is the Lagrangian density expressed in the new coordinates. Rewrite Eq. (9.3) as follows:

$$0 = \int_{V'} \left(\overline{\mathscr{L}}(\overline{x}) - \mathscr{L}(\overline{x})\right) \mathrm{d}^4\overline{x} + \int_{V'} \mathscr{L}(\overline{x})\mathrm{d}^4\overline{x} - \int_V \mathscr{L}(x)\mathrm{d}^4x . \tag{9.4}$$

We've massaged Eq. (9.3) in this way so that the integrand of the first integral in Eq. (9.4) is first order in small quantities; consequently, we can let $V' \to V$ and consistently work to first order. At

[1] There is no shortage of proofs to Noether's theorem, discovered in 1918. We follow that given in Gross.[40]

that point we can substitute $\overline{x}$ for x as a change in dummy variables. Equation (9.4) then becomes to first order

$$0=\int_V\left(\overline{\mathscr{L}}(x)-\mathscr{L}(x)\right)\mathrm{d}^4x+\int_V\left(\mathscr{L}(x)+\partial_\mu\mathscr{L}(x)\delta x^\mu\right)J\mathrm{d}^4x-\int_V\mathscr{L}(x)\mathrm{d}^4x\,, \tag{9.5}$$

where in the second integral we have replaced $\mathscr{L}(\overline{x})$ with its first-order Taylor expansion around x, and where J is the Jacobian determinant of the transformation $\overline{x}\to x$. From Eq. (5.53) and Eq. (9.1), $J=\det\partial\overline{x}^\mu/\partial x^\nu=\det\left(\delta^\mu_\nu+\epsilon^i\partial_\nu\lambda^\mu_i(x)\right)$. Because we want only terms to first order in ϵ, we need keep only the product of the diagonal terms in the Jacobian determinant, which to first order is $J=1+\epsilon^i\partial_\mu\lambda^\mu_i$. Substituting this result for J into Eq. (9.5), we have, keeping terms to first order,

$$0=\int_V\left(\overline{\mathscr{L}}(x)-\mathscr{L}(x)\right)\mathrm{d}^4x+\epsilon^i\int_V\partial_\mu(\lambda^\mu_i\mathscr{L})\mathrm{d}^4x\,. \tag{9.6}$$

Equation (9.6) is equivalent to Eq. (9.3) to first order in small quantities.

The integrand of the first integral in Eq. (9.6) is the *functional variation* $\delta\mathscr{L}(x)$ at point x (see Eq. (D.6)),

$$\delta\mathscr{L}(x)\equiv\overline{\mathscr{L}}(x)-\mathscr{L}(x)=\frac{\partial\mathscr{L}}{\partial\phi_\alpha}\delta\phi_\alpha(x)+\frac{\partial\mathscr{L}}{\partial(\partial_\mu\phi_\alpha)}\delta(\partial_\mu\phi_\alpha(x))\,, \tag{9.7}$$

where the variations $\delta\phi_\alpha$ and $\delta(\partial_\mu\phi_\alpha)$ are similarly defined (Appendix D). We can eliminate $\partial\mathscr{L}/\partial\phi_\alpha$ from Eq. (9.7) using the Lagrange equation for fields, Eq. (D.65). Together with $\delta(\partial_\mu\phi_\alpha)=\partial_\mu(\delta\phi_\alpha)$, Eq. (9.7) is equivalent to

$$\delta\mathscr{L}(x)=\partial_\mu\left(\frac{\partial\mathscr{L}}{\partial(\partial_\mu\phi_\alpha)}\right)\delta\phi_\alpha(x)+\frac{\partial\mathscr{L}}{\partial(\partial_\mu\phi_\alpha)}\partial_\mu(\delta\phi_\alpha(x))=\partial_\mu\left[\frac{\partial\mathscr{L}}{\partial(\partial_\mu\phi_\alpha)}\delta\phi_\alpha(x)\right]\,. \tag{9.8}$$

The variation $\delta\phi_\alpha(x)$ can be written

$$\begin{aligned}\delta\phi_\alpha(x)&\equiv\overline{\phi}_\alpha(x)-\phi_\alpha(x)=\overline{\phi}_\alpha(\overline{x})-\phi_\alpha(x)-(\overline{\phi}_\alpha(\overline{x})-\overline{\phi}_\alpha(x))\\&=\epsilon^i\Omega_{i\alpha}(x)-\epsilon^i\lambda^\mu_i(x)\partial_\mu\overline{\phi}_\alpha=\epsilon^i\left[\Omega_{i\alpha}(x)-\lambda^\mu_i(x)\partial_\mu\phi_\alpha\right]\,,\end{aligned} \tag{9.9}$$

where we've used Eqs. (9.1) and (9.2), and in the last equality we've replaced $\overline{\phi}_\alpha\to\phi_\alpha$ because we're already at first order in ϵ. Combining Eqs. (9.9), (9.8), and (9.6),

$$0=\epsilon^i\int_V\mathrm{d}^4x\partial_\mu\left[\frac{\partial\mathscr{L}}{\partial(\partial_\mu\phi_\alpha)}\left(-\lambda^\nu_i(x)\partial_\nu\phi_\alpha+\Omega_{i\alpha}(x)\right)+\lambda^\mu_i(x)\mathscr{L}\right]\,. \tag{9.10}$$

The ϵ^i are independent parameters and V is arbitrary: The integrand of Eq. (9.10) must vanish for each value of the index i,

$$\partial_\mu J^\mu_i=0\,,\qquad\qquad(i=1,2,\cdots) \tag{9.11}$$

where

$$J^\mu_i\equiv\lambda^\mu_i(x)\mathscr{L}+\frac{\partial\mathscr{L}}{\partial(\partial_\mu\phi_\alpha)}\left[-\lambda^\nu_i(x)\partial_\nu\phi_\alpha+\Omega_{i\alpha}(x)\right]\,. \tag{9.12}$$

Equation (9.11) is a continuity equation for each i. Noether's theorem is that *for each continuous symmetry of* S *there is a conserved current,* J^μ_i prescribed by Eq. (9.12). *Conservation laws for fields are expressed as continuity equations among the components of a tensor field.*

9.2 SPACETIME HOMOGENEITY: ENERGY-MOMENTUM TENSOR

The first thing to try in an application of Noether's theorem is invariance under infinitesimal translations, $x^\mu \to \overline{x}^\mu = x^\mu + \epsilon^\mu$ (homogeneity of spacetime), implying from Eq. (9.1) that $\lambda^\mu_\nu(x) = \delta^\mu_\nu$. In addition, $\Omega_{i\alpha}(x) = 0$ because $\overline{\phi}_\alpha(\overline{x}) \equiv \phi_\alpha(x)$, for any tensor field under a shift in coordinates. We show that *invariance under spacetime translations implies conservation of field energy and momentum.*

From Eqs. (9.11) and (9.12) with[2] $\lambda^\mu_\nu(x) = \delta^\mu_\nu = g_\nu{}^\mu$ and $\Omega_{i\alpha} = 0$,

$$\frac{\partial}{\partial x^\mu}\left[g_\nu{}^\mu \mathscr{L} - (\partial_\nu \phi_\alpha)\frac{\partial \mathscr{L}}{\partial(\partial_\mu \phi_\alpha)}\right] = 0 \,. \qquad (\nu = 0,1,2,3) \tag{9.13}$$

The terms in square brackets comprise the *canonical energy-momentum tensor*,

$$T_\nu{}^\mu \equiv g_\nu{}^\mu \mathscr{L} - (\partial_\nu \phi_\alpha)\frac{\partial \mathscr{L}}{\partial(\partial_\mu \phi_\alpha)} \,. \tag{9.14}$$

It's convenient to raise the index on Eq. (9.14) and work with the contravariant tensor

$$T^{\mu\nu} = g^{\mu\nu}\mathscr{L} - (\partial^\mu \phi_\lambda)\frac{\partial \mathscr{L}}{\partial(\partial_\nu \phi_\lambda)} \,, \tag{9.15}$$

in terms of which Eq. (9.13) is a set of four continuity equations

$$\partial_\nu T^{\mu\nu} = 0 \,. \qquad (\mu = 0,1,2,3) \tag{9.16}$$

Continuity equations imply the existence of conserved quantities (Section 8.3)

$$Q^\mu \equiv \int_\Sigma T^{\mu 0}\mathrm{d}^3 y = \int T^{\mu\nu}\mathrm{d}\Sigma_\nu \qquad (\mu = 0,1,2,3) \tag{9.17}$$

contained within a given SH Σ. The four quantities Q^μ, the *Noether charges*—we'll look at their physical interpretations shortly—transform as the components of a four-vector. To establish that, use the quotient theorem. The strategy is to show that $B_\mu Q^\mu$ is a scalar for some four-vector, B_μ. Let B_μ be a vector with constant (non-spatially varying) components. Contract Eq. (9.16) with B_μ, $\partial_\nu B_\mu T^{\mu\nu} = \partial_\nu C^\nu = 0$, where $C^\nu \equiv B_\mu T^{\mu\nu}$. By Section 8.3, $\int_\Sigma C^0 \mathrm{d}^3 y$ is a scalar. From Eq. (9.17), $B_\mu Q^\mu = \int_\Sigma B_\mu T^{\mu 0}\mathrm{d}^3 y = \int_\Sigma C^0 \mathrm{d}^3 y$, a scalar.[3]

Interpretation of $T^{\mu\nu}$

Noether's theorem provides a recipe for constructing the tensor $T^{\mu\nu}$ for any system of fields and for identifying conserved quantities given only the Lagrangian density $\mathscr{L}$ and metric tensor $g^{\mu\nu}$. The quantity $T^{\mu\nu}$ plays a significant role in GR—it's "one half" of Einstein's field equation! *It's essential that we understand the physical interpretation of* $T^{\mu\nu}$.

From its definition T^{00} is the Hamiltonian density (compare Eq. (9.15) with Eq. (D.72)),

$$T^{00} = -\mathscr{L} + \partial_0 \phi^\lambda \frac{\partial \mathscr{L}}{\partial(\partial_0 \phi^\lambda)} = \mathscr{H} \,,$$

where we've used $\eta^{00} = -1$ and $\partial^0 = -\partial_0$. From Eq. (9.17),

$$Q^0 = \int_\Sigma \mathscr{H}\mathrm{d}^3 x = E \,.$$

The tensor element T^{00} is the energy density of the field.

---

[2]The mixed metric tensor is the Kronecker delta, Eq. (5.10). We write $\lambda^\mu_\nu = g_\nu{}^\mu$ with the contravariant index set off for no reason other than cosmetic: In this way it's the *rows* of $T^{\nu\mu}$ that turn out to satisfy continuity equations, $\partial_\mu T^{\nu\mu} = 0$ (which makes comprehending the content of the energy-momentum tensor easier in my opinion). If we had taken $\lambda^\mu_\nu = g^\mu{}_\nu$ (covariant index set off), the columns would have satisfied continuity equations, $\partial_\mu T^{\mu\nu} = 0$.

[3]The essentials of this argument can be traced to Weyl.[41, p272]

What is the interpretation of the other conserved quantities, Q^i? Invariance of the action under spatial translations in the ith-direction implies a conserved quantity we will call the i-component of the field momentum. Because the time component (of the four-momentum) $cP^0 = E = Q^0$, we identify a four-momentum for the field through the relation $P^\mu = Q^\mu/c$. Under this interpretation, which is standard, we have for the spatial components of the field momentum,

$$P^i = \frac{1}{c}Q^i = \frac{1}{c}\int_\Sigma T^{i0}\mathrm{d}^3x \, .$$

T^{i0}/c *is the momentum density of the field in the* i^{th} *direction*, denoted g^i. Thus, $T^{i0} = g^i c$.

If we integrate Eq. (9.16) over a *finite* three-volume V, we have the balance equation

$$\frac{\mathrm{d}}{\mathrm{d}t}\int_V T^{\mu 0}\mathrm{d}^3x = -c\int_V \partial_i T^{\mu i}\mathrm{d}^3x = -c\int_S T^{\mu i}\mathrm{d}\Sigma_i \, . \tag{9.18}$$

For $\mu = 0$ in Eq. (9.18), we see that the rate of change of field energy in V is balanced by the energy transported across the boundary of V. Thus, cT^{0i} *is the energy current density in the* i^{th} *direction*, denoted S^i; $T^{0i} = S^i/c$. The spatial components of Eq. (9.18) are

$$\frac{\mathrm{d}}{\mathrm{d}t}\int_V \frac{1}{c}T^{i0}\mathrm{d}^3x = -\int_S T^{ij}\mathrm{d}\Sigma_j$$

or that the rate of change of the i-component of momentum in V is balanced by the transport of i-momentum through the boundary of V. The quantity T^{ij} thus represents the rate of change of i-momentum per unit area in the jth direction,

$$T^{ij} = \frac{\Delta P^i/\Delta t}{\Delta\Sigma_j} = \frac{F^i}{\Delta\Sigma_j} \equiv \sigma^{ij} \, ,$$

where F^i is a force (time rate of change of momentum is a force). The spatial components T^{ij} thus tell us the force exerted by the field in the ith-direction across a surface oriented in the jth-direction, what is termed a *stress*. The space-space part of $T^{\mu\nu}$ is the *stress tensor*, σ^{ij}.

There are thus four parts to the energy-momentum tensor,

$$T^{00} = \text{energy density} \equiv w \qquad cT^{0i} = (\text{energy current density})^i \equiv S^i$$

$$\frac{1}{c}T^{i0} = (\text{momentum density})^i \equiv g^i \qquad T^{ij} = \frac{(\text{force})^i}{(\text{area})_j} \equiv \sigma^{ij} \, ,$$

which we can display as a matrix

$$[T^{\mu\nu}] = \left(\begin{array}{c|ccc} w & S^1/c & S^2/c & S^3/c \\ \hline g^1c & \sigma^{11} & \sigma^{12} & \sigma^{13} \\ g^2c & \sigma^{21} & \sigma^{22} & \sigma^{23} \\ g^3c & \sigma^{31} & \sigma^{32} & \sigma^{33} \end{array}\right) . \tag{9.19}$$

From here on, we take for granted the interpretation of the energy-momentum tensor:

$$T = \left(\begin{array}{c|c} \text{energy density} & \text{energy current density} \\ \hline \text{momentum density} & \text{stress tensor} \end{array}\right) .$$

9.3 SPACETIME ISOTROPY: ANGULAR MOMENTUM TENSOR

Next consider what Noether's theorem has to say about invariance under LTs, which implies isotropy of spacetime. For *particles*, invariance of the action under infinitesimal LTs implies conservation of angular momentum along worldlines (Section 7.6). Do fields have angular momentum? (Hint: Photons have spin.) In this section, we examine angular momentum as a consequence of Noether's theorem.

LTs depend on six parameters (Chapter 6, three for rotations, three for boosts). The range of the index i in Eq. (9.1) for infinitesimal LTs is therefore $1 \leq i \leq 6$. Instead of arranging the six parameters as a vector, however, they can be "packaged" in an antisymmetric 4×4 matrix as in Eq. (6.32). To use Noether's theorem, we need to know how coordinates and fields transform under $L^\mu_\nu = \delta^\mu_\nu + \lambda^\mu_\nu$, where $[\lambda^\mu_\nu]$ contains the infinitesimal parameters of the transformation, Eq. (6.31). For coordinates, compare Eq. (9.1) with the variation in coordinates produced by the LT. We identify $\epsilon^i \lambda^\mu_i(x) = \lambda^\mu_\nu x^\nu$ (apologies for notational abuse). For fields, we have to do a little more work.

How do four-vector fields (such as $A^\mu(x)$) transform under the LT? To answer, form the scalar product $\psi(x) \equiv n_\mu \phi^\mu(x)$ where n_μ are the elements of a fixed but otherwise arbitrary four-vector. From the invariance of scalar fields (Section 5.1),

$$\overline{\psi}(\overline{x}) = n_{\nu'} \phi^{\nu'}(\overline{x}) = n_\mu \phi^\mu(x) = \psi(x) \,, \tag{9.20}$$

where $\overline{x}^\mu = L^\mu_\nu x^\nu$. From Eq. (9.20), $n_\mu \phi^\mu(x) = L^{\nu'}_\mu n_{\nu'} \phi^\mu(x) = n_{\nu'} \phi^{\nu'}(\overline{x})$. *Four-vector fields transform as a four-vector when the position coordinates are also transformed,*

$$\phi^{\nu'}(\overline{x}) = L^{\nu'}_\mu \phi^\mu(x) \,. \tag{9.21}$$

Comparing Eq. (9.2) with Eq. (9.21) when we use $L^\mu_\nu = \delta^\mu_\nu + \lambda^\mu_\nu$, we identify $\epsilon^i \Omega_{i\alpha}(x) = \lambda^\nu_\alpha \phi_\nu(x)$.

Let $J^\mu = \epsilon^i J^\mu_i$ be the net current in Eq. (9.12) summed over the six independent parameters of the LT. With $\epsilon^i \lambda^\mu_i(x) \to \lambda^\mu_\nu x^\nu$ and $\epsilon^i \Omega_{i\alpha}(x) \to \lambda^\nu_\alpha \phi_\nu(x)$, we have using Eq. (9.12),

$$J^\mu = \lambda^\mu_\nu x^\nu \mathscr{L} + \frac{\partial \mathscr{L}}{\partial(\partial_\mu \phi_\alpha)} \left[-\lambda^\rho_\nu x^\nu \partial_\rho \phi_\alpha + \lambda^\rho_\alpha \phi_\rho(x) \right] \tag{9.22}$$

Combine Eq. (9.14) (canonical energy-momentum tensor $T_\rho{}^\mu$) with Eq. (9.22) to obtain

$$J^\mu = \lambda^\rho_\nu x^\nu T_\rho{}^\mu + \lambda^\rho_\alpha \phi_\rho \frac{\partial \mathscr{L}}{\partial(\partial_\mu \phi_\alpha)} \,. \tag{9.23}$$

Raise indices here appropriately, with the result

$$J^\mu = \eta_{\tau\nu} \lambda^\nu_\rho \left[x^\rho T^{\tau\mu} + \phi^\tau \frac{\partial \mathscr{L}}{\partial(\partial_\mu \phi_\rho)} \right] \equiv \omega_{\tau\rho} \left[x^\rho T^{\tau\mu} + \phi^\tau \frac{\partial \mathscr{L}}{\partial(\partial_\mu \phi_\rho)} \right] , \tag{9.24}$$

where $\omega_{\tau\rho} = \eta_{\tau\nu} \lambda^\nu_\rho$, Eq. (6.32). Because $\omega_{\tau\rho}$ is antisymmetric, the contraction in Eq. (9.24) picks out the antisymmetric part with respect to (τ, ρ) (Section 5.10.1):

$$J^\mu = \tfrac{1}{2} \omega_{\tau\rho} J^{\rho\tau\mu} \,,$$

where $J^{\rho\tau\mu}$ is the *angular momentum current density tensor*

$$J^{\rho\tau\mu} \equiv x^\rho T^{\tau\mu} - x^\tau T^{\rho\mu} + \phi^\tau \frac{\partial \mathscr{L}}{\partial(\partial_\mu \phi_\rho)} - \phi^\rho \frac{\partial \mathscr{L}}{\partial(\partial_\mu \phi_\tau)} \,. \tag{9.25}$$

Noether's theorem therefore implies six conservation laws (the parameters $\omega_{\tau\rho}$ are indepdent),

$$\partial_\mu J^{\rho\tau\mu} = 0 \,. \qquad (\rho = 0, 1, 2, \tau > \rho) \tag{9.26}$$

By the reasoning of Section 8.3, Eq. (9.26) implies six conserved quantities,

$$J^{\beta\lambda} \equiv \int_\Sigma J^{\beta\lambda 0} \mathrm{d}^3 y = \int J^{\beta\lambda\mu} \mathrm{d}\Sigma_\mu \, . \tag{9.27}$$

The quantity $J^{\beta\lambda}$ is the (antisymmetric) angular momentum tensor for fields.

In seeking an interpretation of the angular momentum current, a natural idea is to take the first group of terms in Eq. (9.25) as representing what can be termed the orbital properties of the field—the spacetime flow of energy and momentum, with the second group representing something else, an intrinsic angular momentum, call it spin:

$$J^{\rho\tau\mu} = \underbrace{x^\rho T^{\tau\mu} - x^\tau T^{\rho\mu}}_{\text{orbital momentum density}} + \underbrace{\phi^\tau \frac{\partial \mathscr{L}}{\partial(\partial_\mu \phi_\rho)} - \phi^\rho \frac{\partial \mathscr{L}}{\partial(\partial_\mu \phi_\tau)}}_{\text{spin density}} \, .$$

Thus, write $J^{\rho\tau\mu}$ as the sum of two terms $J^{\rho\tau\mu} = L^{\rho\tau\mu} + S^{\rho\tau\mu}$, with

$$L^{\rho\tau\mu} \equiv x^\rho T^{\tau\mu} - x^\tau T^{\rho\mu} \tag{9.28}$$

$$S^{\rho\tau\mu} \equiv \phi^\tau \frac{\partial \mathscr{L}}{\partial(\partial_\mu \phi_\rho)} - \phi^\rho \frac{\partial \mathscr{L}}{\partial(\partial_\mu \phi_\tau)} \, . \tag{9.29}$$

Splitting the angular momentum into orbital and spin parts is an appealing idea, but there's no guarantee it's anything other than giving names to the terms: neither $L^{\rho\tau\mu}$ nor $S^{\rho\tau\mu}$ need be separately conserved—it's the total angular momentum $J^{\rho\tau\mu}$ that's conserved, Eq. (9.26). For this idea to be well defined physically, an *independently existing* orbital angular momentum would have to transform like a tensor under the LT, which would be defined by an integral over a SH, $L^{\beta\lambda} \equiv \int L^{\beta\lambda\mu} \mathrm{d}\Sigma_\mu$ (likewise with the spin, $S^{\beta\lambda} \equiv \int S^{\beta\lambda\mu} \mathrm{d}\Sigma_\mu$). Independence of SH, however, would imply a conservation law, $\partial_\mu L^{\beta\lambda\mu} = 0$ (similarly $\partial_\mu S^{\beta\lambda\mu} = 0$). From Eq. (9.26),

$$0 = \partial_\mu J^{\rho\tau\mu} = \partial_\mu L^{\rho\tau\mu} + \partial_\mu S^{\rho\tau\mu} = T^{\tau\rho} - T^{\rho\tau} + \partial_\mu S^{\rho\tau\mu} \, ,$$

where we've differentiated $L^{\rho\tau\mu}$ using Eq. (9.28) and we've used the continuity equations for $T^{\tau\mu}$, Eq. (9.16), implying

$$\partial_\mu S^{\rho\tau\mu} = T^{\rho\tau} - T^{\tau\rho} \equiv 2T^{[\rho\tau]} \, . \tag{9.30}$$

We arrive at an important conclusion: *We can meaningfully speak of a field spin density when the canonical energy-momentum tensor is symmetric; otherwise, the separation into orbital and spin angular momentum cannot be made relativistically invariant.*

Scalar fields

Scalar fields transform under the LT such that $\Omega_{i\alpha}(x) = 0$ in Noether's theorem. Tracing through the steps, for a scalar field $J^{\rho\tau\mu} = L^{\rho\tau\mu}$: *there is no spin part to the angular momentum of a scalar field.* Because the total (conserved) angular momentum is "all orbital" in this case, *the canonical energy-momentum tensor of a scalar field is necessarily symmetric*. Only multi-component fields (like $A^\mu(x)$) can have a nonzero spin density.

Example. The elastic displacement field $\phi(x,t)$ of an isotropic medium is a scalar field. Its Lagrangian density is (see Eq. (D.66)) $\mathscr{L} = -\frac{1}{2}(\partial^\mu \phi)(\partial_\mu \phi)$. Using Eq. (9.15), we find the energy-momentum tensor for this field:

$$T^{\mu\nu} = g^{\mu\nu}\mathscr{L} - (\partial^\mu \phi) \frac{\partial \mathscr{L}}{\partial(\partial_\nu \phi)} = g^{\mu\nu}\mathscr{L} + (\partial^\mu \phi)(\partial^\nu \phi) \, ,$$

which is symmetric. As an exercise, you should be able to show that $T^{00} = \frac{1}{2}\left[(\partial_0\phi)^2 + \sum_i (\partial_i \phi)^2\right]$.

9.4 SYMMETRIC ENERGY-MOMENTUM TENSOR

Is it possible to represent the angular momentum in "$\boldsymbol{r} \times \boldsymbol{p}$" form? We defer for now the *why* behind this question; let's see if it can be done. At issue is whether an equivalent energy-momentum tensor $\theta^{\mu\nu}$ can be found such that $J^{\rho\tau\mu}$ in Eq. (9.25) can be written in the form

$$J_{\boldsymbol{\theta}}^{\rho\tau\mu} \equiv x^\rho \theta^{\tau\mu} - x^\tau \theta^{\rho\mu} . \tag{9.31}$$

What would we require of $\theta^{\mu\nu}$? First, do no harm: The conservation law Eq. (9.26) should be satisfied. From Eq. (9.31),

$$\partial_\mu J_{\boldsymbol{\theta}}^{\rho\tau\mu} = \theta^{\tau\rho} - \theta^{\rho\tau} + x^\rho \partial_\mu \theta^{\tau\mu} - x^\tau \partial_\mu \theta^{\rho\mu} . \tag{9.32}$$

For the right side of Eq. (9.32) to vanish, $\theta^{\mu\nu}$ must 1) be symmetric and 2) have zero divergence, $\partial_\nu \theta^{\mu\nu} = 0$, i.e., $\theta^{\mu\nu}$ should satisfy the conservation law expected of an energy-momentum tensor, Eq. (9.16). Add to this list the requirement that conserved quantities be preserved, the Noether charges Q^μ, Eq. (9.17), and the angular momentum tensor $J^{\beta\lambda}$, Eq. (9.27). As we'll see, the symmetry requirement on $\theta^{\mu\nu}$ is the most difficult; after that, the rest fall into place.

The antisymmetric part of the canonical energy-momentum tensor $T^{[\mu\nu]}$ is equal to the divergence of the spin density, Eq. (9.30); $T^{[\mu\nu]}$ is the *source* of the field spin density. Perhaps we can use the spin current to build a symmetric tensor $\theta^{\mu\nu}$. Let

$$T^{\mu\nu} \to \theta^{\mu\nu} \equiv T^{\mu\nu} + \Delta^{\mu\nu} = T^{(\mu\nu)} + T^{[\mu\nu]} + \Delta^{(\mu\nu)} + \Delta^{[\mu\nu]} , \tag{9.33}$$

where $\Delta^{\mu\nu}$ is to be determined such that $\theta^{\mu\nu}$ is symmetric. Choose $\Delta^{[\mu\nu]}$ so that it cancels $T^{[\mu\nu]}$, where we use Eq. (9.30):

$$\Delta^{[\mu\nu]} = -T^{[\mu\nu]} = -\tfrac{1}{2}\partial_\lambda S^{\mu\nu\lambda} .$$

We are done if we can find a suitable expression for the symmetric part, $\Delta^{(\mu\nu)}$. Let

$$\Delta^{(\mu\nu)} = \tfrac{1}{2}\partial_\lambda \left(S^{\mu\lambda\nu} + S^{\nu\lambda\mu} \right) .$$

Thus, $\Delta^{\mu\nu} = \frac{1}{2}\partial_\lambda \left(S^{\mu\lambda\nu} + S^{\nu\lambda\mu} - S^{\mu\nu\lambda} \right)$ results in a symmetric tensor $\theta^{\mu\nu}$. But wait, there's more. Consider the terms

$$\psi^{\mu\lambda\nu} \equiv \tfrac{1}{2}(S^{\mu\lambda\nu} + S^{\nu\lambda\mu} - S^{\mu\nu\lambda}) , \tag{9.34}$$

which are antisymmetric in the second and third indices: $\psi^{\mu\lambda\nu} = -\psi^{\mu\nu\lambda}$, as follows from the antisymmetry of $S^{\lambda\nu\mu}$, Eq. (9.29). The antisymmetry of $\psi^{\mu\lambda\nu}$ implies that

$$\partial_\nu \theta^{\mu\nu} = \partial_\nu T^{\mu\nu} + \partial_\nu \partial_\lambda \psi^{\mu\lambda\nu} = \partial_\nu T^{\mu\nu}$$

because $\partial_\nu \partial_\lambda$ is symmetric. Thus, $\partial_\nu \theta^{\mu\nu} = 0$ if $\partial_\nu T^{\mu\nu} = 0$. The modification of the canonical energy-momentum tensor in Eq. (9.33) is therefore equivalent to

$$T^{\mu\nu} \to \theta^{\mu\nu} = T^{\mu\nu} + \partial_\lambda \psi^{\mu\lambda\nu} . \tag{9.35}$$

We can add to $T^{\mu\nu}$ the divergence of a third-rank tensor antisymmetric in its latter two indices and achieve the *symmetric energy-momentum tensor*, $\theta^{\mu\nu}$.

What we require of $T^{\mu\nu}$ is that it have zero divergence (Noether's theorem). Beyond that, we're free to modify it so that it's symmetric. The Noether charges Q^μ are invariant under Eq. (9.35). From Eq. (9.17),

$$\begin{aligned}\overline{Q}^\mu \equiv \int_V \theta^{\mu 0} \mathrm{d}^3x = \int_V T^{\mu 0}\mathrm{d}^3 x + \int_V \partial_\lambda \psi^{\mu\lambda 0}\mathrm{d}^3 x = Q^\mu + \int_V \partial_i \psi^{\mu i 0} \mathrm{d}^3 x \\ = Q^\mu + \int_S \psi^{\mu i 0} \mathrm{d}S_i , \end{aligned} \tag{9.36}$$

where we've used $\psi^{\mu 00} = 0$ (antisymmetry). Now let V encompass all of space (implicit in the definition of Q^μ), and assume that the surface terms vanish in Eq. (9.36) because the fields fall off sufficiently rapidly as $|\boldsymbol{x}| \to \infty$; $\overline{Q}^\mu = Q^\mu$.

The same conclusion holds for the angular momentum tensor, but it takes more doing. It can be shown that (Exercise 9.10)

$$J_{\boldsymbol{\theta}}^{\alpha\beta\mu} = J^{\alpha\beta\mu} + \partial_\lambda \left(x^\alpha \psi^{\beta\lambda\mu} - x^\beta \psi^{\alpha\lambda\mu} \right) . \tag{9.37}$$

Thus, under $T^{\mu\nu} \to \theta^{\mu\nu} = T^{\mu\nu} + \partial_\lambda \psi^{\mu\lambda\nu}$, where $\psi^{\mu\lambda\nu}$ is an antisymmetric tensor, the angular momentum current transforms as $J^{\alpha\beta\mu} \to J_{\boldsymbol{\theta}}^{\alpha\beta\mu} = J^{\alpha\beta\mu} + \partial_\lambda G^{\alpha\beta\lambda\mu}$ where $G^{\alpha\beta\lambda\mu} \equiv x^\alpha \psi^{\beta\lambda\mu} - x^\beta \psi^{\alpha\lambda\mu}$ is antisymmetric in (α, β) and (λ, μ). Using Eq. (9.37),

$$\overline{J}^{\alpha\beta} \equiv \int J_{\boldsymbol{\theta}}^{\alpha\beta 0} \mathrm{d}^3x = \int J^{\alpha\beta 0} \mathrm{d}^3x + \int \mathrm{d}^3x \partial_\lambda G^{\alpha\beta\lambda 0} = J^{\alpha\beta} + \int \mathrm{d}^3x \partial_i G^{\alpha\beta i0} = J^{\alpha\beta} + \int G^{\alpha\beta i0} \mathrm{d}S_i \, .$$

As usual, argue that for physical fields the surface terms vanish; $\overline{J}^{\alpha\beta} = J^{\alpha\beta}$.

Thus, it's always possible to find a symmetric energy-momentum tensor $\theta^{\mu\nu}$ by removing the antisymmetric part of the canonical energy-momentum tensor $T^{[\mu\nu]}$, which is the source of the spin angular momentum of the field. The symmetric tensor $\theta^{\mu\nu}$ describes the same total energy-momentum as does $T^{\mu\nu}$, implying that *the total energy-momentum due to field spin is zero.* Field spin only modifies the *distribution* of energy-momentum. The *location* of energy-momentum is important in GR, because energy-momentum is the source of the gravitational field, and in GR the field couples to $\theta^{\mu\nu}$ rather than $T^{\mu\nu}$.

9.5 THE ELECTROMAGNETIC FIELD

9.5.1 Symmetric energy-momentum tensor for the electromagnetic field

What we've developed applies to *any* system of fields; all we require is a Lagrangian density function. Here we specialize to the electromagnetic field. Combining $\mathscr{L} = -F^{\alpha\beta}F_{\alpha\beta}/(4\mu_0)$ (Section 8.9) with Eq. (9.15), we have the canonical energy-momentum tensor: (using Eq. (8.48))

$$\mu_0 T^{\mu\nu} = -\frac{1}{4} g^{\mu\nu} F_{\alpha\beta} F^{\alpha\beta} - (\partial^\mu A_\lambda) F^{\lambda\nu} \, . \tag{9.38}$$

It's straightforward to show from Eq. (9.38) that

$$T^{00} = \frac{1}{2\mu_0} |B|^2 + \frac{\epsilon_0}{2} |E|^2 + \epsilon_0 \left[\boldsymbol{\nabla} \cdot (\phi \boldsymbol{E}) - \phi \boldsymbol{\nabla} \cdot \boldsymbol{E} \right] . \tag{9.39}$$

The components T^{i0} and T^{0i} are more difficult to calculate. It can be shown that

$$T^{i0} = g^i c + \frac{1}{c\mu_0} \left[\boldsymbol{\nabla} \cdot (\boldsymbol{E} A^i) - A^i \boldsymbol{\nabla} \cdot \boldsymbol{E} \right] , \tag{9.40}$$

where $g^i = \epsilon_0 (\boldsymbol{E} \times \boldsymbol{B})^i$, and

$$T^{0i} = \frac{1}{c} S^i + \frac{1}{c\mu_0} (\boldsymbol{\nabla}\phi \times \boldsymbol{B})^i - \frac{\epsilon_0}{c} \frac{\partial \phi}{\partial t} E^i ,$$

with $S^i = (\boldsymbol{E} \times \boldsymbol{B})^i / \mu_0$. That suffices; we won't calculate T^{ij}.

The canonical energy-momentum tensor is thus not symmetric, $T^{0i} \neq T^{i0}$. Moreover, the tensor elements are not what we expect: T^{00} is not the energy density, $T^{i0} \neq g^i c$, and $T^{0i} \neq S^i/c$. Despite these shortcomings, the conserved quantities Q^μ *are* obtained correctly from $T^{\mu 0}$. From Eqs. (9.39) and (9.40), we have, setting $\boldsymbol{\nabla} \cdot \boldsymbol{E} = 0$ for the free field,

$$\begin{aligned} Q^0 &= \int_{\text{all space}} T^{00} \mathrm{d}^3 x = \int_{\text{all space}} \left(\frac{1}{2\mu_0} |B|^2 + \frac{\epsilon_0}{2} |E|^2 \right) \mathrm{d}^3 x \equiv U_{\text{field}} \\ Q^i &= \int_{\text{all space}} T^{i0} \mathrm{d}^3 x = c \int_{\text{all space}} g^i \mathrm{d}^3 x = c P^i_{\text{field}} , \end{aligned}$$

where the additional terms in Eqs. (9.39) and (9.40) convert to surface integrals and vanish at infinity.

We can put $T^{\mu\nu}$ into symmetric form by adding a suitable term, $T^{\mu\nu} \rightarrow \theta^{\mu\nu} = T^{\mu\nu} + \partial_\lambda \psi^{\mu\nu\lambda}$. How to choose ψ? Guess. We can construct ψ out of the field tensor, $F^{\lambda\nu}$, which is antisymmetric. Let's try $\psi^{\mu\lambda\nu} = f A^\mu F^{\lambda\nu}$, where f is an unknown scalar. From Eq. (9.35),

$$\theta^{\mu\nu} = T^{\mu\nu} + f \partial_\lambda \left(A^\mu F^{\lambda\nu} \right) = T^{\mu\nu} + f \left(\partial_\lambda A^\mu \right) F^{\lambda\nu} + f A^\mu \partial_\lambda F^{\lambda\nu} . \tag{9.41}$$

Combining Eqs. (9.38) and (9.41),

$$\theta^{\mu\nu} = g^{\mu\nu} \mathscr{L} - \frac{1}{\mu_0} \partial^\mu A_\lambda F^{\lambda\nu} + f \left(\partial_\lambda A^\mu \right) F^{\lambda\nu} + f A^\mu \partial_\lambda F^{\lambda\nu} .$$

Using $A_\lambda = g_{\lambda\alpha} A^\alpha$ and $\partial_\lambda = g_{\lambda\alpha} \partial^\alpha$,

$$\theta^{\mu\nu} = g^{\mu\nu} \mathscr{L} - \frac{1}{\mu_0} g_{\lambda\alpha} F^{\lambda\nu} \left(\partial^\mu A^\alpha - f \mu_0 \partial^\alpha A^\mu \right) + f A^\mu \partial_\lambda F^{\lambda\nu} .$$

By choosing $f = 1/\mu_0$,

$$\theta^{\mu\nu} = g^{\mu\nu} \mathscr{L} - \frac{1}{\mu_0} g_{\alpha\lambda} F^{\mu\alpha} F^{\lambda\nu} + \frac{1}{\mu_0} A^\mu \partial_\lambda F^{\lambda\nu} , \tag{9.42}$$

where we've used Eq. (8.18). Combining Eq. (9.42) with $\mathscr{L} = -F^{\alpha\beta} F_{\alpha\beta}/(4\mu_0)$, and using $\partial_\nu F^{\mu\nu} = 0$ for free fields, we have the symmetric energy-momentum tensor

$$\theta^{\mu\nu} = -\frac{1}{\mu_0} \left(g_{\alpha\lambda} F^{\mu\alpha} F^{\lambda\nu} + \frac{1}{4} g^{\mu\nu} F_{\alpha\beta} F^{\alpha\beta} \right) . \tag{9.43}$$

Using Eq. (9.43) we find

$$\begin{aligned} \theta^{00} &= \frac{1}{2\mu_0} |B|^2 + \frac{\epsilon_0}{2} |E|^2 = u_{EM} \qquad \theta^{i0} = \theta^{0i} = \frac{1}{c\mu_0} \left(\boldsymbol{E} \times \boldsymbol{B} \right)^i = \frac{S^i}{c} = g^i c \\ \theta^{ij} &= -\epsilon_0 \left(E^i E^j - \tfrac{1}{2} \delta^{ij} |E|^2 \right) - \frac{1}{\mu_0} \left(B^i B^j - \tfrac{1}{2} \delta^{ij} |B|^2 \right) \equiv -\sigma^{ij} . \end{aligned} \tag{9.44}$$

The elements of $\theta^{\mu\nu}$ agree with the interpretation of the energy-momentum tensor (Section 9.2) by reproducing the known energy and momentum density of the electromagnetic field. The quantity σ^{ij} is the *Maxwell stress tensor*. We'll show (next section) that for *free fields*, $\partial_\nu \theta^{\mu\nu} = 0$, i.e., electromagnetic energy-momentum is conserved when there are no charges to steal it. Because of the symmetry of $\theta^{\mu\nu}$, $\boldsymbol{g} = \boldsymbol{S}/c^2$, i.e., the electromagnetic momentum is parallel to the Poynting vector—wherever there's a flow of energy, there's a flow of momentum.

9.5.2 Field-particle interactions

Electromagnetic fields, represented by the field tensor $F^{\mu\nu}$, are generated by the four-current J^μ through the Maxwell equation $\partial_\nu F^{\mu\nu} = \mu_0 J^\mu$, Eq. (8.27). Particles respond to the electromagnetic field through the equation of motion $\mathrm{d}P^\mu/\mathrm{d}\tau = qF^{\mu\nu}U_\nu$, Eq. (8.43). The paradigm is that *fixed particle currents* generate fields ($F^{\mu\nu}$) and *fixed fields* determine the motion of charged particles, P^μ. There must be a more dynamical linkage, however, between particles and fields. The intermediary is the energy-momentum tensor $\theta^{\mu\nu}$. We show that $\theta^{\mu\nu}$ responds to fields and currents through the equation of motion $\partial_\nu\theta^{\mu\nu} = -f^\mu$, where $f^\mu \equiv F^{\mu\alpha}J_\alpha$ is the *Lorentz force-density four-vector* (Eq. (9.48)).[4]

Let's take the divergence of $\theta^{\mu\nu}$ from Eq. (9.43):

$$\begin{aligned}-\mu_0\partial_\nu\theta^{\mu\nu} =&\partial_\nu\left(g_{\alpha\lambda}F^{\mu\alpha}F^{\lambda\nu} + \frac{1}{4}g^{\mu\nu}F_{\alpha\beta}F^{\alpha\beta}\right)\\ =&g_{\alpha\lambda}F^{\mu\alpha}\mu_0J^\lambda + g_{\alpha\lambda}F^{\lambda\nu}\partial_\nu F^{\mu\alpha} + \frac{1}{4}\partial^\mu\left(F_{\alpha\beta}F^{\alpha\beta}\right) ,\end{aligned} \tag{9.45}$$

where we've used Eq. (8.27), $\partial_\nu F^{\lambda\nu} = \mu_0J^\lambda$. To massage this equation further, we note that $F^{\alpha\beta}\partial^\mu F_{\alpha\beta} = F_{\alpha\beta}\partial^\mu F^{\alpha\beta}$. Thus, from Eq. (9.45),

$$-\mu_0\partial_\nu\theta^{\mu\nu} = \mu_0J_\alpha F^{\mu\alpha} + g_{\alpha\lambda}F^{\lambda\nu}\partial_\nu F^{\mu\alpha} + \frac{1}{2}F_{\alpha\beta}\partial^\mu F^{\alpha\beta} . \tag{9.46}$$

Next, we note that $g_{\alpha\lambda}F^{\lambda\nu}\partial_\nu F^{\mu\alpha} = F_{\alpha\beta}\partial^\beta F^{\mu\alpha}$. Equation (9.46) can then be written

$$-\mu_0\partial_\nu\theta^{\mu\nu} = \mu_0J_\alpha F^{\mu\alpha} + \frac{1}{2}\left[F_{\alpha\beta}\left(2\partial^\beta F^{\mu\alpha} + \partial^\mu F^{\alpha\beta}\right)\right] . \tag{9.47}$$

The terms in square brackets vanish, as we now show. The homogeneous Maxwell equations can be written $\partial^\mu F^{\alpha\beta} + \partial^\beta F^{\mu\alpha} + \partial^\alpha F^{\beta\mu} = 0$ ($\mu \neq \alpha \neq \beta$), Eq. (8.21). Thus,

$$\begin{aligned}2\partial^\beta F^{\mu\alpha} + \partial^\mu F^{\alpha\beta} =&\partial^\beta F^{\mu\alpha} + \partial^\beta F^{\mu\alpha} + \partial^\mu F^{\alpha\beta} = \partial^\beta F^{\mu\alpha} - \partial^\mu F^{\alpha\beta} - \partial^\alpha F^{\beta\mu} + \partial^\mu F^{\alpha\beta}\\ =&\partial^\beta F^{\mu\alpha} - \partial^\alpha F^{\beta\mu} .\end{aligned}$$

The terms in square brackets then vanish[5] because $F_{\alpha\beta}\partial^\beta F^{\mu\alpha} = F_{\alpha\beta}\partial^\alpha F^{\beta\mu}$.

The equation of motion for $\theta^{\mu\nu}$ is then, from Eq. (9.47),

$$\partial_\nu\theta^{\mu\nu} = -J_\alpha F^{\mu\alpha} \equiv -f^\mu . \tag{9.48}$$

The force density $f^\mu = J_\alpha F^{\mu\alpha} = (\boldsymbol{J}\cdot\boldsymbol{E}/c, \rho\boldsymbol{E} + \boldsymbol{J}\times\boldsymbol{B}) = (f^0, \boldsymbol{f})$. The time and space components of Eq. (9.48) are

$$\partial_\nu\theta^{0\nu} = \partial_0\theta^{00} + \partial_i\theta^{0i} = \partial_0\theta^{00} + \frac{1}{c}\partial_iS^i = -f^0 = -\frac{1}{c}\boldsymbol{J}\cdot\boldsymbol{E} ,$$

which is Poynting's theorem, $\partial u_{EM}/\partial t + \boldsymbol{\nabla}\cdot\boldsymbol{S} = -\boldsymbol{J}\cdot\boldsymbol{E}$, and

$$\partial_\nu\theta^{i\nu} = \partial_0\theta^{i0} + \partial_j\theta^{ij} = -f^i = -\left(\rho\boldsymbol{E} + \boldsymbol{J}\times\boldsymbol{B}\right)^i . \tag{9.49}$$

Written as a vector equation, Eq. (9.49) is equivalent to

$$\frac{\partial}{\partial t}\boldsymbol{g} + \boldsymbol{\nabla}\cdot\boldsymbol{\theta} = -\left(\rho\boldsymbol{E} + \boldsymbol{J}\times\boldsymbol{B}\right) . \tag{9.50}$$

Equation (9.50) is the balance equation for the field momentum: The rate of loss of field momentum is balanced by the Lorentz force acting on charged particles. The single covariant equation,

[4]The covariant expression for the Lorentz force is $qF^{\mu\nu}U_\nu$, whereas the force density (force per volume) is $F^{\mu\nu}J_\nu$. The four-current is a four-vector density.

[5]Use the antisymmetry of $F_{\alpha\beta}$ and $F^{\mu\alpha}$, and let $\beta \to \alpha$ and $\alpha \to \beta$.

Eq. (9.48), contains Poynting's theorem on energy balance as the time component and momentum balance for its spatial part.

From Eq. (9.48) we see that field energy-momentum is *not* conserved due to the coupling of the field to currents. We expect, however, for the system of fields *and* particles that energy-momentum *is* conserved. Define an energy-momentum tensor for the combined system of fields and particles, $\theta^{\mu\nu} \equiv \theta_F^{\mu\nu} + \theta_M^{\mu\nu}$, where $\theta_F^{\mu\nu}$ is the field energy-momentum tensor, Eq. (9.43), and $\theta_M^{\mu\nu}$ is an energy-momentum tensor for "matter," so that for the combined system energy-momentum is conserved, $\partial_\nu \theta^{\mu\nu} = 0$. Because $\partial_\nu \theta_F^{\mu\nu} = -f^\mu$, Eq. (9.48), total energy-momentum conservation requires

$$\partial_\nu \theta_M^{\mu\nu} = f^\mu \, . \tag{9.51}$$

Equation (9.51) of course begs the question of what is $\theta_M^{\mu\nu}$. We can guess at $\theta_M^{\mu\nu}$ for a point mass m as a generalization of the four-current for a point charge, Eq. (8.12), the transport of four-momentum in the various spacetime directions:

$$\theta_M^{\mu\nu} \equiv c \int_{-\infty}^{\infty} \mathrm{d}\tau P^\mu U^\nu \delta^4(x - z(\tau)) = mc \int_{-\infty}^{\infty} \mathrm{d}\tau U^\mu U^\nu \delta^4(x - z(\tau)) \, . \tag{9.52}$$

We see that $\theta_M^{\mu\nu}$ defined this way is a symmetric tensor. Keep in mind that $\theta_M^{\mu\nu}$ does not follow from Noether's theorem; it's been "cooked up," yet it does the job. The divergence of Eq. (9.52) is (Exercise 9.7)

$$\partial_\nu \theta_M^{\mu\nu} = mc \int_{-\infty}^{\infty} \dot{A}^\mu \delta^4(x - z(\tau)) \mathrm{d}\tau = c \int_{-\infty}^{\infty} q F^{\mu\alpha} U_\alpha \delta^4(x - z(\tau)) \mathrm{d}\tau = F^{\mu\alpha} J_\alpha = f^\mu \, , \tag{9.53}$$

where we've used Eq. (8.43) in the second equality and Eq. (8.12) in the third. Thus, Eq. (9.51) is satisfied; the energy-momentum of particles and fields is conserved.

SUMMARY

- For every continuous symmetry of the action, there's a conserved quantity associated with fields (Noether's theorem). Invariance under spacetime translations leads to energy-momentum conservation, and invariance under LTs leads to conservation of angular momentum.
- Conservation of energy-momentum is expressed in terms of continuity equations of the rows of the energy-momentum tensor, $\partial_\nu T^{\mu\nu} = 0$. The quantities $Q^\mu = \int_\Sigma \mathrm{d}^3x T^{\mu 0}$ are conserved.
- The tensor $T^{\mu\nu}$ must be symmetric to conserve angular momentum. If the canonical energy-momentum tensor is not symmetric, it can be symmetrized by adding to it the divergence of a third-rank antisymmetric tensor, $T^{\mu\nu} \to \theta^{\mu\nu} \equiv T^{\mu\nu} + \partial_\lambda \psi^{\mu\lambda\nu}$. This procedure does not change the conserved quantities Q^μ.
- The energy-momentum tensor has the structure:

$$[T^{\mu\nu}] = \left(\begin{array}{c|ccc} W & S^1/c & S^2/c & S^3/c \\ \hline g^1 c & \sigma^{11} & \sigma^{12} & \sigma^{13} \\ g^2 c & \sigma^{21} & \sigma^{22} & \sigma^{23} \\ g^3 c & \sigma^{31} & \sigma^{32} & \sigma^{33} \end{array} \right) .$$

  where W is the energy energy, S^i is the energy current density, g^i is the momentum density, and σ^{ij} is the stress tensor, $\sigma^{ij} = F^i/(\Delta a)_j$. Symmetry of the tensor implies that $S^i = g^i c^2$.
- For the electromagnetic field interacting with particles, $\partial_\nu \theta^{\mu\nu} = -f^\mu$, where $f^\mu = F^{\mu\nu} J_\nu$ is the Lorentz force-density four-vector. The one tensor equation $\partial_\nu \theta^{\mu\nu} = -f^\mu$ contains as its time component Poynting's theorem on energy balance and for its spatial part the equation for momentum balance in terms of the Maxwell stress tensor.

EXERCISES

9.1 Fill in the steps from Eq. (9.5) to Eq. (9.6).

9.2 Derive Eq. (9.39) for T^{00} using the canonical energy-momentum tensor, Eq. (9.38).

9.3 Show that $\theta^{\mu\nu}$ as given in Eq. (9.43) is symmetric. Show that the covariant version of Eq. (9.43) is given by $-\mu_0\theta_{\mu\nu} = \frac{1}{4}g_{\mu\nu}F^{\alpha\beta}F_{\alpha\beta} + g^{\lambda\beta}F_{\mu\beta}F_{\lambda\nu}$.

9.4 Verify the first line of Eq. (9.44) using Eq. (9.43) for the symmetric energy-momentum tensor.

9.5 Verify the result given in Section 9.5.2 for $f^{\mu} = (\boldsymbol{J}\cdot\boldsymbol{E}/c, \rho\boldsymbol{E} + \boldsymbol{J}\times\boldsymbol{B})$ is correct.

9.6 a. Derive the energy-momentum tensor for the Proca Lagrangian, Eq. (8.52). Using Eq. (9.42) show that the symmetric tensor is given by

$$\theta^{\mu\nu} = \theta_0^{\mu\nu} + \frac{1}{\mu_0}\left(\frac{m_\gamma c}{\hbar}\right)^2\left(A^\mu A^\nu - \frac{1}{2}g^{\mu\nu}A_\alpha A_\alpha\right),$$

where $\theta_0^{\mu\nu}$ is the tensor for $m_\gamma = 0$, Eq. (9.43).

b. From $\theta^{0\mathrm{i}}$, show that the Poynting vector in this theory is

$$\boldsymbol{S} = \frac{1}{\mu_0}\left(\boldsymbol{E}\times\boldsymbol{B} + \left(\frac{m_\gamma c}{\hbar}\right)^2\phi\boldsymbol{A}\right).$$

c. Show from θ^{00} that the energy density is given by

$$W = \frac{1}{2\mu_0}\left(|B|^2 + \left(\frac{m_\gamma c}{\hbar}\right)^2|A|^2\right) + \frac{\epsilon_0}{2}\left(|E|^2 + \left(\frac{m_\gamma c}{\hbar}\right)^2\phi^2\right).$$

d. Show that while $\theta^{\mu\nu}$ has been modified by the inclusion of a massive photon, the form of Eq. (9.48) is preserved. The Lorentz force density is thus unchanged for a massive photon. In particular, Poynting's theorem is still given by $\partial W/\partial t + \boldsymbol{\nabla}\cdot\boldsymbol{S} = -\boldsymbol{J}\cdot\boldsymbol{E}$.

9.7 Derive the first equality in Eq. (9.53). Use the identity given in Exercise 8.1 and integrate by parts.

9.8 Is the canonical energy-momentum tensor for the electromagnetic field, Eq. (9.38), gauge invariant? What about the symmetric tensor, Eq. (9.43)?

9.9 Fill in the steps between Eq. (9.22) and Eq. (9.23).

9.10 Derive Eq. (9.37). Start with Eq. (9.31) and substitute Eq. (9.35). Show that

$$x^\alpha\partial_\lambda\psi^{\beta\lambda\mu} = \partial_\lambda\left(x^\alpha\psi^{\beta\lambda\mu}\right) - \psi^{\beta\alpha\mu}.$$

Use Eq. (9.34) to show that

$$\psi^{\alpha\beta\mu} - \psi^{\beta\alpha\mu} = S^{\alpha\beta\mu}.$$

Use Eq. (9.25) combined with Eq. (9.29). Bob's your uncle.

CHAPTER 10

Relativistic hydrodynamics

RELATIVISTIC hydrodynamics is the study of matter in the *fluid* state,[1] with relativistic flow velocities or with energy densities sufficiently great that they generate gravitational fields. The subject is needed in astrophysics (gamma ray bursts, relativistic jets, and the collapse of stars that produce black holes); it's also used in cosmology. In this chapter we develop the energy-momentum tensor $T_{\mu\nu}$ of a *perfect fluid*—an idealized fluid in which energy dissipation due to viscosity or heat flow is ignored. Classical hydrodynamics is based on *balance equations* for mass, energy, and momentum, concepts that are distinct in nonrelativistic physics, but which are interrelated in relativity, mass-energy, and energy-momentum. We begin with a review of the nonrelativistic theory.

10.1 NONRELATIVISTIC HYDRODYNAMICS

10.1.1 Mass balance

Conservation of mass is expressed as a continuity equation, Eq. (8.9),

$$\frac{\partial \rho}{\partial t} + \boldsymbol{\nabla} \cdot (\rho \boldsymbol{v}) = 0 \,, \tag{10.1}$$

where $\boldsymbol{J} = \rho \boldsymbol{v}$ is the mass current density with ρ the mass density and $\boldsymbol{v}$ the fluid velocity.

10.1.2 Momentum balance

The balance equation for momentum in a volume V is (see Eq. (8.7), general balance equation)

$$\underbrace{\frac{\mathrm{d}}{\mathrm{d}t}\int_V \rho \boldsymbol{v} \mathrm{d}V}_{\text{rate of change of momentum in } V} = \underbrace{-\oint_S \rho \boldsymbol{v}\boldsymbol{v} \cdot \mathrm{d}\boldsymbol{S}}_{\text{momentum flux through } S} + \underbrace{\int_V \rho \boldsymbol{F} \mathrm{d}V}_{\substack{\text{momentum produced by} \\ \text{external forces}}} + \underbrace{\oint_S \boldsymbol{\sigma} \cdot \mathrm{d}\boldsymbol{S}}_{\substack{\text{momentum produced by} \\ \text{short-range internal forces at } S}} \,, \tag{10.2}$$

where we're using dyadic notation: $(\boldsymbol{v}\boldsymbol{v})_{ij} \equiv v_i v_j$. The first integral on the right accounts for the momentum convected through the surface S bounding V. The other two integrals represent *sources* of momentum production: $\boldsymbol{F}$ is an external force (per mass) that couples to the particles within V (such as the gravitational field), and $\boldsymbol{\sigma}$, the stress tensor, represents the effect of short-range internal forces (per area) acting at S. In a fluid there is always a normal component of the surface force (per area), the pressure P, with $\boldsymbol{\sigma} = -P\mathbf{I}$ ($\mathbf{I}$ is the unit dyad). In terms of components,[2] $\sigma_{ij} = -P\delta_{ij}$. This form of the stress tensor ignores the effects of viscous forces—sufficient for our purposes.

[1] A precise definition of fluid is elusive—a system with a sufficiently large number of particles that the dynamics of individual particles cannot be tracked, and in which the *collective* motion of particles can be approximated in terms of a few variables such as mass density ρ or flow velocity $\boldsymbol{v}$.

[2] Because we're dealing with nonrelativistic hydrodynamics, we dispense with covariant notation.

The surface integrals in Eq. (10.2) can be converted into volume integrals through the divergence theorem, giving us the local form of the momentum balance equation, the *Euler equation*:

$$\frac{\partial(\rho \boldsymbol{v})}{\partial t} + \boldsymbol{\nabla} \cdot (\rho \boldsymbol{v}\boldsymbol{v}) = \rho \boldsymbol{F} + \boldsymbol{\nabla} \cdot \boldsymbol{\sigma} \,, \tag{10.3}$$

where the divergence of a tensor has the components $(\boldsymbol{\nabla} \cdot \boldsymbol{\sigma})_k = \partial_i \sigma^i{}_k$. The divergence of the stress tensor (at a point) is the *force density* (at that point) due to internal forces in the fluid.[3]

If we set the external force $\boldsymbol{F} = 0$ in Eq. (10.3) (a *free fluid*), and move the internal force to the left of the equation, we have a conservation law (continuity equation) for fluid momentum

$$\frac{\partial}{\partial t}\boldsymbol{g} + \boldsymbol{\nabla} \cdot (\rho \boldsymbol{v}\boldsymbol{v} - \boldsymbol{\sigma}) = 0 \,,$$

where $\boldsymbol{g} = \rho \boldsymbol{v}$ is the momentum density. Written in terms of components:

$$\frac{\partial}{\partial t} g_i + \partial_k \left(P\delta_{ik} + \rho v_i v_k \right) = 0 \,, \tag{10.4}$$

where we've used $\boldsymbol{\nabla} \cdot (P\mathbf{I}) = \boldsymbol{\nabla} P$ (Exercise 10.4).

The equations of hydrodynamics tend to simplify if we introduce a special time derivative, the *convective derivative* $\mathrm{D}/\mathrm{d}t \equiv \partial/\partial t + \boldsymbol{v} \cdot \boldsymbol{\nabla}$. To an observer moving with the fluid (*Lagrangian observer*), $\mathrm{D}/\mathrm{d}t = \partial/\partial t$. In the lab frame, however, (*Eulerian observer*) a vector field $\boldsymbol{A}(\boldsymbol{r}, t)$ changes in time and space. For small $\mathrm{d}t$ and $\mathrm{d}\boldsymbol{r}$, $\boldsymbol{A}(\boldsymbol{r} + \mathrm{d}\boldsymbol{r}, t + \mathrm{d}t) \approx \boldsymbol{A}(\boldsymbol{r}, t) + \mathrm{d}t \,(\partial \boldsymbol{A}/\partial t) + (\mathrm{d}\boldsymbol{r} \cdot \boldsymbol{\nabla})\, \boldsymbol{A}$; as $\mathrm{d}t \to 0$, we obtain the total time derivative $\mathrm{D}\boldsymbol{A}/\mathrm{d}t = \partial \boldsymbol{A}/\partial t + (\boldsymbol{v} \cdot \boldsymbol{\nabla})\, \boldsymbol{A}$, where $\boldsymbol{v}(\boldsymbol{r}, t) = \mathrm{d}\boldsymbol{r}/\mathrm{d}t$ is the velocity field in the Eulerian reference frame. The goal of relativity theory is to relate the descriptions obtained in different reference frames. It's a simple exercise to show that $U^\mu \partial_\mu = \gamma \mathrm{D}/\mathrm{d}t$; the convective derivative is *almost* a Lorentz invariant. Using the result of Exercise 10.3, Euler's equation can be written

$$\rho \frac{\mathrm{D}\boldsymbol{v}}{\mathrm{d}t} = \rho \boldsymbol{F} + \boldsymbol{\nabla} \cdot \boldsymbol{\sigma} \,. \tag{10.5}$$

10.1.3 Energy balance

In a perfect fluid energy dissipation is ignored, implying that flows are *isentropic*.[4] In hydrodynamics it's usually more convenient to work with specific, "per mass" quantities. Denoting $s = S/m$ as the *specific entropy* (entropy per mass), isentropic flow is such that

$$\frac{\mathrm{D}s}{\mathrm{d}t} = \frac{\partial s}{\partial t} + \boldsymbol{v} \cdot \boldsymbol{\nabla} s = 0 \,. \tag{10.6}$$

We assume that *local thermodynamic equilibrium* applies at any point of the fluid, and thus the state variables of classical thermodynamics (which describe a global equilibrium state, independent of space and time) become field quantities, functions of space and time. If the only form of work available to the fluid is mechanical, the first law of thermodynamics is $\mathrm{d}E = T\mathrm{d}S - P\mathrm{d}V$, where E is the internal energy and T is the absolute temperature. *Enthalpy* $H \equiv E + PV$ plays a role in hydrodynamics;[5] its differential is $\mathrm{d}H = T\mathrm{d}S + V\mathrm{d}P$. Defining $h = \epsilon + P/\rho$ as the *specific enthalpy*, where ϵ is the *specific energy*, we have $\mathrm{d}h = T\mathrm{d}s + \mathrm{d}P/\rho$, where $V/m = 1/\rho$. For isentropic flow ($\mathrm{d}s = 0$), $\boldsymbol{\nabla} h = \boldsymbol{\nabla} P/\rho$, and Euler's equation becomes $\partial \boldsymbol{v}/\partial t + (\boldsymbol{v} \cdot \boldsymbol{\nabla})\boldsymbol{v} = -\boldsymbol{\nabla} h$. The velocity field of the free fluid responds to gradients in enthalpy, $\mathrm{D}\boldsymbol{v}/\mathrm{d}t = -\boldsymbol{\nabla} h$.

Let's derive the energy transport equation (the analog of Poynting's theorem). The energy per volume is $\frac{1}{2}\rho v^2 + \rho\epsilon$.

---

[3]We *should* denote the stress tensor as a mixed tensor density (such as we found in Eq. (9.14)). The divergence of the stress tensor is a force density: If σ is a force per area, its divergence is a force per volume, $\boldsymbol{\nabla} \cdot \boldsymbol{\sigma} = \boldsymbol{f} \implies \partial_i \sigma^i_j = f_j$.

[4]Thermodynamics enters our discussion. Entropy measures the number of microscopic degrees of freedom consistent with a given macroscopic state. Isentropic flow implies no additional transfer of energy to microscopic degrees of freedom.

[5]Enthalpy includes the work required to "make room" for a system by displacing volume V against the pressure P.

Kinetic energy

For the kinetic energy term,

$$\begin{aligned}\frac{1}{2}\frac{\partial \rho v^2}{\partial t} &= \frac{1}{2}v^2\frac{\partial \rho}{\partial t} + \rho\boldsymbol{v}\cdot\frac{\partial \boldsymbol{v}}{\partial t} = -\frac{v^2}{2}\boldsymbol{\nabla}\cdot\rho\boldsymbol{v} + \rho\boldsymbol{v}\cdot\left(-\frac{1}{\rho}\boldsymbol{\nabla}P - (\boldsymbol{v}\cdot\boldsymbol{\nabla})\boldsymbol{v}\right)\\ &= -\frac{v^2}{2}\boldsymbol{\nabla}\cdot\rho\boldsymbol{v} - \boldsymbol{v}\cdot\boldsymbol{\nabla}P - \rho\boldsymbol{v}\cdot(\boldsymbol{v}\cdot\boldsymbol{\nabla})\boldsymbol{v} = -\frac{v^2}{2}\boldsymbol{\nabla}\cdot\rho\boldsymbol{v} - \rho\boldsymbol{v}\cdot\boldsymbol{\nabla}h\\ &\quad + T\rho\boldsymbol{v}\cdot\boldsymbol{\nabla}s - \frac{1}{2}\rho\boldsymbol{v}\cdot\boldsymbol{\nabla}(v^2)\\ &= -\frac{v^2}{2}\boldsymbol{\nabla}\cdot\rho\boldsymbol{v} - \rho\boldsymbol{v}\cdot\boldsymbol{\nabla}(h+\frac{v^2}{2}) + T\rho\boldsymbol{v}\cdot\boldsymbol{\nabla}s\,, \end{aligned} \tag{10.7}$$

where in the first line we've used the continuity equation and Euler's equation, and in the second line $\boldsymbol{\nabla}P = \rho\boldsymbol{\nabla}h - T\rho\boldsymbol{\nabla}s$ and $\boldsymbol{v}\cdot(\boldsymbol{v}\cdot\boldsymbol{\nabla})\boldsymbol{v} = \frac{1}{2}\boldsymbol{v}\cdot\boldsymbol{\nabla}(v^2)$.

Internal energy

For the internal energy, we use that $\mathrm{d}E = T\mathrm{d}S - P\mathrm{d}V$ is equivalent to $\mathrm{d}\epsilon = T\mathrm{d}s + (P/\rho^2)\mathrm{d}\rho$. We then have

$$\frac{\partial \rho\epsilon}{\partial t} = \rho\frac{\partial \epsilon}{\partial t} + \epsilon\frac{\partial \rho}{\partial t} = \rho\left(T\frac{\partial s}{\partial t} + \frac{P}{\rho^2}\frac{\partial \rho}{\partial t}\right) + \epsilon\frac{\partial \rho}{\partial t} = \rho T\frac{\partial s}{\partial t} + h\frac{\partial \rho}{\partial t} = -\rho T\boldsymbol{v}\cdot\boldsymbol{\nabla}s - h\boldsymbol{\nabla}\cdot(\rho\boldsymbol{v})\,, \tag{10.8}$$

where we've used Eqs. (10.1) and (10.6) and $h = \epsilon + P/\rho$.

Total energy

Combining Eqs. (10.7) and (10.8), the entropy gradients cancel (mercifully),

$$\frac{\partial}{\partial t}\left(\frac{1}{2}\rho v^2 + \rho\epsilon\right) = -\left(h+\frac{v^2}{2}\right)\boldsymbol{\nabla}\cdot(\rho\boldsymbol{v}) - \rho\boldsymbol{v}\cdot\boldsymbol{\nabla}\left(h+\frac{v^2}{2}\right) = -\boldsymbol{\nabla}\cdot\left(\rho\boldsymbol{v}\left(h+\frac{v^2}{2}\right)\right)\,,$$

and we obtain a continuity equation for energy

$$\frac{\partial}{\partial t}\left(\frac{1}{2}\rho v^2 + \rho\epsilon\right) + \boldsymbol{\nabla}\cdot\left(\rho\boldsymbol{v}\left(h+\frac{1}{2}v^2\right)\right) = 0\,. \tag{10.9}$$

Energy is conserved—not surprising because we haven't allowed energy dissipation. The energy current density, however, $\rho\boldsymbol{v}(h+\frac{1}{2}v^2)$ involves the transport of *enthalpy*. To see why that's the case, develop a balance equation by integrating Eq. (10.9) over space,

$$\frac{\mathrm{d}}{\mathrm{d}t}\int \mathrm{d}^3x\rho(\epsilon+v^2/2) + \oint \rho\boldsymbol{v}(\epsilon+v^2/2)\cdot\mathrm{d}\boldsymbol{S} = -\oint P\boldsymbol{v}\cdot\mathrm{d}\boldsymbol{S}\,. \tag{10.10}$$

The right side of Eq. (10.10) *resembles* "Joule heating" in Poynting's theorem (Section 9.5.2), but the comparison isn't apt. The quantity $-\boldsymbol{J}\cdot\boldsymbol{E}$ represents an irreversible transfer of energy from the electromagnetic field into the motion of particles, while the right side of Eq. (10.10) represents a reversible conversion of internal energy into work, and is not dissipated. The total energy (kinetic energy and internal energy of the fluid) is conserved when we take into account transport of enthalpy.

10.1.4 Nonrelativistic energy-momentum tensor

Equations (10.4) and (10.9) express the conservation of fluid momentum and energy. We can write these equations in a compact form *resembling* conservation of energy and momentum of general fields, Eq. (9.16), by defining a *tensor-like* quantity (in analogy with Eq. (9.19))

$$T_{\mu\nu} \equiv \left(\begin{array}{c|ccc} w & J_1 & J_2 & J_3 \\ \hline g_1 & T_{11} & T_{12} & T_{13} \\ g_2 & T_{21} & T_{22} & T_{23} \\ g_3 & T_{31} & T_{32} & T_{33} \end{array}\right) , \tag{10.11}$$

where

$$\begin{aligned} T_{00} &= w \equiv \rho\epsilon + \tfrac{1}{2}\rho v^2 \qquad & T_{0i} &= J_i \equiv \rho v_i (h + \tfrac{1}{2}v^2) \\ T_{i0} &= g_i = \rho v_i \qquad & T_{ij} &= P\delta_{ij} + \rho v_i v_j \,. \end{aligned} \tag{10.12}$$

Equations (10.4) and (10.9) can then be written using the notation of Eq. (10.11)

$$\frac{\partial T_{00}}{\partial t} + \frac{\partial T_{0i}}{\partial x^i} = \frac{\partial w}{\partial t} + \frac{\partial J_i}{\partial x^i} = 0 \qquad \frac{\partial T_{i0}}{\partial t} + \frac{\partial T_{ik}}{\partial x^k} = \frac{\partial g_i}{\partial t} + \frac{\partial T_{ik}}{\partial x^k} = 0 \,. \tag{10.13}$$

We say that Eq. (10.11) is "tensor like" because, even though we've used tensor notation, it's *not* a tensor: It doesn't transform as one. We've written the conservation laws this way only to suggest how to get started with a relativistic theory (next section). While the "space-space" part of Eq. (10.11), T_{ij}, is symmetric, $T_{0i} \neq T_{i0}$—there is no "$E = mc^2$" in classical hydrodynamics.

10.2 ENERGY-MOMENTUM TENSOR FOR PERFECT FLUIDS

Assume that a reference frame can be found in which all particles of the fluid are at rest. For such a frame, set all fluid velocities to zero in Eq. (10.11), leaving us with an energy-momentum tensor:

$$\overline{T}^{\mu\nu} = \begin{pmatrix} \rho c^2 & 0 & 0 & 0 \\ 0 & P & 0 & 0 \\ 0 & 0 & P & 0 \\ 0 & 0 & 0 & P \end{pmatrix} , \tag{10.14}$$

where we've let $\rho\epsilon \to \rho c^2$ as the internal energy density, with ρ the proper density.[6] The conservation laws in Eq. (10.13) reduce to $w =$ constant, $P =$ constant, $\partial_\nu \overline{T}^{\mu\nu} = 0$.

Starting with Eq. (10.14), we can obtain the energy-momentum tensor in an IRF through a LT,

$$T^{\mu\nu} = L^\mu_\alpha L^\nu_\beta \overline{T}^{\alpha\beta} \,. \tag{10.15}$$

Because Eq. (10.14) is diagonal, Eq. (10.15) is equivalent to

$$T^{\mu\nu} = \rho c^2 L^\mu_0 L^\nu_0 + P \sum_{i=1}^{3} L^\mu_i L^\nu_i \,. \tag{10.16}$$

Clearly $T^{\mu\nu}$ is symmetric. Using Eq. (6.16) for an arbitrary boost, it's straightforward to show that[7]

$$T^{\mu\nu} = P\eta^{\mu\nu} + \frac{\psi}{c^2} U^\mu U^\nu , \tag{10.17}$$

where $\psi \equiv \rho c^2 + P$ is the *enthalpy density* (enthalpy per volume). The elements of $T^{\mu\nu}$ are

$$\begin{aligned} T^{00} &= -P + \gamma^2\psi = \gamma^2 \rho c^2 + \left(\gamma^2 - 1\right) P = \gamma^2 \left(\rho c^2 + P\beta^2\right) \\ T^{0i} &= T^{i0} = \frac{\psi}{c}\gamma^2 v^i \qquad T^{ij} = P\delta^{ij} + \frac{\psi}{c^2}\gamma^2 v^i v^j \,. \end{aligned} \tag{10.18}$$

[6]At this point, we could let $\rho\epsilon \to \rho c^2 + \rho\overline{\epsilon}$, where $\overline{\epsilon}$ is the internal energy because (for example) the medium is elastic. Traditional thermodynamics ignores the rest-mass energy, whereas for our purposes we're going to ignore $\overline{\epsilon}$.

[7]Equation (10.17) is often given as the *definition* of a relativistic perfect fluid.

Using the results established in Section 9.2, the energy current $S^i \equiv cT^{0i}$ and the momentum density $g^i \equiv T^{i0}/c$ are, from Eq. (10.18),

$$S^i = \gamma^2 \psi v^i \qquad g^i = \gamma^2 \frac{\psi}{c^2} v^i \,. \tag{10.19}$$

Clearly $g^i = S^i/c^2$ holds in this theory, just as in electrodynamics, Eq. (9.44). The factors of γ^2 can be traced to the fact that there is already one factor of γ in $E = m\gamma c^2$ and $U^\mu = c\gamma\,(1, \boldsymbol{\beta})$; the second factor of γ comes from the LT of densities (Section 8.2).

In the relativistic theory of hydrodynamics, *both* the energy and momentum currents involve transport of enthalpy (in the nonrelativistic theory, enthalpy transport contributes only to the energy current). This is because "$E = mc^2$" is missing from the nonrelativistic theory. It's a good habit to watch for where P enters the theory. Pressure has dimensions of energy density; hence P/c^2 has dimensions of mass density. In GR, the "mass-energy" of pressure contributes to gravitation.

10.3 ENERGY-MOMENTUM CONSERVATION

The equations of energy-momentum conservation are obtained from $\partial_\nu T^{\mu\nu} = 0$, Eq. (9.16). Applied to Eq. (10.17), we have

$$\partial^\mu P + \frac{1}{c^2}\partial_\nu \left(\psi U^\mu U^\nu\right) = 0 \,. \tag{10.20}$$

The equation of energy conservation is the time component of Eq. (10.20), $\mu = 0$,

$$-\partial_0 P + \frac{1}{c}\partial_\nu \left(\gamma \psi U^\nu\right) = \frac{\partial}{\partial t}\left(\gamma^2(\rho c^2 + P\beta^2)\right) + \boldsymbol{\nabla} \cdot \left(\gamma^2 \psi \boldsymbol{v}\right) = 0 \,. \tag{10.21}$$

In the low-speed limit, Eq. (10.21) reduces to:

$$\frac{\partial \rho}{\partial t} + \boldsymbol{\nabla} \cdot \left((\rho + P/c^2)\boldsymbol{v}\right) = 0 \,. \qquad (v \ll c) \tag{10.22}$$

For $P \ll \rho c^2$, Eq. (10.22) reduces to Eq. (10.1) (that is, when we neglect the equivalent mass-energy density of the pressure). Both limits, $v \ll c$ and $P \ll \rho c^2$, are obtained by letting $c \to \infty$. The equation for energy conservation thus reduces to mass conservation in the nonrelativistic limit. *In relativistic hydrodynamics, there is no mass conservation equation.* Mass is not conserved, but *energy* is. *The equation for energy conservation is the relativistic generalization of the mass continuity equation.* Note how mass conservation differs from charge conservation. Charge conservation is built into the field equations of electromagnetism.[8] While charges generate the electromagnetic field, they don't carry electromagnetic energy, which is contained in the electromagnetic field. In the theory of relativity, mass both carries energy and generates the gravitational field.

Momentum conservation is obtained from the spatial parts of Eq. (10.20); setting $\mu = i$,

$$\partial^i P + \frac{1}{c}\partial_0 \left(\gamma^2 \psi v^i\right) + \frac{1}{c^2}\boldsymbol{\nabla} \cdot \left(\gamma^2 \psi \boldsymbol{v} v^i\right) = 0 \,. \tag{10.23}$$

Using Eq. (10.21) in Eq. (10.23), it can be shown

$$\frac{\partial \boldsymbol{v}}{\partial t} + (\boldsymbol{v} \cdot \boldsymbol{\nabla})\boldsymbol{v} = -\frac{1}{\gamma^2\left(\rho + P/c^2\right)}\left(\boldsymbol{\nabla} P + \boldsymbol{v}\frac{\partial}{\partial t}(P/c^2)\right) , \tag{10.24}$$

which is the relativistic generalization of the Euler equation. Equation (10.24) reduces to the nonrelativistic Euler equation in the limit $c \to \infty$.

---

[8]Take the divergence of the Ampère-Maxwell equation, and use Gauss's law.

10.4 PARTICLE NUMBER CONSERVATION

While mass is not conserved, the *number* of particles (baryons at a fundamental level, or atoms at reasonable temperatures) *is* conserved. We must put in "by hand" a continuity equation for particle number. Baryons (particles composed of three quarks) can undergo transmutations and thus while the rest mass of a group of baryons may not be conserved, the number of baryons is. Conservation of particle number (baryon or lepton) is a tenet of the standard model of particle physics.

Let n denote the number density of particles in the rest frame. (We're assuming a reference frame in which all particles are at rest.) Define the *number flux four-vector* N^μ by the requirements that in the rest frame, $\overline{N}^0 = nc$ and $\overline{N}^i = 0$. The quantity nc is the number of particles per area per time to cross a spacelike surface in the rest frame (see Fig. 10.1). Under a LT (to an IRF in which the particles all have the same velocity), $N^\mu = L^\mu_\nu \overline{N}^\nu = L^\mu_0 nc = nU^\mu$. We *posit* a continuity equation for the number of particles,

$$\partial_\mu N^\mu = \partial_\mu (nU^\mu) = \partial_t(\gamma n) + \boldsymbol{\nabla} \cdot (n\gamma \boldsymbol{v}) = 0\,. \tag{10.25}$$

Particle number conservation does not follow from Noether's theorem, from a spacetime symmetry; we take it as an experimentally verified aspect of our world.

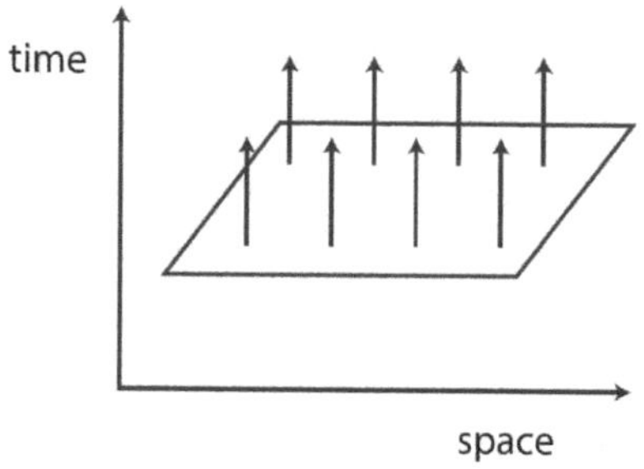

Figure 10.1 $\overline{N}^0 = nc$ is the flux of particles through a spacelike hypersurface.

10.5 COVARIANT EQUATION OF MOTION

Equation (10.24) is the relativistic generalization of the Euler equation. It is, however, not in covariant form—it involves the three-velocity, not the four-velocity. The Euler equation specifies the acceleration of a fluid; a covariant version would involve spacelike quantities because four-acceleration is spacelike and orthogonal to the four-velocity (Section 7.1). What we *do* have in covariant form are the conservation laws, Eq. (10.20). For this purpose it's useful to view $A^\mu \equiv \partial_\nu T^{\mu\nu}$ as the components of a four-vector (components that have magnitude zero because of the conservation laws). We then resolve each of the components A^μ into timelike and spacelike quantities (through the application of a projection operator), $A^\mu = A^\mu_T + A^\mu_S$, which is a Lorentz-invariant procedure. The conservation laws can then be written $0 = \sum_\mu A^\mu_T + \sum_\mu A^\mu_S$. Because timelike and spacelike vectors are linearly independent, we have separately $\sum_\mu A^\mu_T = 0$ and $\sum_\mu A^\mu_S = 0$.

We seek an operator that projects out of an arbitrary four-vector any components lying in the spacelike hypersurface (SH) orthogonal to the four-velocity. A three-vector $\boldsymbol{A}$ can be written as the sum of a projection along a given unit vector $\hat{\boldsymbol{n}}$ and a vector orthogonal to $\hat{\boldsymbol{n}}$, $\boldsymbol{A} = \hat{\boldsymbol{n}}(\hat{\boldsymbol{n}} \cdot \boldsymbol{A}) + \boldsymbol{A}_\perp$. The operator $\mathscr{P}_n \equiv I - \hat{\boldsymbol{n}}(\hat{\boldsymbol{n}}\cdot\ )$ would then project out of $\boldsymbol{A}$ its component orthogonal to $\hat{\boldsymbol{n}}$, $\boldsymbol{A}_\perp = \mathscr{P}_n \boldsymbol{A}$. In the case of four-vectors, take the unit vectors to be in the direction of the four-velocity, $n^\mu = U^\mu/c$. Define the projection operator

$$\mathscr{P}^\mu_\nu \equiv \delta^\mu_\nu + U^\mu U_\nu/c^2 \tag{10.26}$$

that projects out of a four-vector any components orthogonal to U^μ. A plus sign is used in Eq. (10.26) because $U_\mu U^\mu = -c^2$. As can be verified, $\mathscr{P}^\mu_\nu U^\nu = 0$; U^μ (timelike) has no spacelike component. For an arbitrary four-vector B^μ, $\mathscr{P}^\mu_\nu B^\nu = B^\mu + (U_\nu B^\nu)U^\mu/c^2$. We then have that $U_\mu\left(\mathscr{P}^\mu_\nu B^\nu\right) = 0$: Because U^μ is timelike, $\mathscr{P}^\mu_\nu B^\nu$ is spacelike (see Section 4.4.2). The projection of A^ν in the timelike direction is $\left(-U^\mu U_\nu/c^2\right) A^\nu$.

First consider the timelike projection of the conservation laws. The projection of $\partial_\nu T^{\lambda\nu} = 0$ along the direction of U^μ is $(-U^\mu/c^2)U_\lambda \partial_\nu T^{\lambda\nu}$, implying $U_\lambda \partial_\nu T^{\lambda\nu} = 0$. From Eq. (10.20), we have $U^\nu \partial_\nu P + U_\lambda \partial_\nu \left(\psi U^\lambda U^\nu\right)/c^2 = 0$. This equation is equivalent to (using the result of Exercise 10.9)

$$U^\nu \partial_\nu \rho + \frac{1}{c^2}\psi \partial_\nu U^\nu = 0 \,. \tag{10.27}$$

Equation (10.27) is a useful identity relating the compressibility of the fluid to the divergence of the four-velocity. As $c \to \infty$, Eq. (10.27) reduces to Eq. (10.1).

From $\mathscr{P}^\mu_\alpha \partial_\nu T^{\alpha\nu}$ we obtain the projections of $\partial_\nu T^{\alpha\nu} = 0$ onto the directions orthogonal to U^μ,

$$(\rho + P/c^2)U^\nu \partial_\nu U^\mu + \left(\eta^{\mu\nu} + \frac{U^\mu U^\nu}{c^2}\right)\partial_\nu P = 0 \,, \tag{10.28}$$

where we've used Eq. (10.20) and the result of Exercise 10.9. Equation (10.28) is the covariant generalization of the Euler equation. There are four equations implicit in Eq. (10.28), whereas the nonrelativistic Euler equation is among three-vectors. For $v \ll c$ and $P \ll \rho c^2$, the spatial parts of Eq. (10.28) reduce to the nonrelativistic Euler equation. The time component of Eq. (10.28) vanishes as $c \to \infty$.

10.6 LAGRANGIAN DENSITY

What is the Lagrangian density of the perfect fluid? We know that $\mathscr{L}$ should be a Lorentz invariant. What scalar invariants can be built out of the velocity field, U^μ? Let's try $\mathscr{L} = a + bU_\mu U^\mu$, where a and b are unknown scalar fields. Using this form for $\mathscr{L}$ in Eq. (9.15), the energy-momentum tensor obtained from Noether's theorem, we would have $T^{\mu\nu} = \mathscr{L}\eta^{\mu\nu} = \eta^{\mu\nu}a + b\eta^{\mu\nu}U_\nu U^\nu = a\eta^{\mu\nu} + bU^\mu U^\nu$. We obtain agreement with Eq. (10.17), the energy-momentum tensor obtained from hydrodynamics, by choosing $a = P$ and $b = \psi/c^2$. We would then have $\mathscr{L} = P + \psi U_\mu U^\mu/c^2$. However, $U_\mu U^\mu = -c^2$, so that[9]

$$\mathscr{L} = P - \psi = -\rho c^2 \,. \tag{10.29}$$

SUMMARY

- Nonrelativistic hydrodynamics is based on five equations expressing conservation of mass, momentum, and energy, concepts that are distinct in pre-relativistic physics but become interrelated in relativity theory. In relativistic hydrodynamics, there are four continuity equations expressing conservation of energy and momentum. There is no mass conservation law in relativistic hydrodynamics, mass being subsumed by energy.
- The energy-momentum tensor of the perfect fluid[10] is $T^{\mu\nu} = \rho U^\mu U^\nu + P(\eta^{\mu\nu} + U^\mu U^\nu/c^2)$.
- From $T^{\mu\nu}$ we obtain the energy current $S^i = \gamma^2\psi v^i$ where ψ is the enthalpy density in the rest frame, and the momentum density $g^i = \gamma^2\psi v^i/c^2$. In relativistic hydrodynamics energy and momentum transport involve enthalpy. This is where "$E = mc^2$" enters the theory: Pressure has units of energy density and contributes to energy transport; P/c^2 has units of mass density and contributes to momentum transport.

[9] Equation (10.29) would be $\mathscr{L} = -\rho(c^2 + \bar{\epsilon})$ if we included an internal energy function of the medium.
[10] What today we refer to as a perfect fluid, Einstein in his 1916 article called a frictionless adiabatic fluid.[9, p152]

- Conservation of energy-momentum is expressed by the zero divergence of $T^{\mu\nu}$, $\partial_\nu T^{\mu\nu} = 0$. The time component of $\partial_\nu T^{\mu\nu} = 0$, Eq. (10.21), expresses energy conservation and is the relativistic generalization of the mass continuity equation. In relativity, energy is conserved, and energy is tied to mass. The spatial parts of $\partial_\nu T^{\mu\nu} = 0$, Eq. (10.24), are the relativistic generalization of the Euler equation. Conservation of particle number is expressed by Eq. (10.25), a continuity equation for n, the number density in the rest frame. The covariant Euler equation is given in Eq. (10.28).

EXERCISES

10.1 In a gravitational field $\boldsymbol{g}$, there is a force per mass $\boldsymbol{F} = \boldsymbol{g}$ in Euler's equation. Show that Eq. (10.9) is modified to read $\partial(\frac{1}{2}\rho v^2 + \rho\epsilon)/\partial t + \boldsymbol{\nabla}\cdot\left(\rho\boldsymbol{v}(h + v^2/2)\right) = \rho\boldsymbol{v}\cdot\boldsymbol{g}$.

10.2 Show that $\boldsymbol{\nabla}\cdot(\rho\boldsymbol{v}\boldsymbol{v}) = \rho(\boldsymbol{v}\cdot\boldsymbol{\nabla})\boldsymbol{v} + \boldsymbol{v}\boldsymbol{\nabla}\cdot\rho\boldsymbol{v}$. The divergence of a tensor $\mathbf{T}$ is, in this notation, $(\boldsymbol{\nabla}\cdot\mathbf{T})_i \equiv \sum_j \partial_j T_{ij}$.

10.3 Using the results of Exercise 10.2, show that $\partial(\rho\boldsymbol{v})/\partial t + \boldsymbol{\nabla}\cdot(\rho\boldsymbol{v}\boldsymbol{v}) = \rho \mathrm{D}\boldsymbol{v}/\mathrm{D}t$.

10.4 Show that $\boldsymbol{\nabla}\cdot(P\mathbf{I}) = \boldsymbol{\nabla}P$.

10.5 Show that, for any scalar function ϕ, $\rho\mathrm{D}\phi/\mathrm{D}t = (\partial/\partial t)(\rho\phi) + \boldsymbol{\nabla}\cdot(\phi\rho\boldsymbol{v})$.

10.6 Show that $U^\mu\partial_\mu = \gamma\mathrm{D}/\mathrm{D}t$.

10.7 Show that Eq. (10.6) can be combined with Eq. (10.1) to produce an entropy continuity equation, $\partial(\rho s)/\partial t + \boldsymbol{\nabla}\cdot(\rho s\boldsymbol{v}) = 0$.

10.8 Show the vector identity $\boldsymbol{v}\cdot(\boldsymbol{v}\cdot\boldsymbol{\nabla})\,\boldsymbol{v} = \frac{1}{2}\boldsymbol{v}\cdot\boldsymbol{\nabla}\left(v^2\right)$.

10.9 Show that $U_\mu U^\nu \partial_\nu U^\mu = 0$. Start by differentiating $U_\mu U^\mu = -c^2$. Show that this leads to the identity $U_\mu \partial_\nu U^\mu U^\nu = -c^2\partial_\nu U^\nu$.

10.10 For nonrelativistic matter, the pressure is much less than the mass-energy density, $P \ll \rho c^2$. Compute, in Pascals, the value of ρc^2 obtained using the ordinary density of water. Compare with ordinary atmospheric pressure.

10.11 Show, using Eq. (10.17), that $\rho U^\mu = -U_\nu T^{\mu\nu}/c^2$. A covariant definition of the mass current is thus $J^\mu \equiv \rho U^\mu = -U_\nu T^{\mu\nu}/c^2$. Is J^μ conserved? Does $\partial_\mu J^\mu = 0$? Show that $\partial_\mu J^\mu = -P\partial^\nu U_\nu/c^2$ (use the result from Exercise 10.9). Combine this result with Eq. (10.27) to conclude that $\partial_\mu J^\mu = (P/\psi)\, U^\nu \partial_\nu \rho$. The extent to which $\partial_\mu J^\mu \neq 0$ is related to the *compressibility* of the fluid. Mass is energy; if energy can change, mass is not conserved.

10.12 Using Eq. (10.17), show that $U_\mu U_\nu T^{\mu\nu} = \rho c^4$. Thus, $U_\mu U_\nu T^{\mu\nu}$ is a covariant definition of ρ.

10.13 Show that the projection operator defined in Eq. (10.26) obeys the property of a projection operator: $\mathscr{P}^\mu_\nu\left(\mathscr{P}^\nu_\lambda A^\lambda\right) = \mathscr{P}^\mu_\lambda A^\lambda$ for arbitrary four-vector A^λ.

10.14 Show directly that for any four-vector A^ν, the component orthogonal to U^μ, $\mathscr{P}^\mu_\nu A^\nu$, is spacelike.

10.15 Show that $\mathscr{P}^\mu_\alpha \partial_\nu T^{\alpha\nu} = 0$ leads to Eq. (10.28). Use the results of Exercise 10.9.

10.16 Show that by contracting Eq. (10.28) with U_μ, you are led back to the result shown in Exercise 10.9. Why does this make sense?

CHAPTER 11

Equivalence of local gravity and acceleration

IN Section 1.8 we bid *au revoir* to gravity in order to develop spacetime physics as it pertains to non-gravitational phenomena (SR). It's time to return gravity to the fold.

11.1 THE EÖTVÖS EXPERIMENT

Aristotle taught that heavier objects fall in gravity faster than lighter ones—which is true for motion in resistive media. Galileo noted that: "...the difference of speed in moveables of different heaviness is found to be much greater in more resistant mediums."[42, p75] When friction can be minimized, however, objects fall at the same rate: "Yet balls of gold, lead, copper, porphyry, and other heavy materials differ almost insensibly in the inequality of motion through air. ...I came to the opinion that *if one were to remove entirely the resistance of the medium, all materials would descend with equal speed*" (my emphasis).[1] Galileo refuted the Aristotelian theory with a simple argument. Consider two objects of unequal weight, and let them be connected by a string. Under free fall, the nominally faster object would be retarded by the slower, due to the string, and the nominally slower object would be sped up by the faster. But the two objects together make a composite object *heavier* than either of its parts, which now falls *slower* than the heavier of its parts would fall by itself. Galileo: "...from the supposition that the heavier body is moved more swiftly than the less heavy, I conclude that the heavier moves less swiftly."[42, p66]

The resolution is that, in the absence of friction, all objects fall at the same rate. Write Newton's law of motion in the form $\boldsymbol{F} = m_i\boldsymbol{a}$, where the *inertial mass* m_i is the property of matter that resists changes in inertial motion (Section 1.2.1), and the force law in the form $\boldsymbol{F} = m_g\boldsymbol{g}$ (*if* all other forces have been eliminated, such as drag), where the *gravitational mass* m_g is the property of matter that couples to the gravitational field (Section 1.7.1). We customarily take $m_i = m_g \equiv m$, so that $\boldsymbol{a} = \boldsymbol{g}$, *independent of any property of the object.* Acceleration under gravity is independent of *any* attribute of an object! Gravity is a special kind of force.

The two types of mass, however, are logically distinct. What if $m_i \neq m_g$? Objects A and B would *not* accelerate the same under gravity, with $\boldsymbol{a}_A = (m_g/m_i)_A\,\boldsymbol{g}$ and $\boldsymbol{a}_B = (m_g/m_i)_B\,\boldsymbol{g}$. Let's do the experiment: *Does* $\boldsymbol{a}_A = \boldsymbol{a}_B$? The difference in acceleration between objects is characterized by the *Eötvös parameter*, η, a dimensionless quantity defined as the difference in acceleration

[1] Such an experiment was performed by the Apollo astronauts in 1971, where a simultaneously released hammer and feather were observed to strike the lunar surface at the same time.

divided by the average acceleration,

$$\eta \equiv \frac{|\boldsymbol{a}_A - \boldsymbol{a}_B|}{\frac{1}{2}(\boldsymbol{a}_A + \boldsymbol{a}_B)} = 2\frac{|(m_g/m_i)_A - (m_g/m_i)_B|}{(m_g/m_i)_A + (m_g/m_i)_B} .$$

One could measure the time to fall, $t = \sqrt{2h/[g(m_g/m_i)]}$, so that $m_g/m_i = 2h/(gt^2)$, from which $\eta = 2\left|t_B^2 - t_A^2\right| / \left(t_B^2 + t_A^2\right)$.

Another experiment is to compare the oscillation periods of pendulums of the same length made of different materials. Galileo used a lead ball and a cork ball, with the lead ball approximately 100 times as heavy as the cork ball. He watched the two pendulums, released at the same time, and could observe no difference in their motion. The oscillation period $T = 2\pi\sqrt{(m_i/m_g)(l/g)}$ and thus $\eta = 2\left|T_B^2 - T_A^2\right|/(T_B^2 + T_A^2)$. Newton performed this experiment around 1680, and from his data one can estimate $\eta \approx 10^{-3}$. Friedrich Bessel repeated the experiment in 1832 and found $\eta \approx 2 \times 10^{-5}$.

By far the most accurate experiment is that of Loránd Eötvös, who devised a *torsion balance* technique, starting in 1885 and repeated and refined by Eötvös and co-workers until 1920. Eötvös found $\eta < 5 \times 10^{-9}$. The Eötvös experiment has been repeated with ever-increasing precision. In 2008 a group reported $\eta = (0.3 \pm 1.8) \times 10^{-13}$,[43] consistent with $\eta = 0$, or $m_i = m_g$. Einstein took the *identity* $m_i \equiv m_g$ as a fact of experience and used it to far-reaching effect in establishing GR. *The Eötvös experiment provides the foundation for GR, just as the Michelson-Morley experiment provides the foundation for SR.*

The classic Eötvös experiment uses Earth as a laboratory, which as a rotating object allows us to probe the difference between m_g (associated with gravity), and the inertial mass m_i (associated with inertial forces). The experiment consists of a balance suspended in gravity from a torsion fiber (dashed line in Fig. 11.1), having objects A and B at distances l and l' from the fiber. The centrifugal

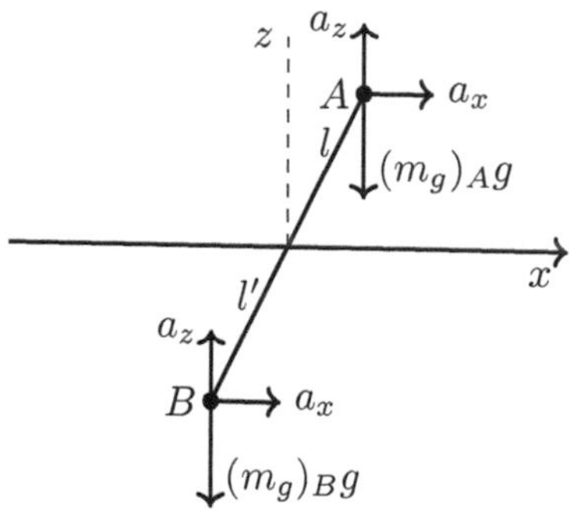

Figure 11.1 Eötvös torsion balance; z is the local vertical and x points from north to south.

force from Earth's rotation has components in the z and x-directions (depending on one's latitude on Earth). Objects A and B, being different, have unequal inertial masses. A torque is produced about the z-axis from the horizontal components of the centrifugal force,

$$\tau_z = (m_i)_B l' a_x - (m_i)_A l a_x . \tag{11.1}$$

The forces in the vertical direction produce a torque about the x-axis, $\tau_x = a_z\left[(m_i)_B l' - (m_i)_A l\right] - g\left[(m_g)_B l' - (m_g)_A l\right]$. In equilibrium there's no rotation about the x-axis; $\tau_x = 0$ implies $l' = l\left[(m_i)_A a_z - (m_g)_A g\right] / \left[(m_i)_B a_z - (m_g)_B g\right]$, which when combined with Eq. (11.1) leads to

$$\tau_z = (m_i)_A a_x l \frac{(m_g/m_i)_A - (m_g/m_i)_B}{(m_g/m_i)_B - a_z/g} .$$

A torque τ_z exists only if $(m_g/m_i)_A \neq (m_g/m_i)_B$. The Eötvös experiment seeks to measure τ_z.

11.2 THE EQUIVALENCE PRINCIPLE

The experimentally verified equivalence of gravitational and inertial mass implies the equivalence of uniform gravity and accelerated reference frames. Consider yourself in an elevator car S at rest in a uniform gravitational field $\boldsymbol{g}$ (left part of Fig. 11.2). A released object accelerates downward with $a = g$. Now consider the same elevator car S' in free space where gravitational fields are negligible, but where the elevator is accelerated with $\boldsymbol{a} = -\boldsymbol{g}$ (right part of Fig. 11.2). A released

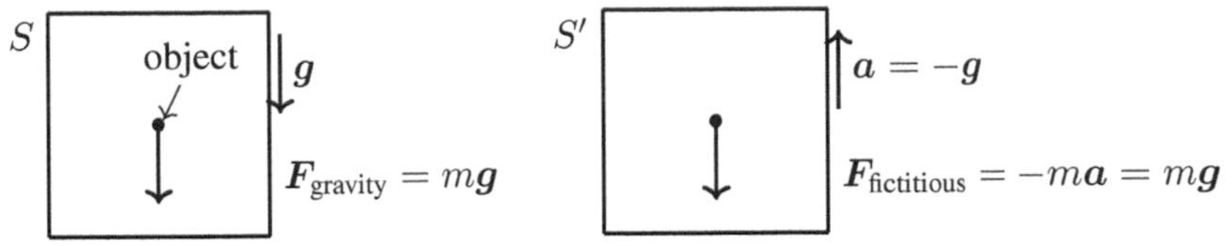

Figure 11.2 Left: Elevator car S stationary in a uniform gravitational field $\boldsymbol{g}$; a released object falls with acceleration g. Right: Elevator car S' in a gravity-free region accelerated upward with magnitude $a = g$; object moves downward with acceleration g.

object in S' accelerates downward with $a = g$: Free objects in linearly accelerated frames have the acceleration of the frame; set $\boldsymbol{F} = 0$ in Eq. (1.9). The object in S moves because of gravity, that in S' because of the reference frame. Are there measurements one could make *in these elevators* that would distinguish a gravitational field from an accelerated frame?

Einstein's *equivalence principle* (EP) says *No*:

> It is impossible to distinguish a uniform gravitational field from an accelerated reference frame; the two are physically equivalent.

The EP provides a framework for understanding physics in gravitational fields if we can first understand it in accelerating reference frames.

The identity $m_i = m_g$ is clearly required for the validity of the EP. Starting from the EP, however, we can go the other way and infer the equality $m_g = m_i$. Assume objects A and B with $(m_g/m_i)_A \neq (m_g/m_i)_B$. In the gravitational field of a source mass M_0, there would be accelerations $a_{A,B} = (GM_0/r^2)(m_g/m_i)_{A,B}$. Step into one of the elevators and let go of the objects. There are two possibilities:

1. $a_A = a_B$, in which case we are in the gravity-free elevator, because all free objects accelerate the same in an accelerated frame.

2. $a_A \neq a_B$, in which case we are in the elevator in a gravitational field.

This thought experiment implies a violation of the EP because a gravitational field *could* be distinguished from an accelerated frame. Taking the EP as valid, we must have $m_g = m_i$ for all objects. It might be thought that the EP and the Eötvös experiment establish only that the *ratio* m_g/m_i is universal, leaving the possibility that m_i is *proportional*, but not equal, to m_g. If $m_g = \alpha m_i$, with α a universal constant, it would be absorbed into the gravitational constant, $G \to G\alpha$, because what's measured would be $G\alpha$.

11.3 TIDAL FORCES AND REFERENCE FRAMES

Note the proviso of the equivalence of *uniform* gravity with acceleration. Strictly speaking, uniform gravitational fields are a convenient fiction.[2] The EP applies in regions of space small enough that the

[2]Add uniform gravitational field to the list of idealizations in physics: massless pulleys, perfectly thermally insulating substances, black-body radiators, parallel-plate capacitor without fringing fields. Idealizations are useful because they can be conceived of as the limiting behavior of actual physical systems.

gravitational field is ostensibly uniform. If the spatial extent of the "elevator" is too large, *tidal forces* become apparent (as we now explain), in which case it *is* possible to distinguish a gravitational field from an accelerated frame. We can "wave away" gravity through a frame transformation, but only locally, not globally. It's an experimental question as to how small a reference frame must be for the effects of gravity to be effectively uniform; we want $\nabla g \cdot \Delta \boldsymbol{r} \ll g$.

11.3.1 Tidal forces

Consider two particles of mass m separated by a distance $\Delta \boldsymbol{r}$ in an *inhomogeneous* gravitational field, $\boldsymbol{g}(\boldsymbol{r})$ (see Fig. 11.3). Because $\boldsymbol{g}$ varies over $\Delta \boldsymbol{r}$, the gravitational forces at $\boldsymbol{r}$ and $\boldsymbol{r}+\Delta \boldsymbol{r}$ are

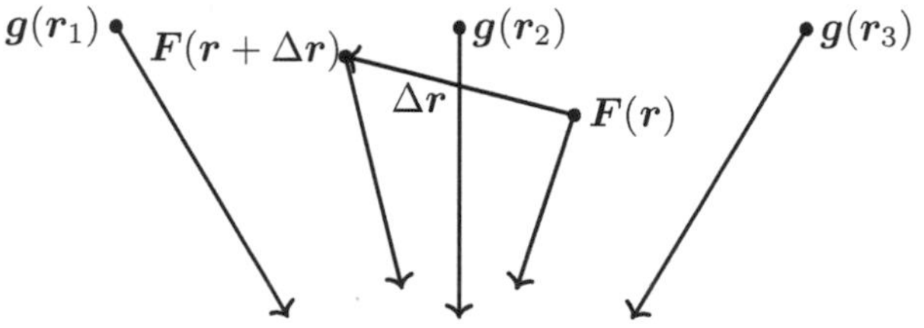

Figure 11.3 Inhomogeneous gravitational field.

not the same. The *difference* in force between the two locations, $\boldsymbol{f}(\boldsymbol{r}, \Delta \boldsymbol{r}) \equiv \boldsymbol{F}(\boldsymbol{r}+\Delta \boldsymbol{r}) - \boldsymbol{F}(\boldsymbol{r})$ is the tidal force. It depends on *two* vectors, a reference position[3] $\boldsymbol{r}$, and $\Delta \boldsymbol{r}$, the distance over which $\boldsymbol{F}(\boldsymbol{r})$ varies appreciably in the vicinity of $\boldsymbol{r}$. In terms of vector components,

$$f_i = F_i(\boldsymbol{r}+\Delta \boldsymbol{r}) - F_i(\boldsymbol{r}) \approx \Delta \boldsymbol{r} \cdot \nabla F_i = m \Delta \boldsymbol{r} \cdot \nabla g_i \,, \tag{11.2}$$

where we've used $F = mg$ and the fact that $|\Delta r|$ is small. Figure 11.4 shows schematically the

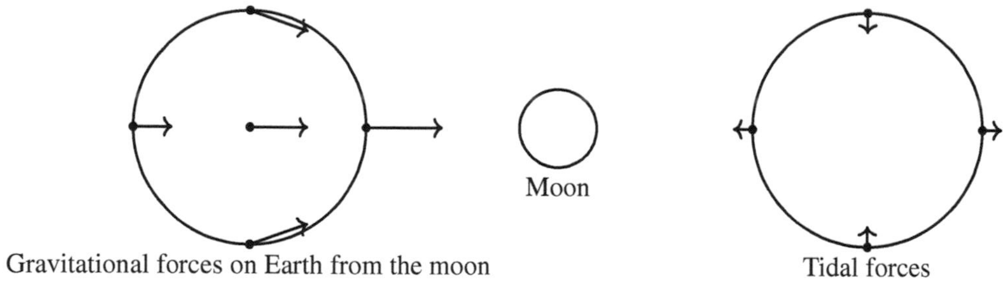

Figure 11.4 Tidal forces on Earth produced by the moon.

tidal forces experienced on Earth, which are what remain of the gravitational field of the moon after we subtract the force on the center of Earth.[4] There's always a reference location in a discussion of tidal forces; here it's the center of Earth. The earth is in *free fall* around the moon, free fall being *motion with no acceleration other than that provided by gravity*. A reference frame in free fall can be treated as an IRF because of the equivalence of inertial and gravitational mass. The tidal force is what we experience of the moon's gravitational field in the rest frame of the earth.

Figure 11.5 shows a coordinate system to analyze tidal forces. At a distance r from the center of Earth, we find, using Eq. (11.2), the tidal force in the radial (z) direction,

$$f_z = \frac{2GMm}{r^3} z + O(z^2/r^4) \,, \tag{11.3}$$

[3]That the tidal force depends on a reference position $\boldsymbol{r}$ indicates that it's a vector field.

[4]There are also tidal forces associated with the sun.

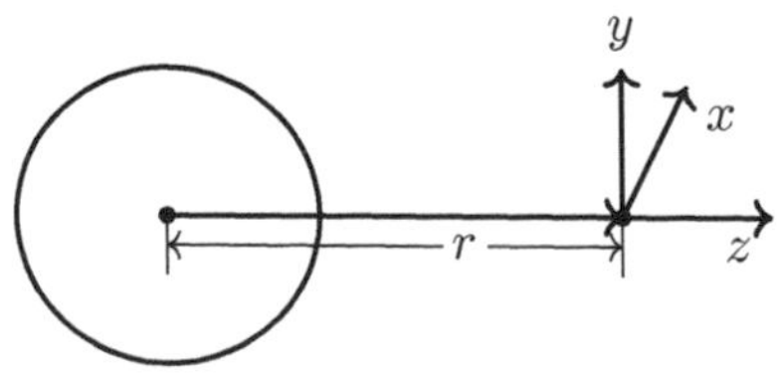

Figure 11.5 Coordinate system for tidal forces at distance r from the center of Earth.

where $|z/r| \ll 1$. Note that $f_z > 0$ ($f_z < 0$) for $z > 0$ ($z < 0$). Tidal forces *stretch* objects in the radial direction. In the lateral directions, tidal forces are *attractive*. The transverse component of the tidal force is given by

$$f_x = -\frac{GMm}{r^3}|x| + O(x^3/r^5) , \tag{11.4}$$

for $|x| \ll r$. Likewise, in the y direction, $f_y \approx -\left(GMm/r^3\right)|y|$.

The tidal force is the difference in forces experienced by neighboring particles. Such particles could be isolated test particles, or they could be infinitesimal mass elements in an extended object. If an object has finite extent in the x, y, z directions, then in an inhomogeneous field all three forces (f_x, f_y, f_z) are present simultaneously. An initially spherical object would be deformed under tidal stresses: stretched in the radial direction and compressed in the transverse directions—what happens to the oceans on Earth.

11.3.2 Tidal acceleration tensor

In terms of the gravitational potential $\Phi(\boldsymbol{r})$ (Section 1.7.2), the tidal force is, using Eq. (11.2),

$$f_i = m\Delta \boldsymbol{r} \cdot \boldsymbol{\nabla} g_i = m\sum_{l=1}^{3}\Delta r_l \frac{\partial g_i}{\partial x_l} = -m\sum_{l=1}^{3}\Delta r_l \frac{\partial^2 \Phi}{\partial x_l \partial x_i} . \tag{11.5}$$

The tidal force is thus a manifestation of the *curvature* of $\Phi(\boldsymbol{r})$. Applying Newton's second law to the difference in forces that make up the tidal force, we have the *relative acceleration* of the masses,

$$f_i = m\frac{\mathrm{d}^2\Delta r_i}{\mathrm{d}t^2} . \tag{11.6}$$

Equating Eqs. (11.5) and (11.6), the equation of motion for the inter-particle separation is

$$\frac{\mathrm{d}^2\Delta r_i}{\mathrm{d}t^2} = -\sum_{l=1}^{3}\frac{\partial^2\Phi}{\partial x_i \partial x_l}\Delta r_l \equiv \sum_{l=1}^{3} a_{il}\Delta r_l . \tag{11.7}$$

Equation (11.7) is the *Newtonian deviation equation.* It can be used (for $\Phi \ll c^2$ and $v \ll c$) to calculate the separation between nearby particles. The quantities $a_{ij} \equiv -\partial^2\Phi/\partial x_i \partial x_j$ form the *tidal acceleration tensor.* We'll derive the relativistic version of Eq. (11.7) in Chapter 14.

11.3.3 Torques due to tidal forces

Tidal forces produce torques. Referring to Fig. 11.5, the infinitesimal torque $\mathrm{d}\boldsymbol{\tau}$ on a mass $\mathrm{d}m$ at location (x, y, z) is, using Eqs. (11.3) and (11.4):

$$\mathrm{d}\boldsymbol{\tau} = \hat{\boldsymbol{x}}(yf_z - zf_y) + \hat{\boldsymbol{y}}(zf_x - xf_z) + \hat{\boldsymbol{z}}(xf_y - yf_x) = \frac{3GM}{r^3}\mathrm{d}m\,(\hat{\boldsymbol{x}}yz - \hat{\boldsymbol{y}}xz) .$$

Integrating over the mass (r is constant), the net torque is

$$\boldsymbol{\tau} = \frac{3GM}{r^3}\left(- \hat{\boldsymbol{x}} I_{yz} + \hat{\boldsymbol{y}} I_{zx} \right) ,$$

where I_{yz}, I_{zx} are the products of inertia. The tidal force thus causes a change in the angular momentum, the spin $\boldsymbol{S}$ of a rigid body, with

$$\frac{\mathrm{d}S_x}{\mathrm{d}t} = -\frac{3GM}{r^3} I_{yz} \qquad \frac{\mathrm{d}S_y}{\mathrm{d}t} = \frac{3GM}{r^3} I_{zx} \qquad \frac{\mathrm{d}S_z}{\mathrm{d}t} = 0 .$$

The spin vector *precesses* about the z-direction. *The rate of change of $\boldsymbol{S}$ is a measure of the tidal force field.*

The Gravity Probe B satellite experiment (GPB), concluded in 2011, measured two hitherto unverified precessional effects predicted by GR, the *geodetic* and *frame-dragging* effects.[44] These effects are quite small, and thus it was essential to eliminate precessions due to tidal forces to the greatest extent possible. The experiment utilized precisely manufactured spherical gyroscopes, with $I_{yz} = I_{zx} \approx 0$ to within a few parts in 10^6. We return to the GPB experiment in Chapter 17.

11.3.4 Freely falling reference frames

Consider a large laboratory falling toward Earth. In the lab depicted in the left portion of Fig. 11.6, the gravitational forces are different at the top and bottom of the laboratory. To an Earth-based

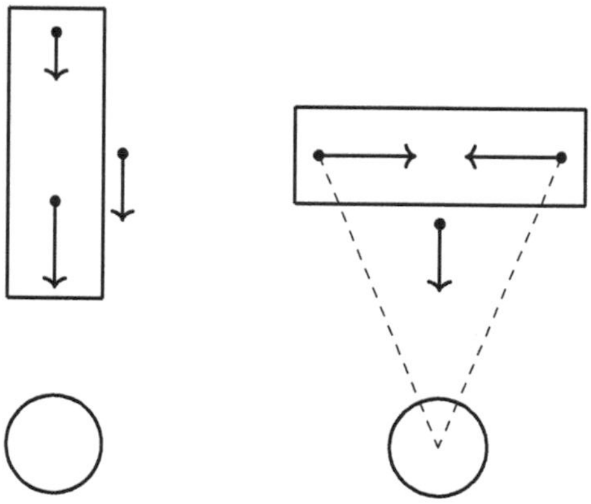

Figure 11.6 Tidal forces are repulsive in the falling laboratory on the left, and attractive in the lab on the right.

observer, the lower particle is accelerating faster than the upper particle. To an observer attached to the center of the laboratory—the *comoving* frame—test particles at those locations would appear to *repel* each other: The upper particle accelerates toward the top of the lab and the lower particle accelerates toward the bottom. Such a frame is *not* an IRF—particles released at rest do not remain at rest. This repulsive force is a fictitious force (Section 1.6.1): It's not due to a physical agency, but is an artifact of the reference frame. In the laboratory depicted in the right part of Fig. 11.6, the particles, which are falling toward the center of the earth, appear to *attract* each other.

If the *size* of the reference frame is restricted so that $\boldsymbol{\nabla} g \cdot \Delta \boldsymbol{r} \ll g$, $\boldsymbol{g}(\boldsymbol{r}) \approx g$ and differential accelerations between nearby particles are eliminated. A freely falling frame of sufficiently limited extent is therefore an IRF. We'll call this a *local* IRF. The transition from SR to GR is the transition from *global* IRFs of unlimited extent to local IRFs.

11.4 WEAK AND STRONG EQUIVALENCE PRINCIPLES

It's instructive to quote from Einstein's 1911 article (my emphasis added):[9, p99]

> In a homogeneous gravitational field (acceleration of gravity g) let there be a stationary system of coordinates S, oriented so that the lines of force of the gravitational field run in the negative direction of the z-axis. In a space free of gravitational fields, let there be a second system of coordinates S', moving with uniform acceleration g in the positive direction of the z-axis. ... Relatively to S, as well as relatively to S', material points which are not subjected to the action of other material points move in keeping with the equations
>
> $$\frac{\mathrm{d}^2x}{\mathrm{d}t^2} = 0, \frac{\mathrm{d}^2y}{\mathrm{d}t^2} = 0, \frac{\mathrm{d}^2z}{\mathrm{d}t^2} = -g \, .$$
>
> For the accelerated system S' this follows directly from Galileo's principle,[5] but for the system S, at rest in a homogeneous gravitational field, from the experience that all bodies in such a field are equally and uniformly accelerated. This experience, of the equal falling of all bodies in the gravitational field, is one of the most universal which the observation of nature has yielded; but in spite of that the law has not found any place in the foundations of our edifice of the physical universe.
>
> But we arrive at a very satisfactory interpretation of this law of experience, if we assume that *the systems S and S' are physically exactly equivalent, that is, if we assume that we may just as well regard the system S as being in a space free from gravitational fields, if we then regard S as uniformly accelerated.*
>
> As long as we restrict ourselves to purely mechanical processes in the realm where Newton's mechanics holds sway, we are certain of the equivalence of the systems S and S'. But this view of ours will not have any deeper significance unless the systems S and S' are equivalent with respect to all physical processes, that is, unless the laws of nature with respect to S are in entire agreement with those with respect to S'. By assuming this to be so, we arrive at a principle, which if it is really true, has great heuristic importance. For by theoretical consideration of processes which take place relatively to a system of reference with uniform acceleration, we obtain information as to the career of processes in a homogeneous gravitational field.

The EP is, as Einstein says, of "great heuristic importance" in that it lets us investigate physics in a gravitational field, physics that we might not understand (such as propagation of light in a gravitational field), by considering the same physics in an accelerated frame. Einstein recognized there might be different *versions* of the EP according to the physics that's equivalent between a frame in a gravitational field and an accelerated frame. Under the overall banner of the EP, refinements have been put forth: the *weak equivalence principle* (WEP) and the *strong equivalence principle* (SEP).

The WEP is simply $m_g = m_i$, that the motion of freely falling test particles is independent of composition and structure.[6] Einstein conjectured that *all* laws of physics, such as electromagnetism and quantum phenomena, would be equivalent between a frame in a uniform gravitational field and an accelerated frame, which is the SEP. The WEP replaces "laws of physics" with "laws of motion of freely falling test particles." The distinction between the SEP and the WEP is useful—it's possible there could be physical effects that violate the SEP and still satisfy the WEP. The name *weak* EP is unfortunate, however, as it might imply that the WEP is lacking in import. To the contrary, the WEP, through Eötvös-type experiments, is one of the most securely established results in all of physics! The WEP is not weak!

[5] By *Galileo's principle*, Einstein means the invariance of acceleration under the GT. S and S' are shown in Fig. 11.2.

[6] A *test particle* is one with no internal structure other than its ability to fall in gravity: electrically neutral; negligible gravitational binding energy compared to its rest mass; negligible angular momentum; and small enough that inhomogeneities of the gravitational field have negligible effect on its motion.

11.5 SPACETIME IS GLOBALLY CURVED, LOCALLY FLAT

The WEP establishes one of the main paradigms of GR, that spacetime is flat locally.[7] In a space free of gravitational fields, a released object floats (left portion of Fig. 11.7).[8] In a freely falling reference

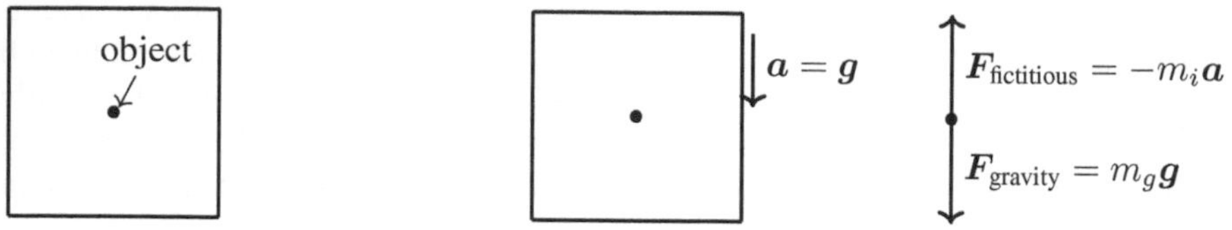

Figure 11.7 Left: Laboratory in free space; released object floats. Right: Freely falling laboratory; released object floats.

frame (right portion of Fig. 11.7) a released object floats; in such a frame the object experiences the force of gravity $F = m_g g$ and the apparent fictitious force, $F = -m_i a$, where $a = g$. The net force is zero (because $m_g = m_i$) and the object floats. *A sufficiently small freely falling reference frame is an IRF*: Gravity disappears! Gravitational effects over a small enough region can be obviated by letting the laboratory fall.

Every freely falling frame of sufficiently small size is a local inertial frame, and hence *is one in which SR applies*, and the spacetime of SR is flat. Thus, in regions of spacetime small enough that gravitational fields are uniform, spacetime is locally flat. A flat *region* of spacetime cannot "cover" all of spacetime, which GR shows is globally curved.

The power of the WEP becomes apparent when combined with the principle of covariance. By the WEP, at any point in an arbitrary gravitational field there is a local inertial frame in which the effects of gravitation are absent. This allows us to write the equations governing any sufficiently small system in a gravitational field if we first know the equations governing it in the *absence* of gravity. We need only write the equations in covariant form. Such equations will be true in the presence of gravity because covariance guarantees that if they're true in one set of coordinates, they're true in all coordinates, and the WEP tells us that there *is* a set of coordinates in which the equations are true—the local IRF at the spacetime point of the system.

11.6 ENERGY COUPLES TO GRAVITY

We established in Chapter 7 the equivalence of mass and energy, $E = mc^2$, where m in this formula is the inertial mass. The equivalence of inertial and gravitational mass would indicate that *energy couples to gravity* as if it had mass, $m_g = E/c^2$ (demonstrated below). Energy is subject to gravitation, and one can have energy without mass (photons).[9]

Gravitational mass of energy

In Fig. 11.8, S is a reference frame at rest in a uniform gravitational field $\boldsymbol{g}$ oriented in the negative z-direction, with instruments K_1, K_2 rigidly attached to the z-axis a distance Δz apart so that the gravitational potential of K_2 exceeds that of K_1 by $g\Delta z$. Instrument K_1 can emit a definite amount of energy E, which K_2 receives. Following the "recipe" of the EP, let S' be the *equivalent* frame that's gravity free but accelerating in the positive z-direction with $a = g$, with identical instruments K_1 and K_2. Let energy E be emitted in S' from K_1 toward K_2 when S' is instantaneously at rest

[7] In Section 14.4.4 we show mathematically that any manifold is locally flat. With the EP we have a physical reason why this must be the case. Flatness is defined as the vanishing of the Riemann tensor, Section 14.4.

[8] Compare Fig. 11.7 with Fig. 11.2.

[9] Energy is more general than mass: Energy is conserved, but not mass (Chapter 10).

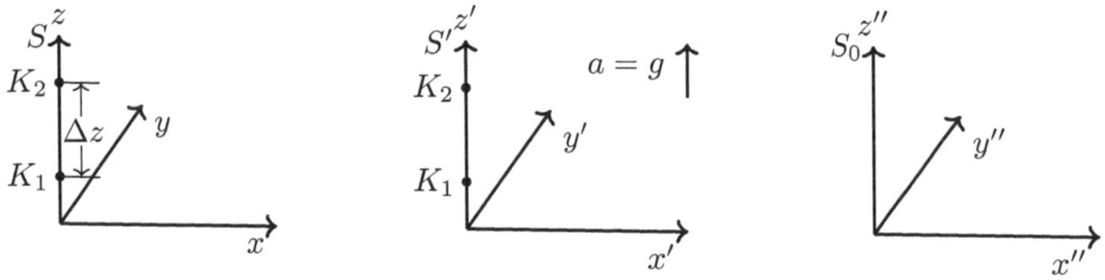

Figure 11.8 Frame S at rest in a uniform gravitational field, S' the equivalent accelerated, gravity-free frame, and S_0 an auxiliary IRF.

with respect to IRF S_0. Radiation arrives at K_2 at time $\Delta t = \Delta z/c$, when K_2 has speed relative to S_0, $v = g\Delta z/c$. Energy is emitted when K_1 is at rest relative to an IRF, and received by K_2 when it has speed $g\Delta z/c$ in an IRF. We can use the LT between these frames.

Use the photon four-momentum $Q^\mu = (E/c)(1, \hat{\boldsymbol{n}})$, where $\hat{\boldsymbol{n}}$ is the direction of propagation (Eq. (7.18)). For a photon traveling in the positive z-direction, we have the LT for a frame also moving in the positive z-direction:

$$\frac{E'}{c}\begin{pmatrix}1\\0\\0\\1\end{pmatrix} = \frac{E}{c}\begin{pmatrix}\gamma & 0 & 0 & -\beta\gamma\\ 0 & 1 & 0 & 0\\ 0 & 0 & 1 & 0\\ -\beta\gamma & 0 & 0 & \gamma\end{pmatrix}\begin{pmatrix}1\\0\\0\\1\end{pmatrix} .$$

Thus, $E' = E\gamma(1-\beta)$. The radiation measured at K_2 does not have energy E, but rather a *smaller* energy E'. At lowest order in β,

$$E' = E(1-\beta) = E\left(1 - g\Delta z/c^2\right) . \tag{11.8}$$

By the EP, the same relation holds in S, which is not accelerated but situated in a gravitational field. In S, we may replace $g\Delta z$ in Eq. (11.8) by the difference in gravitational potential between K_2 and K_1, $\Delta\Phi = g\Delta z$, leaving us with:

$$E' = E - \frac{E}{c^2}\Delta\Phi . \tag{11.9}$$

Equation (11.9) is a statement of energy conservation if we recognize *the photon energy E as having an equivalent gravitational mass E/c^2*. The association of mass with energy, $m = E/c^2$, is the converse of $E = mc^2$, the association of energy with mass. Work is done in traversing the gravitational field, and the energy arriving at K_2 is reduced by an amount $(E/c^2)\Delta\Phi$. If we had reversed the roles of K_2 and K_1, the energy arriving *from* the higher gravitational potential would be *greater* by an amount $(E/c^2)\Delta\Phi$, where E is the energy emitted.

Gravity couples to all forms of energy

Gravitational fields and acceleration cannot be distinguished on the basis of releasing test particles. What about more realistic particles? Nucleons (proton or neutron) participate in the strong force, but electrons don't; it's not unreasonable to ask whether different particles couple to gravity differently. The mass of bound systems—such as nuclei—is *less* than the mass of their parts by the mass equivalent of the *binding energy*,[10] $\Delta m = E_b/c^2$. Gravity must couple to binding energies so that

[10]The binding energy E_b is the energy required to separate a nucleus into its constituent parts. For a nucleus with A nucleons and Z protons, of mass $M(A,Z)$, $E_b(A,Z) \equiv Am_nc^2 + Z(m_p - m_n)c^2 - M(A,Z)c^2$, where m_n (m_p) is the free neutron (proton) mass.

the gravitational mass of the composite system is less than that of its constituent parts. The binding energy, however, depends on the nature of the forces holding the system together—it's a function of the number of nucleons A (which couple to the strong force), and the number of protons Z (which couple to the electromagnetic force).[11] The ratio A/Z is unity in hydrogen, and approximately 2.5 in heavier atoms. Differences in the gravitational force experienced by nuclei of different A/Z ratio (a violation of the SEP) are in principle subject to experimental investigation.

What about gravitationally bound systems, *does the gravitational binding energy contribute equally to the inertial and gravitational masses?* Stated more inclusively: Does gravity couple to *all* forms of energy in an equivalent manner? If the SEP were violated, Earth and Moon, having different gravitational binding energies would have different ratios of inertial to gravitational mass, and hence would have different accelerations toward the sun. Such an effect would eventually lead to a polarization of the moon's orbit around the earth, a phenomenon known as the *Nordvedt effect*.[45] The Lunar Laser Ranging Experiment rules out the Nordvedt effect to high precision and provides experimental support for the SEP.[12]

11.7 GRAVITY AFFECTS TIME

11.7.1 Gravitational frequency shift

We now establish another equivalence associated with gravity, that supplied by the association of energy with frequency through the Planck relation, $E = h\nu$. Gravity couples to energy and energy is related to frequency. Ergo, *gravity affects time*. By applying the Planck formula to Eq. (11.9), the frequency of the received radiation, ν, is related to the frequency of the emitted radiation ν_0 by

$$\nu = \nu_0 \left(1 - \frac{\Delta\Phi}{c^2}\right) , \tag{11.10}$$

where $\Delta\Phi \equiv \Phi_{\text{rec}} - \Phi_{\text{emit}}$. Rewrite Eq. (11.10) as

$$\frac{\nu - \nu_0}{\nu_0} \equiv \frac{\Delta\nu}{\nu_0} = -\frac{\Delta\Phi}{c^2} . \tag{11.11}$$

If light travels "uphill," out of a gravitational potential well, $\Delta\Phi > 0$, and the received frequency is less than that transmitted; it's shifted toward the red end of the spectrum, *redshifted*. For light that travels "downhill," further into a gravitational well, $\Delta\Phi < 0$ and the received frequency is greater than that transmitted; the light is *blueshifted*. The gravitational frequency shift occurs because of 1) energy conservation and 2) energy-mass equivalence; it's not a consequence (yet) of GR, although GR naturally incorporates these effects into the theory.

11.7.2 The Pound-Rebka-Snider experiment

The gravitational frequency shift was verified in the 1959 Pound-Rebka experiment [47] and in the 1964 Pound-Snider experiment [48] which measured the shift in photon energy in either falling or rising through a height of 22.6 m. The expected fractional change in frequency is, from Eq. (11.11),

$$\frac{\Delta\nu}{\nu_0} = \frac{g\Delta z}{c^2} = \frac{9.8\text{m/s}^2 \times 22.6\text{m}}{(3 \times 10^8 \text{m s}^{-1})^2} = 2.46 \times 10^{-15} ,$$

[11]The Weizsäcker formula is an expression for nuclear binding energies:

$$E_b(A, Z) = a_v A - a_s A^{2/3} - a_c \frac{Z^2}{A^{1/3}} - a_a \frac{(A - 2Z)^2}{A} - a_p \frac{1}{A^{1/2}} ,$$

where the parameters $\{a_i\}$ are obtained from a best fit to measured binding energies.

[12]Reflecting arrays were left on the surface of the moon by the astronauts of the *Apollo 11, Apollo 14*, and *Apollo 15* missions. The Lunar Laser Ranging Experiment measures the earth-moon distance using, you guessed it, laser ranging. It's found that a SEP violation parameter (similar to the Eötvös parameter) has the value [46] $\eta = (4.4 \pm 4.5) \times 10^{-4}$.

a small effect. Pound and Rebka measured a fractional frequency shift of $(2.57 \pm 0.26) \times 10^{-15}$, and thus $\Delta\nu_{\text{expt}}/\Delta\nu_{\text{theory}} = 1.05 \pm 0.10$. Pound and Snider improved the agreement between theory and experiment with $\Delta\nu_{\text{expt}}/\Delta\nu_{\text{theory}} = 1.00 \pm 0.01$.

How was such a small effect measured? The experiment used the 14.4 keV photon from the γ decay $^{57}\text{Fe}^* \to {}^{57}\text{Fe} + \gamma$ and allowed it to "fall" through 22.6 m, where they observed its absorption by a ^{57}Fe target. The excited state $^{57}\text{Fe}^*$ has a relatively long lifetime $\approx 10^{-7}$ s, leading to an approximate energy width of the photon, $\Gamma \approx 10^{-8}$eV; Pound and Rebka measured $\Gamma = 1.6 \times 10^{-8}$ eV. The fractional width of the γ-ray is thus $\Gamma/E = 1.1 \times 10^{-12}$. The absorption (by the reverse process, $^{57}\text{Fe} + \gamma \to {}^{57}\text{Fe}^*$) of such a narrow γ-ray would normally be impossible due to the recoil energy of the nucleus, E_R. Using Eq. (P11.1), $E_R \approx 2 \times 10^{-3}$ eV; the recoil energy *far* exceeds the energy width, $E_R \gg \Gamma$. The experiment would not have been possible were it not for the Mössbauer effect, where the nucleus is embedded in a crystal lattice; in this case E_R is absorbed by the crystal so that essentially no energy is lost to recoil, on emission or absorption.

Narrow as it is, the width of the 14.4 keV photon is 500 times *larger* than the effect they were trying to measure, seemingly rendering the experiment impossible. The photon would need to "fall" through a distance of 10 km before $\Delta\Phi/c^2 \approx \Gamma/E$. Pound and Rebka devised a method that allowed the experiment to succeed. If the γ-ray source is imparted a small velocity v, there is a Doppler shift of the spectral line, $(\Delta E/E)_{\text{Doppler}} = \beta$. If $\beta \gtrsim \Gamma/E$, absorption of the γ-ray is impossible. The maximum velocity one could give to the emitter and still have the γ-ray be absorbed is $\beta_{\text{max}} = \Gamma/E \approx 10^{-12}$ or $v = 3 \times 10^{-2}$ cm/s. Pound and Rebka used the Doppler shift to offset the expected gravitational frequency shift. For the configuration where photons are "dropped," we expect a blueshift from the energy gained by the photon. Imparting a slight velocity upward to the emitter would produce a Doppler redshift, and decrease the energy of the photon. The energy arriving at the absorber would then be $E_{\text{arrive}} = E_0 + (\Delta E)_{\text{gravity}} + (\Delta E)_{\text{Doppler}} = E_0 + E_0\left(\Delta\Phi/c^2\right) - \beta E_0$. Maximum absorption occurs for $E_{\text{arrive}} = E_0$, when

$$\beta = \frac{\Delta\Phi}{c^2} = \frac{g\Delta z}{c^2} \,. \tag{11.12}$$

The Doppler shift that offsets the anticipated gravitational blueshift occurs for $v = 7.38 \times 10^{-5}$ cm s^{-1}, a small velocity! Note that Eq. (11.12) holds independent of the photon energy, and hence its gravitational mass; the EP in action.

In the experiment, the ^{57}Fe absorber covered a scintillator, which was viewed by a photomultiplier tube. The γ-ray source was attached to a transducer, which could impart a small velocity. The velocity of the transducer was modulated periodically, with $v = v_0 \cos\omega t$, where v_0 is arbitrary, as long as it's much greater than 7.4×10^{-5} cm/s. Maximum absorption, where Eq. (11.12) is met, corresponds to a *minimum* counting rate from the photomultiplier tube. This minimum signal was correlated with the phase of the transducer, from which they could obtain the velocity. *The experiment showed that the photon has an interaction with the gravitational field in accord with the predictions of the EP.*

11.7.3 Gravitational time dilation

The gravitational frequency shift, Eq. (11.10), presents a conundrum. How do we reconcile the experimentally observed frequency shift with the fact that the number of wave crests reaching the receiver is the same as that emitted by the source? Wave crests are a periodic phenomenon that can be regarded as the ticks of a clock. The gravitational frequency shift has been experimentally verified, and the *number* of ticks (wave crests) is the same at both clocks, and the clocks are identical. We must conclude that the time *interval* between successive ticks is altered by the local gravitational field. *Time is affected by gravity*! Let's let Einstein speak for himself (refer to Fig. 11.8):

> On a superficial consideration, Eq. (11.10) seems to assert an absurdity. If there is constant transmission of light from K_1 to K_2, how can any other number of periods per

> second arrive at K_2 than is emitted at K_1? But the answer is simple. We cannot regard ν_1 or ν_2 simply as frequencies (as the number of periods per second) because we have not yet determined the time in system S. What ν_1 denotes is the number of periods with reference to the time-unit of the clock U in K_1, while ν_2 denotes the number of periods per second with reference to the identical clock in K_2. Nothing compels us to assume that the clocks U in different gravitational potentials must be regarded as going at the same rate.[9, p105]

We can view Eq. (11.10) as stating

$$\nu(r + \Delta r) = \nu(r)\left(1 - \Delta\Phi/c^2\right) , \tag{11.13}$$

where $\Delta\Phi \equiv \Phi(r + \Delta r) - \Phi(r)$. Define a local unit of time as the time between wave crests, $t(r) \equiv 1/\nu(r)$, so that Eq. (11.13) is equivalent to

$$\frac{t(r + \Delta r) - t(r)}{t(r + \Delta r)} = \frac{\Delta\Phi}{c^2} . \tag{11.14}$$

In the limit $\Delta r \to 0$, Eq. (11.14) becomes $\frac{1}{t}\frac{\mathrm{d}t}{\mathrm{d}r} = \frac{1}{c^2}\frac{\mathrm{d}\Phi}{\mathrm{d}r}$, or simply

$$\frac{\mathrm{d}t}{t} = \frac{1}{c^2}\mathrm{d}\Phi . \tag{11.15}$$

Equation (11.15) expresses the change in local time scale with a small change in gravitational potential. Even though Eq. (11.15) cries out to be integrated, $t(\Phi_2) = t(\Phi_1)\exp\left(\Delta\Phi/c^2\right)$, where $\Delta\Phi = \Phi_2 - \Phi_1$, the temptation should be resisted. Why? The argument leading to Eq. (11.8) is based on small velocities and small separations in space; Eq. (11.15) is approximate.[13] We derive in Chapter 17 the result that supplants Eq. (11.15),

$$\mathrm{d}t(r) = \mathrm{d}t(\infty)\sqrt{1 + 2\Phi(r)/c^2} , \tag{11.16}$$

where $\mathrm{d}t(\infty)$ is the time scale where $\Phi = 0$. Equation (11.16) expresses the effect mentioned in Chapter 1: Clocks run slower the lower they are in a gravitational potential.

We cannot observe time dilation locally, at one point in space; the gravitational field affects our time standards the same as it affects the clock being studied. We must compare clocks at *different* points in space. Comparing clocks at locations r_1 and r_2, we have from Eq. (11.16)

$$\frac{\mathrm{d}t_1}{\mathrm{d}t_2} = \sqrt{\frac{1 + 2\Phi(r_1)/c^2}{1 + 2\Phi(r_2)/c^2}} . \tag{11.17}$$

For weak fields ($\Phi/c^2 \ll 1$) and Eq. (11.17) reduces to

$$\frac{\mathrm{d}t_1}{\mathrm{d}t_2} \approx 1 + \left(\Phi_1 - \Phi_2\right)/c^2 = 1 + \Delta\Phi/c^2 , \tag{11.18}$$

the same as Eq. (11.15), $\Delta t/t = \Delta\Phi/c^2$.

---

[13] Einstein recognized the approximate nature of the conclusions reached this way. From his 1911 article:[9, p99] "The relations here deduced, even if the theoretical foundation is sound, are valid only to a first approximation." One must have GR at hand to have a tool capable of incorporating all relevant physics without approximation.

SUMMARY

Because $m_i = m_g$, the effects of a gravitational field cannot be distinguished from a linearly accelerated reference frame over regions of space small enough that tidal forces are negligible. The EP is a cornerstone of GR, along with the principle of covariance. It has several consequences.

- Spacetime is locally flat. Because of the EP, we cannot distinguish between an IRF in the absence of gravity, and a small reference frame freely falling in a gravitational field. Every sufficiently small freely falling frame is locally an IRF, implying that in regions of spacetime small enough that gravitational fields are uniform, spacetime is locally flat.
- Energy couples to gravity. Because of the equivalence of mass and energy, $E = mc^2$, and the equivalence $m_i = m_g$, energy acts as if it has a gravitational mass, $m_g = E/c^2$.
- Photons undergo a frequency shift in a gravitational field, with $\Delta\nu/\nu = -\Delta\Phi/c^2$. Photons are redshifted if they climb out of a gravitational potential well, and blueshifted if they fall into a gravitational potential well.
- Clocks run slower the further they are into a gravitational potential well (gravitational time dilation). Over small changes in Φ, time is altered according to $\Delta t/t = \Delta\Phi/c^2$.

EXERCISES

11.1 In your hand you hold a can of water that's open to the atmosphere at the top. At the same time you puncture a hole in the side of the can at a distance h below the water line, you let go of the can. Describe what happens to the water. Hint: From Bernoulli's principle, the velocity of the water escaping the hole is $v = \sqrt{2gh}$.

11.2 Derive Eq. (11.3). Start with Eq. (11.2), and use the formula for the Newtonian gravitational force at locations r and $r + z$. Assume that $z \ll r$.

11.3 The tidal force considered as a vector field has zero divergence, $\nabla \cdot \boldsymbol{f} = 0$.

a. Show directly from Eqs. (11.3) and (11.4) that $\nabla \cdot \boldsymbol{f} = 0$. Do not differentiate with respect to r (which should be treated as a constant). These formulas are valid only for x, y, z small compared to r.

b. Show from Eq. (11.5) that this result is to be expected. In forming the divergence using Eq. (11.5), you should recognize the Poisson equation—see Table 1.1. Is there a local *source* of tidal forces?

11.4 Compute the curl of the tidal force vector field $\boldsymbol{f}$ using Eqs. (11.3) and (11.4) (treat r as a constant). You should find $\nabla \times \boldsymbol{f} = 0$. Thus we have $\nabla \cdot \boldsymbol{f} = 0$ and $\nabla \times \boldsymbol{f} = 0$. Why doesn't this present a violation of Helmholtz's theorem of vector calculus? (Not treated in this book; look it up!)

11.5 a. Which exerts a stronger gravitational force on the earth, the moon or the sun? Calculate the ratio of the forces acting on the earth, $F_{\text{Sun}}/F_{\text{Moon}}$.

b. Which produces a stronger *tidal* force at the earth, the moon or the sun? Calculate the ratio of $f_{\text{Moon}}/f_{\text{Sun}}$. Use Eq. (11.3).

11.6 A spherical solid moon with mass M_m and radius R orbits a planet with mass M_p at distance r (distance from center of planet to center of moon). Show that if $r = R\,(2M_p/M_m)^{1/3}$, loose rocks on the surface of the moon will become dislodged due to tidal effects. This distance is known as the *Roche limit*. A moon approaching a planet closer than the Roche limit will be disrupted by tidal forces. What is the Roche limit for the earth-moon system? Express your answer in units of the radius of the earth. What is the distance between the earth and the moon, in units of the radius of the earth?

11.7 When a quantum system makes a radiative transition between energy levels E_2 and E_1, the energy of the photon is $E_\gamma = E_2 - E_1$, right? The actual energy of the emitted photon is *reduced* from the nominal value ($E_0 \equiv E_2 - E_1$) by the *recoil energy* of the emitter. In the rest frame of the emitter, the recoil momentum is $p = E_\gamma/c$.

a. Energy conservation requires that $E_0 = E_\gamma + E_R$, where E_R is the kinetic energy of the recoil, $E_R = (\gamma - 1)Mc^2$, with M the mass of the emitter. Show that

$$E_0 - E_\gamma = (\gamma - 1)Mc^2 = Mc^2(\sqrt{1 + (p/Mc)^2} - 1) = Mc^2(\sqrt{1 + (E_\gamma/Mc^2)^2} - 1)\,.$$

b. Show that $E_\gamma = E_0\left(1 + E_0/(2Mc^2)\right) / \left(1 + E_0/(Mc^2)\right)$. Clearly, $E_\gamma < E_0$.

c. Show that $E_R = \left(E_0^2/2Mc^2\right) / \left(1 + E_0/Mc^2\right)$. If $E_0 \ll Mc^2$,

$$E_R = \frac{E_0^2}{2Mc^2} + O\big(E_0^3/(Mc^2)^2\big)\,. \tag{P11.1}$$

11.8 Consider a photon emitted from the sun. Is the photon redshifted or blueshifted when received on Earth? (You choose a frequency of the photon.) What's the total change in gravitational potential from the surface of the sun to the surface of the earth? Calculate the energy shift of the received photon.

CHAPTER 12

Acceleration in special relativity

ONE often hears that SR can't "handle" accelerated motion, and that to treat acceleration GR must be invoked. GR is a theory of gravitation that entails a connection between energy-momentum and the spacetime metric; until that relation is established, we haven't left the realm of SR. What's fair to say is that accelerated motion isn't handled in SR as *naturally* as inertial motion. In this chapter we consider applications of SR to accelerated motion.

We showed in Chapter 3 how the three-acceleration $\boldsymbol{a}$ transforms using the LT, which involves not just the relative velocity $\boldsymbol{v}$ between frames, but also the instantaneous velocity $\boldsymbol{u}$ of a particle in one of the frames. For $\boldsymbol{a}$ parallel to $\boldsymbol{v}$, we have from Eq. (3.34):

$$a = a' \frac{(1 - v^2/c^2)^{3/2}}{(1 + u'v/c^2)^3} . \tag{12.1}$$

SR is based on the equivalence of IRFs, and clearly there is no IRF in which an accelerated particle is always at rest. There is, however, an IRF for which any event on the worldline of a material particle is momentarily at rest, the instantaneous rest frame; see Fig. 7.1. In such a frame, $u' = 0$, and a' is called the *proper acceleration*. It's the device of instantaneous rest frames that allows us to treat accelerated motion in SR.[1]

12.1 LINEAR ACCELERATION

We treat the case of constant linear acceleration *as experienced in the frame of the object*, with $a' = g =$ constant.[2] Equation (12.1) prescribes the equation of motion of such an object in an instantaneous rest frame:

$$\frac{\mathrm{d}u}{\mathrm{d}t} = g\left(1 - u^2/c^2\right)^{3/2} . \tag{12.2}$$

Equation (12.2) can be integrated.[3] For the initial condition of starting at rest, we have the instantaneous speed at time t in "our" IRF:

$$u(t) = \frac{gt}{\sqrt{1 + g^2t^2/c^2}} . \tag{12.3}$$

[1] We used an instantaneous rest frame in Section 11.6 to derive the gravitational mass equivalent of energy, $m = E/c^2$.

[2] We use the symbol g to give it a familiar connotation.

[3] By integrating Eq. (12.2), we are finding the effect of a compound LT made up from a sequence of transformations between instantaneous rest frames. LTs have the group property (Section 4.3) that a LT followed by a LT is itself a LT.

For $t \ll c/g$ the speed varies linearly with time,[4] $u \sim gt$, but then approaches c asymptotically for $t \gg c/g$. Integrating Eq. (12.3), we obtain the equation of the worldline $x = x(t)$,

$$x(t) - x_0 = \frac{c^2}{g}\sqrt{1 + \frac{g^2t^2}{c^2}} - \frac{c^2}{g}, \tag{12.4}$$

where x_0 is the initial position. Equation (12.4) is the equation of a hyperbola:

$$\left(x - x_0 + c^2/g\right)^2 - (ct)^2 = c^4/g^2, \tag{12.5}$$

the asymptotes of which are the lines $x = x_0 - c^2/g \pm ct$. The asymptotes intersect at the focus of the hyperbola P, at $t = 0$ and $x = x_0 - c^2/g$, (see Fig. 12.1). The worldline lies *outside* the future light cone of P. The asymptote is termed an *event horizon*, a boundary in spacetime beyond which signals will never reach the accelerated observer.

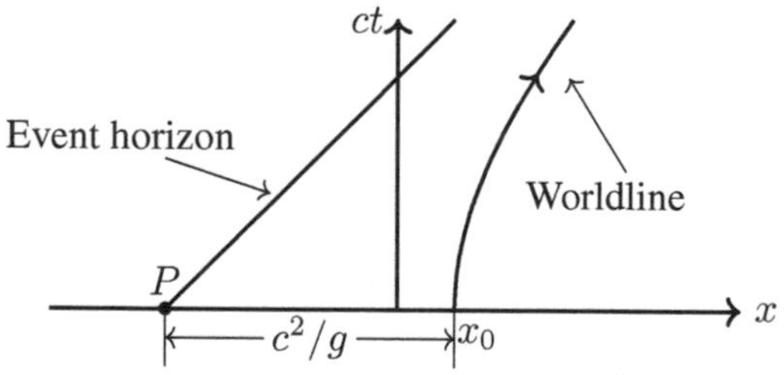

Figure 12.1 Worldline of constant acceleration and event horizon, the future lightline of P.

Accelerated worldlines have some unusual properties in relation to the event horizon. By combining Eqs. (12.3) and (12.4), the instantaneous velocity can be written

$$\beta(t) = \frac{u(t)}{c} = \frac{gt/c}{\sqrt{1 + g^2t^2/c^2}} = \frac{ct}{x(t) - x_0 + c^2/g} = \tan\theta, \tag{12.6}$$

where $\theta < \pi/4$ is shown in Fig. 12.2. In LTs, the angle θ on a spacetime diagram between the

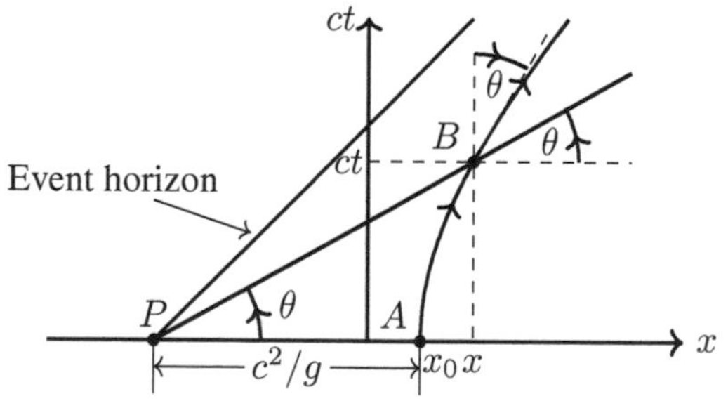

Figure 12.2 The instantaneous speed at B is $\beta = \tan\theta$.

x'-axis (of an IRF at moving at speed β) and the x-axis (of our IRF) is such that $\tan\theta = \beta$ (see Fig. 2.9). The line PB in Fig. 12.2 is therefore the x'-axis of the IRF in which the event B is momentarily at rest. The tangent to the worldline at B, the time axis of the instantaneous rest frame, is also at angle θ relative to the time axis of our IRF. As the speed increases, the time axis (tangent to the worldline) gets continually bent towards the direction of the lightline, with $\theta \to \pi/4$.

[4]For $g \sim 10\,\text{m}\,\text{s}^{-2}$, c/g is about 1 year; $t = c/g$ is the time to accelerate to the speed of light in Newtonian mechanics.

The x'-axis is in general a line of simultaneity (Section 2.4), and thus for the accelerated observer *all events on the line PB in Fig. 12.2 are simultaneous with B*, including P. Because B is chosen arbitrarily, however, we conclude that *P is simultaneous with every event on the worldline*! Such behavior is indicative of a *coordinate singularity*. Infinitely many lines $t' =$ constant pass through P (see Fig. 12.3).[5] The coordinate system breaks down at P. Event horizons and coordinate singularities are features of accelerated reference frames from a spacetime perspective.

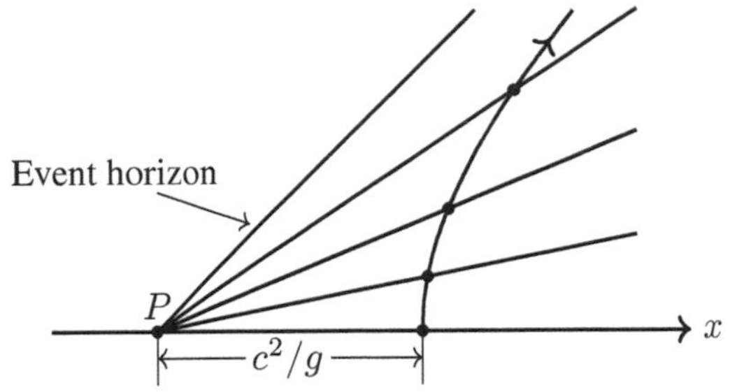

Figure 12.3 Lines of simultaneity. P is simultaneous with every event on the worldline.

There's another remarkable feature about constant linear acceleration: *The distance from the accelerated observer to P is constant*! Figure 12.4 shows the coordinates x and x' assigned to event

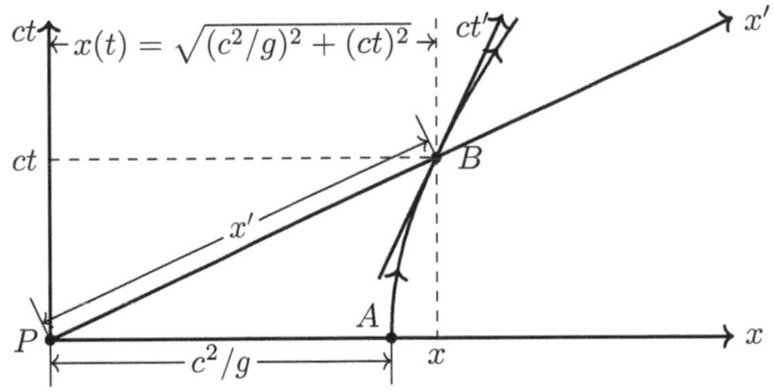

Figure 12.4 In the accelerated frame the distance PB is the same as PA, c^2/g.

B in two IRFs, which are therefore related by a LT. We find $x' = x/\gamma$, which can be seen from the LT:

$$\begin{pmatrix} ct \\ x \end{pmatrix} = \gamma \begin{pmatrix} 1 & \beta \\ \beta & 1 \end{pmatrix} \begin{pmatrix} 0 \\ x' \end{pmatrix} .$$

The instantaneous value of the Lorentz factor is, using Eq. (12.6), $\gamma(t) = \sqrt{1+(gt/c)^2}$. The coordinate x' is thus given by

$$x' = \frac{x}{\gamma} = \frac{\sqrt{(c^2/g)^2 + (ct)^2}}{\sqrt{1+(gt/c)^2}} = \frac{c^2}{g} , \tag{12.7}$$

a constant.

The proper time for the accelerated observer is found by combining Eq. (12.6) with Eq. (7.3):

$$\mathrm{d}\tau = \mathrm{d}t\sqrt{1 - \frac{u^2(t)}{c^2}} = \frac{\mathrm{d}t}{\sqrt{1+g^2t^2/c^2}} . \tag{12.8}$$

[5]We haven't drawn the time axis ct in Fig. 12.3 for simplicity, but it's of course still there. The equations simplify if the initial condition is taken as $x_0 = c^2/g$.

Letting $gt/c = \sinh\theta$, Eq. (12.8) reduces to $\mathrm{d}\tau = c\mathrm{d}\theta/g$, which is trivially integrated:

$$\tau = \frac{c}{g}\theta = \frac{c}{g}\sinh^{-1}(gt/c) \ , \tag{12.9}$$

where we assume $\tau(0) = 0$. Inverting Eq. (12.9), we have the connection between the proper time in the accelerated frame and the time in an IRF,

$$ct = \frac{c^2}{g}\sinh(g\tau/c) \ . \tag{12.10}$$

Equation (12.10) is not the linear relationship we have in SR, $t = \gamma\tau$. Combining Eqs. (12.10) and (12.4), we find

$$x - x_0 = \frac{c^2}{g}\left(\cosh(g\tau/c) - 1\right) \ . \tag{12.11}$$

Equations (12.10) and (12.11) comprise a parameterization of the worldline in terms of τ. In terms of the proper time, the instantaneous velocity is

$$u = (\mathrm{d}x/\mathrm{d}\tau)/(\mathrm{d}t/\mathrm{d}\tau) = c\tanh(g\tau/c) \ .$$

As $g \to 0$, $\tau \to t$, and $x \to x_0 + \frac{1}{2}gt^2$.

Example. An acceleration of 1 $\text{ly}/(\text{year})^2 = 9.5\ \text{m/s}^2$, approximately one Earth "g"! In these units $c = 1$ (one light year per year). Work out the clock time for an observer accelerating with this value of g to travel 10^{10} light years (ly), a sizable fraction of the universe. For large values of x, $\cosh x \approx \sinh x \approx \frac{1}{2}\mathrm{e}^x$. From Eq. (12.11), $\mathrm{e}^\tau \approx 2(\Delta x) = 2\times 10^{10}$, or $\tau \approx 23.7$ years. Not bad! 24 years of constant acceleration takes you across the universe. How much time transpires on an Earth clock? From Eq. (12.10) with $\tau = 23.7$ we have approximately 10^{10} years! Most of the journey is done at a speed, relative to Earth, of almost the speed of light, $u = c\tanh(23.7) \approx c$.

12.1.1 Coordinate system for linearly accelerated observers

A spacetime coordinate system (t', x') can be developed using the worldlines of accelerated particles, just as the worldlines of free particles comprise a coordinate system for Minkowski space. Figure 12.5 shows worldlines associated with different values g_i of the acceleration; these intersect

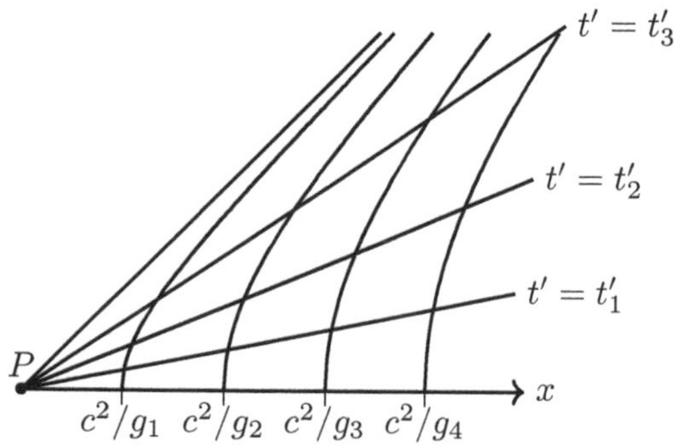

Figure 12.5 Hyperbolas associated with accelerated worldlines are a set of curves equidistant from P; lines of constant slope are lines of simultaneity.

the x-axis at different distances from P, c^2/g_i. Worldlines associated with different values of g thus comprise a set of curves of constant distance to P. The straight lines in Fig. 12.5 are lines of simultaneity. A locus of points equidistant from P is what we require of a *time axis*, just as a locus of points of constant time is what's required of a spatial axis. *Where* in spacetime we place a time axis is more involved than for Minkowski space—the spacetime of SR is homogeneous (Chapter 3), whereas now we have a distinguished feature, the event horizon.

From Eq. (12.6), $\beta\gamma = gt/c$. Because events on the same line connected to P have the same instantaneous speed ($\beta = \tan\theta$), they have the same value of the *product* gt, where, to be clear, t is the time in the IRF. Thus, for events 1 and 2 in Fig. 12.6, $g_1 t_1 = g_2 t_2$. From the invariance of gt on the same line, we infer from Eq. (12.10) the invariance of the product $g\tau$; for events 1 and 2 in Fig. 12.6, $g_1\tau_1 = g_2\tau_2$. From $g_1\tau_1 = g_2\tau_2$ and Eq. (12.10), we have

$$ct_2 = \frac{c^2}{g_2}\sinh(g_2\tau_2/c) = \frac{c^2}{g_2}\sinh(g_1\tau_1/c)\ . \tag{12.12}$$

We'll use this equation momentarily.

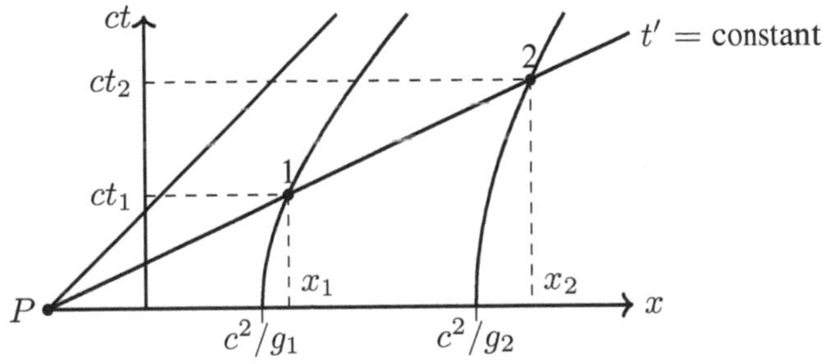

Figure 12.6 Events 1 and 2 are such that $g_1 t_1 = g_2 t_2$ and $g_1\tau_1 = g_2\tau_2$.

We now choose, arbitrarily, worldline 1 in Fig. 12.6 to be the t'-axis; we also place the origin of the (t, x) coordinate system at $t' = 0$ (see Fig. 12.7). Furthermore, we let τ_1 be the *coordinate time*

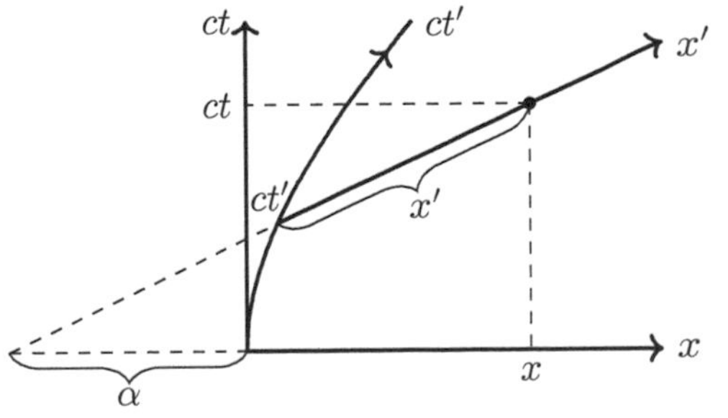

Figure 12.7 Coordinate system for accelerated observer.

t' for all events along the constant-time line from P. In Eq. (12.12) therefore we set $\tau_1 = t'$, giving us a connection between the time coordinates assigned to the same event:

$$ct = \frac{c^2}{g_2}\sinh\left(\frac{g_1}{c^2}ct'\right)\ . \tag{12.13}$$

The coordinate x' is the difference between the invariant distances from P where the worldlines intersect the x-axis: $x' = (c^2/g_2) - (c^2/g_1)$; thus $c^2/g_2 = c^2/g_1 + x'$. Equation (12.13) can therefore be written

$$ct = \left(x' + \frac{c^2}{g_1}\right) \sinh\left(\frac{g_1}{c^2} ct'\right) . \tag{12.14}$$

Let $\alpha \equiv c^2/g_1$ be a *parameter of the coordinate system*—the location of the t'-axis relative to P. For this coordinate system, therefore (that specified by the parameter α), we have a connection between (t', x') in the accelerated system, and t in the IRF:

$$ct = (x' + \alpha) \sinh(ct'/\alpha) . \tag{12.15}$$

The other equation, $x = x(t', x')$, can be obtained, referring to worldline 2 in Fig. 12.6, using Eq. (12.11) with $x_0 = c^2/g_2 - \alpha$, $g_2\tau_2 = g_1\tau_1 = g_1 t'$, and $x' = c^2/g_2 - c^2/g_1$:

$$x = (\alpha + x') \cosh(ct'/\alpha) - \alpha . \tag{12.16}$$

Equations (12.15) and (12.16) are the transformation equations between the uniformly accelerated (primed) system and the IRF (unprimed). The inverse transformation is not difficult to figure out:

$$x' = \sqrt{(x+\alpha)^2 - (ct)^2} - \alpha \qquad ct' = \frac{1}{2}\alpha \ln\left(\frac{x+\alpha+ct}{x+\alpha-ct}\right) . \tag{12.17}$$

The inverse transformation is not a simple matter of swapping primes for unprimes and reversing the velocity; unlike the LT, these transformations are nonlinear.

12.1.2 Metric tensor for linearly accelerated frame

The region to the right of the event horizon can thus be covered by two coordinate grids: the Cartesian grid (t, x) of an inertial observer and the hyperbolic grid (t', x') of accelerated observers. What is the metric tensor for the accelerated coordinate system? We can use Eq. (5.39), the transformation equation for $g_{\mu\nu}$, and the coordinate transformation equations (12.15) and (12.16) to build the Jacobian matrix:

$$g'_{\mu\nu} = A^\alpha_\mu A^\beta_\nu g_{\alpha\beta} , \tag{12.18}$$

where the derivatives (see Eq. (5.24))

$$\begin{aligned} A^0_0 &= \frac{\partial ct}{\partial ct'} = (1 + x'/\alpha)\cosh(ct'/\alpha) \qquad & A^0_1 &= \frac{\partial ct}{\partial x'} = \sinh(ct'/\alpha) \\ A^1_0 &= \frac{\partial x}{\partial ct'} = (1 + x'/\alpha)\sinh(ct'/\alpha) \qquad & A^1_1 &= \frac{\partial x}{\partial x'} = \cosh(ct'/\alpha) . \end{aligned} \tag{12.19}$$

Using the Lorentz metric for the IRF,

$$g_{\alpha\beta} = \begin{pmatrix} -1 & 0 \\ 0 & 1 \end{pmatrix} , \tag{12.20}$$

we find, combining Eqs. (12.18), (12.19), and (12.20), the metric tensor *field*[6] for the accelerated system:

$$g'_{\mu\nu}(x') = \begin{pmatrix} -(1 + x'/\alpha)^2 & 0 \\ 0 & 1 \end{pmatrix} . \tag{12.21}$$

The fact that $g'_{01} = 0$ in Eq. (12.21) implies that the basis vectors in the accelerated coordinate system are orthogonal. Referring to Fig. 12.8, lines from P to the worldline define the x'-axis in the instantaneous rest frame at the point of intersection. We can put a coordinate basis vector e_1 at that point, pointing in the spatial direction. The tangent to the worldline is the time axis in the instantaneous rest frame, and we can attach a timelike basis vector e_0 at the same point. Because the accelerated worldline is obtained from a progression of instantaneous rest frames, the local basis vectors remain orthogonal at each point of the worldline. We have indicated with dashed lines the light cone at each point.

[6]The metric tensor is different at each point in spacetime—a tensor field.

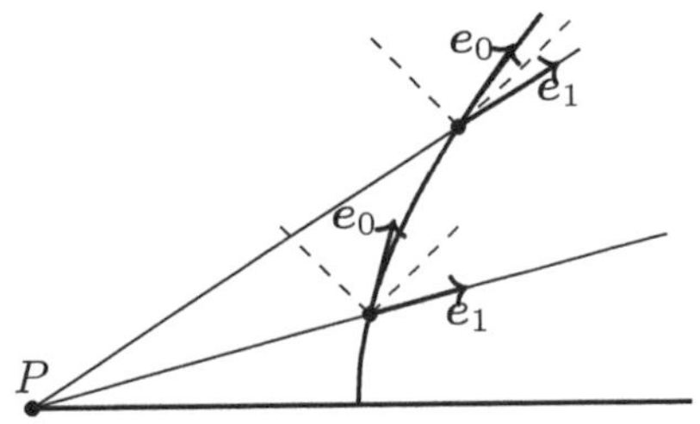

Figure 12.8 Timelike and spacelike basis vectors remain orthogonal along the worldline.

12.1.3 Time dilation for accelerated clocks

What is the implication of the *time* component g_{00} being a function of position, $g_{00} = g_{00}(x)$? From $(\mathrm{d}s)^2 = g'_{\mu\nu}\mathrm{d}x'^{\mu}\mathrm{d}x'^{\nu} = -(1+x'/\alpha)^2(c\mathrm{d}t')^2 + (\mathrm{d}x')^2$, we see that for a stationary clock ($\mathrm{d}x' = 0$) *the proper time depends on the location in the coordinate system*:

$$\mathrm{d}\tau = (1 + x'/\alpha)\mathrm{d}t' = \sqrt{|g_{00}(x')|}\mathrm{d}t' \ . \tag{12.22}$$

Positions are relative to the origin of coordinates, which in this case is at the distance c^2/g_1 from P. Relative to P, $x' = c^2/g - c^2/g_1$. A *positive* coordinate x' labels a worldline with a *weaker* acceleration $g < g_1$; negative x' labels a worldline of stronger acceleration $g > g_1$. The coordinate system, and hence the metric tensor, represents the effects of a spatially varying acceleration field, which, per the EP, we can think of as mimicking a gravitational field where the magnitude of the acceleration is dependent upon the distance from a singular location. That $\sqrt{|g_{00}|}$ becomes *larger* for $x' > 0$ implies that clocks "there" run *faster* than at $x' = 0$. Conversely, relative to a worldline at $x' > 0$, clocks at the origin run *slower*. These effects are analogous to gravitational time dilation, Section 11.7.3. There is a limit to how negative the coordinate x' can become, $x' > -c^2/g_1 = -\alpha$. As $x' \to -\alpha$, $g_{00} \to 0$, and time differences *vanish*: Point P is the intersection of an infinite number of lines with $t' =$ constant.

12.2 TWIN PARADOX

The twin paradox, another of the supposed paradoxes associated with SR, would customarily have been treated in an earlier chapter; it's been deferred to now so we can bring to bear the effects of acceleration. The problem is to predict which of two observers, E, who stays on Earth, or S, who goes off in a spaceship and returns, will be younger when they reunite. S takes off at speed $\beta = 0.8$ for a location $D = 4$ ly from Earth. If S turns around instantly upon reaching D and heads back with the same speed, S returns (by Earth clocks) in $T_E = 2D/\beta = 10$ years. S, however, sees D Lorentz-contracted to $D/\gamma = 2.4$ ly ($\gamma = 5/3$). By spaceship clocks S returns to Earth in $T_S = 2(D/\gamma)/\beta = T_E/\gamma = 6$ years. Apparently S ages less than E: Moving clocks run slow, including the clock of a beating heart. We showed in Section 4.2, however, that time dilation is a *symmetrical* effect: Both inertial observers conclude that a clock in motion runs slow. Each of E or S would conclude that the *other* is younger when they reunite. There is no contradiction, however: S is not a single IRF, whereas E is; there's no symmetry between the observers. As we have framed the problem, S is associated with *two* IRFs connected by an infinite acceleration. Paradox lost.

Figure 12.9 shows the spacetime diagram of the journey, where lines of simultaneity are shown in the E-frame (left) and in the S-frame (right). In S, there are *two* lines of simultaneity connected to event A, one associated with the outbound journey (B) and the other with the inbound journey (C). Thus, there's a discontinuity in the perceived age of E—an unphysical artifact of our assumption that S instantly reverses course and undergoes infinite acceleration. S must instantly "jump ship" to another reference frame with $\beta = -0.8$. Physically, S must experience an acceleration to

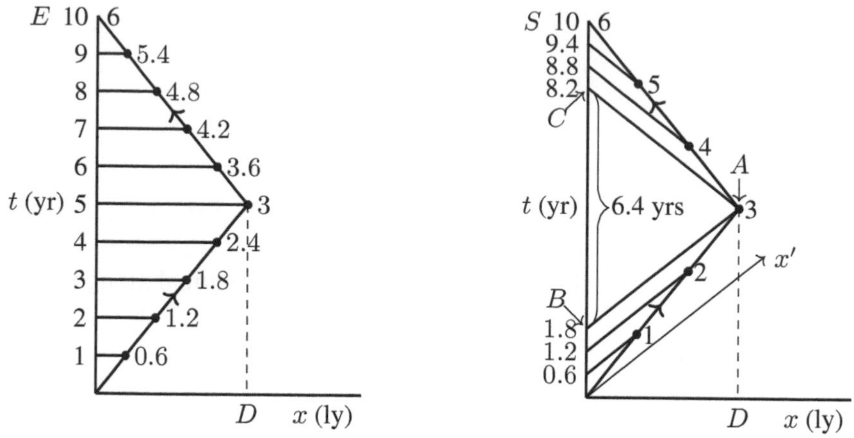

Figure 12.9 S travels to $D = 4$ ly and back at speed $\beta = 0.8$. Lines of simultaneity in the earth frame E (left) and in the spacecraft frame S (right). In S there is a discontinuity in the age of E as S abruptly changes course for the return journey.

turn around, and time slows down in acceleration fields, Eq. (12.22). The "missing" years can be accounted for using what we've developed in this chapter.

Motion in a noninertial frame can be treated using the mechanics of Chapter 7. Start with the *free-particle* Lagrangian, Eq. (7.49),

$$L = \frac{m}{2} g_{\mu\nu} \dot{x}^\mu \dot{x}^\nu = -\frac{m}{2}\left(1 + gx/c^2\right)^2 c^2 \dot{t}^2 + \frac{m}{2}\dot{x}^2 , \tag{12.23}$$

where we've used the metric tensor Eq. (12.21), and where the "dot" notation means a derivative with respect to the proper time of the observer.[7] *The effects of acceleration are contained in the metric*, a theme we'll see in GR. We can treat the observer as a *free* particle in a coordinate system that contains the effects of the acceleration. The metric tensor is physical.

The canonical momenta are obtained from the Lagrangian, $P_\mu \equiv \partial L/\partial \dot{x}^\mu = m g_{\mu\alpha}\dot{x}^\alpha$. Thus,

$$P_0 = m g_{00} \dot{x}^0 = m g_{00} c\dot{t} = -m\left(1 + gx/c^2\right)^2 c\dot{t} \qquad P_1 = m g_{11}\dot{x}^1 = m g_{11}\dot{x} = m\dot{x} . \tag{12.24}$$

The quantity P_0 is a constant of the motion because $\partial L/\partial x^0 = 0$. The norm of the four-momentum is a constant, $P_\alpha P^\alpha = -m^2c^2$, so that $g^{\mu\nu}P_\mu P_\nu = -m^2c^2$. The contravariant metric tensor is the inverse of $g_{\mu\nu}$; from Eq. (12.21), $g^{00} = g_{00}^{-1}$ and $g^{11} = 1$. Thus,

$$g_{00}^{-1}(P_0)^2 + (P_1)^2 = -m^2c^2 , \tag{12.25}$$

a result that can be likened to Eq. (7.16). Using $P_1 = m\dot{x}$ in Eq. (12.25),

$$\dot{x}^2 = -g_{00}^{-1}(x)\left(P_0/m\right)^2 - c^2 . \tag{12.26}$$

Equation (12.26) can be rewritten

$$\mathrm{d}\tau = \frac{\mathrm{d}x}{\sqrt{-g_{00}^{-1}(x)\left(P_0/m\right)^2 - c^2}} = \frac{\mathrm{d}x(1 + gx/c^2)}{\sqrt{\left(P_0/m\right)^2 - c^2\left(1 + gx/c^2\right)^2}} \tag{12.27}$$

[7]Time is a coordinate, $\dot{t} \equiv \mathrm{d}t/\mathrm{d}\tau$. You never had a problem with $\dot{x}$; time is another spacetime coordinate. In IRFs, $\dot{t} = \mathrm{d}t/\mathrm{d}\tau = \gamma$ is a *constant*; in an accelerated frame it's a *dynamical variable*.

which can be integrated,

$$\tau_2 - \tau_1 = -\frac{1}{g}\sqrt{(P_0/m)^2 - c^2(1+gx/c)^2}\Bigg|_{x_1}^{x_2} . \tag{12.28}$$

To make progress we must evaluate P_0. As applied to the motion of S, $\dot{x} = 0$ at the turning point D. Thus from Eq. (12.26), $(P_0/m)^2 = -g_{00}(D)c^2$, which allows us to simplify Eq. (12.28):

$$\tau_2 - \tau_1 = -\frac{c}{g}\sqrt{-g_{00}(D) + g_{00}(x)}\Bigg|_{x_1}^{x_2} = -\sqrt{\frac{2}{g}(D-x) + \frac{1}{c^2}(D^2 - x^2)}\Bigg|_{x_1}^{x_2} . \tag{12.29}$$

Equation (12.29) represents the proper time for a uniformly accelerated observer to travel from $x_1 \to x_2$, where $x_1 < x_2 \leq D$. Now set $x_2 = D$, in which case Eq. (12.29) simplifies further,

$$\Delta\tau = \sqrt{\frac{2}{g}(D-x) + \frac{1}{c^2}(D^2 - x^2)} . \tag{12.30}$$

Equation (12.30) specifies the proper time for constant acceleration between x and D. Clearly the answer depends on the value of g. To make contact with our previous analysis, let g become large. In the limit $g \to \infty$ in Eq. (12.30),

$$\Delta\tau = \frac{1}{c}\sqrt{D^2 - x^2} . \tag{12.31}$$

Let x in Eq. (12.31) denote the point where the acceleration begins, at the point where inertial motion ends, which in the S-frame is $x = D/\gamma$, the Lorentz-contracted length for the traveler in motion at constant speed. From Eq. (12.31),

$$\Delta\tau = \frac{D}{c}\sqrt{1 - \frac{1}{\gamma^2}} = \frac{D\beta}{c} . \tag{12.32}$$

With $D = 4$ ly and $\beta = 0.8$, $\Delta\tau = 3.2$ years, half of the "missing" 6.4 years. The complete turnaround time is $2D\beta/c$.

12.3 ROTATING REFERENCE FRAME

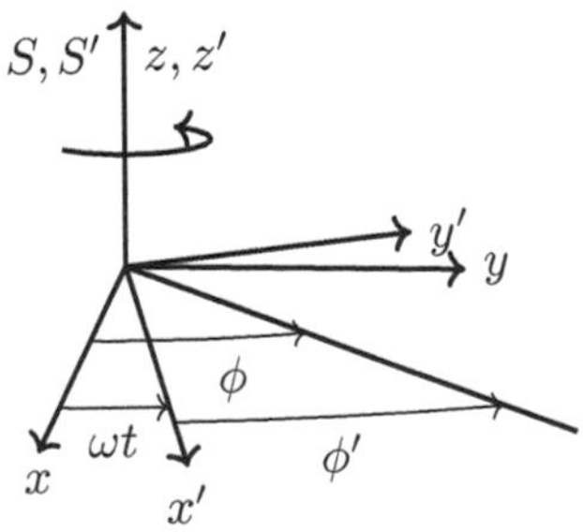

Figure 12.10 Rotating reference frame: $\phi = \omega t + \phi'$.

We now consider a frame S' that rotates at a constant rate ω relative to an inertial frame S, about the z-axis common to both frames (see Fig. 12.10). It's convenient to use cylindrical coordinates, (ρ, ϕ, z), along with the coordinate basis ($\boldsymbol{e}_\rho = \hat{\boldsymbol{\rho}}, \boldsymbol{e}_\phi = \rho\hat{\boldsymbol{\phi}}, \boldsymbol{e}_z = \hat{\boldsymbol{z}}$). We take the coordinates in S' to be related to those in S by

$$t' = t \qquad \rho' = \rho \qquad \phi' = \phi - \omega t \qquad z' = z . \tag{12.33}$$

As we'll see, such a simple coordinate transformation has surprisingly far-reaching consequences. The azimuth angle in the rotating frame is related to that in S by $\phi = \phi' + \omega t$. Taking $t' = t$, however, would seemingly be a throwback to the GT. In accelerated systems, proper time intervals *vary* throughout the system[8] because of the spatial dependence of g_{00}, $\mathrm{d}\tau = \sqrt{|g_{00}|}\mathrm{d}t$. In such systems there will always be a location where $g_{00} = -1$. In the rotating frame that location is the *inertial* frame at $\rho' = 0$. The coordinate transformation Eq. (12.33) leads to a metric tensor for the rotating frame where $g_{00} = g_{00}(\rho')$, Eq. (12.35). Setting $t' = t$ is a kind of bootstrap process: We will be led to a description of time in a rotating frame through the metric tensor brought about by Eq. (12.33). Because time is unambiguous at $\rho = 0$, it becomes a standard coordinate in our analysis. Back to Eq. (12.33), taking $\rho' = \rho$ seems reasonable because coordinates perpendicular to the velocity remain unchanged. However, $\rho' = \rho$ immediately leads to a problem known as *Ehrenfest's paradox*. Lengths are contracted in the direction of motion. We would expect therefore the *circumference* of a rotating disk as observed in S, $C = 2\pi\rho$, to be less than that in the rest frame, $C' = 2\pi\rho'$. Thus we have an apparent contradiction: $\rho' = \rho$ and $2\pi\rho < 2\pi\rho'$. The paradox arises from applying to a system with non-constant velocities, the constant-velocity results of SR, and hence points to the limits of SR. We'll see how the paradox is resolved.

12.3.1 Metric tensor for rotating frame

The spacetime separation in S, $(\mathrm{d}s)^2 = -(c\mathrm{d}t)^2 + (\mathrm{d}\rho)^2 + (\rho\mathrm{d}\phi)^2 + (\mathrm{d}z)^2$, becomes, under the substitutions in Eq. (12.33), the spacetime separation expressed in the rotating-frame coordinates:

$$(\mathrm{d}s)^2 = -(c\mathrm{d}t')^2\left(1 - \frac{\rho'^2\omega^2}{c^2}\right) + \frac{2\omega\rho'^2}{c}\mathrm{d}\phi'\mathrm{d}(ct') + (\mathrm{d}\rho')^2 + (\rho'\mathrm{d}\phi')^2 + (\mathrm{d}z')^2 \,. \tag{12.34}$$

We can then "read off" the metric tensor from Eq. (12.34):

$$[g_{\mu\nu}] = \begin{pmatrix} -1+\rho^2\omega^2/c^2 & 0 & \omega\rho^2/c & 0 \\ 0 & 1 & 0 & 0 \\ \omega\rho^2/c & 0 & \rho^2 & 0 \\ 0 & 0 & 0 & 1 \end{pmatrix} = \begin{pmatrix} -1/\gamma^2 & 0 & \omega\rho^2/c & 0 \\ 0 & 1 & 0 & 0 \\ \omega\rho^2/c & 0 & \rho^2 & 0 \\ 0 & 0 & 0 & 1 \end{pmatrix}, \tag{12.35}$$

where $\gamma \equiv 1/\sqrt{1-\omega^2\rho^2/c^2}$. A coordinate singularity lurks in Eq. (12.35): As $\omega\rho \to c$, $g_{00} \to 0$. The proper time is, from $\mathrm{d}\tau = \sqrt{|g_{00}|}\mathrm{d}t$,

$$\mathrm{d}\tau = \sqrt{1 - \frac{\rho^2\omega^2}{c^2}}\mathrm{d}t = \mathrm{d}t/\gamma \,. \tag{12.36}$$

Time runs slower in a rotating frame the further the distance from the axis of rotation. Do we see a pattern? For linear acceleration, $\mathrm{d}\tau(0) < \mathrm{d}\tau(x)$, Eq. (12.22), where $x > 0$ is the direction of *less* acceleration, and in Eq. (12.36), $\mathrm{d}\tau(\rho) < \mathrm{d}\tau(0)$, where $\rho > 0$ is the direction of *greater* acceleration. *Time runs slower in regions of greater acceleration.*

The metric tensor Eq. (12.35) is not diagonal; $g_{02} \neq 0$. The rotating coordinate system is not *time-orthogonal*; in a rotating frame, time is not orthogonal to the space spanned by spatial coordinates. Using Eq. (5.33), $\boldsymbol{\alpha}_\mu = A^\nu_\mu \boldsymbol{e}_\nu$, we can construct a set of spacetime basis vectors $\{\boldsymbol{\alpha}_\mu\}$ for the rotating frame, where the Jacobian matrix is found by differentiating the coordinate transformation equations (12.33), $A^\nu_\mu \equiv \partial x^\nu/\partial x'^\mu$. The basis vectors found this way are

$$\boldsymbol{\alpha}_0 = \boldsymbol{e}_0 + \frac{\omega}{c}\boldsymbol{e}_2 \qquad \boldsymbol{\alpha}_i = \boldsymbol{e}_i \,. \qquad (i = 1,2,3) \tag{12.37}$$

Even though the time coordinates are equal, $t' = t$, the timelike basis vectors $\boldsymbol{\alpha}_0$ and $\boldsymbol{e}_0$ are not—the "direction" of time for a rotating observer is *skewed* to have a component in the ϕ-direction, the direction of rotation.

[8] As we've seen in Section 12.1.2 and as we will see in GR.

12.3.2 Rest space of a rotating observer

The *rest space* of an observer is defined as the local three-dimensional space orthogonal to the timelike basis vector, i.e., a SH (Section 8.3). To find a set of basis vectors for this space, define the projections of the spacelike vectors $\boldsymbol{\alpha}_i$ onto the direction orthogonal to $\boldsymbol{\alpha}_0$ as $\boldsymbol{\alpha}_{i\perp}$, where

$$\boldsymbol{\alpha}_{i\perp} \equiv \boldsymbol{\alpha}_i - \left(\frac{\boldsymbol{\alpha}_i \cdot \boldsymbol{\alpha}_0}{\boldsymbol{\alpha}_0 \cdot \boldsymbol{\alpha}_0}\right)\boldsymbol{\alpha}_0 = \boldsymbol{\alpha}_i - \frac{g_{i0}}{g_{00}}\boldsymbol{\alpha}_0 \, . \qquad (i = 1, 2, 3) \tag{12.38}$$

We then define the *spatial metric tensor* as

$$\gamma_{ij} \equiv \boldsymbol{\alpha}_{i\perp} \cdot \boldsymbol{\alpha}_{j\perp} = \left(\boldsymbol{\alpha}_i - \frac{g_{i0}}{g_{00}}\boldsymbol{\alpha}_0\right) \cdot \left(\boldsymbol{\alpha}_j - \frac{g_{j0}}{g_{00}}\boldsymbol{\alpha}_0\right) = g_{ij} - \frac{g_{i0}g_{j0}}{g_{00}} \, , \tag{12.39}$$

where γ_{ij} is symmetric. By construction, $\gamma_{00} = \gamma_{0i} = \gamma_{i0} = 0$. If the geometry is such that $g_{0i} = 0$, there is no difference between γ_{ij} and g_{ij}. From Eqs. (12.35) and (12.39),

$$[\gamma_{ij}] = \begin{pmatrix} 1 & 0 & 0 \\ 0 & \rho^2\gamma^2 & 0 \\ 0 & 0 & 1 \end{pmatrix} . \tag{12.40}$$

Using γ_{ij}, the spacetime interval can be written

$$\begin{aligned}(\mathrm{d}s)^2 =& g_{\mu\nu}\mathrm{d}x^\mu \mathrm{d}x^\nu = \gamma_{ij}\mathrm{d}x^i\mathrm{d}x^j + g_{00}\left[(\mathrm{d}x^0)^2 + 2\frac{g_{0i}}{g_{00}}\mathrm{d}x^0\mathrm{d}x^i + \frac{g_{i0}g_{j0}}{g_{00}^2}\mathrm{d}x^i\mathrm{d}x^j\right] \\ \equiv& (\mathrm{d}l)^2 - (\mathrm{d}\hat{t})^2 \, ,\end{aligned} \tag{12.41}$$

where the *spatial line element* is

$$(\mathrm{d}l)^2 \equiv \gamma_{ij}\mathrm{d}x^i\mathrm{d}x^j = (\mathrm{d}\rho)^2 + \gamma^2(\rho\mathrm{d}\phi)^2 + (\mathrm{d}z)^2 \, , \tag{12.42}$$

and where

$$\mathrm{d}\hat{t} \equiv \sqrt{|g_{00}|}\left[\mathrm{d}x^0 + \frac{g_{0i}}{g_{00}}\mathrm{d}x^i\right] = \frac{1}{\gamma}\left[\mathrm{d}x^0 - \omega\rho^2\gamma^2\mathrm{d}\phi/c\right] \tag{12.43}$$

is a time increment orthogonal to the spacelike directions. The local hypersurface of simultaneous events, the rest space, is specified by $\mathrm{d}\hat{t} = 0$.

12.3.3 Tetrad basis for rotating observer

A *tetrad* is a local set of orthonormal vectors[9] $\{\boldsymbol{e}'_\mu(\tau)\}$, where one is timelike and three are spacelike, such that $\boldsymbol{e}'_0(\tau)$ is tangent to the worldline and $\boldsymbol{e}'_\mu(\tau) \cdot \boldsymbol{e}'_\nu(\tau) = \eta_{\mu\nu}$, the Lorentz metric tensor—a metric tensor for a time-orthogonal coordinate system. Define $\boldsymbol{e}'_0 \equiv \boldsymbol{\alpha}_0/\sqrt{|g_{00}|}$; that ensures $\boldsymbol{e}'_0 \cdot \boldsymbol{e}'_0 = -1$. The local rest space, spanned by the vectors $\{\boldsymbol{\alpha}_i\}$, is orthogonal to $\boldsymbol{e}'_0(\tau)$, which can be used to construct the other members of the tetrad. Because the magnitude of $\boldsymbol{\alpha}_{i\perp}$ is $\sqrt{\gamma_{ii}}$ (Eq. (12.39)), we can take one of the spatial vectors as $\boldsymbol{e}'_i = \boldsymbol{\alpha}_{i\perp}/\sqrt{\gamma_{ii}} = (\boldsymbol{\alpha}_i - g_{i0}\boldsymbol{\alpha}_0/g_{00})/\sqrt{\gamma_{ii}}$. Clearly $\boldsymbol{e}'_0 \cdot \boldsymbol{e}'_i = 0$. Another spacelike member of the tetrad is chosen such that $\boldsymbol{e}'_j \cdot \boldsymbol{e}'_i = \delta_{ij}$. The third spacelike vector can be taken as $\boldsymbol{e}'_k = \boldsymbol{e}'_i \times \boldsymbol{e}'_j$. Following this recipe, we find

$$\begin{aligned}\boldsymbol{e}'_0 =& \gamma\boldsymbol{\alpha}_0 = \gamma\left(\boldsymbol{e}_0 + \frac{\omega}{c}\boldsymbol{e}_\phi\right) = \gamma\left(\boldsymbol{e}_0 + \frac{\omega\rho}{c}\hat{\boldsymbol{\phi}}\right) \\ \boldsymbol{e}'_\phi =& \frac{1}{\gamma\rho}\left(\boldsymbol{\alpha}_\phi + \gamma^2\frac{\omega\rho^2}{c}\boldsymbol{\alpha}_0\right) = \gamma\left(\frac{\omega\rho}{c}\boldsymbol{e}_0 + \hat{\boldsymbol{\phi}}\right) \\ \boldsymbol{e}'_\rho =& \boldsymbol{\alpha}_\rho = \boldsymbol{e}_\rho \qquad \boldsymbol{e}'_z = \boldsymbol{\alpha}_z = \boldsymbol{e}_z \, ,\end{aligned} \tag{12.44}$$

[9] We mentioned tetrad basis in Section 5.1.1. The choice of basis for a given vector space is not unique.

where we've used Eq. (12.37). The vectors e'_0 and e'_ϕ are shown in Fig. 12.11. The definition of e'_0 and e'_ϕ is a local LT (Exercise 12.6),

$$\begin{pmatrix} e'_0 \\ e'_\phi \end{pmatrix} = \gamma \begin{pmatrix} 1 & \omega\rho/c \\ \omega\rho/c & 1 \end{pmatrix} \begin{pmatrix} e_0 \\ \hat{\phi} \end{pmatrix} . \tag{12.45}$$

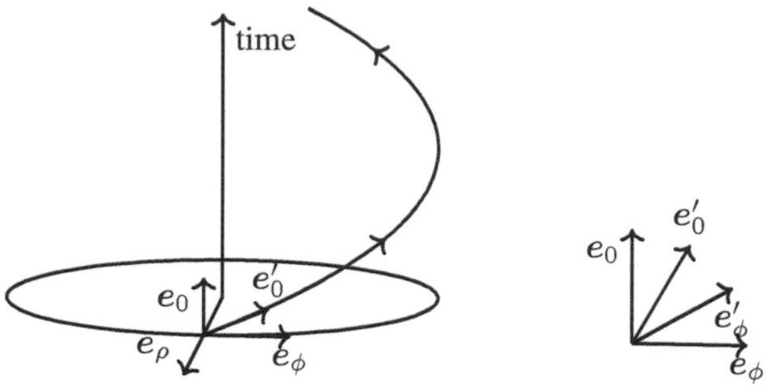

Figure 12.11 Basis vectors e'_0 and e'_ϕ and worldline of a rotating observer (not drawn to scale). The z-direction is not included in this figure.

12.3.4 Simultaneity

The distance from (ρ, ϕ, z) to $(\rho, \phi + \mathrm{d}\phi, z)$ in the rotating frame is, from Eq. (12.43), $\mathrm{d}l = \sqrt{\gamma_{22}}\mathrm{d}\phi = \gamma\rho\mathrm{d}\phi$. Thus the circumference in the rotating frame is $C' = \int_0^{2\pi}\sqrt{\gamma_{22}}\mathrm{d}\phi = 2\pi\gamma\rho$, implying that in S, $C = 2\pi\rho = C'/\gamma$, as we'd expect. Note, however, that $C'/\rho = 2\pi\gamma > 2\pi$, the hallmark of a *curved surface* (Chapter 14)—a development not anticipated in the statement of Ehrenfest's paradox! The rest space of a rotating disk is *curved*; moreover, it has a *negative* curvature because $C'/\rho > 2\pi$. It's not *spacetime* that's curved in this example. We started out with Minkowski space, and a flat spacetime remains so under coordinate transformations such as Eq. (12.33).[10] Our notions of space are radically altered in a rotating frame, and so is the measure of time.

Time is presented to us as the progression of SHs, "one at a time." By writing the spacetime separation in the form of Eq. (12.41), simultaneous events are characterized by $\mathrm{d}\hat{t} = 0$. From Eq. (12.43), simultaneous events are those that are infinitesimally related by

$$g_{00}\mathrm{d}x^0 + g_{0i}\mathrm{d}x^i = 0 , \tag{12.46}$$

which specifies a local plane spanned by the coordinate differentials $\{\mathrm{d}x^\mu\}$.

Clocks may be synchronized ($\hat{t}$ = constant) for events that lie on an *open curve* in spacetime by integrating Eq. (12.46),

$$\Delta x^0 = -\int \frac{g_{0i}}{g_{00}}\mathrm{d}x^i = -\int \frac{g_{02}}{g_{00}}\mathrm{d}\phi = \frac{\gamma^2\omega\rho^2}{c}\Delta\phi . \tag{12.47}$$

This equation gives the time difference (along the t-axis) for events that are simultaneous in the rotating frame. Synchronization of clocks around a *closed* path is not possible because the closed-loop integral does not vanish,[11]

$$\Delta x^0 = -\oint \frac{g_{0i}}{g_{00}}\mathrm{d}x^i \neq 0 , \tag{12.48}$$

[10] The Riemann curvature tensor (Chapter 14) is zero in Minkowski space, and a tensor that vanishes in one reference frame vanishes in another that's connected to it by linear coordinate transformations such as Eq. (12.33).

[11] Equation (12.46) is not an exact differential form.

implying that $\hat{t}$ is not *single-valued.* It's not in general possible to synchronize clocks over the entire reference frame; the exceptional case is when all the terms $g_{0i} = 0$. The impossibility of global clock synchronization is not a property of spacetime, but rather a property of the coordinate system we use to describe it. It's always possible to choose a coordinate system such that *at a point* the metric tensor is diagonal, but it may not be possible to arrange for a global coordinate system in which $g_{0i} = 0$ everywhere.

12.4 THE SAGNAC EFFECT

The Sagnac effect is named after an experiment performed by Georges Sagnac in 1913. The experimental arrangement is indicated in Fig. 12.12. There is a source of light, a beam splitter (BS), three

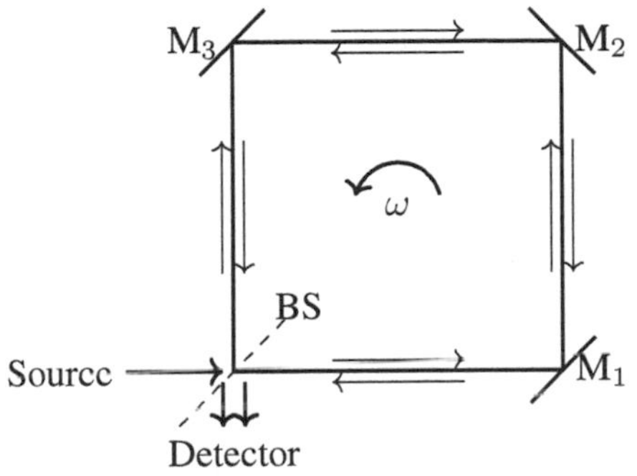

Figure 12.12 Sagnac interferometer. M_1, M_2, M_3 are mirrors, and BS is a beam splitter.

mirrors (M_1, M_2, M_3), and a detector. The beams traverse a closed path in opposite directions, and then recombine. Sagnac found that the interference pattern changes when the apparatus is rotated at a rate ω. In contrast to the Michelson interferometer, which was *rotated* through $\pi/2$ (Section 2.6.1), the Sagnac interferometer is continuously *rotating* at a constant rate.

We can understand the Sagnac effect in terms of the time delay between the beams propagating in opposite directions in a rotating reference frame. In writing the spacetime separation in the rest space as $(\mathrm{d}s)^2 = (\mathrm{d}l)^2 - [\mathrm{d}\hat{t}]^2$ (Eq. (12.41)), the null separation $(\mathrm{d}s)^2 = 0$ implies that light propagates in a rotating frame such that $\mathrm{d}l = \pm \mathrm{d}\hat{t}$. Using Eqs. (12.42) and (12.43), the path of null separation for light propagating in a ring ($\mathrm{d}\rho = 0$, $\mathrm{d}z = 0$) is specified by

$$\gamma \rho \mathrm{d}\phi = \pm \frac{1}{\gamma} \left(\mathrm{d}x^0 - \omega \rho^2 \gamma^2 \mathrm{d}\phi / c \right) ,$$

or

$$\mathrm{d}x^0 = \rho \gamma^2 \mathrm{d}\phi \left(1 \pm \omega\rho/c\right) . \tag{12.49}$$

Integrating Eq. (12.49) over 2π, we have the time for light to propagate around the ring in the two directions

$$T_{\pm} = \frac{2\pi\rho\gamma^2}{c}\left(1 \pm \omega\rho/c\right) = \frac{2\pi\rho/c}{1 \mp \omega\rho/c} = \frac{2\pi\rho}{c \mp \omega\rho} , \tag{12.50}$$

where the upper (lower) sign refers to the time for light to propagate around one circuit of the system in (against) the direction of rotation. The time delay between the beams is thus given by,

$$\Delta T \equiv T_+ - T_- = \frac{4\pi\omega\rho^2\gamma^2}{c^2} = \frac{4A\gamma^2\omega}{c^2} , \tag{12.51}$$

where A is the area enclosed by the path. We've used $A = \pi\rho^2$, the area of a circle. Experiment shows that the effect depends only on the magnitude of the area enclosed by the beams, and not the shape. From the time delay, there's a phase difference $2\pi f \Delta T$ between the beams, where f is the frequency of the light. The number of fringe shifts is then given by $\Delta N = f\Delta T = c\Delta T/\lambda$, or

$$\Delta N = \frac{c}{\lambda}\Delta T = \frac{4\gamma^2 \omega A}{c\lambda} . \tag{12.52}$$

In Sagnac's experiment, $\omega = 14$ rad sec^{-1}, $A = 0.0863$ m^2, $\lambda = 0.436 \times 10^{-6}$ m, and $\gamma \approx 1$. Equation (12.52) predicts a shift of $\Delta N = 0.037$ fringes, in agreement with his findings! Equation (12.52) has subsequently been verified in numerous experiments.

Of note, in 1925 Albert Michelson performed a version of the Sagnac experiment (that Michelson had proposed in 1904, but not performed until 1925), utilizing an interferometer with a *large* area: 1113×2010 feet $= 207836\,\text{m}^2$. The rotation of the interferometer was provided by the *rotation of Earth*! He detected $\Delta N = 0.230 \pm 0.005$ fringe shifts, agreeing with the theoretical value of 0.236 fringe shifts.[49] How did Michelson "stop the earth" in order to get a reference against which to measure the fringe shift? He employed in the same experiment *two* interferometers, where the area enclosed by the second interferometer was too small to give a measurable Sagnac effect. In this way, through clever design of the optics, he was able to measure a displacement between two sets of fringes.

12.5 RELATIVISTIC DESCRIPTION OF SPIN

Spin angular momentum is defined in the theory of relativity as a four-vector S^μ such that $S^0 = 0$ in the rest frame.[12] As a consequence S^μ is orthogonal to U^μ,

$$U^\mu S_\mu = 0 , \tag{12.53}$$

where $S_\mu = (-S^0, \boldsymbol{S})$ (in the rest frame $U^\mu = (c, 0)$ and $S_\mu = (0, \boldsymbol{S})$). As a covariant equation, (12.53) is valid in any IRF[13] and can be taken as an alternate definition of S^μ. In an arbitrary inertial frame,

$$S^0 = -S_0 = \frac{1}{c}\boldsymbol{u} \cdot \boldsymbol{S} . \tag{12.54}$$

The spin four-vector is spacelike and lies in a SH orthogonal to the worldline. Because the direction of U^μ in general changes between t and $t + \mathrm{d}t$, the SH at time t in which $\boldsymbol{S}$ lies, does not in general coincide with that at time $t + \mathrm{d}t$. As a consequence, $\boldsymbol{S}(t)$ has an infinitesimal component out of the hyperplane of $\boldsymbol{S}(t + \mathrm{d}t)$; see Fig. 12.13. The vector $\boldsymbol{S}(t + \mathrm{d}t)$ is the *projection* of $\boldsymbol{S}(t)$ onto

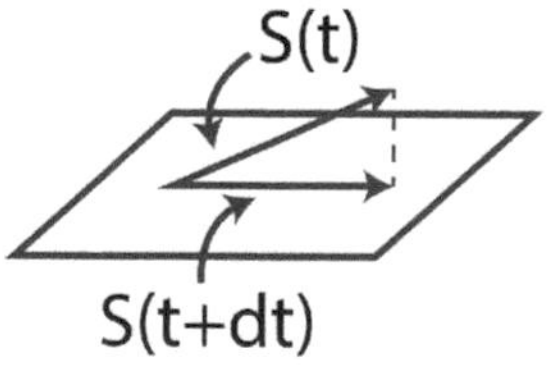

Figure 12.13 The spin vector at time t lies out of the SH at time $t + \mathrm{d}t$.

the hypersurface at time $t + \mathrm{d}t$ (as $\mathrm{d}t \to 0$); $\boldsymbol{S}(t + \mathrm{d}t)$ has the same length as $\boldsymbol{S}(t)$ because $\mathrm{d}t$ is infinitesimal. Thus, *the components of $\boldsymbol{S}$ in the instantaneous rest frame at $t + \mathrm{d}t$ are equal to the components of $\boldsymbol{S}$ in the instantaneous rest frame at time t*, which may be taken as another definition of the spin vector, along with $S^0 = 0$ in a rest frame.

[12] The orbital angular momentum three-vector $\boldsymbol{L}$ is the dual of the spatial parts of the angular momentum tensor $M^{\mu\nu}$, Eq. (7.61), which vanishes in rest frame of a particle.

[13] By the quotient theorem, S^μ comprises a four-vector, when defined by Eq. (12.53).

Our goal is to derive an equation of motion for S^μ; the reader uninterested in the details should skip to Eq. (12.60). Let a particle have velocity $\boldsymbol{u}$ in IRF K. The particle appears instantaneously at rest in IRF K'. Frame K' is related to K by a boost $L(\boldsymbol{u})$. At $t + \mathrm{d}t$ the particle has velocity $\boldsymbol{u} + \mathrm{d}\boldsymbol{u}$ relative to K, and appears instantaneously at rest in IRF K'', which is related to K by a boost $L(\boldsymbol{u}+\mathrm{d}\boldsymbol{u})$. We showed previously that $L(\boldsymbol{u}+\mathrm{d}\boldsymbol{u}) = R(\hat{\boldsymbol{n}}\mathrm{d}\phi)L(\mathrm{d}\boldsymbol{w})L(\boldsymbol{u})$, Eq. (6.53), where the infinitesimal velocity $\mathrm{d}\boldsymbol{w}$ is such that the relativistic addition of velocities $\boldsymbol{u} \oplus \mathrm{d}\boldsymbol{w} = \boldsymbol{u} + \mathrm{d}\boldsymbol{u}$ (Section 3.3) and the rotation is about the direction of $\boldsymbol{u} \times \mathrm{d}\boldsymbol{u}$ through the angle $\mathrm{d}\phi$, Eq. (6.51). A boost followed by a boost (here $L(\mathrm{d}\boldsymbol{w})L(\boldsymbol{u})$) is not a boost when the velocities are non-colinear.

By the properties of the spin vector in its rest frame, we have, referring to the spin in frame K, the equality (written loosely) $L(\boldsymbol{u}+\mathrm{d}\boldsymbol{u})S(t+\mathrm{d}t) = R(\hat{\boldsymbol{n}}\mathrm{d}\phi)L(\boldsymbol{u})S(t)$, where we have to take into account the rotation of the reference frame between t and $t + \mathrm{d}t$. Written in terms of components,

$$S^\alpha(t+\mathrm{d}t) = \left[L(-(\boldsymbol{u}+\mathrm{d}\boldsymbol{u}))R(\hat{\boldsymbol{n}}\mathrm{d}\phi)L(\boldsymbol{u})\right]^\alpha_{\ \nu} S^\nu(t) \,. \tag{12.55}$$

Using the result of Exercise 6.19 with $\boldsymbol{u} \to -\boldsymbol{u}$ and $\mathrm{d}\boldsymbol{u} \to -\mathrm{d}\boldsymbol{u}$, we have the infinitesimal LT that effects the terms in Eq. (12.55)

$$S^\alpha(t+\mathrm{d}t) = \left[I + \frac{\gamma^2}{c^2}(\boldsymbol{u}\times\mathrm{d}\boldsymbol{u})\cdot\boldsymbol{J} + \frac{\gamma^2}{c}\boldsymbol{K}\cdot\mathrm{d}\boldsymbol{u}\right]^\alpha_{\ \nu} S^\nu(t) \,, \tag{12.56}$$

where $\boldsymbol{J}$ and $\boldsymbol{K}$ are vectors comprised of the rotation and boost generators (Section 6.1).

From the time component of Eq. (12.56) we obtain the equation of motion for S^0,

$$\frac{\mathrm{d}S^0}{\mathrm{d}t} = \frac{\gamma^2}{c}\boldsymbol{S}\cdot\frac{\mathrm{d}\boldsymbol{u}}{\mathrm{d}t} \,. \tag{12.57}$$

The spatial components of Eq. (12.56) are

$$\boldsymbol{S}(t+\mathrm{d}t) = \boldsymbol{S}(t) + \frac{\gamma^2}{c}S^0\mathrm{d}\boldsymbol{u} + \frac{\gamma^2}{c^2}\boldsymbol{S}\times(\boldsymbol{u}\times\mathrm{d}\boldsymbol{u}) \,. \tag{12.58}$$

Using Eq. (12.54) to eliminate S^0 from Eq. (12.58), we find the equation of motion for the spatial components

$$\frac{\mathrm{d}}{\mathrm{d}t}\boldsymbol{S} = \frac{\gamma^2}{c^2}\boldsymbol{u}\left(\boldsymbol{S}\cdot\frac{\mathrm{d}\boldsymbol{u}}{\mathrm{d}t}\right) \,. \tag{12.59}$$

By combining Eqs. (12.57) and (12.59), the covariant equation of motion for S^μ is (as can be verified):

$$\frac{\mathrm{d}S^\mu}{\mathrm{d}\tau} = \frac{U^\mu}{c^2}S_\nu\frac{\mathrm{d}U^\nu}{\mathrm{d}\tau} \,. \tag{12.60}$$

Equation (12.60) is a purely *kinematical* prediction of SR: Spin components evolve by virtue of the acceleration of the particle (the Thomas-Wigner rotation), without reference to the underlying *cause* of the acceleration. For a free particle, $\mathrm{d}S^\mu/\mathrm{d}\tau = 0$.

How does the spin vector transform between IRFs? Boost into the instantaneous rest frame, $S' = L(\boldsymbol{u})S$. Using Eq. (6.16), we find

$$\boldsymbol{S}' = \boldsymbol{S} - \frac{\gamma}{(1+\gamma)c^2}\boldsymbol{u}(\boldsymbol{u}\cdot\boldsymbol{S}) \,. \tag{12.61}$$

The inverse transformation is

$$\boldsymbol{S} = \boldsymbol{S}' + \frac{\gamma^2}{(1+\gamma)c^2}\boldsymbol{u}(\boldsymbol{u}\cdot\boldsymbol{S}') \,. \tag{12.62}$$

Note the minus sign difference between Eqs. (12.61) and (12.62) and also that there is a single factor of γ in Eq. (12.61).

12.6 COVARIANT SPIN DYNAMICS

Equation (12.60) is a kinematic equation of motion for an accelerated spin without regard to the forces causing the acceleration. It would be natural at this point to consider Thomas precession and spin-orbit coupling in atomic physics—topics we omit in favor of brevity.[14] If we *were*, however, to consider these phenomena, we would see that to introduce the coupling of a spin to the electromagnetic field typically involves a series of nonrelativistic approximations. To sort through these approximations and to make them relativistically correct is a daunting task. Is there a systematic way to derive a covariant equation of motion for spins that accounts for the causes of acceleration? There is, and it's instructive to see the process: Simply "write down" the answer by examining all the possibilities that could go into a covariant equation.[15]

We want a covariant equation of motion for the four-vector S^α. The left side of the equation should simply be $\mathrm{d}S^\alpha/\mathrm{d}\tau$. The right side must then be a four-vector as well. How many ways can we construct a four-vector that builds in the relevant physics of the fields and the acceleration? Such an equation should be linear in the spin and in the fields, "riffing" the nonrelativistic physics of Larmor precession. The equation can also involve U^α and $\mathrm{d}U^\alpha/\mathrm{d}\tau$, such as occurs in Eq. (12.60). Higher time derivatives are assumed absent. Using S^α, the electromagnetic field tensor $F^{\alpha\beta}$, U^α, and $\mathrm{d}U^\alpha/\mathrm{d}\tau$ as building blocks, we can construct the four-vectors

$$F^{\alpha\beta}S_\beta \qquad (S_\lambda F^{\lambda\mu}U_\mu)\,U^\alpha \qquad \left(S_\beta\frac{\mathrm{d}U^\beta}{\mathrm{d}\tau}\right)U^\alpha\,.$$

That's it. Other possibilities ($F^{\alpha\beta}U_\beta S_\lambda U^\lambda$, $U_\lambda F^{\lambda\mu}U_\mu S^\alpha$, or $S_\lambda F^{\lambda\mu}U_\mu \mathrm{d}U^\alpha/\mathrm{d}\tau$) either vanish, are higher order in the fields, or reduce to multiples of the three above. Our candidate equation is then

$$\frac{\mathrm{d}S^\alpha}{\mathrm{d}\tau} = A_1F^{\alpha\beta}S_\beta + \frac{A_2}{c^2}\,(S_\lambda F^{\lambda\mu}U_\mu)\,U^\alpha + \frac{A_3}{c^2}\left(S_\beta\frac{\mathrm{d}U^\beta}{\mathrm{d}\tau}\right)U^\alpha\,, \tag{12.63}$$

where A_1, A_2, and A_3 are unknown constants. Equation (12.53) provides an important constraint that must be built in. Differentiating Eq. (12.53),

$$\frac{\mathrm{d}}{\mathrm{d}\tau}(U_\alpha S^\alpha) = U_\alpha\frac{\mathrm{d}S^\alpha}{\mathrm{d}\tau} + S^\alpha\frac{\mathrm{d}U^\alpha}{\mathrm{d}\tau} = 0\,. \tag{12.64}$$

Contract Eq. (12.63) with U_α,

$$U_\alpha\frac{\mathrm{d}S^\alpha}{\mathrm{d}\tau} = A_1U_\alpha F^{\alpha\beta}S_\beta - A_2S_\lambda F^{\lambda\mu}U_\mu - A_3S_\beta\frac{\mathrm{d}U^\beta}{\mathrm{d}\tau}\,, \tag{12.65}$$

where we've used $U_\alpha U^\alpha = -c^2$. Combine Eqs. (12.64) and (12.65) to obtain

$$0 = (1-A_3)\,S_\beta\frac{\mathrm{d}U^\beta}{\mathrm{d}\tau} + (A_1+A_2)\,U_\alpha F^{\alpha\beta}S_\beta\,, \tag{12.66}$$

where we've used $S_\beta \mathrm{d}U^\beta/\mathrm{d}\tau = S^\alpha \mathrm{d}U_\alpha/\mathrm{d}\tau$ and the antisymmetry of $F^{\alpha\beta}$. The choices $A_3 = 1$ and $A_2 = -A_1$ cry out to be made, reducing Eq. (12.63) to one unknown,

$$\frac{\mathrm{d}S^\alpha}{\mathrm{d}\tau} = A_1\left[F^{\alpha\beta}S_\beta - \frac{1}{c^2}\,(S_\lambda F^{\lambda\mu}U_\mu)\,U^\alpha\right] + \frac{1}{c^2}\left(S_\beta\frac{\mathrm{d}U^\beta}{\mathrm{d}\tau}\right)U^\alpha\,. \tag{12.67}$$

Examine Eq. (12.67) in the rest frame. The time component is identically zero (Exercise 12.12). The spatial components of Eq. (12.67) in the rest frame reduce to

$$\frac{\mathrm{d}\boldsymbol{S}}{\mathrm{d}t} = A_1(\boldsymbol{S}\times\boldsymbol{B})\,, \tag{12.68}$$

[14] The woods are lovely, dark and deep, but I have promises to keep, and miles to go before I sleep.—Robert Frost

[15] We often tell students that the way to solve differential equations is to guess. Here we're guessing at the form of the differential equation itself!

where we've used Eq. (8.26). To make contact with Larmor precession, we identify $A_1 = gq/(2m)$ where g is the *g-factor*. We then have the covariant equation of motion,

$$\frac{\mathrm{d}S^\alpha}{\mathrm{d}\tau} = \frac{1}{c^2}\left(S_\beta \frac{\mathrm{d}U^\beta}{\mathrm{d}\tau}\right)U^\alpha + \frac{gq}{2m}\left[F^{\alpha\beta}S_\beta - \frac{1}{c^2}\left(S_\lambda F^{\lambda\mu}U_\mu\right)U^\alpha\right]. \tag{12.69}$$

If the fields vanish, Eq. (12.69) reduces to Eq. (12.60). If we now combine Eq. (12.69) with Eq. (8.43), $\mathrm{d}U^\alpha/\mathrm{d}\tau = (q/m)F^{\alpha\beta}U_\beta$, we obtain the *Bargmann-Michel-Telegdi equation* [50]

$$\frac{\mathrm{d}S^\alpha}{\mathrm{d}\tau} = \frac{gq}{2m}F^{\alpha\beta}S_\beta + \frac{q}{mc^2}\left(1 - \frac{g}{2}\right)\left(S_\beta F^{\beta\lambda}U_\lambda\right)U^\alpha. \tag{12.70}$$

Our definition of spin (four-vector S^μ with $S^0 = 0$ in its rest frame) *describes* how experimentally measured quantities can be treated in the theory of relativity. It does not, however, *explain* the phenomenon of spin. It's often said that spin is a relativistic effect because the Dirac equation of relativistic quantum mechanics describes the class of particles (fermions) with spin $\frac{1}{2}\hbar$. However, the nonrelativistic limit of the Dirac equation (the Pauli equation) describes the interaction of spin-$\frac{1}{2}$ particles with the electromagnetic field. Spin is a quantum effect. Relativity provides an elegant framework for describing (and hence understanding) spin, but does not explain it; indeed it can have no intuitive explanation—it's new!

SUMMARY

This chapter has examined several applications of SR to accelerated motion.

- Constant linear acceleration leads to hyperbolic motion in spacetime. The asymptotes of the hyperbola form an event horizon, a boundary in spacetime beyond which events cannot send signals to the worldline of the accelerated observer. An event horizon is associated with a coordinate singularity—in this case infinitely many lines of t = constant pass through a single point. The coordinate system is ill-defined at that point.

- A coordinate system can be developed using the worldlines of linearly accelerated observers, the metric tensor for which is given by Eq. (12.21). The effects of acceleration are contained in the metric tensor. The proper time $\mathrm{d}\tau = \sqrt{|g_{00}(x)|}\mathrm{d}t$ then depends on the location relative to the event horizon, or, said differently, on the value of the acceleration. Clocks in regions of greater acceleration run slower than in regions of lesser acceleration. Carried to an extreme, time differences vanish at the event horizon.

- The metric tensor in a rotating frame is given by Eq. (12.35). In this system, time slows down toward the perimeter of a rotating disk, where the acceleration is greatest. Because $g_{02} \neq 0$, the metric tensor describes a non-time-orthognal reference frame, where the direction of time is skewed to have a component in the direction of rotation. The timelike basis vector in the rotating frame, $\boldsymbol{\alpha}_0$, is a combination of the timelike and spacelike basis vectors in the inertial reference frame.

- The rest space of an observer is the local three-dimensional hypersurface orthogonal to the timelike basis vector. We defined a new set of spacelike basis vectors orthogonal to the timelike basis vector, $\boldsymbol{\alpha}_{i\perp} = \boldsymbol{\alpha}_i - (g_{i0}/g_{00})\boldsymbol{\alpha}_0$. The metric tensor obtained from the $\boldsymbol{\alpha}_{i\perp}$, $\gamma_{ij} \equiv \boldsymbol{\alpha}_{i\perp} \cdot \boldsymbol{\alpha}_{j\perp}$, the spatial metric tensor, $\gamma_{ij} = g_{ij} - g_{i0}g_{0j}/g_{00}$. The spacetime separation can be written as $(\mathrm{d}s)^2 = (\mathrm{d}l)^2 - (\mathrm{d}\hat{t})^2$, the difference between a spatial line element spanning the rest space, $(\mathrm{d}l)^2 = \gamma_{ij}\mathrm{d}x^i\mathrm{d}x^j$, and a time increment $\mathrm{d}\hat{t}$ orthogonal to the spacelike directions, $\mathrm{d}\hat{t} = \sqrt{|g_{00}|}\left(\mathrm{d}x^0 + g_{0i}\mathrm{d}x^i/g_{00}\right)$. The rest space is specified by $\mathrm{d}\hat{t} = 0$.

- A tetrad is a set of four orthonormal basis vectors, $\{\boldsymbol{e}'_\mu(\tau)\}$, one timelike and three spacelike, that satisfy: $\boldsymbol{e}'_0(\tau)$ is a unit vector tangent to the worldline, and $\boldsymbol{e}'_\mu(\tau) \cdot \boldsymbol{e}'_\nu(\tau) = \eta_{\mu\nu}$.

- The Sagnac experiment is a rotating interferometer where the light beams traverse closed paths in opposite directions and then recombine. There is a time delay between the beams, $\Delta T = 4A\gamma^2\omega^2/c^2$, where A is the area enclosed by the beams.
- Spin in relativity is a four-vector S^μ such that the time component vanishes in the rest frame of the particle. Its spatial part $\boldsymbol{S}$ lies in a SH orthogonal to the worldline.
- Equation (12.60) is a covariant equation of motion for S^μ given the acceleration of the particle, $\mathrm{d}U^\mu/\mathrm{d}\tau$. It's a relativistic kinematic effect, that the product of non-colinear boosts produces a rotation of the coordinate axes, regardless of the nature of the forces causing the acceleration.

EXERCISES

12.1 A large parallel plate capacitor is charged to a potential difference of 10^6 V. The plates are 5 cm apart. How long does it take an electron to travel from the negative to the positive plate, starting from rest? Compute both the relativistic and nonrelativistic results.

12.2 Show using Eq. (12.17) that all points on the event horizon have the coordinates $x' = -\alpha$ and $t' \to \infty$. Hint: Set $x = -\alpha + ct$.

12.3 Derive Eq. (12.28).

12.4 Derive Eq. (12.34). In passing to Eq. (12.35), why are there no factors of 2 in the metric tensor?

12.5 Derive Eq. (12.37). Build the Jacobian matrix using the coordinate transformation Eq. (12.33).

12.6 Show that the transformation in Eq. (12.45) meets the requirement of a Lorentz transformation, Eq. (4.12).

12.7 Derive Eqs. (12.57) and (12.58) from Eq. (12.56).

12.8 Derive Eq. (12.59).

12.9 Show that the time derivative of S^0, as given by Eq. (12.54), reduces to Eq. (12.57) when you make use of Eq. (12.59).

12.10 Show that the time component of Eq. (12.60) reduces to Eq. (12.57). Hint: You will encounter terms proportional to $\mathrm{d}\gamma/\mathrm{d}\tau$. Do not evaluate $\mathrm{d}\gamma/\mathrm{d}\tau$. Show, using Eq. (12.54) that all terms proportional to $\mathrm{d}\gamma/\mathrm{d}\tau$, when combined, vanish.

12.11 Derive the spin transformation equations (12.61) and (12.62). Use Eq. (12.54) when needed and that $S^{0'} = 0$.

12.12 Show that the time component of Eq. (12.67) vanishes identically in the rest frame of the spin.

12.13 Show that

$$S_\lambda F^{\lambda\mu} U_\mu = -\frac{\gamma}{c^2}(\boldsymbol{u}\cdot\boldsymbol{S})(\boldsymbol{E}\cdot\boldsymbol{u}) + \gamma\boldsymbol{S}\cdot(\boldsymbol{E}+\boldsymbol{u}\times\boldsymbol{B}) \,.$$

Use Eq. (12.54) to eliminate the time component.

12.14 Show that the time and space components of $F^{\alpha\beta}S_\beta$ are given by

$$F^{0\beta}S_\beta = \frac{1}{c}(\boldsymbol{E}\cdot\boldsymbol{S}) \qquad F^{i\beta}S_\beta = \frac{E^i}{c^2}(\boldsymbol{u}\cdot\boldsymbol{S}) + (\boldsymbol{S}\times\boldsymbol{B})^i \,.$$

CHAPTER 13

Tensors on manifolds

THE treatment of tensors in Chapter 5 implicitly presumes a flat geometry, that spacetime can be covered by a single coordinate system. A coordinate grid covering Minkowski space can be constructed from the worldlines of free particles—straight lines at constant speed, always and everywhere. When it comes to gravitation, one realizes that global inertial frames are not possible: We can't get away from the matter of the universe(!); force-free motion has only local, approximate validity. As we've seen (Chapter 11), uniform gravitational fields are equivalent to IRFs and thus *uniform gravitational fields are equivalent to no gravitational field.* Real gravitational fields are *inhomogeneous* and lead to relative accelerations between neighboring observers (Section 11.3). The effective interaction brought about by inhomogeneous gravitational fields implies a metrical relation between neighboring points in spacetime involving a metric tensor $g_{\mu\nu}(x)$ that *varies* in spacetime, a *tensor field,* such as occurs in accelerated systems (Chapter 12). We'll see that a non-constant metric field implies a curved geometry (Chapter 14).

A relativistic theory of gravitation requires a spacetime more general than Minkowski space, one that can be flat locally, approximating local inertial frames, but which is not flat globally. This need is met by the mathematical structure of a *manifold.* To get on the same page with GR, we must acquire the mathematical tools of tensors on manifolds. Whereas Chapter 5 presumes a fixed space-time metric (Lorentz metric), in GR we will be *solving* for the metric field $g_{\mu\nu}(x)$ that represents gravitation. We can benefit therefore from a more general perspective on tensors. Such a program, however, represents a decided jump in mathematical sophistication—welcome to GR.

This chapter introduces manifolds and tensor fields. *It could be skipped on a first reading* (read it when you need it). It may seem we're becoming unduly immersed in points of mathematical *finery,* which experience in physics shows can typically be overlooked. The theory of manifolds, however, is a subject where it pays to be precise; a little bit of rigor goes a long way.

13.1 MANIFOLDS

13.1.1 Definitions

A manifold of dimension n is a set of points M with the property that subsets of M can be placed in local correspondence with points[1] of $\mathbb{R}^n$. A basic concept is a *chart,* a pair (U_α, ϕ_α) consisting of an open subset $U_\alpha \subset M$ and a *homeomorphic mapping*[2] $\phi_\alpha : U_\alpha \to \mathbb{R}^n$ such that the image

[1]As we'll discuss, points of M are in local correspondence with points of $\mathbb{R}^n$ in such a way that M inherits the smoothness property of $\mathbb{R}^n$. According to Robert Geroch:[51] "The idea, then, is to isolate, from all the rich structure of $\mathbb{R}^n$ (e.g., its metric structure, its vector-space structure, its topological structure), that one small bit of structure we call 'smoothness'." Summaries of manifolds are given in Wald [37, Chapter 2] and in Hawking and Ellis.[38, Chapter 2]

[2]A map $\phi : U \to \mathbb{R}^n$ is one-to-one (or injective) if no two distinct points of U are associated with the same point of $\mathbb{R}^n$. A *homeomorphism* is a continuous one-to-one map ϕ with a continuous inverse mapping, ϕ^{-1}. A subset O of $\mathbb{R}^n$ is *open* if for any $x \in O$ there is a number $\epsilon > 0$ such that whenever $d(x, y) < \epsilon$, y is also in O, where $d(x, y)$ is the usual

of U_α under ϕ_α, $O_\alpha \equiv \phi_\alpha(U_\alpha) \subset \mathbb{R}^n$, is open in $\mathbb{R}^n$ (see Fig. 13.1). A chart (U_α, ϕ_α) specifies a continuous one-to-one correspondence between a set of points $U_\alpha \subset M$ and a set of points $O_\alpha \subset \mathbb{R}^n$. For $p \in U_\alpha$, because $\phi_\alpha(p) \in \mathbb{R}^n$, the chart defines n real-valued functions, *coordinate functions* $x^i : U_\alpha \to \mathbb{R}$, $i = 1, \cdots, n$, so that $\phi_\alpha(p) = (x^1(p), \cdots, x^n(p))$. The numbers $x^i(p)$ are the *coordinates* of p. A chart thus specifies a *coordinate system*[3] in $\mathbb{R}^n$ for a neighborhood of $p \in M$. Back to what is a manifold, a manifold is a set M together with a *collection* of charts $\{(U_\alpha, \phi_\alpha)\}$ (an *atlas*) such that several conditions are met. We require that the subsets $\{U_\alpha\}$ *cover* M, that $M = \bigcup_\alpha U_\alpha$ is the union of subsets. Each $p \in M$ is contained in *at least* one subset[4] U_α. Figure 13.1 indicates that charts U_α, U_β on M overlap, which they must: Unless M is flat it cannot

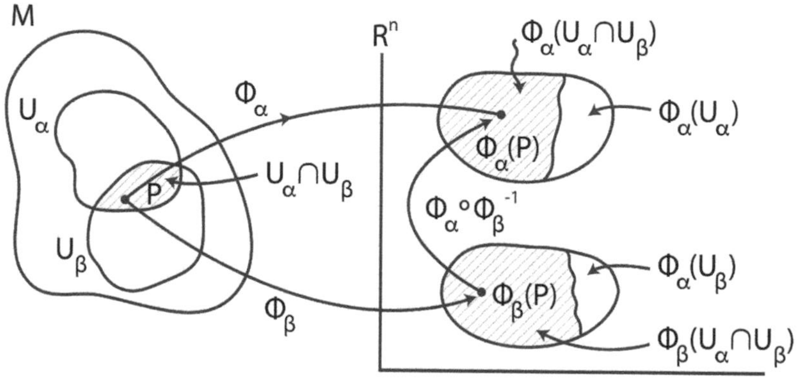

Figure 13.1 Subsets of an n-dimensional manifold M are mapped into subsets of $\mathbb{R}^n$. The coordinates of the overlap of subsets on M are related by the transition map $\phi_\alpha \circ \phi_\beta^{-1}$.

be covered by a single chart.[5] Our definition so far features continuity, but not differentiability (which implies continuity). A map ϕ of an open set $O \subset \mathbb{R}^n$ to an open set $O' \subset \mathbb{R}^m$ is of class C^r if the coordinates $(x'^1, \cdots, x'^m)$ of the image point $\phi(p) \in O'$ are r-times continuously differentiable functions of the coordinates $(x^1, \cdots, x^n)$ of $p \in O$. If a map is C^r for all $r \geq 0$, it is said to be C^∞, or *smooth*. *Smoothness* is a statement about differentiability. We restrict ourselves to smooth manifolds where all mappings are C^∞. We want the charts of M to overlap smoothly. Two charts are *compatible* if: 1) $\phi_\alpha(U_\alpha \cap U_\beta)$ and $\phi_\beta(U_\alpha \cap U_\beta)$ are open in $\mathbb{R}^n$, and 2) the *transition maps* (coordinate transformations) $\phi_\beta \circ \phi_\alpha^{-1} : \phi_\alpha(U_\alpha \cap U_\beta) \to \phi_\beta(U_\alpha \cap U_\beta)$, and their inverses $\phi_\alpha \circ \phi_\beta^{-1} : \phi_\beta(U_\alpha \cap U_\beta) \to \phi_\alpha(U_\alpha \cap U_\beta)$ are smooth. We can now define manifold as we'll use it: a pair (M, C) of a set M and an atlas C of charts on M such that the charts cover M and any two charts in C are compatible.[6]

Euclidean distance in $\mathbb{R}^n$. A set is open if every point in the set is surrounded by points that are also in the set. We have invoked subsets of $\mathbb{R}^n$ as being open because we know what Euclidean distance means. To say that a subset of a *manifold* is open requires that a *topology* be defined on it. A topology retains the notion of *nearness* without having to say *how* close. Without delving into the intricacies of topology, we can say that a subset S of M is open if, for all $p \in S$, there is a chart (U, ϕ) such that $p \in U$ and $U \subset S$. For our purposes, the domain of every chart is open.

[3]Note that no one coordinate system associated with a manifold is distinguished at the expense of others; all concepts developed in the theory of manifolds are therefore "coordinate-system neutral"—well suited for the theory of relativity.

[4]Every point of a manifold M must be in at least one chart, yet we require that no point of M be in an *infinite* number of charts. This requirement is met by imposing on M the (not very restrictive) property of *paracompactness*.

[5]A sphere for example (a two-dimensional manifold) cannot be mapped in its entirety onto $\mathbb{R}^2$ in a continuous one-to-one manner; at least two coordinate systems are required.

[6]The fine print. Two additional conditions are imposed on manifolds. First, manifolds are required to be *Hausdorff*. A manifold M is Hausdorff if for distinct points $p, q \in M$ there exist disjoint open sets $U_\alpha, U_\beta \subset M$ such that $p \in U_\alpha$ and $q \in U_\beta$. We can therefore distinguish points, a reasonable requirement. The Hausdorff property does not follow from the definition of manifold and must be stipulated; non-Hausdorff manifolds are pathological. The second requirement is that

Example. Examples of manifolds

(1) n-sphere, S^n

Let M consist of $(n+1)$ real numbers $(x^1, \cdots, x^{n+1})$ such that $\sum_{i=1}^{n+1}(x^i)^2 = 1$, the n-sphere S^n. Define subsets of M, $U_i^+ \equiv \{(x^1, \cdots, x^{n+1}) : x^i > 0\}$, the "upper half" with respect to coordinate x^i, and $U_i^- \equiv \{(x^1, \cdots, x^{n+1}) : x^i < 0\}$, the "lower half." Shown in Fig. 13.2 is U_1^+, the portion of the 1-sphere (a circle) corresponding to points $x^1 > 0$. Define the projections $\phi_i^\pm : U_i^\pm \to \mathbb{R}^n$ that eliminate x^i from the list $(x^1, \cdots, x^{n+1})$

$$\phi_i^+(x^1, \cdots, x^{n+1}) = (x^1, \cdots, x^{i-1}, x^{i+1}, \cdots, x^{n+1}) = \phi_i^-(x^1, \cdots, x^{n+1}) .$$

The $n+1$ pairs $(U_i^\pm, \phi_i^\pm)$ are compatible charts on M and cover M. We thus have an n-dimensional manifold, the n-sphere S^n. For $n = 1$ the charts provide one-to-one mappings between half-circles (subsets of M) and lines (see Fig. 13.2). In this way, S^1 consists of points that locally "look" like lines. For $n = 2$, M is the unit sphere; the charts provide one-to-one mappings onto planes. S^2 consists of points that locally look like planes.

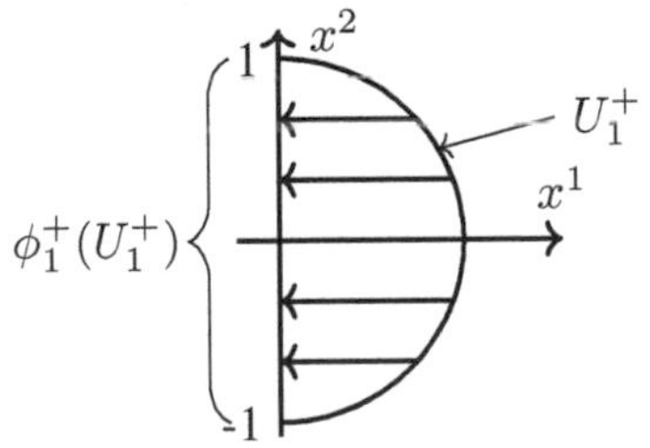

Figure 13.2 One-dimensional manifold U_1^+ consisting of points $(x^1)^2 + (x^2)^2 = 1$ such that $x_1 > 0$. U_1^+ locally looks like a portion of the real line because open subsets of U_1^+ can be placed in one-to-one correspondence with open subsets of $\mathbb{R}$.

(2) Product manifolds

Let M, M' be manifolds of dimensions n, n'. The *product manifold* $M \times M'$ is a new manifold of dimension $(n + n')$. Let (U, ϕ) be a chart on M and (U', ϕ') a chart on M', i.e., for $p \in U$ and $p' \in U'$, $\phi(p) = (x^1, \cdots, x^n)$ and $\phi'(p') = (y^1, \cdots, y^{n'})$. Associate with U, U' the set $V \equiv U \times U'$ where $(p, p') \in V$. Define a map $\phi : V \to \mathbb{R}^{n+n'}$ as $\phi(p, p') = (x^1, \cdots, x^n, y^1, \cdots, y^{n'})$. We thus have a chart (V, ϕ) on $M \times M'$. Because M, M' are manifolds, $M \times M'$ is a manifold. As examples, $\mathbb{R}^1 \times S^1$ is a cylinder and $S^1 \times S^1$ is a torus. The manifold $\mathbb{R}^{n_1} \times \mathbb{R}^{n_2}$ is $\mathbb{R}^{n_1+n_2}$. (By the way, $\mathbb{R}^n$ is a manifold.)

manifolds be *paracompact*. A collection $\{O_\alpha\}$ of open sets of M is said to be an *open cover* of a subset $A \subset M$ if the union of these sets contains A. A *finite* cover contains a finite number of sets. A *refinement* of a cover is a new cover such that every set in the new cover is a subset of some set in the old cover. A cover is *locally finite* if every point $p \in M$ has an open neighborhood that intersects only finitely many sets in the cover. M is paracompact if every open cover has a locally finite open refinement.[52] A connected Hausdorff manifold M is paracompact if and only if M can be covered by a locally finite, countable family of charts, and a *partition of unity* exists, which allows integration on M to be defined. Examples of non-Hausdorff and nonparacompact manifolds are given by Geroch in [53, p100].

13.1.2 Maps on manifolds

We now consider mappings between manifolds.[7] We start with *functions*[8] on M, $f : M \to \mathbb{R}$, an assignment of a real number to each point $p \in M$. A function is *smooth* if for all charts (U, ϕ) of M, the composite function $f \circ \phi^{-1} : \mathbb{R}^n \to \mathbb{R}$ is smooth;[9] see Fig. 13.3. The set of all smooth functions at $p \in M$ is denoted $\mathcal{F}_p$. For $\{x^i\}$ the coordinates of $p \in M$, the effect of the composite map $f \circ \phi^{-1}$ can be written $f(x^i)$: $f(x^i)$ represents the value of f at the point $p \in M$ associated with coordinates x^i.

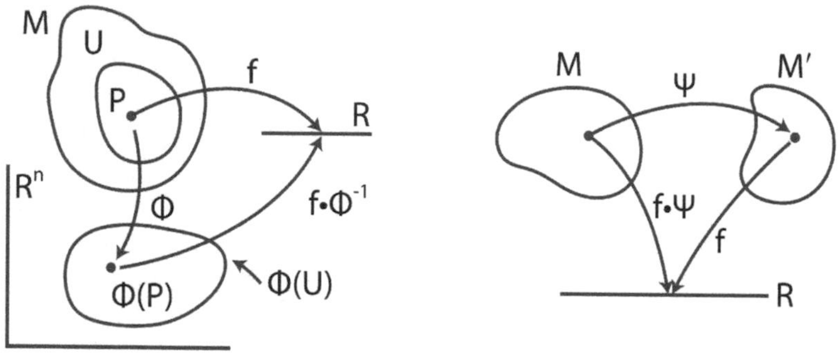

Figure 13.3 Left: A function $f : M \to \mathbb{R}$ is smooth if $f \circ \phi^{-1} : \mathbb{R}^n \to \mathbb{R}$ is smooth. Right: A map $\psi : M \to M'$ is smooth if $f \circ \psi : M \to \mathbb{R}$ is smooth.

Smooth functions define smooth mappings between manifolds. Let $\psi : M \to M'$ be a mapping between manifolds M, M' of dimensions n, n'. For f a function on M', $f \circ \psi$ is a function[10] on M. The map $\psi : M \to M'$ is smooth[11] if for every smooth function f on M', $f \circ \psi$ is a smooth function on M. In terms of charts, for (U, ϕ) on M and (V, μ) on M', the map ψ is smooth if $\mu \circ \psi \circ \phi^{-1}$ is smooth. A smooth function is thus a special case of a smooth mapping, one for which the target manifold M' is $\mathbb{R}$. The definition of smoothness applies to maps of the form $\psi : O \to M'$, where O is an open subset of M (the *restriction* of ψ to O, $\psi|_O$). *An open subset of M is a manifold in its own right* because it inherits the charts of M whose domains are subsets of O. A *diffeomorphism* $\psi : M \to M'$ is a one-to-one map having a smooth inverse, in which case M and M' are said to be *diffeomorphic*.[12] A diffeomorphism requires $n' = n$; the dimensions of the manifolds must be equal. A diffeomorphism is a manifold analog of isomorphism between vector spaces. Diffeomorphic manifolds are identical in terms of their smoothness properties.

13.1.3 Curves on manifolds

A *smooth curve* on a manifold M is a smooth mapping $\gamma : I \to M$ from an open interval[13] $I \subset \mathbb{R}$ to M; see Fig. 13.4. The image of the parameter $t \in I$ under γ is denoted $\gamma(t)$. Note that the "curve" is the *map* and not the set of image points in M: The same image can be produced by different maps.

[7] Manifolds are equipped with mappings $\phi : M \to \mathbb{R}^n$. Here we consider mappings between manifolds M and M'.

[8] We've already invoked functions on manifolds through the coordinate functions $x^i : U_\alpha \to \mathbb{R}$, $i = 1, \cdots, n$.

[9] Reread the definition of C^∞, which is in terms of mappings between $\mathbb{R}^n$. Smoothness of functions on M is defined in terms of functions on $\mathbb{R}^n$ where we know what such things mean. Manifolds "import" the smoothness property of $\mathbb{R}^n$.

[10] The function $f \circ \psi$ is referred to as the pullback of f; see Section 13.2.2.1.

[11] Smoothness of mappings between manifolds is determined by the smoothness of mappings between $\mathbb{R}^n$ and $\mathbb{R}^{n'}$.

[12] A homeomorphism is a one-to-one map having a continuous inverse. A diffeomorphism is a one-to-one map having a smooth (differentiable) inverse.

[13] As an open subset of $\mathbb{R}$, I inherits the manifold structure of $\mathbb{R}$.

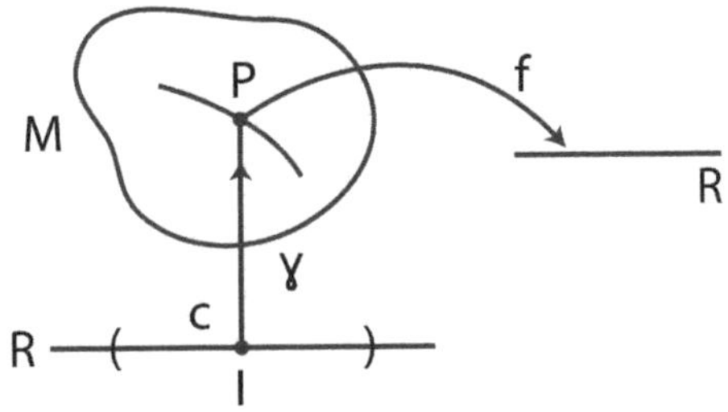

Figure 13.4 A curve γ on M is a mapping $\gamma : I \to M$ from an open interval $I \subset \mathbb{R}$ to M.

13.2 VECTOR AND TENSOR FIELDS

13.2.1 Vectors as operators, tangent space

Referring to Fig. 13.4, let $p \in M$ be such that $\gamma(c) = p$ for $c \in I$. For a function f on M, the composite map $f \circ \gamma : \mathbb{R} \to \mathbb{R}$ is a function on $\mathbb{R}$. For $u \in I$, how do the numbers $f \circ \gamma(u) \equiv f(\gamma(u))$ vary for u near c? The quantity $[f(\gamma(u)) - f(\gamma(c))]$ is a measure of how close $\gamma(u) \in M$ is to p for u near c. The *directional derivative* on M,

$$\left.\frac{\mathrm{d}(f \circ \gamma)}{\mathrm{d}u}\right|_c \equiv \lim_{s \to 0} \frac{1}{s} \left[f(\gamma(c+s)) - f(\gamma(c))\right] , \tag{13.1}$$

specifies the rate of change of $f : M \to \mathbb{R}$ along the points specified by γ. Contrast Eq. (13.1) with the directional derivative on $\mathbb{R}^n$, $\mathrm{D}_{\boldsymbol{v}}$, the derivative of a function ϕ in the direction of a vector $\boldsymbol{v}$,

$$\mathrm{D}_{\boldsymbol{v}}\phi(\boldsymbol{x}) \equiv \lim_{h \to 0} \frac{1}{h} \left[\phi(\boldsymbol{x} + h\boldsymbol{v}) - \phi(\boldsymbol{x})\right] = v^i(\partial\phi/\partial x^i) = \boldsymbol{v} \cdot \boldsymbol{\nabla}\phi(\boldsymbol{x}) ,$$

where $(v^1, \cdots, v^n)$ are the components of $\boldsymbol{v}$ with respect to the coordinate axes in $\mathbb{R}^n$. On M no "axes" are distinguished. In $\mathbb{R}^n$, $\mathrm{D}_{\boldsymbol{v}}$ is based on the direction specified by $\boldsymbol{v}$. On M there is no sense of *global* direction.[14] *Local* direction on M is specified by $\gamma(u)$. In $\mathbb{R}^n$ a vector is a displacement from "here" to "there." *On a manifold the notion of finite displacement loses its meaning.* In $\mathbb{R}^n$ there is a one-to-one correspondence between vectors and directional derivatives, $\boldsymbol{v} \leftrightarrow \mathrm{D}_{\boldsymbol{v}}$. We break that correspondence on manifolds and *define* vectors at a point $p \in M$, *tangent vectors*, as directional derivatives at that point. Tangent vectors are "here" on M, precisely at one point; hence the term *tangent*.

Tangent vectors have a vector-space structure—the *tangent space*. The tangent space at $p \in M$, the set of all directional derivatives at p, is denoted $T_p(M)$. For fixed γ, the directional derivative $\boldsymbol{t}_\gamma(f) \equiv \mathrm{d}(f \circ \gamma)/\mathrm{d}u|_p$ is an *operator* that maps functions $f \in \mathcal{F}_p$ onto numbers, $\boldsymbol{t}_\gamma : \mathcal{F}_p \to \mathbb{R}$. To cement the relation between directional derivatives and vectors, we require that $\boldsymbol{t}$ possess the properties customarily associated with derivatives, that for fixed γ and for $f, g \in \mathcal{F}_p$ and $\alpha, \beta \in \mathbb{R}$:

1. $\boldsymbol{t}$ is linear: $\boldsymbol{t}(\alpha f + \beta g) = \alpha\boldsymbol{t}(f) + \beta\boldsymbol{t}(g)$;
2. $\boldsymbol{t}$ satisfies the *derivation property* (product rule of calculus): $\boldsymbol{t}(fg) = f(p)\boldsymbol{t}(g) + g(p)\boldsymbol{t}(f)$.

Both points are easily checked. The linearity property establishes $T_p(M)$ as a vector space.

A point $p \in M$, with chart (U, ϕ), has coordinates $(x^1(p), \cdots, x^n(p))$ under $\phi : U \to \mathbb{R}^n$. A *coordinate curve*[15] γ^i through p is specified by $\phi^{-1} : \mathbb{R}^n \to M$ when all coordinates are held fixed

[14] For an arrow moving "west" on Earth, its direction on one side of the earth is opposite to that on the other side.

[15] Make sure you understand the distinction between coordinate functions and coordinate curves.

except the ith, $\gamma^i(u) \equiv \phi^{-1}(x^1(p), \cdots, x^{i-1}(p), x^i(p) + u, x^{i+1}(p), \cdots, x^n(p))$. The directional derivative associated with γ^i at p defines the partial derivative of f with respect to x^i, $\mathrm{d}(f \circ \gamma^i)/\mathrm{d}u|_{u=0} \equiv \partial f/\partial x^i|_p \equiv (\partial_i|_p)f$. When $f = x^j(p)$ (the jth coordinate function), $\partial_i x^j = \delta^j_i$.

The partial derivatives $\{\partial_i|_p\}_{i=1}^n$ form a basis for $T_p(M)$ (the *coordinate basis*), and thus *the dimension of $T_p(M)$ is the same as that of M*. To be a basis the set $\{\partial_i|_p\}_{i=1}^n$ must be linearly independent and it must span $T_p(M)$. If the set was *not* linearly independent, we could find nonzero numbers V^j such that $V^j\partial_j|_p = 0$; applied to x^k, $V^j\partial_j x^k = V^j\delta^k_j = V^k = 0$ for $k = 1, \cdots, n$, a contradiction. For the second requirement it must be shown that any $\boldsymbol{t} \in T_p(M)$ can be expressed as a linear combination

$$\boldsymbol{t} = a^i \frac{\partial}{\partial x^i} \equiv a^i \boldsymbol{e}_i \,. \qquad (a_i \in \mathbb{R}) \tag{13.2}$$

We omit a proof of Eq. (13.2); see [29, p52]. Equation (13.2) makes precise what we said in Section 5.1.2, that basis vectors are locally tangent to coordinate curves.

Figure 13.5 shows a two-dimensional manifold, M. Through $p \in M$ pass two coordinate curves,

Figure 13.5 Tangent space $T_p(M)$ attached to point p of manifold M. $T_p(M)$ is spanned by basis vectors $\boldsymbol{e}_u$ and $\boldsymbol{e}_v$. Each point has its own tangent space, e.g., $T_q(M)$.

u and v. The vectors tangent to the curves at p, $\boldsymbol{e}_u$, $\boldsymbol{e}_v$, are a basis for $T_p(M)$. Tangent vectors are not "in" the manifold, they are in $T_p(M)$ which is said to be attached to M at p. A point $q \in M \neq p$ has its own distinct tangent space, $T_q(M)$. Figure 13.5 shows a two-dimensional manifold so that it may be visualized. A manifold, however, has its own existence; it should not be thought of as embedded in a higher-dimensional Euclidean space.

If in one chart $p \in M$ has coordinates $\{x^i\}$ and $T_p(M)$ has the basis $\{\partial/\partial x^i\}$, then in an overlapping chart p has coordinates $\{y^j\}$ and $T_p(M)$ has the basis $\{\partial/\partial y^j\}$. The same vector $\boldsymbol{t} \in T_p(M)$ can be expressed in either basis, $\boldsymbol{t} = a^i(\partial/\partial x^i) = b^j(\partial/\partial y^j)$. Using Eq. (13.2), the coefficients a^i and b^j are related by (x^i is a coordinate function)

$$\boldsymbol{t}\left(x^i\right) = a^j \frac{\partial x^i}{\partial x^j} = a^j \delta^i_j = a^i = \boldsymbol{t}\left(x^i\right) = b^j \frac{\partial x^i}{\partial y^j} \,. \tag{13.3}$$

Equation (13.3) is the transformation law for contravariant vector components, Eq. (5.35). The basis vectors of $T_p(M)$ transform the "other" way, $\partial/\partial x^i = (\partial y^j/\partial x^i)\partial/\partial y^j$, Eq. (5.33). *Contravariant vectors at $p \in M$ are elements of the tangent space $T_p(M)$.*

13.2.2 Differentials as operators, cotangent space

If vectors on manifolds are directional derivatives, what would *dual vectors* on manifolds be? Dual vectors map vectors onto numbers (Appendix C). A dual vector $\boldsymbol{\omega}$ at $p \in M$ would be a mapping $\boldsymbol{\omega} : T_p(M) \to \mathbb{R}$. The set of all dual vectors at point p is the *cotangent space*, denoted $T^*_p(M)$. What mathematical objects act on derivatives to produce numbers? Those that are defined to do so! To develop that idea requires that we learn more about mappings between manifolds.

13.2.2.1 Pullback of functions, pushforward of vectors

For M, N manifolds of dimensions m, n, let $\phi : M \to N$ be a smooth map. Associated with ϕ are two other maps,[16] the *pullback* ϕ^* and the *pushforward* ϕ_*. Let $f : N \to \mathbb{R}$ be a smooth function. Then $f \circ \phi : M \to \mathbb{R}$ is a function on M (see the right part of Fig. 13.3), which can be formalized as an operator, the *pullback* of f, $\phi^* f \equiv f \circ \phi$ (see Fig. 13.6). For $p \in M$, let $q = \phi(p) \in N$. The

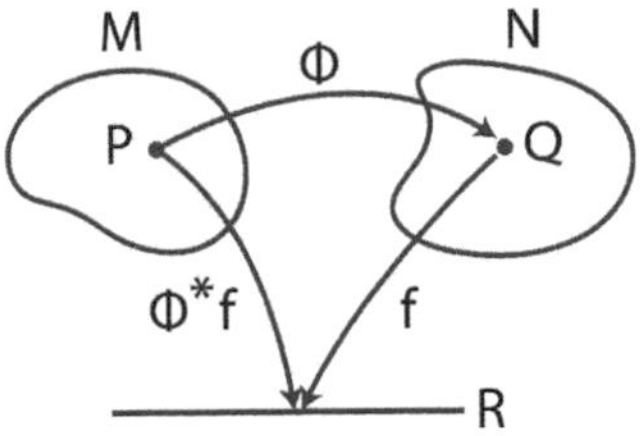

Figure 13.6 Pullback $\phi^* f \equiv f \circ \phi$ is a function on M for f a function on N.

pullback "pulls" f (defined on N) to a function defined on M, $\phi^* f$, such that

$$(\phi^* f)(p) \equiv f(\phi(p)) . \tag{13.4}$$

The pullback is a mapping between the space of functions, $\phi^* f : \mathcal{F}_q \to \mathcal{F}_p$.

The pullback effects another operation, the *pushforward* of a vector $\boldsymbol{t} \in T_p(M)$ as a mapping between tangent spaces $\phi_* \boldsymbol{t} : T_p(M) \to T_{\phi(p)}(N)$. The pushforward "pushes" $\boldsymbol{t} \in T_p(M)$ to $\phi_* \boldsymbol{t} \in T_{\phi(p)}(N)$, such that for all $f \in \mathcal{F}_q$,

$$(\phi_* \boldsymbol{t}) f \equiv \boldsymbol{t}(\phi^* f) = \boldsymbol{t}(f \circ \phi) . \tag{13.5}$$

It can be shown that $\phi_* \boldsymbol{t}$ meets the requirements of a tangent vector: linearity and the derivation property.[29, p56] The mapping $\phi : M \to N$ is an association between the *points* of manifolds; it effects (in math-speak *induces*) the operations ϕ^* and ϕ_* *that map functions and vectors between manifolds*. These operations are indicated schematically in Fig. 13.7.

$$\mathcal{F}_p \xleftarrow{\quad \phi^* f \quad} \mathcal{F}_{\phi(p)} \qquad T_p(M) \xrightarrow{\quad \phi_* \boldsymbol{t} \quad} T_{\phi(p)}(N)$$

Figure 13.7 Operations of pullback of a function and pushforward of a vector.

It helps to see how this works with coordinates. Let $\{x^i\}_{i=1}^m$ be the coordinates of $p \in M$ and $\{y^a\}_{a=1}^n$ be the coordinates of $\phi(p) \in N$. An arbitrary $\boldsymbol{t} \in T_p(M)$ can be expressed in the coordinate basis $t^i(\partial/\partial x^i)|_p$, Eq. (13.2). Because $\phi_* \boldsymbol{t} \in T_q(N)$, we can write $\phi_* \boldsymbol{t} = \beta^a(\partial/\partial y^a)|_q$. From Eq. (13.5) (using the chain rule),

$$(\phi_* \boldsymbol{t}) f = \beta^a \frac{\partial f}{\partial y^a} = \boldsymbol{t}(f \circ \phi) = t^i \frac{\partial (f \circ \phi)}{\partial x^i} = t^i \frac{\partial y^a}{\partial x^i} \frac{\partial f}{\partial y^a} . \tag{13.6}$$

The components of $\phi_* \boldsymbol{t}$ are thus related to those of $\boldsymbol{t}$ by $\beta^a = (\partial y^a / \partial x^i) t^i$; the action of $\phi_* \boldsymbol{t}$ is the $n \times m$ Jacobian matrix $\partial y^a / \partial x^i$. Equation (13.6) has the form of Eq. (5.35) (transformation law for the contravariant components of a vector), except that Eq. (13.6) is more general: It connects contravariant vectors on manifolds of unequal dimensions. Because Eq. (13.6) holds for all f,

$$\phi_* \boldsymbol{t} = t^i \frac{\partial y^a}{\partial x^i} e_a , \tag{13.7}$$

where $\{e_a \equiv \partial / \partial y^a\}$ is the coordinate basis in the target space.

[16] The notation for pullbacks (ϕ^*, upper star) and pushforwards (ϕ_*, lower star) is fairly standard, but not universal.

13.2.2.2 Differential of a map

In the special case $N = \mathbb{R}$, ϕ is a function $f : M \to \mathbb{R}$. The pushforward under f is, by Eq. (13.5), $(f_*\boldsymbol{t})g = \boldsymbol{t}(f^*g) = \boldsymbol{t}(g \circ f)$, where $g : \mathbb{R} \to \mathbb{R}$. By definition $f_*\boldsymbol{t} : T_p(M) \to T_c(\mathbb{R})$ where $c \equiv f(p)$. Because $T_c(\mathbb{R})$ is one-dimensional, $f_*\boldsymbol{t} = a(\mathrm{d}/\mathrm{d}u)|_c$, where a is a constant. To evaluate a choose $g(u) = u$, the identity function, which implies that $a = \boldsymbol{t}(f)$. The pushforward of a vector $\boldsymbol{t} \in T_p(M)$ under a function $f : M \to \mathbb{R}$ *defines* a linear map known as the *differential of* f, $\mathbf{d}\boldsymbol{f} : T_p(M) \to \mathbb{R}$, such that

$$\mathbf{d}\boldsymbol{f}(\boldsymbol{t}) \equiv \boldsymbol{t}(f) \,. \tag{13.8}$$

The quantity $\mathbf{d}\boldsymbol{f}$ therefore brings to the party what we sought for the dual of a tangent vector, a mapping from tangent vectors to numbers.[17] The cotangent space $T_p^*(M)$ is the set of all differentials $\mathbf{d}\boldsymbol{f}$ at p. For f the coordinate function x^i and $\boldsymbol{t}$ the tangent to the jth coordinate curve, $\partial_j|_p$, the action of $\mathbf{d}\boldsymbol{x}^i$ (the differential of the coordinate function x^i) on $\partial_j|_p$ is, by Eq. (13.8):

$$\mathbf{d}\boldsymbol{x}^i(\partial_j|_p) = (\partial_j|_p)\, x^i = \frac{\partial x^i}{\partial x^j} = \delta^i_j \,. \tag{13.9}$$

The set of operators $\{\mathbf{d}\boldsymbol{x}^i|_p\}_{i=1}^m$ *are dual to the basis of* $T_p(M)$, $\{\partial_j|_p\}_{j=1}^m$, *and form a basis for* $T_p^*(M)$. Compare Eq. (13.9) with Eq. (5.48) (or Eq. (C.1)).

Let's look at $\mathbf{d}\boldsymbol{f}$ in terms of coordinates. Let $f : M \to \mathbb{R}$ and let $\{x^i\}$ be the coordinates of $p \in M$. For $\boldsymbol{t} \in T_p(M)$, $\boldsymbol{t} = t^i(\partial_i|_p)$. From Eq. (13.8), $(\mathbf{d}\boldsymbol{f})\boldsymbol{t} = t^i\partial_i f$, where $t^i = \boldsymbol{t}(x^i) = (\mathbf{d}\boldsymbol{x}^i)\boldsymbol{t}$, also by Eq. (13.8). Hence, $\mathbf{d}\boldsymbol{f}(\boldsymbol{t}) = (\partial_i f)\mathbf{d}\boldsymbol{x}^i(\boldsymbol{t})$. Because this relation holds for any $\boldsymbol{t}$,

$$\mathbf{d}\boldsymbol{f} = (\partial_i f)\mathbf{d}\boldsymbol{x}^i \,. \tag{13.10}$$

While Eq. (13.10) looks familiar, $\mathbf{d}\boldsymbol{x}^i$ is an *operator*, a *symbol* for the map $\mathbf{d}\boldsymbol{x}^i(\partial/\partial x^j) = \delta^i_j$; see Eq. (13.9). For another set of coordinates at p, $\{y^j\}$ (those obtained from an overlapping chart), we would have another basis for $T_p^*(M)$, $\{\mathbf{d}\boldsymbol{y}^j\}$, related to the basis vectors $\{\mathbf{d}\boldsymbol{x}^i\}$ by

$$\mathbf{d}\boldsymbol{y}^j = \frac{\partial y^j}{\partial x^i}\mathbf{d}\boldsymbol{x}^i \equiv A^j_{i'}\mathbf{d}\boldsymbol{x}^{i'} \,. \tag{13.11}$$

Equation (13.11) has the same form as Eq. (5.34) (transformation equation for the dual basis vectors). The *components* of $\mathbf{d}\boldsymbol{f}$, $\partial_i f$, transform as in Eq. (5.38), the transformation equation for the covariant components of a vector, $\partial/\partial y^j = (\partial x^i/\partial y^j)\partial/\partial x^i$. *Covariant vectors at* $p \in M$ *are elements of the cotangent space* $T_p^*(M)$.

13.2.3 Tensor fields

We can now define tensors on manifolds. Vector spaces $T_p(M)$, $T_p^*(M)$ exist at every point of M. Tensors of type (k,l) can be constructed *at each point* $p \in M$ as multilinear functions on products of T_p and T_p^*, $\mathbf{T}^k_l|_p : (T_p^*)^k \times (T_p)^l \to \mathbb{R}$ (see Section 5.5). A *tensor field* is an assignment of a tensor to every point $p \in M$, where the tensors all have the same index structure. We can anticipate an issue with tensor fields. Whereas the points of Minkowski space are *isomorphic* with those of $\mathbb{R}^4$, the spacetime of GR—a four-dimensional manifold—has points only *locally* in correspondence with those of $\mathbb{R}^4$ and does not possess the global vector-space structure of $\mathbb{R}^4$. Comparing tensors from different points of M, for example for computing derivatives, is therefore problematic. If our goal is to represent physics on the spacetime manifold, e.g., in terms of partial differential equations, we must find a way of establishing a relationship between tensors at separate points of M. A means for doing that is presented in Chapter 14 which comes from equipping the manifold with an additional

[17] Equation (13.8) is a special case of the natural pairing between vectors and duals discussed in Appendix C, Eq. (C.2).

structure, the *affine connection*, leading to one way of taking derivatives, the covariant derivative. In this chapter we restrict ourselves to what can be said about tensor fields using just the properties of smooth manifolds. We will be led to another derivative, the Lie derivative.

A *vector field* is an assignment of a vector to every point of a manifold. A vector selected from the tangent space at every point, $\boldsymbol{v}|_p \in T_p(M)$, is a contravariant vector field, the *tangent field*. For $f : M \to \mathbb{R}$, $\boldsymbol{v}|_p(f)$ is a number for each $p \in M$. The tangent field is said to be smooth if for each smooth function f, $\boldsymbol{v}(f)$ is smooth. A covariant vector field $\boldsymbol{\omega}$ is smooth if for each smooth tangent field $\boldsymbol{v}$, the function $\boldsymbol{\omega}(\boldsymbol{v})$ is smooth. A tensor field $\mathbf{T}^k_l$ is smooth if for smooth covariant vector fields $\boldsymbol{\omega}^1, \cdots, \boldsymbol{\omega}^k$ and smooth tangent fields $\boldsymbol{v}_1, \cdots, \boldsymbol{v}_l$, $\mathbf{T}^k_l(\boldsymbol{\omega}^1, \cdots, \boldsymbol{\omega}^k, \boldsymbol{v}_1, \cdots, \boldsymbol{v}_l)$ is a smooth function. Note the progression: from functions on M, contravariant vector fields; from contravariant vector fields, covariant vector fields; from both, tensor fields.

The *metric field* $\mathbf{g}$ is an assignment[18] to every $x \in M$ of the tensor $\mathbf{g}|_x = g_{\mu\nu}(x)\boldsymbol{e}^\mu \otimes \boldsymbol{e}^\nu$ where the components $g_{\mu\nu}(x) = \mathbf{g}|_x(\boldsymbol{e}_\mu, \boldsymbol{e}_\nu)$ are smooth functions (with $\{\boldsymbol{e}_\mu\}$ the coordinate basis of $T_x(M)$ and $\{\boldsymbol{e}^\mu\}$ the coordinate basis for $T^*_x(M)$). Einstein's field equation is a set of differential equations for $g_{\mu\nu}(x)$. At every point p of a manifold M equipped with a metric, a basis can be found (a $\boldsymbol{g}$-orthonormal basis, Section 5.6), $\{\boldsymbol{v}_i\}_{i=1}^n \in T_p(M)$ such that $\mathbf{g}(\boldsymbol{v}_i, \boldsymbol{v}_j) = 0$ for $i \neq j$ and $\mathbf{g}(\boldsymbol{v}_i, \boldsymbol{v}_i) = \pm 1$. A *Riemannian manifold* has *signature* $(+\cdots+)$ for the terms $\mathbf{g}(\boldsymbol{v}_i, \boldsymbol{v}_i)$, while a manifold with signature $(-+++)$ is called *Lorentzian*. Once a metric field has been introduced, *the signature is the same at every point of a connected manifold.*[19]

13.3 INTEGRAL CURVES, CONGRUENCES, AND FLOWS

Every point of a curve $\gamma(t)$ has a tangent vector, the directional derivative at that point. Is the converse true? Let $\boldsymbol{X}$ be a tangent field on M. A curve $\gamma : I \to M$ is the *integral curve* of $\boldsymbol{X}$ if, for each $t \in I \subset \mathbb{R}$, the tangent field of $\gamma(t)$ coincides with $\boldsymbol{X}|_{\gamma(t)}$. An integral curve "threads" a smooth vector field seen as a collection of arrows; Fig. 13.8. Integral curves always exist. If $\gamma(t)$ has coordinates $x^i(t)$ and $\boldsymbol{X}$ has components X^i, finding the integral curve associated with $\boldsymbol{X}$ reduces to solving a set of coupled first-order differential equations,

$$\frac{\mathrm{d}}{\mathrm{d}t}x^i = X^i(x^1(t), \cdots, x^n(t)) \,. \qquad (i = 1, \cdots, n) \tag{13.12}$$

Such systems of differential equations have unique solutions for prescribed starting values.[20]

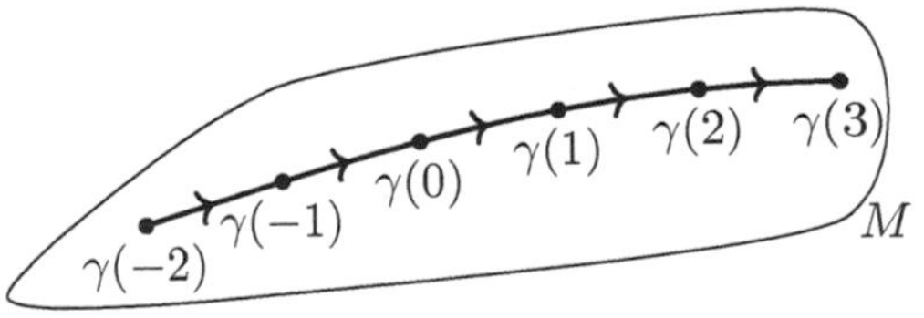

Figure 13.8 Integral curve $\gamma(t)$ threads a given vector field on M.

[18]It would be possible to have more than one metric field on a manifold. There are *bi-metric* theories of gravity in which the spacetime manifold has two metric tensors instead of one.

[19]More fine print. A manifold M is *connected* if the only subsets of M that are simultaneously open and closed are the empty set and M itself.[29, p46] The set of points on which the metric field has any given signature is open—the components $g_{\mu\nu}(x)$ are continuous functions, and because a metric must be invertible to be a metric (Section 5.6), if the metric has a given signature at a point $p \in M$, there is an open subset O of M containing p such that $g_{\mu\nu}$ has the same signature at each point of O. The set of points of M at which $g_{\mu\nu}$ has some *other* signature is also open, and thus the set of points at which $g_{\mu\nu}$ has the given signature is also closed. (A set of points is closed if its complement is open.) These statements imply that the metric field on a *connected* manifold has the same signature at every point.

[20]Existence and uniqueness of the solutions of ordinary differential equations is proved in books on differential equations.

Example. Examples of integral curves

1. Let $\boldsymbol{X} = x\partial_x + y\partial_y$ be a vector field ($\boldsymbol{X}$ assigns to the function f the number $x\partial_x f + y\partial_y f$). Equation (13.12) implies $\mathrm{d}x/\mathrm{d}t = x$ and $\mathrm{d}y/\mathrm{d}t = y$. The integral curve passing through (a, b) at $t = 0$ is $\gamma(t) = (ae^t, be^t)$.

2. Let $\boldsymbol{X} = -y\partial_x + x\partial_y$ be a vector field. Implied by Eq. (13.12) is the pair of differential equations $\mathrm{d}x/\mathrm{d}t = -y$ and $\mathrm{d}y/\mathrm{d}t = x$. The integral curve passing through (a, b) at $t = 0$ is $\gamma(t) = (a\cos t - b\sin t, a\sin t + b\cos t)$.

Integral curves associated with the same vector field cannot cross.[21] Every smooth vector field $\boldsymbol{X}$ has a unique *family* of curves having the property that precisely one curve passes through each $p \in M$ such that the tangent to the curve at p is $\boldsymbol{X}|_p$. Because an integral curve passes through every point p, they "fill" the manifold. A non-intersecting, manifold-filling set of curves is referred to as a *congruence*. The congruence can be regarded as the manifold itself.

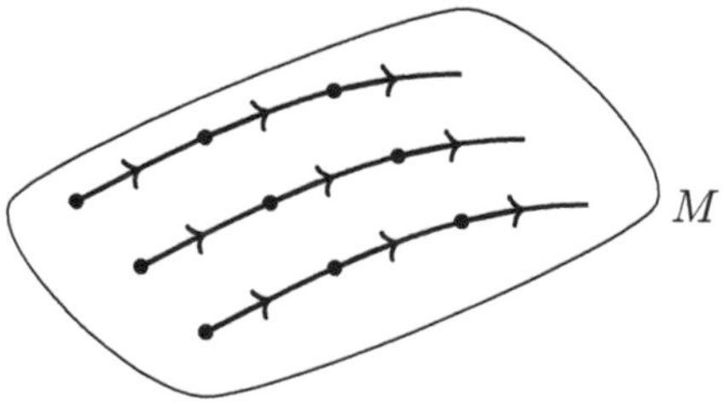

Figure 13.9 A manifold filling set of curves is a congruence.

Two integral curves can coincide if one is a reparameterization of the other. Let $\gamma : I \to M$ be an integral curve of $\boldsymbol{X}$ with $I = (a, b)$. Let c be a real number with $I' \equiv (a + c, b + c)$. Then $\gamma'(t) \equiv \gamma(t - c)$ is also an integral curve for $\boldsymbol{X}$ with $t \in I'$.[29, p121] We can state this idea formally: Let $\alpha : I' \to I$ be a diffeomorphism mapping the interval I' to the interval I (open subsets of $\mathbb{R}$ are manifolds). The reparameterized curve $\gamma' = \gamma \circ \alpha : I' \to M$. If 0 is in both I and I', the starting value of γ', $\gamma'(0)$ is the point $\gamma(-c)$ of M (see Fig. 13.10). The starting value has thus been shifted under a diffeomorphism.

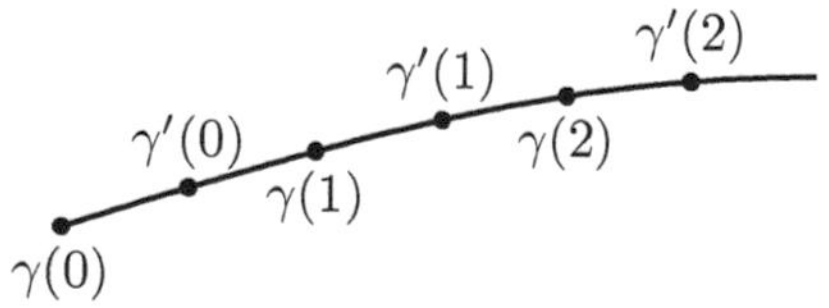

Figure 13.10 Two parameterizations of the same integral curve.

That idea is quite powerful and leads to a useful concept. Consider the integral curves of a vector field as the flow lines of a fluid (as in Fig. 13.9), where the "motion" of a point on the curve is provided by a diffeomorphism *of the manifold to itself.* The *flow* of a vector field $\boldsymbol{X}$ is the continuous sequence of mappings $\phi_t : \mathbb{R} \times M \to M$, $t \in \mathbb{R}$, such that for fixed t, $\phi_t : M \to M$ is the diffeomorphism $\phi_t(\gamma(0)) \to \gamma(t)$, where $\gamma(t)$ is the integral curve associated with $\boldsymbol{X}$ that passes

[21] The argument is the same as why in classical mechanics, trajectories in phase space cannot cross: Each is obtained from first-order differential equations, the solutions of which are unique.

through $\gamma(0)$. In other words, ϕ_t "pushes" a point along the integral curve by an amount[22] controlled by the parameter t. The set of maps $\{\phi_t | t \in \mathbb{R}\}$ comprises a group: for $t, s \in \mathbb{R}$, $\phi_t \circ \phi_s = \phi_{t+s}$, the identity map is ϕ_0, and the inverse $(\phi_t)^{-1} = \phi_{-t}$. The flow ϕ_t is a one-parameter group of diffeomorphisms[23] of M associated with vector field $\boldsymbol{X}$.

A congruence can be considered a mapping of a manifold onto itself. Through each point $p \in M$ there passes one curve of the family. Each curve is associated with a one-parameter group of diffeomorphisms, ϕ_t, such that for sufficiently small[24] t, each $p \in M$ is mapped to the point $q = \phi_t(p)$ lying at a parameter distance t along the curve (see Fig. 13.11).

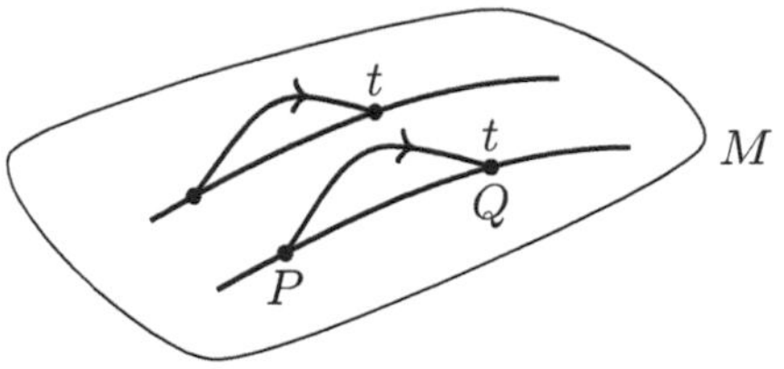

Figure 13.11 Each $p \in M$ is mapped under ϕ_t to $q = \phi_t(p)$ lying at parameter distance t along an integral curve.

13.4 MAPPINGS OF TENSORS

The pullback operation applies to dual vectors[25] with $\phi^*\boldsymbol{\omega} : T^*_{\phi(p)}(N) \to T^*_p(M)$, a mapping between cotangent spaces such that for $\boldsymbol{\omega} \in T^*_{\phi(p)}(N)$ and for all $\boldsymbol{t} \in T_p(M)$,

$$(\phi^*\boldsymbol{\omega})\boldsymbol{t} \equiv \boldsymbol{\omega}(\phi_*\boldsymbol{t}) \,. \tag{13.13}$$

Compare Eq. (13.13) with Eq. (13.4). For x^i the coordinates of $p \in M$ and y^a the coordinates of $\phi(p) \in N$, because $\phi^*\boldsymbol{\omega} \in T^*_p(M)$, $\phi^*\boldsymbol{\omega} = f_i\boldsymbol{e}^i$, where $\boldsymbol{e}^i = \mathbf{d}\boldsymbol{x}^i$. Thus, for $\boldsymbol{t} = t^i\boldsymbol{e}_i$ (with $\boldsymbol{e}_i = \partial/\partial x^i$), and for $\boldsymbol{\omega} = \omega_b\boldsymbol{e}^b$ ($\boldsymbol{e}^b = \mathbf{d}\boldsymbol{y}^b$), we have from Eq. (13.13),

$$(\phi^*\boldsymbol{\omega})\,\boldsymbol{t} = f_i\boldsymbol{e}^i\left(t^j\boldsymbol{e}_j\right) = f_it^i = \boldsymbol{\omega}\left(\phi_*\boldsymbol{t}\right) = \omega_b\boldsymbol{e}^b\left(t^i\frac{\partial y^a}{\partial x^i}\boldsymbol{e}_a\right) = \omega_at^i\frac{\partial y^a}{\partial x^i} \,, \tag{13.14}$$

where we've used Eqs. (13.9) and (13.7). The components of $\phi^*\boldsymbol{\omega}$ (i.e., f_i) are thus related to those of $\boldsymbol{\omega}$ (i.e., ω_a) by $f_i = \omega_a(\partial y^a/\partial x^i)$, a generalization of the transformation law for covariant vector components, Eq. (5.38). Because Eq. (13.14) holds for all $\boldsymbol{t}$,

$$\phi^*\boldsymbol{\omega} = \omega_a\frac{\partial y^a}{\partial x^i}\mathbf{d}\boldsymbol{x}^i \,. \tag{13.15}$$

The pullback operation commutes with the differential of a function,

$$\phi^*\left(\mathbf{d}\boldsymbol{f}\right) = \mathbf{d}(\phi^*\boldsymbol{f}) \,, \tag{13.16}$$

which follows from $[\phi^*(\mathbf{d}\boldsymbol{f})]\boldsymbol{t} = \mathbf{d}\boldsymbol{f}(\phi_*\boldsymbol{t}) = (\phi_*\boldsymbol{t})f = \boldsymbol{t}(\phi^*f) = \mathbf{d}(\phi^*\boldsymbol{f})\boldsymbol{t}$, where we've used Eqs. (13.13), (13.8), (13.5), and (13.8) again. When expressed in terms of coordinates, Eq. (13.16) is equivalent to the chain rule for partial derivatives.

---

[22] We're sweeping under the rug that the parameter values of the integral curve can be taken from all of $\mathbb{R}$. A vector field $\boldsymbol{X}$ is said to be *complete* if its integral curve is defined by the entire real line. If $\boldsymbol{X}$ is not complete, its flow is defined only locally and the parameter t must be considered sufficiently small.

[23] See [37, p18] or [38, p27].

[24] The proviso for small t is to avoid potential problems of "running out of manifold."

[25] Functions (scalar fields) can be considered rank-zero covariant vectors, just as *points* can be considered rank-zero contravariant vectors.

For diffeomorphisms ($\phi^{-1} : N \to M$ is one-one), we can use ϕ^{-1} to extend ϕ^* and ϕ_*. The pullback under ϕ^{-1} of a function $f \in \mathcal{F}_p$ is such that $\left((\phi^{-1})^* f\right)\phi(p) \equiv \left(f \circ \phi^{-1}\right)\phi(p) = f(p)$ for $p \in M$. Hence, $(\phi^{-1})^* f$ is effectively a *pushfoward* on functions, $\mathcal{F}_p \to \mathcal{F}_{\phi(p)}$, $\left(\phi^{-1}\right)^* f \equiv \phi_* f$. The pushforward of a vector under ϕ^{-1}, $\left((\phi^{-1})_* \boldsymbol{t}\right) f \equiv \boldsymbol{t}\left((\phi^{-1})^* f\right)$ for $\boldsymbol{t} \in T_{\phi(p)}(N)$ for all $f \in \mathcal{F}_p$, is effectively the pullback under ϕ, $\left(\phi^{-1}\right)_* \boldsymbol{t} \equiv \phi^* \boldsymbol{t}$, with $(\phi^* \boldsymbol{t})\, f = \boldsymbol{t}\,(\phi_* f)$. The pullback of a dual vector under ϕ^{-1}, $\left((\phi^{-1})^* \boldsymbol{\omega}\right) \boldsymbol{t} \equiv \boldsymbol{\omega}\left((\phi^{-1})_* \boldsymbol{t}\right)$ for $\boldsymbol{\omega} \in T_p^*(M)$ for all $\boldsymbol{t} \in T_{\phi(p)}(N)$ is effectively a pushforward under ϕ, $\left(\phi^{-1}\right)^* \boldsymbol{\omega} \equiv \phi_* \boldsymbol{\omega}$, with $(\phi_* \boldsymbol{\omega})\, \boldsymbol{t} = \boldsymbol{\omega}\,(\phi^* \boldsymbol{t})$. These operations are illustrated in Fig. 13.12 (compare Fig. 13.12 with Fig. 13.7).

$$T_p^*(M) \xleftarrow{\;\phi^*\boldsymbol{\omega}\;} T_{\phi(p)}^*(N) \qquad \mathcal{F}_p \xrightarrow{\;(\phi^{-1})^* f\;} \mathcal{F}_{\phi(p)} \qquad T_p(M) \xleftarrow{\;(\phi^{-1})_* \boldsymbol{t}\;} T_{\phi(p)}(N)$$

Figure 13.12 Pullback of a dual vector, pushforward of a function, and pullback of a vector.

The pushforward of a type (k, l) tensor under a diffeomorphism is defined such that

$$\left(\phi_* \mathbf{T}_l^k\right)\left(\boldsymbol{\omega}^1, \cdots, \boldsymbol{\omega}^k, \boldsymbol{t}_1, \cdots, \boldsymbol{t}_l\right)|_{\phi(p)} \equiv \mathbf{T}_l^k\left(\phi^* \boldsymbol{\omega}^1, \cdots, \phi^* \boldsymbol{\omega}^k, (\phi^{-1})_* \boldsymbol{t}_1, \cdots, (\phi^{-1})_* \boldsymbol{t}_l\right)|_p\,,$$

for $\boldsymbol{\omega}^i \in T_{\phi(p)}^*(N)$ and $\boldsymbol{t}_i \in T_{\phi(p)}(N)$. The operator $(\phi^{-1})_*$ is effectively the pullback, for $\boldsymbol{\omega}^i \in T_p^*(M)$ and $\boldsymbol{t}_i \in T_p(M)$,

$$\left(\left(\phi^{-1}\right)_* \mathbf{T}_l^k\right)\left(\boldsymbol{\omega}^1, \cdots, \boldsymbol{\omega}^k, \boldsymbol{t}_1, \cdots, \boldsymbol{t}_l\right)|_p \equiv \mathbf{T}_l^k\left((\phi^{-1})^* \boldsymbol{\omega}^1, \cdots, (\phi^{-1})^* \boldsymbol{\omega}^k, \phi_* \boldsymbol{t}_1, \cdots, \phi_* \boldsymbol{t}_l\right)|_{\phi(p)}\,.$$

13.5 THE LIE DERIVATIVE

We're now in a position to define a derivative of tensor fields. A manifold can be covered by a non-intersecting family of integral curves associated with a vector field $\boldsymbol{v}$, where each curve is generated by a group of diffeomorphisms, ϕ_t with $t \in \mathbb{R}$ (Section 13.3). The pushforward associated with ϕ_t, ϕ_{t*}, would map a tensor at $p \in M$, $\mathbf{T}|_p$, to the point $\phi_t(p)$. How does $\phi_{t*}\mathbf{T}$ compare with the tensor that's "already there," $\mathbf{T}|_{\phi_t(p)}$?

The *Lie derivative*[26] $\pounds_{\boldsymbol{v}}$ of tensor field $\mathbf{T}$ with respect to tangent field $\boldsymbol{v}$ is defined in terms of the difference between the pullback of $\mathbf{T}|_{\phi_t(p)}$ and $\mathbf{T}|_p$ as $t \to 0$,

$$\left(\pounds_{\boldsymbol{v}} \mathbf{T}\right)_p \equiv \lim_{t \to 0} \frac{1}{t}\left[(\phi_{-t})_* \mathbf{T}|_{\phi_t(p)} - \mathbf{T}|_p\right]\,, \tag{13.17}$$

where $\phi_{-t} \equiv \left(\phi_t\right)^{-1}$. *The Lie derivative measures the rate of change of* $\mathbf{T}$ *relative to the contravariant vector field* $\boldsymbol{v}$. This idea is illustrated in Fig. 13.13 where we've used an arrow to represent a

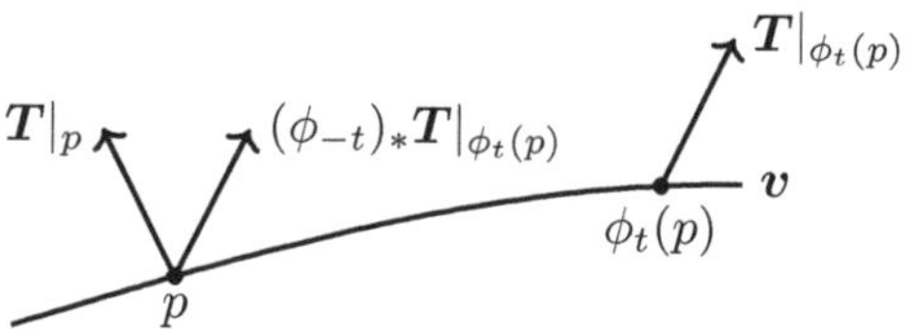

Figure 13.13 Lie derivative of a tensor field $\mathbf{T}$ is obtained from the difference between the tensor at p, $\mathbf{T}|_p$, with that at $\phi_t(p)$, $\mathbf{T}|_{\phi_t(p)}$, by pulling it back to p.

[26]Named after the mathematician Sophus Lie, pronounced Lee. It's traditionally denoted $\pounds_{\boldsymbol{v}}$, a notation we adhere to.

generic tensor field $\mathbf{T}$. As with any derivative, $\pounds_{\boldsymbol{v}}$ is linear, $\pounds_{\boldsymbol{v}}(\lambda\mathbf{S}+\mu\mathbf{T}) = \lambda\pounds_{\boldsymbol{v}}\mathbf{S}+\mu\pounds_{\boldsymbol{v}}\mathbf{T}$, and it obeys the product rule, $\pounds_{\boldsymbol{v}}(\mathbf{S}\otimes\mathbf{T}) = (\pounds_{\boldsymbol{v}}\mathbf{S})\otimes\mathbf{T}+\mathbf{S}\otimes(\pounds_{\boldsymbol{v}}\mathbf{T})$. Additionally, it commutes with contractions:[27] $\pounds_{\boldsymbol{v}}\left(T^{\mu}_{\mu}\right) = \delta^{\mu}_{\nu}\pounds_{\boldsymbol{v}}T^{\nu}_{\mu}$. It's also *type preserving*: For $\mathbf{T}$ a type (k,l) tensor, $\pounds_{\boldsymbol{v}}\mathbf{T}$ is a type (k,l) tensor. The Lie derivative is one of several ways to define derivatives on manifolds (see Chapter 14). Its advantage is that it uses only the intrinsic properties of smooth manifolds; its disadvantage is that it requires a vector field $\boldsymbol{v}$ to be specified.[28]

The Lie derivative of various objects is found by working out their pullbacks. Let $p\in M$ have coordinates x^{μ}. The coordinates of $\phi_t(p)$ are, for infinitesimal t, generated by the vector $\boldsymbol{v}$ at p,

$$\phi_t(x^{\mu}) \equiv \overline{x}^{\mu} = x^{\mu}+tv^{\mu}+O(t^2)\,. \qquad (t\to 0) \tag{13.18}$$

The pullback of a function at $\overline{x}^{\mu}$ is, from Eq. (13.4), $(\phi_t^* f)(x^{\mu}) = f(x^{\mu}+tv^{\mu}) \approx f(x^{\mu})+tv^{\mu}\partial_{\mu}f$. Using Eq. (13.17), the Lie derivative of a scalar field is the directional directive,

$$\pounds_{\boldsymbol{v}}f = v^{\mu}\partial_{\mu}f\,. \tag{13.19}$$

Evaluating the pullback of tensors is more involved. Key to understanding why is that the diffeomorphism ϕ_t is an active transformation:[29] The point p *with its local coordinate system* x^{μ} is sent to $\phi_t(p)$, which has a different system of coordinates, $\overline{x}^{\mu}$ (see Eq. (13.18)). Because tensors are expressed with respect to the basis vectors underlying these coordinates, we have to transform the basis vectors. The pushforward of a contravariant vector field $\boldsymbol{u}$ under ϕ_{-t} is, using Eq. (13.7),

$$\begin{aligned}\phi_{-t*}\left(\boldsymbol{u}|_{\phi_t(p)}\right) =& u^a(x^{\lambda}+tv^{\lambda})\frac{\partial x^{\mu}}{\partial\overline{x}^a}\boldsymbol{e}_{\mu} = \left(u^a(x)+tv^{\lambda}\partial_{\lambda}u^a\right)\left(\delta^{\mu}_a - t\partial_a v^{\mu}\right)\boldsymbol{e}_{\mu}+O(t^2)\\ =& \boldsymbol{u}|_p + t\left[v^{\lambda}\partial_{\lambda}u^{\mu} - u^a\partial_a v^{\mu}\right]\boldsymbol{e}_{\mu}+O(t^2)\,,\end{aligned}$$

where we've kept terms to first order in t. Using Eq. (13.17),

$$\pounds_{\boldsymbol{v}}\boldsymbol{u} = \left(v^{\lambda}\partial_{\lambda}u^{\mu} - u^{\lambda}\partial_{\lambda}v^{\mu}\right)\boldsymbol{e}_{\mu} \equiv [\boldsymbol{v},\boldsymbol{u}]\,, \tag{13.20}$$

where $\boldsymbol{e}_{\mu}\equiv\partial_{\mu}$. Equation (13.20) defines the *commutator* (*Lie bracket*) of the vector fields $\boldsymbol{u}$ and $\boldsymbol{v}$, *which is itself a contravariant vector field* (Lie derivative is type preserving). In component form, $(\pounds_{\boldsymbol{v}}\boldsymbol{u})^{\mu} = v^{\lambda}\partial_{\lambda}u^{\mu} - u^{\lambda}\partial_{\lambda}v^{\mu}$. Note that we need to know *all* components of $\boldsymbol{v}$ and $\boldsymbol{u}$ to evaluate just one component of the derivative, $(\pounds_{\boldsymbol{v}}\boldsymbol{u})^{\mu}$. The Lie bracket is antisymmetric: $[\boldsymbol{v},\boldsymbol{u}] = -[\boldsymbol{u},\boldsymbol{v}]$. In particular, $[\boldsymbol{v},\boldsymbol{v}]=0$, which makes sense: the rate of change of $\boldsymbol{v}$ relative to itself is zero.

The pullback of a dual vector is, using Eqs. (13.15) and (13.18),

$$\begin{aligned}\phi_t^*\boldsymbol{\omega} &= \omega_a(x^{\lambda}+tv^{\lambda})\frac{\partial\overline{x}^a}{\partial x^{\mu}}\boldsymbol{e}^{\mu} = \left(\omega_a(x)+tv^{\lambda}\partial_{\lambda}\omega_a\right)\left(\delta^a_{\mu}+t\partial_{\mu}v^a\right)\boldsymbol{e}^{\mu}+O(t^2)\\ &= \boldsymbol{\omega}|_p + t\left[\omega_a\partial_{\mu}v^a + v^{\lambda}\partial_{\lambda}\omega_{\mu}\right]\boldsymbol{e}^{\mu}+O(t^2)\,.\end{aligned}$$

Thus, from Eq. (13.17),

$$\pounds_{\boldsymbol{v}}\boldsymbol{\omega} = [v^{\lambda}\partial_{\lambda}\omega_{\mu}+\omega_{\lambda}\partial_{\mu}v^{\lambda}]\boldsymbol{e}^{\mu} \equiv \{\boldsymbol{\omega},\boldsymbol{v}\}\,, \tag{13.21}$$

the *anticommutator* of $\boldsymbol{\omega}$ and $\boldsymbol{v}$. In component form, $(\pounds_{\boldsymbol{v}}\boldsymbol{\omega})_{\mu} = v^{\lambda}\partial_{\lambda}\omega_{\mu}+\omega_{\lambda}\partial_{\mu}v^{\lambda}$.

The pullback of $\mathbf{g} = g_{ab}\boldsymbol{e}^a\otimes\boldsymbol{e}^b$ is $\phi_t^*\mathbf{g}|_{\phi_t(p)} = g_{\mu\nu}(x^{\lambda}+tv^{\lambda})\partial_a\overline{x}^{\mu}\partial_b\overline{x}^{\nu}\boldsymbol{e}^a\otimes\boldsymbol{e}^b$, from which it's straightforward to show that

$$(\pounds_{\boldsymbol{v}}\mathbf{g})_{ab} = v^{\lambda}\partial_{\lambda}g_{ab}+g_{a\lambda}\partial_b v^{\lambda}+g_{\lambda b}\partial_a v^{\lambda}\,. \tag{13.22}$$

---

[27] The three properties of the Lie derivative mirror the fundamental operations on tensors at a point of addition, outer product, and contraction. The Lie derivative interacts with tensors preserving these fundamental properties.

[28] The Lie derivative also does not reduce to the partial derivative on flat spaces; it generalizes the directional derivative.

[29] The distinction between active and passive transformations is discussed in Section 4.1.1.

The analysis can be repeated for a tensor of type (k, l) with the result that there is a plus sign for every covariant index and a negative sign for every contravariant index

$$(\pounds_v \mathbf{T})^{a_1 \cdots a_k}_{b_1 \cdots b_l} = v^\lambda \partial_\lambda T^{a_1 \cdots a_k}_{b_1 \cdots b_l} - \sum_{n=1}^{k} T^{a_1 \cdots a_{n-1} \alpha a_{n+1} a \cdots a_k}_{b_1 \cdots b_l} \partial_\alpha v^{a_n} + \sum_{m=1}^{l} T^{a_1 \cdots a_k}_{b_1 \cdots b_{m-1} \alpha b_{m+1} \cdots b_l} \partial_{b_m} v^\alpha \,. \tag{13.23}$$

13.6 SUBMANIFOLDS, EMBEDDINGS, AND HYPERSURFACES

13.6.1 Submanifolds

A k-dimensional *submanifold* S of an n-dimensional manifold M $(1 \le k \le n)$ is a subset of M that is itself a manifold. Roughly speaking, a k-dimensional submanifold is a k-dimensional surface in M that locally looks like $\mathbb{R}^k$, as indicated in Fig. 13.14. Precisely speaking, $S \subset M$ is a k-dimensional submanifold if, for each point $p \in S$, there is an n-chart (U, ϕ) of M such that for $p \in U$, $\phi(U \cap S)$ consists of the points of $\phi(U)$ with $x^{k+1} = \cdots = x^n = 0$. That is, for each point p of S, there is an n-chart containing p such that S intersects U at those points (of U) that are assigned the value zero for $(n - k)$ coordinates. The number $(n - k)$ is the *co-dimension* of S. A submanifold is thus a manifold.[30] For $k = n$, S is an open subset of M; for $k = n - 1$, S is a hypersurface[31] in M.

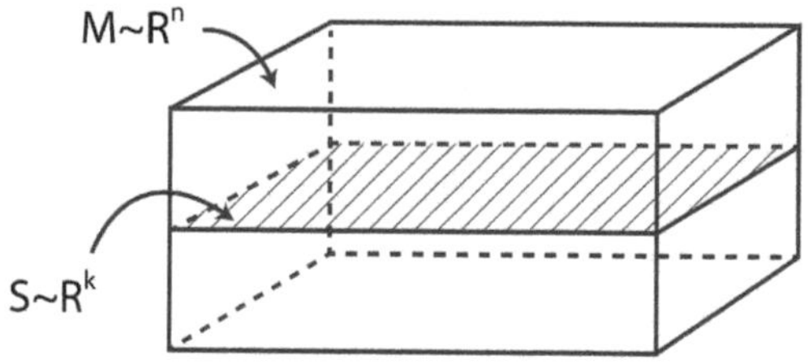

Figure 13.14 k-dimensional submanifold S of an n-dimensional manifold M.

Example. Examples of submanifolds

1. Let S be a subset of $\mathbb{R}^n$ with $\sum_{i=1}^{n} (x^i)^2 = 1$. Then S is a $(n-1)$-dimensional submanifold of $\mathbb{R}^n$; it's the manifold S^{n-1} defined in Section 13.1.1. Note that the charts used to define S^{n-1} simply eliminate one coordinate, which the definition of submanifold sets to zero.

2. Let M_1 and M_2 be manifolds of dimension n_1 and n_2, and let $M = M_1 \times M_2$ be the product manifold. Fix a point $p_1 \in M_1$, and let S consist of all points of M of the form (p_1, p_2) with $p_2 \in M_2$. Then S is a submanifold of M of dimension n_2. Consider the cylinder $S^1 \times \mathbb{R}^1$. Fix a point of S^1; $\mathbb{R}^1$ is a one-dimensional submanifold of the cylinder.

---

[30] *Proof*: S is a subset of M, so S is a set. If (U, ϕ) is a chart on M with U consisting of points having $(n-k)$ coordinates zero, then $(x^1, \cdots, x^k)$ are coordinate functions on $U \cap S$. But $U \cap S$ is a subset of S and $(x^1, \cdots, x^k)$ are coordinates on this subset; therefore we have a k-chart. A k-dimensional submanifold of M is thus a k-dimensional manifold; it inherits the smoothness of M through its charts.

[31] Hypersurfaces were introduced in Section 8.3.

13.6.2 Embeddings

Let S and M be manifolds of dimensions k and n, with $1 \leq k \leq n$. Referring to Fig. 13.15, an *embedding* of S in M (with S *embedded* in M) is a smooth map[32] $\psi : S \to M$ if: 1) ψ is one-to-one, and 2) for $p \in S$ there is a neighborhood $U \subset S$ so that for $\psi(p) \in M$ there is a chart (V, μ) with $\mu : V \to \mathbb{R}^n$ and $\mu(\psi(p)) = (y^1, \cdots, y^n)$, the quantities $x^i \equiv y^i \circ \psi|_U$ $(i = 1, \cdots, k)$ are coordinates for p. The manifold S is an embedded submanifold of M if the *identity map* is an

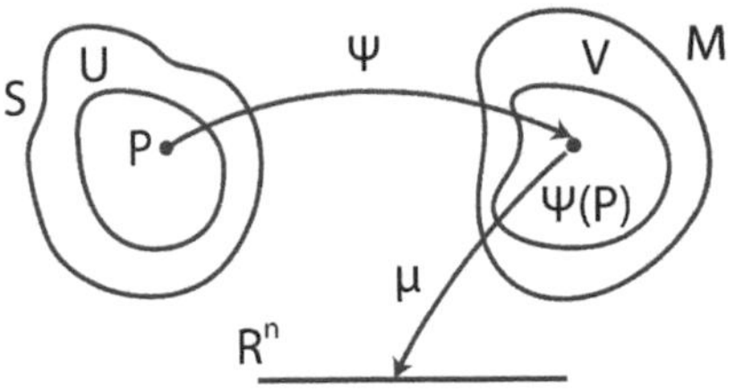

Figure 13.15 Embedding of S in M under the mapping ψ.

embedding.[33] *An embedded submanifold cannot intersect itself* because embeddings are one-to-one. Curves *can* intersect themselves, however ($\gamma(t_1) = \gamma(t_2)$ for $t_1 \neq t_2$). One-dimensional embedded submanifolds are curves, but not the converse; not all curves are embedded submanifolds. If the requirement that ψ is one-to-one is omitted but the requirement of obtaining coordinates of S from those of M is retained, ψ is said to be an *immersion* (with S *immersed* in M).[34] In what follows we work with embedded submanifolds.

13.6.3 Coordinate slices

A *coordinate slice* of dimension k in an n-dimensional manifold M (a k-slice) are points in $U \subset M$ with coordinates $(x^1, \cdots, x^n)$ such that $(n - k)$ of the coordinates are held fixed: for $p \in U$, $\{x^i(p) = c^i\}_{i=k+1}^n$, where the constants c^i determine the slice. A slice specifies a submanifold.[29, p42] Figure 13.16 shows a one-dimensional coordinate slice of $\mathbb{R}^3$.

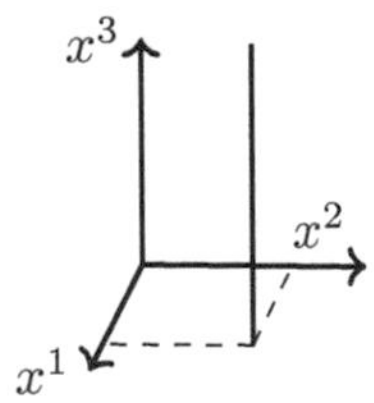

Figure 13.16 One-dimensional coordinate slice of $\mathbb{R}^3$: x^1 and x^2 are held fixed.

13.6.4 Tangents and normals to submanifolds

Let S be an embedded k-dimensional submanifold of an n-dimensional manifold M specified by the identity map $\psi : S \to M$. A curve $\gamma : I \to S$ passing through $p \in S$ is also a curve in M through

32 The embedding is the map $\psi : S \to M$.

33 For S an embedded submanifold of M, a point of S, which is also a point of M, is taken by the identity map to that point of M. The identity map is one-to-one and smooth because S is a smooth manifold.

34 There are several inequivalent definitions of embeddings and immersions in the literature; beware. We have adopted definitions that guarantee the greatest smoothness; differential geometry is the study of smoothness. What we have defined as embeddings are also referred to in the literature as *imbeddings*.

p. Vectors tangent to curves on M passing through p are elements of the n-dimensional vector space $T_p(M)$. Vectors tangent to curves on S (passing through p) are elements of a k-dimensional subspace of $T_p(M)$, $T_p(S)$. A contravariant vector $\boldsymbol{u}$ at $p \in S$ with components $(u^1, \cdots, u^k)$, is, under the embedding, a contravariant vector at $p \in M$ with components $(u^1, \cdots, u^k, 0, \cdots, 0)$, which we can indicate with the pushforward map $\psi_* : T_p(S) \rightarrow T_{\psi(p)}(M)$, where $\psi(p) \equiv p$. Vectors in M of the form $(u^1, \cdots, u^k, 0, \cdots, 0)$ are said to be *tangents* to S. Let $\boldsymbol{\omega}$ be a covariant vector at $p \in M$, an element of the cotangent space $T_p^*(M)$. Such a vector is said to be *normal* to S (at p) if $\omega_i u^i = 0$ ($i = 1, \cdots, n$) for *all* tangents to S (at p), which is possible only if $\boldsymbol{\omega}$ has the form $(0, \cdots, 0, \omega_{k+1}, \cdots, \omega_n)$. Normal vectors to S (at p) therefore form an $(n-k)$-dimensional subspace of $T_p^*(M)$. Under the pullback $\psi^* : T^*_{\psi(p)}(M) \rightarrow T_p^*(S)$, $\boldsymbol{\omega}$ is normal to S if its pullback is the zero vector of $T_p^*(S)$, $\psi^*(\boldsymbol{\omega}) = 0$. At any point p of a k-dimensional submanifold of M *there is a k-dimensional subspace of $T_p(M)$ and an $(n-k)$-dimensional subspace of $T_p^*(M)$.*

Example. For Minkowski space, consider the three-dimensional submanifold S obtained by setting the time coordinate to zero, $x^0 = 0$. Corresponding to contravariant vectors in S, (x^1, x^2, x^3), are the tangent vectors in M, $(0, x^1, x^2, x^3)$. Vectors normal to S are of the form $\boldsymbol{\omega} = (\omega_1, 0, 0, 0)$. The pullback of $\boldsymbol{\omega}$ is the zero vector $(0, 0, 0)$.

13.6.5 Metric submanifolds, hypersurfaces, and induced metrics

Could a vector be both tangent to, *and* normal to a submanifold S? The question can be answered if we equip M with a metric field $\mathbf{g}$. For a metric $\mathbf{g}$ on M, the pullback ψ^* associated with the embedding $\psi : S \rightarrow M$ defines a symmetric tensor field on S, $h_{ab}|_p \equiv (\psi^* g_{ab})|_{\psi(p)}$. If $h_{ab}|_p$ is nondegenerate at all $p \in S$, h_{ab} is a metric field on S, the *induced metric*,[35] where for $\boldsymbol{u}, \boldsymbol{v} \in T_p(S)$, $\psi^*\mathbf{g}(\boldsymbol{u}, \boldsymbol{v})|_p = \mathbf{g}(\psi_*\boldsymbol{u}, \psi_*\boldsymbol{v})|_{\psi(p)}$ (see Section 13.4). In such a case, S is said to be a *metric submanifold* relative to the background metric $\mathbf{g}$ on M. *A submanifold S is a metric submanifold if and only if at each point p of $\psi(S)$, the only vector $\boldsymbol{v}$ tangent to, and normal to S is $\boldsymbol{v} = 0$.*[36] For a metric submanifold there are at each point of $\psi(S) \in M$ two sets of vectors (tangents and normals to S) having no vector in common except the zero vector. There is therefore a vector space at each point of $\psi(S)$ comprised of basis vectors that are either tangent to, or normal to S, but not both.

For S a submanifold of a Riemannian manifold[37] M (under an embedding $\psi : S \rightarrow M$), let there be a contravariant vector $\boldsymbol{v}$ tangent to S with components v^i. Use the metric tensor to lower the index, $v_j = g_{ji}v^i$. If the covariant vector $\boldsymbol{v}$ were normal to S, we would have $v_i v^i = g_{ij}v^i v^j = 0$. But there cannot be *nonzero* vectors satisfying this condition for a positive-definite metric. *Every submanifold of a Riemannian manifold is a metric submanifold.* The induced metric is positive definite: $(\psi^*\mathbf{g})(\boldsymbol{u}, \boldsymbol{v})|_p = \mathbf{g}(\psi_*\boldsymbol{u}, \psi_*\boldsymbol{v})|_{\psi(p)} > 0$ for $\boldsymbol{u}, \boldsymbol{v} \in T_p(S)$.

The same cannot be said of a Lorentzian manifold. Consider the hypersurfaces S of a Lorentzian manifold M. We know there is a vector n_α normal to all vectors tangent to S. Because the metric $\mathbf{g}$ on M is indefinite, there are three possibilities for the inner product: $n^\alpha n_\alpha > 0$, $n^\alpha n_\alpha = 0$, and $n^\alpha n_\alpha < 0$. If $n^\alpha n_\alpha = 0$, we have a nonzero vector that is both tangent to, and normal to S (a *null*

---

[35] A metric must be invertible (nondegenerate) to be a metric, Appendix C.

[36] *Proof*: First assume that h_{ab} is not invertible (and hence S is not a metric submanifold), that for all $v^b \in T_p(S)$ there is some *nonzero* $u^a \in T_p(S)$ such that $h_{ab}u^a v^b = 0$ (the null space of h_{ab} is non-empty). By definition, $h_{ab}u^a v^b = (\psi^* g_{ab})u^a v^b = g_{ab}\psi_*(u^a)\psi_*(v^b)$. The pushforward $\psi_* v^b \equiv w^b$ is tangent to S (the pushforward of u^a is also a nonzero tangent to S). Thus, $g_{ab}\psi_*(u^a)w^b = 0$ for all w^b. The covariant version of $\psi_*(u^a)$, $g_{ab}\psi_*(u^a)$, is therefore normal to S. Hence, if h_{ab} is not invertible, there are nonzero vectors both tangent to, and normal to S. The converse is straightforward. Let there be a nonzero vector w^a tangent to S with $w_a = g_{ab}w^b$ normal to S. Then, $0 = w_a w^a = g_{ab}w^a w^b = g_{ab}\psi_* u^a \psi_* u^b = (\psi^* g_{ab})u^a u^b$; u^a, the vector on S such that $w^a = \psi_* u^a$, is in the null space of $\psi^* g_{ab}$. Hence, h_{ab} is not invertible and S is not a metric submanifold.

[37] Riemannian and Lorentzian manifolds are equipped with metric fields—Section 13.2.3.

hypersurface); by the above theorem, a metric does not exist on S because $\psi^*\mathbf{g}$ is not invertible. If $n^\alpha n_\alpha > 0$ (*timelike hypersurface*), the induced metric $\mathbf{g}(\psi_*(\boldsymbol{u}), \psi_*(\boldsymbol{v}))$ (for $\boldsymbol{u}, \boldsymbol{v} \in T_p(S)$) is indefinite to preserve the indefinite character of $\mathbf{g}$. For the same reason, if $n^\alpha n_\alpha < 0$ (*spacelike hypersurface*), the induced metric is positive definite. These results are summarized in Table 13.1.

Table 13.1 Hypersurfaces of a Lorentzian manifold.

$n_\alpha n^\alpha$	Induced metric
>0	indefinite, timelike hypersurface
$=0$	not defined, null hypersurface
<0	positive definite, spacelike hypersurface

13.7 DIFFERENTIAL FORMS AND EXTERIOR DIFFERENTIATION

13.7.1 Differential forms

We defined p-forms in Chapter 5 as elements of the wedge-product space $\wedge^p V^*$ (over the dual of the n-dimensional vector space V), the space of all totally antisymmetric type $(0, p)$ tensors, $0 \le p \le n$ (Section 5.10.2). We now define a *differential p-form* as a p-form *field* at every point of an n-dimensional manifold. (To save writing, we'll refer to differential p-forms simply as p-forms.) A 0-form field is a smooth function on M, and a 1-form field is a covariant vector field. For $\{x^i\}$ the coordinates of $x \in M$, the set of differentials $\{\mathbf{d}x^i\}$ (a basis for $T_x^*(M)$) is a basis for 1-forms: Any 1-form $\boldsymbol{\omega}$ can be expressed as $\omega_i \mathbf{d}x^i$, where the coefficients ω_i are smooth functions. Denote the vector space of p-forms at $x \in M$ as $\wedge^p T_x^*(M)$. An element $\boldsymbol{\omega} \in \wedge^p T_x^*(M)$ can be constructed from *basis p-forms*,

$$\boldsymbol{\omega} = \frac{1}{p!}\omega_{i_i \cdots i_p}(x)\mathbf{d}x^{i_1} \wedge \cdots \wedge \mathbf{d}x^{i_p} = \sum_{i_1 < \cdots < i_p} \omega_{i_1 \cdots i_p}(x)\mathbf{d}x^{i_1} \wedge \cdots \wedge \mathbf{d}x^{i_p}\,, \tag{13.24}$$

where the coefficients $\omega_{i_1 \cdots i_p}$ are smooth functions and antisymmetric in all indices. Differential forms on $\mathbb{R}^3$ are listed in Table 13.2.

Table 13.2 Differential forms on $\mathbb{R}^3$.

0-form	$\boldsymbol{\omega} = F(x, y, z)$
1-form	$\boldsymbol{\omega} = A(x, y, z)\mathbf{d}x + B(x, y, z)\mathbf{d}y + C(x, y, z)\mathbf{d}z$
2-form	$\boldsymbol{\omega} = A(x, y, z)\mathbf{d}y \wedge \mathbf{d}z + B(x, y, z)\mathbf{d}z \wedge \mathbf{d}x + C(x, y, z)\mathbf{d}x \wedge \mathbf{d}y$
3-form	$\boldsymbol{\omega} = F(x, y, z)\mathbf{d}x \wedge \mathbf{d}y \wedge \mathbf{d}z$

Example. The wedge product between 1-forms is a 2-form: $(A\mathbf{d}x + B\mathbf{d}y + C\mathbf{d}z) \wedge (D\mathbf{d}x + E\mathbf{d}y + F\mathbf{d}z) = (AE - BD)\mathbf{d}x\mathbf{d}y + (BF - CE)\mathbf{d}y\mathbf{d}z + (CD - AF)\mathbf{d}z\mathbf{d}x$, where we have suppressed the functional dependence of the coefficients and we have written $\mathbf{d}x\mathbf{d}y$ for $\mathbf{d}x \wedge \mathbf{d}y$, etc. The wedge product between a 1-form and a 2-form is a 3-form: $(A\mathbf{d}x + B\mathbf{d}y + C\mathbf{d}z) \wedge (P\mathbf{d}y\mathbf{d}z + Q\mathbf{d}z\mathbf{d}x + R\mathbf{d}x\mathbf{d}y) = (AP + BQ + CR)\mathbf{d}x\mathbf{d}y\mathbf{d}z$. There are no higher p-forms on $\mathbb{R}^3$.

13.7.2 Exterior derivative

The *exterior derivative*, $\mathbf{d}$, is a mapping between wedge-product spaces *at the same point of the manifold* $\mathbf{d} : \wedge^p T_x^*(M) \to \wedge^{p+1} T_x^*(M)$ having the following properties.[38] For $\boldsymbol{\omega} \in \wedge^p T_x^*(M)$ and $\boldsymbol{\eta} \in \wedge^q T_x^*(M)$ (p and q forms),

(1) $\mathbf{d}(\boldsymbol{\omega} + \boldsymbol{\eta}) = \mathbf{d}\boldsymbol{\omega} + \mathbf{d}\boldsymbol{\eta}$. (linearity)

(2) $\mathbf{d}(\boldsymbol{\omega} \wedge \boldsymbol{\eta}) = \mathbf{d}\boldsymbol{\omega} \wedge \boldsymbol{\eta} + (-1)^p \boldsymbol{\omega} \wedge \mathbf{d}\boldsymbol{\eta}$. (*antiderivation* property)

(3) $\mathbf{d}(\mathbf{d}\boldsymbol{\omega}) = 0$ for any $\boldsymbol{\omega}$. ($\mathbf{d} \circ \mathbf{d} \equiv \mathbf{d}^2 = 0$ "Poincaré lemma")

(4) $\mathbf{d}\boldsymbol{f} = (\partial_i f)\mathbf{d}\boldsymbol{x}^i$. (Exterior derivative of a 0-form is a 1-form)

The fourth requirement reproduces Eq. (13.10). There is precisely one expression for $\mathbf{d}$ meeting these requirements (for $\boldsymbol{\omega}$ a p-form),[39]

$$\mathbf{d}\boldsymbol{\omega} = \frac{1}{p!}\frac{\partial \omega_{\mu_1 \cdots \mu_p}}{\partial x^\nu}\mathbf{d}\boldsymbol{x}^\nu \wedge \mathbf{d}\boldsymbol{x}^{\mu_1} \wedge \cdots \wedge \mathbf{d}\boldsymbol{x}^{\mu_p} = \frac{\partial \omega_{|\mu_1 \cdots \mu_p|}}{\partial x^\nu}\mathbf{d}\boldsymbol{x}^\nu \wedge \mathbf{d}\boldsymbol{x}^{\mu_1} \wedge \cdots \wedge \mathbf{d}\boldsymbol{x}^{\mu_p} \,, \tag{13.25}$$

where $|\mu_1 \cdots \mu_p|$ means that the indices are ordered with $\mu_1 < \cdots < \mu_p$. Note the "extra" $\mathbf{d}\boldsymbol{x}^\nu$ in Eq (13.25); $\mathbf{d}\boldsymbol{\omega}$ is a $(p+1)$-form (compare with Eq. (13.24)). We can see that $\mathbf{d}^2 = 0$ using Eq. (13.25): $\mathbf{d}^2\boldsymbol{\omega} = \partial^2 \omega_{|\mu_1 \cdots \mu_p|}/\partial x^\alpha \partial x^\beta \mathbf{d}\boldsymbol{x}^\alpha \wedge \mathbf{d}\boldsymbol{x}^\beta \wedge \mathbf{d}\boldsymbol{x}^{\mu_1} \wedge \cdots \wedge \mathbf{d}\boldsymbol{x}^{\mu_p} = 0$. The 2-form $\mathbf{d}\boldsymbol{x}^\alpha \wedge \mathbf{d}\boldsymbol{x}^\beta$ is antisymmetric in (α, β), but the second derivatives are symmetric (if $\partial_\alpha \partial_\beta = \partial_\beta \partial_\alpha$); the contraction between a symmetric and antisymmetric tensor is zero (Section 5.10.1). The Lie derivative (which applies to any tensor field) is a generalization of the directional derivative; the exterior derivative (which applies only to differential forms) generalizes the gradient.

The exterior derivative of a 1-form is a 2-form. For $\boldsymbol{\omega} = \omega_i \mathbf{d}\boldsymbol{x}^i$, from Eq. (13.25),

$$\mathbf{d}\boldsymbol{\omega} = (\partial_j \omega_i)\mathbf{d}\boldsymbol{x}^j \wedge \mathbf{d}\boldsymbol{x}^i = \sum_{j<i} (\partial_j \omega_i - \partial_i \omega_j)\mathbf{d}\boldsymbol{x}^j \wedge \mathbf{d}\boldsymbol{x}^i \,,$$

which in $\mathbb{R}^3$ we recognize as the components of the curl, $\epsilon^i{}_{jk}(\boldsymbol{\nabla} \times \boldsymbol{\omega})_i = (\mathbf{d}\boldsymbol{\omega})_{jk}$, where $(\boldsymbol{\nabla} \times \boldsymbol{\omega})_i = \epsilon_i{}^{lm}\partial_l \omega_m$. The exterior derivative of a 2-form is a 3-form. For a 2-form in $\mathbb{R}^3$, $\boldsymbol{F} = F_1 \mathbf{d}\boldsymbol{x}^2 \wedge \mathbf{d}\boldsymbol{x}^3 + F_2 \mathbf{d}\boldsymbol{x}^3 \wedge \mathbf{d}\boldsymbol{x}^1 + F_3 \mathbf{d}\boldsymbol{x}^1 \wedge \mathbf{d}\boldsymbol{x}^2$, the exterior derivative is, from Eq. (13.25),

$$\mathbf{d}\boldsymbol{F} = \left(\frac{\partial F_1}{\partial x^1} + \frac{\partial F_2}{\partial x^2} + \frac{\partial F_3}{\partial x^3}\right)\mathbf{d}\boldsymbol{x}^1 \wedge \mathbf{d}\boldsymbol{x}^2 \wedge \mathbf{d}\boldsymbol{x}^3 \,,$$

which we recognize as the divergence, $\boldsymbol{\nabla} \cdot \boldsymbol{F} = (\mathbf{d}\boldsymbol{F})_{123}$. The classic results of vector calculus, $\boldsymbol{\nabla} \times \boldsymbol{\nabla} f = 0$ and $\boldsymbol{\nabla} \cdot (\boldsymbol{\nabla} \times \boldsymbol{\omega}) = 0$, are instances of $\mathbf{d}^2 = 0$.

Let's see what kind of fun we can have by combining the exterior derivative with the Hodge star operator (Section 5.10.6). For a 1-form $\boldsymbol{\alpha}$, $*\boldsymbol{\alpha}$ is an $(n-1)$-form, and thus $\mathbf{d}(*\boldsymbol{\alpha})$ is an n-form. It can be shown that (the Hodge operator involves the metric tensor)

$$\mathbf{d}(*\boldsymbol{\alpha}) = \frac{1}{\sqrt{|g|}}\frac{\partial(\sqrt{|g|}\alpha^\nu)}{\partial x^\nu}\boldsymbol{\epsilon} \,, \tag{13.26}$$

where $\boldsymbol{\epsilon} \equiv \sqrt{|g|}\mathbf{d}\boldsymbol{x}^1 \wedge \cdots \wedge \mathbf{d}\boldsymbol{x}^n$ is the *volume form*, a generalization of the invariant volume element, Eq. (5.57). (Thus, the volume form requires that a metric be defined on the manifold.) The dual of Eq. (13.26) produces a 0-form, a scalar field

$$*\mathbf{d}(*\boldsymbol{\alpha}) = (-1)^{n-}\frac{1}{\sqrt{|g|}}\frac{\partial}{\partial x^\nu}(\sqrt{|g|}\alpha^\nu) \,, \tag{13.27}$$

[38]The exterior derivative $\mathbf{d}\boldsymbol{\alpha}$ of a differential form field $\boldsymbol{\alpha}$ occurs in the version of Stokes's theorem on manifolds, Eq. (13.34), and for that reason the exterior derivative is a generalization of the gradient.

[39]The uniqueness of Eq. (13.25) is shown in [54, p21] and in [55, p17].

the *covariant divergence*. There are other combinations of $\mathbf{d}$ and $*$ that generalize the differential operators of vector calculus.

13.7.3 Maxwell's equations

From Eq. (8.19), $F_{\mu\nu}$ is a second-rank antisymmetric covariant tensor. The associated differential 2-form is, using Eq. (13.24):

$$\boldsymbol{F} = -\frac{E_1}{c}\mathbf{d}\boldsymbol{x}^0\wedge\mathbf{d}\boldsymbol{x}^1 - \frac{E_2}{c}\mathbf{d}\boldsymbol{x}^0\wedge\mathbf{d}\boldsymbol{x}^2 - \frac{E_3}{c}\mathbf{d}\boldsymbol{x}^0\wedge\mathbf{d}\boldsymbol{x}^3 + B^3\mathbf{d}\boldsymbol{x}^1\wedge\mathbf{d}\boldsymbol{x}^2 + B^2\mathbf{d}\boldsymbol{x}^3\wedge\mathbf{d}\boldsymbol{x}^1 + B^1\mathbf{d}\boldsymbol{x}^2\wedge\mathbf{d}\boldsymbol{x}^3 .$$

The exterior derivative of $\boldsymbol{F}$ is then a 3-form. From Eq. (13.25),

$$\begin{aligned}\mathbf{d}\boldsymbol{F} = &-\frac{1}{c}\frac{\partial E_1}{\partial x^\mu}\mathbf{d}\boldsymbol{x}^\mu\wedge\mathbf{d}\boldsymbol{x}^0\wedge\mathbf{d}\boldsymbol{x}^1 - \frac{1}{c}\frac{\partial E_2}{\partial x^\nu}\mathbf{d}\boldsymbol{x}^\nu\wedge\mathbf{d}\boldsymbol{x}^0\wedge\mathbf{d}\boldsymbol{x}^2 - \frac{1}{c}\frac{\partial E_3}{\partial x^\lambda}\mathbf{d}\boldsymbol{x}^\lambda\wedge\mathbf{d}\boldsymbol{x}^0\wedge\mathbf{d}\boldsymbol{x}^3 \qquad (13.28)\\ &+\frac{\partial B^3}{\partial x^\rho}\mathbf{d}\boldsymbol{x}^\rho\wedge\mathbf{d}\boldsymbol{x}^1\wedge\mathbf{d}\boldsymbol{x}^2 + \frac{\partial B^2}{\partial x^\sigma}\mathbf{d}\boldsymbol{x}^\sigma\wedge\mathbf{d}\boldsymbol{x}^3\wedge\mathbf{d}\boldsymbol{x}^1 + \frac{\partial B^1}{\partial x^\phi}\mathbf{d}\boldsymbol{x}^\phi\wedge\mathbf{d}\boldsymbol{x}^2\wedge\mathbf{d}\boldsymbol{x}^3 .\end{aligned}$$

Expanding the terms in Eq. (13.28), it can be shown that

$$\mathbf{d}\boldsymbol{F} = 0 . \qquad (13.29)$$

The homogeneous Maxwell equations are subsumed into one equation involving a differential form, $\mathbf{d}\boldsymbol{F} = 0$. We see that a gauge transformation consists of modifying $\boldsymbol{F}$ such that $\boldsymbol{F} \to \boldsymbol{F}' = \boldsymbol{F} + \mathbf{d}\boldsymbol{A}$, where $\boldsymbol{A}$ is any 1-form. Under such a transformation $\mathbf{d}\boldsymbol{F}' = 0$ because $\mathrm{d}^2 = 0$ for any $\boldsymbol{A}$.

The dual of $\boldsymbol{F}$, $*\boldsymbol{F}$, is a 2-form (in Minkowski space). We find (using Eq. (5.74) with $|g| = 1$)

$$*\boldsymbol{F} = B^1\mathbf{d}\boldsymbol{x}^0\wedge\mathbf{d}\boldsymbol{x}^1 + B^2\mathbf{d}\boldsymbol{x}^0\wedge\mathbf{d}\boldsymbol{x}^2 + B^3\mathbf{d}\boldsymbol{x}^0\wedge\mathbf{d}\boldsymbol{x}^3 + \frac{1}{c}E_3\mathbf{d}\boldsymbol{x}^1\wedge\mathbf{d}\boldsymbol{x}^2 + \frac{1}{c}E_2\mathbf{d}\boldsymbol{x}^3\wedge\mathbf{d}\boldsymbol{x}^1 + \frac{1}{c}E^1\mathbf{d}\boldsymbol{x}^2\wedge\mathbf{d}\boldsymbol{x}^3 .$$

The exterior derivative of $*\boldsymbol{F}$ is a 3-form. From Eq. (13.25),

$$\begin{aligned}\mathbf{d}(*\boldsymbol{F}) = &\frac{\partial B^1}{\partial x^\nu}\mathbf{d}\boldsymbol{x}^\nu\wedge\mathbf{d}\boldsymbol{x}^0\wedge\mathbf{d}\boldsymbol{x}^1 + \frac{\partial B^2}{\partial x^\mu}\mathbf{d}\boldsymbol{x}^\mu\wedge\mathbf{d}\boldsymbol{x}^0\wedge\mathbf{d}\boldsymbol{x}^2 + \frac{\partial B^3}{\partial x^\lambda}\mathbf{d}\boldsymbol{x}^\lambda\wedge\mathbf{d}\boldsymbol{x}^0\wedge\mathbf{d}\boldsymbol{x}^3 \qquad (13.30)\\ &+\frac{1}{c}\frac{\partial E_3}{\partial x^\sigma}\mathbf{d}\boldsymbol{x}^\sigma\wedge\mathbf{d}\boldsymbol{x}^1\wedge\mathbf{d}\boldsymbol{x}^2 + \frac{1}{c}\frac{\partial E_2}{\partial x^\rho}\mathbf{d}\boldsymbol{x}^\rho\wedge\mathbf{d}\boldsymbol{x}^3\wedge\mathbf{d}\boldsymbol{x}^1 + \frac{1}{c}\frac{\partial E_1}{\partial x^\phi}\mathbf{d}\boldsymbol{x}^\phi\wedge\mathbf{d}\boldsymbol{x}^2\wedge\mathbf{d}\boldsymbol{x}^3 .\end{aligned}$$

Expanding the terms in Eq. (13.30), we find

$$\mathbf{d}(*\boldsymbol{F}) = \mu_0(*\boldsymbol{J}) , \qquad (13.31)$$

where the dual of the 1-form J_μ is a 3-form. We find, using Eq. (5.76),

$$*\boldsymbol{J} = -J_1\mathbf{d}\boldsymbol{x}^0\wedge\mathbf{d}\boldsymbol{x}^2\wedge\mathbf{d}\boldsymbol{x}^3 - J_2\mathbf{d}\boldsymbol{x}^0\wedge\mathbf{d}\boldsymbol{x}^3\wedge\mathbf{d}\boldsymbol{x}^1 - J_3\mathbf{d}\boldsymbol{x}^0\wedge\mathbf{d}\boldsymbol{x}^1\wedge\mathbf{d}\boldsymbol{x}^2 + \rho c\mathbf{d}\boldsymbol{x}^1\wedge\mathbf{d}\boldsymbol{x}^2\wedge\mathbf{d}\boldsymbol{x}^3 . \qquad (13.32)$$

The inhomogeneous Maxwell equations are equivalent to a single equation among differential forms, $\mathbf{d}(*\boldsymbol{F}) = \mu_0(*\boldsymbol{J})$. Equations (13.29) and (13.31) are equivalent to the tensor form of Maxwell's equations, Eqs. (8.22) and (8.27).

13.8 INTEGRATION ON MANIFOLDS

13.8.1 Orientable manifolds

We introduced in Section 5.10.5 the *orientation* of an n-dimensional vector space V as a nonzero element θ of $\wedge^n V$. An *ordered basis* for V, $\{e_i\}_{i=1}^n$, is in the orientation specified by θ if for

$e_1 \wedge \cdots \wedge e_n = \alpha\theta$, $\alpha > 0$. There are precisely two possibilities: $\alpha > 0$ or $\alpha < 0$; $\alpha \neq 0$ because basis vectors are linearly independent. Orientation on a manifold requires, at each $p \in M$, an orientation provided by the basis vectors of $T_p(M)$ or $T_p^*(M)$. In what follows, we use the basis of the cotangent space. A manifold M is *orientable* if it's possible to assign a *nowhere vanishing* orientation on M, i.e., if the sign of the orientation is the same at all points of M. Manifolds without orientation exist, the Möbius strip being a famous example.

Two charts of an n-manifold M, (ϕ_α, U_α) and (ϕ_β, U_β), with overlapping domains are *orientation compatible* if the Jacobian determinant of the transition map is positive. Let $\boldsymbol{\omega}_\alpha \equiv \boldsymbol{e}_\alpha^1 \wedge \cdots \wedge \boldsymbol{e}_\alpha^n$ be a basis vector of $\wedge^n T_p^*(M)$, where the set $\{\boldsymbol{e}_\alpha^i\}_{i=1}^n$ is the coordinate basis of $T_p^*(M)$ associated with (ϕ_α, U_α). Let $\{\boldsymbol{e}_\beta^j\}_{j=1}^n$ be the coordinate basis of $T_p^*(M)$ associated with (ϕ_β, U_β). The two sets of basis vectors are related by a linear transformation $\boldsymbol{e}_\alpha^i = A_j^i \boldsymbol{e}_\beta^j$, where the A_j^i are elements of the Jacobian matrix associated with the transition map, Eq. (13.11). Under the transformation,

$$\begin{aligned}\boldsymbol{\omega}_\alpha &= \boldsymbol{e}_\alpha^1 \wedge \cdots \wedge \boldsymbol{e}_\alpha^n = A_{j_1}^1 \cdots A_{j_n}^n \boldsymbol{e}_\beta^{j_1} \wedge \cdots \wedge \boldsymbol{e}_\beta^{j_n} \\ &= A_{j_1}^1 \cdots A_{j_n}^n \varepsilon^{j_1 \cdots j_n} \boldsymbol{e}_\beta^1 \wedge \cdots \wedge \boldsymbol{e}_\beta^n = \det A_j^i \boldsymbol{e}_\beta^1 \wedge \cdots \wedge \boldsymbol{e}_\beta^n \equiv J \boldsymbol{\omega}_\beta ,\end{aligned} \tag{13.33}$$

where we've used Eqs. (5.82) and (5.53). The transition maps must have $J > 0$ for the orientation to be well defined. A manifold is orientable if, in addition to the criteria established in Section 13.1, its charts are all orientation compatible.

An n-dimensional manifold M is orientable if and only if it admits a non-vanishing n-form. Let $\boldsymbol{\alpha}$ be a nowhere-vanishing n-form on M, i.e., for each local chart there is a function $f \neq 0$ such that $\boldsymbol{\alpha} = f \boldsymbol{e}^1 \wedge \cdots \wedge \boldsymbol{e}^n$. It follows that $\boldsymbol{\alpha}(e_1, \cdots, e_n) = f \neq 0$, where $\{e_i\}_{i=1}^n$ are coordinate basis vectors of $T_x(M)$. We can always find a chart for each point $x \in M$ so that $f(x) > 0$ (otherwise replace the coordinate x^1 with $-x^1$). Let (ϕ_α, U_α) and (ϕ_β, U_β) be two such charts, so that $\boldsymbol{\alpha} = f \boldsymbol{e}_\alpha^1 \wedge \cdots \wedge \boldsymbol{e}_\alpha^n = g \boldsymbol{e}_\beta^1 \wedge \cdots \wedge \boldsymbol{e}_\beta^n$, where $f, g > 0$, i.e., $\boldsymbol{\alpha}$ is a geometric object, independent of chart. From Eq. (13.33), we must have $fJ = g$. Hence, $J > 0$ because $f, g > 0$. To prove the converse requires that M be paracompact.[56, p63]

13.8.2 Manifold with boundary

A *manifold with boundary* is a manifold with an "edge" to it. To describe that edge, define $\mathbb{R}_+^n \equiv \{(x^1, \cdots, x^n) | x^n \geq 0\}$, i.e., $\mathbb{R}_+^n$ is $\mathbb{R}^n$ with the nth coordinate non-negative; it can be considered the "upper half" of $\mathbb{R}^n$ and is sometimes denoted $\frac{1}{2}\mathbb{R}^n$. (One could similarly define $\mathbb{R}_-^n$.) The hemisphere $x^2 + y^2 + z^2 = 1$ with $z \geq 0$ is a two-dimensional manifold with boundary. The edge of the hemisphere ($z = 0$) is in correspondence with the circle $x^2 + y^2 = 1$, while the rest of the hemisphere (the *interior* with $z > 0$) is in correspondence with the open disk $x^2 + y^2 < 1$.

An n-dimensional manifold with boundary M is defined as in Section 13.1, except for the modified requirement that the image of every chart map is an open subset of $\mathbb{R}^n$ *or* $\mathbb{R}_+^n$. The *boundary* of M, denoted ∂M, is the set of points of M that are mapped (by the chart maps of M) into $(x^1, \cdots, x^{n-1}, 0)$. The boundary ∂M is an $(n-1)$-dimensional manifold *without* boundary; it's also an $(n-1)$-dimensional submanifold of M.[40] For example, the closed ball $B^n(1) = \{x \in \mathbb{R}^n | \, |x| \leq 1\}$ is a smooth manifold with boundary. Its boundary is $\partial B^n(1) = S^{n-1}$, the $(n-1)$-sphere.

Let M be an oriented n-manifold with boundary. The boundary ∂M is an oriented $(n-1)$-dimensional submanifold of M. The orientation on ∂M, the *induced orientation*, is defined as follows. Suppose that near the boundary M is described by $x^n \geq 0$. The orientation on ∂M is specified by the differential form $\delta \boldsymbol{d}x^1 \wedge \cdots \wedge \boldsymbol{d}x^{n-1}$ where $\delta = \pm 1$ is determined by the requirement $\delta(-\boldsymbol{d}x^n) \wedge \boldsymbol{d}x^1 \wedge \cdots \wedge \boldsymbol{d}x^{n-1} = \boldsymbol{d}x^1 \wedge \cdots \wedge \boldsymbol{d}x^n$. Here $-\boldsymbol{d}x^n$ is outwardly pointing to

[40] Let $p \in \partial M$ and let (U, ϕ) be a chart of M near p that maps into $\mathbb{R}_+^n$. Then $(U \cap \partial M, \phi)$ is a chart that maps to $\{(x^1, \cdots, x^n) | x^n = 0\}$, i.e., an open subset of $\mathbb{R}^{n-1}$.

the manifold with boundary. It may be instructive to review the choice of orientations described in Section 5.11 so that the orientation of the boundary matches that of the embedding space.

13.8.3 Stokes's theorem

To establish notation, consider an n-manifold M and a subset $N \subset M$. The *complement* of N is the set $M - N \equiv \{x \in M | x \notin N\}$. N is said to be *closed* if its complement is open. The *closure* of N, denoted $\overline{N}$, is defined as the intersection of all closed sets that contain N. (The intersection of an arbitrary collection of closed sets is closed, the union of a finite number of closed sets is closed.) The *interior* of N, denoted $\text{int}(N)$, is defined as the union of all *open* sets contained within N. The *boundary* of N, ∂N, is the set of all points that lie in $\overline{N}$ but not in $\text{int}(N)$. Note that $\text{int}(N) \equiv N - \partial N$ is an n-dimensional manifold without boundary.

We now state the *generalized Stokes's theorem*.[41] Let M be an n-dimensional oriented manifold with boundary, and let $\boldsymbol{\alpha}$ be an $(n-1)$-form field on M. Then

$$\int_{\text{int}(M)} \mathbf{d}\boldsymbol{\alpha} = \int_{\partial M} \boldsymbol{\alpha} \, . \tag{13.34}$$

Note the correct number of "d's" in Eq. (13.34): $\boldsymbol{\alpha}$ is an $(n-1)$-form and comes to the party with $(n-1)$ differentials; $\mathbf{d}\boldsymbol{\alpha}$ is an n-form and has n differentials.[42]

Example. On a two-dimensional manifold M, let there be the 1-form $\boldsymbol{\phi} = \phi_i \mathbf{d}\boldsymbol{x}^i$. Using Eq. (13.25), $\mathbf{d}\boldsymbol{\phi}$ is the 2-form

$$\mathbf{d}\boldsymbol{\phi} = (\partial \phi_i / \partial x^j) \mathbf{d}\boldsymbol{x}^j \wedge \mathbf{d}\boldsymbol{x}^i = \left(\frac{\partial \phi_2}{\partial x^1} - \frac{\partial \phi_1}{\partial x^2} \right) \mathbf{d}\boldsymbol{x}^1 \wedge \mathbf{d}\boldsymbol{x}^2 \, .$$

From Stokes's theorem, Eq. (13.34),

$$\int_M \mathbf{d}\boldsymbol{\phi} = \int_M \left(\frac{\partial \phi_2}{\partial x^1} - \frac{\partial \phi_1}{\partial x^2} \right) \mathbf{d}\boldsymbol{x}^1 \wedge \mathbf{d}\boldsymbol{x}^2 = \int_{\partial M} \boldsymbol{\phi} = \int_{\partial M} \phi_i \mathbf{d}\boldsymbol{x}^i \, ,$$

from which we recognize the customary vector-calculus form of Stokes's theorem, $\int (\boldsymbol{\nabla} \times \boldsymbol{\phi}) \cdot \mathrm{d}\boldsymbol{S} = \oint \boldsymbol{\phi} \cdot \mathrm{d}\boldsymbol{l}$. The orientation of ∂M must match that of M.

Stokes's theorem equates two integrals, *each over a manifold without boundary*. We have, however, yet to define integration on manifolds! The integral of a continuous n-form field $\boldsymbol{\alpha} = \alpha(x^1, \cdots, x^n) \mathbf{d}\boldsymbol{x}^1 \wedge \cdots \wedge \mathbf{d}\boldsymbol{x}^n$ over the domain of the chart (U, ϕ) of an orientable n-manifold is defined as

$$\int_U \boldsymbol{\alpha} \equiv \int_{\phi(U)} \alpha(x^1, \cdots, x^n) \mathrm{d}x^1 \cdots \mathrm{d}x^n \, , \tag{13.35}$$

where the integral on the right is the standard Riemann multiple integral in $\mathbb{R}^n$. It might seem that the integral defined this way is dependent on the choice of chart. From Eq. (13.33) we see that an n-form on an n-manifold transforms in just the same way as integrals transform in $\mathbb{R}^n$ (see Eq. (5.52)). Equation (13.35) provides a coordinate-independent definition of integral over a single chart of M.

To extend the integral in Eq. (13.35) to all of M, sum over the charts. To do so relies on the paracompactness property of M. Paracompact manifolds are (by definition) covered by *locally finite* atlases, $\{(U_\alpha, \phi_\alpha)\}$, which means that each point of M is covered by a *finite* number of open sets

[41] Proven in [57, p124] and in [54, p64].

[42] The fact that Stokes's theorem on manifolds is formulated in terms of differential forms is one of the primary reasons to study the exterior derivative.

U_α. On a paracompact manifold, one can always find a *partition of unity* subordinate to the charts $\{U_\alpha\}$, a set of smooth functions $\{g_\alpha(x)\}$ such that (x denotes $x^1, \cdots, x^n$): 1) $0 \le g_\alpha(x) \le 1$ for each α; 2) the *support* of each function g_α (the closure of the set where g_α is nonvanishing) is contained in the corresponding U_α; 3) $\sum_\alpha g_\alpha(x) = 1$ at every point of $x \in M$ (the sum here is a finite sum because only finitely many g_α are nonvanishing at any point x).[52, p272] Multiply the sum property by an arbitrary function: $f(x) = \sum_\alpha f(x)g_\alpha(x) \equiv \sum_\alpha f_\alpha(x)$, where $f_\alpha(x) = f(x)g_\alpha(x)$ is the "fraction" of $f(x)$ that has support on the set U_α. The advantage of decomposing $f(x)$ into a sum over the functions $f_\alpha(x)$ is that f_α is zero outside of U_α. The integral of each f_α is defined by a relation analogous to Eq. (13.35). With these technicalities, the integral of an n-form $\boldsymbol{\alpha}$ over an oriented n-manifold is

$$\int_M \boldsymbol{\alpha} = \sum_\beta \int_{\phi_\beta(U_\beta)} g_\beta(x^1_\beta, \cdots, x^n_\beta)\alpha(x^1_\beta, \cdots, x^n_\beta)\mathrm{d}x^1_\beta \cdots \mathrm{d}x^n_\beta \,, \tag{13.36}$$

where $\{x^i_\beta\}$ are the local coordinates associated with U_β.[37, Appendix B][38, p26]

SUMMARY

This chapter introduced the properties of manifolds and tensor fields, concepts that are used in GR.

- **Manifold**: An n-dimensional manifold M is a set of points such that open subsets can be put into correspondence with open subsets of $\mathbb{R}^n$. A chart on M is a pair (U_α, ϕ_α) consisting of $U_\alpha \subset M$ and a mapping $\phi_\alpha : U_\alpha \to O_\alpha \subset \mathbb{R}^n$. For $p \in U_\alpha$, $\phi_\alpha(p)$ specifies coordinates $x^i(p) \in \mathbb{R}^n$ ($i = 1, \cdots, n$), which is the job of a manifold: establish coordinate systems for each $p \in M$. The domains of the charts cover the manifold, $M = \bigcup_\alpha U_\alpha$, and meet other criteria discussed in Section 13.1. Functions on manifolds, $f : M \to \mathbb{R}$, and mappings between them, $\phi : M \to N$, are defined in Section 13.1.2.

- **Tangent space**: Vectors at each point $p \in M$, $\boldsymbol{t}$, are directional derivatives evaluated at that point. As derivatives, they act on functions to produce numbers, $\boldsymbol{t}(f)$ = a number. They form a vector space, the tangent space $T_p(M)$, which has the same dimension as M. Vectors (directional derivatives) are not "in" the manifold M; they "touch" M at one point and are said to be tangent to M. The tangent space $T_p(M)$ is the set of all tangents at $p \in M$. The coordinate basis for $T_p(M)$ is the set of partial derivatives with respect to the coordinate curves $\{\boldsymbol{e}_i \equiv \partial/\partial x^i|_p\}_{i=1}^n$. Contravariant vectors are elements of the tangent space.

- **Cotangent space**: The cotangent space $T_p^*(M)$ is the dual space to $T_p(M)$, the set of all mappings $\boldsymbol{\omega} : T_p(M) \to \mathbb{R}$; it requires some fancy footwork to define. Associated with a mapping ϕ between points of manifolds $\phi : M \to N$ are two other mappings, the pullback ϕ^* and the pushforward, ϕ_*. For a function $f : N \to \mathbb{R}$, the pullback of f under ϕ, $\phi^* f$, is a function on M such that $(\phi^* f)p = f(\phi(p))$. The pushforward of a vector $\boldsymbol{t} \in T_p(M)$ is a mapping between tangent spaces, $\phi_* \boldsymbol{t} : T_p(M) \to T_{\phi(p)}(N)$. It maps a tangent vector defined on M, $\boldsymbol{t}$, to one defined on N, $\phi_*\boldsymbol{t}$, such that $(\phi_*\boldsymbol{t})f \equiv \boldsymbol{t}(\phi^* f) = \boldsymbol{t}(f \circ \phi)$, where f is defined on N. For the special case of $N = \mathbb{R}$, the pushforward of a vector $\boldsymbol{t} \in T_p(M)$ under a function $f : M \to \mathbb{R}$ defines a linear map known as the differential of f, $\mathbf{d}\boldsymbol{f} : T_p(M) \to \mathbb{R}$, such that $\mathbf{d}\boldsymbol{f}(\boldsymbol{t}) = \boldsymbol{t}(f)$. The cotangent space $T_p^*(M)$ is the set of all mappings $\mathbf{d}\boldsymbol{f}$ on $T_p(M)$. If f is the coordinate function x^i and $\boldsymbol{t}$ is the tangent to the jth coordinate curve, $\partial_j|_p$, the action of $\mathbf{d}\boldsymbol{x}^i$ on $\partial_j|_p$ is $\mathbf{d}\boldsymbol{x}^i(\partial_i) = \partial_i(x^j) = \delta^i_j$. The operators $\{\mathbf{d}\boldsymbol{x}^i|_p\}_{i=1}^m$ are dual to the basis of $T_p(M)$, $\{\partial_j|_p\}_{j=1}^m$, and form a basis for $T_p^*(M)$. For any $\mathbf{d}\boldsymbol{f}$ we have an expansion in the basis, $\mathbf{d}\boldsymbol{f} = (\partial_i f)\mathbf{d}\boldsymbol{x}^i$. The quantity $\mathbf{d}\boldsymbol{x}^i$ is not a number, but a symbol for the linear map $\mathbf{d}\boldsymbol{x}^i(\partial/\partial x^j) = \delta^i_j$. Covariant vectors are elements of the cotangent space.

- **Tensor field**: Tensors at a point $p \in M$ are multilinear maps over products of vector spaces, $\mathbf{T}^k_l|_p : (T^*_p)^k \times (T_p)^l \to \mathbb{R}$. A tensor field is the assignment of a tensor $\mathbf{T}^k_l|_p$ to each $p \in M$. The tangent field $\boldsymbol{v}$ is the selection of a vector $\boldsymbol{v}|_p \in T_p(M)$ from each $p \in M$. For $f : M \to \mathbb{R}$, $\boldsymbol{v}|_p(f)$ is a number at each $p \in M$. A tangent field is smooth if for each smooth function f, $\boldsymbol{v}(f)$ is smooth. A covariant vector field $\boldsymbol{\omega}$ is smooth if for each smooth tangent field $\boldsymbol{v}$, the function $\boldsymbol{\omega}(\boldsymbol{v})$ is smooth. A tensor field $\mathbf{T}^k_l$ is smooth if for all smooth covariant vector fields $\boldsymbol{\omega}^1, \cdots, \boldsymbol{\omega}^k$ and smooth tangent fields $\boldsymbol{v}_1, \cdots, \boldsymbol{v}_l$, $\mathbf{T}^k_l(\boldsymbol{\omega}^1, \cdots, \boldsymbol{\omega}^k, \boldsymbol{v}_1, \cdots, \boldsymbol{v}_l)$ is a smooth function.

- **Metric signature**: The metric field $\mathbf{g}$ assigns to every $x \in M$ the covariant tensor $\mathbf{g}|_x = g_{\mu\nu}(x)\boldsymbol{e}^\mu \otimes \boldsymbol{e}^\nu$ where the components $g_{\mu\nu}(x) = \mathbf{g}|_x(\boldsymbol{e}_\mu, \boldsymbol{e}_\nu)$ are smooth functions (with $\{\boldsymbol{e}_\mu\}$ the coordinate basis of $T_x(M)$ and $\{\boldsymbol{e}^\mu\}$ the coordinate basis for $T^*_x(M)$). Because directional derivatives generalize the displacement vectors of flat space, the infinitesimal distance "squared" in the neighborhood of p is built up out of $T_p(M) \times T_p(M)$. When $\mathbf{g}$ acts on a pair of vectors from T_p, it produces a number associated with p. At every point $p \in M$, a $\mathbf{g}$-orthonormal basis can be found, $\{\boldsymbol{v}_i\}_{i=1}^n \in T_p(M)$ such that $\mathbf{g}(\boldsymbol{v}_i, \boldsymbol{v}_j) = 0$ for $i \neq j$ and $\mathbf{g}(\boldsymbol{v}_i, \boldsymbol{v}_i) = \pm 1$. A Riemannian manifold has signature $(+\cdots+)$ for the diagonal terms $\mathbf{g}(\boldsymbol{v}_i, \boldsymbol{v}_i)$, while a manifold with signature $(-+++)$ is called Lorentzian. The signature is the same at every point of a connected manifold.

- **Integral curve**: An integral curve is a curve for which its tangent at point p coincides with a contravariant vector field $\boldsymbol{v}$ at that point, $\boldsymbol{v}|_p$. Precisely one integral curve passes through each $p \in M$. Thus they cannot intersect and "fill" a manifold (a congruence). The integral curves of a vector field can be considered the flow lines of a fluid, where the "motion" of a point on the curve is generated by a diffeomorphism of the manifold to itself. The flow of a vector field is a parameterized set of mappings $\phi_t : \mathbb{R} \times M \to M, t \in \mathbb{R}$, such that for fixed t, $\phi_t : M \to M$, and for fixed $p \in M$, $\phi_t(p) : \mathbb{R} \to M$ is a curve that passes through p.

- **Lie derivative**: A congruence is a mapping of the manifold onto itself. For small t, each $p \in M$ is mapped to the point $\phi_t(p)$ lying at the parameter t further along the curve. Starting from $\mathbf{T}|_p$, how does the pushforward tensor $\phi_{t*}\mathbf{T}$ differ from the tensor that's "already there," $\mathbf{T}|_{\phi_t(p)}$? The Lie derivative $\pounds_{\boldsymbol{v}}$ of tensor field $\mathbf{T}$ with respect to vector field $\boldsymbol{v}$ is defined as the difference between the pullback of $\mathbf{T}|_{\phi_t(p)}$ and $\mathbf{T}|_p$ as $t \to 0$, $(\pounds_{\boldsymbol{v}})_p = \lim_{t\to 0}\left(\phi_{-t*}\mathbf{T}|_{\phi_t(p)} - \mathbf{T}|_p\right)/t$. For scalar fields f, $\pounds_{\boldsymbol{v}} f = v^\mu \partial_\mu f$ is the directional derivative along $\boldsymbol{v}$. For tensor fields, $\pounds_{\boldsymbol{v}}\mathbf{T}$ measures the rate of change of $\mathbf{T}$ relative to $\boldsymbol{v}$. The Lie derivative applies to any tensor field, but requires a vector field $\boldsymbol{v}$.

- **Differential forms**: A differential p-form is a tensor field of antisymmetric type $(0, p)$ tensors. The main difference between differential p-forms and p-forms (Section 5.10.2) is that the expansion coefficients $\omega_{\mu_1,\cdots,\mu_p}(x)$ are smooth functions and the basis vectors are obtained from the local cotangent space.

- **Exterior derivative**: The exterior derivative $\mathbf{d}$ is a mapping between wedge-product spaces at the same point of the manifold, $\mathbf{d} : \wedge^p T^*_x(M) \to \wedge^{p+1} T^*_x(M)$. Combinations of the exterior derivative with the Hodge star operator provide generalizations of the classic differential operators of vector calculus. The exterior derivative is a generalization of the gradient; it applies only to differential forms.

- **Stokes's theorem**: Integrals on n-dimensional oriented manifolds are given as integrals over n-form fields, which by their transformation properties are naturally set up to provide a chart-independent definition of integral. For N an n-dimensional oriented manifold with boundary and $\boldsymbol{\alpha}$ an $(n-1)$-form on the $(n-1)$-dimensional boundary ∂N, Stokes's theorem is $\int_{\text{int}(N)} \mathbf{d}\boldsymbol{\alpha} = \int_{\partial N} \boldsymbol{\alpha}$, where int$(N)$ is the interior of N.

EXERCISES

13.1 a. Let $\boldsymbol{X} = \partial_x$ be a vector field on $\mathbb{R}^2$. Find the integral curve associated with $\boldsymbol{X}$ that passes through (a,b) at $t=0$. A: $\gamma(t) = (a+t, b)$.

b. Let $\boldsymbol{X} = x\partial_x - y\partial_y$ be a vector field on $\mathbb{R}^2$. Find the integral curve associated with $\boldsymbol{X}$ that passes through (a,b) at $t=0$. A: $\gamma(t) = (ae^t, be^{-t})$.

13.2 Does the Lie derivative transform like a tensor? Consider the Lie derivative of a vector component, $\pounds_{\boldsymbol{v}} A^\mu = v^\lambda \partial_\lambda A^\mu - A^\lambda \partial_\lambda v^\mu$.

13.3 Take the exterior derivative of $\mathrm{d}(*\boldsymbol{F}) = \mu_0(*\boldsymbol{J})$ (Eq. (13.31), with $*\boldsymbol{J}$ given in Eq. (13.32)), and show that it reproduces the continuity equation

$$\left(\frac{\partial \rho}{\partial t} + \boldsymbol{\nabla}\cdot\boldsymbol{J}\right) \mathbf{d}\boldsymbol{x}^0 \wedge \mathbf{d}\boldsymbol{x}^1 \wedge \mathbf{d}\boldsymbol{x}^2 \wedge \mathbf{d}\boldsymbol{x}^3 = 0\,.$$

13.4 Show that the exterior derivative of a p-form $\boldsymbol{\omega}$ transforms as a tensor.

13.5 Show that when expressed in terms of coordinates, Eq. (13.16) is equivalent to the chain rule for partial derivatives.

13.6 For $\phi : M \to N$ a map between manifolds, and for $\boldsymbol{\omega}_1, \boldsymbol{\omega}_2 \in T_p^*(N)$ show that

$$\phi^*\left(\boldsymbol{\omega}_1 \otimes \boldsymbol{\omega}_2\right) = \phi^*\boldsymbol{\omega}_1 \otimes \phi^*\boldsymbol{\omega}_2\,.$$

Hint: Invent vectors $\boldsymbol{t}_1, \boldsymbol{t}_2 \in T_p(M)$ for $\phi^*(\boldsymbol{\omega}_1 \otimes \boldsymbol{\omega}_2)$ to act on. For the mathematically inclined, the pullback operation ϕ^* is a *homomorphism*, a map that preserves structure between algebraic objects. Show that $\phi^*\left(\boldsymbol{\omega}_1 \wedge \boldsymbol{\omega}_2\right) = \phi^*\boldsymbol{\omega}_1 \wedge \phi^*\boldsymbol{\omega}_2$.

13.7 Show, for $\phi : M \to M'$ a mapping between manifolds M and M', and for $\mathbf{A}$ a differential form on M', that the exterior derivative commutes with the pullback operator, $\mathbf{d}(\phi^*\mathbf{A}) = \phi^*(\mathbf{dA})$.

13.8 Show that the Lie derivative commutes with the exterior derivative, that for $\boldsymbol{\omega}$ a p-form field and $\boldsymbol{v}$ a contravariant vector field, $\mathrm{d}\left(\pounds_{\boldsymbol{v}}\boldsymbol{\omega}\right) = \pounds_{\boldsymbol{v}}\left(\mathrm{d}\boldsymbol{\omega}\right)$.

13.9 From the definition of the commutator Eq. (13.20) and the action of a tangent vector on a function f, Eq. (13.3), show that the commutator of vector fields $\boldsymbol{u}$ and $\boldsymbol{v}$ acting on a scalar field f can be written $[\boldsymbol{u}, \boldsymbol{v}]\, f = \boldsymbol{u}\left(\boldsymbol{v}(f)\right) - \boldsymbol{v}\left(\boldsymbol{u}(f)\right)$.

13.10 Show for $\boldsymbol{u}$, $\boldsymbol{v}$ smooth vector fields and f a smooth function on a manifold M:

a. $\pounds_{(f\boldsymbol{v})}\boldsymbol{u} = f\pounds_{\boldsymbol{v}}\boldsymbol{u} - \boldsymbol{v}\pounds_{\boldsymbol{u}} f\,.$

b. $\left(\pounds_{\boldsymbol{u}}\pounds_{\boldsymbol{v}} - \pounds_{\boldsymbol{v}}\pounds_{\boldsymbol{u}}\right) f = \left(\pounds_{\boldsymbol{u}}\boldsymbol{v}\right) f = \pounds_{(\pounds_{\boldsymbol{u}}\boldsymbol{v})} f.$

CHAPTER 14

Differential geometry

THE central content of GR is the Einstein field equation, $G_{\mu\nu} = \left(8\pi G/c^4\right) T_{\mu\nu}$, a relation between two tensors: $G_{\mu\nu}$, which characterizes the local *curvature* of spacetime, and $T_{\mu\nu}$, which specifies the local density and flux of energy-momentum. The energy-momentum tensor $T_{\mu\nu}$ is treated in Chapters 9 and 10, and we officially take up Einstein's equation in Chapter 15.

In this chapter we develop the mathematics of curvature. Our intuitive notions of curvature are based on experience with two-dimensional surfaces embedded in three-dimensional space. The *extrinsic* curvature is described in terms of quantities available in the space in which the surface is embedded, such as radius of curvature. Spacetime is not embedded in a higher-dimensional geometry as far as we know. The *intrinsic* curvature of a manifold is specified in terms of its attributes without reference to an embedding space. To speak of the curvature of four-dimensional spacetime we require its intrinsic curvature. As we'll see, intrinsic curvature is defined in terms of two interrelated mathematical concepts: *parallel transport* and the *covariant derivative*. The covariant derivative is yet another derivative on manifolds, in addition to the Lie and exterior derivatives.

14.1 COVARIANT DIFFERENTIATION

14.1.1 Is the partial derivative of a tensor, a tensor?

There are two ways to answer that. *First*: How does the derivative of a tensor transform? Equation (5.60) (reproduced here) shows the result of differentiating the transformation law for contravariant vector components, $T^{\lambda'} = A^{\lambda'}_{\rho} T^{\rho}$:

$$\frac{\partial T^{\lambda'}}{\partial x^{\alpha'}} = \frac{\partial x^{\beta}}{\partial x^{\alpha'}} \frac{\partial}{\partial x^{\beta}} \left(A^{\lambda'}_{\rho} T^{\rho} \right) = A^{\beta}_{\alpha'} A^{\lambda'}_{\rho} \frac{\partial T^{\rho}}{\partial x^{\beta}} + A^{\beta}_{\alpha'} \left(\frac{\partial A^{\lambda'}_{\rho}}{\partial x^{\beta}} \right) T^{\rho} . \tag{14.1}$$

The partial derivative does not transform as a tensor (unless $\partial_\beta A^{\lambda'}_{\rho} \equiv 0$, such as in SR). If it weren't for the inhomogeneous terms in Eq. (14.1) we'd conclude that the partial derivative of a type $(1,0)$ tensor is a type $(1,1)$ tensor. For tensors of arbitrary rank, if not for those inhomogeneous terms, the derivative of a type (k,l) tensor would be a type $(k,l+1)$ tensor. Physics is chock-full of equations involving derivatives, and we want tensors to help us write equations in covariant form (independent of coordinate system), yet partial derivatives of tensors are not in general tensors. We must find a *generalized derivative* that transforms as a tensor on manifolds yet reduces to the partial derivative on flat spaces.[1] That need is met by the covariant derivative, defined below.

[1]There are combinations of derivatives that *do* transform as tensors. The generalization of the curl to arbitrary dimensions, $\partial_\mu A_\nu - \partial_\nu A_\mu$, is such an example. The electromagnetic field tensor, Eq. (8.18), is therefore defined on manifolds the same as in flat spacetime.

The *second* way is to look at the definition of derivative,

$$\frac{\partial T^{\rho}}{\partial x^{\beta}} = \lim_{\mathrm{d}x^{\beta}\to 0} \frac{T^{\rho}(x+\mathrm{d}x^{\beta}) - T^{\rho}(x)}{\mathrm{d}x^{\beta}} . \tag{14.2}$$

The numerator in Eq. (14.2) is not in general a vector! We're comparing (subtracting) vectors from different points, yet the transformation properties of tensors are defined *at a point*. Each point of a manifold is equipped with a tangent space, and there is no natural way of identifying the "same" vector from different vector spaces. On flat spaces we're taught to "slide" a parallel copy of a vector from one point to another to form the difference (see Fig. 14.1). How do we ensure that the

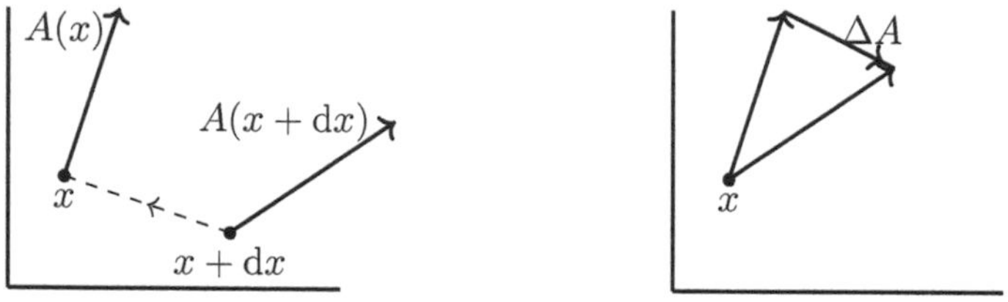

Figure 14.1 Sliding vectors on a flat space to join them at a common base.

vector remains unchanged as it's "transported"? We could demand that the vector components be unchanged in the process. That would work in Cartesian coordinates, where the basis vectors have a fixed orientation, but not in polar coordinates where the basis vectors are a function of position. The mere fact that the difference between vectors could be defined in one coordinate system but not in another tells us that the numerator in Eq. (14.2) is not a vector. *We must agree on a way of subtracting tensors from different points of a manifold that can be done in a covariant manner.* Didn't we do that with the Lie derivative, Eq. (13.17)? In that case "transport" is along a curve associated with a vector field and hence the answer depends on the choice of vector field.[2] The Lie derivative does not reduce to the partial derivative on a flat geometry, a feature we require. The scheme we develop in this chapter is called parallel transport and is defined in Eq. (14.45). The modification of Eq. (14.2) is given in Eq. (14.48). From here on we use the symbol ∇ to denote the covariant derivative (what we said we would do in Section 5.3). To understand the geometric interpretation of ∇ requires that we understand the process of parallel transport, yet parallel transport is defined in terms of ∇. To get started, ∇ is defined by a set of requirements.

14.1.2 Requirements on ∇

The derivative operator ∇ on a manifold M is defined an abstract mapping at a point $p \in M$ from type (k, l) tensors to type $(k, l+1)$ tensors,[3] $\nabla : \mathcal{T}_l^k \to \mathcal{T}_{l+1}^k$. For tensor $\mathbf{T}$ with components $T^{i_1\cdots i_k}_{j_1\cdots j_l}$, we indicate the *components* of $\nabla\mathbf{T}$, $(\nabla T)^{i_1\cdots i_k}_{\mu j_1\cdots j_l}$ (with an extra index μ), as $\nabla_\mu T^{i_1\cdots i_k}_{j_1\cdots j_l}$ even though the notation is misleading: ∇_μ is not a covariant vector; it indicates the extra index that ∇ brings to the party.[4] The index on ∇_μ is (as we'll see) related to the variation of the tensor in the direction of the coordinate curve x^μ.

The covariant derivative ∇ is defined to satisfy five requirements:[37, p31]

[2]What about the exterior derivative, Eq. (13.25)? It applies to tensors only of a *particular index structure*, differential forms that are antisymmetric in all indices.

[3]The covariant derivative thus appears similar to the exterior derivative, a mapping between wedge-product spaces at a point of the manifold, $\mathbf{d} : \wedge^p T_x^*(M) \to \wedge^{p+1} T_x^*(M)$, *spaces that already exist*. The covariant derivative *creates* a new tensor field, for which there are many possibilities.

[4]What we're writing as $\nabla_\mu T^{i_1\cdots i_k}_{j_1\cdots j_l}$ (components of the tensor field produced by the covariant derivative), is denoted in other books as $T^{i_1\cdots i_k}_{j_1\cdots j_l;\mu}$. I've avoided that notation—difficult for students to see that semicolon from the back of the room.

(1) ∇ is linear, $\nabla(\alpha\mathbf{S}+\beta\mathbf{T})=\alpha\nabla\mathbf{S}+\beta\nabla\mathbf{T}$; (for tensors $\mathbf{S}$, $\mathbf{T}$ and $\alpha,\beta\in\mathbb{R}$)

(2) ∇ satisfies the product rule, $\nabla(\mathbf{S}\otimes\mathbf{T})=(\nabla\mathbf{S})\otimes\mathbf{T}+\mathbf{S}\otimes(\nabla\mathbf{T})$;

(3) ∇ commutes with contractions, $\nabla_\mu(T^\lambda{}_{\lambda\rho})=(\nabla T)^\lambda{}_{\mu\lambda\rho}$;

(4) For any scalar field ϕ, $\nabla_\mu\phi=\partial_\mu\phi$;

(5) For any scalar field ϕ, $\nabla_{[\mu}\nabla_{\nu]}\phi=0$. (no torsion property)

The first three would be expected of any derivative; the Lie derivative satisfies these requirements (Section 13.5). The analog of property (4) for the Lie derivative is Eq. (13.19): $\pounds_{\boldsymbol{v}}\phi=v^\mu\partial_\mu\phi$. Requirement (5), the *no-torsion* property, is discussed in Section 14.3. We now have several equivalent ways of writing the directional derivative of scalar fields ϕ along a vector field $\boldsymbol{v}$: tangent vector Eq. (13.2), Lie derivative Eq. (13.19), and now covariant derivative. By property (4),

$$\boldsymbol{v}(\phi)=v^\lambda\partial_\lambda\phi=\pounds_{\boldsymbol{v}}\phi=v^\lambda\nabla_\lambda\phi\,. \tag{14.3}$$

14.1.3 Uniqueness, not

It's useful (essential, actually) to examine the *uniqueness* of operators meeting these criteria.[5] Assume one has two operators ∇ and ∇' on the same manifold M meeting the five requirements. By property (4), ∇ and ∇' must agree in their action on scalar fields, $(\nabla'_\mu-\nabla_\mu)\phi=(\partial_\mu-\partial_\mu)\phi=0$. What about tensors? Let $\boldsymbol{\omega}$ be a covariant vector field on M. The operation $(\nabla'_\mu-\nabla_\mu)\omega_\nu$ effects a linear map from dual vectors to type $(0,2)$ tensors; the operator $(\nabla'_\mu-\nabla_\mu)$ is therefore a type $(1,2)$ tensor,[6] the components of which we denote $C^\alpha_{\ \mu\nu}$. For any two derivative operators ∇' and ∇, there is a tensor $C^\alpha_{\ \mu\nu}$ such that for all ω_ν,

$$\nabla_\mu\omega_\nu=\nabla'_\mu\omega_\nu-C^\alpha_{\ \mu\nu}\omega_\alpha\,. \tag{14.4}$$

Equation (14.4) combined with property (5) implies that[7]

$$C^\alpha_{\ [\mu\nu]}=0 \tag{14.5}$$

i.e., $C^\alpha_{\ \mu\nu}$ is symmetric[8] in μ,ν. Equation (14.5) need not hold if requirement (5) is dropped.

We can use Eq. (14.4) to infer the action of $(\nabla'_\mu-\nabla_\mu)$ on contravariant vector fields. We know that $(\nabla'_\mu-\nabla_\mu)(\omega_\nu t^\nu)=0$ because $\omega_\nu t^\nu$ is a scalar field. Applying property (2) (product rule),

$$\begin{aligned}0=(\nabla'_\mu-\nabla_\mu)(\omega_\nu t^\nu)&=\omega_\nu(\nabla'_\mu-\nabla_\mu)t^\nu+t^\nu(\nabla'_\mu-\nabla_\mu)\omega_\nu\\&=\omega_\nu(\nabla'_\mu-\nabla_\mu)t^\nu+t^\nu C^\alpha_{\ \mu\nu}\omega_\alpha\\&=\omega_\nu\left[(\nabla'_\mu-\nabla_\mu)t^\nu+t^\alpha C^\nu_{\ \mu\alpha}\right]\,,\end{aligned} \tag{14.6}$$

where we've used Eq. (14.4) and renamed indices. We conclude from Eq. (14.6) that

$$\nabla_\mu t^\nu=\nabla'_\mu t^\nu+C^\nu_{\ \mu\alpha}t^\alpha\,. \tag{14.7}$$

---

[5]Uniqueness is not normally of interest to physicists, yet it's important in this case. While we're at it then, what about existence? A connected Hausdorff manifold admits a derivative operator if and only if it's paracompact.[53, p100]

[6]This point follows from the quotient theorem. A nice explanation is given in [37, p33].

[7]*Proof*: Set $\omega_\nu=\nabla_\nu\phi=\nabla'_\nu\phi$ in Eq. (14.4) (property (4)): $\nabla_\mu\nabla_\nu\phi=\nabla'_\mu\nabla'_\nu\phi-C^\alpha_{\ \mu\nu}\nabla_\alpha\phi$. Antisymmetrize over μ, ν, invoke property (5), and thus $C^\alpha_{\ [\mu\nu]}\nabla_\alpha\phi=0$. Because this relation holds for any ϕ, Eq. (14.5) follows.

[8]We'll refer to $C^\alpha_{\ \beta\gamma}$ as a symmetric tensor instead of "symmetric in its lower indices." Symmetry or antisymmetry pertain to one type of index, covariant or contravariant, but not both; see Section 5.7.

Compare Eqs. (14.7) and (14.4); the only effective difference is a minus sign. Using Eqs. (14.4) and (14.7), it's straightforward to infer for a general tensor field that

$$\nabla_\mu T^{\alpha_1\cdots\alpha_k}_{\beta_1\cdots\beta_l} = \nabla'_\mu T^{\alpha_1\cdots\alpha_k}_{\beta_1\cdots\beta_l} + \sum_{i=1}^{k} C^{\alpha_i}_{\mu\nu} T^{\alpha_1\cdots\alpha_{i-1}\nu\alpha_{i+1}\cdots\alpha_k}_{\beta_1\cdots\beta_l} - \sum_{j=1}^{l} C^{\nu}{}_{\mu\beta_j} T^{\alpha_1\cdots\alpha_k}_{\beta_1\cdots\beta_{j-1}\nu\beta_{j+1}\cdots\beta_l} \,. \tag{14.8}$$

The difference between derivative operators is completely characterized by a tensor field $C^\alpha_{(\mu\nu)}$. Conversely, if ∇' is a derivative operator, and $C^\alpha_{\mu\nu}$ is a symmetric tensor field, then ∇ as defined by Eq. (14.8) is also a derivative operator.

Derivative operators on manifolds are never unique: One can choose any type $(1,2)$ tensor field $C^\alpha_{(\mu\nu)}$ and get another derivative. The collection of derivative operators on a manifold *almost* has the structure of a vector space. What's lacking for a complete analogy with vector spaces is that there's no "origin," no natural "zero derivative operator." By selecting (arbitrarily) a derivative operator ∇ as distinguished, there is then a one-to-one correspondence between the collection of derivative operators on M and the vector space of tensors $C^\alpha_{(\mu\nu)}$. Linear spaces lacking a unique origin are called *affine spaces*.

14.1.4 Connection coefficients: Bootstrapping from ∂_μ

Manifolds have plenty of derivative operators, with any one easily obtainable from another. Could we see just one? We know they exist, we understand their non-uniqueness properties, but we don't yet know their form. Reaching for our bootstraps, take ∇' in Eq. (14.8) to be the partial derivative, $\nabla'_\mu = \partial_\mu$. Partial derivatives exist on manifolds (Section 13.2.1). In a particular chart with coordinates $\{x^\mu\}$, we have $\{\partial/\partial x^\mu\}$ and $\{\mathbf{d}x^\mu\}$ as the coordinate bases. For any smooth tensor field $T^{\alpha_1\cdots\alpha_k}_{\beta_1\cdots\beta_l}$ in that coordinate system, we can form partial derivatives, $\partial_\mu T^{\alpha_1\cdots\alpha_k}_{\beta_1\cdots\beta_l}$. The partial derivative satisfies the five conditions listed above. The fifth property in particular is the equality of mixed partial derivatives, which holds for all smooth tensor fields, not just scalar fields. Of course, there's a glaring problem here: We're not entitled to set $\nabla'_\mu = \partial_\mu$! The covariant derivative is supposed to be a mapping between tensor fields, but the partial derivative of a tensor is not a tensor, Eq. (14.1). So, while ∂_μ meets the five conditions, it does not fulfill its mission as a mapping between tensor fields. Not to worry. Invent a set of three-index symbols $\{\Gamma^\alpha_{\beta\gamma}\}$, the *connection coefficients*, $(n)^3$ *functions* on an n-manifold to be determined so that, emulating Eq. (14.7), the quantity

$$\nabla_\mu t^\nu \equiv \partial_\mu t^\nu + \Gamma^\nu_{\mu\alpha} t^\alpha \tag{14.9}$$

does meet the required properties of ∇. That is, the quantities $\{\Gamma^\nu_{\mu\alpha}\}$ are crafted so that $\nabla_\mu t^\nu$ as given by Eq. (14.9) transforms as the element of a type $(1,1)$ tensor, in addition to meeting the five criteria.[9] Note that the covariant derivative of the vector component t^ν requires *all other components* t^α, a feature it shares with the Lie derivative (Section 13.5).

If we demand that $\nabla_\mu t^\nu$ as defined in Eq. (14.9) transforms as a type $(1,1)$ tensor, we can infer how the connection coefficients must transform. We "want" $\nabla_{\mu'} t^{\nu'} = A^\mu_{\mu'} A^{\nu'}_\nu \nabla_\mu t^\nu$. Thus,

$$\nabla_{\mu'} t^{\nu'} \overset{\text{Eq.(14.9)}}{\overset{\downarrow}{=}} \partial_{\mu'} t^{\nu'} + \Gamma^{\nu'}_{\mu'\alpha'} t^{\alpha'} \overset{\text{want}}{\overset{\downarrow}{=}} A^\mu_{\mu'} A^{\nu'}_\nu \nabla_\mu t^\nu \overset{\text{Eq.(14.9)}}{\overset{\downarrow}{=}} A^\mu_{\mu'} A^{\nu'}_\nu \left(\partial_\mu t^\nu + \Gamma^\nu_{\mu\alpha} t^\alpha\right) . \tag{14.10}$$

Use Eq. (14.1) to effect a partial cancellation:

$$\Gamma^{\nu'}_{\mu'\alpha'} t^{\alpha'} = A^\mu_{\mu'} A^{\nu'}_\nu \Gamma^\nu_{\mu\alpha} t^\alpha - A^\mu_{\mu'} \left(\partial_\mu A^{\nu'}_\alpha\right) t^\alpha \,. \tag{14.11}$$

---

[9] We'll see that for flat spaces the quantities $\Gamma^\alpha_{\beta\gamma}$ vanish, in which case ∇_μ reduces to the partial derivative, ∂_μ. In Section 14.2 we look at the geometric meaning of the connection coefficients.

On the left side of Eq. (14.11) substitute $t^{\alpha'} = A^{\alpha'}_{\alpha} t^{\alpha}$ (Eq. (5.35)); the equation that remains is true for each t^{α} and thus α is a free index. Contract with $A^{\alpha}_{\rho'}$ and use the orthogonality property Eq. (5.27), $A^{\alpha}_{\rho'} A^{\alpha'}_{\alpha} = \delta^{\alpha'}_{\rho'}$. What remains is the transformation equation:

$$\Gamma^{\nu'}_{\mu'\rho'} = A^{\alpha}_{\rho'} A^{\mu}_{\mu'} A^{\nu'}_{\nu} \Gamma^{\nu}_{\mu\alpha} - A^{\alpha}_{\rho'} A^{\mu}_{\mu'} \partial_{\mu} A^{\nu'}_{\alpha} . \tag{14.12}$$

Any set of quantities $\{\Gamma^{\nu}_{\mu\rho}\}$ *that transform as in* Eq. (14.12) *is a valid set of connection coefficients.* They do not, however, comprise the elements of a tensor (why they're called *symbols*).[10] In Eq. (14.9) *a sum of two non-tensorial objects is a tensor.* The connection coefficients are sometimes given the Zen-like notation $\{{}^{\lambda}_{\mu\nu}\}$ to emphasize that they're not tensors. Just to be clear, in passing from Eq. (14.7) to Eq. (14.9), the symmetric tensor $C^{\alpha}_{\beta\gamma}$ has been replaced by a non-tensorial set of quantities $\Gamma^{\alpha}_{\beta\gamma}$ which turn out to be symmetric in the lower indices (Section 14.1.5).

Example. Show explicitly that $\nabla_{\mu} t^{\nu}$ transforms as a type $(1,1)$ tensor. From Eqs. (14.9), (14.1), and (14.12), (and using orthogonality, $A^{\alpha}_{\alpha'} A^{\alpha'}_{\sigma} = \delta^{\alpha}_{\sigma}$)

$$\begin{aligned}
\nabla_{\mu'} t^{\nu'} &= \partial_{\mu'} t^{\nu'} + \Gamma^{\nu'}_{\mu'\alpha'} t^{\alpha'} = A^{\mu}_{\mu'} A^{\nu'}_{\nu} \partial_{\mu} t^{\nu} + A^{\beta}_{\mu'} (\partial_{\beta} A^{\nu'}_{\rho}) t^{\rho} + \left(A^{\nu'}_{\nu} A^{\mu}_{\mu'} A^{\alpha}_{\alpha'} \Gamma^{\nu}_{\mu\alpha} - A^{\rho}_{\alpha'} A^{\beta}_{\mu'} (\partial_{\beta} A^{\nu'}_{\rho}) \right) A^{\alpha'}_{\sigma} t^{\sigma} \\
&= A^{\mu}_{\mu'} A^{\nu'}_{\nu} \left(\partial_{\mu} t^{\nu} + A^{\alpha}_{\alpha'} A^{\alpha'}_{\sigma} \Gamma^{\nu}_{\mu\alpha} t^{\sigma} \right) + A^{\beta}_{\mu'} (\partial_{\beta} A^{\nu'}_{\rho}) \left(t^{\rho} - A^{\rho}_{\alpha'} A^{\alpha'}_{\sigma} t^{\sigma} \right) = A^{\mu}_{\mu'} A^{\nu'}_{\nu} \left(\partial_{\mu} t^{\nu} + \Gamma^{\nu}_{\mu\alpha} t^{\alpha} \right) \\
&= A^{\mu}_{\mu'} A^{\nu'}_{\nu} \nabla_{\mu} t^{\nu} .
\end{aligned}$$

Equation (14.9) specifies the form of ∇ acting on a contravariant vector component. We can infer that for a covariant vector by forming the scalar field $\phi = \omega_{\nu} t^{\nu}$ and using property (4). Using Eq. (14.9), we require that

$$\nabla_{\mu} \omega_{\nu} = \partial_{\mu} \omega_{\nu} - \Gamma^{\alpha}_{\mu\nu} \omega_{\alpha} \tag{14.13}$$

for the product rule to be satisfied, analogous to Eq. (14.4). Using Eqs. (14.9) and (14.13), it's straightforward to derive the analog of Eq. (14.8):[11]

$$\nabla_{\mu} T^{\alpha_1 \cdots \alpha_k}_{\beta_1 \cdots \beta_l} = \partial_{\mu} T^{\alpha_1 \cdots \alpha_k}_{\beta_1 \cdots \beta_l} + \sum_{i=1}^{k} \Gamma^{\alpha_i}_{\mu\nu} T^{\alpha_1 \cdots \alpha_{i-1} \nu \alpha_{i+1} \cdots \alpha_k}_{\beta_1 \cdots \beta_l} - \sum_{j=1}^{l} \Gamma^{\nu}_{\mu\beta_j} T^{\alpha_1 \cdots \alpha_k}_{\beta_1 \cdots \beta_{j-1} \nu \beta_{j+1} \cdots \beta_l} . \tag{14.14}$$

There's a $+$ $(-)$ sign for every contravariant (covariant) index, *opposite* to the Lie derivative in its assignment of plus and minus signs, Eq. (13.23).

Equation (14.14) indicates that ∇_{μ} *is completely specified by the connection coefficients* $\{\Gamma^{\alpha}_{\mu\gamma}\}$. Connection coefficients, however, are not prescribed in the definition of manifold (Chapter 13); they must be *imposed* as an additional structure.[12] The covariant derivative is thus defined *with respect to a particular set of connection coefficients*. The message here is that we have to *choose* the connection coefficients. It would be possible to have more than one type of connection on a manifold, so long as Eq. (14.12) is satisfied for each. Suppose we have two covariant derivatives on a manifold, ∇_{μ} and $\widetilde{\nabla}_{\mu}$, each defined as in Eq. (14.9) with their own connections $\Gamma^{\lambda}_{\mu\nu}$ and $\widetilde{\Gamma}^{\lambda}_{\mu\nu}$ which separately satisfy Eq. (14.12). The *difference* $\Gamma^{\lambda}_{\mu\nu} - \widetilde{\Gamma}^{\lambda}_{\mu\nu}$ is a tensor. Subtract the versions

---

[10] This point could have been anticipated (by a very prescient reader!): In flat space, the connection coefficients vanish. If the $\{\Gamma^{\lambda}_{\mu\nu}\}$ were the elements of a tensor, then $\Gamma^{\lambda}_{\mu\nu} = 0$ in one coordinate system would imply they vanish in all coordinate systems. Because $\Gamma^{\lambda}_{\mu\nu} \neq 0$ in general coordinate systems, the connection coefficients are not be a tensor.

[11] To derive Eq. (14.14) form a scalar field $\phi = T^{\mu_1 \cdots \mu_k}_{\nu_1 \cdots \nu_l} A^{\nu_1}_{(1)} \cdots A^{\nu_l}_{(l)} B^{(1)}_{\mu_1} \cdots B^{(k)}_{\mu_k}$, where $A_{(i)}$, $1 \leq i \leq l$ ($B^{(j)}$, $1 \leq j \leq k$) are a collection of contravariant (covariant) vectors, and apply the rules for ∇ that have already been established.

[12] Analogous to imposing an inner product on a vector space; the axioms of vector spaces do not include an inner product.

of Eq. (14.9) for each derivative and use Eq. (14.7): $(\nabla_\mu - \widetilde{\nabla}_\mu)t^\nu = (\Gamma^\nu{}_{\mu\alpha} - \widetilde{\Gamma}^\nu{}_{\mu\alpha})t^\alpha = C^\nu{}_{\mu\alpha}t^\alpha$, implying that

$$\widetilde{\Gamma}^\lambda_{\mu\nu} = \Gamma^\lambda_{\mu\nu} - C^\lambda_{\mu\nu} \,. \tag{14.15}$$

There is thus a family of connection coefficients on a manifold differing by type $(1,2)$ tensors. The quantities $\Gamma^\alpha_{\beta\gamma}$ are called the *affine connection.*

14.1.5 Torsion tensor

By Eq. (14.15) the connection coefficients share the symmetry of $C^\lambda_{\mu\nu}$, i.e., $\Gamma^\alpha_{[\alpha\beta]} = 0$. What if the no-torsion requirement is dropped? Set $\omega_\nu = \nabla_\nu\phi = \nabla'_\nu\phi$ in Eq. (14.4): $\nabla_\mu\nabla_\nu\phi = \nabla'_\mu\nabla'_\nu\phi - C^\alpha_{\mu\nu}\nabla_\alpha\phi$. Antisymmetrize over μ and ν, and assume ∇' is torsion free, but ∇ is not. In that case $\nabla_{[\mu}\nabla_{\nu]}\phi = -C^\alpha_{[\mu\nu]}\nabla_\alpha\phi$. The *torsion tensor* is defined $T^\alpha_{\mu\nu} \equiv 2C^\alpha_{[\mu\nu]}$, implying that

$$\left(\nabla_\mu\nabla_\nu - \nabla_\nu\nabla_\mu\right)\phi = -T^\alpha_{\mu\nu}\nabla_\alpha\phi \,. \tag{14.16}$$

Substituting Eq. (14.9) in Eq. (14.16), we find

$$T^\alpha_{\mu\nu} = \Gamma^\alpha_{\mu\nu} - \Gamma^\alpha_{\nu\mu} = 2\Gamma^\alpha_{[\mu\nu]} \,. \tag{14.17}$$

The torsion tensor is zero in standard GR (for physical reasons explained in Section 14.3). Only $n^2(n+1)/2$ elements of the set $\{\Gamma^\alpha_{\mu\nu}\}$ are independent when $\Gamma^\alpha_{[\mu\nu]} = 0$.

14.1.6 Christoffel symbols and metric compatibility

Equation (14.8) indicates that operators ∇, ∇' satisfying the requirements in Section 14.1.2 are not uniquely specified in their action on tensor fields; the requirements allow covariant derivatives to differ by tensors $C^\alpha_{(\beta\gamma)}$. Equation (14.14) shows the action ∇_μ on tensor fields when we've "bootstrapped" from the partial derivative, $\nabla'_\mu = \partial_\mu$; we can define ∇_μ this way if we can find connection coefficients $\Gamma^\alpha_{\beta\gamma}$ satisfying Eq. (14.12). The connection coefficients, however, are not unique; they can also differ by a tensor $C^\alpha_{(\beta\gamma)}$, Eq. (14.15). How to choose the connection coefficients?

We've been concerned with the general issue of defining derivatives on manifolds. We require, however, not the most general manifolds but (for use in GR) those equipped with metric fields. That allows us to single out a set of connection coefficients. We show for $g_{\alpha\beta}$ a given metric field on M, *there is a unique torsion-free derivative satisfying* $\nabla_\mu g_{\alpha\beta} = 0$. We start by noting from Eq. (14.8) a result specific to torsion-free covariant derivatives of a metric field,

$$\left(\nabla_\mu - \nabla'_\mu\right)g_{\alpha\beta} = -C^\nu{}_{\mu\alpha}g_{\nu\beta} - C^\nu{}_{\mu\beta}g_{\nu\alpha} = -\left(C_{\beta\mu\alpha} + C_{\alpha\mu\beta}\right) = -2C_{(\alpha\beta)\mu} \,, \tag{14.18}$$

where we've lowered an index[13] and used the symmetry of $C_{\alpha\beta\gamma}$ in its second and third indices (no-torsion requirement). Uniqueness of $\nabla_\mu g_{\alpha\beta}$ is implied by $C_{(\alpha\beta)\mu} = 0$, which combined with the no-torsion requirement implies $C^\alpha_{\beta\gamma} = 0$. If we can find a form of ∇_μ such that $\nabla_\mu g_{\alpha\beta} = 0$ implies $C^\alpha_{\beta\gamma} = 0$ for all $g_{\alpha\beta}$, we're done.

Let's see what we can learn from Eq. (14.18) by equating

$$\nabla'_\mu g_{\alpha\beta} = 2C_{(\alpha\beta)\mu} \,. \tag{14.19}$$

Equation (14.19) can be solved formally. A tensor satisfying Eq. (14.19) is (check it!)

$$C^\alpha_{\mu\nu} = \frac{1}{2}g^{\alpha\rho}\left(\nabla'_\mu g_{\nu\rho} + \nabla'_\nu g_{\mu\rho} - \nabla'_\rho g_{\mu\nu}\right) \,. \tag{14.20}$$

[13] Thus, Eq. (14.18) applies to the metric tensor and not a general second-rank covariant tensor.

Equation (14.20) represents the unique solution of Eq. (14.19). To show uniqueness, assume another solution of Eq. (14.19) exists, $\widetilde{C}^{\alpha}_{\ \mu\nu}$. Form the difference $D^{\alpha}_{\ \mu\nu} \equiv \widetilde{C}^{\alpha}_{\ \mu\nu} - C^{\alpha}_{\ \mu\nu}$ and show that $D = 0$. Directly from Eq. (14.19) we have that $D_{(\alpha\mu)\nu} = 0$ (show this), i.e., $D_{\alpha\mu\nu}$ is antisymmetric in its first two indices, $D_{\alpha\mu\nu} = -D_{\mu\alpha\nu}$. By the torsion-free requirement, $D_{\alpha\mu\nu}$ is symmetric in its second and third indices, $D_{\alpha\mu\nu} = D_{\alpha\nu\mu}$. By permuting indices $D_{\alpha\mu\nu} = 0$: $D_{\alpha\mu\nu} = -D_{\mu\alpha\nu} = -D_{\mu\nu\alpha} = D_{\nu\mu\alpha} = D_{\nu\alpha\mu} = -D_{\alpha\nu\mu} = -D_{\alpha\mu\nu}$. Now use Eq. (14.14) to write

$$\nabla'_{\mu} g_{\alpha\beta} = \partial_{\mu} g_{\alpha\beta} - \widetilde{\Gamma}^{\nu}_{\ \mu\alpha} g_{\nu\beta} - \widetilde{\Gamma}^{\nu}_{\ \mu\beta} g_{\nu\alpha} \,. \tag{14.21}$$

Combine Eq. (14.21) with Eq. (14.20); it can be shown that (using the symmetry of $\widetilde{\Gamma}^{\alpha}_{\ \mu\nu}$)[14]

$$C^{\alpha}_{\ \mu\nu} = \frac{1}{2} g^{\alpha\rho} \left(\partial_{\mu} g_{\nu\rho} + \partial_{\nu} g_{\mu\rho} - \partial_{\rho} g_{\mu\nu}\right) - \widetilde{\Gamma}^{\alpha}_{\ \mu\nu} \,. \tag{14.22}$$

If we take as *the* connection coefficients the terms $\widetilde{\Gamma}^{\alpha}_{\ \mu\nu} = \Gamma^{\alpha}_{\ \mu\nu}$, where

$$\Gamma^{\alpha}_{\ \mu\nu} \equiv \frac{1}{2} g^{\alpha\rho} \left(\frac{\partial g_{\nu\rho}}{\partial x^{\mu}} + \frac{\partial g_{\mu\rho}}{\partial x^{\nu}} - \frac{\partial g_{\mu\nu}}{\partial x^{\rho}} \right) , \tag{14.23}$$

then $C^{\alpha}_{\ \beta\gamma} = 0$ from Eq. (14.22), implying $\nabla'_{\mu} g_{\alpha\beta} = 0$ from Eq. (14.19) and uniqueness.

The terms in Eq. (14.23) are the *Christoffel symbols*.[15,16] You should memorize Eq. (14.23); you'll end up memorizing it if you use it enough. The Christoffel symbols have the property that

$$g_{\beta\nu} \Gamma^{\nu}_{\ \mu\alpha} + g_{\alpha\nu} \Gamma^{\nu}_{\ \mu\beta} = \partial_{\mu} g_{\alpha\beta} \,, \tag{14.24}$$

and hence $\nabla_{\mu} g_{\alpha\beta} = 0$. One can verify (laboriously) that the Christoffel symbols transform as required by Eq. (14.12). Note that the Christoffel symbols are symmetric in the lower two indices.

Example. Evaluate the Christoffel symbols in plane polar coordinates. The metric tensor in this coordinate system is given by

$$[g_{ij}] = \begin{pmatrix} 1 & 0 \\ 0 & r^2 \end{pmatrix} \qquad [g^{ij}] = \begin{pmatrix} 1 & 0 \\ 0 & r^{-2} \end{pmatrix} .$$

Of the 2^3 possible Christoffel symbols in this coordinate system, three are nonzero:

$$\Gamma^{\theta}_{\ r\theta} = \Gamma^{\theta}_{\ \theta r} = \frac{1}{r} \qquad \Gamma^{r}_{\ \theta\theta} = -r \tag{14.25}$$

(and thus $\Gamma^{r}_{\ rr} = \Gamma^{r}_{\ \theta r} = \Gamma^{r}_{\ r\theta} = \Gamma^{\theta}_{\ rr} = \Gamma^{\theta}_{\ \theta\theta} = 0$). Let's derive some of these results using Eq. (14.23),

$$\begin{aligned}
\Gamma^{r}_{\ r\theta} = \Gamma^{r}_{\ \theta r} &= \tfrac{1}{2} g^{rr} \left[\frac{\partial g_{r\theta}}{\partial r} + \frac{\partial g_{rr}}{\partial \theta} - \frac{\partial g_{r\theta}}{\partial r} \right] + \tfrac{1}{2} g^{r\theta} \left[\frac{\partial g_{\theta\theta}}{\partial r} + \frac{\partial g_{\theta r}}{\partial \theta} - \frac{\partial g_{r\theta}}{\partial \theta} \right] = 0 \\
\Gamma^{r}_{\ \theta\theta} &= \tfrac{1}{2} g^{rr} \left[\frac{\partial g_{r\theta}}{\partial \theta} + \frac{\partial g_{r\theta}}{\partial \theta} - \frac{\partial g_{\theta\theta}}{\partial r} \right] + \tfrac{1}{2} g^{r\theta} \left[\frac{\partial g_{\theta\theta}}{\partial \theta} + \frac{\partial g_{\theta\theta}}{\partial \theta} - \frac{\partial g_{\theta\theta}}{\partial \theta} \right] \\
&= -\tfrac{1}{2} g^{rr} \frac{\partial g_{\theta\theta}}{\partial r} = -\tfrac{1}{2}(1)(2r) = -r \,.
\end{aligned}$$

---

[14] Equation (14.22) is in the form of Eq. (14.15), $C^{\lambda}_{\ \mu\nu} = \Gamma^{\lambda}_{\ \mu\nu} - \widetilde{\Gamma}^{\lambda}_{\ \mu\nu}$.

[15] The Christoffel symbols are derived in another context in Appendix D, Eq. (D.23).

[16] A manifold with the Christoffel symbols as the connection coefficients is said to be *metrically connected*.

Example. The Christoffel symbols associated with the Lorentz metric are all zero.

A derivative operator is *metric compatible* if $\nabla_\mu g_{\alpha\beta} = 0$. We've just shown there is precisely one torsion-free covariant derivative compatible with a given metric. We still have the flexibility to choose the metric, which is what Einstein's field equation does (determine the metric field). We now have a complete algebraic specification of the covariant derivative, Eq. (14.14) combined with Eq. (14.23). We're lacking a geometric interpretation, but that will come shortly (Section 14.3).

14.1.7 Comparison of derivatives on manifolds

In Chapter 13 we introduced the Lie and the exterior derivative. Table 14.1 compares the three types of derivatives defined on manifolds. In this section we show how the action of these derivatives can be expressed in terms of the covariant derivative.

Table 14.1 Derivatives on manifolds.

	Lie derivative	Exterior derivative	Covariant derivative
Generalization of	Directional derivative	Gradient	Gradient
Requires	Contravariant vector field	Nothing	Connection coefficients
Applies to	All tensor fields	Differential forms	All tensor fields
Index structure of result	Type preserving	Adds one index	Adds one index

The action of the Lie derivative on scalar fields can be expressed in terms of the covariant derivative, Eq. (14.3): $\pounds_{\boldsymbol{v}}\phi = v^\lambda \nabla_\lambda \phi$. Combined with the result of Exercise 13.10, we have for vector fields $\boldsymbol{u}$, $\boldsymbol{v}$, and scalar field ϕ,

$$\begin{aligned}(\pounds_{\boldsymbol{u}}\boldsymbol{v})\,\phi &= (\pounds_{\boldsymbol{u}}\pounds_{\boldsymbol{v}} - \pounds_{\boldsymbol{v}}\pounds_{\boldsymbol{u}})\,\phi = \pounds_{\boldsymbol{u}}\left(v^\lambda\nabla_\lambda\phi\right) - \pounds_{\boldsymbol{v}}\left(u^\lambda\nabla_\lambda\phi\right)\\ &= u^\beta\nabla_\beta\left(v^\lambda\nabla_\lambda\phi\right) - v^\beta\nabla_\beta\left(u^\lambda\nabla_\lambda\phi\right)\\ &= u^\beta v^\lambda\left(\nabla_\beta\nabla_\lambda - \nabla_\lambda\nabla_\beta\right)\phi + \left(u^\beta\nabla_\beta v^\lambda - v^\beta\nabla_\beta u^\lambda\right)\nabla_\lambda\phi\,. \end{aligned} \tag{14.26}$$

The second line of Eq. (14.26) follows because the Lie derivative is type preserving. For the same reason, the left side of Eq. (14.26) represents a tangent vector acting on ϕ; thus $(\pounds_{\boldsymbol{u}}\boldsymbol{v})\phi = (\pounds_{\boldsymbol{u}}v^\lambda)\nabla_\lambda\phi$. Invoking the no-torsion property in Eq. (14.26),

$$\pounds_{\boldsymbol{u}}v^\lambda = u^\beta\nabla_\beta v^\lambda - v^\beta\nabla_\beta u^\lambda\,. \tag{14.27}$$

For a covariant vector, $\pounds_{\boldsymbol{v}}\omega_\beta$, use the trick that the Lie derivative satisfies the product rule: $\pounds_{\boldsymbol{v}}(u^\beta\omega_\beta) = u^\beta\pounds_{\boldsymbol{v}}\omega_\beta + \omega_\beta\pounds_{\boldsymbol{v}}u^\beta$. Thus,

$$\begin{aligned} u^\beta\pounds_{\boldsymbol{v}}\omega_\beta &= \pounds_{\boldsymbol{v}}\left(u^\beta\omega_\beta\right) - \omega_\beta\pounds_{\boldsymbol{v}}u^\beta = v^\lambda\nabla_\lambda\left(u^\beta\omega_\beta\right) - \omega_\beta\left(v^\lambda\nabla_\lambda u^\beta - u^\lambda\nabla_\lambda v^\beta\right)\\ &= u^\beta\left(v^\lambda\nabla_\lambda\omega_\beta + \omega_\lambda\nabla_\beta v^\lambda\right)\,, \end{aligned}$$

where we have used Eq. (14.27). Because u^β is arbitrary,

$$\pounds_{\boldsymbol{v}}\omega_\beta = v^\lambda\nabla_\lambda\omega_\beta + \omega_\lambda\nabla_\beta v^\lambda\,. \tag{14.28}$$

Using Eqs. (14.27) and (14.28), it's straightforward to derive the general result

$$\pounds_{\boldsymbol{v}} T^{\mu_1 \cdots \mu_k}_{\nu_1 \cdots \nu_l} = v^\lambda \nabla_\lambda T^{\mu_1 \cdots \mu_k}_{\nu_1 \cdots \nu_l} + \sum_{i=1}^{l} T^{\mu_1 \cdots \mu_k}{}_{\nu_1 \cdots \nu_{i-1} \alpha \nu_{i+1} \cdots \nu_l} \nabla_{\nu_i} v^\alpha - \sum_{j=1}^{k} T^{\mu_1 \cdots \mu_{j-1} \alpha \mu_{j+1} \cdots \mu_k}_{\nu_1 \cdots \nu_l} \nabla_\alpha v^{\mu^j} . \tag{14.29}$$

Equations (13.23), (14.8), (14.14), and (14.29) have definite similarities.

An implication of formulas such as Eqs. (14.27)–(14.29), which connect $\pounds_{\boldsymbol{v}}$ with combinations of covariant derivatives is that the right side of each equation is *independent* of the choice of connection coefficients (so long as they're torsion free), because the left side certainly is. Take Eq. (14.28). Allegedly (using Eqs. (14.4) and (14.7)),

$$\begin{aligned} v^\lambda \nabla'_\lambda \omega_\beta + \omega_\lambda \nabla'_\beta v^\lambda &\stackrel{?}{=} v^\lambda \left(\nabla_\lambda \omega_\beta + C^\alpha_{\lambda\beta} \omega_\alpha \right) + \omega_\lambda \left(\nabla_\beta v^\lambda - C^\lambda_{\beta\alpha} v^\alpha \right) \\ &= v^\lambda \nabla_\lambda \omega_\beta + \omega_\lambda \nabla_\beta v^\lambda + \left[v^\lambda \omega_\alpha C^\alpha_{\lambda\beta} - \omega_\lambda v^\alpha C^\lambda_{\beta\alpha} \right] . \end{aligned}$$

The terms in square brackets vanish through index substitutions (check it!). A similar conclusion is obtained from the totally antisymmetric combination of indices $\nabla_{[\mu} \omega_{\nu_1 \cdots \nu_n]}$. Using Eq. (14.8),

$$\nabla_\mu \omega_{\nu_1 \cdots \nu_n} = \nabla'_\mu \omega_{\nu_1 \cdots \nu_n} - \sum_{j=1}^{n} C^\nu_{\mu\nu_j} \omega_{\nu_1 \cdots \nu_{j-1} \nu \nu_{j+1} \cdots \nu_n} . \tag{14.30}$$

Antisymmetrizing over all indices in Eq. (14.30) and using Eq. (14.5), we have $\nabla_{[\mu} \omega_{\nu_1 \cdots \nu_n]} = \nabla'_{[\mu} \omega_{\nu_1 \cdots \nu_n]}$, i.e., $\nabla_{[\mu} \omega_{\nu_1 \cdots \nu_n]}$ is independent of derivative operator. From Eqs. (13.25) and (14.14), $\nabla_{[\mu} \omega_{\nu_1 \cdots \nu_n]}$ is the same as the exterior derivative of an n-form:

$$(\mathrm{d}\omega)_{\mu\nu_1 \cdots \nu_n} = \partial_{[\mu} \omega_{\nu_1 \cdots \nu_n]} = \nabla_{[\mu} \omega_{\nu_1 \cdots \nu_n]} , \tag{14.31}$$

further underscoring that $\nabla_{[\mu} \omega_{\nu_1 \cdots \nu_n]}$ is independent of our choice of ∇.

14.1.8 Covariant divergence

In Chapter 13 we obtained the covariant divergence through a combination of the Hodge star operator and the exterior derivative, Eq. (13.27). We now derive the same result using the covariant derivative. To do so, we require the formula for the derivative of a determinant, Eq. (C.7). Applying Eq. (C.7) to the determinant of the covariant metric tensor (g) is particularly simple because, from Eq. (5.18), $\left(g^{-1}\right)_{ij} = g^{ij}$. Hence, from Eq. (C.7),

$$\mathrm{d}g = g g^{ji} \mathrm{d}g_{ij} . \tag{14.32}$$

The total differential of any element g_{ij} is $\mathrm{d}g_{ij} = (\partial g_{ij} / \partial x^l) \mathrm{d}x^l$; likewise for the determinant, $\mathrm{d}g = (\partial g / \partial x^l) \mathrm{d}x^l$. From Eq. (14.32), then, we have the intermediate result

$$\frac{1}{g} \frac{\partial g}{\partial x^l} = g^{ji} \frac{\partial g_{ij}}{\partial x^l} . \tag{14.33}$$

Next, take the expression for Γ^i_{jk}, Eq. (14.23), set $k = i$ and sum,

$$\Gamma^i_{ji} = \frac{1}{2} g^{il} \left[\frac{\partial g_{li}}{\partial x^j} + \frac{\partial g_{lj}}{\partial x^i} - \frac{\partial g_{ji}}{\partial x^l} \right] = \frac{1}{2} g^{il} \frac{\partial g_{li}}{\partial x^j} , \tag{14.34}$$

where the second and third terms in Eq. (14.34) cancel. Combining Eqs. (14.33) and (14.34), we obtain the useful result

$$\Gamma^i_{ji} = \frac{1}{2g} \frac{\partial g}{\partial x^j} = \frac{\partial}{\partial x^j} \ln \sqrt{g} .$$

Finally, from the equation for $\nabla_j A^k$, Eq. (14.9), set $j = k$ and sum:

$$\nabla_k A^k = \frac{\partial A^k}{\partial x^k} + A^i \Gamma^k_{ki} = \frac{\partial A^k}{\partial x^k} + A^i \frac{\partial}{\partial x^i} \ln \sqrt{g} = \frac{1}{\sqrt{g}} \frac{\partial}{\partial x^i} \left(\sqrt{g} A^i\right) , \tag{14.35}$$

the same as Eq. (13.27).

Example. Spherical coordinates. The metric tensor for spherical coordinates is given in Eq. (5.5). The determinant is therefore $g = r^4 \sin^2\theta$ and hence $\sqrt{g} = r^2 \sin\theta$. A generic vector in spherical coordinates is $\boldsymbol{A} = A^r \boldsymbol{e}_r + A^\theta \boldsymbol{e}_\theta + A^\phi \boldsymbol{e}_\phi$. The covariant divergence is, using Eq. (14.35),

$$\begin{aligned}\frac{1}{\sqrt{g}} \frac{\partial}{\partial x^i} \left(\sqrt{g} A^i\right) =& \frac{1}{r^2 \sin\theta} \left[\frac{\partial}{\partial r} \left(r^2 \sin\theta A^r\right) + \frac{\partial}{\partial \theta} \left(r^2 \sin\theta A^\theta\right) + \frac{\partial}{\partial \phi} \left(r^2 \sin\theta A^\phi\right) \right] \\ =& \frac{1}{r^2} \frac{\partial}{\partial r} \left(r^2 A^r\right) + \frac{1}{\sin\theta} \frac{\partial}{\partial \theta} \left(\sin\theta A^\theta\right) + \frac{\partial A^\phi}{\partial \phi} = \boldsymbol{\nabla} \cdot \boldsymbol{A} ,\end{aligned}$$

the usual expression for the divergence in spherical coordinates.

14.2 WHAT DO THE CONNECTION COEFFICIENTS TELL US?

Consider the total differential of a vector field $\boldsymbol{t} = t^\alpha(x^1, \cdots, x^n)\boldsymbol{e}_\alpha$ (without specifying what "d" means):

$$\mathrm{d}\boldsymbol{t} = \mathrm{d}t^\alpha \boldsymbol{e}_\alpha + t^\alpha \mathrm{d}\boldsymbol{e}_\alpha = \left(\frac{\partial t^\alpha}{\partial x^\beta} \mathrm{d}x^\beta \right) \boldsymbol{e}_\alpha + t^\alpha \left(\frac{\partial \boldsymbol{e}_\alpha}{\partial x^\beta} \mathrm{d}x^\beta \right) . \tag{14.36}$$

The first term is a linear combination of basis vectors; we know what that means. The second term involves derivatives of vectors—the very quantity we're trying to formulate with the covariant derivative. We *expect* a change $\boldsymbol{e}_\beta \to \boldsymbol{e}_\beta + \mathrm{d}\boldsymbol{e}_\beta$ under $x^\alpha \to x^\alpha + \mathrm{d}x^\alpha$ because coordinate basis vectors are tangent to coordinate curves (Section 13.2.1). We know that the spacetime manifold M is locally flat (Section 11.5). Thus, in a sufficiently small neighborhood of $p \in M$ there is a local inertial frame, the basis vectors of which are *constants*; call them $\{\boldsymbol{e}^0_\beta\}$. The coordinate basis $\{\boldsymbol{e}_\alpha\}$ in a neighborhood of $p \in M$ can be expressed in the basis $\{\boldsymbol{e}^0_\beta\}$, $\boldsymbol{e}_\alpha(x) = A^{\beta'}_\alpha(x)\boldsymbol{e}^0_{\beta'}$. Differentiating this formula (the $\boldsymbol{e}^0_{\beta'}$ are constants), $\partial_\mu \boldsymbol{e}_\alpha = (\partial_\mu A^{\beta'}_\alpha)\boldsymbol{e}^0_{\beta'} = (\partial_\mu A^{\beta'}_\alpha)A^\rho_{\beta'}\boldsymbol{e}_\rho$, where we've inverted the basis transformation, $\boldsymbol{e}^0_{\beta'} = A^\rho_{\beta'}\boldsymbol{e}_\rho$. Thus *there is a three-index symbol*, call it $\gamma^\rho_{\mu\alpha} \equiv (\partial_\mu A^{\beta'}_\alpha)A^\rho_{\beta'}$, effecting the derivative of $\boldsymbol{e}_\alpha$ along the coordinate curve x^μ:

$$\partial_\mu \boldsymbol{e}_\alpha = \gamma^\rho_{\mu\alpha} \boldsymbol{e}_\rho , \qquad (= \Gamma^\rho_{\mu\alpha} \boldsymbol{e}_\rho) \tag{14.37}$$

which we note involves all other vectors of the basis. We'll show that $\gamma^\nu_{\alpha\mu} = \Gamma^\nu_{\alpha\mu}$, the Christoffel symbols.[17] While we invoked a physical argument to get to this point (spacetime manifold is locally flat), it's not necessary to do so. We'll reach the same conclusion once we define *geodesic curves* on a manifold, which in turn require parallel transport and the covariant derivative. It's all connected!

We first show that $\gamma^\rho_{\alpha\beta}$ is symmetric. In a coordinate basis, $\boldsymbol{e}_\alpha = \partial_\alpha$ (Section 13.2.1), and partial derivatives commute, $\partial_\mu \partial_\alpha = \partial_\alpha \partial_\mu$. Thus, $\partial_\mu \boldsymbol{e}_\alpha = \partial_\alpha \boldsymbol{e}_\mu$. This symmetry implies, using Eq. (14.37), $\gamma^\rho_{\mu\alpha}\boldsymbol{e}_\rho = \gamma^\rho_{\alpha\mu}\boldsymbol{e}_\rho$, i.e., $\gamma^\rho_{\mu\alpha}$ is symmetric in its lower indices. We now show that $\gamma^\rho_{\alpha\nu} = \Gamma^\rho_{\alpha\nu}$. Using Eq. (14.37), $\mathrm{d}\boldsymbol{e}_\alpha = (\partial_\mu \boldsymbol{e}_\alpha)\mathrm{d}x^\mu = \gamma^\rho_{\alpha\mu}\boldsymbol{e}_\rho \mathrm{d}x^\mu$. Form the differential of $g_{\alpha\beta} \equiv \boldsymbol{e}_\alpha \cdot \boldsymbol{e}_\beta$, Eq. (5.3): $\mathrm{d}g_{\alpha\beta} = \boldsymbol{e}_\alpha \cdot \mathrm{d}\boldsymbol{e}_\beta + \mathrm{d}\boldsymbol{e}_\alpha \cdot \boldsymbol{e}_\beta = (\gamma^\rho_{\mu\beta} g_{\alpha\rho} + \gamma^\rho_{\alpha\mu} g_{\beta\rho})\mathrm{d}x^\mu$. The partial derivative of $g_{\alpha\beta}$ can

[17] Many books *define* the connection coefficient with Eq. (14.37); Eq. (14.23) is then obtained when metric compatibility is imposed. We've inverted that order, postulating the connection in Eq. (14.9), arriving at Eq. (14.23) through metric compatibility, and then motivating Eq. (14.37) as reasonable.

thus be written $\partial_\mu g_{\alpha\beta} = \gamma^\rho_{\mu\beta} g_{\alpha\rho} + \gamma^\rho_{\alpha\mu} g_{\beta\rho}$ (compare with Eq. (14.24)). Combine this result with Eq. (14.21) (erase primes and tildes):

$$\nabla_\mu g_{\alpha\beta} = (\gamma^\rho_{\mu\beta} - \Gamma^\rho_{\mu\beta}) g_{\alpha\rho} + (\gamma^\rho_{\alpha\mu} - \Gamma^\rho_{\mu\alpha}) g_{\rho\beta} \,. \tag{14.38}$$

With an appeal to uniqueness, metric compatibility ($\nabla_\mu g_{\alpha\beta} = 0$) requires $\gamma^\rho_{\alpha\beta} = \Gamma^\rho_{\alpha\beta}$, Eq. (14.23).

With $\gamma^\rho_{\mu\alpha} = \Gamma^\rho_{\mu\alpha}$ in Eq. (14.37), isolate $\Gamma^\rho_{\mu\alpha}$ using the orthogonality properties of the basis vectors. Let the dual basis vector $\boldsymbol{e}^\beta$ act on Eq. (14.37),

$$\Gamma^\beta_{\mu\alpha} = \boldsymbol{e}^\beta \left(\partial_\mu \boldsymbol{e}_\alpha \right) \,. \tag{14.39}$$

The connection coefficients describe how basis vectors change along coordinate curves; $\Gamma^\beta_{\mu\alpha}$ is the rate of change of $\boldsymbol{e}_\alpha$ along the coordinate curve associated with x^μ, projected onto $\boldsymbol{e}^\beta$.

Example. In polar coordinates $\Gamma^r_{\theta\theta} = -r$, Eq. (14.25). The rate of change of $\boldsymbol{e}_\theta = r\hat{\boldsymbol{\theta}}$ in the θ direction is $\mathrm{d}\boldsymbol{e}_\theta = -r\hat{\boldsymbol{r}}\mathrm{d}\theta$ and thus $\Gamma^r_{\theta\theta} = -r$ (projection onto the r direction).

The connection "connects" the bases of tangent spaces $T_x(M)$ and $T_{x+\mathrm{d}x}(M)$. The total change in basis vectors between a point of the manifold having coordinates x^α and a point with coordinates $x^\alpha + \mathrm{d}x^\alpha$ is found from Eq. (14.37) by summing over all directions:

$$\mathrm{d}\boldsymbol{e}_\alpha = (\partial_\mu \boldsymbol{e}_\alpha)\mathrm{d}x^\mu = \left(\Gamma^\rho_{\mu\alpha} \boldsymbol{e}_\rho\right) \mathrm{d}x^\mu \,. \tag{14.40}$$

Equation (14.40) can be "psyched out" by noting that the infinitesimal changes $\mathrm{d}\boldsymbol{e}_\alpha$ are: 1) proportional to the differentials $\mathrm{d}x^\mu$; and 2) $\mathrm{d}\boldsymbol{e}_\alpha$ can be represented in the basis set $\{\boldsymbol{e}_\rho\}$. Make the proportionality into an equality by inventing a three-index proportionality coefficient $\gamma^\rho_{\mu\alpha}$, which, as we've shown, is the same as $\Gamma^\rho_{\mu\alpha}$.

Example. Illustrate Eq. (14.40) in polar coordinates using the Christoffel symbols:

$$\begin{aligned}\mathrm{d}\boldsymbol{e}_r &= \Gamma^k_{rj} \boldsymbol{e}_k \mathrm{d}x^j = \Gamma^r_{rj} \boldsymbol{e}_r \mathrm{d}x^j + \Gamma^\theta_{rj} \boldsymbol{e}_\theta \mathrm{d}x^j = \Gamma^r_{rr} \boldsymbol{e}_r \mathrm{d}r + \Gamma^r_{r\theta} \boldsymbol{e}_r \mathrm{d}\theta + \Gamma^\theta_{rr} \boldsymbol{e}_\theta \mathrm{d}r + \Gamma^\theta_{r\theta} \boldsymbol{e}_\theta \mathrm{d}\theta \\ &= \Gamma^\theta_{r\theta} \boldsymbol{e}_\theta \mathrm{d}\theta = \frac{1}{r} \boldsymbol{e}_\theta \mathrm{d}\theta = \frac{1}{r}\left(r\hat{\boldsymbol{\theta}}\right) \mathrm{d}\theta = \hat{\boldsymbol{\theta}} \mathrm{d}\theta \,,\end{aligned}$$

where we've used Eq. (14.25). For the other basis vector we have, keeping only the nonzero terms,

$$\mathrm{d}\boldsymbol{e}_\theta = \Gamma^r_{\theta\theta} \boldsymbol{e}_r \mathrm{d}\theta + \Gamma^\theta_{\theta r} \boldsymbol{e}_\theta \mathrm{d}r = -r\boldsymbol{e}_r \mathrm{d}\theta + \frac{1}{r} \boldsymbol{e}_\theta \mathrm{d}r = -r\hat{\boldsymbol{r}}\mathrm{d}\theta + \hat{\boldsymbol{\theta}}\mathrm{d}r \,, \tag{14.41}$$

where we've used $\boldsymbol{e}_r = \hat{\boldsymbol{r}}$ and $\boldsymbol{e}_\theta = r\hat{\boldsymbol{\theta}}$. Using $\mathrm{d}\boldsymbol{e}_\theta = \mathrm{d}(r\hat{\boldsymbol{\theta}}) = r\mathrm{d}\hat{\boldsymbol{\theta}} + \hat{\boldsymbol{\theta}}\mathrm{d}r$, we see that Eq. (14.41) is the same as $\mathrm{d}\hat{\boldsymbol{\theta}} = -\mathrm{d}\theta\hat{\boldsymbol{r}}$, a familiar result.

How do the dual basis vectors $\{\boldsymbol{e}^\alpha\}$ change under $x^\beta \to x^\beta + \mathrm{d}x^\beta$? From Eq. (C.1), $\boldsymbol{e}_\alpha\left(\boldsymbol{e}^\beta\right) = \delta^\beta_\alpha$; hence, $\mathrm{d}\left(\boldsymbol{e}_\alpha(\boldsymbol{e}^\beta)\right) = 0$, or $\boldsymbol{e}_\alpha\left(\mathrm{d}\boldsymbol{e}^\beta\right) = -\boldsymbol{e}^\beta\left(\mathrm{d}\boldsymbol{e}_\alpha\right)$. Using Eq. (14.40), $\boldsymbol{e}_\alpha\left(\mathrm{d}\boldsymbol{e}^\beta\right) = -\boldsymbol{e}^\beta\left(\Gamma^\sigma_{\alpha\rho}\boldsymbol{e}_\sigma \mathrm{d}x^\rho\right) = -\Gamma^\sigma_{\alpha\rho}\delta^\beta_\sigma \mathrm{d}x^\rho = -\Gamma^\beta_{\alpha\rho}\mathrm{d}x^\rho$, implying that

$$\mathrm{d}\boldsymbol{e}^\beta = -\Gamma^\beta_{\rho\alpha} \boldsymbol{e}^\alpha \mathrm{d}x^\rho \,. \tag{14.42}$$

Note that neither $\mathrm{d}\boldsymbol{e}_\alpha$ nor $\mathrm{d}\boldsymbol{e}^\beta$ are tensor quantities.

Returning to Eq. (14.36), we have, using Eq. (14.40)

$$\mathrm{d}\boldsymbol{t} = \left(\partial_\beta t^\alpha + t^\lambda \Gamma^\alpha_{\lambda\beta}\right) \mathrm{d}x^\beta \boldsymbol{e}_\alpha = \nabla_\beta t^\alpha \mathrm{d}x^\beta \boldsymbol{e}_\alpha \,. \tag{14.43}$$

The derivative $\nabla_\beta t^\alpha$ is *the rate of change of* t^α *along coordinate curve* x^β *taking into account the change in basis vectors as the coordinate grid changes.* The quantity $\mathrm{d}t^\alpha = \nabla_\beta t^\alpha \mathrm{d}x^\beta$ is the total change of t^α summed over coordinate differentials $\mathrm{d}x^\beta$. The total change in the *vector* $\boldsymbol{t}$ is found by summing over the directions $\boldsymbol{e}_\alpha$, $\mathrm{d}\boldsymbol{t} = \mathrm{d}t^\alpha \boldsymbol{e}_\alpha = \nabla_\beta t^\alpha \mathrm{d}x^\beta \mathrm{d}\boldsymbol{e}_\alpha$, Eq. (14.43).

Equation (14.43) is problematic, however—it mixes coordinate differentials and basis vectors, quantities which belong to different spaces (Chapter 13). The problem is obviated if $\mathrm{d}\boldsymbol{t}$ is taken to be a *tensor*, with $\mathrm{d}x^\beta$ a basis vector of the dual space, $T_x^*(M)$. The covariant derivative $\nabla \boldsymbol{t}$, *by definition* a type $(1,1)$ tensor field, can be written:

$$\nabla \boldsymbol{t} = (\nabla \boldsymbol{t})^\alpha{}_\beta \, \boldsymbol{e}^\beta \otimes \boldsymbol{e}_\alpha \equiv \nabla_\beta t^\alpha \boldsymbol{e}^\beta \otimes \boldsymbol{e}_\alpha = \left(\partial_\beta t^\alpha + t^\sigma \Gamma^\alpha_{\beta\sigma}\right) \boldsymbol{e}^\beta \otimes \boldsymbol{e}_\alpha \,. \tag{14.44}$$

Concordance between Eqs. (14.44) and (14.36) (considered now a type $(1,1)$ tensor field) is achieved if we let $\mathrm{d}x^\beta \to \boldsymbol{e}^\beta$ and we write[18] Eq. (14.40) as $\mathrm{d}\boldsymbol{e}_\alpha \equiv \Gamma^\rho_{\beta\alpha} \boldsymbol{e}^\beta \otimes \boldsymbol{e}_\rho$.

14.3 PARALLEL TRANSPORT AND GEODESIC CURVES

At this point we know quite a bit about the covariant derivative. What we don't know, however, is how it solves the problem posed in Section 14.1. Let's do that, and then explore further uses of ∇_μ.

14.3.1 Covariant directional derivative

Consider a curve $\gamma : I \to M$, parameterized by $\lambda \in I \subset \mathbb{R}$ (see Fig. 14.2), the points of which

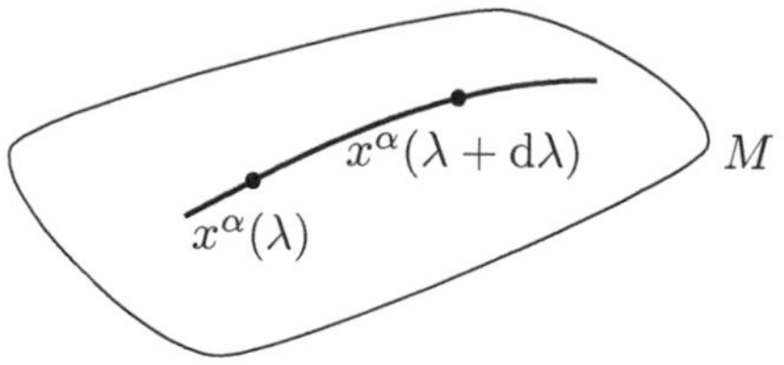

Figure 14.2 The points of $\gamma(\lambda)$ are characterized by coordinates $x^\alpha(\lambda)$.

are associated with coordinates $x^\alpha(\lambda)$. Let U^β denote the components of the tangent vector to the curve, $U^\beta = \mathrm{d}x^\beta/\mathrm{d}\lambda$. The change in a vector field $\boldsymbol{A} = A^\mu \boldsymbol{e}_\mu$ *along the curve*, the *covariant directional derivative*,[19] is found by setting $\mathrm{d}x^\beta = U^\beta \mathrm{d}\lambda$ in Eq. (14.43),

$$\frac{\mathrm{D}}{\mathrm{d}\lambda}\boldsymbol{A} \equiv U^\beta \nabla_\beta A^\alpha \boldsymbol{e}_\alpha \,.$$

The rate of change of A^μ along $\gamma(\lambda)$ is

$$\frac{\mathrm{D}}{\mathrm{d}\lambda}A^\mu = U^\alpha \nabla_\alpha A^\mu = \frac{\mathrm{d}x^\alpha}{\mathrm{d}\lambda}\left(\frac{\partial A^\mu}{\partial x^\alpha} + \Gamma^\mu_{\alpha\beta} A^\beta\right) = \frac{\mathrm{d}}{\mathrm{d}\lambda}A^\mu + U^\alpha \Gamma^\mu_{\alpha\rho} A^\rho \,. \tag{14.45}$$

14.3.2 Parallel transport

A vector $\boldsymbol{v}$ given at every point of $\gamma(\lambda)$ with tangent vector U^β is said to be *parallel transported*[20] if the covariant directional derivative vanishes at all points of $\gamma(\lambda)$:

$$U^\mu \nabla_\mu v^\alpha = U^\mu \left(\partial_\mu v^\alpha + \Gamma^\alpha_{\mu\rho} v^\rho\right) = 0 \,. \tag{14.46}$$

[18] Similarly, Eq. (14.42) should be written $\mathrm{d}\boldsymbol{e}^\beta = -\Gamma^\beta_{\rho\alpha} \boldsymbol{e}^\rho \otimes \boldsymbol{e}^\alpha$.

[19] Some books define the covariant derivative as what we're calling the covariant directional derivative, and use the notation $\nabla_U \equiv U^\beta \nabla_\beta$ where ∇_β is what we have defined as the covariant derivative. Beware.

[20] Parallel transport (a noun) is defined by Eq. (14.46). We have, in making a verb out of parallel transport, referred to a vector as being parallel transported. Some books use *parallely transported* as the verb form of parallel transport.

Equation (14.46)—a set of coupled first-order differential equations—specifies a criterion by which vectors remain constant relative to a curve, which we indicated in Section 14.1 that we would need. The magnitude of vectors is constant under parallel transport: $U^{\mu}\nabla_{\mu}\left(A^{\alpha}A_{\alpha}\right) = U^{\mu}\nabla_{\mu}\left(g_{\alpha\beta}A^{\alpha}A^{\beta}\right) = g_{\alpha\beta}U^{\mu}\nabla_{\mu}\left(A^{\alpha}A^{\beta}\right) = g_{\alpha\beta}\left(A^{\alpha}U^{\mu}\nabla_{\mu}A^{\beta} + A^{\beta}U^{\mu}\nabla_{\mu}A^{\alpha}\right) = 2A_{\alpha}U^{\mu}\nabla_{\mu} A^{\alpha} = 0$, where we've used metric compatibility. Moreover, the inner product of two vectors each parallel transported along the same curve is constant: $U^{\mu}\nabla_{\mu}\left(A_{\alpha}B^{\alpha}\right) = A_{\alpha}U^{\mu}\nabla_{\mu}B^{\alpha} + B_{\beta}U^{\mu}\nabla_{\mu}A^{\beta} = 0$. Equation (14.46) provides a criterion for the *parallelity* of a vector along $\gamma(\lambda)$ (with respect to ∇).[21] The definition of parallel transport extends to tensors of arbitrary rank,

$$U^{\mu}\nabla_{\mu}T^{i_1\cdots i_k}_{j_1\cdots j_l} = 0\,. \tag{14.47}$$

It's a trivial observation that the parallel transport of zero is zero; yet that implies *tensor equations are preserved under parallel transport* (see discussion in Section 5.1.9).

Example. The nonzero Christoffel symbols in polar coordinates are $\Gamma^{\theta}_{r\theta} = \Gamma^{\theta}_{\theta r} = r^{-1}$ and $\Gamma^{r}_{\theta\theta} = -r$, Eq. (14.25). Using Eq. (14.45), we have the differential equations for parallel transport

$$\frac{\mathrm{d}A^r}{\mathrm{d}\lambda} + \Gamma^{r}_{\theta\theta}A^{\theta}\frac{\mathrm{d}\theta}{\mathrm{d}\lambda} = 0 \qquad \frac{\mathrm{d}A^{\theta}}{\mathrm{d}\lambda} + \Gamma^{\theta}_{r\theta}A^{r}\frac{\mathrm{d}\theta}{\mathrm{d}\lambda} + \Gamma^{\theta}_{\theta r}A^{\theta}\frac{\mathrm{d}r}{\mathrm{d}\lambda} = 0\,.$$

Let the curve for parallel transport be the circle $r = R$, in which case:

$$\frac{\mathrm{d}A^r}{\mathrm{d}\theta} = RA^{\theta} \qquad \frac{\mathrm{d}A^{\theta}}{\mathrm{d}\theta} = -\frac{1}{R}A^r\,.$$

The solution of these coupled differential equations is, for initial conditions $A^r(\theta_0)$ and $A^{\theta}(\theta_0)$,

$$\begin{aligned} A^r(\theta) &= A^r(\theta_0)\cos(\theta-\theta_0) + RA^{\theta}(\theta_0)\sin(\theta-\theta_0) \\ A^{\theta}(\theta) &= A^{\theta}(\theta_0)\cos(\theta-\theta_0) - \frac{1}{R}A^r(\theta_0)\sin(\theta-\theta_0)\,. \end{aligned}$$

The magnitude of a vector is constant, $g_{\alpha\beta}A^{\alpha}(\theta)A^{\beta}(\theta) = g_{\alpha\beta}A^{\alpha}(\theta_0)A^{\beta}(\theta_0)$, yet the same is not true of the individual components. Consider $A^{\theta}(\theta_0) = 0$ (see Fig. 14.3). In that case,

$$A^r(\theta) = A^r(\theta_0)\cos(\theta-\theta_0) \qquad A^{\theta}(\theta) = -\frac{1}{R}A^r(\theta_0)\sin(\theta-\theta_0)$$

and A^{θ} does not remain zero. The vector tangent to the curve is $U^{\mu} = (0,1)$ and thus $U_{\mu} = (0, R^2)$. The inner product between $\boldsymbol{U}$ and $\boldsymbol{A}$ varies over the circle, $U_{\mu}A^{\mu} = -RA^r(\theta_0)\sin(\theta-\theta_0)$.

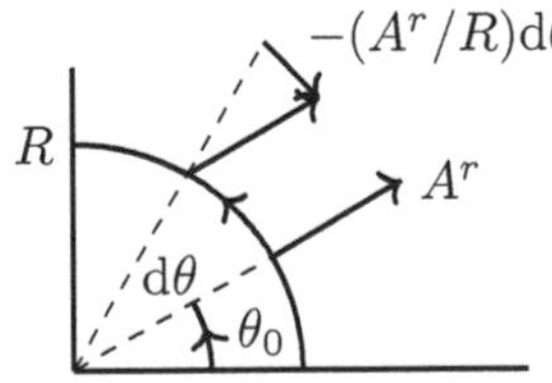

Figure 14.3 Parallel transport of vector with $A^{\theta}(\theta_0) = 0$.

---

[21] *Parallelity* seems not to be an official word of the English language. But it should be (or perhaps *parallelness*). Paralelity: The property of being parallel. This word is intended to be different from the closely related official word *parallelism*.

Example. Parallel transport on the 2-sphere of radius R. The nonzero elements of the metric tensor are $g_{\theta\theta} = R^2$ and $g_{\phi\phi} = R^2 \sin^2\theta$, Eq. (5.5), and the nonzero Christoffel symbols are $\Gamma^\theta_{\phi\phi} = -\sin\theta\cos\theta$ and $\Gamma^\phi_{\theta\phi} = \Gamma^\phi_{\phi\theta} = \cot\theta$, Exercise 14.4. The equations of parallel transport are, from Eq. (14.45),

$$\frac{\mathrm{d}A^\theta}{\mathrm{d}\lambda} - \sin\theta\cos\theta A^\phi \frac{\mathrm{d}\phi}{\mathrm{d}\lambda} = 0 \qquad \frac{\mathrm{d}A^\phi}{\mathrm{d}\lambda} + \cot\theta\left(A^\theta\frac{\mathrm{d}\phi}{\mathrm{d}\lambda} + A^\phi\frac{\mathrm{d}\theta}{\mathrm{d}\lambda}\right) = 0\,.$$

Let the curve for parallel transport be the circle $\theta = \theta_0$, in which case

$$\frac{\mathrm{d}A^\theta}{\mathrm{d}\phi} = \sin\theta_0\cos\theta_0 A^\phi \qquad \frac{\mathrm{d}A^\phi}{\mathrm{d}\phi} = -\cot\theta_0 A^\theta\,.$$

The solution of these coupled differential equations is

$$A^\theta(\phi) = A^\theta_0\cos\alpha\phi + A^\phi_0\sin\theta_0\sin\alpha\phi \qquad A^\phi(\phi) = A^\phi_0\cos\alpha\phi - A^\theta_0\sin\alpha\phi/\sin\theta_0\,,$$

where $\alpha \equiv \cos\theta_0$ and A^θ_0, A^ϕ_0 are the initial conditions at $\phi = 0$. The vector tangent to the circle is $U^\mu = (0,1)$ and thus $U_\mu = (0, R^2\sin^2\theta_0)$. The inner product $U_\mu A^\mu = R^2\sin^2\theta_0 A^\phi(\phi) = R^2\sin^2\theta_0(A^\phi_0\cos\alpha\phi - A^\theta_0\sin\alpha\phi/\sin\theta_0)$. The parallel-transported vector *rotates* relative to $\boldsymbol{U}$ (except when $\theta_0 = \pi/2$, in which case $U_\mu A^\mu = R^2 A^\phi_0 =$ constant). Consider $A^\phi_0 = 0$, $A^\theta_0 = -1 \implies U_\mu A^\mu = R^2\sin\theta_0\sin\alpha\phi$. At $\phi = 0$, A^μ is orthogonal to the curve; at $\phi = \pi/(2\cos\theta_0)$, it's aligned with the tangent.

14.3.3 Parallel transport and the covariant derivative

Armed with the knowledge of parallel transport, we can complete the picture of the covariant derivative. Let $\delta A^\mu \equiv -\Gamma^\mu_{\beta\nu}A^\beta \mathrm{d}x^\nu$ denote the change in A^μ that occurs in parallel transport through $\mathrm{d}x^\nu$ (see Fig. 14.4). The covariant derivative is almost the same as the customary definition of derivative,

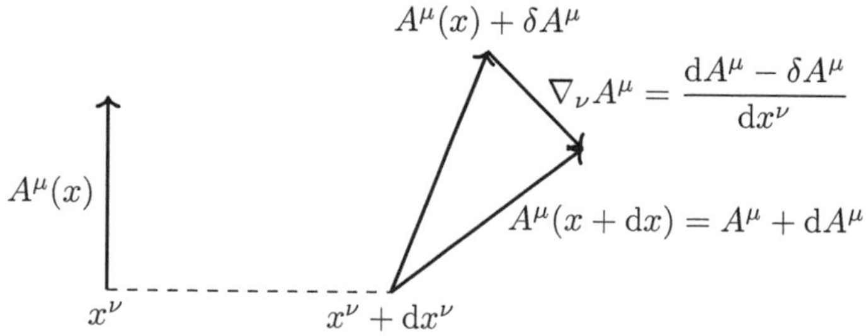

Figure 14.4 $A^\mu(x)$ is parallel transported from x^ν to $x^\nu + \mathrm{d}x^\nu$, picking up an extra component δA^μ. The covariant derivative is the difference $A^\mu(x + \mathrm{d}x^\nu) - (A^\mu(x) + \delta A^\mu)$.

Eq. (14.2), except it's the parallel-transported version of the vector that's subtracted at $x^\nu + \mathrm{d}x^\nu$:

$$\begin{aligned}\nabla_\nu A^\mu &= \lim_{\mathrm{d}x^\nu\to 0}\frac{A^\mu(x+\mathrm{d}x) - (A^\mu(x) + \delta A^\mu)}{\mathrm{d}x^\nu} \\ &= \lim_{\mathrm{d}x^\nu\to 0}\frac{A^\mu(x+\mathrm{d}x) - A^\mu(x)}{\mathrm{d}x^\nu} + \Gamma^\mu_{\alpha\beta}A^\beta\frac{\mathrm{d}x^\alpha}{\mathrm{d}x^\nu} = \partial_\nu A^\mu + \Gamma^\mu_{\nu\beta}A^\beta\,,\end{aligned} \tag{14.48}$$

in agreement with Eq. (14.9) ($\mathrm{d}x^\alpha/\mathrm{d}x^\nu = \delta^\alpha_\nu$).

14.3.4 No-torsion property

We can use parallel transport to understand the geometric significance of the no-torsion requirement. Let P be a point with coordinates x^μ. Let points Q and S have coordinates $x^\mu + \mathrm{d}x^\mu$ and $x^\mu + \mathrm{d}y^\mu$, and let $\boldsymbol{X} = \mathrm{d}x^\mu \boldsymbol{e}_\mu$ and $\boldsymbol{Y} = \mathrm{d}y^\mu \boldsymbol{e}_\mu$ be infinitesimal vectors. Parallel transport $\boldsymbol{X}$ along the line PS to obtain the vector ST. Similarly, parallel transport $\boldsymbol{Y}$ along the line PQ to obtain vector QR. Points T and R will coincide if $\mathrm{d}y^\mu + \mathrm{d}x^\mu - \Gamma^\mu_{\nu\lambda}\mathrm{d}x^\lambda \mathrm{d}y^\nu = \mathrm{d}x^\mu + \mathrm{d}y^\mu - \Gamma^\mu_{\lambda\nu}\mathrm{d}y^\nu \mathrm{d}x^\lambda$, or if

$$(\Gamma^\mu_{\lambda\nu} - \Gamma^\mu_{\nu\lambda})\, \mathrm{d}x^\lambda \mathrm{d}y^\nu = 0\,. \tag{14.49}$$

For Eq. (14.49) to hold, i.e., for the parallelogram in Figure 14.5 to close, we must have $\Gamma^\mu_{\nu\lambda} = \Gamma^\mu_{\lambda\nu}$. If $\Gamma^\mu_{\alpha\beta}$ were not symmetric, and the parallelogram did not close, the manifold wouldn't be flat even on a small scale, which in the context of GR is a violation of the equivalence principle (Section 11.5). Standard GR requires a torsion-free spacetime manifold.

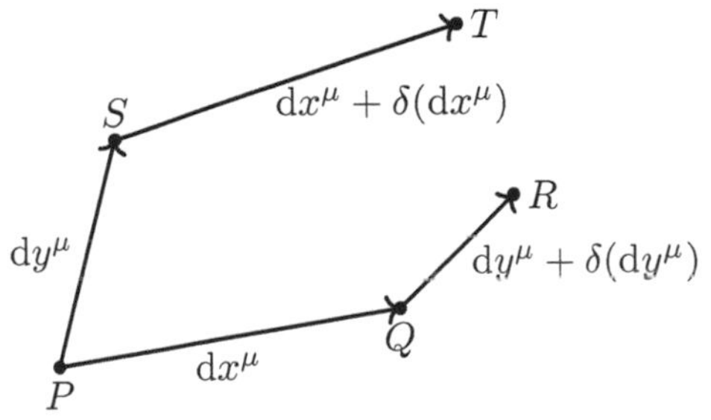

Figure 14.5 Parallelogram does not close unless the manifold is torsion free.

14.3.5 Geodesic curves

Geodesics are a class of curves of particular importance to GR. *Geodesics are curves along which the tangent vector $\boldsymbol{U}$ is parallel transported*:

$$U^\mu \nabla_\mu U^\alpha = 0\,. \tag{14.50}$$

The tangent to any curve "follows" the curve, but not all curves are such that the tangent at one point of the curve is obtained from parallel transport of the tangent from any other point. A vector $\boldsymbol{V}$ parallel transported *along a geodesic* maintains a constant inner product with $\boldsymbol{U}$.

Using Eq. (14.45), we have the differential equation for the coordinates $x^\alpha(\lambda)$ of the geodesic,

$$\frac{\mathrm{d}^2 x^\alpha}{\mathrm{d}\lambda^2} + \Gamma^\alpha_{\mu\rho}\frac{\mathrm{d}x^\mu}{\mathrm{d}\lambda}\frac{\mathrm{d}x^\rho}{\mathrm{d}\lambda} = 0\,. \tag{14.51}$$

Equation (14.51), *the geodesic equation*, is a system of coupled second-order ordinary differential equations for the functions $x^\mu(\lambda)$, and as such has unique solutions[22] for given $p \in M$ and tangent vector $U^\beta \in T_p(M)$. Similar to Newtonian mechanics where a unique solution of the equation of motion exists for given initial position and velocity, *a unique geodesic passes through any point of a manifold with a given tangent vector*. Note that the summation over $U^\alpha U^\beta$ in Eq. (14.51) is symmetric in (α, β), and thus an *antisymmetric* component of $\Gamma^\mu_{\alpha\beta}$ would not contribute to Eq. (14.51).

[22]Existence and uniqueness of the solutions of ordinary differential equations is proved in books on differential equations, such as [58, p85] or [59, p22].

There is another sense in which geodesics are unique. Suppose operators ∇ and ∇' exist on M having identical geodesic curves, i.e., for all smooth curves $\gamma : I \to M$, γ is a geodesic of ∇ if and only if it's a geodesic of ∇'. Then[23] $\nabla' = \nabla$. A derivative operator determines a family of geodesics; conversely, a covariant derivative is characterized by its geodesics.

Example. Geodesics on the 2-sphere. The nonzero Christoffel symbols are $\Gamma^{\theta}_{\phi\phi} = -\sin\theta\cos\theta$ and $\Gamma^{\phi}_{\theta\phi} = \Gamma^{\phi}_{\phi\theta} = \cot\theta$ (Exercise 14.4). The differential equations for the geodesic are, from Eq. (14.51)

$$\frac{\mathrm{d}^2\theta}{\mathrm{d}\lambda^2} - \sin\theta\cos\theta\left(\frac{\mathrm{d}\phi}{\mathrm{d}\lambda}\right)^2 = 0 \qquad \frac{\mathrm{d}^2\phi}{\mathrm{d}\lambda^2} + 2\cot\theta\frac{\mathrm{d}\phi}{\mathrm{d}\lambda}\frac{\mathrm{d}\theta}{\mathrm{d}\lambda} = 0\,.$$

Clearly, $\theta = \pi/2$, $\phi = a + b\lambda$ is a solution to both differential equations; another is $\phi = c$, $\theta = d + e\lambda$, where (a, b, c, d, e) are constants. Any *great circle* is a geodesic.

What parameterizations leave the form of the geodesic equation invariant? Change to a new parameter $\tau = \tau(\lambda)$:

$$\frac{\mathrm{d}}{\mathrm{d}\lambda} = \frac{\mathrm{d}\tau}{\mathrm{d}\lambda}\frac{\mathrm{d}}{\mathrm{d}\tau} \qquad \frac{\mathrm{d}^2}{\mathrm{d}\lambda^2} = \left(\frac{\mathrm{d}\tau}{\mathrm{d}\lambda}\right)^2\frac{\mathrm{d}^2}{\mathrm{d}\tau^2} + \frac{\mathrm{d}^2\tau}{\mathrm{d}\lambda^2}\frac{\mathrm{d}}{\mathrm{d}\tau}\,. \tag{14.52}$$

Substitute Eq. (14.52) in Eq. (14.51):

$$\frac{\mathrm{d}^2x^\nu}{\mathrm{d}\tau^2} + \Gamma^{\nu}_{\alpha\beta}\frac{\mathrm{d}x^\alpha}{\mathrm{d}\tau}\frac{\mathrm{d}x^\beta}{\mathrm{d}\tau} = -\frac{\mathrm{d}^2\tau}{\mathrm{d}\lambda^2}\left(\frac{\mathrm{d}\tau}{\mathrm{d}\lambda}\right)^{-2}\frac{\mathrm{d}x^\nu}{\mathrm{d}\tau}\,. \tag{14.53}$$

For Eq. (14.53) to look like Eq. (14.51) we require $\mathrm{d}^2\tau/\mathrm{d}\lambda^2 = 0$ or that $\tau(\lambda) = a\lambda + b$, where (a, b) are constants: We can shift the origin of $\lambda \in I \subset \mathbb{R}$ and we can rescale λ, which amounts to a scaling of the tangent vector, $\boldsymbol{U} \to \boldsymbol{U}' = \boldsymbol{U}/a$. A parameterization that preserves the form of Eq. (14.51) is an *affine parameterization.* It was used in Section 7.5 to show the equivalence of the two forms of the free-particle Lagrangian, Eqs. (7.46) and (7.49).

Geodesics have the property of being the straightest possible curves,[24] which we expect would extremize the distance between points (see Appendix D). Distance is a *metrical* concept. The length of a curve $\gamma : I \to M$, with parameterization $\lambda \in I$, having tangent field U^μ, passing through $p, q \in M$ with coordinates $x^\mu(\lambda_1)$ and $x^\mu(\lambda_2)$, for M having metric field $g_{\mu\nu}$, is defined as:[25]

$$S(p, q) = \int_{\lambda_1}^{\lambda_2}\sqrt{g_{\mu\nu}\frac{\mathrm{d}x^\mu}{\mathrm{d}\lambda}\frac{\mathrm{d}x^\nu}{\mathrm{d}\lambda}}\mathrm{d}\lambda \equiv \int_{\lambda_1}^{\lambda_2}\sqrt{g_{\mu\nu}U^\mu U^\nu}\mathrm{d}\lambda\,. \tag{14.54}$$

As shown in Appendix D, the path of stationary action for a free particle is the same as the geodesic curve when the connection coefficients are given by the Christoffel symbols (set $V = 0$ in Eq. (D.22)). The derivatives of the metric tensor that emerge in a variational calculation reflect the fact that in varying the path around the geodesic we must in general take into account variations in the metric field. The geodesic differential equation (14.51) is the generalization of what we have in SR for the motion of free particles: $\mathrm{d}U^\mu/\mathrm{d}\tau = \mathrm{d}^2x^\mu/\mathrm{d}\tau^2 = 0$. The version of Newton's first law of motion in GR is that *free particles follow geodesic curves on the spacetime manifold.*

---

[23] *Proof*: Contract Eq. (14.7) with U^μ and set $t^\nu = U^\nu$: $U^\mu\nabla_\mu U^\nu = U^\mu\nabla'_\mu U^\nu + C^{\nu}_{\mu\alpha}U^\mu U^\alpha$. By assumption $U^\mu\nabla_\mu U^\nu = 0$ and $U^\mu\nabla'_\mu U^\nu = 0$. Thus $C^{\nu}_{\mu\alpha}U^\mu U^\alpha = 0$, implying $C^{\nu}_{\mu\alpha} = 0$, and hence that $\nabla' = \nabla$ from Eq. (14.8).

[24] Hermann Weyl characterized a geodesic as a curve that "preserves its direction unchanged."[41, p115]

[25] Equation (14.54) presumes $g_{ij}U^iU^j > 0$, true for a Riemannian manifold or for spacelike separated events on a Lorentzian manifold. For timelike separated events, $\sqrt{g_{ij}U^iU^j}$ is replaced with $\sqrt{-g_{ij}U^iU^j}$; Section 4.4.1.

14.3.6 Maxwell equations on a manifold

We can now ask what is the form of Maxwell's equations on a manifold? By the equivalence principle, if we can write equations in covariant form in the absence of gravitation, then they are true in a gravitational field (Section 11.5). On manifolds we must replace partial derivatives by the covariant derivative, $\partial_\nu \to \nabla_\nu$. Is that all we need to do—rewrite Maxwell's equations using the covariant derivative? Basically, yes. Maxwell's equations in Minkowski space are given by Eqs. (8.21) and (8.27). The field tensor on a manifold is still given by Eq. (8.18), $F_{\mu\nu} = \partial_\mu A_\nu - \partial_\nu A_\mu$—the extra terms introduced by the covariant derivative cancel (Exercise 14.17). The homogeneous Maxwell equations, (8.21), are unchanged for the same reason (Exercise 14.18). The generalization of Eq. (8.27) (inhomogeneous Maxwell equations) on a manifold is, using Eq. (14.35),

$$\nabla_\beta F^{\alpha\beta} = \frac{1}{\sqrt{-g}}\partial_\lambda\left(\sqrt{-g}F^{\alpha\lambda}\right) = \mu_0 J^\alpha \,. \tag{14.55}$$

The equation for charge conservation is $\nabla_\mu J^\mu = \mu_0^{-1}\nabla_\mu\nabla_\nu F^{\mu\nu} = 0$ ($\nabla_\mu\nabla_\nu$ is symmetric in (μ,ν) and $F^{\mu\nu}$ is antisymmetric), so that $\nabla_\mu J^\mu = \partial_\lambda(\sqrt{-g}J^\lambda) = 0$.

The equation of motion for a charged particle on flat spacetime is Eq. (8.43), $m\mathrm{d}U^\mu/\mathrm{d}\tau = qF^{\mu\nu}U_\nu$. On a manifold the ordinary derivative of U^μ is replaced by the directional covariant derivative along the worldline, Eq. (14.45):

$$\frac{\mathrm{D}U^\mu}{\mathrm{d}\tau} = U^\alpha\nabla_\alpha U^\mu = \frac{\mathrm{d}U^\mu}{\mathrm{d}\tau} + \Gamma^\mu_{\alpha\rho}U^\alpha U^\rho = \frac{q}{m}F^{\mu\nu}U_\nu \,. \tag{14.56}$$

In the absence of the electromagnetic field, Eq. (14.56) becomes the geodesic equation (14.51). The motion of a free particle has zero four-acceleration along the worldline, $\mathrm{D}U^\mu/d\tau = 0$. We haven't stated so explicitly, but the four-acceleration is given by $\mathcal{A}^\mu = U^\alpha\nabla_\alpha U^\mu$.

To convert a covariant equation valid in flat spacetime to one valid in arbitrary coordinates:

(1) Replace partial derivatives with covariant derivatives, $\partial_\mu \to \nabla_\mu$;

(2) Replace ordinary derivatives with the directional covariant derivative, $\mathrm{d}/\mathrm{d}\tau \to \mathrm{D}/\mathrm{d}\tau$;

(3) Replace the metric tensor for flat spacetime, $\eta_{\mu\nu} \to g_{\mu\nu}$.

14.4 THE RIEMANN TENSOR

We can now define the intrinsic curvature of a manifold, which is described by a fourth-rank tensor.

14.4.1 Commutator of covariant derivatives

In the following we need the *second* covariant derivative. We start by forming the total differential of the derivative operator, considered as a type $(1,1)$ tensor field, $\mathbf{T} \equiv \nabla_j T^k e^j e_k$:

$$\begin{aligned}\mathrm{d}\mathbf{T} &= \mathrm{d}(\nabla_j T^k)e^j e_k + \nabla_j T^k e^j \mathrm{d}e_k + \nabla_j T^k e_k \mathrm{d}e^j \\ &= \mathrm{d}(\nabla_j T^k)e^j e_k + \nabla_j T^k e^j \Gamma^m_{kn} e_m \mathrm{d}x^n - \nabla_j T^k e_k \Gamma^j_{mn} e^m \mathrm{d}x^n \,,\end{aligned} \tag{14.57}$$

where we've used Eqs. (14.40) and (14.42). The total differential of $\nabla_j T^k$ is, using Eq. (14.9),

$$\begin{aligned}\mathrm{d}(\nabla_j T^k) &= \frac{\partial}{\partial x^l}\left(\nabla_j T^k\right)\mathrm{d}x^l = \frac{\partial}{\partial x^l}\left(\frac{\partial T^k}{\partial x^j} + \Gamma^k_{jp}T^p\right)\mathrm{d}x^l \\ &= \left[\frac{\partial^2 T^k}{\partial x^l \partial x^j} + \Gamma^k_{jp}\frac{\partial T^p}{\partial x^l} + T^p\frac{\partial \Gamma^k_{jp}}{\partial x^l}\right]\mathrm{d}x^l \,.\end{aligned} \tag{14.58}$$

The terms in square brackets in Eq. (14.58) are therefore an explicit expression for $\partial_l(\nabla_j T^k)$. Combining Eqs. (14.58) and (14.57),

$$\mathrm{d}\mathbf{T} = \left[\partial_l(\nabla_j T^k) + \Gamma^k_{lm}\nabla_j T^m - \Gamma^m_{lj}\nabla_m T^k\right] \mathrm{d}x^l \boldsymbol{e}^j \boldsymbol{e}_k \equiv \nabla_l \nabla_j T^k \boldsymbol{e}^j \boldsymbol{e}_k \mathrm{d}x^l \,. \tag{14.59}$$

The terms in square brackets in Eq. (14.59) define the second covariant derivative, $\nabla_l\nabla_j T^k$. The second covariant derivative transforms as a type $(1,2)$ tensor, $\nabla_{l'}\nabla_{j'}T^{k'} = A^{k'}_{\alpha} A^{\beta}_{l'} A^{\lambda}_{j'} \nabla_\beta \nabla_\lambda T^\alpha$.

Covariant derivatives do not commute, $\nabla_\mu\nabla_\nu T^\alpha \neq \nabla_\nu\nabla_\mu T^\alpha$. Using Eq. (14.59),

$$\begin{aligned}
(\nabla_\mu\nabla_\nu - \nabla_\nu\nabla_\mu) T^\alpha &= \partial_\mu(\nabla_\nu T^\alpha) + \Gamma^\alpha_{\beta\mu}\nabla_\nu T^\beta - \Gamma^\beta_{\nu\mu}\nabla_\beta T^\alpha - [\partial_\nu(\nabla_\mu T^\alpha) \\
&\quad + \Gamma^\alpha_{\beta\nu}\nabla_\mu T^\beta - \Gamma^\beta_{\mu\nu}\nabla_\beta T^\alpha] \\
&= \partial_\mu\left(\frac{\partial T^\alpha}{\partial x^\nu} + \Gamma^\alpha_{\nu\rho}T^\rho\right) + \Gamma^\alpha_{\beta\mu}\nabla_\nu T^\beta - \partial_\nu\left(\frac{\partial T^\alpha}{\partial x^\mu} + \Gamma^\alpha_{\mu\rho}T^\rho\right) - \Gamma^\alpha_{\beta\nu}\nabla_\mu T^\beta \\
&= \left[(\partial_\mu\Gamma^\alpha_{\nu\rho} + \Gamma^\alpha_{\beta\mu}\Gamma^\beta_{\nu\rho}) - (\partial_\nu\Gamma^\alpha_{\mu\rho} + \Gamma^\alpha_{\beta\nu}\Gamma^\beta_{\mu\rho})\right] T^\rho \\
[\nabla_\mu, \nabla_\nu] T^\alpha &\equiv R^\alpha_{\ \rho\mu\nu} T^\rho \,,
\end{aligned} \tag{14.60}$$

where Eq. (14.60) uniquely defines a fourth-order tensor field

$$R^\alpha_{\ \rho\mu\nu} \equiv \left(\frac{\partial\Gamma^\alpha_{\nu\rho}}{\partial x^\mu} + \Gamma^\alpha_{\mu\beta}\Gamma^\beta_{\nu\rho}\right) - \left(\frac{\partial\Gamma^\alpha_{\mu\rho}}{\partial x^\nu} + \Gamma^\alpha_{\nu\beta}\Gamma^\beta_{\mu\rho}\right) , \tag{14.61}$$

the *Riemann curvature tensor field.*[26] The Riemann tensor involves derivatives of the Christoffel symbols and hence second derivatives of the metric tensor. The second group of terms in Eq. (14.61) follows from the first under the interchange $\mu \leftrightarrow \nu$. Thus $R^\alpha_{\ \rho\mu\nu}$ is antisymmetric in the last two indices, $R^\alpha_{\ \rho\mu\nu} = -R^\alpha_{\ \rho\nu\mu}$, as it must be—the commutator on the left side of Eq. (14.60) is antisymmetric in μ and ν. The Riemann tensor transforms as a type $(1,3)$ tensor, $R^{i'}_{\ j'k'l'} = A^{i'}_{\alpha} A^{\beta}_{j'} A^{\gamma}_{k'} A^{\delta}_{l'} R^\alpha_{\ \beta\gamma\delta}$.

A manifold for which $R^\alpha_{\ \rho\mu\nu} = 0$ everywhere is said to be *flat*. A necessary and sufficient condition for the metric tensor to contain constant elements is the vanishing of the Riemann tensor.[60, p25] Either condition, a metric tensor with constant elements or $R^\alpha_{\ \beta\gamma\delta} = 0$ can be taken as the hallmark of a flat geometry.

14.4.2 Parallel transport around a closed curve

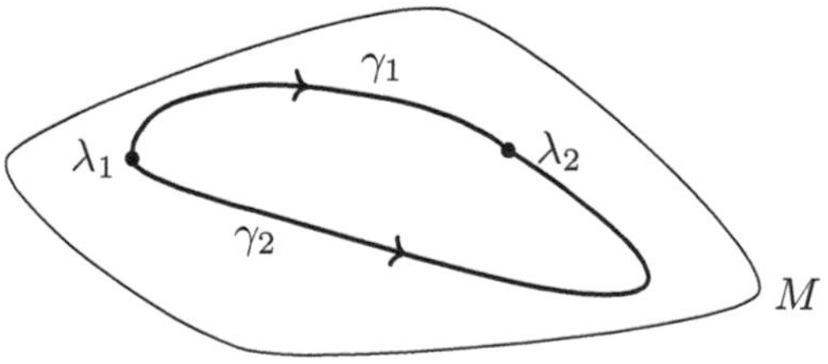

Figure 14.6 Two curves connecting points λ_1, λ_2 on M.

Parallel transport is defined with respect to prescribed curves. By integrating Eq. (14.45) along a curve $\gamma(\lambda)$ starting at $\lambda = \lambda_1$, we obtain the parallel-transported value of A^μ at $\lambda = \lambda_2$, $A^\mu_\parallel$:

$$A^\mu_\parallel(\gamma(\lambda_2)) \equiv A^\mu(\gamma(\lambda_1)) - \int_{\gamma(\lambda_1)}^{\gamma(\lambda_2)} \Gamma^\mu_{\alpha\beta} A^\beta(\lambda) \frac{\mathrm{d}x^\alpha}{\mathrm{d}\lambda} \mathrm{d}\lambda \,. \qquad (\mu = 0,1,2,3) \tag{14.62}$$

[26]Equation (14.60) should be compared with Eq. (14.16). The no-torsion property of the manifold $[\nabla_\mu, \nabla_\nu]\phi = 0$ refers to scalar fields, while Eq. (14.60) refers to the non-commutativity of the covariant derivative acting on vector fields.

To calculate $A_{\|}^{\mu}$ we must know the value of the other vector components A^{β} as well as the Christoffel symbols at each point of $\gamma(\lambda)$ for $\lambda_1 \leq \lambda \leq \lambda_2$. The result is *path dependent*:[27] Referring to Fig. 14.6, $A_{\|}^{\mu}(\lambda_2)$ obtained on γ_1 is not the same as that obtained on γ_2.

Apply Eq. (14.62) to a *closed curve* C (for example join $\gamma_1(\lambda)$ in Fig. 14.6, $\lambda_1 \leq \lambda \leq \lambda_2$, to $\gamma_2(\lambda)$, $\lambda_2 \geq \lambda \geq \lambda_1$): $A_{\|}^{\mu}(\lambda_1) = A^{\mu}(\lambda_1) - \oint_C \Gamma^{\mu}_{\alpha\beta} A^{\beta} \mathrm{d}x^{\alpha}$. In general, $A_{\|}^{\mu}(\lambda_1) \neq A^{\mu}(\lambda_1)$. Unless the manifold is flat, *a vector parallel transported around a closed curve produces a vector different from the initial vector*. This effect is illustrated in Fig. 14.7. Define the *difference* between

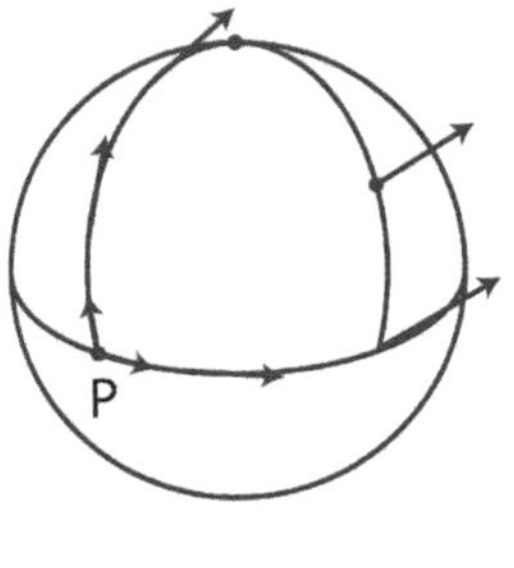

Figure 14.7 Parallel transport of a vector around a closed curve consisting of three segments of great circles (geodesics).

the starting value, $A_0^{\mu}(\lambda)$, and the value of $A_{\|}^{\mu}$ after parallel transport around a closed curve:

$$\Delta A^{\mu} \equiv A_{\|}^{\mu}(\lambda) - A_0^{\mu}(\lambda) = -\oint_C \Gamma^{\mu}_{\alpha\beta} A^{\beta} \mathrm{d}x^{\alpha} \,. \tag{14.63}$$

The extent to which $\Delta A^{\mu} \neq 0$ provides another way to characterize curvature;[28] in essence ΔA^{μ} is a "curvature meter." As we show, ΔA^{μ} is related to the Riemann tensor for an infinitesimal closed curve. The reader uninterested in the details should skip to Eq. (14.70).

Consider a point p with coordinates x_p^{μ} and a closed curve C surrounding p. Figure 14.8 shows

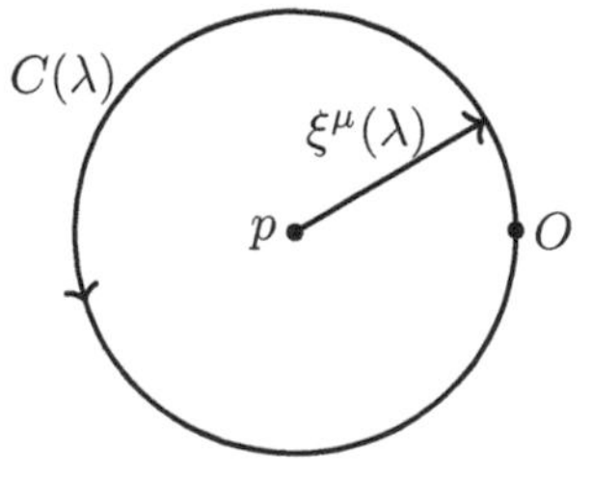

Figure 14.8 Parallel transport around a closed circle $C(\lambda)$.

such a curve as a circle, but no conclusions we reach will depend on C being a circle. Let the points of C have coordinates $x^{\mu}(\lambda)$, which we can express relative to the coordinates of p, $x^{\mu}(\lambda) = x_p^{\mu} + \xi^{\mu}(\lambda)$. Equation (14.63) can then be written as an integral over $\mathrm{d}\xi^{\mu}$ because x_p^{μ} is a constant,

$$\Delta A^{\mu} = -\oint_C \Gamma^{\mu}_{\alpha\beta} A^{\alpha} \mathrm{d}\xi^{\beta} \,. \tag{14.64}$$

[27]For parallel transport to be independent of path, $\Gamma^{\mu}_{\alpha\beta} A^{\beta} \mathrm{d}x^{\alpha}$ must be an exact differential, which occurs only if the Riemann tensor vanishes, i.e., the manifold is flat. We omit a proof—this chapter is long enough already.

[28]$\Delta A^{\mu} = 0$ only for a flat space; $\Delta A^{\mu} \neq 0$ on a curved space.

Equation (14.64) is an *integral equation*; it cannot be integrated directly because the unknown A^α is part of the integrand. We can, however, set up an iterative method of solution, one that can be implemented to any desired level of accuracy. Approximate Eq. (14.64) by setting $A^\alpha = A_0^\alpha$ in the integral, and thus $A^\mu(\lambda) \approx A_0^\mu - A_0^\alpha \int \Gamma^\mu_{\alpha\beta} \mathrm{d}\xi^\beta$. Substitute this result back into Eq. (14.64):

$$A^\mu(\lambda) \approx A_0^\mu - \int \Gamma^\mu_{\alpha\beta} \left(A_0^\alpha - A_0^\gamma \int \Gamma^\alpha_{\gamma\rho} \mathrm{d}\xi^\rho \right) \mathrm{d}\xi^\beta = A_0^\mu - A_0^\alpha \int \Gamma^\mu_{\alpha\beta} \mathrm{d}\xi^\beta + A_0^\gamma \int \Gamma^\mu_{\alpha\beta} \left(\int \Gamma^\alpha_{\gamma\rho} \mathrm{d}\xi^\rho \right) \mathrm{d}\xi^\beta . \tag{14.65}$$

Equation (14.65) represents two terms in a *Neumann series*.[29] The iterative process can be repeated indefinitely, but we stop at second order in ξ, which we presume to be small. We now Taylor expand the Christoffel symbols $\Gamma^\mu_{\alpha\beta}(\lambda)$ around their values at x_p^μ,

$$\Gamma^\mu_{\alpha\beta}(\lambda) \approx \left(\Gamma^\mu_{\alpha\beta} \right)_p + \left(\partial_\sigma \Gamma^\mu_{\alpha\beta} \right)_p \xi^\sigma . \tag{14.66}$$

Substitute Eq. (14.66) into the first integral on the right of Eq. (14.65),

$$\oint \Gamma^\mu_{\alpha\beta} \mathrm{d}\xi^\beta \approx \left(\partial_\sigma \Gamma^\mu_{\alpha\beta} \right)_p \oint \xi^\sigma \mathrm{d}\xi^\beta , \tag{14.67}$$

where we've extended the integral to the closed curve and we've used $\oint \mathrm{d}\xi^\beta = 0$. For the second term on the right of Eq. (14.65) we can, consistent to second order in ξ, keep only the zeroth-order values for the Christoffel coefficients. Thus,

$$\oint \Gamma^\mu_{\alpha\beta} \left(\int \Gamma^\alpha_{\gamma\rho} \mathrm{d}\xi^\rho \right) \mathrm{d}\xi^\beta \approx \left(\Gamma^\mu_{\alpha\beta} \Gamma^\alpha_{\gamma\rho} \right)_p \oint \left(\int \mathrm{d}\xi^\rho \right) \mathrm{d}\xi^\beta = \left(\Gamma^\mu_{\alpha\beta} \Gamma^\alpha_{\gamma\rho} \right)_p \oint \xi^\rho \mathrm{d}\xi^\beta . \tag{14.68}$$

Combine Eqs. (14.68) and (14.67) with Eq. (14.65), and we have to lowest order in ξ

$$\Delta A^\mu = -A_0^\alpha \left(\partial_\sigma \Gamma^\mu_{\alpha\beta} - \Gamma^\mu_{\gamma\beta} \Gamma^\gamma_{\alpha\sigma} \right)_p \oint \xi^\sigma \mathrm{d}\xi^\beta . \tag{14.69}$$

Because $\oint \mathrm{d}(\xi^\sigma \xi^\beta) = 0$, it follows that $\oint \xi^\sigma \mathrm{d}\xi^\beta = -\oint \xi^\beta \mathrm{d}\xi^\sigma$, and thus

$$\oint \xi^\sigma \mathrm{d}\xi^\beta = \frac{1}{2} \oint (\xi^\sigma \mathrm{d}\xi^\beta - \xi^\beta \mathrm{d}\xi^\sigma) \equiv \oint \mathrm{d}\Sigma^{\sigma\beta} \equiv S^{\sigma\beta}$$

defines an antisymmetric tensor, $S^{\sigma\beta} = -S^{\beta\sigma}$. We know from Section 5.11 that a differential element of surface area is an antisymmetric tensor, $\mathrm{d}\Sigma^{\alpha\beta} \equiv (\mathrm{d}\boldsymbol{x}_{(1)} \wedge \mathrm{d}\boldsymbol{x}_{(2)})^{\alpha\beta}$. It's straightforward to show from Eq. (14.69) and the antisymmetry of $S^{\alpha\beta}$ that

$$\Delta A^\mu = -\frac{1}{2} A_0^\alpha \left[\left(\partial_\sigma \Gamma^\mu_{\alpha\beta} + \Gamma^\mu_{\gamma\sigma} \Gamma^\gamma_{\alpha\beta} \right) - \left(\partial_\beta \Gamma^\mu_{\alpha\sigma} + \Gamma^\mu_{\gamma\beta} \Gamma^\gamma_{\alpha\sigma} \right) \right]_p S^{\sigma\beta} = -\frac{1}{2} R^\mu{}_{\alpha\sigma\beta} A_0^\alpha S^{\sigma\beta} , \tag{14.70}$$

where we've used Eq. (14.61). The quantity $S^{\sigma\beta}$ is the projection of the area enclosed by the curve onto the plane spanned by ξ^σ and ξ^β.

Example. Let $x = r\cos\theta$ and $y = r\sin\theta$; then

$$\frac{1}{2} \oint (x dy - y dx) = \frac{1}{2} \int_0^{2\pi} d\theta r^2 (\cos^2\theta + \sin^2\theta) = \pi r^2 .$$

---

[29]Neumann series occur in the solution of Fredholm integral equations of the second kind. Such equations occur frequently in physics, notably in quantum and electromagnetic scattering.

We now have two interpretations of the Riemann tensor. From Eq. (14.60) it's related to the commutator of the covariant derivative. The non-commutativity of derivatives is related to the fact that the coordinate basis vectors in an arbitrary coordinate system change along different coordinate curves in a way that they have no correlation with each other. Equation (14.70) provides a more qualitative picture: The change in a vector parallel transported around a small closed curve is proportional to the area enclosed by the curve and to the starting value of the vector, $\Delta A \propto [A_0 \times (\text{area})]$. The proportionality factor is the Riemann tensor.

14.4.3 Geodesic deviation

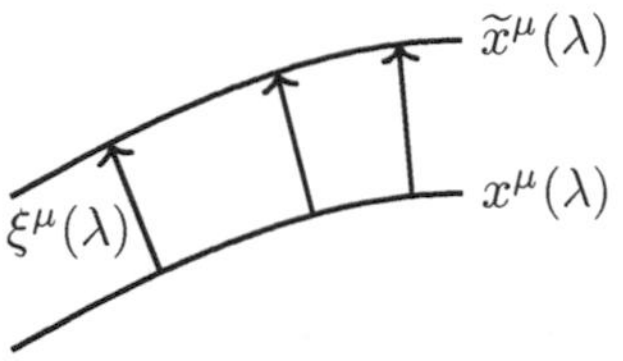

Figure 14.9 Deviation vector $\xi^\mu(\lambda)$ between two nearby geodesic curves.

There is a third manifestation of curvature that can be quantified in terms of the Riemann tensor, the rate at which neighboring geodesic curves either diverge from or approach each other. Consider a geodesic curve $x^\mu(\lambda)$. Let there be another, nearby geodesic $\widetilde{x}^\mu(\lambda)$, also parameterized by λ. Let the separation between the curves be given by the vector $\xi^\mu(\lambda) \equiv \widetilde{x}^\mu(\lambda) - x^\mu(\lambda)$; see Fig. 14.9. We wish to obtain a differential equation for $\xi^\mu(\lambda)$. Both $\widetilde{x}^\mu$ and x^μ satisfy Eq. (14.51),

$$\frac{\mathrm{d}^2 x^\mu}{\mathrm{d}\lambda^2} + \Gamma^\mu_{\alpha\beta}\frac{\mathrm{d}x^\alpha}{\mathrm{d}\lambda}\frac{\mathrm{d}x^\beta}{\mathrm{d}\lambda} = 0 \qquad \frac{\mathrm{d}^2 \widetilde{x}^\mu}{\mathrm{d}\lambda^2} + \widetilde{\Gamma}^\mu_{\alpha\beta}\frac{\mathrm{d}\widetilde{x}^\alpha}{\mathrm{d}\lambda}\frac{\mathrm{d}\widetilde{x}^\beta}{\mathrm{d}\lambda} = 0\,, \tag{14.71}$$

where $\widetilde{\Gamma}^\mu_{\alpha\beta}$ is obtained from the coordinates $\widetilde{x}^\mu(\lambda)$. To obtain an equation for $\xi^\mu(\lambda)$, subtract the equations in (14.71), substituting $\widetilde{x}^\mu = x^\mu + \xi^\mu$, and, as in Eq. (14.66), take to first order in ξ^α, $\widetilde{\Gamma}^\mu_{\alpha\beta} \approx \Gamma^\mu_{\alpha\beta} + (\partial_\sigma \Gamma^\mu_{\alpha\beta})\xi^\sigma$. To first order in ξ we find:

$$\frac{\mathrm{d}^2 \xi^\mu}{\mathrm{d}\lambda^2} + \Gamma^\mu_{\alpha\beta}\left[\frac{\mathrm{d}x^\alpha}{\mathrm{d}\lambda}\frac{\mathrm{d}\xi^\beta}{\mathrm{d}\lambda} + \frac{\mathrm{d}\xi^\alpha}{\mathrm{d}\lambda}\frac{\mathrm{d}x^\beta}{\mathrm{d}\lambda}\right] + \left(\partial_\sigma \Gamma^\mu_{\alpha\beta}\right)\xi^\sigma \frac{\mathrm{d}x^\alpha}{\mathrm{d}\lambda}\frac{\mathrm{d}x^\beta}{\mathrm{d}\lambda} = 0\,. \tag{14.72}$$

We will be aided in the interpretation of Eq. (14.72) if we stop and define the *second covariant directional derivative.*

The second covariant directional derivative is, using Eq. (14.45),

$$\begin{aligned}\frac{\mathrm{D}^2 A^\mu}{\mathrm{d}\lambda^2} &= \frac{\mathrm{D}}{\mathrm{d}\lambda}\left(\frac{\mathrm{D}A^\mu}{\mathrm{d}\lambda}\right) = \frac{\mathrm{d}}{\mathrm{d}\lambda}\left(\frac{\mathrm{D}A^\mu}{\mathrm{d}\lambda}\right) + \frac{\mathrm{d}x^\alpha}{\mathrm{d}\lambda}\Gamma^\mu_{\alpha\rho}\frac{\mathrm{D}A^\rho}{\mathrm{d}\lambda} \\ &= \frac{\mathrm{d}}{\mathrm{d}\lambda}\left(\frac{\mathrm{d}A^\mu}{\mathrm{d}\lambda} + \frac{\mathrm{d}x^\beta}{\mathrm{d}\lambda}\Gamma^\mu_{\beta\sigma}A^\sigma\right) + \frac{\mathrm{d}x^\alpha}{\mathrm{d}\lambda}\Gamma^\mu_{\alpha\rho}\left(\frac{\mathrm{d}A^\rho}{\mathrm{d}\lambda} + \frac{\mathrm{d}x^\gamma}{\mathrm{d}\lambda}\Gamma^\rho_{\gamma\psi}A^\psi\right)\,.\end{aligned}$$

Equation (14.72) can be written in terms of $\mathrm{D}^2/\mathrm{d}\lambda^2$ by adding and subtracting the appropriate terms. For example, let $\ddot{\xi}^\mu \to (\mathrm{d}/\mathrm{d}\lambda)(\dot{\xi}^\mu + \dot{x}^\beta \Gamma^\mu_{\beta\sigma}\xi^\sigma) - (\mathrm{d}/\mathrm{d}\lambda)(\dot{x}^\beta \Gamma^\mu_{\beta\sigma}\xi^\sigma)$, where the dot indicates a derivative with respect to λ. When this is done, we obtain

$$\frac{\mathrm{D}^2}{\mathrm{d}\lambda^2}\xi^\mu + R^\mu{}_{\beta\sigma\alpha}\xi^\sigma \frac{\mathrm{d}x^\alpha}{\mathrm{d}\lambda}\frac{\mathrm{d}x^\beta}{\mathrm{d}\lambda} = 0\,. \tag{14.73}$$

Equation (14.73) is the equation of *geodesic deviation.* It describes how geodesics deviate from each other (either approaching or separating), according to the curvature described by the Riemann tensor. Because free particles follow geodesics, Eq. (14.73) describes the relative *acceleration* of two free particles on a curved manifold, an acceleration due entirely to the geometry. In Section 11.3.2, we described the Newtonian deviation between particles under the influence of tidal forces, Eq. (11.7). In the Newtonian paradigm, accelerations are caused by forces; here we see that the relative acceleration of particles in free fall is associated with the geometry.

14.4.4 Normal coordinates: Local flatness

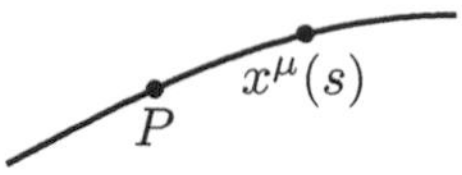

Figure 14.10 Point $x^\mu(s)$ on geodesic curve where s is the arc length from P.

Consider a point P on a geodesic curve. Let another point on the curve have coordinates $x^\mu(s)$, where s is the arc length from P; see Fig. 14.10. Form the Taylor expansion of $x^\mu(s)$,

$$x^\mu(s) = x^\mu_P + s\dot{x}^\mu + \frac{1}{2}s^2\ddot{x}^\mu + \frac{1}{6}s^3\dddot{x}^\mu + O(s^4)\,, \tag{14.74}$$

where the dot indicates a derivative with respect to s, evaluated at P. Because $x^\mu(s)$ is on the geodesic, we have from Eq. (14.51) $\ddot{x}^\mu = -\Gamma^\mu_{\alpha\beta}\dot{x}^\alpha\dot{x}^\beta$. Evaluate the third derivative as the derivative of the second derivative:

$$\begin{aligned}\dddot{x}^\mu &= -\frac{\mathrm{d}}{\mathrm{d}s}(\Gamma^\mu_{\alpha\beta}\dot{x}^\alpha\dot{x}^\beta) = -\Gamma^\mu_{\alpha\beta}(\dot{x}^\alpha\ddot{x}^\beta + \ddot{x}^\alpha\dot{x}^\beta) - \dot{x}^\alpha\dot{x}^\beta\partial_\sigma(\Gamma^\mu_{\alpha\beta})\dot{x}^\sigma \\ &= \left[2\Gamma^\mu_{\alpha\rho}\Gamma^\rho_{\beta\sigma} - \partial_\sigma\Gamma^\mu_{\alpha\beta}\right]\dot{x}^\sigma\dot{x}^\alpha\dot{x}^\beta \equiv S^\mu_{\alpha\beta\sigma}\dot{x}^\sigma\dot{x}^\alpha\dot{x}^\beta\,,\end{aligned}$$

where we've used $\ddot{x}^\mu = -\Gamma^\mu_{\alpha\beta}\dot{x}^\alpha\dot{x}^\beta$ and $\mathrm{d}(\Gamma^\mu_{\alpha\beta})/\mathrm{d}s = \partial_\sigma(\Gamma^\mu_{\alpha\beta})\dot{x}^\sigma$. Substitute back into Eq. (14.74), and we have the Taylor expansion *along the geodesic*,

$$x^\mu(s) = x^\mu_P + s\dot{x}^\mu - \frac{1}{2}s^2\Gamma^\mu_{\alpha\beta}\dot{x}^\alpha\dot{x}^\beta + \frac{1}{6}s^3 S^\mu_{\alpha\beta\sigma}\dot{x}^\alpha\dot{x}^\beta\dot{x}^\sigma + O(s^4)\,. \tag{14.75}$$

Now define a new coordinate,

$$y^\mu(s) \equiv x^\mu(s) - x^\mu_P + \frac{1}{2}s^2\Gamma^\mu_{\alpha\beta}\dot{x}^\alpha\dot{x}^\beta - \frac{1}{6}s^3 S^\mu_{\alpha\beta\sigma}\dot{x}^\alpha\dot{x}^\beta\dot{x}^\sigma\,. \tag{14.76}$$

By combining Eqs. (14.75) and (14.76), the new coordinates, the *normal coordinates* (also called *Riemannian coordinates*), are *linear* in s along the geodesic,

$$y^\mu(s) = s\dot{x}^\mu + O(s^4)\,, \tag{14.77}$$

where $\dot{x}^\mu$ is a constant, the tangent vector that singles out the geodesic passing through x^μ_P.

In terms of these coordinates, *geodesics are straight lines* (for small s); $\mathrm{d}^2y^\mu/\mathrm{d}s^2 = O(s^2)$ from Eq. (14.77). As we now show, Christoffel symbols evaluated in normal coordinates *vanish* if the $\{\Gamma^\alpha_{\mu\nu}\}$ are symmetric. From Eq. (14.76), we have to lowest order

$$y^\mu = x^\mu - x^\mu_P + \frac{1}{2}\Gamma^\mu_{\alpha\beta}(x^\alpha - x^\alpha_P)(x^\beta - x^\beta_P)\,,$$

where we've approximated $\dot{x} \approx (x - x_P)/s$. The transformation properties between the coordinates x^μ and y^μ are described by

$$A^\alpha_\beta = \frac{\partial x^\alpha}{\partial y^\beta}\Big|_P = \delta^\alpha_\beta \qquad \overline{A}^\alpha_\beta = \frac{\partial y^\alpha}{\partial x^\beta}\Big|_P = \delta^\alpha_\beta \qquad \partial_\lambda\overline{A}^\alpha_\beta\big|_P = \frac{1}{2}\left(\Gamma^\alpha_{\lambda\beta} + \Gamma^\alpha_{\beta\lambda}\right)\big|_P\,;$$

see Eqs. (5.24) and (5.25). Using Eq. (14.12), the Christoffel symbols $\overline{\Gamma}^{\lambda}_{\mu\nu}$ in normal coordinates have the value

$$\overline{\Gamma}^{\lambda}_{\mu\nu} = \delta^{\lambda}_{\beta}\delta^{\alpha}_{\mu}\delta^{\tau}_{\nu}\Gamma^{\beta}_{\alpha\tau} - \frac{1}{2}\delta^{\alpha}_{\mu}\delta^{\tau}_{\nu}\left(\Gamma^{\lambda}_{\alpha\tau} + \Gamma^{\lambda}_{\tau\alpha}\right) = \frac{1}{2}\left(\Gamma^{\lambda}_{\mu\nu} - \Gamma^{\lambda}_{\nu\mu}\right) = \Gamma^{\lambda}_{[\mu\nu]} \,. \tag{14.78}$$

Through any point on a manifold passes a unique geodesic curve with a given tangent vector, and now we see that a coordinate system exists at a point on a geodesic where the connection coefficients vanish if they are symmetric—Eq. (14.78). This result is significant for two reasons. From a practical standpoint, because we have found a local frame in which all the Christoffel symbols vanish, any results proved in that frame involving tensors are valid in any other frame. From the physics perspective, the existence of normal coordinates provides the mathematical justification for the picture presented by the equivalence principle that spacetime is locally flat (re-read the argument motivating Eq. (14.37)). The converse of the Christoffel symbols vanishing at a point is that the first derivatives of the metric vanish at that point—see Eq. (14.24). In the neighborhood of this point, the manifold is *locally flat* and we have a *local inertial frame*, where SR holds sway. Locally the spacetime manifold looks like Minkowski space. At any point P, $g_{\mu\nu}|_P = \eta_{\mu\nu}$.

14.4.5 Symmetries of the Riemann tensor

From Eq. (14.61), $R^{\alpha}{}_{\rho\mu\nu}$ is antisymmetric in the last two indices,

$$R^{\alpha}{}_{\rho\mu\nu} = -R^{\alpha}{}_{\rho\nu\mu} \,. \tag{14.79}$$

Thus, the Riemann tensor vanishes if any two of the latter pair of indices (μ, ν) are equal. For example, $R^{\alpha}{}_{\rho 11} = 0$. If we lower the first index, $R_{\alpha\rho\mu\nu} = g_{\alpha\beta}R^{\beta}{}_{\rho\mu\nu}$, it turns out that the tensor is antisymmetric in the first two indices as well,

$$R_{\alpha\rho\mu\nu} = -R_{\rho\alpha\mu\nu} \,. \tag{14.80}$$

The Riemann tensor vanishes if any two of the first pair of indices are equal, $R_{11\mu\nu} = 0$. Equation (14.79) is manifestly obvious but Eq. (14.80) is not. Start from a scalar field $\phi = A_\beta B^\beta = g_{\beta\alpha}A^\alpha B^\beta$ and apply Eq. (14.60),

$$\begin{aligned}
[\nabla_\mu, \nabla_\nu]\,\phi &= [\nabla_\mu, \nabla_\nu]\, g_{\alpha\beta}A^\alpha B^\beta = g_{\alpha\beta}\,[\nabla_\mu, \nabla_\nu]\, A^\alpha B^\beta \\
&= g_{\alpha\beta}\left(A^\alpha\,[\nabla_\mu, \nabla_\nu]\, B^\beta + ([\nabla_\mu, \nabla_\nu]\, A^\alpha)B^\beta\right) \\
&= g_{\alpha\beta}\left(A^\alpha R^{\beta}{}_{\rho\mu\nu}B^\rho + R^{\alpha}{}_{\rho\mu\nu}A^\rho B^\beta\right) = R_{\alpha\rho\mu\nu}A^\alpha B^\rho + R_{\beta\rho\mu\nu}A^\rho B^\beta \\
0 &= \left(R_{\alpha\rho\mu\nu} + R_{\rho\alpha\mu\nu}\right) A^\alpha B^\rho \,,
\end{aligned} \tag{14.81}$$

where we've taken $g_{\alpha\beta}$ through the commutator ($\nabla_\rho g_{\alpha\beta} = 0$). The left side of Eq. (14.81) vanishes by the no-torsion property. Because A^α and B^ρ are arbitrary, Eq. (14.80) holds. A consequence of Eqs. (14.79) and (14.80) is that $R_{\alpha\beta\gamma\delta} = R_{\beta\alpha\delta\gamma}$. A third symmetry relation is (see Exercise 14.13):

$$R_{\alpha\beta\gamma\delta} = R_{\gamma\delta\alpha\beta} \,. \tag{14.82}$$

The final symmetry is the *cyclic relation* among the lower indices:

$$R^{\alpha}{}_{\rho\mu\nu} + R^{\alpha}{}_{\mu\nu\rho} + R^{\alpha}{}_{\nu\rho\mu} = 0 \,. \tag{14.83}$$

This relation follows from Eq. (14.61) under a cyclic permutation of the three lower indices. Equation (14.83) holds if the first index is lowered, $R_{\alpha\rho\mu\nu} + R_{\alpha\mu\nu\rho} + R_{\alpha\nu\rho\mu} = 0$.

There are n^4 elements of $R^{\alpha}{}_{\rho\mu\nu}$ on an n-dimensional space. The number of *independent* elements, however, is reduced by the symmetries of the Riemann tensor. As we now show, there are

$n^2(n^2-1)/12$ independent elements. Let $m \equiv \binom{n}{2} = n(n-1)/2$. Antisymmetry in the first and second pairs of indices, Eqs. (14.79) and (14.80), would imply that only m^2 elements of the Riemann tensor can be chosen independently. That number, however, is reduced because there are $\binom{m}{2} = m(m-1)/2$ ways of satisfying Eq. (14.82) included in m^2. Additionally, Eq. (14.83) provides $\binom{n}{4}$ equations of constraint. The number of independent components is

$$\binom{n}{2}^2 - \binom{m}{2} - \binom{n}{4} = \frac{1}{12}n^2(n^2-1)\,. \tag{14.84}$$

Table 14.2 summarizes the number of independent elements versus the dimension of the space.

Table 14.2 Number of independent elements of the Riemann tensor.

Dimension	n^4	$n^2(n^2-1)/12$
2	16	1
3	81	6
4	256	20

Example. Riemann tensor on a 2-sphere. By Eq. (14.84) there is only one independent element, $R_{\theta\phi\theta\phi}$; all other nonzero terms are related to $R_{\theta\phi\theta\phi}$ by symmetry. From Eq. (14.61),

$$\begin{aligned} R^{\theta}{}_{\phi\theta\phi} &= \frac{\partial}{\partial\theta}\Gamma^{\theta}_{\phi\phi} + \Gamma^{\theta}_{\theta\beta}\Gamma^{\beta}_{\phi\phi} - \frac{\partial}{\partial\phi}\Gamma^{\theta}_{\theta\phi} - \Gamma^{\theta}_{\phi\beta}\Gamma^{\beta}_{\theta\phi} \\ &= \frac{\partial}{\partial\theta}\Gamma^{\theta}_{\phi\phi} - \frac{\partial}{\partial\phi}\Gamma^{\theta}_{\theta\phi} + \Gamma^{\theta}_{\theta\theta}\Gamma^{\theta}_{\phi\phi} + \Gamma^{\theta}_{\theta\phi}\Gamma^{\phi}_{\phi\phi} - \Gamma^{\theta}_{\phi\theta}\Gamma^{\theta}_{\theta\phi} - \Gamma^{\theta}_{\phi\phi}\Gamma^{\phi}_{\theta\phi}\,. \end{aligned}$$

By the results of Exercise 14.4,

$$R^{\theta}{}_{\phi\theta\phi} = \frac{\partial}{\partial\theta}\Gamma^{\theta}_{\phi\phi} - \Gamma^{\theta}_{\phi\phi}\Gamma^{\phi}_{\theta\phi} = \frac{\partial}{\partial\theta}(-\sin\theta\cos\theta) - (-\sin\theta\cos\theta)\cot\theta = \sin^2\theta\,.$$

Lowering the index, $R_{\theta\phi\theta\phi} = g_{\theta\beta}R^{\beta}{}_{\phi\theta\phi} = g_{\theta\theta}R^{\theta}{}_{\phi\theta\phi} + g_{\theta\phi}R^{\phi}{}_{\phi\theta\phi} = R^{\theta}{}_{\phi\theta\phi} = \sin^2\theta$. The nonzero elements are thus given by $R_{\theta\phi\theta\phi} = -R_{\theta\phi\phi\theta} = R_{\phi\theta\phi\theta} = -R_{\phi\theta\theta\phi}$. If we wanted to find $R^{\phi}{}_{\theta\phi\theta}$ we would raise an index

$$R^{\phi}{}_{\theta\phi\theta} = g^{\phi\alpha}R_{\alpha\theta\phi\theta} = g^{\phi\phi}R_{\phi\theta\phi\theta} = \frac{1}{\sin^2\theta}\sin^2\theta = 1\,.$$

Note that $R^{\phi}{}_{\theta\phi\theta} \neq R_{\phi\theta\phi\theta}$. The sphere is not flat—the Riemann tensor does not vanish.

14.4.6 The Bianchi identity

We now prove a relation that turns out to be of crucial importance in GR, the *Bianchi identity*:

$$\nabla_\lambda R_{\alpha\beta\gamma\delta} + \nabla_\delta R_{\alpha\beta\lambda\gamma} + \nabla_\gamma R_{\alpha\beta\delta\lambda} = 0\,, \tag{14.85}$$

which involves a cyclic permutation among the indices that label derivatives, $(\lambda, \gamma, \delta)$. Equation (14.85) is sometimes referred to as the *second Bianchi identity*, with the first Bianchi identity being the cyclic relation, Eq. (14.83). While Eq. (14.83) follows trivially from the definition of the Riemann tensor, proving Eq. (14.85) is considerably more involved.

We first prove a relation involving the *third covariant derivative*. Start with Eq. (14.60) and lower the index,[30]

$$[\nabla_\mu, \nabla_\nu]T_\alpha = R_{\alpha\rho\mu\nu}T^\rho \, . \tag{14.86}$$

Let $T_\alpha = A_{\alpha\beta}v^\beta$, where $A_{\alpha\beta}$ is a second-rank tensor and v^β is a vector; the contravariant version is $T^\alpha = A^\alpha{}_\beta v^\beta$. Substitute into Eq. (14.86): $[\nabla_\mu, \nabla_\nu](A_{\alpha\beta}v^\beta) = R_{\alpha\rho\mu\nu}A^\rho{}_\beta v^\beta$, or, equivalently,

$$A_{\alpha\beta}[\nabla_\mu, \nabla_\nu]v^\beta + ([\nabla_\mu, \nabla_\nu]A_{\alpha\beta})v^\beta = R_{\alpha\rho\mu\nu}A^\rho{}_\beta v^\beta \, . \tag{14.87}$$

Make use of Eq. (14.60) in the first term in Eq. (14.87): $A_{\alpha\beta}R^\beta{}_{\rho\mu\nu}v^\rho + ([\nabla_\mu, \nabla_\nu]A_{\alpha\beta})v^\beta = R_{\alpha\rho\mu\nu}A^\rho{}_\beta v^\beta$. By relabeling indices we have $([\nabla_\mu, \nabla_\nu]A_{\alpha\beta})v^\beta = R_{\alpha\rho\mu\nu}A^\rho{}_\beta v^\beta - A_{\alpha\lambda}R^\lambda{}_{\beta\mu\nu}v^\beta$. Because this equation applies for all v^β,

$$[\nabla_\mu, \nabla_\nu]A_{\alpha\beta} = R_{\alpha\rho\mu\nu}A^\rho{}_\beta - R^\lambda{}_{\beta\mu\nu}A_{\alpha\lambda} \, . \tag{14.88}$$

Equation (14.88) defines the action of the commutator on a type $(0, 2)$ tensor. Now set $A_{\alpha\beta} = \nabla_\beta A_\alpha$ in Eq. (14.88); the desired result is

$$[\nabla_\mu, \nabla_\nu]\nabla_\beta A_\alpha = R_{\alpha\rho\mu\nu}\nabla_\beta A^\rho - R^\lambda{}_{\beta\mu\nu}\nabla_\lambda A_\alpha \, . \tag{14.89}$$

Take the covariant derivative of Eq. (14.60),

$$\nabla_\lambda([\nabla_\mu, \nabla_\nu]A_\alpha) = \nabla_\lambda(R_{\alpha\rho\mu\nu}A^\rho) = R_{\alpha\rho\mu\nu}\nabla_\lambda A^\rho + (\nabla_\lambda R_{\alpha\rho\mu\nu})A^\rho \, . \tag{14.90}$$

Permute the indices (λ, μ, ν) in Eq. (14.90) cyclically and add the equations. The left-hand side (lhs) of the resulting equation will involve a sum of commutators of the form $\nabla_\lambda[\nabla_\mu, \nabla_\nu]A_\alpha$. These terms can be converted into a sum of commutators of the form $[\nabla_\mu, \nabla_\nu]\nabla_\lambda A_\alpha$ through the following identity among commutators:

$$\nabla_\lambda[\nabla_\mu, \nabla_\nu] + \nabla_\mu[\nabla_\nu, \nabla_\lambda] + \nabla_\nu[\nabla_\lambda, \nabla_\mu] = [\nabla_\mu, \nabla_\nu]\nabla_\lambda + [\nabla_\nu, \nabla_\lambda]\nabla_\mu + [\nabla_\lambda, \nabla_\mu]\nabla_\nu \, . \tag{14.91}$$

Apply Eq. (14.89) to the terms in Eq. (14.91). The lhs is then given by

$$\text{lhs} = [R_{\alpha\rho\mu\nu}\nabla_\lambda + R_{\alpha\rho\nu\lambda}\nabla_\mu + R_{\alpha\rho\lambda\mu}\nabla_\nu]\, A^\rho - \left(R^\rho{}_{\lambda\mu\nu} + R^\rho{}_{\mu\nu\lambda} + R^\rho{}_{\nu\lambda\mu}\right)\nabla_\rho A_\alpha \, . \tag{14.92}$$

The terms in parentheses in Eq. (14.92) vanish by Eq. (14.83). The right side is given by

$$\text{rhs} = [R_{\alpha\rho\mu\nu}\nabla_\lambda + R_{\alpha\rho\nu\lambda}\nabla_\mu + R_{\alpha\rho\lambda\mu}\nabla_\nu]\, A^\rho + (\nabla_\lambda R_{\alpha\rho\mu\nu} + \nabla_\mu R_{\alpha\rho\nu\lambda} + \nabla_\nu R_{\alpha\rho\lambda\mu})\, A^\rho \, . \tag{14.93}$$

After canceling common terms between Eqs. (14.92) and (14.93), we are left with the Bianchi identity, Eq. (14.85). Note that Eq. (14.85) is a relation among *five* indices—lots of room for mischief.

14.5 THE RICCI TENSOR AND SCALAR FIELD

14.5.1 The Ricci tensor

The *Ricci tensor* is a type $(0, 2)$ tensor obtained from a contraction (the *trace*) of the Riemann tensor. The contraction $R^\alpha{}_{\alpha\mu\nu} = 0$: Starting from $R^\alpha{}_{\rho\mu\nu} = g^{\alpha\sigma}R_{\sigma\rho\mu\nu}$, it follows that $R^\alpha{}_{\alpha\mu\nu} = g^{\rho\sigma}R_{\sigma\rho\mu\nu} = -g^{\sigma\rho}R_{\rho\sigma\mu\nu} = -R^\sigma{}_{\sigma\mu\nu}$, and hence $R^\alpha{}_{\alpha\mu\nu} = 0$, where we've used Eq. (14.80). Because $R^\alpha{}_{\rho\mu\alpha} = -R^\alpha{}_{\rho\alpha\mu}$, Eq. (14.79), *there is only one independent contraction of the Riemann*

[30] We can lower the index because of metric compatibility.

tensor. We define the *Ricci tensor* as the contraction of the first and third indices of the Riemann tensor,[31]

$$R_{\mu\nu} \equiv R^{\rho}{}_{\mu\rho\nu} = g^{\rho\sigma} R_{\sigma\mu\rho\nu} \,. \tag{14.94}$$

The Ricci tensor is symmetric. Use Eq. (14.83): $R^{\alpha}{}_{\rho\mu\nu} + R^{\alpha}{}_{\mu\nu\rho} + R^{\alpha}{}_{\nu\rho\mu} = 0$. Set $\alpha = \rho$ and sum,

$$R^{\rho}{}_{\rho\mu\nu} + R^{\rho}{}_{\mu\nu\rho} + R^{\rho}{}_{\nu\rho\mu} = 0 \,. \tag{14.95}$$

The first term vanishes, and for the second, $R^{\rho}{}_{\mu\nu\rho} = -R^{\rho}{}_{\mu\rho\nu} = -R_{\mu\nu}$; thus, Eq. (14.95) establishes that $R_{\nu\mu} = R_{\mu\nu}$. There are $n(n+1)/2$ independent elements for $R_{\mu\nu}$ on an n-dimensional space.

14.5.2 The curvature scalar

The *curvature scalar* is the scalar field obtained from the trace of the Ricci tensor,

$$R \equiv R^{\alpha}{}_{\alpha} = g^{\alpha\mu} g^{\rho\sigma} R_{\sigma\mu\rho\alpha} = g^{\alpha\mu} R_{\alpha\mu} \,. \tag{14.96}$$

Yes, the same symbol R is used to denote the Riemann tensor, the Ricci tensor, and the curvature scalar. The meaning should be clear from the number of indices. The quantity R is a scalar *field*—not in general a number—and is a complicated expression involving first and second derivatives of the metric tensor in such a way as to obtain a scalar. See Exercise 14.16.

Example. The three independent elements of the Ricci tensor on the unit sphere are, contracting over the elements of the Riemann tensor (see example on page 264):

$$\begin{aligned}
R_{\theta\theta} &= R^{\rho}{}_{\theta\rho\theta} = R^{\theta}{}_{\theta\theta\theta} + R^{\phi}{}_{\theta\phi\theta} = g^{\phi\phi} R_{\phi\theta\phi\theta} = 1 \\
R_{\theta\phi} &= R^{\rho}{}_{\theta\rho\phi} = R^{\theta}{}_{\theta\theta\phi} + R^{\phi}{}_{\theta\phi\phi} = 0 \\
R_{\phi\phi} &= R^{\rho}{}_{\phi\rho\phi} = R^{\theta}{}_{\phi\theta\phi} + R^{\phi}{}_{\phi\phi\phi} = g^{\theta\theta} R_{\theta\phi\theta\phi} = \sin^2\theta \,.
\end{aligned}$$

To obtain the curvature scalar, we must raise an index, $R^{\alpha}{}_{\beta} = g^{\alpha\rho} R_{\rho\beta}$,

$$\begin{aligned}
R^{\theta}{}_{\theta} &= g^{\theta\rho} R_{\rho\theta} = g^{\theta\theta} R_{\theta\theta} + g^{\theta\phi} R_{\phi\theta} = 1 \\
R^{\phi}{}_{\theta} &= g^{\phi\theta} R_{\theta\theta} + g^{\phi\phi} R_{\phi\theta} = 0 \\
R^{\phi}{}_{\phi} &= g^{\phi\theta} R_{\theta\phi} + g^{\phi\phi} R_{\phi\phi} = 1 \,.
\end{aligned}$$

The curvature scalar is $R = R^{\theta}{}_{\theta} + R^{\phi}{}_{\phi} = 2$. If we had kept the radius of the sphere r in our calculations, we would have obtained $R = 2/r^2$.

14.5.3 The Weyl tensor

On an n-dimensional manifold, the *traceless part* of the Ricci tensor is defined as

$$S_{\mu\nu} \equiv R_{\mu\nu} - \frac{1}{n} g_{\mu\nu} R \,,$$

where $R = R^{\alpha}{}_{\alpha}$ is the Ricci scalar. It is said to be traceless, because, as is easily shown,[32] $S^{\alpha}{}_{\alpha} = 0$. We could therefore write $R_{\mu\nu} = n^{-1} g_{\mu\nu} R + S_{\mu\nu}$; we've split off its traceless part. $S_{\mu\nu}$ is symmetric.

[31]The Ricci tensor is defined in some books as the contraction of the first and last indices of the Riemann tensor, introducing another source of minus sign ambiguity, in addition to that of the metric tensor in SR. Beware.

[32]The trick is to note that $g^{\mu}_{\nu} = \delta^{\mu}_{\nu}$ and thus $g^{\alpha}_{\alpha} = n$.

The *Ricci decomposition* is a way of expressing the Riemann tensor valid for $n \geq 3$:

$$R_{\alpha\beta\gamma\delta} = -\frac{2}{(n-2)(n-1)} R g_{\alpha[\gamma} g_{\delta]\beta} + \frac{2}{(n-2)} \left(g_{\alpha[\gamma} R_{\delta]\beta} - g_{\beta[\gamma} R_{\delta]\alpha} \right) + C_{\alpha\beta\gamma\delta} \tag{14.97}$$

where the "remainder" $C_{\alpha\beta\gamma\delta}$ is the *Weyl curvature tensor*.[60, p90] The Weyl tensor has the same symmetries as the Riemann tensor: Eqs. (14.79), (14.80), (14.82), and (14.83).

As can be verified using Eq. (14.97), all contractions of $C_{\alpha\beta\gamma\delta}$ vanish, $C^{\alpha}{}_{\alpha\gamma\delta} = C^{\alpha}{}_{\beta\alpha\delta} = C^{\alpha}{}_{\beta\gamma\alpha} = 0$. The Weyl tensor is thus the traceless part of the Riemann tensor. Einstein's equation involves (as we'll see) the Ricci tensor—the trace of the Riemann tensor. It's convenient to remove from the Riemann tensor its traceless part, the Weyl tensor. *No part of the Weyl tensor is contained in the Ricci tensor.* The Riemann tensor has the property that $R^{\alpha}{}_{\alpha\gamma\delta} = 0$ (page 265), and thus the *additional* contraction properties of the Weyl tensor ($C^{\alpha}{}_{\beta\alpha\delta} = C^{\alpha}{}_{\beta\gamma\alpha} = 0$) imply that it has fewer independent elements than the Riemann tensor. From the antisymmetry property of the last two indices, $C^{\alpha}{}_{\beta\gamma\alpha} = 0$ is not independent of $C^{\alpha}{}_{\beta\alpha\delta} = 0$, and thus, to count the extra number of equations of constraint, it suffices to consider $C^{\alpha}{}_{\beta\alpha\delta} = 0$. The quantity $C^{\alpha}{}_{\beta\alpha\delta}$ is symmetric in (β, δ) (because $R_{\beta\delta}$ is symmetric). Thus, there are $n(n+1)/2$ equations of constraint implied by $C^{\alpha}{}_{\beta\alpha\delta} = 0$ not accounted for by Eq. (14.84). The Weyl tensor has, using Eq. (14.84),

$$\frac{1}{12}n^2(n^2-1) - \frac{1}{2}n(n+1) = \frac{1}{12}n(n+1)(n+2)(n-3)$$

independent elements. For $n = 3$, there are zero independent elements, and thus *the Weyl tensor vanishes in three dimensions*. For $n = 4$, there are 10 independent elements of the Weyl tensor. If $R_{\mu\nu} = 0$, then $R_{\alpha\beta\gamma\delta} = C_{\alpha\beta\gamma\delta} \neq 0$. The vanishing of the Ricci tensor ("Ricci flat") is not sufficient to say that a space is flat—the Riemman tensor need not be zero. In Einstein's equation, $R_{\alpha\beta}$ and R are determined by the energy-momentum in spacetime; the Weyl tensor therefore represents a part of the Riemann tensor that's unconstrained by the physics of spacetime.

14.6 THE EINSTEIN TENSOR

We now define the *Einstein curvature tensor*, the importance of which will not be apparent until we consider GR:

$$G_{\mu\nu} \equiv R_{\mu\nu} - \frac{1}{2} g_{\mu\nu} R \, . \tag{14.98}$$

Because $R_{\mu\nu}$ is symmetric, so is $G_{\mu\nu}$. A key feature of $G_{\mu\nu}$ is that its covariant divergence vanishes,

$$\nabla_\nu G^{\mu\nu} = 0 \, . \tag{14.99}$$

Equation (14.99) is crucially important for GR, and it's worthwhile to derive it. Start from the Bianchi identity, Eq. (14.85),

$$\nabla_\nu R^{\alpha\beta}{}_{\gamma\mu} + \nabla_\gamma R^{\alpha\beta}{}_{\mu\nu} + \nabla_\mu R^{\alpha\beta}{}_{\nu\gamma} = 0 \, , \tag{14.100}$$

where we've raised the two indices that are not permuted. Set $\alpha = \gamma$ and sum:

$$\nabla_\nu R^{\gamma\beta}{}_{\gamma\mu} + \nabla_\gamma R^{\gamma\beta}{}_{\mu\nu} + \nabla_\mu R^{\gamma\beta}{}_{\nu\gamma} = 0 \, .$$

In the first term, we recognize the Ricci tensor, $R^{\gamma\beta}{}_{\gamma\mu} = R^{\beta}{}_{\mu}$, and in the last term we have $R^{\gamma\beta}{}_{\nu\gamma} = -R^{\gamma\beta}{}_{\gamma\nu} = -R^{\beta}{}_{\nu}$. Thus, $\nabla_\nu R^{\beta}{}_{\mu} + \nabla_\gamma R^{\gamma\beta}{}_{\mu\nu} - \nabla_\mu R^{\beta}{}_{\nu} = 0$. Now set $\beta = \nu$ and sum:

$$\nabla_\nu R^{\nu}{}_{\mu} + \nabla_\gamma R^{\gamma\nu}{}_{\mu\nu} - \nabla_\mu R^{\nu}{}_{\nu} = 0 \, .$$

In the middle term, interchange the indices in both sets of indices, $R^{\gamma\nu}{}_{\mu\nu} = R^{\nu\gamma}{}_{\nu\mu} = R^{\gamma}{}_{\mu}$, and in the final term we recognize the curvature scalar $R^{\nu}{}_{\nu} = R$. Thus,[33]

$$\nabla_\nu R^{\nu}{}_{\mu} + \nabla_\gamma R^{\gamma}{}_{\mu} - \nabla_\mu R = 0\,. \tag{14.101}$$

The first two terms in Eq. (14.101) are identical. Thus, buried in the Bianchi identity is the relation

$$\nabla_\nu \left(R^{\nu}{}_{\mu} - \frac{1}{2}\delta^{\nu}_{\mu} R \right) = \nabla_\nu \left(R^{\nu}{}_{\mu} - \frac{1}{2} g^{\nu}_{\mu} R \right) \equiv \nabla_\nu G^{\nu}{}_{\mu} = 0\,,$$

where we have used $g^{\nu}_{\mu} = \boldsymbol{e}^{\nu} \cdot \boldsymbol{e}_{\mu} = \delta^{\nu}_{\mu}$. Raise the index μ and we have Eq. (14.99). As a symmetric tensor, $G_{\mu\nu}$ would have $n(n+1)/2$ independent components, but because of the zero-divergence condition, that number is reduced by n, so that $G_{\mu\nu}$ has $n(n-1)/2$ independent components. Table 14.3 summarizes the number of independent elements in the curvature tensors we've defined, the Riemann, Ricci, Weyl, and Einstein tensors.

Table 14.3 Number of independent elements in the curvature tensors.

n	$R^{\alpha}{}_{\beta\gamma\delta}$	$R_{\mu\nu}$	$C_{\alpha\beta\gamma\delta}$	$G_{\mu\nu}$
2	1	3	Not defined	1
3	6	6	0	3
4	20	10	10	6

14.7 ISOMETRIES, KILLING VECTORS, AND CONSERVATION LAWS

Symmetry involves 1) a transformation and 2) something invariant under the transformation (Section 4.1.2). Under a diffeomorphism $\phi : M \to M$ of a manifold M to itself (Section 13.3), we can compare a tensor $\mathbf{T}|_{p\in M}$ with the pullback of the tensor field at $\phi(p)$, $\phi^*\mathbf{T}|_{\phi(p)}$ (Section 13.5). If $\phi^*\mathbf{T} = \mathbf{T}$, ϕ is a *symmetry transformation* of the tensor field $\mathbf{T}$. The Lie derivative thus serves as a symmetry detector: A tensor field $\mathbf{T}$ is invariant along the vector field $\boldsymbol{v}$ if $\pounds_{\boldsymbol{v}}\mathbf{T} = 0$.

Of interest are mappings that preserve the metric tensor (for reasons we'll discuss shortly). A diffeomorphism $\phi : M \to M$ is an *isometry* if it carries a metric field $\mathbf{g}$ to itself. If the local group of diffeomorphisms ϕ_t generated by a vector field $\boldsymbol{v}$ is a group of isometries,[34] $\boldsymbol{v}$ is said to be a *Killing field*.[35] For a Killing field $\boldsymbol{v}$, the Lie derivative of the metric tensor is zero,

$$(\pounds_{\boldsymbol{v}}\mathbf{g})_{\alpha\beta} = v^\lambda \nabla_\lambda g_{\alpha\beta} + g_{\alpha\lambda}\nabla_\beta v^\lambda + g_{\beta\lambda}\nabla_\alpha v^\lambda = \nabla_\beta v_\alpha + \nabla_\alpha v_\beta = 0\,,$$

where we've used Eq. (14.29) and metric compatibility. A Killing field $\boldsymbol{v}$ satisfies *Killing's equation*

$$\nabla_\alpha v_\beta + \nabla_\beta v_\alpha = 0\,. \tag{14.102}$$

Equation (14.102) is a system of first-order differential equations for the vectors v^μ that generate isometries.

[33]Note that in the reduction of Eq. (14.100) to (14.101), we've passed from an equation involving five indices to three to one. Every contraction lowers the rank of a tensor by two.

[34]That is, for each t, ϕ_t is an isometry.

[35]Named after the mathematician Wilhelm Killing.

Example. Consider $\mathbb{R}^2$ with Cartesian coordinates, so that $g_{ij} = \delta_{ij}$. From Eq. (14.102) the differential equations for the Killing vectors are $\partial_j v^i + \partial_i v^j = 0$, the solution of which can be written

$$\begin{pmatrix} v^1 \\ v^2 \end{pmatrix} = \begin{pmatrix} a \\ b \end{pmatrix} + \theta \begin{pmatrix} 0 & 1 \\ -1 & 0 \end{pmatrix} \begin{pmatrix} x^1 \\ x^2 \end{pmatrix} ,$$

where (a, b, θ) are constants. The isometries are therefore translations and rotations. Under $x \to x' = x + a + y\theta$, and $y \to y' = y + b - x\theta$, $(\mathrm{d}x')^2 + (\mathrm{d}y')^2 = (\mathrm{d}x)^2 + (\mathrm{d}y)^2$ for infinitesimal θ.

There is a connection between Killing vectors and the Riemann tensor. From Exercise 14.12, $\nabla_\mu \nabla_\nu \xi_\lambda - \nabla_\nu \nabla_\mu \xi_\lambda = -R^\alpha_{\ \lambda\mu\nu} \xi_\alpha$. From Killing's equation, however, $\nabla_\mu \xi_\lambda = -\nabla_\lambda \xi_\mu$, and hence $\nabla_\nu \nabla_\mu \xi_\lambda = -\nabla_\nu \nabla_\lambda \xi_\mu$. We thus have specifically for Killing vectors, $\nabla_\mu \nabla_\nu \xi_\lambda + \nabla_\nu \nabla_\lambda \xi_\mu = -R^\alpha_{\ \lambda\mu\nu} \xi_\alpha$. Permute the indices $(\mu\nu\lambda)$ cyclically. Add the $(\mu\nu\lambda)$ equation to the $(\nu\lambda\mu)$ equation and subtract from it the $(\lambda\mu\nu)$ equation. The result is $2\nabla_\nu \nabla_\lambda \xi_\mu = -(R^\alpha_{\ \lambda\mu\nu} + R^\alpha_{\ \mu\nu\lambda} - R^\alpha_{\ \nu\lambda\mu})\xi_\alpha$. Use Eq. (14.83) to conclude that, for a Killing vector

$$\nabla_\nu \nabla_\lambda \xi_\mu = R^\alpha_{\ \nu\lambda\mu} \xi_\alpha \, . \tag{14.103}$$

Second derivatives of Killing vectors thus occur as linear combinations of Killing vectors. From uniqueness theorems of second-order differential equations, Killing fields are completely determined by their values at a point, $\xi^\mu|_p$, and by the values of their first derivatives at a point, $L_{\mu\nu} \equiv \nabla_\mu \xi_\nu|_p$. If we're given $(\xi^\mu, L_{\mu\nu})$ at $p \in M$, then their value at any other point is determined by integrating Killing's equation. From Eq. (14.102) $L_{\mu\nu} = -L_{\nu\mu}$, and thus on an n-dimensional manifold there are $n + \binom{n}{2} = n(n+1)/2$ initial values $(\xi^\mu, L_{\mu\nu})$, and, accordingly, a maximum of $n(n+1)/2$ linearly independent Killing vectors. Manifolds that possess the maximum number of Killing vectors are called *maximally symmetric spaces.*

Example. Isometries of Minkowski space
The Killing equation in Minkowski space is $\partial_\mu \xi_\nu + \partial_\nu \xi_\mu = 0$, the solutions of which are $\xi_\alpha = \epsilon_\alpha + \omega_{\alpha\beta} x^\beta$, where the ϵ_α are constants and $\omega_{\alpha\beta}$ is an antisymmetric matrix with constant elements. There are 10 independent Killing fields ξ_α specified by the four parameters ϵ_α and the six independent elements of $\omega_{\alpha\beta}$, the maximum number in four dimensions—Minkowski space is a maximally symmetric space. The four parameters ϵ_α specify translations and the remaining six specify Lorentz transformations, three for rotations and three for boosts; Section 6.2. The Lorentz metric is thus preserved under translations and Lorentz transformations. The matrix $\omega_{\alpha\beta}$ is the generator of Lorentz transformations, Eq. (6.32).

As discussed in Chapter 9, for every continuous symmetry of the action integral there is a corresponding conservation law. We now take this to a deeper level: The isometries of the spacetime manifold have physical significance in that they imply the existence of conserved quantities. For ξ^α a Killing vector field and γ a geodesic curve with tangent field u^β, then $\xi_\alpha u^\alpha$ is constant along γ. The proof is simple: $u^\beta \nabla_\beta (\xi_\alpha u^\alpha) = u^\beta u^\alpha \nabla_\beta \xi_\alpha + \xi_\alpha u^\beta \nabla_\beta u^\alpha = 0$. The second term vanishes by Eq. (14.50) (geodesic), and the first term vanishes by Killing's equation—$u^{(\alpha} u^{\beta)} \nabla_{[\beta} \xi_{\alpha]} = 0$. The quantity $\xi_\alpha u^\alpha$ is thus constant along a geodesic. Because freely falling particles and photons travel on geodesics, *there is a conserved quantity for every Killing vector field.* For Minkowski space with its 10 isometries, there are four spacetime translations corresponding to conservation of four-momentum, three rotations corresponding to conservation of angular momentum, and three boosts, which are tied to the relativistic center of mass theorem (all shown in Section 7.6).

14.8 MAXIMALLY SYMMETRIC SPACES

A manifold may not have the maximum number of Killing vectors. From Exercise 14.20,

$$[\nabla_\mu, \nabla_\nu]\nabla_\lambda \xi_\rho = -R^\sigma{}_{\lambda\mu\nu}\nabla_\sigma\xi_\rho - R^\sigma{}_{\rho\mu\nu}\nabla_\lambda\xi_\sigma \,. \tag{14.104}$$

Equation (14.104) is valid for general tensors $\nabla_\lambda\xi_\rho$, and is not specific to Killing vectors. Expand out the left side of Eq. (14.104),

$$\nabla_\mu\nabla_\nu\nabla_\lambda\xi_\rho - \nabla_\nu\nabla_\mu\nabla_\lambda\xi_\rho = -R^\sigma{}_{\lambda\mu\nu}\nabla_\sigma\xi_\rho - R^\sigma{}_{\rho\mu\nu}\nabla_\lambda\xi_\sigma \,. \tag{14.105}$$

Now substitute Eq. (14.103) into Eq. (14.105), and we have a result specific to Killing vectors

$$\nabla_\mu\left(R^\alpha{}_{\nu\lambda\rho}\xi_\alpha\right) - \nabla_\nu\left(R^\alpha{}_{\mu\lambda\rho}\xi_\alpha\right) = -R^\sigma{}_{\lambda\mu\nu}\nabla_\sigma\xi_\rho - R^\sigma{}_{\rho\mu\nu}\nabla_\lambda\xi_\sigma \,. \tag{14.106}$$

Expand out the derivatives on the left side of Eq. (14.106), and it's straightforward to show

$$\left(\nabla_\mu R^\alpha{}_{\nu\lambda\rho} - \nabla_\nu R^\alpha{}_{\mu\lambda\rho}\right)\xi_\alpha = \left[-R^\alpha{}_{\nu\lambda\rho}\delta^\beta_\mu + R^\alpha{}_{\mu\lambda\rho}\delta^\beta_\nu + R^\alpha{}_{\lambda\mu\nu}\delta^\beta_\rho - R^\alpha{}_{\rho\mu\nu}\delta^\beta_\lambda\right]\nabla_\beta\xi_\alpha \,. \tag{14.107}$$

There are thus metric-dependent restrictions on ξ_α and $\nabla_\beta\xi_\alpha$ that would prevent a given manifold from possessing the maximum number of independent Killing vectors.

Under what conditions does a manifold have the maximum number of Killing vectors? For such spaces, we want there to be no restrictions on the values of ξ_α and $\nabla_\lambda\xi_\alpha$. This will be the case if the left side of Eq. (14.107) vanishes,

$$\nabla_\mu R^\alpha{}_{\nu\lambda\rho} = \nabla_\nu R^\alpha{}_{\mu\lambda\rho} \,. \tag{14.108}$$

This condition would permit any value of ξ_α. The right side of Eq. (14.107) must then also vanish, which we can write in the form $\Delta^{\alpha\beta}{}_{\nu\lambda\rho\mu}\nabla_\beta\xi_\alpha = 0$. Because $\nabla_\beta\xi_\alpha$ is antisymmetric (Killing's equation), the vanishing of the right side of Eq. (14.107) requires that $\Delta^{[\alpha\beta]}{}_{\nu\lambda\rho\mu} = 0$. Thus, the terms in square brackets in Eq. (14.107) must be symmetric in (α, β):

$$-R^\alpha{}_{\nu\lambda\rho}\delta^\beta_\mu + R^\alpha{}_{\mu\lambda\rho}\delta^\beta_\nu + R^\alpha{}_{\lambda\mu\nu}\delta^\beta_\rho - R^\alpha{}_{\rho\mu\nu}\delta^\beta_\lambda = -R^\beta{}_{\nu\lambda\rho}\delta^\alpha_\mu + R^\beta{}_{\mu\lambda\rho}\delta^\alpha_\nu + R^\beta{}_{\lambda\mu\nu}\delta^\alpha_\rho - R^\beta{}_{\rho\mu\nu}\delta^\alpha_\lambda \,. \tag{14.109}$$

Set $\beta = \mu$ in Eq. (14.109) and sum; we find

$$-nR^\alpha{}_{\nu\lambda\rho} + R^\alpha{}_{\nu\lambda\rho} + R^\alpha{}_{\lambda\rho\nu} - R^\alpha{}_{\rho\lambda\nu} = -R^\alpha{}_{\nu\lambda\rho} + R^\mu{}_{\mu\lambda\rho}\delta^\alpha_\nu + R^\mu{}_{\lambda\mu\nu}\delta^\alpha_\rho - R^\mu{}_{\rho\mu\nu}\delta^\alpha_\lambda \,. \tag{14.110}$$

On the left of Eq. (14.110) we can set $R^\alpha{}_{\rho\lambda\nu} = -R^\alpha{}_{\rho\nu\lambda}$, in which case we can use the cyclic relation to eliminate the three terms with cyclic permutations of $(\nu\lambda\rho)$. On the right side, we can set $R^\mu{}_{\mu\lambda\rho} = 0$ (Section 14.5.1), and the other terms we recognize as the Ricci tensor. Thus, we obtain a requirement on the curvature tensor for a maximally symmetric space,

$$(n-1)R^\alpha{}_{\nu\lambda\rho} = R_{\nu\rho}\delta^\alpha_\lambda - R_{\nu\lambda}\delta^\alpha_\rho \,. \tag{14.111}$$

Lower α and use the fact that $\delta^\beta_\sigma = g^\beta_\sigma$; Eq. (14.111) is equivalent to

$$(n-1)R_{\alpha\nu\lambda\rho} = R_{\nu\rho}g_{\alpha\lambda} - R_{\nu\lambda}g_{\alpha\rho} \,. \tag{14.112}$$

The Riemann tensor is antisymmetric in the first two indices. From Eq. (14.112) we therefore require $R_{\nu\rho}g_{\alpha\lambda} - R_{\nu\lambda}g_{\alpha\rho} = -R_{\alpha\rho}g_{\nu\lambda} + R_{\alpha\lambda}g_{\nu\rho}$. Now raise ν on this equation, $R^\nu{}_\rho g_{\alpha\lambda} - R^\nu{}_\lambda g_{\alpha\rho} = -R_{\alpha\rho}\delta^\nu_\lambda + R_{\alpha\lambda}\delta^\nu_\rho$. Contract over ν and λ, and we have for the Ricci tensor of a maximally symmetric space

$$nR_{\alpha\rho} = Rg_{\alpha\rho} \,, \tag{14.113}$$

where $R = R^{\nu}{}_{\nu}$. Combining Eqs. (14.113) and (14.112), we have

$$R_{\alpha\nu\lambda\rho} = \frac{R}{n(n-1)} \left(g_{\alpha\lambda} g_{\nu\rho} - g_{\alpha\rho} g_{\nu\lambda} \right) . \tag{14.114}$$

Equation (14.114) is used in the relativistic theory of cosmology.

What can we say about the curvature scalar, R? Combining Eqs. (14.99) (secretly the Bianchi identity) and (14.113), we have for each μ:

$$\nabla_{\nu} \left(R^{\mu\nu} - \frac{1}{2} g^{\mu\nu} R \right) = \left(\frac{1}{n} - \frac{1}{2} \right) g^{\mu\nu} \nabla_{\nu} R = \left(\frac{1}{n} - \frac{1}{2} \right) \partial^{\mu} R = 0 , \tag{14.115}$$

where we've used metric compatibility and that the action of ∇_{ν} on a scalar field is the same as ∂_{ν}. For $n \geq 3$, we conclude from Eq. (14.115) that $R =$ constant, i.e., *a maximally symmetric space in three or more dimensions is one of constant, homogeneous curvature.*

SUMMARY

Chapter 14 is a foundational mathematical chapter, along with Chapters 5 and 13. In Chapter 5 we introduced tensors on flat spaces, in Chapter 13 we put tensors on manifolds, and in this chapter we considered derivatives of tensors and how they relate to curvature.

- **Covariant derivative:** The covariant derivative ∇_{μ} transforms as a tensor and reduces to ∂_{μ} on a flat space. ∇_{μ} gives the rate of change of a tensor in the coordinate direction x^{μ} taking into account the change in tensor components due to changes in basis vectors in an arbitrary coordinate system. We discussed ∇_{μ} in two ways, algebraic and geometric.
  - Equation (14.9) is an algebraic definition: $\nabla_{\mu} T^{\lambda} = \partial_{\mu} T^{\lambda} + \Gamma^{\lambda}_{\mu\nu} T^{\nu}$, where the connection coefficients $\{\Gamma^{\lambda}_{\mu\nu}\}$ transform according to Eq. (14.12); they are not tensors. Equation (14.9) combines two non-tensorial objects in such a way as to transform as a tensor. Standard GR is restricted to connection coefficients symmetric in the lower indices, otherwise spacetime could not be locally flat, contrary to the prediction of the equivalence principle. The Christoffel symbols $\Gamma^{\lambda}_{\mu\nu} = \frac{1}{2} g^{\lambda\rho} (\partial_{\mu} g_{\nu\rho} + \partial_{\nu} g_{\mu\rho} - \partial_{\rho} g_{\mu\nu})$ are a unique set of connection coefficients symmetric in the lower indices and compatible with the metric tensor, $\nabla_{\rho} g_{\mu\nu} = 0$.
  - The connection coefficients connect the basis vectors of the tangent spaces at nearby points of the manifold, $T_x(M)$ and $T_{x+\mathrm{d}x}(M)$. The change $\boldsymbol{e}_{\alpha} \to \boldsymbol{e}_{\alpha} + \mathrm{d}\boldsymbol{e}_{\alpha}$ under $x^{\beta} \to x^{\beta} + \mathrm{d}x^{\beta}$ is $\mathrm{d}\boldsymbol{e}_{\alpha} = \Gamma^{\mu}_{\alpha\beta} \mathrm{d}x^{\beta} \boldsymbol{e}_{\mu}$. The covariant derivative captures the total change in a vector: $\mathrm{d}(A^{\alpha} \boldsymbol{e}_{\alpha}) = \mathrm{d}A^{\alpha} \boldsymbol{e}_{\alpha} + A^{\alpha} \mathrm{d}\boldsymbol{e}_{\alpha} \equiv (\nabla_{\beta} A^{\alpha}) \mathrm{d}x^{\beta} \boldsymbol{e}_{\alpha}$. Along a parametrized curve $x^{\mu}(\lambda)$, $\mathrm{d}x^{\beta} = (\mathrm{d}x^{\beta}/\mathrm{d}\lambda)\mathrm{d}\lambda \equiv u^{\beta} \mathrm{d}\lambda$, where u^{β} is the tangent vector to the curve. The directional covariant derivative is $\mathrm{D}A^{\mu}/\mathrm{d}\lambda = u^{\alpha} \nabla_{\alpha} A^{\mu}$. Parallel transport is the requirement that $\mathrm{D}A^{\mu} = u^{\alpha} \nabla_{\alpha} A^{\mu} \mathrm{d}\lambda = 0$ along a curve. Under parallel transport, the change in vector components due to the change in basis vectors is $\delta A^{\mu} \equiv -\Gamma^{\mu}_{\alpha\beta} A^{\alpha} \mathrm{d}x^{\beta}$. This permits a geometric understanding of ∇_{μ} provided by Eq. (14.48). Parallel transport is path dependent (unless the space is flat): The vector that results from parallel transport along a given path depends on the path taken.
- **Geodesic curves:** A geodesic is the straightest possible curve on a manifold, defined so that the tangent to the curve is parallel transported along itself, $u^{\alpha} \nabla_{\alpha} u^{\mu} = 0$. The geodesic equation $\mathrm{d}^2 x^{\mu} + \Gamma^{\mu}_{\alpha\beta} \mathrm{d}x^{\alpha} \mathrm{d}x^{\beta} = 0$ is a system of coupled second-order differential equations. Geodesics extremize the distance between fixed points. Free particles follow geodesics. The geodesic equation is the Euler-Lagrange equation for the free-particle Lagrangian $L =$

$\sqrt{g_{\mu\nu}\dot{x}^\mu\dot{x}^\nu}$. An affine parameterization, $\tau = a\lambda + b$, where (a,b) are constants, leaves the geodesic equation invariant. There is a unique geodesic through any point on a manifold with a given tangent vector.

- **Curvature:** The extent to which covariant derivatives do not commute provides an algebraic specification of the Riemann curvature tensor, $[\nabla_\mu, \nabla_\nu]T^\alpha \equiv R^\alpha{}_{\rho\mu\nu}T^\rho$, where $R^\alpha{}_{\rho\mu\nu}$ is given in Eq. (14.61). A manifold is flat if $R^\alpha{}_{\rho\mu\nu} = 0$ everywhere, or equivalently if the components of the metric tensor are all constant. A geometric interpretation of curvature is the extent to which a vector parallel transported around a close curve changes, $\Delta A^\mu = -\frac{1}{2}R^\mu{}_{\alpha\sigma\beta}A_0^\alpha S^{\sigma\beta}$, where A_0^α is the initial value of the vector and $S^{\sigma\beta}$ is an antisymmetric tensor that contains the area of the loop. Another manifestation of curvature is the extent to which two nearby geodesic curves, separated by the vector ξ^μ, either approach or diverge from each other as specified by the geodesic deviation equation, $\mathrm{D}^2\xi^\mu + R^\mu{}_{\beta\sigma\alpha}\xi^\sigma \mathrm{d}x^\alpha \mathrm{d}x^\beta = 0$. Nearby free particles either approach or diverge from each other, giving rise to an apparent acceleration that's caused not by forces, but by the geometry of spacetime.

- **Ricci tensor and scalar field:** The Ricci tensor $R_{\mu\nu} = R^\rho{}_{\mu\rho\nu}$ is the contraction of the first and third indices of the Riemann tensor. The Ricci tensor is symmetric, $R_{\mu\nu} = R_{\nu\mu}$. The curvature scalar field is the contraction over the Ricci tensor, $R = R^\alpha{}_\alpha = g^{\alpha\beta}R_{\alpha\beta}$.

- **Einstein tensor:** The Einstein tensor $G_{\mu\nu} = R_{\mu\nu} - \frac{1}{2}g_{\mu\nu}R$ is a symmetric tensor and has zero covariant divergence, $\nabla_\nu G^{\mu\nu} = 0$. This property is a consequence of the Bianchi identity.

- **Killing vector fields:** A Killing field $\boldsymbol{\xi}$ is a vector field along which the Lie derivative of the metric tensor vanishes, $\pounds_{\boldsymbol{\xi}} g_{\mu\nu} = 0$. The components of the Killing field obey Killing's equation $\nabla_\mu\xi_\nu + \nabla_\nu\xi_\mu = 0$. There is a conserved quantity for every Killing field. The maximum number of Killing vectors on an n-dimensional manifold is $n(n+1)/2$. Maximally symmetric spaces (maximum number of Killing fields) are characterized by constant curvature, R.

EXERCISES

14.1 We showed for a torsion-free covariant derivative that $\pounds_{\boldsymbol{u}} v^\lambda = u^\beta \nabla_\beta v^\lambda - v^\beta \nabla_\beta u^\lambda$, Eq. (14.27). We also showed, however, that $\pounds_{\boldsymbol{u}} v^\lambda = u^\beta\partial_\beta v^\lambda - v^\beta \partial_\beta u^\lambda$, Eq. (13.20). Show how these equations are consistent under the no-torsion proviso. Hint: What is $u^{[\beta}v^{\alpha]}\Gamma^\lambda_{\beta\alpha}$?

14.2 Show that $\nabla_\mu \omega_\nu$ transforms as a type $(0,2)$ tensor. Use $\partial_\alpha(A^\rho_{\lambda'}A^{\lambda'}_\tau) = 0$ and hence that $A^\rho_{\lambda'}\partial_\alpha A^{\lambda'}_\tau = -A^{\lambda'}_\tau \partial_\alpha A^\rho_{\lambda'}$.

14.3 Show that the torsion tensor $T^\lambda_{\mu\nu} \equiv \Gamma^\lambda_{\mu\nu} - \Gamma^\lambda_{\nu\mu}$ is indeed a tensor. Use Eq. (14.12).

14.4 Derive the Christoffel symbols on the surface of a sphere. Show that

$$\begin{aligned} &\Gamma^\theta_{\theta\theta} = \Gamma^\theta_{\theta\phi} = \Gamma^\theta_{\phi\theta} = 0 \qquad && \Gamma^\theta_{\phi\phi} = -\sin\theta\cos\theta \\ &\Gamma^\phi_{\theta\theta} = \Gamma^\phi_{\phi\phi} = 0 && \Gamma^\phi_{\theta\phi} = \Gamma^\phi_{\phi\theta} = \cot\theta\,. \end{aligned}$$

14.5 Show explicitly that $\nabla_\sigma g_{\mu\nu} = 0$ on the surface of the sphere. Use Eq. (14.14) and the results from Exercise 14.4. For example, show that $\nabla_\theta g_{\phi\phi} = 0$. Repeat for all the indices.

14.6 Show that for the determinant of the metric tensor g, $\nabla_\mu g = 0$. The quantity g is a scalar density (Section 5.2), not a scalar; it's a mistake to assume $\nabla_\mu g = \partial_\mu g$.

a. The direct way is to use the property of determinants that g is an alternating sum of n-tuple products of elements of the metric tensor (Section 5.8.2); apply ∇_μ to this definition of g and use $\nabla_\mu g_{\alpha\beta} = 0$ for all (α, β).

b. Another way is to make use of the covariant derivative of tensor densities, not covered in this chapter, but which is presented here. For a tensor density of weight w of any index structure, $\mathfrak{T}^{\alpha_1\alpha_2\cdots}_{\beta_1\beta_2\cdots}$, we can write $\mathfrak{T}^{\alpha_1\cdots}_{\beta_1\cdots} = \sqrt{g}^{w}\sqrt{g}^{-w}\mathfrak{T}^{\alpha_1\cdots}_{\beta_1\cdots}$. Thus,

$$\nabla_\mu \mathfrak{T}^{\alpha_1\cdots}_{\beta_1\cdots} = \nabla_\mu\left(\sqrt{g}^{w}\sqrt{g}^{-w}\mathfrak{T}^{\alpha_1\cdots}_{\beta_1\cdots}\right) = \left(\nabla_\mu \sqrt{g}^{w}\right)\sqrt{g}^{-w}\mathfrak{T}^{\alpha_1\cdots}_{\beta_1\cdots} + \sqrt{g}^{w}\nabla_\mu\left(\sqrt{g}^{-w}\mathfrak{T}^{\alpha_1\cdots}_{\beta_1\cdots}\right).$$

The first term vanishes by the result of the first part of this exercise. Make use of the fact that $\sqrt{g}^{-w}\mathfrak{T}^{\alpha_1\cdots}_{\beta_1\cdots}$ transforms as an absolute tensor (Section 5.2) and apply the rule for the covariant derivative of a general tensor, Eq. (14.14). Show that the covariant derivative of a tensor density of weight w is

$$\nabla_\mu \mathfrak{T}^{\alpha_1\cdots}_{\beta_1\cdots} = \Big(\text{usual terms if } \mathfrak{T}^{\alpha_1\cdots}_{\beta_1\cdots} \text{ were a tensor, Eq.(14.14)}\Big) - w\Gamma^{\alpha}_{\mu\alpha}\mathfrak{T}^{\alpha_1\cdots}_{\beta_1\cdots}. \quad \text{(P14.1)}$$

Use $\partial_\mu g = 2g\Gamma^{\alpha}_{\mu\alpha}$, Section 14.1.8. Use Eq. (P14.1) to show that $\nabla_\mu g = 0$. What's the weight w of g?

14.7 Using the result of Exercise 14.6, show that:

a. $\partial_\mu\left(\sqrt{-g}g^{\alpha\beta}\right) = \sqrt{-g}\left(\Gamma^{\sigma}_{\mu\sigma}g^{\alpha\beta} - \Gamma^{\alpha}_{\mu\sigma}g^{\sigma\beta} - \Gamma^{\beta}_{\mu\sigma}g^{\alpha\sigma}\right)$. Hint: $\nabla_\mu\left(\sqrt{-g}g^{\alpha\beta}\right) = 0$.

b. For $\mathfrak{T}^\alpha$ a tensor density of weight $w = 1$, $\nabla_\alpha \mathfrak{T}^\alpha = \partial_\alpha \mathfrak{T}^\alpha$.

14.8 Derive the Christoffel symbols for the rotating coordinate system discussed in Section 12.3.1. You will need to work out the inverse metric, $g^{\mu\nu}$. You should find that there are eight nonzero Christoffel symbols for this system. Show that:

$$\Gamma^{\rho}_{tt} = -\frac{\omega^2}{c^2}\rho \qquad \Gamma^{\rho}_{t\phi} = \Gamma^{\rho}_{\phi t} = -\frac{\omega}{c}\rho \qquad \Gamma^{\rho}_{\phi\phi} = -\rho$$

$$\Gamma^{\phi}_{t\rho} = \Gamma^{\phi}_{\rho t} = \frac{\omega}{\rho c} \qquad \Gamma^{\phi}_{\rho\phi} = \Gamma^{\phi}_{\phi\rho} = \frac{1}{\rho}.$$

14.9 Show that the condition for parallel transport of a covariant vector is given by $\mathrm{d}A_\mu = \Gamma^{\rho}_{\mu\sigma}A_\rho \mathrm{d}x^\sigma$.

14.10 Show that the analog of Eq. (14.70) for a covariant vector is $\Delta A_\mu = \frac{1}{2}A_{\lambda,0}R^{\lambda}{}_{\mu\beta\sigma}S^{\beta\sigma}$.

14.11 Can geodesic curves cross? Integral curves (Section 13.3) cannot intersect. Are geodesic curves, integral curves?

14.12 Show that $[\nabla_\mu, \nabla_\nu]T_\lambda = -R^{\alpha}{}_{\lambda\mu\nu}T_\alpha$. Note how this equation differs from Eq. (14.86).

14.13 Show that $R_{\alpha\beta\gamma\delta} = R_{\gamma\delta\alpha\beta}$. Write the cyclic relation Eq. (14.83) for each of the four values of the first index. Now add these equations and use the symmetries already established.

14.14 Verify Eq. (14.91). This is an identity involving commutators—you don't have to invoke any properties of the covariant derivative.

14.15 Show that the Riemann tensor can be written in terms of derivatives of the metric tensor,

$$R_{\mu\nu\rho\sigma} = \frac{1}{2}\left[\partial_\nu\partial_\rho g_{\mu\sigma} - \partial_\mu\partial_\rho g_{\nu\sigma} - \partial_\nu\partial_\sigma g_{\mu\rho} + \partial_\mu\partial_\sigma g_{\nu\rho}\right] + g_{\alpha\beta}\left[\Gamma^{\alpha}_{\nu\rho}\Gamma^{\beta}_{\mu\sigma} - \Gamma^{\alpha}_{\mu\rho}\Gamma^{\beta}_{\nu\sigma}\right].$$

You may find it useful to first show that

$$g_{\mu\alpha}\partial_\rho\Gamma^\alpha_{\nu\sigma} = \frac{1}{2}\partial_\rho(\partial_\nu g_{\mu\sigma} + \partial_\sigma g_{\mu\nu} - \partial_\mu g_{\nu\sigma}) - \Gamma^\alpha_{\nu\sigma}\partial_\rho g_{\mu\alpha}\,.$$

Then use Eq. (14.24).

14.16 Show from the result of Exercise 14.15 that the Ricci scalar is given by

$$R = \partial^\sigma\partial^\mu g_{\mu\sigma} - g^{\sigma\nu}\partial^\rho\partial_\rho g_{\nu\sigma} + g^{\sigma\nu}g^{\rho\mu}g_{\alpha\beta}\left[\Gamma^\alpha_{\nu\rho}\Gamma^\beta_{\mu\sigma} - \Gamma^\alpha_{\mu\rho}\Gamma^\beta_{\nu\sigma}\right]$$

From the complexity of this result, we should not underestimate the richness of the Ricci scalar field!

14.17 Show that the "covariant curl" $\nabla_\alpha A_\beta - \nabla_\beta A_\alpha = \partial_\alpha A_\beta - \partial_\beta A_\alpha$. Use Eq. (14.13).

14.18 Show that the "cyclical divergence," where the indices are permuted is a tensor, i.e., show for $F_{\alpha\beta}$ an antisymmetric tensor that $\nabla_\alpha F_{\beta\gamma} + \nabla_\beta F_{\gamma\alpha} + \nabla_\gamma F_{\alpha\beta} = \partial_\alpha F_{\beta\gamma} + \partial_\beta F_{\gamma\alpha} + \partial_\gamma F_{\alpha\beta}$. Use Eq. (14.14). The homogeneous Maxwell equations have this property.

14.19 Suppose that the only symmetry property of the Riemann tensor was Eq. (14.79), $R_{\alpha\beta\gamma\delta} = -R_{\alpha\beta\delta\gamma}$ (which would be the case if the connection coefficients were not symmetric). How many independent elements would the Riemann tensor have? Answer: $n^3(n-1)/2$, which in four dimensions is 96.

14.20 Show that $[\nabla_\rho, \nabla_\sigma]T_{\mu\nu} = -R^\lambda{}_{\mu\rho\sigma}T_{\lambda\nu} - R^\lambda{}_{\nu\rho\sigma}T_{\mu\lambda}$. This is an instance of a more general result known as the *Ricci identity*,

$$[\nabla_\mu, \nabla_\nu]\,T^{\alpha_1\cdots\alpha_p}_{\beta_1\cdots\beta_q} = \sum_{r=1}^{p} R^{\alpha_r}{}_{\sigma\mu\nu}T^{\alpha_1\cdots\alpha_{r-1}\sigma\alpha_{r+1}\cdots\alpha_p}_{\beta_1\cdots\beta_q} - \sum_{k=1}^{q} R^\sigma{}_{\beta_k\mu\nu}T^{\alpha_1\cdots\alpha_p}_{\beta_1\cdots\beta_{k-1}\sigma\beta_{k+1}\cdots\beta_p}\,.$$

Equation (14.60) is a special case of Ricci's identity.

14.21 The equation of a great circle in polar coordinates is $\cot\theta = \beta\sin(\phi-\alpha)$, where α, β are constants. Using this equation, compute $\ddot\theta$. Show that

$$\ddot\theta - \sin\theta\cos\theta\dot\phi^2 + \beta\cos(\phi-\alpha)\sin^2\theta\left(\ddot\phi + 2\cot\theta\dot\phi\dot\theta\right) = 0\,.$$

Argue that for this equation to be satisfied for all α and β, the geodesic equations on a sphere must be satisfied.

14.22 Show that

$$\nabla_\nu T^{\mu\nu} = \frac{1}{\sqrt{g}}\partial_\nu(\sqrt{g}T^{\mu\nu}) + \Gamma^\mu_{\nu\lambda}T^{\lambda\nu}\,.$$

Is there a "Gauss's law" for $T^{\mu\nu}$ for a curved manifold? What if $T^{\mu\nu}$ is antisymmetric?

14.23 Show that the Killing equations on the unit sphere are given by

$$\partial_\theta\xi^\theta = 0 \qquad \partial_\phi\xi^\theta + \sin^2\theta\,\partial_\theta\xi^\phi = 0 \qquad \partial_\phi\xi^\phi + \cot\theta\,\xi^\theta = 0\,.$$

Use the results of Exercise 14.4. Show that the solutions to these equations are

$$\xi^\theta = A\sin(\phi+\phi_0) \qquad \xi^\phi = A\cot\theta\cos(\phi+\phi_0) + B\,,$$

where A, B, and ϕ_0 are constants. How many linearly independent Killing vectors are there in this case? Is this a maximally symmetric space? Find values of A, B, and ϕ_0 such that the components of the three killing vectors are given by

$$\xi^{(1)} = (\sin\phi, \cot\theta\cos\phi) \qquad \xi^{(2)} = (\cos\phi, -\cot\theta\sin\phi) \qquad \xi^{(3)} = (0,1)\,.$$

CHAPTER 15

General relativity

15.1 INTRODUCTION

OUR exposition has brought together several theories of space, time, and motion. In pre-relativistic physics space and time are absolute, existing independently of the contents of the universe, and of each other. An instance of "now" is the same for every observer, at any point in the universe. Space has the manifold structure of $\mathbb{R}^3$. Coordinates assigned to points x^i are arbitrary, but the distance between neighboring points, $(\mathrm{d}s)^2 = a_{ij}\mathrm{d}x^i\mathrm{d}x^j$—where the metric $a = \text{diag}(1, 1, 1)$, is independent of coordinate system and describes an intrinsic property of space. The Christoffel symbols vanish and the geodesic equation is $\mathrm{d}^2x^i/\mathrm{d}t^2 = 0$. Free particles follow straight lines at constant speed, $x^i = b^i + v^it$ (Newton's first law). The Riemann tensor vanishes, and geodesics do not deviate from each other. Einstein, in SR, modified the Newtonian framework by showing that space and time do not have a separate existence, but together form an absolute spacetime that has the manifold structure of $\mathbb{R}^4$. Coordinates x^μ assigned to events are arbitrary, but the distance between events, $(\mathrm{d}s)^2 = \eta_{\alpha\beta}\mathrm{d}x^\alpha\mathrm{d}x^\beta$ with $\eta = \text{diag}(-1, 1, 1, 1)$, is independent of observer and describes an intrinsic property of spacetime. The metric is constant, but with Lorentz signature. The Christoffel symbols vanish and geodesics are described by $\mathrm{d}^2x^\mu/\mathrm{d}\lambda^2 = 0$. Worldlines of free particles are straight, $x^\mu(\lambda) = a^\mu + b^\mu\lambda$, a geometric feature that all inertial observers agree upon. The Riemann tensor vanishes, geodesics do not deviate, and we can establish a single coordinate system that covers all of spacetime.

In GR, however (as we'll see), Einstein showed that spacetime is not the passive, fixed backdrop that it is in SR, but rather is physical, determined by the presence of matter, energy, and momentum. Spacetime is modeled as a four-dimensional manifold with a metric field $g_{\mu\nu}(x)$ prescribed by the (yet-to-be-established) Einstein field equation. In the framework of GR, gravity is not a force in the usual sense, but is a manifestation of the curvature of spacetime. Whereas in Newtonian mechanics tidal forces cause a relative acceleration between neighboring particles, in GR the same acceleration is accounted for by geodesic deviation on a curved manifold. Just as a fictitious force can be transformed away by changing to an IRF, the force of gravity can be transformed away (locally) by changing to a freely falling reference frame. By the equivalence principle, a freely falling frame is equivalent to an inertial frame in the absence of gravity, and is one in which the laws of SR apply. At any point P of a manifold we can find coordinates in which the Christoffel symbols vanish, $(\Gamma^\lambda_{\mu\nu})_P = 0$ (Section 14.4.4). At P then, $(\partial_\sigma g_{\mu\nu})_P = 0$, Eq. (14.24), and $g_{\mu\nu}|_P = \eta_{\mu\nu}$. Before we write down Einstein's equation for the metric field—and a truly new equation can only be written down, not derived from something more fundamental—we show how gravitational effects can be associated with the spacetime metric tensor.

15.2 WEAK, STATIC GRAVITY

Suppose that in the neighborhood of a point in four-dimensional spacetime, the metric field has the property

$$g_{\mu\nu}(x) = \eta_{\mu\nu} + h_{\mu\nu}(x) \,, \tag{15.1}$$

where $|h_{\mu\nu}(x)| \ll 1$, i.e., the Lorentz metric is *perturbed* by a tensor field $h_{\mu\nu}(x)$. Clearly $h_{\mu\nu} = h_{\nu\mu}$. We show that for a time-independent perturbation and for nonrelativistic speeds, the geodesic equation takes the form of the Newtonian equation of motion in a gravitational field.

Let's write the geodesic equation not in terms of the proper time, as in Eq. (14.51), but in terms of the coordinate time $x^0 = ct$. Under the change of variables $\tau = \tau(t)$, the geodesic equation is, from Eq. (14.53),

$$\frac{\mathrm{d}^2x^\nu}{\mathrm{d}t^2} + \Gamma^\nu_{\alpha\lambda}\frac{\mathrm{d}x^\alpha}{\mathrm{d}t}\frac{\mathrm{d}x^\lambda}{\mathrm{d}t} = -\left(\frac{\mathrm{d}\tau}{\mathrm{d}t}\right)^2\frac{\mathrm{d}^2t}{\mathrm{d}\tau^2}\frac{\mathrm{d}x^\nu}{\mathrm{d}t} = \left(\frac{\mathrm{d}\tau}{\mathrm{d}t}\right)^{-1}\frac{\mathrm{d}^2\tau}{\mathrm{d}t^2}\frac{\mathrm{d}x^\nu}{\mathrm{d}t} \,, \tag{15.2}$$

where we've used the chain rule

$$\frac{\mathrm{d}^2t}{\mathrm{d}\tau^2} = \frac{\mathrm{d}}{\mathrm{d}\tau}\left(\frac{\mathrm{d}t}{\mathrm{d}\tau}\right) = \frac{\mathrm{d}t}{\mathrm{d}\tau}\frac{\mathrm{d}}{\mathrm{d}t}\left(\frac{\mathrm{d}\tau}{\mathrm{d}t}\right)^{-1} = -\left(\frac{\mathrm{d}\tau}{\mathrm{d}t}\right)^{-3}\frac{\mathrm{d}^2\tau}{\mathrm{d}t^2} \,.$$

From $-c^2(\mathrm{d}\tau)^2 = (\mathrm{d}s)^2 = g_{\mu\nu}\mathrm{d}x^\mu\mathrm{d}x^\nu$ and Eq. (15.1),

$$(\mathrm{d}\tau)^2 = -\frac{1}{c^2}g_{\mu\nu}\mathrm{d}x^\mu\mathrm{d}x^\nu = -\frac{1}{c^2}(\eta_{\mu\nu} + h_{\mu\nu})\mathrm{d}x^\mu\mathrm{d}x^\nu \,. \tag{15.3}$$

Expand out the terms in Eq. (15.3) and divide by $(\mathrm{d}t)^2$:

$$\left(\frac{\mathrm{d}\tau}{\mathrm{d}t}\right)^2 = 1 - h_{00} - \frac{2}{c}h_{0i}\frac{\mathrm{d}x^i}{\mathrm{d}t} - \frac{1}{c^2}(\eta_{ij} + h_{ij})\frac{\mathrm{d}x^i}{\mathrm{d}t}\frac{\mathrm{d}x^j}{\mathrm{d}t} \,. \tag{15.4}$$

Assuming nonrelativistic speeds, we have to lowest order in small quantities,

$$\frac{\mathrm{d}\tau}{\mathrm{d}t} \approx \sqrt{1 - h_{00}} \approx 1 - \frac{1}{2}h_{00} \,. \tag{15.5}$$

Using Eq. (15.5),

$$\frac{\mathrm{d}^2\tau}{\mathrm{d}t^2} \approx -\frac{1}{2}c\partial_0 h_{00} \,. \tag{15.6}$$

Using Eqs. (15.5) and (15.6), the term on the right of Eq. (15.2) is to lowest order

$$\left(\frac{\mathrm{d}\tau}{\mathrm{d}t}\right)^{-1}\frac{\mathrm{d}^2\tau}{\mathrm{d}t^2} \approx \frac{-\frac{1}{2}c\partial_0 h_{00}}{1 - \frac{1}{2}h_{00}} \approx -\frac{1}{2}c\partial_0 h_{00} \,. \tag{15.7}$$

The time and space components of Eq. (15.2) are, under the assumptions we've made:

$$\begin{aligned}
\Gamma^0_{00} + \frac{2}{c}\Gamma^0_{0j}\frac{\mathrm{d}x^j}{\mathrm{d}t} + \frac{1}{c^2}\Gamma^0_{ij}\frac{\mathrm{d}x^i}{\mathrm{d}t}\frac{\mathrm{d}x^j}{\mathrm{d}t} &= -\frac{1}{2}\partial_0 h_{00} \\
\frac{1}{c^2}\frac{\mathrm{d}^2x^i}{\mathrm{d}t^2} + \Gamma^i_{00} + \frac{2}{c}\Gamma^i_{0j}\frac{\mathrm{d}x^j}{\mathrm{d}t} + \frac{1}{c^2}\Gamma^i_{jk}\frac{\mathrm{d}x^j}{\mathrm{d}t}\frac{\mathrm{d}x^k}{\mathrm{d}t} &= -\frac{1}{2c}\partial_0 h_{00}\frac{\mathrm{d}x^i}{\mathrm{d}t} \,.
\end{aligned} \tag{15.8}$$

Meanwhile, what about the Christoffel symbols? Substituting Eq. (15.1) into Eq. (14.23),

$$\Gamma^\lambda_{\mu\nu} = \frac{1}{2}\eta^{\lambda\rho}\left(\partial_\mu h_{\rho\nu} + \partial_\nu h_{\rho\mu}\right) - \frac{1}{2}\partial^\lambda h_{\mu\nu} + O(h^2) \,. \tag{15.9}$$

The Christoffel symbols thus start at first order in the perturbation—consistent with what we know must hold: $\Gamma^{\lambda}_{\mu\nu} \to 0$ as $h_{\alpha\beta} \to 0$. From Eq. (15.9) we find $\Gamma^{0}_{00} = -\frac{1}{2}\partial_0 h_{00}$. The time component in Eq. (15.8) thus starts at *second order* in small quantities and can be ignored.[1] From Eq. (15.9),

$$\Gamma^{i}_{00} = \partial_0 h_{i0} - \frac{1}{2}\partial^i h_{00} \qquad \Gamma^{i}_{0j} = \frac{1}{2}\left(\partial_0 h_{ij} + \partial_j h_{i0} - \partial_i h_{0j}\right) . \tag{15.10}$$

We now set $\partial_0 h_{\mu\nu} = 0$; our goal is to consider a time-independent perturbation to the geodesic equation. Using Eq. (15.10), the spatial part in Eq. (15.8) is

$$\frac{1}{c^2}\frac{\mathrm{d}^2 x^i}{\mathrm{d}t^2} = \frac{1}{2}\partial^i h_{00} - \frac{1}{c}\left(\partial_j h_{i0} - \partial_i h_{0j}\right)\frac{\mathrm{d}x^j}{\mathrm{d}t} , \tag{15.11}$$

where we've kept second-order terms on the right of Eq. (15.11); we'll wave them away shortly. These terms look like a velocity-dependent force associated with a rotating coordinate system; we'll see them again when we consider *frame dragging* in Section 18.5. We'll refer to a coordinate system as *nonrotating* if $\partial_j h_{i0} - \partial_i h_{j0} = 0$. Thus, in a nonrotating coordinate system for small velocities in a static field, Eq. (15.11) is

$$\frac{\mathrm{d}^2 x^i}{\mathrm{d}t^2} = \frac{c^2}{2}\partial^i h_{00} . \tag{15.12}$$

Compare Eq. (15.12) with the Newtonian equation of motion, $m_i \boldsymbol{a} = -m_g \boldsymbol{\nabla}\Phi$, where Φ is the gravitational potential. With $m_g = m_i$, $\boldsymbol{a} = -\boldsymbol{\nabla}\Phi$, or $a^i = -\partial^i \Phi$. Consistency with Newtonian gravitation is achieved if we equate[2]

$$h_{00} = -\frac{2}{c^2}\Phi + \text{constant} . \tag{15.13}$$

The constant can be set to zero: h_{00} vanishes as $\Phi \to 0$. We noted previously (Section 1.7.3) that the gravitational potential has the dimension of velocity squared; Eq. (15.13) is consistent with the starting assumption of $|h_{00}|$ small, or $|\Phi| \ll c^2$. Combining Eq. (15.13) with Eq. (15.1),

$$g_{00} \approx -\left(1 + \frac{2}{c^2}\Phi\right) . \tag{15.14}$$

Thus, the geodesic equation, *an equation of motion for free particles* in an arbitrary coordinate system, has the form of the Newtonian equation of motion of a particle in a gravitational field (nominally not a free particle) when the spacetime metric is perturbed from its value in SR.

15.3 THE EINSTEIN FIELD EQUATION

15.3.1 Motivation

Equation (15.14) illustrates a connection between gravitation, a *physical effect*, and the metric tensor, a mathematical concept. It represents an evolution in Einstein's view on the physicality of spacetime. Whereas in SR he held that spacetime is independent (absolute) in its physical properties, having a physical effect, but not in itself influenced by physical conditions, in GR he contends that physical conditions *determine* the spacetime metric. In doing so, he "breathes life" into the metric tensor as a physical quantity. Let's let Einstein speak for himself on how gravity is represented by the spacetime metric tensor:

[1] The metric perturbation $h_{\mu\nu}$ is assumed small, and we're assuming nonrelativistic speeds.

[2] What about the other terms, $h_{\mu\nu}$ for $\mu \neq 0, \nu \neq 0$? This method of analysis can only determine h_{00} in the limit of weak gravity and nonrelativistic speeds—there is but one Newtonian field equation for gravity.

> ··· it follows that the quantities $g_{\sigma\tau}$ are to be regarded from the physical standpoint as the quantities which describe the gravitational field in relation to the chosen system of reference. For, if we now assume the special theory of relativity to apply to a certain four-dimensional region with the coordinates properly chosen, then the $g_{\sigma\tau}$ have the values given by $\eta_{\sigma\tau}$. A free material point then moves, relatively to this system, with uniform motion in a straight line. Then if we introduce new spacetime coordinates x^{μ}, by means of any substitution we choose, the $g_{\sigma\tau}$ in this new system will no longer be constants, but functions of space and time. At the same time the motion of the free material point will present itself in the new coordinates as a curvilinear nonuniform motion, and the law of this motion will be independent of the nature of the moving particle. We shall therefore interpret this motion as a motion under the influence of a gravitational field connected with the spacetime variability of $g_{\sigma\tau}$.[9, p120]

Let's analyze this argument. In IRFs the worldlines of free particles are straight and the metric tensor $g_{\mu\nu} = \eta_{\mu\nu}$ consists of constants. In terms of a new set of coordinates obtained under arbitrary coordinate transformations ("substitutions"), worldlines of the same particles would not be straight, $g_{\mu\nu}$ would contain functions of space and time, yet the description of motion would not depend on inherent features of the particles. The free-fall motion of particles is likewise independent of material composition (Eötvös-type experiments).[3] We can therefore associate nonuniform worldlines of otherwise free particles subject to gravitational fields with the spacetime variability of the metric tensor—basically the equivalence principle argument repeated with arbitrary reference frames instead of elevators. Einstein is saying that what we customarily associate with the manifestations of a gravitational force is indistinguishable from the manifestations of the variability of the spacetime metric field. Gravity is not a force in the usual sense of the word.

The fact of experience revealed by Eötvös-type experiments—free-fall worldlines independent of the nature of the particle, depending only on initial conditions—cries out to be modeled with geodesic curves,[4] and suggests strongly that free-fall motion is controlled by the geometry of the spacetime manifold. What laws might govern the gravitational field, seen as a geometric property of spacetime? Harken back to pre-relativistic gravitation theory: The Newtonian gravitational potential is specified by the Poisson equation, $\nabla^2\Phi = 4\pi G\rho(\boldsymbol{r})$, where $\rho(\boldsymbol{r})$ is the local mass density function (Table 1.1). In regions of space where $\rho(\boldsymbol{r})$ is negligible, Φ satisfies the Laplace equation, $\nabla^2\Phi = 0$. Of course, $\nabla^2\Phi = 0$ does not imply $\Phi = 0$; there are boundary conditions on Φ controlled by nearby masses. Only if $\Phi = 0$ everywhere can we say that gravitation is absent.[5] In a spacetime with $\Phi = 0$ everywhere, free-particle worldlines are specified by $\mathrm{d}^2x^{\mu}/\mathrm{d}\lambda^2 = 0$, and the Riemann tensor vanishes everywhere.[6] Thus, to associate gravitation with spacetime geometry, the correspondence $\Phi \leftrightarrow R_{\alpha\beta\gamma\delta}$ is suggested. What property of the spacetime manifold would the Poisson equation correspond to? Loosely speaking, the Laplacian operator involves second derivatives, and thus we might expect a correspondence between $\nabla^2\Phi$ and a contraction of the Riemann tensor, the Ricci tensor. Let's examine the form of the Ricci tensor using Eq. (15.1) for the metric tensor. Ordinarily, to find the Ricci tensor we must first have the Riemann tensor, which can be a daunting task to evaluate. In this case, however, the Christoffel symbols begin at first order in the perturbation $h_{\mu\nu}$, and thus products of Christoffel symbols can be ignored to lowest order. Thus, from Eq. (14.61),

$$R^{\alpha}_{\ \rho\mu\nu} = \partial_{\mu}\Gamma^{\alpha}_{\nu\rho} - \partial_{\nu}\Gamma^{\alpha}_{\mu\rho} + O(h^2)\,. \tag{15.15}$$

[3] Free fall is defined on page 188.

[4] Through any point of a manifold there is a unique geodesic for given tangent vector.

[5] The Newtonian gravitational potential represents the work done per mass in bringing a test mass from a distant region in which gravity vanishes into a region where a gravitational field exists, that produced by other masses.

[6] A necessary and sufficient condition for the metric tensor to consist of constant elements is that the Riemann tensor vanish.[60, p25]

Set $\alpha = \mu$ in Eq. (15.15) and sum:

$$R_{\rho\nu} = \partial_\mu \Gamma^\mu_{\nu\rho} - \partial_\nu \Gamma^\mu_{\mu\rho} + O(h^2) . \tag{15.16}$$

In particular, $R_{00} \approx \partial_\mu \Gamma^\mu_{00} - \partial_0 \Gamma^\mu_{\mu 0}$ from Eq. (15.16), and, because we're considering a time-independent perturbation, set time derivatives to zero. Thus, $R_{00} \approx \partial_i \Gamma^i_{00}$. Using Eqs. (15.10) and (15.13), we have for R_{00} under the approximation of weak, static gravity:

$$R_{00} \approx -\frac{1}{2}\partial_i \partial^i h_{00} = \frac{1}{c^2}\nabla^2 \Phi .$$

We thus have the correspondence $c^2 R_{00} \leftrightarrow \nabla^2 \Phi$ in the weak-field limit. On the right side of the Poisson equation is the mass density function, which through the equivalence of mass and energy ($E = mc^2$) represents an energy density. The Poisson equation can therefore be written:

$$R_{00} \approx \frac{1}{c^2}\nabla^2 \Phi = \frac{4\pi G}{c^2}\rho = \frac{4\pi G}{c^4}\left(\rho c^2\right) = \frac{4\pi G}{c^4} T_{00} , \tag{15.17}$$

where $\rho c^2 = T_{00}$ is an element of $T_{\mu\nu}$, the energy-momentum tensor for a perfect fluid, Eq. (10.17).

15.3.2 Writing it down

Equation (15.17) is not covariant, but it suggests a covariant generalization. Einstein proposed (in October 1915) a simple proportionality between $R_{\mu\nu}$ and $T_{\mu\nu}$,

$$R_{\mu\nu} = \kappa T_{\mu\nu} , \qquad \text{(wrong)} \tag{15.18}$$

where κ is a constant to be determined.[7,8] He quickly realized, however (November 1915), that the covariant divergence of the field equation must be zero. In flat spacetime, conservation of energy-momentum of fields is associated with $\partial_\nu T^{\mu\nu} = 0$, Eq. (9.16); in curved spacetime we should have $\nabla_\nu T^{\mu\nu} = 0$. The left side of Eq. (15.18) must therefore have zero covariant divergence (which it does not). He was led to propose as a relativistic field theory of gravitation,

$$R_{\mu\nu} - \frac{1}{2} g_{\mu\nu} R = \kappa T_{\mu\nu} . \qquad \text{(right)} \tag{15.19}$$

We showed in Section 14.6 that the combination of terms in Eq. (15.19) (now called the Einstein tensor $G_{\mu\nu} \equiv R_{\mu\nu} - \frac{1}{2} g_{\mu\nu} R$) has zero divergence, $\nabla_\nu G^{\mu\nu} = 0$. Equation (15.19) is the *Einstein field equation*, the foundational equation for gravitational phenomena. We show that $\kappa = 8\pi G/c^4$ below. As noted in Section 14.6, the Einstein tensor has six independent components in four-dimensional spacetime, yet the metric tensor has 10. That leaves four degrees of freedom—just enough to permit a free choice of spacetime coordinate system (see Section 18.1.1). Table 15.1 compares GR with Newtonian gravity.

15.3.3 Alternate form of the field equation

There is an equivalent way of writing Einstein's equation that's often more convenient. Start with Eq. (15.19) and raise an index,

$$R^\mu_{\ \nu} - \frac{1}{2}\delta^\mu_\nu R = \kappa T^\mu_{\ \nu} , \tag{15.20}$$

[7] Equation (15.18) should not be underestimated. Einstein struggled for 10 years in his attempts to find a relativistic field theory of gravity. You can't get there from here: Equation (15.18) is fairly "outside the box" as compared with SR; getting to Eq. (15.18) required Einstein's deep insights into the nature of space, time, and gravity.

[8] Don't we already have the proportionality constant in Eq. (15.17)? It turns out $\kappa = 4\pi G/c^4$ is off by a factor of two.

Table 15.1 Comparison of Newtonian gravitation with general relativity.

	Newtonian gravitation	General relativity ($\Lambda = 0$)
Field equation	$\nabla^2\Phi = 4\pi G\rho$	$R_{\mu\nu} - \frac{1}{2}g_{\mu\nu}R = \kappa T_{\mu\nu}$
Vacuum equation	$\nabla^2\Phi = 0$	$R_{\mu\nu} = 0$
No gravitational field	$\Phi = 0$	$R_{\alpha\beta\gamma\delta} = 0$

where we've used $g^{\mu}_{\nu} = \delta^{\mu}_{\nu}$. Now set $\nu = \mu$ in Eq. (15.20) and sum: $R - \frac{1}{2}4R = \kappa T$, where $R = R^{\mu}{}_{\mu}$, Eq. (14.96), $4 = \delta^{\mu}_{\mu}$, and where we've defined a scalar field associated with the energy-momentum tensor, $T \equiv T^{\mu}{}_{\mu}$. Thus, $R = -\kappa T$. Combine this result with Eq. (15.19), and we have an equivalent form of the field equation,

$$R_{\mu\nu} = \kappa\left(T_{\mu\nu} - \frac{1}{2}g_{\mu\nu}T\right) . \tag{15.21}$$

Clearly when $T_{\mu\nu} = 0$, Eq. (15.21) becomes the *vacuum equation*, $R_{\mu\nu} = 0$.

15.3.4 Determining the proportionality constant: $\kappa = 8\pi G/c^4$

To evaluate the constant κ, we make use of the energy-momentum tensor for a perfect fluid, Eq. (10.17) with $\eta_{\mu\nu} \to g_{\mu\nu}$: $T_{\mu\nu} = (\rho + P/c^2)U_\mu U_\nu + Pg_{\mu\nu}$. To construct the scalar $T = T^{\mu}_{\mu}$, raise an index on $T_{\mu\nu}$ and sum:

$$T = T^{\mu}_{\mu} = \left(\rho + \frac{P}{c^2}\right)U^{\mu}U_{\mu} + P\delta^{\mu}_{\mu} = \left(\rho + \frac{P}{c^2}\right)(-c^2) + 4P = -\rho c^2 + 3P , \tag{15.22}$$

where we've used $U_\alpha U^\alpha = -c^2$ (Section 7.1). We then have

$$T_{\mu\nu} - \frac{1}{2}g_{\mu\nu}T = \left(\rho + \frac{P}{c^2}\right)U_\mu U_\nu + \frac{1}{2}g_{\mu\nu}\left(\rho c^2 - P\right) . \tag{15.23}$$

In the nonrelativistic limit $U_0 = -c$, and thus

$$T_{00} - \frac{1}{2}g_{00}T = \frac{1}{2}\rho c^2 + \frac{3}{2}P . \tag{15.24}$$

As part of the "nonrelativistic limit" we assume that $\rho c^2 \gg P$. This approximation is almost always valid; we ignore the pressure term here. Combining Eqs. (15.24) and (15.21), we have in the weak-field limit

$$R_{00} \approx \frac{1}{c^2}\nabla^2\Phi = \kappa\left(T_{00} - \frac{1}{2}g_{00}T\right) = \frac{1}{2}\kappa\rho c^2 = \frac{4\pi G\rho}{c^2} . \tag{15.25}$$

With $\kappa = 8\pi G/c^4$, Einstein's equation agrees with Poisson's equation in its domain of validity.[9]

With κ determined, Einstein's equation is completely specified. Once specified, there is no "wiggle room"—either its predictions agree with the results of experiment or they do not. Determining κ by requiring that Eq. (15.19) reproduce the Poisson equation for weak fields is standard practice

[9]Einstein's equation is sometimes written with the value of $\kappa = -8\pi G/c^4$. There are two reasons for this. One is the "other" choice of Lorentz metric, $\eta = \text{diag}(1, -1, -1, -1)$. However, even with this metric one also sees κ as given by $\kappa = 8\pi G/c^4$. The definition of the Ricci tensor is not standard. Whereas we have taken the Ricci tensor as the contraction of the first and third indices of the Riemann tensor, $R_{\alpha\beta} = R^{\rho}_{\ \alpha\rho\beta}$, the Ricci tensor is also defined as the contraction over the first and last indices, $R_{\mu\nu} = R^{\alpha}_{\ \mu\nu\alpha}$, leading to another source of minus sign. Beware!

in theoretical physics, "nailing down" unknown parameters by seeking agreement with previously established theories in their domains of validity. This method does not guarantee that Eq. (15.19) is correct for *all* gravitational phenomena, it merely asserts that the theory is not obviously wrong in the nonrelativistic limit. Even for weak fields, however, GR predicts *new* phenomena not contained in the Newtonian theory (precession of orbits, deflection of light, etc.—see Chapter 17). Whether Eq. (15.19) allows us to interpret phenomena for strong gravitational effects remains to be seen. In the 100 years of its existence, GR *has successfully met every test.* Tomorrow, however, discoveries may be made that are not consistent with the framework of GR, in which case modifications to the theory will be required.

15.3.5 The cosmological constant

Speaking of wiggle room, Einstein later realized (in 1917) that the field equation could be modified and still satisfy the Bianchi identity. Because $\nabla_\nu g^{\mu\nu} = 0$, a term proportional to the metric tensor, $\Lambda g_{\mu\nu}$ could be added to Eq. (15.19), where Λ is a constant:

$$R_{\mu\nu} - \frac{1}{2}g_{\mu\nu}R + \Lambda g_{\mu\nu} = \frac{8\pi G}{c^4}T_{\mu\nu} \,. \tag{15.26}$$

The parameter Λ must have dimensions of $(\text{length})^{-2}$ so that $1/\sqrt{|\Lambda|}$ is a length. Current estimates are $\Lambda \approx 10^{-52}\ \text{m}^{-2}$, so that $1/\sqrt{|\Lambda|} \approx 10^{26}$ m (10^{10} ly). Thus, Λ is associated with length scales comparable to the size of the universe—hence the term *cosmological constant.* Experimental tests of GR on the scale of the solar system ignore Λ.

15.4 LAGRANGIAN FORMULATION

Einstein's field equation, which one can say is the result of inspired guesswork, can be derived from Hamilton's principle.[10] One requires a Lagrangian density of the gravitational field, $\mathscr{L}_g$, such that the action integral $S = \int_V \mathscr{L}_g \text{d}^4x$, where V is a region of spacetime bounded by a hypersurface is stationary under suitable variations, $\delta S = 0$. The gravitational field is associated with derivatives of the metric tensor; we vary S by varying the metric[11] $g^{\mu\nu} \to g^{\mu\nu} + \delta g^{\mu\nu}$, where $\delta g^{\mu\nu} = 0$ on the boundary of V. Having a method for *deriving* Einstein's equation provides a framework for including the coupling of physical fields to gravity. Of course, Hamilton's principle only reproduces that which is already known, but the more ways you have of looking at a problem the better.[12] It turns out that having a variational approach to Einstein's equation is highly significant.

15.4.1 Vacuum equation: $R_{\mu\nu} = 0$

How do we find $\mathscr{L}_g$? We require that it be a scalar density, so it should be a function of scalars made up out of the metric tensor and its derivatives. Effectively the only scalar we can obtain from the Riemann tensor is the Ricci scalar field,[13] R. The mathematician David Hilbert took $\mathscr{L}_g = \sqrt{-g}R$,

[10]Variational calculus and Hamilton's principle are reviewed in Appendix D.

[11]A change in coordinates induces a change in metric tensor, and S, a scalar, must be the same in all coordinate systems.

[12]From Richard Feynman:[32] "Theories of the known, which are described by different physical ideas, may be equivalent in all their predictions and hence scientifically indistinguishable. However, they are not psychologically identical when trying to move from that base to the unknown. For different views suggest different kinds of modifications which might be made and hence are not equivalent in the hypotheses one generates from them in one's attempt to understand what is not yet understood."

[13]The scalar field $R^\alpha_{\ \alpha}$ is the only invariant that can be obtained from the Riemann tensor made up of the metric tensor $g_{\alpha\beta}$ as well as its first and second derivatives and that's linear in the second derivatives.[41, Appendix II] If one tried any other scalar fields such as $R^{\mu\nu}R_{\mu\nu}$ or $R^{\alpha\beta\gamma\delta}R_{\alpha\beta\gamma\delta}$, one would obtain equations involving fourth-order derivatives of the metric, outside of the customary class of equations in physics.

which[14] (as we'll show) leads to the Einstein vacuum equation under $\delta S = 0$. We can include Λ by replacing $R \to R - 2\Lambda$ (see Eq. (15.26)). The action integral for the gravitational field (the *Hilbert action*) is

$$S = \int_V (R - 2\Lambda)\sqrt{-g}\, \mathrm{d}^4 x \,. \tag{15.27}$$

The quantity $R = g^{\rho\nu} R_{\rho\nu}$, where $R_{\rho\nu} = R^{\alpha}{}_{\rho\alpha\nu}$ is (from the definition of the Riemann tensor),

$$R_{\rho\nu} = \partial_\mu \Gamma^\mu_{\nu\rho} - \partial_\nu \Gamma^\mu_{\mu\rho} + \Gamma^\mu_{\mu\beta}\Gamma^\beta_{\nu\rho} - \Gamma^\mu_{\nu\beta}\Gamma^\beta_{\mu\rho} \,. \tag{15.28}$$

Thus, to vary S we require $\delta g^{\alpha\beta}$, $\delta\sqrt{-g}$, and $\delta R_{\mu\nu}$. The first two are straightforward. From $g^{\alpha\beta} g_{\beta\gamma} = \delta^\alpha_\gamma$,

$$\delta g^{\rho\alpha} = -g^{\rho\nu} g^{\alpha\beta} \delta g_{\nu\beta} \,. \tag{15.29}$$

Using Eq. (14.32),

$$\delta\sqrt{-g} = \frac{1}{2}\sqrt{-g} g^{\alpha\beta} \delta g_{\beta\alpha} = -\frac{1}{2}\sqrt{-g} g_{\alpha\beta} \delta g^{\alpha\beta} \,. \tag{15.30}$$

The third type of variation $\delta R_{\mu\nu}$ entails a fairly involved calculation. The reader who is uninterested in the details should skip to Eq. (15.39).

The variation in Christoffel symbols is, using Eq. (14.23),

$$\delta\Gamma^\lambda_{\mu\nu} = \frac{1}{2} g^{\lambda\rho} \left(\partial_\mu \delta g_{\rho\nu} + \partial_\nu \delta g_{\rho\mu} - \partial_\rho \delta g_{\mu\nu}\right) + \frac{1}{2}\delta(g^{\lambda\rho}) \left(\partial_\mu g_{\rho\nu} + \partial_\nu g_{\rho\mu} - \partial_\rho g_{\mu\nu}\right) \,. \tag{15.31}$$

Combining Eqs. (15.29) and (15.31), it's straightforward to show that

$$\delta\Gamma^\lambda_{\mu\nu} = -g^{\lambda\alpha}(\delta g_{\alpha\beta})\Gamma^\beta_{\mu\nu} + \frac{1}{2} g^{\lambda\rho} \left(\partial_\mu \delta g_{\rho\nu} + \partial_\nu \delta g_{\rho\mu} - \partial_\rho \delta g_{\mu\nu}\right) \,. \tag{15.32}$$

Next we make use of the covariant derivative of terms like

$$\nabla_\mu \delta g_{\rho\nu} = \partial_\mu \delta g_{\rho\nu} - \Gamma^\sigma_{\mu\rho} \delta g_{\sigma\nu} - \Gamma^\sigma_{\mu\nu} \delta g_{\rho\sigma} \,, \tag{15.33}$$

where we've used Eq. (14.14). It can then be shown that for the terms in Eq. (15.32),

$$\partial_\mu \delta g_{\rho\nu} + \partial_\nu \delta g_{\rho\mu} - \partial_\rho \delta g_{\mu\nu} = \nabla_\mu \delta g_{\rho\nu} + \nabla_\nu \delta g_{\rho\mu} - \nabla_\rho \delta g_{\mu\nu} + 2\Gamma^\sigma_{\mu\nu} \delta g_{\rho\sigma} \,. \tag{15.34}$$

Combining Eqs. (15.34) and (15.32),

$$\delta\Gamma^\lambda_{\mu\nu} = \frac{1}{2} g^{\lambda\rho} \left[\nabla_\mu \delta g_{\rho\nu} + \nabla_\nu \delta g_{\rho\mu} - \nabla_\rho \delta g_{\mu\nu}\right] \,. \tag{15.35}$$

From Section 14.1.4 we know that $\delta\Gamma^\lambda_{\mu\nu}$, the difference between two connections, is a tensor, and because $\delta\Gamma^\lambda_{\mu\nu}$ is a tensor, we can form *its* covariant derivative

$$\nabla_\kappa(\delta\Gamma^\lambda_{\mu\nu}) = \partial_\kappa(\delta\Gamma^\lambda_{\mu\nu}) + \Gamma^\lambda_{\kappa\sigma}\delta\Gamma^\sigma_{\mu\nu} - \Gamma^\sigma_{\kappa\mu}\delta\Gamma^\lambda_{\sigma\nu} - \Gamma^\sigma_{\kappa\nu}\delta\Gamma^\lambda_{\mu\sigma} \,, \tag{15.36}$$

where we've used Eq. (14.14). Using Eq. (15.36), we arrive at an intermediate result:

$$\nabla_\kappa(\delta\Gamma^\lambda_{\mu\lambda}) - \nabla_\lambda(\delta\Gamma^\lambda_{\mu\kappa}) = \partial_\kappa(\delta\Gamma^\lambda_{\mu\lambda}) - \partial_\lambda(\delta\Gamma^\lambda_{\mu\kappa}) + \Gamma^\lambda_{\kappa\sigma}\delta\Gamma^\sigma_{\mu\lambda} - \Gamma^\sigma_{\kappa\mu}\delta\Gamma^\lambda_{\sigma\lambda} - \Gamma^\lambda_{\lambda\sigma}\delta\Gamma^\sigma_{\mu\kappa} + \Gamma^\sigma_{\lambda\mu}\delta\Gamma^\lambda_{\sigma\kappa} \,. \tag{15.37}$$

We have set $\nu = \lambda$ in Eq. (15.36) and then subtracted from it the version of Eq. (15.36) where we first let $\kappa \to \lambda$ and then $\nu \to \kappa$. Comparing Eq. (15.37) with Eq. (15.28), it's straightforward to show that (*Palatini's identity*)

$$\delta R_{\mu\kappa} = \nabla_\lambda(\delta\Gamma^\lambda_{\mu\kappa}) - \nabla_\kappa(\delta\Gamma^\lambda_{\mu\lambda}) \,. \tag{15.38}$$

[14] See Section 5.2 for making integrals relativistically invariant with the inclusion of $\sqrt{-g}$.

To vary the Hilbert action, Eq. (15.27), we need to form the variation

$$\delta\left([g^{\rho\nu}R_{\rho\nu}-2\Lambda]\sqrt{-g}\right)=(R-2\Lambda)\delta\sqrt{-g}+\sqrt{-g}\left(g^{\rho\nu}\delta R_{\rho\nu}+R_{\rho\nu}\delta g^{\rho\nu}\right). \tag{15.39}$$

The $\delta R_{\rho\nu}$ term in Eq. (15.39) leads to a four-divergence. Using Eq. (15.38),

$$\begin{aligned} g^{\rho\nu}\delta R_{\rho\nu} =& \nabla_\lambda(g^{\rho\nu}\delta\Gamma^\lambda_{\rho\nu})-\nabla_\nu(g^{\rho\nu}\delta\Gamma^\lambda_{\rho\lambda})=\frac{1}{\sqrt{-g}}\partial_\sigma(\sqrt{-g}g^{\rho\nu}\delta\Gamma^\sigma_{\rho\nu})-\frac{1}{\sqrt{-g}}\partial_\sigma(\sqrt{-g}g^{\rho\sigma}\delta\Gamma^\lambda_{\rho\lambda}) \\ =& \frac{1}{\sqrt{-g}}\partial_\sigma\left(\sqrt{-g}[g^{\rho\nu}\delta\Gamma^\sigma_{\rho\nu}-g^{\rho\sigma}\delta\Gamma^\lambda_{\rho\lambda}]\right)\equiv\frac{1}{\sqrt{-g}}\partial_\sigma\left(\sqrt{-g}W^\sigma\right), \end{aligned}$$

where we've used Eq. (14.35) and we've taken $g^{\rho\nu}$ inside the covariant derivative. Thus, $\sqrt{-g}g^{\rho\nu}\delta R_{\rho\nu}=\partial_\sigma\left(\sqrt{-g}W^\sigma\right)$, a four-divergence. The integral of $\partial_\sigma\left(\sqrt{-g}W^\sigma\right)$ over V becomes an integral of W^σ on the hypersurface bounding V, which vanishes because $\delta g^{\mu\nu}=0$ on the boundary. The variation $\delta R_{\rho\nu}$ therefore *does not make a contribution to the variational problem*; the Ricci tensor is effectively a constant for the purposes of doing variational calculations. Thus,

$$\delta S=\int_V\left[(R-2\Lambda)\frac{\delta\sqrt{-g}}{\delta g^{\mu\nu}}+\sqrt{-g}R_{\mu\nu}\right]\delta g^{\mu\nu}\,\mathrm{d}^4x. \tag{15.40}$$

Using Eq. (15.30) in Eq. (15.40), we have, comparing with Eq. (15.27)

$$\delta S=\int_V\frac{\delta\mathscr{L}_g}{\delta g^{\mu\nu}}\delta g^{\mu\nu}\mathrm{d}^4x=\int_V\left[R_{\alpha\beta}-\frac{1}{2}g_{\alpha\beta}(R-2\Lambda)\right]\delta g^{\alpha\beta}\sqrt{-g}\mathrm{d}^4x. \tag{15.41}$$

The requirement $\delta S=0$ under variations of the metric thus results in the Einstein vacuum equation.

15.4.2 Matter fields

Can we obtain the full field equation from a variational treatment? We enlarge the scope of the action $S=\int\mathscr{L}\,\mathrm{d}^4x$ to include the Lagrangian of matter fields,[15] $\mathscr{L}_m$, such that the total Lagrangian $\mathscr{L}=\mathscr{L}_g+\alpha\mathscr{L}_m$, where α is a constant. Invariance of S implies that

$$\delta S=\int_V\left[\frac{\delta\mathscr{L}_g}{\delta g^{\mu\nu}}+\alpha\frac{\delta\mathscr{L}_m}{\delta g^{\mu\nu}}\right]\delta g^{\mu\nu}\mathrm{d}^4x=0. \tag{15.42}$$

We "want" the terms in square brackets to be Einstein's equation, (15.26). That can be achieved by *defining* the energy-momentum tensor density such that

$$\sqrt{-g}T_{\mu\nu}\equiv-\frac{\alpha}{\kappa}\frac{\delta\mathscr{L}_m}{\delta g^{\mu\nu}}. \tag{15.43}$$

We can take the derivative in Eq. (15.43) by writing $\mathscr{L}_m=\sqrt{-g}\left(\mathscr{L}_m/\sqrt{-g}\right)$, with the result:

$$\sqrt{-g}T_{\mu\nu}=g_{\mu\nu}\mathscr{L}_m-2\frac{\delta(\mathscr{L}_m/\sqrt{-g})}{\delta g^{\mu\nu}}, \tag{15.44}$$

where we've used Eq. (15.30) and we've set[16] $\alpha=2\kappa$. We see from Eq. (15.44) that among the fields $\{\Phi^\alpha\}$ represented by $\mathscr{L}_m(\Phi^\alpha,\nabla_\mu\Phi^\alpha)$ we must include the metric field $g_{\mu\nu}$ as well. *The description of a physical system must include the metric field $g_{\mu\nu}$ as a fundamental field.*

Equation (15.44) is not in the form of Eq. (9.15), the form of $T_{\mu\nu}$ from Noether's theorem, which satisfies conservation laws $\partial_\nu T^{\mu\nu}=0$, but is not necessarily symmetric (Section 9.4).[17] While $T_{\mu\nu}$

[15]In GR, nongravitational fields are referred to as "matter" fields, which include the electromagnetic field.

[16]Taking $\alpha=2\kappa$ reproduces known energy-momentum tensors.

[17]The energy-momentum tensor must be symmetric to conserve angular momentum; Section 9.3.

as defined by Eq. (15.44) is symmetric, does it satisfy $\nabla_\nu T^{\mu\nu} = 0$? Until that's been established, Eq. (15.44) is not an acceptable definition of $T_{\mu\nu}$. To check this point, we adopt the approach of Noether's theorem, that the action be invariant under a change in coordinates. Technically, however, that's difficult: We must connect a general coordinate transformation with a variation in the metric. The two are of course related: A variation in the metric is tantamount to a change of coordinates.

We consider an infinitesimal coordinate transformation $x^\mu \to x'^\mu = x^\mu + \xi^\mu(x)$, where $\xi^\mu(x)$ is an infinitesimal smooth vector field. As a type $(0,2)$ tensor, $g_{\mu\nu}$ transforms as

$$\begin{aligned} g'_{\mu\nu}(x') =& \frac{\partial x^\rho}{\partial x'^\mu}\frac{\partial x^\sigma}{\partial x'^\nu} g_{\rho\sigma}(x) = [\delta^\rho_\mu - \partial_\mu \xi^\rho(x)][\delta^\sigma_\nu - \partial_\nu \xi^\sigma(x)] g_{\rho\sigma}(x) \\ =& g_{\mu\nu}(x) - g_{\mu\sigma}(x)\partial_\nu \xi^\sigma(x) - g_{\rho\nu}(x)\partial_\mu \xi^\rho(x) + O(\xi)^2 , \end{aligned} \tag{15.45}$$

where we've inverted $\partial x'^\mu/\partial x^\rho = \delta^\mu_\rho + \partial_\rho \xi^\mu$ to first order in ξ to obtain $\partial x^\rho / \partial x'^\mu$. A variation is an infinitesimal change in functional form *at a single point*, Eq. (D.6). Thus,

$$\begin{aligned} \delta g_{\mu\nu}(x) \equiv& g'_{\mu\nu}(x) - g_{\mu\nu}(x) = [g'_{\mu\nu}(x') - g_{\mu\nu}(x)] - [g'_{\mu\nu}(x') - g'_{\mu\nu}(x)] \qquad (15.46) \\ =& [g'_{\mu\nu}(x') - g_{\mu\nu}(x)] - \xi^\sigma \partial_\sigma g'_{\mu\nu}(x) = [g'_{\mu\nu}(x') - g_{\mu\nu}(x)] - \xi^\sigma \partial_\sigma g_{\mu\nu}(x) + O(\xi^2) . \end{aligned}$$

Use Eq. (15.45) in Eq. (15.46):

$$\begin{aligned} \delta g_{\mu\nu}(x) =& - g_{\mu\sigma}(x)\partial_\nu \xi^\sigma(x) - g_{\rho\nu}(x)\partial_\mu \xi^\rho(x) - \xi^\sigma \partial_\sigma g_{\mu\nu}(x) \\ =& - \pounds_{\boldsymbol{\xi}} g_{\mu\nu} = -(\nabla_\mu \xi_\nu + \nabla_\nu \xi_\mu) , \end{aligned} \tag{15.47}$$

where we've used Eqs. (13.22) and (14.29).

From Eq. (15.42), the variation in the nongravitational action is (with $\alpha = 2\kappa$)

$$\begin{aligned} \delta S_m =& 2\kappa \int_V \frac{\delta \mathscr{L}_m}{\delta g^{\mu\nu}} \delta g^{\mu\nu} \mathrm{d}^4 x = -\frac{1}{2}\int_V T_{\mu\nu}\delta g^{\mu\nu}\sqrt{-g}\mathrm{d}^4x \\ =& \frac{1}{2}\int_V T^{\mu\nu}\delta g_{\mu\nu}\sqrt{-g}\mathrm{d}^4x = \int_V T^{\mu\nu}\nabla_\mu \xi_\nu \sqrt{-g}\mathrm{d}^4 x , \end{aligned} \tag{15.48}$$

where we've used Eqs. (15.29), (15.43) and (15.47). Then, using $T^{\mu\nu}\nabla_\mu \xi_\nu = \nabla_\mu (T^{\mu\nu}\xi_\nu) - \xi_\nu \nabla_\mu T^{\mu\nu}$, Eq. (15.48) becomes

$$\delta S_m = \int_V \nabla_\mu (T^{\mu\nu}\xi_\nu)\sqrt{-g}\mathrm{d}^4x - \int_V \xi_\nu \nabla_\mu T^{\mu\nu}\sqrt{-g}\mathrm{d}^4x . \tag{15.49}$$

The first integral in Eq. (15.49) vanishes: It contains a covariant divergence which can be converted into a surface integral and the functions $\xi^\mu(x)$ vanish on the boundary. Having $\delta S_m = 0$ thus requires the second integral in Eq. (15.49) to vanish. Because the ξ^μ are arbitrary,

$$\nabla_\mu T^{\mu\nu} = 0 . \tag{15.50}$$

The energy-momentum tensor defined by Eq. (15.43) *has vanishing covariant divergence if the action for matter fields is a scalar.*[18]

We can now formulate the gravitational field equations for systems where gravity couples to matter fields. Take $\mathscr{L} = (R - 2\Lambda)\sqrt{-g} + 2\kappa\mathscr{L}_m$; this generates from $\delta S = 0$ Einstein's equation in the form $G_{\mu\nu} + \Lambda g_{\mu\nu} = \kappa T_{\mu\nu}$, where $T_{\mu\nu}$ is given by Eq. (15.43). Note that the coupling between spacetime curvature and energy-matter fields occurs without *explicit* specification of the fields themselves; all we require is that there is an energy-momentum tensor associated with the field. The geometry thus does not distinguish between physically different types of fields if they have the same energy-momentum distribution. This is the general relativistic version of the equivalence principle. We now show that Eq. (15.43) gives expected results for $T_{\mu\nu}$.

[18]The connection between $\nabla_\mu T^{\mu\nu} = 0$ and conservation laws is discussed in Section 18.9.

15.4.2.1 Electromagnetic field

For the electromagnetic field,[19] $\mathscr{L}_F/\sqrt{-g} = -F^{\alpha\beta}F_{\alpha\beta}/(4\mu_0)$ (Section 8.9.2). From Eq. (15.44),

$$-\mu_0 T_{\mu\nu} = \frac{1}{4}g_{\mu\nu}F^{\alpha\beta}F_{\alpha\beta} - \frac{1}{2}\frac{\delta(F^{\alpha\beta}F_{\alpha\beta})}{\delta g^{\mu\nu}} . \tag{15.51}$$

To take the derivative in Eq. (15.51), note that $F^{\alpha\beta}F_{\alpha\beta} = g^{\alpha\lambda}g^{\beta\sigma}F_{\lambda\sigma}F_{\alpha\beta}$. Then,

$$\frac{\delta(F^{\alpha\beta}F_{\alpha\beta})}{\delta g^{\mu\nu}} = F_{\lambda\sigma}F_{\alpha\beta}\left(g^{\alpha\lambda}\delta^{\beta}_{\mu}\delta^{\sigma}_{\nu} + g^{\beta\sigma}\delta^{\alpha}_{\mu}\delta^{\lambda}_{\nu}\right) = -2g^{\alpha\lambda}F_{\mu\alpha}F_{\lambda\nu} . \tag{15.52}$$

Combining Eqs. (15.52) and (15.51),

$$-\mu_0 T_{\mu\nu} = \frac{1}{4}g_{\mu\nu}F^{\alpha\beta}F_{\alpha\beta} + g^{\alpha\lambda}F_{\mu\alpha}F_{\lambda\nu} , \tag{15.53}$$

the symmetric energy-momentum tensor for the electromagnetic field, $\theta_{\mu\nu}$ (Exercise 9.3).

15.4.2.2 Perfect fluid

Equation (15.44) reproduces $T_{\mu\nu}$ for the perfect fluid. To *show* this is more difficult than the previous example: We need to vary the metric subject to the constraints of constant entropy and particle number, and to do that requires some thermodynamics.

In Chapter 10 we found it convenient to work with per-mass or per-volume quantities. Here it will be convenient to work with *per-particle* quantities. The number density in the rest frame is $n = N/V$. The enthalpy per particle $w \equiv H/N = (E + PV)/N = (\rho c^2 + P)/n$, where $E = \rho c^2 V$ is the rest-mass energy contained in V. An exact rewrite of the first law of thermodynamics ($\mathrm{d}E = T\mathrm{d}S - P\mathrm{d}V$) for fixed N is $\mathrm{d}(\rho c^2) = nT\mathrm{d}s + w\mathrm{d}n$, where $s \equiv S/N$ is the entropy per particle. We can therefore take ρ as a function of the independent variables s and n, $\rho = \rho(s, n)$, and hence the total differential $\mathrm{d}\rho = (\partial\rho/\partial s)_n\mathrm{d}s + (\partial\rho/\partial n)_s\mathrm{d}n$. Comparing with the first law of thermodynamics, we have the desired thermodynamic identity

$$\left(\frac{\partial\rho}{\partial n}\right)_s = \frac{w}{c^2} = \frac{1}{n}\left(\rho + P/c^2\right) . \tag{15.54}$$

We define the particle current vector density[20] $N^\mu \equiv nU^\mu\sqrt{-g}$. Contracting this quantity with itself, $N^\mu N_\mu = n^2U^\mu U_\mu(-g) = n^2c^2g$, and thus the total density (in the rest frame) is

$$n = \frac{1}{c}\sqrt{g_{\alpha\beta}N^\alpha N^\beta/g} . \tag{15.55}$$

We vary n with respect to the metric, keeping the particle number fixed, $\delta N^\mu = 0$,

$$\begin{aligned}\delta n =& \frac{1}{2nc^2}\left[\frac{\delta g_{\alpha\beta}}{g}N^\alpha N^\beta - \frac{g_{\alpha\beta}}{g^2}N^\alpha N^\beta\delta g\right] = -\frac{n}{2}\left[\delta g_{\alpha\beta}\frac{U^\alpha U^\beta}{c^2} + \frac{\delta g}{g}\right]\\ =& \frac{n}{2}\left[\frac{U_\alpha U_\beta}{c^2} + g_{\alpha\beta}\right]\delta g^{\alpha\beta} .\end{aligned} \tag{15.56}$$

We can now apply Eq. (15.44) to the ideal fluid, for which $\mathscr{L} = -\rho c^2$, Eq. (10.29):

$$T_{\mu\nu} = -g_{\mu\nu}\rho c^2 + 2c^2\frac{\delta\rho}{\delta g^{\mu\nu}} . \tag{15.57}$$

[19]Note the factor of $\sqrt{-g}$, something we should have been including all along; $\sqrt{-g} = 1$ in SR.

[20]Differs from the definition in Section 10.4 by the factor of $\sqrt{-g}$.

The variation of ρ, holding the entropy fixed, is obtained from Eqs. (15.54) and (15.56):

$$\delta\rho = \left(\frac{\partial\rho}{\partial n}\right)_s \delta n = \frac{w}{c^2}\delta n = \frac{1}{2}(\rho + P/c^2)\left(\frac{U_\alpha U_\beta}{c^2} + g_{\alpha\beta}\right)\delta g^{\alpha\beta} . \tag{15.58}$$

Combining Eqs. (15.57) and (15.58),

$$T_{\mu\nu} = Pg_{\mu\nu} + (\rho + P/c^2)U_\mu U_\nu , \tag{15.59}$$

the same as Eq. (10.17) except for the replacement $\eta_{\mu\nu} \to g_{\mu\nu}$.

15.4.2.3 *Is Eq. (15.43) cheating?*

One might think that *defining* $T_{\mu\nu}$ in terms of $\mathscr{L}_m$, Eq. (15.43), is facile because it's *guaranteed* to reproduce the form of Einstein's field equation. We've just given two examples where the "recipe" of Eq. (15.44) reproduces previously established results[21] for $T_{\mu\nu}$, and it leads to $\nabla_\nu T^{\mu\nu} = 0$. Carrying it a step further, combining Eqs. (15.43) and (15.42), we have that (requiring $\delta S = 0$),

$$\kappa T_{\mu\nu} = \frac{1}{\sqrt{-g}}\frac{\delta\mathscr{L}_g}{\delta g^{\mu\nu}} = G_{\mu\nu} + \Lambda g_{\mu\nu} , \tag{15.60}$$

where the second equality is from Eq. (15.41). Equation (15.60) is a change in emphasis from how GR is usually presented, that energy-momentum, as represented by $T_{\mu\nu}$, causes the curvature of spacetime through $G_{\mu\nu} = \kappa T_{\mu\nu}$ (setting $\Lambda = 0$). The interpretation of Eq. (15.60) is that the extent to which $G_{\mu\nu} \neq 0$, *is* the energy-momentum of non-gravitational fields, $\kappa T_{\mu\nu} = G_{\mu\nu}$. The usual interpretation of Gauss's law, $\boldsymbol{\nabla}\cdot\boldsymbol{E} = \rho/\epsilon_0$ is that ρ controls the divergence of $\boldsymbol{E}$. Turning it around, as in Eq. (15.60), charge does not *cause* a non-zero divergence of $\boldsymbol{E}$, charge *is* the non-vanishing divergence of the electric field; charge is quantified by the extent to which $\boldsymbol{\nabla}\cdot\boldsymbol{E} \neq 0$. In this view, "matter" (energy-momentum) does not *cause* the non-vanishing of $G_{\mu\nu}$, matter *is* the non-vanishing of $G_{\mu\nu}$; matter is described by $G_{\mu\nu} \neq 0$.

15.5 DUST

In GR, *dust* is a *pressure-free perfect fluid*; such a system is of interest because an exact solution of Einstein's field equation can be found. A pressureless "fluid" can be visualized as a collection of particles that move in an orderly fashion and interact with each other only gravitationally.[22]

For the energy-momentum tensor of dust, set $P = 0$ in Eq. (15.59), $T_{\mu\nu} = \rho U_\mu U_\nu$, and thus the field equation $G_{\mu\nu} = \kappa\rho U_\mu U_\nu$. Because $\nabla_\nu G^{\mu\nu} = 0$, the equation of motion for dust is buried in

$$\nabla_\nu(\rho U^\mu U^\nu) = 0 , \tag{15.61}$$

Equation (15.61) is equivalent to both[23]

$$\begin{aligned} \nabla_\nu\left(\rho U^\nu\right) &= 0 \\ U^\nu\nabla_\nu U^\mu &= 0 . \end{aligned} \tag{15.62}$$

The first relation in Eq. (15.62) is the generalization of Eq. (10.25) (conservation of particle number) to GR. The second is the equation for a geodesic, Eq. (14.50).

[21] Hawking and Ellis give two other examples—the Klein-Gordon field and a charged scalar fied.[38, pp67–70]

[22] Think of an ideal gas having no pressure—it would have no temperature; there would be no randomness to the motion of the atoms. Pressure arises from short-range internal forces between the atoms of a fluid; Section 10.1.2.

[23] First show that $U_\alpha\nabla_\nu U^\alpha = 0$. Start from $\nabla_\nu(U^\alpha U_\alpha) = 0$ (because $U^\alpha U_\alpha = -c^2$); use metric compatibility to show that $U^\alpha\nabla_\nu U_\alpha = U_\alpha\nabla_\nu U^\alpha$. Then expand Eq. (15.61): $\nabla_\nu(\rho U^\mu U^\nu) = U^\mu\nabla_\nu(\rho U^\nu) + \rho U^\nu\nabla_\nu U^\mu = 0$. Contract this equation with U_μ to conclude that $\nabla_\nu(\rho U^\nu) = 0$ and hence that $U^\nu\nabla_\nu U^\mu = 0$.

These results are significant for the theory, although it might not appear so at first sight. It would be natural to surmise that of course free particles follow geodesic curves with zero four-acceleration. Bear in mind that Eq. (15.62) is a consequence of the general relativistic field equation. Each particle brings with it the energy-momentum that determines the local curvature of spacetime that in turn determines the motion, and hence the energy-momentum. The theory provides a self-consistent mechanism that embodies Wheeler's adage: Matter tells space how to curve, space tells matter how to move. Contrast with the electromagnetic field, where the field equations, i.e., Maxwell's equations, do not determine the equation of motion. If you follow the steps leading to the covariant equation of motion for a charged particle, Eq. (8.43), it relies upon the form of the Lagrangian for a charged particle, which in turn relies on the Lorentz force having been put in "by hand"; Section D.5. There's nothing like that in GR: *The field equations determine the equation of motion.* The distinction between gravitational and electromagnetic fields lies in the nonlinear nature of the Einstein equations. This discussion is continued in Section 18.9.

SUMMARY

- Einstein's field equation is $R_{\mu\nu} - \frac{1}{2}g_{\mu\nu}R + \Lambda g_{\mu\nu} = \kappa T_{\mu\nu}$, where $\kappa = 8\pi G/c^4$ and Λ is a constant. The value of κ is obtained by requiring that it reproduce the Poisson equation in the limit of weak, static gravity. An alternate way to write the Einstein equation is $R_{\mu\nu} = \kappa(T_{\mu\nu} - \frac{1}{2}g_{\mu\nu}T) + \Lambda g_{\mu\nu}$ where $T = T^{\lambda}_{\lambda}$. The unknown parameter Λ has dimension (length)$^{-2}$; it's sufficiently small that $1/\sqrt{\Lambda}$ is comparable to the size of the universe. Tests of GR on the scale of the solar system ignore Λ.
- The field equation can be derived by varying the action $S = \int \mathscr{L} \mathrm{d}^4 x$ with respect to the metric, where $\mathscr{L} = (R - 2\Lambda)\sqrt{-g} + 2\kappa\mathscr{L}_m$, with $\mathscr{L}_m$ the Lagrangian density for nongravitational fields. The fact that there is the same factor of κ for any $\mathscr{L}_m$ is the general relativistic version of the equivalence principle: Gravity couples to all fields in the same manner.

EXERCISES

15.1 Show that Eq. (15.26) can be written

$$R_{\mu\nu} = \kappa\left(T_{\mu\nu} - \frac{1}{2}g_{\mu\nu}T\right) + \Lambda g_{\mu\nu} \,.$$

Hint: Show that $R = 4\Lambda - \kappa T$.

15.2 Show that $T_{\mu\nu}\delta g^{\mu\nu} = -T^{\alpha\beta}\delta g_{\alpha\beta}$. Use Eq. (15.29).

15.3 Suppose in Eq. (15.24) we did not assume $\rho c^2 \gg P$. Show that the Poisson equation would then be given by $\nabla^2\Phi = 4\pi G(\rho + 3P/c^2)$. Even in Newtonian gravity there is an implied "$E = mc^2$" because pressure has the dimension of energy density and P/c^2 has the dimension of mass density.

15.4 Show that $\partial_\nu\left(\sqrt{-g}g^{\rho\nu}\right) = -\sqrt{-g}g^{\alpha\beta}\Gamma^{\rho}_{\alpha\beta}$. There are several parts to showing this result. You'll need Eq. (14.33), $\partial_\lambda g = g g^{\alpha\beta}\partial_\lambda g_{\alpha\beta}$. You'll also need $\partial_\lambda g^{\alpha\beta} = -g^{\alpha\beta}g^{\rho\gamma}\partial_\lambda g_{\beta\gamma}$, Eq. (15.29). Next take the definition of $\Gamma^{\alpha}_{\mu\nu}$ and contract with $g^{\mu\nu}$.

15.5 Show that $\partial_\mu\left(\sqrt{-g}g^{\rho\nu}\right) = -\sqrt{-g}\left(\Gamma^{\rho}_{\mu\lambda}g^{\lambda\nu} + \Gamma^{\nu}_{\mu\lambda}g^{\rho\lambda} - \Gamma^{\alpha}_{\mu\alpha}g^{\rho\nu}\right)$. Use Eq. (14.33), the fact that $\nabla_\mu g^{\rho\nu} = 0$, $\nabla_\mu g_{\alpha\beta} = 0$, and Eq. (14.14). Does this expression agree with the result of Exercise 15.4 when you set $\mu = \nu$ and sum? Extra credit if you recall Exercise 14.7.

15.6 In deriving the Einstein equation from a variational principle (Section 15.4), we varied the action with respect to $g^{\mu\nu}$ and obtained the field equation with covariant indices, $G_{\mu\nu} + \Lambda g_{\mu\nu} = \kappa T_{\mu\nu}$ when we defined $T_{\mu\nu}$ as in Eq. (15.44). Suppose we had chosen instead to vary with respect to $g_{\mu\nu}$? Show that the field equation is obtained with contravariant indices by varying the action with respect to $g_{\mu\nu}$ if we define

$$\sqrt{-g}T^{\mu\nu} = 2\frac{\delta \mathscr{L}_m}{\delta g_{\mu\nu}} = g^{\mu\nu}\mathscr{L}_m + 2\frac{\delta(\mathscr{L}_m/\sqrt{-g})}{\delta g_{\mu\nu}} .$$

This result can be obtained either from the variational principle or by starting from Eq. (15.44) and applying Eq. (15.29).

15.7 We found in Section 15.4.1 that starting from the Hilbert action for the gravitational field ($S = \int (R - 2\Lambda)\sqrt{-g}d^4x$), in forming the variation δS, the variation $\delta R_{\rho\nu}$ is equivalent to a four-divergence ($\sqrt{-g}g^{\rho\nu}\delta R_{\rho\nu} = \partial_\sigma(\sqrt{-g}W^\sigma)$), and makes no contribution to the vacuum equation. Instead of having to discover this in a variational calculation, is it possible to explicitly display the four-divergence from the outset? Show that

$$\sqrt{-g}R = \sqrt{-g}g^{\rho\nu}R_{\rho\nu} = \sqrt{-g}g^{\rho\nu}\left(\Gamma^\mu_{\alpha\rho}\Gamma^\alpha_{\mu\nu} - \Gamma^\mu_{\nu\rho}\Gamma^\alpha_{\mu\alpha}\right) + \partial_\sigma\left(\sqrt{-g}V^\sigma\right) . \qquad \text{(P15.1)}$$

a. Contract Eq. (15.28) with $\sqrt{-g}g^{\rho\nu}$,

$$\sqrt{-g}g^{\rho\nu}R_{\rho\nu} = \sqrt{-g}g^{\rho\nu}\partial_\mu\Gamma^\mu_{\nu\rho} - \sqrt{-g}g^{\rho\nu}\partial_\nu\Gamma^\mu_{\mu\rho} + \sqrt{-g}g^{\rho\nu}\left(\Gamma^\mu_{\mu\beta}\Gamma^\beta_{\nu\rho} - \Gamma^\mu_{\nu\beta}\Gamma^\beta_{\mu\rho}\right) .$$

b. To the two terms involving derivatives, apply the product rule of calculus,

$$\partial_\mu\left(\sqrt{-g}g^{\rho\nu}\Gamma^\mu_{\nu\rho}\right) = \sqrt{-g}g^{\rho\nu}\partial_\mu\Gamma^\mu_{\nu\rho} + \Gamma^\mu_{\nu\rho}\partial_\mu\left(\sqrt{-g}g^{\rho\nu}\right) .$$

Show that we then have the intermediate result

$$\begin{aligned}\sqrt{-g}g^{\rho\nu}R_{\rho\nu} =& \Gamma^\mu_{\mu\rho}\partial_\nu\left(\sqrt{-g}g^{\rho\nu}\right) - \Gamma^\mu_{\nu\rho}\partial_\mu\left(\sqrt{-g}g^{\rho\nu}\right) + \sqrt{-g}g^{\rho\nu}\left(\Gamma^\mu_{\mu\beta}\Gamma^\beta_{\nu\rho} - \Gamma^\mu_{\nu\beta}\Gamma^\beta_{\mu\rho}\right) \\ &+ \partial_\sigma\left(\sqrt{-g}V^\sigma\right) ,\end{aligned} \qquad \text{(P15.2)}$$

where we have the four-divergence of $V^\sigma \equiv g^{\rho\nu}\Gamma^\sigma_{\nu\rho} - g^{\rho\sigma}\Gamma^\mu_{\mu\rho}$.

c. Show that Eq. (P15.1) follows from Eq. (P15.2) using the results of Exercises 15.4 and 15.5. Thus, instead of $\sqrt{-g}R$ as the Lagrangian density, we have an equivalent Lagrangian,[24] which we'll denote as $\mathcal{L} \equiv \sqrt{-g}g^{\rho\nu}\left(\Gamma^\mu_{\alpha\rho}\Gamma^\alpha_{\mu\nu} - \Gamma^\mu_{\nu\rho}\Gamma^\alpha_{\mu\alpha}\right)$. The significance of $\mathcal{L}$ is that it contains only first derivatives of the metric tensor, whereas R contains second derivatives (Exercise 14.16). We'll use this fact in Section 18.9.

[24] Two Lagrangians can differ by the derivative of a function that vanishes on the boundary of the variation domain and yield the same equation of motion.

CHAPTER 16

The Schwarzschild metric

THE Schwarzschild metric is a static, spherically symmetric solution of the Einstein vacuum equation $R_{\mu\nu} = 0$, such as would apply exterior to a spherical, nonrotating mass[1] where the mass density is presumed small enough that $T_{\mu\nu} \approx 0$ (and $\Lambda = 0$). The Schwarzschild metric is one of a handful of exact solutions of the Einstein equation, and is undoubtedly the most important. There are small but measurable effects that occur in Schwarzschild spacetime, and such measurements constitute a crucial part of the extent to which GR has been quantitatively tested.

16.1 STATIC, SPHERICALLY SYMMETRIC SPACETIME METRICS

16.1.1 Static vs. stationary spacetimes

A spacetime is said to be *stationary* if it has a timelike Killing field, ξ^μ. Along a Killing field the metric tensor is invariant; a timelike Killing vector is the requirement that *the metric is preserved under time translations.* A spacetime is said to be *static* if it has a timelike Killing field ξ^μ that's *hypersurface orthogonal.*[2] Static spacetimes are stationary, but stationary spacetimes are not static.

The difference between the two is that stationary spacetimes have metric tensors that are time-translation symmetric, $t \to t +$ constant, whereas for static spacetimes the metric is also time-reflection invariant, $t \to -t$.[37, p120] For a static spacetime, $g_{0i} = 0$, i.e., "time" (timelike Killing vector) has no projection onto spatial axes (such as found in a rotating coordinate system, Section 12.3.1). The metric for a static spacetime therefore has the form

$$(\mathrm{d}s)^2 = -U(x^1,x^2,x^3)(\mathrm{d}x^0)^2 + \sum_{ij=1}^{3} g_{ij}(x^1,x^2,x^3)\mathrm{d}x^i\mathrm{d}x^j ,$$

where x^i are hypersurface coordinates. The Schwarzschild solution presumes a static spacetime.

16.1.2 Spherical symmetry

Having defined static spacetime, a spherically symmetric spacetime is one whose metric remains invariant under rotations. We should therefore build the metric out of rotational invariants, which for coordinates $\boldsymbol{x}$ and their differentials $\mathrm{d}\boldsymbol{x}$ are $\boldsymbol{x}\cdot\boldsymbol{x}$, $\boldsymbol{x}\cdot\mathrm{d}\boldsymbol{x}$, and $\mathrm{d}\boldsymbol{x}\cdot\mathrm{d}\boldsymbol{x}$. A spherically symmetric, static spacetime interval has the form

$$(\mathrm{d}s)^2 = -F(r)(\mathrm{d}x^0)^2 + D(r)(\boldsymbol{x}\cdot\mathrm{d}\boldsymbol{x})^2 + C(r)\mathrm{d}\boldsymbol{x}\cdot\mathrm{d}\boldsymbol{x} ,$$

[1] We can conceive of the region exterior to a star, except real stars rotate.

[2] A vector field ξ^μ is hypersurface orthogonal if it satisfies $\xi_{[\alpha}\nabla_\beta\xi_{\gamma]} = 0$.[37, p436]

where F, D, and C are functions of $r \equiv \sqrt{\boldsymbol{x} \cdot \boldsymbol{x}}$. In spherical coordinates the invariants are r^2, $r\mathrm{d}r$, and $(\mathrm{d}r)^2 + r^2((\mathrm{d}\theta)^2 + \sin^2\theta(\mathrm{d}\phi)^2)$. With these substitutions,

$$(\mathrm{d}s)^2 = -F(r)(\mathrm{d}x^0)^2 + \left[D(r)r^2 + C(r)\right](\mathrm{d}r)^2 + C(r)r^2((\mathrm{d}\theta)^2 + \sin^2\theta(\mathrm{d}\phi)^2) .$$

If we now redefine the radial coordinate $\widetilde{r} \equiv r\sqrt{C(r)}$, we can write

$$(\mathrm{d}s)^2 = -A(\widetilde{r})(\mathrm{d}x^0)^2 + B(\widetilde{r})(\mathrm{d}\widetilde{r})^2 + \widetilde{r}^2((\mathrm{d}\theta)^2 + \sin^2\theta(\mathrm{d}\phi)^2 ,$$

where A and B are yet other functions. At this point we simply erase the tilde, and start with the form of the Schwarzschild metric,

$$(\mathrm{d}s)^2 = -A(r)(c\mathrm{d}t)^2 + B(r)(\mathrm{d}r)^2 + r^2\left((\mathrm{d}\theta)^2 + \sin^2\theta(\mathrm{d}\phi)^2\right) . \tag{16.1}$$

The angular terms are often denoted $(\mathrm{d}\Omega)^2 \equiv (\mathrm{d}\theta)^2 + \sin^2\theta(\mathrm{d}\phi)^2$. In matrix form,

$$[g_{\alpha\beta}] = \begin{pmatrix} -A(r) & 0 & 0 & 0 \\ 0 & B(r) & 0 & 0 \\ 0 & 0 & r^2 & 0 \\ 0 & 0 & 0 & r^2\sin^2\theta \end{pmatrix} . \tag{16.2}$$

If $A = B = 1$ we recover the Lorentz metric in spherical coordinates. The task is to find functions[3] A and B such that $R_{\mu\nu} = 0$. To get the Ricci tensor we require the Christoffel symbols.[4]

16.2 RICCI TENSOR FOR THE SCHWARZSCHILD METRIC

We reproduce the formula for the Christoffel symbols,

$$\Gamma^{\alpha}_{\mu\nu} = \frac{1}{2}g^{\alpha\rho}\left(\partial_\mu g_{\nu\rho} + \partial_\nu g_{\mu\rho} - \partial_\rho g_{\mu\nu}\right) . \tag{14.23}$$

For an n-dimensional space, there would be n^3 possible connection coefficients (64 for $n = 4$). That number is reduced because of symmetry in the lower indices: There are $n^2(n+1)/2$ independent Christoffel symbols (40 for $n = 4$). The Schwarzschild metric is diagonal, however, which reduces the number of calculations we must do. Equation (14.23) shows that for an orthogonal coordinate system ($g_{\alpha\beta} = 0$ for $\beta \neq \alpha$), $\Gamma^{\alpha}_{\mu\nu} = 0$ if $\alpha \neq \mu \neq \nu$. How many ways are there for the indices to *not all be different*? For each α, there are n terms $\Gamma^{\alpha}_{\mu\mu}$ and $n-1$ independent terms $\Gamma^{\alpha}_{\alpha\nu}$ where $\nu \neq \alpha$. Thus, there are $n(2n-1)$ *possibly* nonzero terms (28 for $n = 4$). Table 16.1 shows the Christoffel symbols associated with the Schwarzschild metric; only nine are nonzero.

We reproduce the formula for the elements of the Ricci tensor,

$$R_{\rho\nu} = \partial_\mu\Gamma^{\mu}_{\nu\rho} - \partial_\nu\Gamma^{\mu}_{\mu\rho} + \Gamma^{\mu}_{\mu\beta}\Gamma^{\beta}_{\nu\rho} - \Gamma^{\mu}_{\nu\beta}\Gamma^{\beta}_{\mu\rho} . \tag{15.28}$$

Using Table 16.1 we find from direct calculations that $R_{\rho\nu} = 0$ for $\rho \neq \nu$. For the diagonal terms we find:

$$R_{tt} = \partial_r\Gamma^{r}_{tt} + \Gamma^{r}_{tt}\left[\Gamma^{r}_{rr} + \Gamma^{\theta}_{\theta r} + \Gamma^{\phi}_{\phi r} - \Gamma^{t}_{rt}\right] = \frac{A''}{2B} - \frac{A'}{4B}\left(\frac{B'}{B} + \frac{A'}{A}\right) + \frac{A'}{rB} ; \tag{16.3}$$

[3]Equations (16.1) and (16.2) are the form of static, spherically symmetric spacetime metrics. They become *the* Schwarzschild metric when we find $A(r)$ and $B(r)$ such that $R_{\mu\nu} = 0$. For brevity we'll refer to Eqs. (16.1) and (16.2) as the Schwarzschild metric until such a time when A and B are so determined.

[4]There is a recipe involved in GR that you'll get to know: $g_{\alpha\beta} \to \Gamma^{\alpha}_{\beta\gamma} \to R^{\alpha}_{\beta\gamma\delta} \to R_{\mu\nu} \to R \to G_{\mu\nu}$.

Table 16.1 Christoffel symbols $\Gamma^{\alpha}_{\mu\nu}$ for the Schwarzschild metric.

$\underline{\alpha = t}$	$\underline{\alpha = r}$
$\Gamma^t_{tt} = \Gamma^t_{rr} = \Gamma^t_{\theta\theta} = \Gamma^t_{\phi\phi} = \Gamma^t_{t\theta} = \Gamma^t_{t\phi} = 0$	$\Gamma^r_{rt} = \Gamma^r_{r\theta} = \Gamma^r_{r\phi} = 0$
$\Gamma^t_{tr} = \frac{1}{2} g^{tt}\partial_r g_{tt} = \dfrac{A'}{2A}$	$\Gamma^r_{rr} = \frac{1}{2} g^{rr}\partial_r g_{rr} = \dfrac{B'}{2B}$
	$\Gamma^r_{tt} = -\frac{1}{2} g^{rr}\partial_r g_{tt} = \dfrac{A'}{2B}$
	$\Gamma^r_{\theta\theta} = -\frac{1}{2} g^{rr}\partial_r g_{\theta\theta} = -\dfrac{r}{B}$
	$\Gamma^r_{\phi\phi} = -\frac{1}{2} g^{rr}\partial_r g_{\phi\phi} = -\dfrac{r}{B}\sin^2\theta$
$\underline{\alpha = \theta}$	$\underline{\alpha = \phi}$
$\Gamma^\theta_{\theta\theta} = \Gamma^\theta_{tt} = \Gamma^\theta_{rr} = \Gamma^\theta_{\theta t} = \Gamma^\theta_{\theta\phi} = 0$	$\Gamma^\phi_{\phi\phi} = \Gamma^\phi_{tt} = \Gamma^\phi_{rr} = \Gamma^\phi_{\theta\theta} = \Gamma^\phi_{\phi t} = 0$
$\Gamma^\theta_{\phi\phi} = -\frac{1}{2} g^{\theta\theta}\partial_\theta g_{\phi\phi} = -\sin\theta\cos\theta$	$\Gamma^\phi_{\phi\theta} = \frac{1}{2} g^{\phi\phi}\partial_\theta g_{\phi\phi} = \cot\theta$
$\Gamma^\theta_{\theta r} = \frac{1}{2} g^{\theta\theta}\partial_r g_{\theta\theta} = r^{-1}$	$\Gamma^\phi_{\phi r} = \frac{1}{2} g^{\phi\phi}\partial_r g_{\phi\phi} = r^{-1}$

$$\begin{aligned} R_{rr} =& - \partial_r\left(\Gamma^t_{rt} + \Gamma^\theta_{\theta r} + \Gamma^\phi_{\phi r}\right) + \Gamma^t_{rt}\left(\Gamma^r_{rr} - \Gamma^t_{rt}\right) + \Gamma^\theta_{\theta r}\left(\Gamma^r_{rr} - \Gamma^\theta_{\theta r}\right) + \Gamma^\phi_{\phi r}\left(\Gamma^r_{rr} - \Gamma^\phi_{\phi r}\right) \\ =& - \frac{A''}{2A} + \frac{A'}{4A}\left(\frac{B'}{B} + \frac{A'}{A}\right) + \frac{B'}{Br}\,; \end{aligned} \tag{16.4}$$

$$\begin{aligned} R_{\theta\theta} =& - \partial_\theta\Gamma^\phi_{\phi\theta} + \partial_r\Gamma^r_{\theta\theta} + \Gamma^r_{\theta\theta}\left(\Gamma^t_{rt} + \Gamma^r_{rr}\right) - \left(\Gamma^\phi_{\theta\phi}\right)^2 \\ =& 1 - \frac{1}{B} + \frac{r}{2B}\left(\frac{B'}{B} - \frac{A'}{A}\right)\,; \end{aligned} \tag{16.5}$$

$$\begin{aligned} R_{\phi\phi} =& \partial_\theta\Gamma^\theta_{\phi\phi} + \partial_r\Gamma^r_{\phi\phi} + \Gamma^r_{\phi\phi}\left(\Gamma^t_{rt} + \Gamma^r_{rr}\right) - \Gamma^\theta_{\phi\phi}\Gamma^\phi_{\theta\phi} \\ =& \sin^2\theta\left[1 - \frac{1}{B} + \frac{r}{2B}\left(\frac{B'}{B} - \frac{A'}{A}\right)\right] = \sin^2\theta R_{\theta\theta}\,. \end{aligned} \tag{16.6}$$

16.3 THE VACUUM SOLUTION

We seek solutions of the Einstein vacuum equation, $R_{\mu\nu} = 0$. From Eqs. (16.3)–(16.6):

$$R_{tt} = \frac{A''}{2B} - \frac{A'}{4B}\left(\frac{B'}{B} + \frac{A'}{A}\right) + \frac{A'}{rB} = 0 \qquad R_{rr} = -\frac{A''}{2A} + \frac{A'}{4A}\left(\frac{B'}{B} + \frac{A'}{A}\right) + \frac{B'}{Br} = 0$$
$$R_{\theta\theta} = 1 - \frac{1}{B} + \frac{r}{2B}\left(\frac{B'}{B} - \frac{A'}{A}\right) = 0\,.$$

We have three second-order, coupled, nonlinear differential equations for two functions, $A(r)$ and $B(r)$. By the method of solution-by-staring-at-the-equations, the terms $BR_{tt}/A + R_{rr}$ lead us to

$$\frac{A'}{A} + \frac{B'}{B} = 0\,. \tag{16.7}$$

Equation (16.7) is equivalent to $\mathrm{d}(AB)/\mathrm{d}r = 0$, which implies that

$$AB = \lambda = \text{constant}\,. \tag{16.8}$$

The functions A and B are thus inversely proportional. If we now combine Eq. (16.7) with $R_{tt} = 0$, that equation becomes $rA'' + 2A' = 0$, which is equivalent to $\mathrm{d}(r^2 A')/\mathrm{d}r = 0$, the solution of which is

$$A(r) = -\frac{k}{r} + a \,, \tag{16.9}$$

where a and k are constants. What about $R_{\theta\theta} = 0$? When combined with Eqs. (16.7) and (16.8), $R_{\theta\theta} = 0$ is equivalent to $\mathrm{d}(rA)/\mathrm{d}r = \lambda$, the solution of which is equivalent to Eq. (16.9).

The constants are determined by the requirement that as $r \to \infty$ we recover the Lorentz metric. A metric having that property is said to be *asymptotically flat*. We require that $A(r) \to 1$ and $B(r) \to 1$ as $r \to \infty$, forcing $\lambda = 1$ in Eq. (16.8) and $a = 1$ in Eq. (16.9). Thus, $B(r) = (A(r))^{-1}$. For a weak, static perturbation on the spacetime metric we found previously that

$$g_{00} \approx -\left(1 - \frac{2GM}{rc^2}\right) . \tag{15.14}$$

From Eq. (16.1), $g_{00} = -A$ and thus we choose $k = 2GM/c^2$, where M is the mass outside of which the metric form in Eq. (16.1) applies. The quantity

$$r_S \equiv \frac{2GM}{c^2} \tag{16.10}$$

is the *Schwarzschild radius*. For the sun, $r_S \approx 3$ km; for Earth $r_S \approx 1$ cm.

The *Schwarzschild metric* for empty spacetime exterior to a spherical mass M is thus

$$(\mathrm{d}s)^2 = -\left(1 - (r_S/r)\right)(c\mathrm{d}t)^2 + \frac{(\mathrm{d}r)^2}{1 - (r_S/r)} + r^2\,(\mathrm{d}\Omega)^2 \,. \tag{16.11}$$

The Lorentz metric is recovered as $r_S \to 0$, i.e., as $M \to 0$. We see that the metric becomes singular at $r = r_S$ and at $r = 0$. The first is an example of a *coordinate singularity*, where the (t, r, θ, ϕ) coordinate system breaks down at $r = r_S$. Non-flat manifolds cannot be covered by a single coordinate system—coordinate systems fail to provide unique labels for spacetime points *somewhere*. In Chapter 17 we present another coordinate system where the metric is well defined at $r = r_S$. At $r = 0$, however, GR itself breaks down: $r = 0$ is a true singularity—hopefully a quantum theory of gravity will ameliorate that conclusion. The Schwarzschild metric applies to systems where the radius R bounding M exceeds r_S, such as the sun with $R = 7 \times 10^5$ km $\gg r_S = 3$ km. As long as $r > R > r_S$, we can apply the Schwarzschild metric, where the assumption of $T_{\mu\nu} = 0$ remains valid. The other limit, where $R < r_S$ defines a black hole. A black hole has its mass entirely contained within r_S.

16.4 BIRKHOFF'S THEOREM

What if we were to relax the condition that the functions $A(r)$, $B(r)$ in the Schwarzschild metric be time-independent? Consider the time-dependent generalization of Eq. (16.2):

$$[g_{\alpha\beta}] = \begin{pmatrix} -A(r,t) & 0 & 0 & 0 \\ 0 & B(r,t) & 0 & 0 \\ 0 & 0 & r^2 & 0 \\ 0 & 0 & 0 & r^2 \sin^2\theta \end{pmatrix} . \tag{16.12}$$

You know the drill: Find the Christoffel symbols associated with the metric and then find the solutions to $R_{\mu\nu} = 0$.

The Christoffel symbols generated by Eq. (16.12) are the same as in Table 16.1 with the exception of the three nonzero symbols:

$$\Gamma^t_{tt} = \frac{1}{2} g^{tt} \partial_t g_{tt} = \frac{\dot{A}}{2A} \qquad \Gamma^t_{rr} = -\frac{1}{2} g^{tt} \partial_t g_{rr} = \frac{\dot{B}}{2A} \qquad \Gamma^r_{rt} = \frac{1}{2} g^{rr} \partial_t g_{rr} = \frac{\dot{B}}{2B} \,,$$

where the dot indicates a partial time derivative. One can show that $R_{\theta\theta}$ and $R_{\phi\phi}$ are the same as in Eqs. (16.5) and (16.6). The other components of the Ricci tensor are:

$$\begin{aligned} R_{tt} &= [R_{tt}] - \partial_t \Gamma^r_{rt} + \Gamma^r_{rt}\left(\Gamma^t_{tt} - \Gamma^r_{rt}\right) \\ &= \frac{A''}{2B} - \frac{A'}{4B}\left(\frac{B'}{B} + \frac{A'}{A}\right) + \frac{A'}{rB} - \frac{\ddot{B}}{2B} + \frac{\dot{B}}{4B}\left(\frac{\dot{A}}{A} + \frac{\dot{B}}{B}\right) ; \end{aligned} \tag{16.13}$$

$$\begin{aligned} R_{rr} &= [R_{rr}] + \partial_t \Gamma^t_{rr} + \Gamma^t_{rr}\left(\Gamma^t_{tt} - \Gamma^r_{rt}\right) \\ &= -\frac{A''}{2A} + \frac{A'}{4A}\left(\frac{B'}{B} + \frac{A'}{A}\right) + \frac{B'}{Br} + \frac{\ddot{B}}{2A} - \frac{\dot{B}}{4A}\left(\frac{\dot{A}}{A} + \frac{\dot{B}}{B}\right) ; \end{aligned} \tag{16.14}$$

$$R_{tr} = \partial_r \Gamma^r_{rt} - \partial_t \Gamma^r_{rr} + \Gamma^r_{rt}\left(\Gamma^t_{rt} + \Gamma^\theta_{\theta r} + \Gamma^\phi_{\phi r}\right) - \Gamma^r_{tt}\Gamma^t_{rr} = \frac{\dot{B}}{Br} , \tag{16.15}$$

where $[R_{tt}]$, $[R_{rr}]$ denote the terms given in Eqs. (16.3) and (16.4). Note that there is an off-diagonal term R_{tr}, Eq. (16.15), associated with the time-dependent metric.

It follows from Eqs. (15.13) and (15.14) that for $R_{tt} = R_{rr} = 0$,

$$\frac{B}{A} R_{tt} + R_{rr} = \frac{1}{r}\left(\frac{A'}{A} + \frac{B'}{B}\right) = \frac{1}{ABr}\frac{\partial}{\partial r}(AB) = 0 , \tag{16.16}$$

so that $AB = \lambda(t)$, where λ is possibly a function of t. From Eq. (16.15), in order for $R_{tr} = 0$ we must have

$$\dot{B} = 0 . \tag{16.17}$$

Combining Eqs. (16.16) and (16.17) with Eq. (16.13) and rearranging, we have

$$R_{tt} = \frac{1}{r}\frac{\partial}{\partial r}(A + rA') - \ddot{B} = 0 . \tag{16.18}$$

Using Eq. (16.16) in Eq. (16.5), the equation for $R_{\theta\theta} = 0$ is equivalent to $B^2 = B - rB'$. Using $AB = \lambda$ together with this result, we find that $A + rA' = \lambda$, which when combined with Eq. (16.18) implies that $\ddot{B} = 0$.

With $\dot{B} = \ddot{B} = 0$, Eqs. (16.13) and (16.14) revert to Eqs. (16.3) and (16.4). We can satisfy the vacuum equation with $A(r,t) = \lambda(t)A(r)$ and $B(r,t) = (A(r))^{-1}$, where $A(r)$ is the same as in Eq. (16.11). Any time dependence in $A(r,t)$ can be "waved away" by defining a new time coordinate $\mathrm{d}t' = \sqrt{\lambda}\mathrm{d}t$. At this point, we're back to the static Schwarzschild metric, Eq. (16.1). Thus, we have *Birkhoff's theorem* that *a spherically symmetric solution of the vacuum field equation is necessarily static*, with a metric given by the Schwarzschild solution, Eq. (16.11). The gravitational field outside a star that is radially pulsating, collapsing or expanding such that it maintains spherical symmetry, is static. This is analogous to the result in Newtonian theory that the gravitational field outside a spherically symmetric body is as if the entire mass is concentrated at the center: All that matters is the total mass, not how it is arranged (so long as spherical symmetry is maintained). Birkhoff's theorem is also analogous to the result in electrodynamics that a pulsating spherical charge (a charge monopole) emits no electromagnetic radiation. Gravitational radiation requires a time-dependent mass quadrupole to emit radiation, a mode of oscillation that is not spherically symmetric.

16.5 SPATIAL GEOMETRY OF THE SCHWARZSCHILD METRIC

We now look at the geometry associated with the spatial part of the Schwarzschild metric. In Section 12.3.2 we gave a procedure for finding the spatial metric tensor in terms of basis vectors orthogonal to the time direction. For a static spacetime with $g_{0i} = 0$ such a procedure is unnecessary. A time slice (t = constant) through the four-dimensional geometry of the Schwarzschild metric can be obtained by setting $\mathrm{d}t = 0$ in Eq. (16.11). This gives us the metric for a three-dimensional manifold

$$(\mathrm{d}s)^2 = \left(1 - \frac{r_S}{r}\right)^{-1} (\mathrm{d}r)^2 + r^2 \left((\mathrm{d}\theta)^2 + \sin^2\theta(\mathrm{d}\phi)^2\right) . \tag{16.19}$$

The metric in Eq. (16.19) is positive definite (spacelike hypersurface, Table 13.1). Because the Schwarzschild metric is time-independent we can speak of the coordinates (r, θ, ϕ) defined in one time slice as having an enduring permanence that describes the *same* space of a time slice at a different time. This splitting of spacetime into space and time is possible in any spacetime where the metric is time-independent.

We can visualize the geometry described by Eq. (16.19) by *embedding* part of it in a three-dimensional Euclidean space. Consider the slice obtained by setting $\mathrm{d}\theta = 0$ and $\theta = \pi/2$,

$$(\mathrm{d}s)^2 = \left(1 - \frac{r_S}{r}\right)^{-1} (\mathrm{d}r)^2 + r^2(\mathrm{d}\phi)^2 . \tag{16.20}$$

The trick to embedding a two-dimensional geometry in three dimensions is to find a Euclidean metric that yields the same distance relations as the original metric. Starting with cylindrical coordinates, $(\mathrm{d}s)^2 = (\mathrm{d}r)^2 + (\mathrm{d}z)^2 + r^2(\mathrm{d}\phi)^2$. On a two-dimensional surface $z = z(r, \phi)$,

$$(\mathrm{d}s)^2 = (\mathrm{d}r)^2 + (\mathrm{d}z)^2 + r^2(\mathrm{d}\phi)^2 = (\mathrm{d}r)^2 \left[1 + \left(\frac{\mathrm{d}z}{\mathrm{d}r}\right)^2\right] + r^2(\mathrm{d}\phi)^2 . \tag{16.21}$$

Equation (16.21) will agree with Eq. (16.20) if $1 + (\mathrm{d}z/\mathrm{d}r)^2 = (1 - (r_S/r))^{-1}$ or if

$$\left(\frac{\mathrm{d}z}{\mathrm{d}r}\right)^2 = \frac{r_S}{r - r_S} . \tag{16.22}$$

Equation (16.22) is readily integrated to yield

$$z(r) = 2\sqrt{r_S\,(r - r_S)} . \tag{16.23}$$

The surface described by Eq. (16.23) is known as *Flamm's parabaloid.*

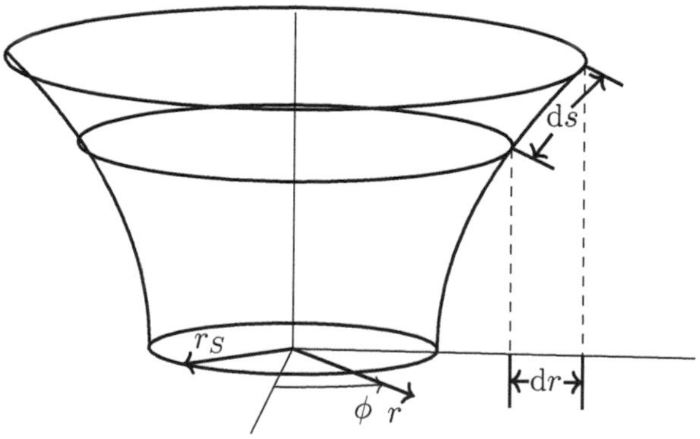

Figure 16.1 Embedding of the 2-surface with metric given by Eq. (16.20) in three-dimensional Euclidean space. The distance $\mathrm{d}s$ is not $\mathrm{d}r$, rather $\mathrm{d}s = \sqrt{g_{rr}}\mathrm{d}r$.

This surface is drawn in Fig. 16.1. The first thing to note is that the radial coordinate is defined only for $r \geq r_S$. As $r \to \infty$ (or $r_S \to 0$), Eq. (16.19) goes over to $(\mathrm{d}s)^2 = (\mathrm{d}r)^2 + r^2(\mathrm{d}\Omega)^2$, and the radial distance is given by the radial coordinate, r. As $r \to r_S$, however, changes in radial *distance* are given by $\sqrt{g_{rr}}\mathrm{d}r$. One should not confuse coordinates with distances.[5] Flamm's paraboloid

[5] On the street where I live, the addresses (*coordinates*) of the houses are 832, 850, 866, 878, and 898, even though the *distance* between houses is the same.

should not be confused with a gravitational potential well. The coordinate z in Eq. (16.21) has no physical reality: It's introduced for *visualizing* the geometry described by the metric Eq. (16.20); the surface $z(r, \phi)$ has the property that distances measured within it match the distances given by the Schwarzschild metric for the same change in radial coordinate, r. No particle could have a worldline on Flamm's paraboloid—a spacelike surface. Actual particles move along timelike trajectories.

SUMMARY

- The Schwarzschild metric,

$$[g_{\mu\nu}] = \begin{pmatrix} -(1-r_S/r) & 0 & 0 & \\ 0 & (1-r_S/r)^{-1} & 0 & 0 \\ 0 & 0 & r^2 & 0 \\ 0 & 0 & 0 & r^2\sin^2\theta \end{pmatrix},$$

where $r_S = 2GM/c^2$ is the Schwarzschild radius, is a solution of the Einstein vacuum equation $R_{\mu\nu} = 0$. It applies to regions of spacetime where space is spherically symmetric and where $T_{\mu\nu} = 0$. M is the mass outside of which the metric applies.

- Any spherically symmetric solution of the vacuum field equations must be static (Birkhoff's theorem). The spacetime surrounding a time-dependent but spherically symmetric mass is governed by the Schwarzschild metric.

- A metric is asymptotically flat when at large distances (from the mass source) the metric becomes indistinguishable from the Lorentz metric.

EXERCISES

16.1 In computing the Ricci tensor we run into the terms $\Gamma^{\mu}_{\mu\beta}\Gamma^{\beta}_{\nu\rho} - \Gamma^{\mu}_{\nu\beta}\Gamma^{\beta}_{\mu\rho}$. Show that for the Christoffel symbols in Table 22.1,

$$\begin{aligned}\Gamma^{\mu}_{\mu\beta}\Gamma^{\beta}_{\rho\nu} - \Gamma^{\mu}_{\nu\beta}\Gamma^{\beta}_{\mu\rho} =& \delta_{\rho,t}\delta_{\nu,t}\Gamma^{r}_{tt}\left[\Gamma^{r}_{rr} + \Gamma^{\theta}_{\theta r} + \Gamma^{\phi}_{\phi r} - \Gamma^{t}_{rt}\right] \\ &+\delta_{\rho,r}\delta_{\nu,r}\left[\Gamma^{t}_{rt}\left(\Gamma^{r}_{rr} - \Gamma^{t}_{rt}\right) + \Gamma^{\theta}_{\theta r}\left(\Gamma^{r}_{rr} - \Gamma^{\theta}_{\theta r}\right) + \Gamma^{\phi}_{\phi r}\left(\Gamma^{r}_{rr} - \Gamma^{\phi}_{r\phi}\right)\right] \\ &+\delta_{\rho,\theta}\delta_{\nu,\theta}\left[\Gamma^{r}_{\theta\theta}\left(\Gamma^{r}_{rt} + \Gamma^{r}_{rr}\right) - \left(\Gamma^{\phi}_{\phi\theta}\right)^2\right] \\ &+\delta_{\rho,\phi}\delta_{\nu,\phi}\left[\Gamma^{r}_{\phi\phi}\left(\Gamma^{t}_{rt} + \Gamma^{r}_{rr}\right) - \Gamma^{\theta}_{\phi\phi}\Gamma^{\phi}_{\theta\phi}\right] .\end{aligned}$$

16.2 Likewise, in computing $R_{\mu\nu}$ we run into the terms $\partial_\mu\Gamma^{\mu}_{\nu\rho} - \partial_\nu\Gamma^{\mu}_{\mu\rho}$. Show for the Christoffel symbols in Table 22.1 that

$$\begin{aligned}\partial_\mu\Gamma^{\mu}_{\nu\rho} - \partial_\nu\Gamma^{\mu}_{\mu\rho} =& \delta_{\nu,t}\delta_{\rho,t}\partial_r\Gamma^{r}_{tt} - \delta_{\nu,r}\delta_{\rho,r}\partial_r\left[\Gamma^{t}_{tr} + \Gamma^{\theta}_{\theta r} + \Gamma^{\phi}_{\phi r}\right] \\ &+\delta_{\nu,\theta}\delta_{\rho,\theta}\left[-\partial_\theta\Gamma^{\phi}_{\phi\theta} + \partial_r\Gamma^{r}_{\theta\theta}\right] + \delta_{\nu,\phi}\delta_{\rho,\phi}\left[\partial_\theta\Gamma^{\theta}_{\phi\phi} + \partial_r\Gamma^{r}_{\phi\phi}\right] .\end{aligned}$$

From the results of this and the previous problem we see that $R_{\nu\rho}$ is diagonal for the static Schwarzschild metric.

16.3 Show directly from (22.4) that $R_{\mu\nu} = 0$ for $\mu \neq \nu$ for the static Schwarzschild metric. Pick one off-diagonal term, e.g., $R_{r\theta}$, and use (22.4) to show that it vanishes.

16.4 Verify the results given for the diagonal terms $R_{\alpha\alpha}$. Pick one term, e.g., R_{tt}, and show from (22.4) that the result given is correct.

16.5 Show that (22.9) implies that $AB =$ constant.

16.6 Show that (22.11) follows from $R_{tt} = 0$ combined with (22.9).

16.7 What is the magnitude of the Schwarzschild radius for the Sun, in km?

16.8 Show that (22.18) is correct. Using $\Gamma^{\alpha}_{\mu\nu}$ associated with the time-dependent Schwarzschild metric, show that $\partial_r \Gamma^r_{rt} = \partial_t \Gamma^r_{rr}$ and $\Gamma^r_{rt}\Gamma^t_{rt} = \Gamma^r_{tt}\Gamma^t_{rr}$.

16.9 Consider the spatial part of the metric for a rotating coordinate system, (18.47),

$$[\gamma_{ij}] = \begin{pmatrix} 1 & 0 & 0 \\ 0 & \gamma^2\rho^2 & 0 \\ 0 & 0 & 1 \end{pmatrix} ,$$

where $i = \rho, \phi, z$ and $\gamma = 1/\sqrt{1-(\omega\rho)^2/c^2}$. Show that the maximum possible number of nonzero, independent Christoffel symbols for this metric is 15. Show that only two are nonzero,

$$\Gamma^{\rho}_{\phi\phi} = -\frac{1}{2} g^{\rho\rho}\partial_\rho g_{\phi\phi} = -\rho\gamma^4 \qquad \Gamma^{\phi}_{\phi\rho} = \frac{1}{2} g^{\phi\phi}\partial_\rho g_{\phi\phi} = \frac{\gamma^2}{\rho} .$$

16.10 Calculate R_{ij} for the spatial geometry associated with the rotating disk (previous problem). Show that the nonzero terms are

$$R_{\rho\rho} = -\partial_\rho \Gamma^{\phi}_{\phi\rho} - \left(\Gamma^{\phi}_{\rho\phi}\right)^2 = -3\gamma^4\frac{\omega^2}{c^2} \qquad R_{\phi\phi} = \partial_\rho \Gamma^{\rho}_{\phi\phi} - \Gamma^{\phi}_{\phi\rho}\Gamma^{\rho}_{\phi\phi} = -3\gamma^6\frac{\omega^2\rho^2}{c^2} .$$

16.11 Calculate the curvature scalar for the spatial geometry of the rotating disk (previous problem). Show that

$$R = R^{\rho}_{\rho} + R^{\phi}_{\phi} = -6\gamma^4\frac{\omega^2}{c^2} = -\frac{6\omega^2}{c^2(1-\rho^2\omega^2/c^2)^2} .$$

This is not a space of constant curvature.

16.12 Show that the Schwarzschild metric (22.13) takes the isotropic form (22.22) with A and B given by (22.24) through the change of variable given in (22.23).

16.13 Consider the spatial part of the Schwarzschild metric, (22.25). Calculate the elements of the Ricci tensor for this metric. Show that

$$R^r_r = -\frac{r_S}{r^3} \qquad R^{\theta}_{\theta} = \frac{r_S}{2r^3} \qquad R^{\phi}_{\phi} = \frac{r_S}{2r^3} .$$

What is the curvature scalar for this three-dimensional manifold?

16.14 How many Killing vectors does the Schwarzschild metric have? Write down the 10 Killing equations; use (19.22), $\pounds_\xi g_{\mu\nu} = 0$, so you don't have to get into the Christoffel symbols. First work out the Killing equations for the (θ, ϕ), (ϕ, ϕ), and (θ, θ) coordinates. Compare your results with those of Problem 20.34. To preserve the symmetry of the sphere, we must take $\xi^r = 0$. The components ξ^θ and ξ^ϕ are thus those of the unit sphere, which have three free parameters. You should find that the remaining Killing equations imply that $\xi^0 =$ constant. There are thus four Killing vectors associated with the Schwarzschild metric.

CHAPTER 17

Physical effects of Schwarzschild spacetime

WE consider the physical effects associated with Schwarzschild spacetime, noting in particular the predictions of the theory that have been experimentally tested.

17.1 GEODESICS IN SCHWARZSCHILD SPACETIME

The differential equation for a geodesic is given by Eq. (14.51), which we reproduce here:

$$\frac{\mathrm{d}^2 x^\mu}{\mathrm{d}\tau^2} + \Gamma^\mu_{\alpha\beta}\frac{\mathrm{d}x^\alpha}{\mathrm{d}\tau}\frac{\mathrm{d}x^\beta}{\mathrm{d}\tau} = 0\,. \tag{14.51}$$

The Christoffel symbols for the Schwarzschild metric are given in Table 16.1. With this information we can calculate the worldlines of freely falling particles exterior to a spherical mass.

17.1.1 Angular coordinates

Let's first consider the angular variables.[1] Using Eq. (14.51) and the nonzero entries in Table 16.1, the geodesic equation for ϕ is

$$\ddot{\phi} + 2\Gamma^\phi_{\phi\theta}\dot{\phi}\dot{\theta} + 2\Gamma^\phi_{\phi r}\dot{\phi}\dot{r} = \ddot{\phi} + 2\cot\theta\dot{\phi}\dot{\theta} + \frac{2}{r}\dot{\phi}\dot{r} = 0\,, \tag{17.1}$$

where the dot signifies a derivative with respect to τ, and where the factors of two account for $\Gamma^\phi_{\phi\theta} = \Gamma^\phi_{\theta\phi}$ and $\Gamma^\phi_{\phi r} = \Gamma^\phi_{r\phi}$. Equation (17.1) is equivalent to

$$\frac{1}{r^2\sin^2\theta}\frac{\mathrm{d}}{\mathrm{d}\tau}\left(r^2\sin^2\theta\dot{\phi}\right) = 0\,.$$

Thus there is a constant of the motion having the dimensions of angular momentum per mass

$$r^2\sin^2\theta\dot{\phi} = l = \text{constant}\,. \tag{17.2}$$

The geodesic equation for θ is

$$\ddot{\theta} + 2\Gamma^\theta_{\theta r}\dot{\theta}\dot{r} + \Gamma^\theta_{\phi\phi}(\dot{\phi})^2 = \ddot{\theta} + \frac{2}{r}\dot{\theta}\dot{r} - \sin\theta\cos\theta\dot{\phi}^2 = 0\,. \tag{17.3}$$

[1] Reminiscent of quantum mechanics, where we solve "once and for all" the angular part of the wave function (the spherical harmonics, $Y_{lm}(\theta,\phi)$) for a spherically symmetric potential.

Equation (17.3) is equivalent to

$$\frac{1}{r^2}\frac{\mathrm{d}}{\mathrm{d}\tau}\left(r^2\dot{\theta}\right) = \sin\theta\cos\theta\dot{\phi}^2 \,. \tag{17.4}$$

If $\theta = \pi/2$ in Eq. (17.4), we have $r^2\dot{\theta} =$ constant. Then if $\dot{\theta} = 0$ at one time, $\dot{\theta} = 0$ for all times. By spherical symmetry, we're free to orient the coordinate system so that $\theta = \pi/2$ and $\dot{\theta} = 0$ at an instant of time. We take, without loss of generality, $\dot{\theta} = 0$ and $\theta = \pi/2$. *Trajectories are planar in the vicinity of a spherical mass.* With $\theta = \pi/2$ in Eq. (17.2), we have $r^2\dot{\phi} =$ constant—seemingly Kepler's second law. The spatial geometry associated with the Schwarzschild metric is not Euclidean, however, and we cannot speak of "areas swept out"; yet we have formally $r^2\dot{\phi} =$ constant.[2]

17.1.2 Time coordinate

The geodesic equation for t is

$$\ddot{t} + 2\Gamma^t_{tr}\dot{t}\dot{r} = \ddot{t} + \frac{A'}{A}\dot{t}\dot{r} = \frac{1}{A}\frac{\mathrm{d}}{\mathrm{d}\tau}\left(\dot{t}A\right) = 0 \,, \tag{17.5}$$

where $\dot{t} = \mathrm{d}t/\mathrm{d}\tau$ is the derivative of the *coordinate* t with respect to τ. In SR, $\dot{t} = \gamma$; here we leave $\dot{t}$ unspecified, as a differential equation for one of the spacetime coordinates. Equation (17.5) implies another constant of the motion,

$$\dot{t}A = \dot{t}\left(1 - \frac{r_S}{r}\right) = k = \text{constant} \,, \tag{17.6}$$

where k is dimensionless. For $r \gg r_S$, $\dot{t} \to$ constant. In SR (free-particle motion for $r \gg r_S$), $\dot{t} = \gamma$ and $E = \gamma mc^2$. We may interpret k as $E/(mc^2)$, where E is the total energy including the gravitational potential energy along a timelike geodesic.

17.1.3 Radial coordinate

The geodesic equation for r is $\ddot{r} + \Gamma^r_{rr}\dot{r}^2 + c^2\Gamma^r_{tt}\dot{t}^2 + \Gamma^r_{\theta\theta}\dot{\theta}^2 + \Gamma^r_{\phi\phi}\dot{\phi}^2 = 0$. Setting $\dot{\theta} = 0$, $\theta = \pi/2$, making use of the results in Table 16.1, and using Eqs. (17.6) and (17.2), we have

$$\ddot{r} + \frac{B'}{2B}\dot{r}^2 + \frac{A'}{2A^2B}c^2k^2 - \frac{l^2}{Br^3} = 0 \,, \tag{17.7}$$

where $B = (1 - r_S/r)^{-1}$. Equation (17.7) is equivalent to

$$\frac{1}{2\dot{r}B}\frac{\mathrm{d}}{\mathrm{d}\tau}\left(\dot{r}^2B - \frac{k^2c^2}{A} + \frac{l^2}{r^2}\right) = 0 \,.$$

We thus have a fourth constant of the motion,

$$\dot{r}^2B - \frac{k^2c^2}{A} + \frac{l^2}{r^2} = \text{constant} \,. \tag{17.8}$$

This constant is straightforward to evaluate because it's related to the spacetime separation: $-c^2(\mathrm{d}\tau)^2 = g_{\mu\nu}\mathrm{d}x^\mu\mathrm{d}x^\nu$. Divide by $(\mathrm{d}\tau)^2$, use the Schwarzschild metric and $\dot{\theta} = 0$; we have $-c^2 = g_{\mu\nu}\dot{x}^\mu\dot{x}^\nu = -Ac^2\dot{t}^2 + B\dot{r}^2 + r^2\dot{\phi}^2$, which is the same as Eq. (17.8). The constant in Eq. (17.8) is then $-c^2$ for material particles. For photons, however, $(\mathrm{d}s)^2 = 0$, and the constant is zero. We'll represent both cases with $-c^2\alpha$ where $\alpha = 1$ for particles and $\alpha = 0$ for photons.

[2] Kepler's first law (elliptical orbits) is not obeyed in general, and only approximately in the vicinity of the sun.

Equation (17.8) is equivalent to

$$\frac{1}{2}\dot{r}^2 - \alpha\frac{GM}{r} + \frac{l^2}{2r^2} - \frac{l^2GM}{c^2r^3} = \frac{1}{2}c^2(k^2 - \alpha) \equiv \mathscr{E} \,. \tag{17.9}$$

Equation (17.9) has the form of the nonrelativistic equation of motion in the effective potential

$$V_{\text{eff}}(r) \equiv -\alpha\frac{GM}{r} + \frac{l^2}{2r^2} - \frac{l^2GM}{c^2r^3} \,. \tag{17.10}$$

The first two terms in Eq. (17.10) are the Newtonian potential function for a particle gravitationally interacting with M. The third is relativistic in origin; it vanishes as $c \to \infty$. This term is dominant for small r. If we write Eq. (17.9) as $\frac{1}{2}m\dot{r}^2 + mV_{\text{eff}}(r) = m\mathscr{E}$, it resembles the Newtonian statement of energy conservation, with the conserved quantity $\frac{1}{2}mc^2(k^2 - \alpha)$. If we write the energy of a particle ($\alpha = 1$) as $E = mc^2 + E_N$, where E_N is the Newtonian total energy, $k = E/(mc^2) = 1 + E_N/(mc^2)$. Equation (17.9) can then be written $\frac{1}{2}m\dot{r}^2 + mV_{\text{eff}}(r) = E_N(1 + E_N/(2mc^2))$, which is consistent with Newtonian energy conservation when $|E_N| \ll mc^2$. Equation (17.9) is *almost* in the form of the Newtonian equation for energy conservation, but not exactly.

17.1.4 Orbit equation

Often what's wanted is not $r(\tau)$ but the closely related quantity $r(\phi)$, the *orbit equation*, the differential equation for which can be obtained from $\mathrm{d}r/\mathrm{d}\phi = (\mathrm{d}r/\mathrm{d}\tau)(\mathrm{d}\tau/\mathrm{d}\phi) = \dot{r}/\dot{\phi}$. As noted in Section 7.4, it's easier to work with the variable $u \equiv r^{-1}$. Using Eq. (17.2),

$$\frac{\mathrm{d}u}{\mathrm{d}\phi} = -\frac{1}{r^2}\frac{\mathrm{d}r}{\mathrm{d}\phi} = -\frac{1}{r^2}\frac{\dot{r}}{\dot{\phi}} = -\frac{1}{l}\dot{r} \,.$$

Thus, $\mathrm{d}u/\mathrm{d}\phi$ is proportional to $\mathrm{d}r/\mathrm{d}\tau$ because l is constant. Using Eq. (17.9) we have

$$\left(\frac{\mathrm{d}u}{\mathrm{d}\phi}\right)^2 = 2\frac{\mathscr{E}}{l^2} + \frac{2\alpha GM}{l^2}u - u^2 + \frac{2GM}{c^2}u^3 \,. \tag{17.11}$$

By differentiating Eq. (17.11), we obtain a second-order differential equation:

$$\frac{\mathrm{d}^2u}{\mathrm{d}\phi^2} + u = \alpha\frac{GM}{l^2} + 3\frac{GM}{c^2}u^2 \,. \tag{17.12}$$

Both Eqs. (17.11) and (17.12) are used as the equations governing orbits.

The Newtonian version of Eq. (17.12) ($\alpha = 1$ and $c \to \infty$) is

$$\frac{\mathrm{d}^2u}{\mathrm{d}\phi^2} + u = \frac{GM}{l^2} \,, \qquad \text{(nonrelativistic)} \tag{17.13}$$

the solution of which is $u = GM/l^2 + A\cos(\phi - \phi_0)$, where A and ϕ_0 are constants. Setting $\phi_0 = 0$, the nonrelativistic orbit equation is

$$r = u^{-1} = (GM/l^2 + A\cos\phi)^{-1} \equiv \frac{p}{1 + \epsilon\cos\phi} \,, \tag{17.14}$$

where $p = l^2/(GM)$ is the distance at $\phi = \pi/2$, and the eccentricity $\epsilon = (r_{\text{max}} - r_{\text{min}})/(r_{\text{max}} + r_{\text{min}})$. The distance $a = \frac{1}{2}(r_{\text{max}} + r_{\text{min}}) = p/(1 - \epsilon^2)$ is the *semimajor axis*; thus $p = a(1 - \epsilon^2)$.

Comparing Eqs. (17.12) and (17.13), it might appear that the only difference between the relativistic and nonrelativistic theories—a difference achieved through considerable theoretical development—is the single extra term in Eq. (17.12). Such an observation is misleading, however. Whereas in nonrelativistic mechanics r represents the actual radial distance, in GR r is a *coordinate*; the distance is obtained from the metric tensor.

17.2 PARTICLE TRAJECTORIES

Understanding the allowed types of particle trajectories[3] is greatly aided by familiarity with the effective potential function, Eq. (17.10) with $\alpha = 1$. It's useful to write V_{eff} in dimensionless form. In terms of $y \equiv r/r_S$, Eq. (17.10) can be written

$$\frac{1}{c^2}V_{\text{eff}}(y) = \frac{1}{2}\left[-\frac{1}{y} + \frac{\xi^2}{y^2} - \frac{\xi^2}{y^3}\right] , \tag{17.15}$$

where $\xi \equiv lc/(2GM)$ is a dimensionless parameter characterizing the angular momentum of the trajectory. Figure 17.1 shows V_{eff}/c^2 for $\xi = 2.25$.

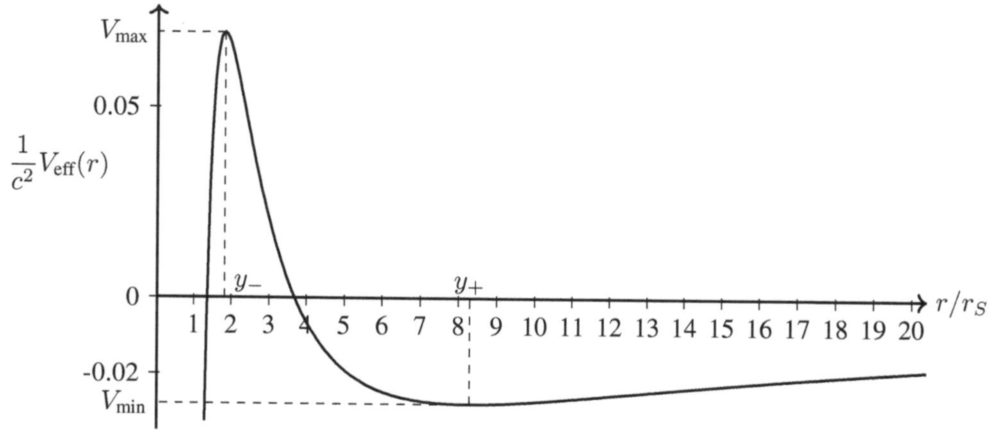

Figure 17.1 Effective gravitational potential for particles ($\xi = 2.25$).

As $y \to 0$, the y^{-1}, y^{-3} terms in V_{eff} diverge to negative infinity, while the y^{-2} term (the *centrifugal barrier*) diverges to positive infinity. The y^{-3} term dominates for small y, but whether the function has local extrema depends on the value of ξ. Taking the derivative of Eq. (17.15), $V'_{\text{eff}} = c^2(y^2 - 2y\xi^2 + 3\xi^2)/(2y^4)$. The roots of $y^2 - 2y\xi^2 + 3\xi^2 = 0$ are

$$y_\pm \equiv \xi^2 \pm \sqrt{\xi^4 - 3\xi^2} \, . \tag{17.16}$$

Real and distinct roots of Eq. (17.16) require $\xi > \sqrt{3}$. For $\xi = \sqrt{3}$, there are inflection points at $y = 6$ and $y = 3$. Thus, for $\xi \leq \sqrt{3}$, an inwardly directed particle spirals towards $y \to 0$. Of course, the Schwarzschild metric is defined only for $r > r_S$ ($y > 1$). The *zeros* of $V_{\text{eff}}(y)$ occur at the roots of $y^2 - \xi^2 y + \xi^2 = 0$, the two values $(\xi^2 \pm \sqrt{\xi^4 - 4\xi^2})/2$. Real and distinct zeros require $\xi > 2$. Thus, only for $\xi > 2$ does $V_{\text{eff}}(y)$ have the shape indicated in Fig. 17.1 where the maximum at $y = y_-$ has a positive value. The maximum and minimum values of the potential are

$$\frac{1}{c^2}V_{\text{eff}}(y_\pm) = \frac{1}{54\xi^2}y_\mp\,(y_\mp - 6) \, . \tag{17.17}$$

In general, $y_+ \geq 3$ for $\xi \geq \sqrt{3}$, with $y_+ \sim 2\xi^2$ for $\xi \gg \sqrt{3}$. The other root satisfies the inequality $\frac{3}{2} \leq y_- \leq 3$ for $\sqrt{3} \leq \xi < \infty$, with the lower bound attained for $\xi \to \infty$. For any ξ, the roots satisfy $y_+ y_- = 3\xi^2$. For $\xi \gg \sqrt{3}$, $V_{\text{max}} \sim \frac{2}{27}\xi^2$ and $V_{\text{min}} \sim \frac{1}{24}\xi^{-2}$.

[3]Trajectory or orbit? An orbit is the gravitationally curved trajectory of an object, so in principle there is no difference between the two terms, and we'll use them interchangeably. Yet, orbit tends to connote closed, regularly repeating orbits, such as the periodic elliptical path of planets in the solar system. In GR, closed orbits are rare, with non-repeating trajectories the norm.

The form of $V_{\text{eff}}(y)$ (Fig. 17.1) indicates the kinds of particle trajectories that can occur. Depending on the energy $\mathscr{E}$—see Eq. (17.9), we can have *bound orbits* for $V_{\text{min}} < \mathscr{E} < 0$, where a particle oscillates between the turning points (a special case is a circular orbit for $\mathscr{E} = V_{\text{min}}$). We'll see that we don't get closed orbits as in Newtonian mechanics (Keplerian ellipses); rather the line of apsides precesses (see Section 7.4). *Scattering orbits* occur for $0 < \mathscr{E} < V_{\text{max}}$, unbound trajectories where a particle comes in from infinity, scatters from V_{eff}, and returns to infinity. For $\mathscr{E} > V_{\text{max}}$, we have a new type of orbit—*plunge orbits*, where a particle spirals toward $y \to 1$. Plunge orbits for $l \neq 0$ are a feature exclusive to GR; in Newtonian mechanics such orbits are precluded (the centrifugal barrier becomes indefinitely large as $y \to 0$). In the following we analyze two special cases: The radial plunge orbit ($l = 0$) and the circular orbit. We return to bound orbits in Section 17.5.

17.2.1 Radial plunge trajectory

A particle inwardly directed toward M with constant ϕ has zero angular momentum, $l = 0$, Eq. (17.2). In this case, we cannot use Eq. (17.12), which presumes $l \neq 0$. From Eq. (17.9), however, with $l = 0$ and $\alpha = 1$,

$$\dot{r}^2 - \frac{2GM}{r} = c^2(k^2 - 1) \,. \tag{17.18}$$

Differentiating Eq. (17.18),

$$\ddot{r} = -\frac{GM}{r^2} \,, \tag{17.19}$$

seemingly the same as Newtonian gravity! This development should be not too surprising: By setting $l = 0$ we have "wiped out" the part of the effective potential that's unique to GR. It should be kept in mind, however, that r is a *coordinate* (not a distance) and that the derivative in Eq. (17.19) is with respect to τ, the proper time, not the Newtonian absolute time. Appearances can be deceiving.

The constant k is determined by the initial conditions. From Eq. (17.18),

$$c^2(k^2 - 1) = \dot{r}_0^2 - \frac{2GM}{r_0} \,, \tag{17.20}$$

where r_0 is the initial coordinate and $\dot{r}_0$ is the initial coordinate speed. For a particle released from rest at $r = r_0$,

$$k^2 = 1 - \frac{2GM}{r_0 c^2} = 1 - \frac{r_S}{r_0} \,. \tag{17.21}$$

Because $k^2 > 0$, Eq. (17.21) applies for $r_0 > r_S$. Combining Eqs. (17.21) and (17.18),

$$\dot{r}^2 = 2GM\left(\frac{1}{r} - \frac{1}{r_0}\right) \,. \qquad \text{(particle released from rest)} \tag{17.22}$$

Equation (17.22) has the same form as the Newtonian equation of energy conservation for a particle in a gravitational field. Because $\dot{r}^2 > 0$, Eq. (17.22) is defined only for $r < r_0$, i.e., for a radially in-falling particle.

The proper time for an observer starting from rest at $r = r_0$ can be calculated from Eq. (17.22),

$$\dot{r} = \frac{\mathrm{d}r}{\mathrm{d}\tau} = -\sqrt{2GM\left(\frac{1}{r} - \frac{1}{r_0}\right)} \,, \tag{17.23}$$

where we've taken the negative square root to account for an inwardly directed particle. From Eq. (17.23) we have, assuming $\tau = 0$ at $r = r_0$,

$$\tau(r) = \frac{1}{\sqrt{2GM}} \int_r^{r_0} \mathrm{d}r \sqrt{\frac{r r_0}{r_0 - r}} \,. \tag{17.24}$$

The integral can be evaluated,

$$\tau(r) = \frac{r_0^{3/2}}{\sqrt{2GM}} \left[\cos^{-1}\left(\sqrt{\frac{r}{r_0}}\right) + \sqrt{\frac{r}{r_0}}\sqrt{1-\frac{r}{r_0}} \right] . \tag{17.25}$$

The most important conclusion to draw from Eq. (17.25) is that, for finite r_0, *the particle reaches the Schwarzschild radius in finite proper time.*

In terms of the *coordinate* time, however, it takes an *infinite* amount of time to reach the Schwarzschild radius! We can eliminate the proper time:

$$\frac{\mathrm{d}t}{\mathrm{d}r} = \frac{\mathrm{d}t}{\mathrm{d}\tau}\frac{\mathrm{d}\tau}{\mathrm{d}r} = \frac{\dot{t}}{\dot{r}} = -\frac{1}{(1-r_S/r)}\frac{1}{c}\sqrt{\frac{r(r_0-r_S)}{r_S(r_0-r)}} , \tag{17.26}$$

where we've used Eqs. (17.6), (17.21) and (17.22). Assuming $t(r_0) = 0$,

$$t(r) = \frac{\sqrt{r_0 - r_S}}{\sqrt{2GM}} \int_r^{r_0} \frac{r^{3/2}}{(r-r_S)\sqrt{r_0 - r}} \mathrm{d}r . \tag{17.27}$$

The integral in Eq. (17.27) diverges as $r \to r_S$. Because $r_S < r < r_0$, we have the inequality

$$\int_r^{r_0} \frac{r^{3/2}}{(r-r_S)\sqrt{r_0-r}} \mathrm{d}r > \frac{r_S^{3/2}}{\sqrt{r_0}} \int_r^{r_0} \frac{\mathrm{d}r}{r - r_S} = \frac{r_S^{3/2}}{\sqrt{r_0}} \ln\left(\frac{r_0 - r_S}{r - r_S}\right) ,$$

and the right side diverges as $r \to r_S$.

We thus have a *qualitative* difference: To an observer at infinity (where $\mathrm{d}\tau = \mathrm{d}t$) a particle falling toward the origin never reaches the Schwarzschild radius; an observer comoving with the particle, however, does not find anything peculiar as it reaches the Schwarzschild radius in finite proper time. This kind of discrepancy is a symptom of the coordinate singularity in the Schwarzschild metric noted in Section 16.3. We can appreciate what's happening here by considering the *coordinate speed*, $|\mathrm{d}r/\mathrm{d}t|$. To simplify matters, let $r_0 \to \infty$. We have from Eq. (17.26) in this limit,

$$\frac{1}{c}\left|\frac{\mathrm{d}r}{\mathrm{d}t}\right| = \sqrt{r_S}\frac{r - r_S}{r^{3/2}} . \qquad (r_0 \to \infty)$$

In Schwarzschild coordinates, a particle released from rest at infinity reaches a maximum coordinate speed of $\approx 0.385c$ at $r = 3r_S$, and then declines to zero as $r \to r_S$; it "never gets there." Contrast with Newtonian mechanics where a particle falling from rest at infinity has speed

$$\frac{1}{c}\left|\frac{\mathrm{d}r}{\mathrm{d}t}\right|_N = \sqrt{\frac{r_S}{r}} \qquad \text{(Newtonian mechanics)}$$

at position r. The Newtonian speed monotonically goes to c as $r \to r_S$.

17.2.2 Circular orbits: $r \geq 3r_S$

Setting $\mathrm{d}^2u/\mathrm{d}\phi^2 = 0$ and $\alpha = 1$ in Eq. (17.12), $u = GM/l^2 + (3GM/c^2)u^2$. Solving for l^2,

$$l^2 = \frac{GMr^2}{r - 3GM/c^2} , \tag{17.28}$$

where $r = u^{-1}$ is the radius of the orbit. We cannot have circular orbits for $r < \frac{3}{2}r_S$ (because $l^2 > 0$). As we now show, the minimum *stable* circular orbit occurs for $r = 3r_S$. From Eq. (17.28), the radii of circular orbits can be given in terms of the angular momentum, $r = r(l)$. In terms of

$\xi = lc/(2GM)$ we find that $r(\xi) = r_S y_\pm(\xi)$, where $y_\pm$ are given in Eq. (17.16). Only the orbits given by $r = r_S y_+(\xi)$ are stable because $V''_{\text{eff}}(y_+) > 0$ ($V''_{\text{eff}}(y_-) < 0$). Because in general $y_+ \geq 3$, the smallest stable circular orbit occurs at $r = 3r_S$. The existence of a minimum stable circular orbit has interesting astrophysical consequences in terms of *accretion disks* surrounding massive objects.

The constant k can be found by combining Eq. (17.28) with Eq. (17.9):

$$k^2 = \frac{(1 - r_S/r)^2}{1 - \frac{3}{2}r_S/r} \; . \tag{17.29}$$

For bound orbits we require $\mathscr{E} < 0$, or that $0 < k^2 < 1$, which from Eq. (17.29) is satisfied for $r > 2r_S$. In general, $k^2 \leq 8/9$ for $r \geq 3r_S$, and thus the energy of a circular orbit is such that $-1/2 < \mathscr{E}/c^2 \leq -1/18$.

17.3 RADIAL NULL GEODESICS: KRUSKAL COORDINATES

We claimed in Section 16.3 that coordinate systems exist in which no singularity occurs at $r = r_S$. To see that's the case, consider an invariant formed from the Riemann tensor (of which there are many). The *Kretschmann scalar* is the contraction[4] $K \equiv R^{\alpha\beta\gamma\delta}R_{\alpha\beta\gamma\delta}$. For the Schwarzschild metric, one finds [61, p332] $K = 48G^2M^2/(c^4r^6)$. This value of K is specific to the Schwarzschild metric.[5] The salient point is that K is not singular at $r = r_S$. Thus, *coordinate systems exist in which there is no singularity at* $r = r_S$: K is a scalar and has the same value in all coordinate systems.

Such coordinate systems can be developed using radial null geodesics.[6] With $l = 0$ and $\alpha = 0$ in Eq. (17.9), $\dot{r}^2 = c^2k^2$, which, combined with Eq. (17.6), is equivalent to $\dot{r}^2 = c^2\dot{t}^2(1 - r_S/r)^2$. We can eliminate the proper time,[7]

$$\left(\frac{\mathrm{d}t}{\mathrm{d}r}\right)^2 = \left(\frac{\mathrm{d}t}{\mathrm{d}\tau}\frac{\mathrm{d}\tau}{\mathrm{d}r}\right)^2 = \left(\frac{\dot{t}}{\dot{r}}\right)^2 = \frac{1}{c^2\left(1 - r_S/r\right)^2} \; . \tag{17.30}$$

Integrate Eq. (17.30) (keeping the integral indefinite):

$$ct = \pm \int \frac{r\mathrm{d}r}{r - r_S} = \pm\left[r + r_S \ln\left((r/r_S) - 1\right)\right] + \text{constant} \; .$$

Define a new coordinate

$$r^* \equiv r + r_S \ln\left((r/r_S) - 1\right) \; , \tag{17.31}$$

so that in terms of this coordinate the radial null geodesics are straight lines:

$$ct = \pm r^* + \text{constant} \; , \tag{17.32}$$

where the plus (minus) sign indicates outgoing (incoming) light rays. Define new coordinates:[8]

$$u \equiv ct - r^* \qquad\qquad v \equiv ct + r^* \; . \tag{17.33}$$

[4]The Kretschmann scalar is analogous to the invariant we obtained for the electromagnetic field $F^{\mu\nu}F_{\mu\nu}$, Section 8.8.

[5]To derive the value of K, it's helpful to use the result of Exercise 14.15 (together with $g_{\alpha\beta}$ and the Christoffel symbols associated with the Schwarzschild metric).

[6]The "trouble" with the Schwarzschild metric at $r = r_S$ involves the time and radial coordinates. Restrict our attention to the two-dimensional space associated with $\mathrm{d}\Omega = 0$.

[7]We've used this step in deriving the orbit equation, and in deriving the coordinate time for radial plunge orbits for particles. Equation (17.30) could appear problematic because $\mathrm{d}\tau = 0$ for photons. Null geodesics can be described by coordinates $x^\alpha(\lambda)$ as functions of an affine parameter λ (Section 14.3.5), with the null vector $\mathrm{d}x^\alpha/\mathrm{d}\lambda$ tangent to the worldline.

[8]From Eq. (17.32), (u, v) are *constant* on null trajectories; each null trajectory is labeled by values of (u, v).

In terms of these coordinates,[9] from Eq. (16.11) for $\mathrm{d}\Omega = 0$:

$$(\mathrm{d}s)^2 = -(1 - r_S/r)\mathrm{d}u\mathrm{d}v \,, \tag{17.34}$$

where now r is defined implicitly through Eq. (17.31): $r^* = (v-u)/2$. Using Eqs. (17.31) and (17.33), it's straightforward to show

$$1 - \frac{r_S}{r} = \frac{r_S}{r}\mathrm{e}^{r^*/r_S}\mathrm{e}^{-r/r_S} = \frac{r_S}{r}\mathrm{e}^{-r/r_S}\mathrm{e}^{(v-u)/(2r_S)} \,. \tag{17.35}$$

Eliminate the exponentials of u and v through another coordinate transformation:

$$U \equiv -\mathrm{e}^{-u/(2r_S)} \qquad V \equiv \mathrm{e}^{v/(2r_S)} \,. \tag{17.36}$$

Using Eq. (17.36), Eq. (17.34) can be written:

$$(\mathrm{d}s)^2 = -\frac{4r_S^3}{r}\mathrm{e}^{-r/(2r_S)}\mathrm{d}U\mathrm{d}V \,. \tag{17.37}$$

We're now free to "unfactorize" the product of differentials through another coordinate transformation (our last), basically the inverse of Eq. (17.33). Let

$$\begin{aligned} T &\equiv \frac{1}{2}(V+U) = \sqrt{(r/r_S)-1}\,\mathrm{e}^{r/(2r_S)}\sinh(ct/(2r_S)) \\ X &\equiv \frac{1}{2}(V-U) = \sqrt{(r/r_S)-1}\,\mathrm{e}^{r/(2r_S)}\cosh(ct/(2r_S)) \,. \end{aligned} \tag{17.38}$$

The (T, X) coordinate system is known as the *Kruskal-Szekeres* coordinate system.[10][62][63] It's straightforward to show that $-(\mathrm{d}T)^2 + (\mathrm{d}X)^2 = -\mathrm{d}U\mathrm{d}V$. The net effect of these transformations is that we have a coordinate system in which the Schwarzschild metric takes the form

$$(\mathrm{d}s)^2 = \frac{4r_S^3}{r}\mathrm{e}^{-r/(2r_S)}\left(-(\mathrm{d}T)^2 + (\mathrm{d}X)^2\right) + r^2(\mathrm{d}\Omega)^2 \,, \tag{17.39}$$

which has no singularity at $r = r_S$. Kruskal-Szekeres coordinates are a "gateway" into an analysis of black holes.

17.4 GRAVITATIONAL DEFLECTION OF LIGHT

The gravitational potential for photons follows from Eq. (17.10) for $\alpha = 0$. In dimensionless form,

$$\frac{1}{c^2}V_{\text{eff}}(y) = \frac{\xi^2}{2}\left[\frac{1}{y^2} - \frac{1}{y^3}\right] \,; \tag{17.40}$$

see Fig. 17.2. Taking the derivative of Eq. (17.40), $V'_{\text{eff}} = c^2\xi^2(3-2y)/(2y^4)$. There is a single root at $y = 3/2$ (for any ξ), where the potential has the value $V_{\text{eff}}(y = 3/2) = 2c^2\xi^2/27$. The potential goes to zero at $y = 1$, $V_{\text{eff}}(y = 1) = 0$. There are no stable, bound orbits for photons. Photons either scatter from V_{eff} or, for sufficiently high energy, spiral into M. A circular orbit of $r = 3r_S/2$ is possible for $\mathscr{E}/l^2 = c^4/(54G^2M^2) = 2/(27r_S^2)$, but it's not stable.

The differential equation for photon orbits follows from Eq. (17.9) with $\alpha = 0$:

$$\frac{\mathrm{d}^2u}{\mathrm{d}\phi^2} + u = 3\frac{GM}{c^2}u^2 \,. \tag{17.41}$$

Let's first examine the case of no scattering: $M = 0$ in Eq. (17.41). The solutions to $\mathrm{d}^2u/\mathrm{d}\phi^2 + u = 0$ are $u = A\cos(\phi + \delta)$, where A and δ are constants. For $\delta = 0$, $u = r_0^{-1}\cos\phi$ describes the photon path as the line specified by $r_0 = r\cos\phi$, where r_0 is the point of closest approach to the nominal scattering center at $\phi = 0$. The asymptotes, where $u \to 0$ ($r \to \infty$), are given by $\phi = \pm\pi/2$. The geometry of "no scattering" is shown in Fig. 17.3.

[9] To show Eq. (17.34), solve Eq. (17.33) for (ct, r^*) in terms of (u, v), and note that $\mathrm{d}r^* = \mathrm{d}r/(1 - r_S/r)$.
[10] Published independently by Kruskal and Szekeres in 1960.

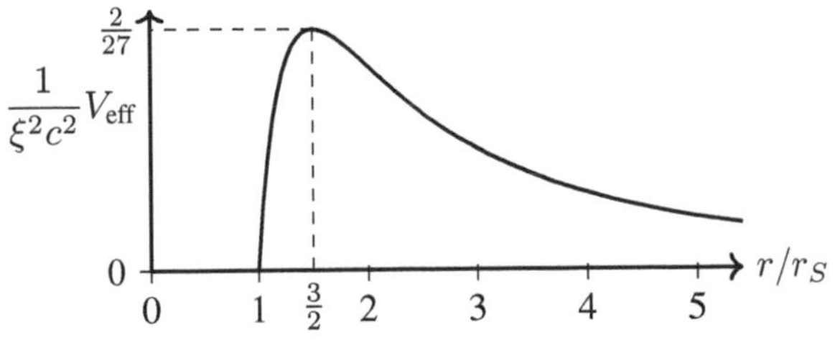

Figure 17.2 Gravitational potential for photons.

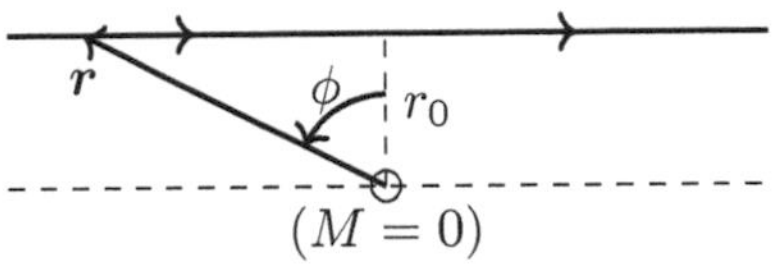

Figure 17.3 Geometry of no scattering.

17.4.1 Small-angle scattering

The nonlinear term $3GMu^2/c^2$ in Eq. (17.41) has the dimension of $(\text{length})^{-1}$. Because $u \leq r_0^{-1}$ (r_0 distance of closest approach), $GM/(c^2 r_0)$ sets a numerical scale for the nonlinear term. For M the solar mass and R the solar radius as the distance of closest approach, $GM/(c^2 R) \approx 2 \times 10^{-6}$, suggesting a perturbative approach to solving Eq. (17.41). Define $\lambda \equiv GM/(c^2 r_0)$ as a small dimensionless parameter.[11] Next, assume that the solution to Eq. (17.41) can be written as a *perturbation expansion* $u(\phi) = u_0(\phi) + u_1(\phi) + u_2(\phi) + \cdots$ where $u_0(\phi) = r_0^{-1} \cos\phi$ is the solution corresponding to no scattering, and $u_i(\phi), i \geq 1$, are unknown functions.[12] Substituting into Eq. (17.41), we have through second order,

$$\frac{\mathrm{d}^2 u_0}{\mathrm{d}\phi^2} + u_0 + \frac{\mathrm{d}^2 u_1}{\mathrm{d}\phi^2} + u_1 + \frac{\mathrm{d}^2 u_2}{\mathrm{d}\phi^2} + u_2 + \cdots = 3\lambda r_0 (u_0 + u_1 + \cdots)^2 .$$

This equation implies a hierarchy of coupled differential equations for the unknown functions. Through second order,

$$\begin{aligned} \frac{\mathrm{d}^2 u_1}{\mathrm{d}\phi^2} + u_1 &= 3\lambda r_0 u_0^2 = \frac{3\lambda}{r_0} \cos^2\phi \\ \frac{\mathrm{d}^2 u_2}{\mathrm{d}\phi^2} + u_2 &= 6\lambda r_0 u_0 u_1 = 6\lambda \cos\phi u_1 = \frac{6\lambda^2}{r_0} \cos\phi (1 + \sin^2\phi) , \end{aligned} \tag{17.42}$$

where we've used that the particular solution of the first-order equation is $u_1 = \lambda \left(1 + \sin^2\phi\right)/r_0$ (check it!). The solution to the second-order equation in Eq. (17.42) is given in Exercise 17.2. The orbit equation for photons, through second order in λ, is:

$$u(\phi) = \frac{1}{r_0}\left(\cos\phi + \lambda(1 + \sin^2\phi) + \frac{3}{4}\lambda^2 (5\phi \sin\phi - \sin^2\phi \cos\phi) + O(\lambda^3)\right) . \tag{17.43}$$

[11]Note that λ is the dimensionless parameter introduced in Section 1.7.2 as a measure of the relativistic importance of gravity, Eq. (1.11). Here we see that this parameter (however small) leads to the gravitational bending of light.

[12]Perturbation theory is a method for obtaining approximate solutions to problems by using the solutions to simpler problems (obtained from the original problem as some parameter is allowed to vanish). Hence u_1 is based on u_0, u_2 is based on u_0 and u_1, etc. It's a method based on *hope*: While it can be quite successful (plan on success!), there's no guarantee that a given perturbation scheme will work, and one can get into trouble this way. See Section 17.4.3.

The distance of closest approach, $r_1 \equiv (u(\phi = 0))^{-1}$, is no longer equal to r_0, but is numerically quite close. From Eq. (17.43), $r_1 = r_0(1 + \lambda + O(\lambda^3))^{-1} = r_0(1 - \lambda + \lambda^2 + O(\lambda^3))$.

What asymptotes are implied by Eq. (17.43)? Assume[13] $u \to 0$ as $\phi \to \pm(\pi/2 + \delta)$. From Eq. (17.43) with $u = 0$,

$$\sin\delta = \lambda\left(1 + \cos^2\delta\right) + \frac{3}{4}\lambda^2\left(5(\frac{\pi}{2} + \delta)\cos\delta + \cos^2\delta\sin\delta\right) + O(\lambda^3) . \tag{17.44}$$

From Eq. (17.44) it's consistent to take δ small if λ is small. Treating δ as small in Eq. (17.44),

$$\delta = 2\lambda + \frac{15\pi}{8}\lambda^2 + O(\lambda^3) . \tag{17.45}$$

The deflection angle $\Delta \equiv 2\delta$ (see Fig. 17.4). Thus, the deflection angle predicted by GR for the sun is, to lowest order,[14]

$$\Delta = \frac{4GM}{Rc^2} = 1.75 \text{ arcseconds} . \tag{17.46}$$

Equation (17.46) is one of the central predictions of GR, and measuring Δ is of the utmost importance to assess the validity of the theory.[15] Measurements of the gravitational deflection by the sun yield the result [64]

$$\frac{\Delta_{\text{observed}}}{\Delta_{\text{theory}}} = 0.9998 \pm 0.0008 . \tag{17.47}$$

Thus there is impressive agreement between theory and experiment. The gravitational bending of light by typical stars is small. Consider however that for a *galaxy* (perhaps 10^{11} solar masses) the deflection of light would be strong. The gravitational field of one galaxy can be used (*gravitational lensing*) to image distant galaxies.

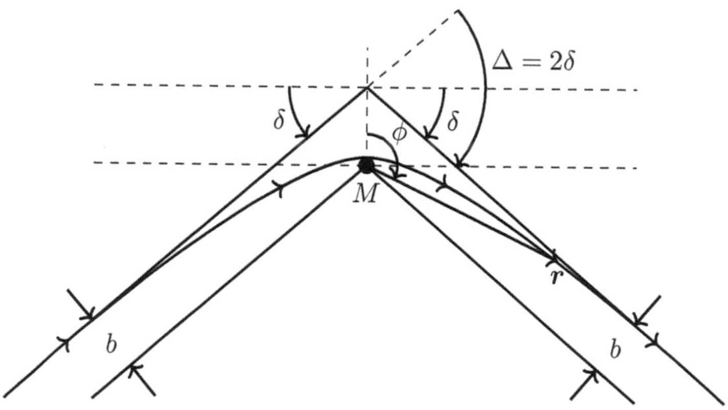

Figure 17.4 Geometry of gravitational deflection of light.

[13] Note that Eq. (17.43) is even in the variable ϕ.

[14] There are 60 arcminutes/degree and 60 arcseconds/arcminute. 1 arcsecond = $2\pi/(360 \times 3600) = 4.848 \times 10^{-6}$ rad.

[15] The traditional method for measuring Δ consists of photographing a star field around the sun during a total eclipse, and then comparing with photographs of the same star field taken later. The smallness of the deflection angle pushes this technique to its limits. Better results are obtained using radio telescopes where, because the sun is not very bright at radio frequencies, measurements can be made on radio sources near the sun at any time. The most accurate measurements use interferometry to measure the relative separation between radio sources, as one of them passes behind the sun. Radio interferometry (at GHz frequencies) provides much greater angular resolution than obtainable with optical instruments.

17.4.2 Arbitrary deflection angle

A perturbative approach to solving Eq. (17.41) relies on the scattering angle being small. What if the scattering is large? In that case it's better to start with the first-order differential equation for the orbit Eq. (17.11) (with $\alpha = 0$),

$$\left(\frac{\mathrm{d}u}{\mathrm{d}\phi}\right)^2 = \frac{2\mathscr{E}}{l^2} - u^2(1 - r_S u) \equiv \frac{1}{b^2} - u^2(1 - r_S u) , \tag{17.48}$$

where $b = l/\sqrt{2\mathscr{E}}$ is the *impact parameter*—see Fig. 17.4 ($\mathscr{E} > 0$ for scattering orbits). Photon orbits depend on a single parameter $b^2 = l^2/(2\mathscr{E})$, in contrast to particle orbits which depend on two independent parameters, l and $\mathscr{E}$. The distance of closest approach is the point in the orbit u_1 where $\mathrm{d}u/\mathrm{d}\phi = 0$, where $u_1^2(1-r_S u_1) = 1/b^2$; it's also the turning point defined by $V_{\text{eff}}(r_1 \equiv u_1^{-1}) = \mathscr{E}$, where $\dot{r} = 0$ (show this). All photon scattering orbits are such that $4b^2 > 27r_S^2$ (show this).

The turning angle is found by integrating Eq. (17.48) from asymptote to asymptote,[16]

$$\Delta\phi = 2\int_0^{u_1} \mathrm{d}u \left[\frac{1}{b^2} - u^2 + r_S u^3\right]^{-1/2} \equiv \frac{2}{\sqrt{r_S}} \int_0^{u_1} \frac{\mathrm{d}u}{\sqrt{(u-u_1)(u-u_2)(u-u_3)}} , \tag{17.49}$$

where u_1, u_2, and u_3 are the roots of the cubic equation $r_S u^3 - u^2 + 1/b^2 = 0$. Referring to Fig. 17.4, $\Delta\phi = (\pi/2 + \delta) - (-\pi/2 - \delta) = \pi + 2\delta = \pi + \Delta$, where Δ is the total scattering angle. Expressions for u_1, u_2, u_3 are given in Appendix E, where it's shown that u_3 is always negative. The root u_1, associated with the turning point in the photon trajectory (see Fig. 17.5), is the smallest positive root. The integral in Eq. (17.49) can be expressed as an incomplete elliptic integral of the first kind (Section E.1), which are numerically tabulated. As $4b^2 \to 27r_S^2$ and a circular orbit becomes possible, u_1 and u_2 coalesce to a common value associated with the circular orbit, $2/(3r_S)$. In this limit the integral in Eq. (17.49) diverges: A circular orbit represents infinite scattering.

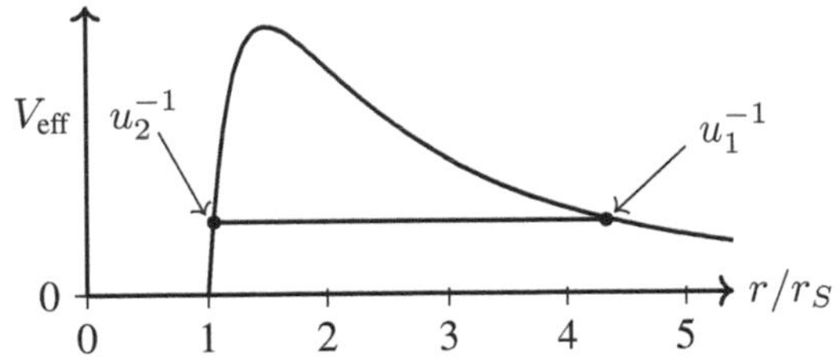

Figure 17.5 Locations of u_1, u_2 for a given $\mathscr{E}$; u_1 locates the turning point.

17.4.3 Note on perturbation theory

A perturbation treatment of Eq. (17.49) provides an instructive lesson in perturbation theory: Choosing the right zeroth-order solution. The zeroth-order solution of Eq. (17.41), $u = b^{-1}\cos\phi$ (no scattering), used in Eq. (17.49) also leads to no scattering ($\Delta = 0$) if we set $r_S = 0$. Can we approximate the integrand in Eq. (17.49) for small r_S? Let's try an expansion in powers of r_S:

$$\left(\frac{1}{b^2} - u^2 + r_S u^3\right)^{-1/2} = \frac{1}{\sqrt{b^{-2} - u^2}}\left(1 - \frac{1}{2}\frac{r_S u^3}{b^{-2} - u^2} + \cdots\right) .$$

That idea leads to trouble. The limit of integration in Eq. (17.49), $u_1 > b^{-1}$, causing the leading term in such an expansion to be singular. From Eq. (E.6), $u_1 = b^{-1}(1 + r_S/(2b) + O(r_S/b)^2)$. Even if the

[16] The problem is symmetric in ϕ; hence the factor of two in Eq. (17.49).

limit of integration could be fixed, the integrals in succeeding terms in the expansion would diverge. One can integrate over a square-root singularity, but not over $(b^{-2} - u^2)^{-3/2}$ at the next order in the perturbation expansion. We need another idea. We're trying to develop a scheme for calculating small modifications to the no-scattering photon trajectory that occurs in the nonrelativistic theory $(c \to \infty)$. While photons are never nonrelativistic, we're referring to what remains of V_{eff} when we make the r^{-3} term from GR disappear by formally letting $c \to \infty$. We note from Eq. (17.10) with $\alpha = 0$ that the gravitational potential for photons, when we let $c \to \infty$ still involves the centrifugal barrier that diverges as $r \to 0$. For finite c, the height of the potential is proportional to c^4 ($V_{\text{max}} \sim \xi^2 c^2 \sim c^4$). By formally letting c get large we force V_{eff} into having a large peak at $r = \frac{3}{2} r_S$, which then rapidly plummets to zero at $r = r_S$, all as $r_S \to 0$ as $c \to \infty$. From Eq. (E.7), the root $u_2 = r_S^{-1}(1 - (r_S/b)^2 + O(r_S/b)^4)$ for small r_S. As $c \to \infty$, $u_2^{-1} \to 0$ and thus u_2^{-1} presents itself as a natural small parameter, and we have the following way to expand Eq. (17.49):

$$\begin{aligned}\Delta\phi &= \frac{2}{\sqrt{r_S u_2}} \int_0^{u_1} \frac{du}{\sqrt{(u_1 - u)(u - u_3)}} \frac{1}{\sqrt{1 - u/u_2}} \\ &= \frac{2}{\sqrt{r_S u_2}} \int_0^{u_1} \frac{du}{\sqrt{(u_1 - u)(u - u_3)}} \left(1 + \frac{u}{2u_2} + \cdots\right) .\end{aligned}$$

This expansion does not suffer from the maladies discussed above. In particular, the limit of integration is correct at every order of perturbation theory.

It's straightforward to evaluate the integrals:

$$\begin{aligned}\int_0^{u_1} \frac{du}{\sqrt{(u_1 - u)(u - u_3)}} &= 2\psi \\ \int_0^{u_1} \frac{u du}{\sqrt{(u_1 - u)(u - u_3)}} &= \sqrt{-u_1 u_3} + (u_1 + u_3)\psi ,\end{aligned} \tag{17.50}$$

where

$$\sin^2 \psi \equiv \frac{u_1}{u_1 - u_3} = \frac{1}{2} + \frac{1}{2\sqrt{3}} \tan(\theta/6) . \tag{17.51}$$

We've used the explicit expressions for the roots given by Eq. (E.2) to evaluate the right side of Eq. (17.51), where θ is defined in Eq. (E.3). By writing $\psi = \pi/4 + \delta$, $\sin^2(\pi/4 + \delta) = \frac{1}{2} + \frac{1}{2}\sin 2\delta$; hence from Eq. (17.51) $\sin 2\delta = \frac{1}{\sqrt{3}} \tan(\theta/6)$. From Eq. (E.4), $\theta \approx 3\sqrt{3}(r_S/b)$ for small r_S. We conclude that for small r_S/b,

$$\psi = \frac{\pi}{4} + \frac{r_S}{4b} + O(r_S/b)^2 . \tag{17.52}$$

From Appendix E, for small $x \equiv r_S/b$, $u_1 + u_3 = x/b + O(x^3)$, $r_S u_2 = 1 - x^2 + O(x^4)$, and $u_1 u_3 = -(1/b^2)(1 + x^2 + O(x^4))$. From Eq. (17.50),

$$\Delta\phi = \frac{2}{\sqrt{r_S u_2}} \left(2\psi + \frac{1}{2u_2}(\sqrt{-u_1 u_3} + (u_1 + u_3)\psi) + \cdots\right) . \tag{17.53}$$

Expanding all quantities in Eq. (17.53) to first order in $1/c^2$, we have

$$\Delta\phi = \pi + \frac{4GM}{c^2 b} + O(1/c^4) ,$$

the same as Eq. (17.46) when we use $\Delta = \Delta\phi - \pi$.

17.5 APSIDAL PRECESSION

We now examine bound orbits, the differential equation for which is Eq. (17.12),

$$\frac{d^2u}{d\phi^2} + u = \frac{1}{p} + 3\frac{GM}{c^2}u^2 , \tag{17.12}$$

where $p = l^2/(GM) = a(1-\epsilon^2)$, with a the semimajor axis of the orbit and ϵ the eccentricity. The orbit of Mercury exhibits the largest effects due to GR within the solar system (closest to the sun), and we'll focus our attention on that case. With $\epsilon = 0.2056$ and $a = 5.79 \times 10^{10}$ m for Mercury, and with M the solar mass, $3GM/(c^2p) \approx 5 \times 10^{-8}$. The effects of GR are small and we can apply the perturbation methods previously discussed.[17]

We therefore approximate the orbit equation as the sum of the Keplerian orbit, Eq. (17.14), and an unknown function, $u_1(\phi)$: $u(\phi) = p^{-1}(1 + \epsilon\cos\phi) + u_1(\phi)$. Substituting this guess into Eq. (17.12), we obtain a differential equation for $u_1(\phi)$,

$$\frac{d^2u_1}{d\phi^2} + u_1 = \frac{\lambda}{p}(1 + \epsilon\cos\phi)^2 , \tag{17.54}$$

where $\lambda \equiv 3GM/(c^2p)$. The particular solution to Eq. (17.54) is (check it!)

$$u_1(\phi) = \frac{\lambda}{p}\left[1 + \epsilon\phi\sin\phi + \frac{\epsilon^2}{2}\left(1 - \frac{1}{3}\cos 2\phi\right)\right] .$$

The orbit equation, correct to first order in the small quantity λ, is thus

$$u(\phi) = \frac{1}{p}\left[1 + \epsilon\cos\phi + \lambda\left(1 + \epsilon\phi\sin\phi + \frac{\epsilon^2}{2}(1 - \frac{1}{3}\cos 2\phi)\right)\right] + O(\lambda^2) . \tag{17.55}$$

Let's rewrite Eq. (17.55),

$$u(\phi) = \frac{1}{p}\left[1 + \lambda\left(1 + \frac{\epsilon^2}{2}(1 - \frac{1}{3}\cos 2\phi)\right) + \epsilon\left(\cos\phi + \lambda\phi\sin\phi\right)\right] . \tag{17.56}$$

The terms in parentheses proportional to λ in Eq. (17.56) represent a small, periodic alteration of the orbit that repeats every two orbital periods. These terms would be difficult to detect experimentally. The *secular term*, however, $\phi\sin\phi$, represents an observable *nonperiodic* modification of the orbit.[18] Consider that for $\lambda \ll 1$, $\cos(\phi(1-\lambda)) = \cos\phi\cos\lambda\phi + \sin\lambda\phi\sin\phi \approx \cos\phi + \lambda\phi\sin\phi$, precisely the terms we have in Eq. (17.56). We thus take as the approximate solution[19] to Eq. (17.12), for orbits such that $\lambda \ll 1$:

$$u(\phi) \approx \frac{1}{p}\left[1 + \epsilon\cos(\phi(1-\lambda))\right] . \tag{17.57}$$

Equation (17.57) describes a precessing ellipse[20]—see Fig. 7.5. The angle through which the line of apsides advances in one period is defined by $u(2\pi + \delta) = u(0)$, implying that $\delta(1-\lambda) = 2\pi\lambda$, or that $\delta = 2\pi\lambda + O(\lambda^2)$. The precession angle is thus

$$\delta = 2\pi\lambda = \frac{6\pi GM}{c^2p} = \frac{3\pi r_S}{p} = \frac{6\pi GM}{c^2a(1-\epsilon^2)} \text{ rad/period} . \tag{17.58}$$

---

[17]We're fortunate to live in an environment where the effects of GR are small and we can speak of approximate Keplerian orbits. Otherwise our understanding of physics would have taken considerably longer to achieve than it did historically.

[18]There is a secular term at second order in the orbit equation for photons—Eq. (17.43). There are no bound, stable orbits for photons, and hence the magnitudes of these terms have no chance to build up in time.

[19]Equation (17.57) reduces to the Keplerian orbit for $\lambda = 0$. The errors incurred upon adopting Eq. (17.57) as the solution to Eq. (17.12) are on the same order as the terms in parentheses in Eq. (17.56) proportional to λ.

[20]The Kepler problem was treated in Chapter 7 using SR.

The angle specified by Eq. (17.58) is small. For Mercury, $\delta = 5.019 \times 10^{-7}$ rad/period. It's customary to report the precession angle as the total angle through which the orbit precesses in 100 years. Using Kepler's third law, $T = 2\pi\sqrt{a^3/GM}$, the period of Mercury is 7.6×10^6 s, or about 88 days. In 100 years there are 415.2 orbits of Mercury around the Sun. The precession angle per century as predicted by GR for Mercury is therefore

$$\Delta\phi = 42.98 \text{ arcseconds/century} . \tag{17.59}$$

This number, $43''$/cy, is now famous. The measured precession of the perihelion of Mercury is $574''$ per century.[65, p316] That number can largely be accounted for by the fact that Mercury (and the other planets) do not respond to *exactly* a $1/r$ gravitational potential from the sun; the total gravitational force also includes interactions with the other massive objects of the solar system (planets, asteroids, etc.). When these interactions are included in a calculation of the orbit of Mercury (based entirely on Newtonian mechanics), all but $43''$ can be accounted for—just the amount predicted by GR! The "missing" $43''$ was known to 19th-century astronomers. Le Verrier, whose calculations on the perturbations of the orbit of Uranus had led to the discovery of Neptune in 1846, proposed in 1859 that another planet, Vulcan, must exist inside the orbit of Mercury to account for the observed precession. It was never found. The natural explanation of the residual precession, beyond what's accounted for by Newtonian mechanics, is one of the triumphs of GR. The perihelion advances of the other planets predicted by GR (in arcseconds per century: 8.62 for Venus, 3.84 for Earth, 1.35 for Mars) are in excellent agreement with observations.[66]

We calculated the small-angle deflection of light in two ways: perturbative treatments of the solutions of Eq. (17.41) and of the integral in Eq. (17.49). Let's do the same for particle orbits. From Eq. (17.11) with $\alpha = 1$, the angle through one period of a bound orbit is given by

$$\Delta\phi = 2\int_{u_3}^{u_2} \mathrm{d}u \left[\frac{2\mathscr{E}}{l^2} + \frac{2GM}{l^2}u - u^2 + r_S u^3\right]^{-1/2} = \frac{2}{\sqrt{r_S}}\int_{u_3}^{u_2} \frac{\mathrm{d}u}{\sqrt{(u-u_3)(u-u_2)(u-u_1)}} , \tag{17.60}$$

where $u_3 < u_2 < u_1$ are the factors of the cubic polynomial in Eq. (17.60), where u_3 and u_2 locate the turning points of the motion—see Fig. 17.6. Expressions for u_i are given in Section E.2.

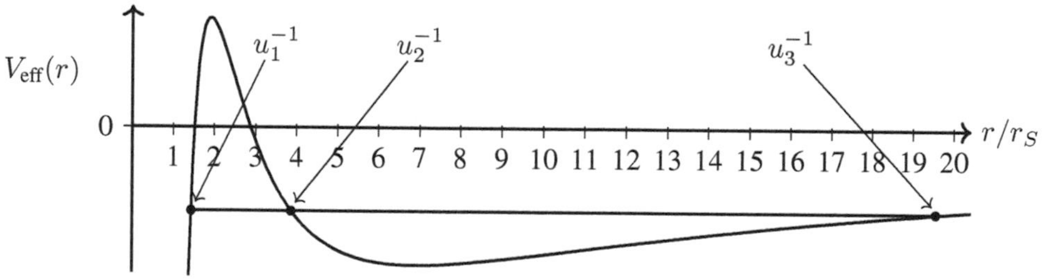

Figure 17.6 Locations of u_1, u_2, u_3 for given $\mathscr{E}$; u_2 and u_3 locate the turning points.

In Newtonian mechanics we would "lose" the u^3 term in Eq. (17.60) ($c \to \infty$):

$$\Delta\phi_N = 2\int_{u_-}^{u_+} \mathrm{d}u \left[\frac{2\mathscr{E}}{l^2} + \frac{2GM}{l^2}u - u^2\right] \equiv 2\int_{u_-}^{u_+} \frac{\mathrm{d}u}{\sqrt{(u_+ - u)(u - u_-)}} . \tag{17.61}$$

We needn't write down expressions for $u_\pm$ because the value of the integral in Eq. (17.61) is the same regardless of $u_\pm$:

$$\int_{u_-}^{u_+} \frac{\mathrm{d}u}{\sqrt{(u_+ - u)(u - u_-)}} = \pi . \tag{17.62}$$

In Newtonian mechanics, a bound orbit closes on itself with $\Delta\phi_N = 2\pi$.

How to approximate the integrand in Eq. (17.60) for small r_S? We've already seen that movie, in Section 17.4. For large c, $u_1 \approx r_S^{-1}(1 - 1/\xi^2)$, Eq. (E.14), and hence u_1^{-1} is a natural small parameter. We thus expand Eq. (17.60) as

$$\begin{aligned}\Delta\phi =& \frac{2}{\sqrt{r_S u_1}} \int_{u_3}^{u_2} \frac{du}{\sqrt{(u-u_3)(u_2-u)}} \frac{1}{\sqrt{1-u/u_1}} \\ =& \frac{2}{\sqrt{r_S u_1}} \int_{u_3}^{u_2} \frac{du}{\sqrt{(u-u_3)(u_2-u)}} \left(1 + \frac{u}{2u_1} + \cdots\right) . \end{aligned} \tag{17.63}$$

Equation (17.62) provides the integral at lowest order. The integral at the next order in perturbation theory is straightforward to evaluate, with the result

$$\int_{u_3}^{u_2} \frac{u du}{\sqrt{(u-u_3)(u_2-u)}} = \frac{\pi}{2}(u_2 + u_3) . \tag{17.64}$$

Combining Eqs. (17.62)–(17.64),

$$\Delta\phi = \frac{2\pi}{\sqrt{r_S u_1}} \left(1 + \frac{u_2 + u_3}{4u_1} + \cdots\right) . \tag{17.65}$$

Expanding all quantities in Eq. (17.65) to first order in $1/c^2$,

$$\Delta\phi = 2\pi\left(1 + \frac{3}{4\xi^2}\right) = 2\pi\left(1 + \frac{3(GM)^2}{c^2 l^2}\right) = 2\pi\left(1 + \frac{3GM}{c^2 p}\right) ,$$

where we've made use of Eq. (E.14) and the angular momentum of the orbit, $l^2 = GMp$. The advance of the perihelion in one period is thus $6\pi GM/(c^2 p)$, the same as Eq. (17.58).

17.6 GRAVITATIONAL TIME DELAY

In Section 17.4 we computed the geodesic taken by photons, but not the *time* to traverse a given trajectory. Light is *delayed* as its path is gravitationally bent near a massive object—see Fig. 17.7, something that can be measured from radar signals reflected from planets such as Venus in the vicinity of the sun.

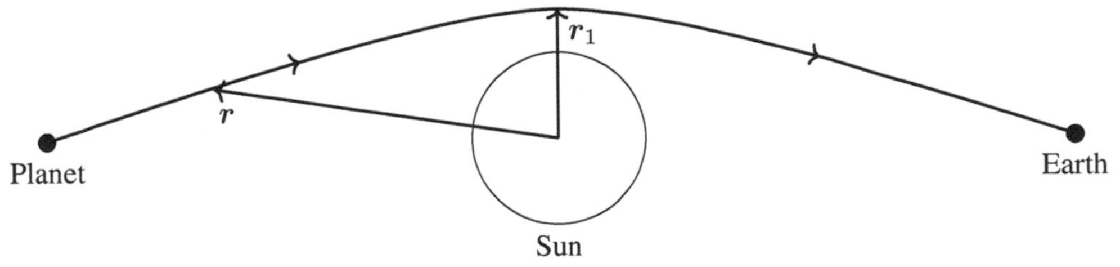

Figure 17.7 Light is delayed in time relative to the direct path between planet and Earth.

To calculate the time delay, start with

$$\left(\frac{dr}{dt}\right)^2 = \left(\frac{dr}{d\tau}\frac{d\tau}{dt}\right)^2 = \left(\frac{\dot{r}}{\dot{t}}\right)^2 = (1 - r_S/r)^2 \left(c^2 - \frac{l^2}{k^2 r^2}(1 - r_S/r)\right) , \tag{17.66}$$

where $\dot{r}^2$ is obtained from Eq. (17.9) with $\alpha = 0$ and $\dot{t}$ is from Eq. (17.6). We can evaluate the constant k at the point of closest approach r_1 where $\dot{r} = 0$. From Eq. (17.9) with $\alpha = 0$,

$$\frac{l^2}{k^2} = \frac{c^2 r_1^2}{1 - r_S/r_1} . \tag{17.67}$$

Combining Eqs. (17.66) and (17.67),

$$\frac{\mathrm{d}r}{\mathrm{d}t} = \pm c(1 - r_S/r)\left[1 - \left(\frac{r_1}{r}\right)^2 \frac{1 - r_S/r}{1 - r_S/r_1}\right]^{1/2} . \tag{17.68}$$

For the sun $r > r_1 > r_S$, $(r_S/r) < 1$ and $(r_S/r_1) < 1$. Expand Eq. (17.68) to first order in r_S and integrate: The time from r to r_1, $t(r, r_1)$, is given by

$$ct(r, r_1) = \int_{r_1}^{r} \frac{r\mathrm{d}r}{\sqrt{r^2 - r_1^2}} \left[1 + \frac{r_S}{r} + \frac{r_1 r_S}{2r(r + r_1)} + O(r_S^2)\right] . \tag{17.69}$$

The integrals in Eq. (17.69) are straightforward:

$$\int_{r_1}^{r} \frac{r\mathrm{d}r}{\sqrt{r^2 - r_1^2}} = \sqrt{r^2 - r_1^2}$$

$$\int_{r_1}^{r} \frac{\mathrm{d}r}{\sqrt{r^2 - r_1^2}} = \ln\left((r/r_1) + \sqrt{(r/r_1)^2 - 1}\right)$$

$$\int_{r_1}^{r} \frac{\mathrm{d}r}{\sqrt{r^2 - r_1^2}} \frac{1}{(r + r_1)} = \frac{1}{r_1}\sqrt{\frac{r - r_1}{r + r_1}} . \tag{17.70}$$

Combining Eqs. (17.69) and (17.70), the propagation time from r to r_1, to first order in $1/c^2$, is

$$ct(r, r_1) = \sqrt{r^2 - r_1^2} + r_S \ln\left((r/r_1) + \sqrt{(r/r_1)^2 - 1}\right) + \frac{r_S}{2}\sqrt{\frac{r - r_1}{r + r_1}} + O(r_S^2) . \tag{17.71}$$

The leading term in Eq. (17.71) represents the distance traveled by light in the absence of gravity.[21] The other terms are positive and represent the extra propagation time caused by the bending of the photon trajectory. The *gravitational time delay* (also known as the *Shapiro time delay*) is defined as $\Delta t \equiv t(r, r_1) - \sqrt{r^2 - r_1^2}/c$. The time delay for a signal sent to a planet (or spacecraft) at distance r_p from the sun, passing the sun at distance r_1, from Earth at distance r_E from the sun, and back, is

$$\Delta t = \frac{2r_S}{c}\left[\ln\left((r_E/r_1) + \sqrt{(r_E/r_1)^2 - 1}\right) + \frac{1}{2}\sqrt{\frac{r_E - r_1}{r_E + r_1}}\right.$$
$$\left. + \ln\left((r_p/r_1) + \sqrt{(r_p/r_1)^2 - 1}\right) + \frac{1}{2}\sqrt{\frac{r_p - r_1}{r_p + r_1}}\right] \approx \frac{2r_S}{c}\left[1 + \ln\left(\frac{4r_E r_p}{r_1^2}\right)\right] . \tag{17.72}$$

What is measured on Earth is actually the proper time $\Delta\tau = \sqrt{1 - r_S/r_E}\Delta t$, but we can ignore the distinction because $r_S/r_E \approx 2 \times 10^{-8}$. The time delay is as large as it can be when the planet and Earth are on opposite sides of the Sun, in *superior conjunction*. Figure 17.8 shows time delay measurements from Venus over a 600-day period as it swept through superior conjunction.[67] It's important to appreciate the difficulty of the measurements shown in Fig. 17.8. The round-trip time for a radar signal to Venus and back (in superior conjunction) is 1720 s, or about 29 minutes. To detect 200 μs out of 1720 s implies an accuracy of 1 part in 10^7; 1% accuracy requires 1 part in 10^9.

17.7 PARAMETERIZED POST-NEWTONIAN FRAMEWORK

The parameterized post-Newtonian (PPN) framework is a formalism designed to put gravitational physics on a more experimentally oriented footing.[68] The method posits a *parametrized* metric,

[21] Draw a Euclidean right triangle.

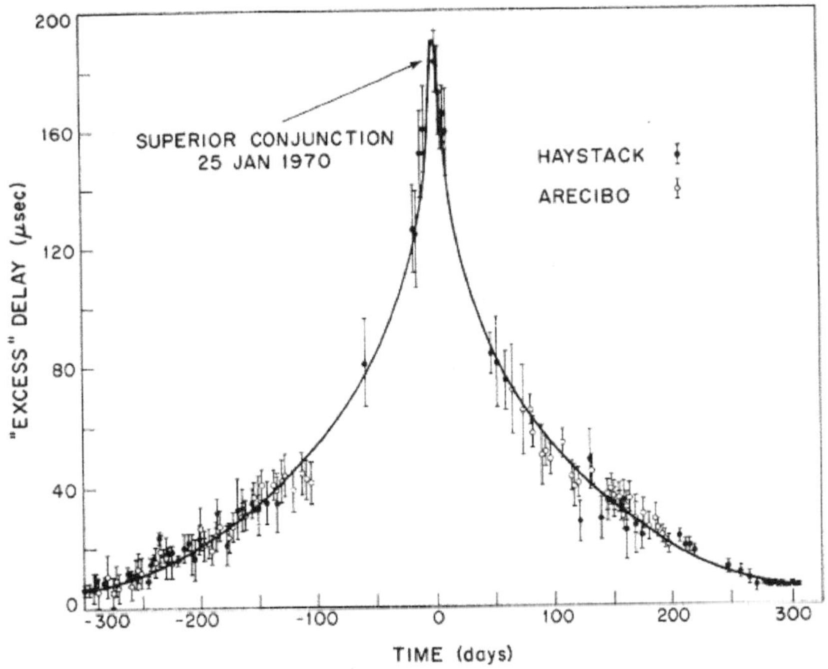

Figure 17.8 Gravitational time delay of radar signals reflected from Venus. Reprinted figure with permission from [67]. Copyright 1971 by the American Physical Society. Solid curve is the prediction of GR.

which in its simplest form is

$$(\mathrm{d}s)^2 = -\left[1 - \frac{2GM}{rc^2} + 2(\beta - \gamma)\left(\frac{GM}{rc^2}\right)^2\right](c\mathrm{d}t)^2 + \left(1 + 2\gamma\frac{GM}{rc^2}\right)(\mathrm{d}r)^2 + r^2(\mathrm{d}\Omega)^2 \,, \tag{17.73}$$

where β and γ are dimensionless parameters whose values are to be found from experiment.[22] The predictions of Eq. (17.73) agree with those of GR for $\beta = \gamma = 1$. What values of β and γ have been obtained from measurements? The most accurate determination of γ is from the *Cassini* spacecraft mission. From the time delay of signals sent to and from the satellite as it went through solar conjunction, a result of $\gamma = 1+(2.1\pm2.3)\times10^{-5}$ was found.[69] The most accurate determination of β is from the lunar laser ranging experiment, where a value of $\beta = 1 + (1.2 \pm 1.1) \times 10^{-4}$ has been found.[46] These measurements confirm the predictions of GR to high accuracy.

17.8 THE GLOBAL POSITIONING SYSTEM

The Global Positioning System (GPS) is a collection of 24 satellites carrying atomic clocks in circular orbits about the earth with 12-hour periods, arranged in six orbital planes, with each plane inclined at 55° to the equatorial plane of Earth.[70] The satellites are positioned so that at least four are always above the local horizon, from almost any point on Earth. Each satellite broadcasts coded information on the positions and times of transmission events. A GPS receiver solves four simultaneous equations, $|\boldsymbol{r} - \boldsymbol{r}_i| = c(t - t_i)$, $i = 1, 2, 3, 4$, between the spacetime coordinates of the four satellites to infer the spacetime coordinates on Earth. The GPS provides a rare *practical*

[22] They are not the same symbols used in SR.

application of GR, how time is affected by gravity—the GPS would quickly become useless if relativistic effects were not taken into account, as we now discuss.

How fast does a GPS satellite travel (orbital period $P = 12$ hours)? Using Kepler's third law, $r^3 = GMP^2/(4\pi^2)$ (with M the mass of Earth), we find $r = 2.661 \times 10^4$ km $= 4.18R_E$, where R_E denotes the radius of Earth. Relative to the center of Earth, the speed of the satellite is $v_{\text{Sat}} = 2\pi r/P = 3870$ m s^{-1}, or $\beta_{\text{Sat}} \equiv v_{\text{Sat}}/c = 1.29 \times 10^{-5}$. The satellite speed should be contrasted with the "orbital" speed of the surface of the earth (relative to the center of Earth), which at the equator travels with a speed of $v_{\text{E}} = 464$ m s^{-1}, or $\beta_{\text{E}} = 1.55 \times 10^{-6}$.

We now want to compare the satellite's proper time with that at the surface of Earth. From $-(c\mathrm{d}\tau)^2 = g_{\mu\nu}\mathrm{d}x^\mu \mathrm{d}x^\nu$ and the Schwarzschild metric:

$$(\mathrm{d}\tau)^2 = (1 - r_S/r)(\mathrm{d}t)^2 - \frac{(\mathrm{d}r)^2}{c^2(1 - r_S/r)} - \frac{r^2}{c^2}\left[(\mathrm{d}\theta)^2 + (\sin\theta \mathrm{d}\phi)^2\right] .$$

For simplicity set $\theta = \pi/2$. For a circular orbit, $\mathrm{d}r = 0$. Thus,

$$\left(\frac{\mathrm{d}\tau}{\mathrm{d}t}\right)^2 = 1 - \frac{r_S}{r} - \beta^2 , \tag{17.74}$$

where $\beta = r\mathrm{d}\phi/(c\mathrm{d}t)$. Equation (17.74) generalizes time dilation from SR, moving clocks run slow, Eq. (7.3), to include a general-relativistic effect, clocks are slow in a gravitational potential, and which also generalizes what we obtained from the equivalence principle, Eq. (11.18), to include a special-relativistic effect. We can factor out $\mathrm{d}t$ by applying Eq. (17.74) to two locations and taking a ratio

$$\frac{\left(\dfrac{\mathrm{d}\tau_{\text{Sat}}}{\mathrm{d}t}\right)}{\left(\dfrac{\mathrm{d}\tau_{\text{E}}}{\mathrm{d}t}\right)} = \frac{\mathrm{d}\tau_{\text{Sat}}}{\mathrm{d}\tau_{\text{E}}} = \sqrt{\frac{1 - \dfrac{r_S}{r_{\text{Sat}}} - \beta^2_{\text{Sat}}}{1 - \dfrac{r_S}{R_E} - \beta^2_{\text{E}}}} . \tag{17.75}$$

Because the terms in the square root in Eq. (17.75) are small, of order 10^{-10}, we can approximate it with a Taylor expansion. To first order,

$$\frac{\mathrm{d}\tau_{\text{Sat}}}{\mathrm{d}\tau_{\text{E}}} \approx 1 + \frac{1}{2}\left(\frac{r_S}{R_E} - \frac{r_S}{r_{\text{Sat}}}\right) - \frac{1}{2}\left(\beta^2_{\text{Sat}} - \beta^2_{\text{E}}\right) . \tag{17.76}$$

The first term in Eq. (17.76) is the general-relativistic effect, that by being *higher* in the gravitational potential of the earth, clocks on satellites *speed up* relative to Earth clocks, while the second term is the special-relativistic effect that satellite clocks in motion *slow down* relative to Earth clocks. Putting in numbers,

$$\begin{aligned}
\frac{\mathrm{d}\tau_{\text{Sat}}}{\mathrm{d}\tau_{\text{E}}} &= 1 + \frac{1}{2}\frac{r_S}{R_E}\left(1 - \frac{1}{4.18}\right) - \frac{1}{2}\left(1.66 \times 10^{-10} - 2 \times 10^{-12}\right) \\
&= 1 + 5.28 \times 10^{-10} - 0.82 \times 10^{-10} \\
&= 1 + 4.46 \times 10^{-10} .
\end{aligned} \tag{17.77}$$

The gravitational effect is larger than that from SR.

The net effect is that the satellite clock gains 0.446 ns every second relative to a clock on Earth due to fundamental physics. In one day the satellite clock gains $86,400 \times 4.46 \times 10^{-10} = 38.5$ μs on the earth clock. Is that significant? Light travels 11.4 km in 38 μs, thus rendering the GPS completely inaccurate. To maintain an accuracy on the ground of 10 m, satellite clocks must be accurate to within 33.3 ns. With the satellite clock gaining 0.446 ns per second, the GPS would be unable to meet this level of accuracy within minutes if relativistic effects were ignored. Of course, the operators of the GPS *do* take into account relativistic effects, as well as other sources of error to keep the system highly accurate.

17.9 SPIN PRECESSION I: GEODETIC EFFECT

Geodetic precession is the change that occurs in the orientation of a spacelike vector as it's transported around a closed timelike geodesic in the curved spacetime of a static mass source. It was discovered theoretically by Willem de Sitter in 1916, within months of the publication of GR; hence the effect is also called *de Sitter precession*. In this section we work out the geodetic effect as another testable prediction of GR. It was announced in 2011 that the *Gravity Probe B* (GPB) experiment had measured the geodetic precession of a gyroscope carried aboard a satellite in Earth orbit in good agreement with GR. There is another, smaller, effect due to the earth's rotation, the *Lense-Thirring* effect. The GPB experiment also measured the Lense-Thirring effect, as we discuss in the next chapter. Geodetic precession differs from Thomas precession (mentioned, but not developed, in Section 12.6); whereas Thomas precession results from an *accelerated* spin, geodetic precession occurs during free fall (no force other than gravity) in curved spacetime.

Spin is a four-vector S^μ with $S^0 = 0$ in the rest frame (Section 12.5), implying that it's orthogonal to the four-velocity, $S^\mu U_\mu = 0$, Eq. (12.53). For a free particle in flat spacetime, $\mathrm{d}S^\mu/\mathrm{d}\tau = 0$, Eq. (12.60). For a free-fall particle in *curved* spacetime, $\mathrm{D}S^\mu = 0$, or, from Eq. (14.45):

$$\frac{\mathrm{d}S^\mu}{\mathrm{d}\tau} = -\Gamma^\mu_{\alpha\beta} S^\alpha U^\beta \,. \tag{17.78}$$

Assume, as in the GPB experiment, that the spin is carried in a circular orbit around the source mass. Thus, $U^r = U^\theta = 0$ and we can set $\theta = \pi/2$. Using Table 16.1 we find from Eq. (17.78),

$$\frac{\mathrm{d}S^0}{\mathrm{d}\tau} = -\Gamma^t_{rt} U^0 S^r \qquad \frac{\mathrm{d}S^r}{\mathrm{d}\tau} = -U^0 S^0 \Gamma^r_{tt} - U^\phi S^\phi \Gamma^r_{\phi\phi} \tag{17.79}$$

$$\frac{\mathrm{d}S^\theta}{\mathrm{d}\tau} = 0 \qquad \frac{\mathrm{d}S^\phi}{\mathrm{d}\tau} = -\Gamma^\phi_{r\phi} S^r U^\phi \,. \tag{17.80}$$

The equation of motion for S^0 in Eq. (17.79) is equivalent to that for S^ϕ in Eq. (17.80) (Exercise 17.4). Using the orthogonality condition, we can eliminate $S^0 U^0 = -g_{\phi\phi} U^\phi S^\phi / g_{00}$ from the equation for S^r:

$$\frac{\mathrm{d}S^r}{\mathrm{d}\tau} = U^\phi S^\phi \left(\frac{g_{\phi\phi}}{g_{00}} \Gamma^r_{tt} - \Gamma^r_{\phi\phi} \right) \,. \tag{17.81}$$

For a circular orbit, $U^\mu = (c\dot{t}, 0, 0, \Omega \dot{t})$, where $\Omega = \mathrm{d}\phi/\mathrm{d}t$ is the constant coordinate angular speed. Using Table 16.1 and converting from proper time to coordinate time, we have the coupled differential equations

$$\frac{\mathrm{d}S^r}{\mathrm{d}t} = \Omega r \left(1 - \frac{3r_S}{2r}\right) S^\phi \qquad \frac{\mathrm{d}S^\phi}{\mathrm{d}t} = -\frac{\Omega}{r} S^r \,.$$

These coupled first-order differential equations are equivalent to uncoupled second-order differential equations:

$$\frac{\mathrm{d}^2}{\mathrm{d}t^2} \left\{ \begin{array}{c} S^r \\ S^\phi \end{array} \right\} = -\Omega^2 \left(1 - \frac{3r_S}{2r}\right) \left\{ \begin{array}{c} S^r \\ S^\phi \end{array} \right\} \equiv -(\Omega')^2 \left\{ \begin{array}{c} S^r \\ S^\phi \end{array} \right\} , \tag{17.82}$$

where we note that $\Omega' < \Omega$. Assume that at $t = 0$ the spin is oriented in the radial direction, $S^r(t = 0) = S$, where S is the magnitude of the spin, implying $S^\theta = 0$ and $S^\phi(t = 0) = 0$. With this initial condition,

$$S^r(t) = S \cos \Omega' t \qquad S^\phi(t) = -\left[\Omega/(\Omega' r)\right] S \sin \Omega' t \,.$$

The spin vector rotates at the rate Ω'. After one period of the orbit, i.e., after a time $2\pi/\Omega$, the phase of the spin differs from that of the orbit by $(2\pi/\Omega)(\Omega - \Omega')$. The precession angle per period is thus

$$\frac{2\pi}{\Omega}(\Omega - \Omega') = 2\pi \left(1 - \frac{\Omega'}{\Omega}\right) = 2\pi \left(1 - \sqrt{1 - \frac{3r_S}{2r}}\right) \approx \frac{3\pi GM}{c^2 r} \,. \tag{17.83}$$

The GPB satellite had a nearly-circular orbit of semimajor axis 7027.4 km.[23] For this orbit Eq. (17.83) gives an angle of 1.227 mas per period, where a "mas" is a milliarcsecond (1 mas $= 4.848 \times 10^{-9}$ rad). It's customary to quote this number in mas per year. Using Kepler's third law, we infer that the orbital period was 5863 s (about 98 minutes), and hence there are 5382.8 orbits per year. The expected geodetic precession angle from GR is then 6606.1 mas/yr. The measured value was 6601.8 ± 18.3 mas/yr, confirming the result from GR to within 0.3%.[71]

Gravity Probe B experiment

The GPB experiment measured the precession effects predicted by GR for a freely spinning gyroscope. To do so, the experimental design had to minimize any precession due to the torques produced by tidal forces (Section 11.3.3). The gyroscopes were fabricated to be almost perfectly spherical and homogeneous. The quartz-sphere gyroscopes, 3.8 cm in diameter, were spherical to within less than 40 atomic layers and had density variations $\Delta\rho/\rho \sim 10^{-6}$. The differences in moments of inertia about any axis were $\Delta I/I \sim 10^{-6}$. How to measure small changes in the spin-axis orientation of a near-perfect sphere without any markings on it? The answer lies in superconductivity. The gyroscopes were coated with a layer of niobium 1270 nm thick. Niobium becomes a superconductor for temperatures below 9 K. *Rotating* superconductors produce a magnetic field, the *London moment*.[72, p213] A magnetic field is generated by surface currents created because of the difference in rotation speeds between superconducting electrons and that of the material lattice. Any change in this magnetic field (due to the gyroscope precessing) can be detected with an induction loop around the gyroscope. The gyros were kept at a temperature of 2 K and spun at about 72 Hz (4300 RPM). The satellite maintained its orientation by constantly keeping a telescope (on the satellite) pointed at a particular star.

17.10 WEIGHT OF AN AT-REST OBSERVER

So far we've considered free fall along the geodesics of Schwarzschild spacetime. The worldline of an observer "at rest," however, with constant coordinates (r, θ, ϕ) is *not* a geodesic. Let's calculate the four-acceleration in curved spacetime, the rate of change of U^μ in the direction of U^μ (covariant directional derivative) $\mathcal{A}^\mu = U^\alpha \nabla_\alpha U^\mu$. The four-velocity of an observer at rest is $U^\mu = (U^0, 0, 0, 0)$. Using $U_\mu U^\mu = -c^2$, $U^0 = c/\sqrt{-g_{00}} = c/\sqrt{1 - r_S/r}$, which is constant for an observer at rest. Thus,

$$\mathcal{A}^\mu = \frac{\mathrm{d}U^\mu}{\mathrm{d}\tau} + \Gamma^\mu_{\alpha\beta} U^\alpha U^\beta = \frac{\mathrm{d}U^\mu}{\mathrm{d}\tau} + \Gamma^\mu_{00} (U^0)^2 . \tag{17.84}$$

The only nonzero Christoffel symbol of the type Γ^μ_{00} is $\Gamma^r_{00} = A'/(2B)$ (Table 16.1), and thus the only nonzero component of $\mathcal{A}^\mu$ is $\mathcal{A}^r = \Gamma^r_{00}(U^0)^2 = GM/r^2$, the Newtonian acceleration! The magnitude of the four-acceleration is $\sqrt{\mathcal{A}_\mu \mathcal{A}^\mu} = \sqrt{g_{rr}}\mathcal{A}^r = (GM/r^2)/\sqrt{1 - r_S/r}$. For $r \gg r_S$ the acceleration is the Newtonian value. Don't get too near r_S.

An observer at rest is clearly not in free fall, and experiences a nonzero four-acceleration, the "force of gravity." Weight is the force we experience in a gravitational field, $F^\mu = m\mathcal{A}^\mu$, an *inertial force* that must be supplied to *prevent* us from being in a state of free fall.[24] A geodesic is the shortest path in spacetime between fixed endpoints. As we've seen even in SR, a worldline that's "straight" by virtue of maintaining constant spatial coordinates is not the shortest path in spacetime.

[23] The orbit of the GPB satellite was very nearly circular. The eccentricity was $\epsilon = 0.0014$, with a semimajor axis $a = 7027.4$ km. The orbit was also very nearly directly over the poles of Earth, with an inclination of 90.007°.

[24] Richard Feynman was once asked in a popular lecture if he could design an anti-gravity machine. Feynman jokingly replied that the best anti-gravity machine he knew of was the chair the questioner was sitting in, because it "supports your behind above the surface of the earth."

SUMMARY

We considered the physical effects predicted by the Schwarzschild spacetime metric, which are generally small (in the environment of the earth and the solar system), yet which are measurable. Such tests validate GR to high accuracy.

- Apsidal precession: GR shows that gravitationally bound orbits do not close on themselves. Instead the apsidal line advances in the direction of the orbit. GR predicts an apsidal precession for the orbit of Mercury of 43″ per century, in precise agreement with astronomical observations (beyond what can be accounted for through perturbations from other planets).

- Gravitational deflection of light: GR predicts that the path followed by light will be bent in a gravitational field. For light rays just passing the surface of the sun, GR predicts a deflection angle of 1.75″ in excellent agreement with measured values.

- Gravitational time delay: As the path of light is bent in the gravitationally modified spacetime of a large mass, the time for light to propagate on that path is delayed with respect to the propagation time in the absence of gravitation. Measurements of the gravitational time delay agree well with the predictions of GR (see Fig. 17.8).

- Gravitational frequency shift: The Pound-Rebka-Snider experiment (Section 11.7.2) confirmed the gravitational redshift or blueshift of photons either rising or falling in a gravitational field. Closely related is time dilation in a gravitational field, an effect necessary to keep the GPS working accurately—time slows down in a gravitational potential relative to far-away clocks.

- Parameterized Post-Newtonian framework: The PPN framework specifies the spacetime metric in terms of experimentally measurable parameters. The simplest PPN theory has two parameters, β and γ, both of which have the value unity in the Schwarzschild metric. Experimentally it's found that $\gamma = 1 + (2.1 \pm 2.3) \times 10^{-5}$ and $\beta = 1 + (1.2 \pm 1.4) \times 10^{-4}$.

- Geodetic precession: The GPB experiment confirmed to high accuracy the prediction of geodetic spin precession (the change in the orientation of a spacelike vector as it is transported around a timelike geodesic in curved spacetime).

- Worldlines of at-rest observers (constant values of spatial coordinates) are not geodesics, and hence are accelerated. "Standing still" is not the shortest path in spacetime. The acceleration experienced by a stationary observer is the force (weight) that must be supplied to prevent a state of free fall. The Schwarzschild metric predicts the Newtonian gravitational acceleration.

EXERCISES

17.1 For a radial particle trajectory, if we don't make the assumption of starting from rest, show that:

a. The equation of motion would be given by $\dot{r}^2 = \dot{r}_0^2 + 2GM\left(\frac{1}{r} - \frac{1}{r_0}\right)$.

b. The value of k would be given by $k^2 = 1 - \frac{r_S}{r_0} + \frac{\dot{r}_0^2}{c^2}$.

c. The energy of the trajectory would be given by $\mathscr{E} = \frac{1}{2}\dot{r}_0^2 - \frac{GM}{r_0}$.

17.2 Verify that the solution to the second-order equation in Eq. (17.42) is

$$u_2(\phi) = \frac{3\lambda^2}{4r_0}\left(5\phi\sin\phi - \sin^2\phi\cos\phi\right) .$$

17.3 Derive Eq. (17.69).

17.4 In Section 17.9 the equivalence is asserted of the equation of motion for S^0 in Eq. (17.79) and that for S^ϕ in Eq. (17.80). Let's show this.

a. From the orthogonality condition $S^\mu U_\mu = 0$, show that for a circular orbit $S^0 = -g_{\phi\phi}U^\phi S^\phi/(g_{00}U^0)$. For a circular orbit, all components of the four-velocity are constant as are the elements of the metric tensor. Thus there is a constant proportionality between S^ϕ and S^0.

b. Show, using the result from part *a* and Eq. (17.79), that

$$\frac{\mathrm{d}S^\phi}{\mathrm{d}\tau} = \frac{g_{00}}{g_{\phi\phi}}\frac{(U^0)^2}{U^\phi}\Gamma^t_{rt}S^r .$$

Show that for this result to be equivalent to Eq. (17.80), the following identity must hold

$$\frac{\Gamma^t_{rt}}{\Gamma^\phi_{r\phi}} = -\left(\frac{U^\phi}{U^0}\right)^2\frac{g_{\phi\phi}}{g_{00}} .$$

c. Show from $U_\mu U^\mu = -c^2$ that for a circular orbit

$$-\frac{g_{\phi\phi}}{g_{00}}\left(\frac{U^\phi}{U^0}\right)^2 = 1 + \frac{1}{g_{00}}\left(\frac{c}{U^0}\right)^2 .$$

d. The equivalence of the equations of motion for S^0 and S^ϕ thus requires that the following be an identity

$$\frac{\Gamma^t_{rt}}{\Gamma^\phi_{r\phi}} = 1 + \frac{1}{g_{00}}\left(\frac{c}{U^0}\right)^2 .$$

Verify that this is indeed an identity using the Christoffel symbols for the Schwarzschild metric, $U^0 = c\dot{t}$, and the value of k^2 for a circular orbit, Eq. (17.29).

CHAPTER 18

Linearized gravity

In this chapter we consider the linearized version of Einstein's field equation, where the spacetime metric field is assumed to be slightly perturbed away from the Lorentz metric. We already considered such a procedure in Section 15.2, where we treated the static case for nonrelativistic speeds for the purposes of identifying the constant κ. Here we undertake a more systematic analysis of the linearized field equation and its numerous consequences.

18.1 LINEARIZED FIELD EQUATION

Assume a spacetime metric field $g_{\mu\nu}(x^\alpha) = \eta_{\mu\nu} + h_{\mu\nu}(x^\alpha)$ comprised of a small perturbation to the Lorentz metric, $|h_{\mu\nu}(x)| \ll 1$. The contravariant version is $g^{\mu\nu} = \eta^{\mu\nu} - h^{\mu\nu}$; in that way $g_{\mu\nu}g^{\nu\lambda} = \delta^\lambda_\mu + O(h^2)$. For $h_{\mu\nu} \equiv 0$, $G_{\mu\nu} = 0$, so we expect that $G_{\mu\nu}$ vanishes in a smooth way as $h_{\mu\nu} \to 0$; our job is to "dig out" the part of $G_{\mu\nu}$ that's first order in $h_{\mu\nu}$. The right side of Einstein's equation does not vanish, however, as $h_{\mu\nu} \to 0$; $T_{\mu\nu} \neq 0$ on flat spacetime. There is then something of a juggling act here: It would seem one cannot consistently take the limit $h_{\mu\nu} \to 0$ on both sides of Einstein's equation. The resolution is that we must rely on κ to set a smallness scale. We'll find solutions to the linearized field equation where $h_{\mu\nu}$ is proportional to κ and that are small everywhere. As we use $T_{\mu\nu}$ at lowest order, it's useful to think of $h_{\mu\nu}(x)$ as a field *on* flat spacetime, like the electromagnetic field $F_{\mu\nu}(x)$. Indeed, we'll find strong similarities between the perturbed metric field in linearized gravity and the electromagnetic field on Minkowski spacetime.

18.1.1 Inhomogeneous wave equation

The Christoffel symbols at lowest order in $h_{\mu\nu}$ are given by Eq. (15.9):

$$\Gamma^\lambda_{\mu\nu} = \frac{1}{2}\left[\eta^{\lambda\rho}\left(\partial_\mu h_{\rho\nu} + \partial_\nu h_{\rho\mu}\right) - \partial^\lambda h_{\mu\nu}\right] + O(h^2) . \tag{15.9}$$

Because $\Gamma^\lambda_{\mu\nu} \sim O(h)$, the products of Christoffel symbols in the Riemann tensor, Eq. (14.61), can be ignored. In the linear approximation the elements of the Ricci tensor are given by Eq. (15.16): $R_{\rho\nu} = \partial_\mu \Gamma^\mu_{\nu\rho} - \partial_\nu \Gamma^\mu_{\mu\rho} + O(h^2)$. Combining Eqs. (15.9) and (15.16), and working out the indicated contractions,

$$R_{\rho\nu} = \frac{1}{2}\left[-\partial_\mu\partial^\mu h_{\rho\nu} - \partial_\nu\partial_\rho h + \partial^\lambda\partial_\rho h_{\lambda\nu} + \partial^\mu\partial_\nu h_{\mu\rho}\right] + O(h_{\mu\nu})^2 , \tag{18.1}$$

where $h \equiv h^\alpha_\alpha$ is the scalar field associated with $h_{\mu\nu}$. The symmetry $R_{\rho\nu} = R_{\nu\rho}$ is preserved in this approximation, Eq. (18.1). At this point we drop the "$+O(h^2)$" symbols, it being understood that equality means "to first order in $h_{\mu\nu}$." The Ricci scalar field obtained from Eq. (18.1) is:

$$R = -\partial_\mu\partial^\mu h + \partial^\lambda\partial^\rho h_{\rho\lambda} . \tag{18.2}$$

Using Eqs. (18.1) and (18.2), we have the Einstein tensor at lowest order

$$R_{\rho\nu} - \frac{1}{2}g_{\rho\nu}R = \frac{1}{2}\left[-\partial_\mu\partial^\mu h_{\rho\nu} - \partial_\nu\partial_\rho h + \eta_{\rho\nu}\partial_\mu\partial^\mu h - \eta_{\rho\nu}\partial^\lambda\partial^\sigma h_{\sigma\lambda} + \partial^\lambda\partial_\rho h_{\lambda\nu} + \partial^\mu\partial_\nu h_{\mu\rho}\right] . \tag{18.3}$$

Remarkably, Eq. (18.3) can be simplified through a change of variables. Let $\psi_{\rho\nu} \equiv h_{\rho\nu} - \frac{1}{2}\eta_{\rho\nu}h$, where $\psi_{\rho\nu}(x)$ is a new field, the "trace-reversed" metric perturbation. If we know $\psi_{\rho\nu}$, then we know $h_{\rho\nu}$ through

$$h_{\rho\nu} = \psi_{\rho\nu} - \frac{1}{2}\eta_{\rho\nu}\psi , \tag{18.4}$$

where $\psi \equiv \psi^\sigma_\sigma = -h$. Combining Eqs. (18.3) and (18.4), Einstein's equation at linear order is

$$\partial^\mu\partial_\mu\psi_{\rho\nu} + \left(-\partial^\lambda\partial_\rho\psi_{\lambda\nu} - \partial^\mu\partial_\nu\psi_{\mu\rho} + \eta_{\rho\nu}\partial_\lambda\partial_\sigma\psi^{\sigma\lambda}\right) = -2\kappa T_{\rho\nu} , \tag{18.5}$$

where the terms in parentheses have been grouped together for a reason that will be explained shortly—they can be made to disappear!

Equation (18.5) can be simplified further through an *infinitesimal coordinate transformation*,

$$x^\mu \to x'^\mu = x^\mu + \xi^\mu , \tag{18.6}$$

where ξ^μ is an unknown vector field of the same order of smallness as $h_{\mu\nu}$. As with any coordinate transformation, we need the Jacobian matrices, Eqs. (5.24) and (5.25): $\partial x'^\mu/\partial x^\nu = \delta^\mu_\nu + \partial_\nu\xi^\mu$ and $\partial x^\nu/\partial x'^\lambda = \delta^\nu_\lambda - \partial_\lambda\xi^\nu$, where the latter equation is defined so that Eq. (5.27) holds to first order. The metric tensor in the new coordinate system is, to linear order, $g'_{\mu\nu} = g_{\mu\nu} - \partial_\nu\xi_\mu - \partial_\mu\xi_\nu$, where[1] $\xi_\mu = \eta_{\mu\nu}\xi^\nu$. Using $g_{\mu\nu} = \eta_{\mu\nu} + h_{\mu\nu}$, we have to linear order,

$$h'_{\mu\nu} = h_{\mu\nu} - \partial_\nu\xi_\mu - \partial_\mu\xi_\nu . \tag{18.7}$$

By raising an index, Eq. (18.7) implies that $h'^\lambda_{\ \nu} = h^\lambda_{\ \nu} - \partial^\lambda\xi_\nu - \partial_\nu\xi^\lambda$. Setting $\lambda = \nu$ and summing,

$$h' = h - \partial^\lambda\xi_\lambda - \partial_\lambda\xi^\lambda = h - 2\partial^\lambda\xi_\lambda . \tag{18.8}$$

From Eqs. (18.7) and (18.8) we obtain the transformation of $\psi_{\mu\nu}$ under $x^\mu \to x^\mu + \xi^\mu$

$$\psi'_{\mu\nu} = \psi_{\mu\nu} - \partial_\nu\xi_\mu - \partial_\mu\xi_\nu + \eta_{\mu\nu}\partial^\lambda\xi_\lambda . \tag{18.9}$$

Now raise both indices in Eq. (18.9) and take the divergence; it's straightforward to show that

$$\partial_\nu\psi'^{\mu\nu} = \partial_\nu\psi^{\mu\nu} - \partial_\nu\partial^\nu\xi^\mu . \tag{18.10}$$

We now come to the key point: If we choose ξ^μ so that $\partial_\nu\partial^\nu\xi^\mu = \partial_\nu\psi^{\mu\nu}$, we have

$$\partial_\nu\psi'^{\mu\nu} = 0 . \tag{18.11}$$

With Eq. (18.11) satisfied, each of the terms in parentheses in Eq. (18.5) vanishes in the transformed coordinate system (show this). In this coordinate system, the *linearized field equation is an inhomogeneous wave equation,*[2]

$$\partial_\mu\partial^\mu\psi'_{\rho\nu} = -2\kappa T'_{\rho\nu} , \tag{18.12}$$

where $T'_{\rho\nu}$ is the energy-momentum tensor in the transformed coordinate system. At this point we can drop the primes from Eq. (18.12), which, as a tensor equation, holds in any coordinate system so long as Eq. (18.11) is satisfied. After Eq. (18.12) has been solved, we have the elements of the perturbed metric tensor using Eq. (18.4),

$$g_{\mu\nu} = \eta_{\mu\nu} + h_{\mu\nu} = \eta_{\mu\nu}\left(1 - \tfrac{1}{2}\psi\right) + \psi_{\mu\nu} . \tag{18.13}$$

---

[1] See Eq. (15.45). When raising or lowering the index on first-order quantities, we must use η the zeroth-order metric.

[2] Thus, in the transformed coordinate system $G_{\mu\nu}$ is related to the wave equation operator $G_{\mu\nu} = -\frac{1}{2}\partial_\lambda\partial^\lambda\psi_{\mu\nu}$.

18.1.2 Gauge invariance

Equation (18.6) is a *gauge transformation.*[3] In electrodynamics, the four-potential A^μ satisfies an inhomogeneous wave equation related to the sources of the fields, $\partial_\mu \partial^\mu A^\alpha = -\mu_0 J^\alpha$, Eq. (8.16), when the Lorenz condition $\partial_\mu A^\mu = 0$ is met, Eq (8.17). The Lorenz condition can always be satisfied through a gauge transformation, $A^\mu \to A^\mu + \partial^\mu \chi$, Eq. (8.20), where the gauge scalar field χ satisfies the inhomogeneous wave equation, $\partial_\alpha \partial^\alpha \chi = -\partial_\beta A^\beta$ (Section 8.5). In linearized gravity, the tensor field $\psi_{\mu\nu}$ satisfies an inhomogeneous wave equation related to its sources, $\partial_\mu \partial^\mu \psi_{\rho\nu} = -2\kappa T_{\rho\nu}$, Eq. (18.12), when the analogous Lorenz condition is met, $\partial_\nu \psi^{\mu\nu} = 0$, Eq. (18.11). Equation (18.11) can always be satisfied if the gauge vector field ξ^μ satisfies the inhomogeneous wave equation, $\partial_\nu \partial^\nu \xi^\mu = \partial_\nu \psi^{\mu\nu}$. These results are summarized in Table 18.1.

In electrodynamics, the field tensor $F^{\mu\nu} = \partial^\mu A^\nu - \partial^\nu A^\mu$, Eq. (8.18), is invariant under $A^\mu \to A^\mu + \partial^\mu \chi$, and the theory is said to be gauge invariant. What's invariant under the gauge transformation in GR? It's the *linearized Riemann tensor.* From Eq. (14.61), we have at linear order $R^\alpha_{\ \rho\mu\nu} = \partial_\mu \Gamma^\alpha_{\nu\rho} - \partial_\nu \Gamma^\alpha_{\mu\rho}$. Using the linearized Christoffel symbols, Eq. (15.9), the linearized Riemann tensor is

$$R_{\alpha\rho\mu\nu} = \tfrac{1}{2}\left[\partial_\alpha\left(\partial_\nu h_{\mu\rho} - \partial_\mu h_{\nu\rho}\right) - \partial_\rho\left(\partial_\nu h_{\alpha\mu} - \partial_\mu h_{\alpha\nu}\right)\right] . \qquad \text{(linearized)} \qquad (18.14)$$

It's straightforward to show that Eq. (18.14) is invariant under the transformation in Eq. (18.7).

Table 18.1 Comparison of Linearized Gravity with Electrodynamics.

	Linearized Gravity	Electrodynamics
Field	$\psi_{\mu\nu}$ perturbed metric tensor	A^μ four-potential
Conservation law(s)	$\partial_\nu T^{\mu\nu} = 0$	$\partial_\nu J^\nu = 0$
Field equation	$\partial^\mu \partial_\mu \psi_{\rho\nu} - \partial^\lambda \partial_\rho \psi_{\lambda\nu} - \partial^\mu \partial_\nu \psi_{\mu\rho}$ $+\eta_{\rho\nu} \partial_\lambda \partial_\sigma \psi^{\sigma\lambda} = -2\kappa T_{\rho\nu}$	$\partial_\nu\left(\partial^\mu A^\nu - \partial^\nu A^\mu\right) = \mu_0 J^\mu$
Gauge transformation	$x^\mu \to x^\mu + \xi^\mu(x)$	$A^\mu \to A^\mu + \partial^\mu \chi(x)$
Invariant under the gauge transformation	linearized Riemann tensor	field tensor $F^{\mu\nu}$
Lorenz condition	$\partial_\nu \psi^{\mu\nu} = 0$	$\partial_\mu A^\mu = 0$
Gauge function	$\partial_\nu \partial^\nu \xi^\mu = \partial_\nu \psi^{\mu\nu}$ vector gauge function	$\partial_\nu \partial^\nu \chi = -\partial_\mu A^\mu$ scalar gauge function
Simplified field equation	$\partial_\mu \partial^\mu \psi_{\rho\nu} = -2\kappa T_{\rho\nu}$	$\partial_\mu \partial^\mu A^\alpha = -\mu_0 J^\alpha$
Retarded solution	$\psi_{\rho\nu}(t, \boldsymbol{x}) =$ $\frac{\kappa}{2\pi} \int \frac{T_{\rho\nu}(t - \lvert\boldsymbol{x} - \boldsymbol{y}\rvert/c, \boldsymbol{y})}{\lvert\boldsymbol{x} - \boldsymbol{y}\rvert} d^3 y$	$A^\alpha(t, \boldsymbol{x}) =$ $\frac{\mu_0}{4\pi} \int \frac{J^\alpha(t - \lvert\boldsymbol{x} - \boldsymbol{y}\rvert/c, \boldsymbol{y})}{\lvert\boldsymbol{x} - \boldsymbol{y}\rvert} d^3 y$

[3] $G_{\mu\nu}$ has six independent elements in four dimensions (Section 15.3.2), while $T_{\mu\nu}$ has 10; thus there is room for four independent degrees of freedom associated with $g_{\mu\nu}$. Those degrees of freedom are represented by the vector field ξ^μ.

18.1.3 Retarded solution

In the linearized theory the perturbation to the Lorentz metric $\psi_{\mu\nu}(x)$ satisfies a wave equation generated by $\kappa T_{\mu\nu}(x)$. The first step in solving an inhomogeneous differential equation is to find the Green function associated with that differential equation. We omit that step because of length restrictions (even though it's instructive to see in a relativistic context), and because we can simply write down the answer because the same equation is encountered in electrodynamics (a relativistic field theory), and the same math solves the same equations. The electromagnetic four-potential $A^\mu(x)$ satisfies an inhomogeneous wave equation generated by the four-current $J^\mu(x)$ (Eq. (8.16)), the solution of which can be expressed as a *retarded field* (see Table 18.1).[4] Without further ado, the retarded solution of Eq. (18.12) is[5]

$$\psi_{\rho\nu}(t, \boldsymbol{x}) = \frac{\kappa}{2\pi} \int \frac{T_{\rho\nu}\left(t - |\boldsymbol{x} - \boldsymbol{y}|/c, \boldsymbol{y}\right)}{|\boldsymbol{x} - \boldsymbol{y}|} \mathrm{d}^3 y \,, \tag{18.15}$$

where the integration is over the spatial location of sources, $\boldsymbol{y}$, with $T_{\rho\nu}$ evaluated at the *retarded time* $t_r \equiv t - |\boldsymbol{x} - \boldsymbol{y}|/c$ on the past light cone of the field point, $\boldsymbol{x}$. See Fig. 18.1. Equation (18.15) is the general solution to the problem of linearized gravity.

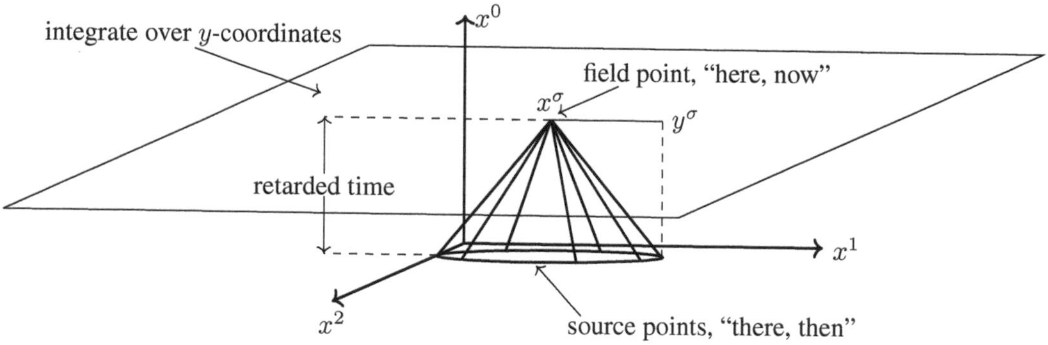

Figure 18.1 Source at $(x^0 - |\boldsymbol{x} - \boldsymbol{y}|, \boldsymbol{y})$ is causally related to the field at $(x^0, \boldsymbol{x})$.

18.2 STATIC SOURCE

As the simplest application of Eq. (18.15), consider a *static* mass distribution $\rho(\boldsymbol{r})$, where $T_{00} = \rho c^2$ and $T_{\mu\nu} = 0$ for $\mu \neq 0$, $\nu \neq 0$. In that case $\psi_{\mu\nu} = 0$ for $\mu \neq 0$, $\nu \neq 0$, with[6]

$$\psi_{00}(\boldsymbol{x}) = \frac{4G}{c^2} \int \frac{\rho(\boldsymbol{y})}{|\boldsymbol{x} - \boldsymbol{y}|} d^3 y \equiv -\frac{4}{c^2}\Phi(\boldsymbol{x}) \,, \tag{18.16}$$

where Φ is the Newtonian gravitational potential; Table 1.1. The assumption that ψ_{00} is small is equivalent to the requirement of weak gravity, $|\Phi| \ll c^2$. The scalar $\psi = \psi^\lambda_\lambda = \eta^{00}\psi_{00} = -\psi_{00}$, and thus from Eq. (18.13), $g_{00} = \eta_{00}(1 + \frac{1}{2}\psi_{00}) + \psi_{00} = -1 + \frac{1}{2}\psi_{00} = -(1 + 2\Phi/c^2)$ and $g_{ii} = 1 + \frac{1}{2}\psi_{00} = 1 - 2\Phi/c^2$. The spacetime geometry is described by

$$(\mathrm{d}s)^2 = -\left(1 + \frac{2}{c^2}\Phi\right)(c\mathrm{d}t)^2 + \left(1 - \frac{2}{c^2}\Phi\right)(\mathrm{d}\sigma)^2 \,, \tag{18.17}$$

[4] There are also advanced fields, which have interesting uses, e.g., in the Wheeler-Feynman theory, but which we ignore here. Retaining the advanced solutions produces a *time-symmetric* theory; including only the retarded solution breaks time-reversal symmetry. A direction of time is selected by working solely with retarded fields.

[5] As shown in books on electrodynamics, Eq. (18.15) solves Eq. (18.12) *and* satisfies the Lorenz condition, Eq. (18.11). Equation (18.11) was assumed in the derivation of Eq. (18.12); consistency requires that Eq. (18.15) satisfy Eq. (18.11).

[6] We can ignore time retardation for static sources.

where $(\mathrm{d}\sigma)^2$ denotes the spatial line element. While Eq. (18.17) has the form of the Schwarzschild metric for weak gravity ($|\Phi| \ll c^2$); Eq. (18.17) is more general than the Schwarzschild metric: The potential Φ in Eq. (18.17) is the solution to the Poisson equation for *any* static mass distribution, not necessarily spherically symmetric; moreover it's not a *vacuum* solution, it's not restricted to the exterior of the mass distribution. Note that the metric in Eq. (18.17) is time-reversal symmetric.[7]

18.3 FAR FROM A SLOWLY VARYING SOURCE

Assume that the field point located by vector $\boldsymbol{r}$ is far removed from the source—in relation to its size—indicated schematically in Fig. 18.2. That is, $r \equiv |\boldsymbol{r}| \gg y \equiv |\boldsymbol{y}|$ for any source point located

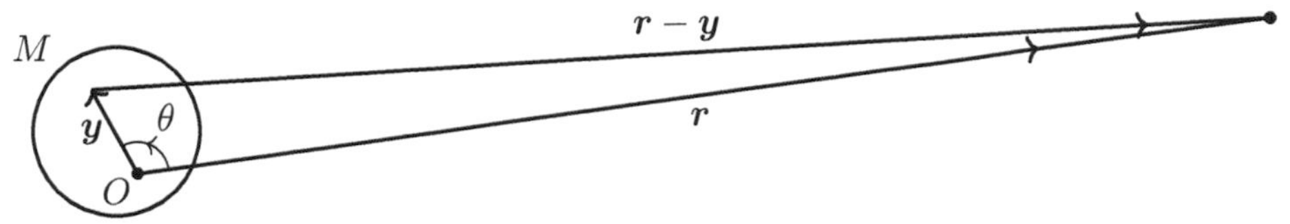

Figure 18.2 Geometry for Eq. (18.21). Mass is shown as spherical for simplicity.

by vector $\boldsymbol{y}$. Under these conditions we can obtain an approximate solution to Eq. (18.15) involving its *multipole expansion*. Start with

$$\begin{aligned}\frac{1}{|\boldsymbol{r}-\boldsymbol{y}|} &= \frac{1}{r}\frac{1}{\sqrt{1-(2y/r)\cos\theta+(y/r)^2}} = \frac{1}{r}\sum_{n=0}^{\infty}\left(\frac{y}{r}\right)^n P_n(\hat{\boldsymbol{r}}\cdot\hat{\boldsymbol{y}}) \\ &= \frac{1}{r}\left[1+\frac{1}{r}\hat{\boldsymbol{r}}\cdot\boldsymbol{y}+\frac{1}{2r^2}\left(3\hat{\boldsymbol{r}}_k\hat{\boldsymbol{r}}_j-\delta_{jk}\right)y^jy^k+\cdots\right], \end{aligned} \tag{18.18}$$

where we've used the generating function for Legendre polynomials [73, p784] and that $\cos\theta = \hat{\boldsymbol{r}}\cdot\hat{\boldsymbol{y}} = \hat{\boldsymbol{r}}_i\hat{\boldsymbol{y}}^i$, with $\hat{\boldsymbol{r}} = \boldsymbol{r}/r$ and $\hat{\boldsymbol{y}} = \boldsymbol{y}/y$ unit vectors. We also require the expansion

$$|\boldsymbol{r}-\boldsymbol{y}| = r - \hat{\boldsymbol{r}}\cdot\boldsymbol{y} - \frac{1}{2r}\left(\hat{\boldsymbol{r}}_j\hat{\boldsymbol{r}}_k-\delta_{jk}\right)y^jy^k+\cdots \equiv r-\Delta(\boldsymbol{y},\boldsymbol{r}), \qquad (r\gg|\boldsymbol{y}|) \tag{18.19}$$

where $\Delta = r - |\boldsymbol{r}-\boldsymbol{y}| = \hat{\boldsymbol{r}}\cdot\boldsymbol{y} + y^jy^k\left(\hat{\boldsymbol{r}}_j\hat{\boldsymbol{r}}_k-\delta_{jk}\right)+\cdots$ is a measure of the size of the source. The retarded time between $\boldsymbol{r}$ and $\boldsymbol{y}$ is characterized by Δ: $t_r = t - r/c + \Delta(\boldsymbol{y},\boldsymbol{r})/c$. The expansion Eq. (18.19) occurs in the time argument of $T_{\mu\nu}(t_r,\boldsymbol{y})$. We can expand around a *fixed* retarded time $r_t^0 \equiv t - r/c$ (that at the origin):

$$T_{\mu\nu}(t-r/c+\Delta/c,\boldsymbol{y}) = \sum_{n=0}^{\infty}\frac{1}{n!}\left(\frac{\Delta}{c}\right)^n\left[\frac{\partial^n T_{\mu\nu}(t,\boldsymbol{y})}{\partial t^n}\right]_{\Delta=0}. \tag{18.20}$$

A derivative with respect to time implies a characteristic time scale t_c over which $T_{\mu\nu}$ varies, with $\partial/\partial t \sim t_c^{-1}$. Because Δ is a measure of the size of the source, the product $\Delta\partial/\partial t$ represents a characteristic *speed* with which the source varies. The series in Eq. (18.20) converges rapidly when $|\Delta\partial/\partial t| \ll c$. Combining Eqs. (18.18) and (18.20) with Eq. (18.15), we have the general expansion probing the variation of the source in space and time:

$$\begin{aligned}\psi_{\mu\nu}(t,\boldsymbol{r}) = \frac{\kappa}{2\pi r}\int \mathrm{d}^3y &\left(\sum_{n=0}^{\infty}\frac{1}{n!}\left(\frac{\Delta(\boldsymbol{y},\boldsymbol{r})}{c}\right)^n\left[\frac{\partial^n T_{\mu\nu}(t,\boldsymbol{y})}{\partial t^n}\right]_{\Delta=0}\right) \\ &\times\left[1+\frac{1}{r}\hat{\boldsymbol{r}}\cdot\boldsymbol{y}+\frac{1}{2r^2}\left(3\hat{\boldsymbol{r}}_k\hat{\boldsymbol{r}}_j-\delta_{jk}\right)y^jy^k+\cdots\right]. \end{aligned} \tag{18.21}$$

[7]The distinction between static and stationary spacetimes is introduced in Section 16.1.1.

Equation (18.21) is clearly an intricate expression from which many details can be inferred. We consider the simplest case of the lowest order in both expansions, appropriate for the observation point far removed from a nonrelativistic source:

$$\psi_{\rho\nu}(t, \boldsymbol{r}) = \frac{4G}{c^4 r} \int \mathrm{d}^3 y T_{\rho\nu}(t - r/c, \boldsymbol{y}) \,. \tag{18.22}$$

Using $T_{00} = T^{00}$, $T_{0i} = -T^{0i} = -T^{i0}$, $T_{ij} = T^{ij}$, and the interpretation of $T^{\mu\nu}$ (Section 9.2),

$$\int T_{00} \mathrm{d}^3 y = c^2 [M]_{t-r/c} = \text{energy of source}$$

$$\int T_{0i} \mathrm{d}^3 y = -c \left[P^i\right]_{t-r/c} = -c \times \text{momentum of source in } i \text{ direction}$$

$$\int T_{ij} \mathrm{d}^3 y = \text{integrated internal stress of source in } i, j \text{ directions.}$$

For an isolated source, M and P^i are constants, and, because it's always possible to work in a reference frame where the momentum is zero, we take

$$\psi_{00} = \frac{4GM}{c^2 r} \qquad \psi_{0i} = \psi_{i0} = 0 \,. \tag{18.23}$$

Note that ψ_{00} in Eq. (18.23) is the monopole approximation to ψ_{00} in Eq. (18.16).

Evaluating ψ_{ij} relies on a trick. Consider $\partial_k(T^{ik} y^l) = T^{ik}\delta^l_k + y^l \partial_k T^{ik} = T^{il} - y^l \partial_0 T^{i0}$, where we've used the conservation law $\partial_\mu T^{i\mu} = 0$. Thus,

$$\int T^{il} \mathrm{d}^3 y = \int \partial_k (y^l T^{ik}) \mathrm{d}^3 y + \int y^l \partial_0 T^{i0} \mathrm{d}^3 y = \frac{1}{2c} \frac{\mathrm{d}}{\mathrm{d}t} \int (y^l T^{i0} + y^i T^{l0}) \mathrm{d}^3 y \,, \tag{18.24}$$

where the integral involving the divergence can be "waved away" (use Gauss's theorem and that for a localized source $T^{ij} = 0$ on the surface), and we have symmetrized with respect to the indices (i, l). Apply the same trick to $\partial_k (T^{ik} y^j y^l) = T^{il} y^j + T^{ij} y^l - y^j y^l \partial_0 T^{i0}$ and for the same reason conclude that

$$\int (T^{il} y^j + T^{ij} y^l) \mathrm{d}^3 y = \frac{1}{c} \frac{\mathrm{d}}{\mathrm{d}t} \int y^j y^l T^{i0} \mathrm{d}^3 y \,. \tag{18.25}$$

Combining Eqs. (18.24) and (18.25) (with $j \to l, l \to i, i \to 0$),

$$\int T^{il} \mathrm{d}^3 y = \frac{1}{2c^2} \frac{\mathrm{d}^2}{\mathrm{d}t^2} \int y^i y^l T^{00} \mathrm{d}^3 y \,. \tag{18.26}$$

Combining Eqs. (18.26) and (18.22),

$$\psi_{ij}(t, r) = \frac{2G}{c^6 r} \left[\frac{\mathrm{d}^2}{\mathrm{d}t^2} I^{ij}(t) \right]_{t-r/c} \,, \tag{18.27}$$

where

$$I^{ij} \equiv \int y^i y^j T^{00} \mathrm{d}^3 y \tag{18.28}$$

is the second mass moment of the source. Equation (18.27) is used in the description of long-wavelength (low frequency) gravitational waves.

18.4 GRAVITOMAGNETISM: STATIONARY SOURCES

We consider a nonrelativistic *stationary* source, one that's invariant under time translations (Section 16.1.1), so that $\partial_0 T^{\mu\nu} = 0$—the particles of the source are in steady-state motion. For this type of source, time retardation is immaterial and Eq. (18.15) is equivalent to:

$$\psi_{\mu\nu}(\boldsymbol{x}) = \frac{4G}{c^4} \int \frac{T_{\mu\nu}(\boldsymbol{y})}{|\boldsymbol{x}-\boldsymbol{y}|} \mathrm{d}^3 y \,. \tag{18.29}$$

Using the energy-momentum tensor in Eq. (10.17), we have for nonrelativistic speeds: $T^{00} = \rho c^2 + O(\beta^2)$, $T^{i0} = (\rho c^2 + P)\beta^i + O(\beta^3)$, and $T^{ij} = P\delta_{ij} + (\rho c^2 + P)\beta^i\beta^j + O(\beta^4)$. Assuming reasonable pressures ($P \ll \rho c^2$), $T^{0i} \sim T^{00}\beta^i$ and $T^{ij} \sim T^{00}\beta^i\beta^j$. We can therefore approximate $T^{\mu\nu}$ schematically as:

$$T^{\mu\nu} \approx T^{00} \left(\begin{array}{c|ccc} 1 & \beta^1 & \beta^2 & \beta^3 \\ \hline \beta^1 & & & \\ \beta^2 & & \beta^i\beta^j & \\ \beta^3 & & & \end{array} \right) .$$

If we further approximate $T^{\mu\nu}$ by keeping only terms up to $O(\beta)$, there are four independent elements, $T^{0\alpha}$. This approximation permits a correspondence between linearized gravity and electromagnetism. The electromagnetic four-potential for a stationary source (no time retardation) is[8]

$$A^\alpha(\boldsymbol{x}) = \frac{\mu_0}{4\pi} \int \frac{J^\alpha(\boldsymbol{y})}{|\boldsymbol{x}-\boldsymbol{y}|} \mathrm{d}^3 y \,.$$

While there are always four components of J^α (four-vector), in linearized gravity there are four independent elements to $T^{\mu\nu}$ at this level of approximation, namely $T^{0\nu}$.

Substituting $T_{00} = \rho c^2$ in Eq. (18.29), we have $\psi_{00}(\boldsymbol{x}) = -4\Phi(\boldsymbol{x})/c^2$, the same as Eq. (18.16). Using $T_{0i} = -T^{0i} = -\rho c v^i$ in Eq. (18.29),

$$\psi_{0i} = -\frac{4G}{c^3} \int \frac{(\rho v^i)(\boldsymbol{y})}{|\boldsymbol{x}-\boldsymbol{y}|} \mathrm{d}^3 y \equiv \frac{A_{g,i}}{c} \,,$$

where we have defined a *gravitational analog* of the vector potential,[9] $A_{g,i} = c\psi_{0i}$. We take the space-space terms $\psi_{ij} = 0$ because the stress tensor T_{ij} is second-order in β. In this case the scalar field $\psi = -\psi_{00}$. Using Eq. (18.13), the spacetime geometry associated a nonrelativistic stationary source is described by

$$(\mathrm{d}s)^2 = -\left(1 + \frac{2\Phi}{c^2}\right)(c\mathrm{d}t)^2 + \frac{2}{c} A_{g,i} \mathrm{d}x^0 \mathrm{d}x^i + \left(1 - \frac{2\Phi}{c^2}\right)(\mathrm{d}\sigma)^2 \,. \tag{18.30}$$

The extra term in Eq. (18.30), compared with Eq. (18.17), reflects the motion of the source.[10] The effect of source motion can be interpreted as a *gravitational* vector potential produced by a mass current, akin to the magnetic vector potential produced by charge currents. We define the *gravitational four-potential* (compare with Eq. (8.15)),

$$A_g^\mu \equiv \left(\frac{4}{c}\Phi, \boldsymbol{A}_g\right) ,$$

[8]Derived in any book on electromagnetism; it's also the analog of Eq. (18.29).

[9]The vector potential in electromagnetism has the dimensions of energy/(charge speed). For the associated quantity in gravitation let charge become mass, in which case the gravitational vector potential has the dimension of speed. The quantity $\psi_{\mu\nu}$ in Eq. (18.29) is dimensionless; thus we have defined $A_{g,i} = c\psi_{0i}$.

[10]The metric tensor for a rotating frame, Eq. (12.34), features a $\mathrm{d}x^0\mathrm{d}x^i$ term.

where Φ is the gravitational potential, Eq. (18.16), and (with $\boldsymbol{J} = \rho\boldsymbol{v}$ the mass current density)

$$\boldsymbol{A}_g = -\frac{4G}{c^2}\int \frac{\boldsymbol{J}(\boldsymbol{y})}{|\boldsymbol{x}-\boldsymbol{y}|}\mathrm{d}^3y \,. \tag{18.31}$$

The analogy with electromagnetism is strong. We find from Eqs. (18.16) and (18.31)[11]

$$\nabla^2\Phi = 4\pi G\rho \qquad \nabla^2\boldsymbol{A}_g = \frac{16\pi G}{c^2}\boldsymbol{J}\,. \tag{18.32}$$

Compared with the Poisson equations in electromagnetism for the scalar and vector potentials, we have the correspondences

$$\epsilon_0 \leftrightarrow -\frac{1}{4\pi G} \qquad \mu_0 \leftrightarrow -\frac{16\pi G}{c^2}\,. \tag{18.33}$$

The minus sign here can be traced to the fact that masses always attract, but like charges repel. The analogy can be taken further by defining the *gravitoelectric field*, $\boldsymbol{E}_g = -\boldsymbol{\nabla}\Phi$, and the *gravitomagnetic field*, $\boldsymbol{B}_g = \boldsymbol{\nabla}\times\boldsymbol{A}_g$. With the identity $\boldsymbol{\nabla}(1/|\boldsymbol{x}-\boldsymbol{y}|) = -(\boldsymbol{x}-\boldsymbol{y})/|\boldsymbol{x}-\boldsymbol{y}|^3$, it's straightforward to show that $\boldsymbol{E}_g$ and $\boldsymbol{B}_g$ reduce to the usual expressions for the electrostatic and magnetostatic fields when the identifications in Eq. (18.33) are used and the symbols ρ and $\boldsymbol{J}$ are interpreted as charge and current densities. Note that it's the potentials $(\Phi, \boldsymbol{A})$ that couple to the spacetime metric in Eq. (18.30), and not the *fields*.

The fields $\boldsymbol{E}_g$, $\boldsymbol{B}_g$ satisfy the *gravitational Maxwell equations*,

$$\begin{aligned} \boldsymbol{\nabla}\cdot\boldsymbol{E}_g &= -4\pi G\rho & \qquad \boldsymbol{\nabla}\cdot\boldsymbol{B}_g &= 0 \\ \boldsymbol{\nabla}\times\boldsymbol{E}_g &= 0 & \qquad \boldsymbol{\nabla}\times\boldsymbol{B}_g &= -\frac{16\pi G}{c^2}\boldsymbol{J}\,. \end{aligned}$$

To show that the analog of Ampere's law holds, we must show that $\boldsymbol{\nabla}\cdot\boldsymbol{A}_g = 0$, which follows from Eq. (18.31). The gravitoelectric field $\boldsymbol{E}_g$ is the Newtonian gravitational field $\boldsymbol{g}$ that we started with in Table 1.1. The gravitomagnetic field, however, $\boldsymbol{B}_g$ is new to GR and predicts an additional type of gravitational acceleration due to mass sources in motion.

18.5 FRAME DRAGGING

What is the motion of a particle in the *gravitoelectromagnetic field*? A charged particle accelerates in the electromagnetic field with $\boldsymbol{a} = (q/m)(\boldsymbol{E}+\boldsymbol{v}\times\boldsymbol{B})$. Under the correspondence $q \leftrightarrow m$, that the "charge" of a particle in a gravitational field is its mass, we would expect $\boldsymbol{a} = \boldsymbol{E}_g + \boldsymbol{v}\times\boldsymbol{B}_g$. In GR a free-fall worldline is a geodesic. The geodesic equation requires the Christoffel symbols associated with Eq. (18.30); these are given in Table 18.2. Note that $\Gamma^i_{jk} = 0$ if the spatial indices are

Table 18.2 Christoffel symbols associated with spacetime metric Eq. (18.30).

$\Gamma^0_{00} = 0$	$\Gamma^i_{00} = -\frac{1}{4}\partial^i\psi_{00} = \frac{1}{c^2}\partial^i\Phi$
$\Gamma^0_{0i} = -\frac{1}{4}\partial_i\psi_{00} = \frac{1}{c^2}\partial_i\Phi$	$\Gamma^i_{0j} = \frac{1}{2}\left[\partial_j\psi^i_0 - \partial^i\psi_{0j}\right] = \frac{1}{2c}\left[\partial_jA^i - \partial^iA_j\right]$
$\Gamma^0_{ij} = -\frac{1}{2}\left[\partial_i\psi_{0j} + \partial_j\psi_{0i}\right]$	$\Gamma^i_{jk} = \frac{1}{c^2}\left[\delta_{jk}\partial^i\Phi - \delta^i_k\partial_j\Phi - \delta^i_j\partial_k\Phi\right]$

[11]Use the standard identity $\nabla^2(1/|\boldsymbol{x}-\boldsymbol{y}|) = -4\pi\delta^{(3)}(\boldsymbol{x}-\boldsymbol{y})$.

all different. This is because the "space-space" part of the metric tensor is diagonal (see comments in Section 16.2).

The linearized geodesic equation for nonrelativistic speeds is worked out in Section 15.2, where it's shown that the time component is second order in small quantities. The spatial part of Eq. (15.8) is, to first order in the velocities, $\ddot{x}^i + c^2\Gamma^i_{00} + 2c\Gamma^i_{0j}\dot{x}^j = 0$. Using Γ^i_{00} and Γ^i_{0j} from Table 18.2,

$$\ddot{x}^i + \partial^i\Phi + (\partial_j A^i - \partial^i A_j)\dot{x}^j = 0\,. \tag{18.34}$$

Let's work out the components of $\boldsymbol{v} \times \boldsymbol{B} = \boldsymbol{v} \times \boldsymbol{\nabla} \times \boldsymbol{A}$,

$$\begin{aligned}(\boldsymbol{v} \times \boldsymbol{B})^i &= \epsilon^i{}_{jk} v^j B^k = \epsilon^i{}_{jk}\epsilon^k{}_{lm} v^j \partial^l A^m = (\delta^i_l \delta_{jm} - \delta^i_m \delta_{jl}) v^j \partial^l A^m \\ &= v^j \left(\partial^i A_j - \partial_j A^i\right) = -2c\Gamma^i_{0j}\dot{x}^j\,.\end{aligned} \tag{18.35}$$

Combining Eqs. (18.34) and (18.35),

$$\ddot{x}^i = (\boldsymbol{E}_g)^i + (\boldsymbol{v} \times \boldsymbol{B}_g)^i\,.$$

We have the remarkable result that *a slow-moving particle in the gravitational field of a stationary source experiences an acceleration entirely analogous to the Lorentz force on a charged particle.*

An inertial reference frame (free-fall) would therefore *appear to be rotating*—free-fall motion would appear to be subject to a Coriolis force.[12] A rotating mass is said to *drag* the reference frame with it—*frame dragging*. Below we calculate $\boldsymbol{B}_g$ for a slowly rotating mass. It might be thought that frame dragging only occurs in the vicinity of a rotating mass. The gravitomagnetic field occurs for *any* slow, steady motion of the mass source: The vector potential $\boldsymbol{A}_g$ is associated with any mass current $\boldsymbol{J}$, Eq. (18.31). From our treatment of the electromagnetic field (Chapter 8), while there are exceptional reference frames in which $\boldsymbol{E}$ or $\boldsymbol{B}$ vanish, there is always some other frame in which $\boldsymbol{E}$ and $\boldsymbol{B}$ are both present; both are aspects of the single entity, the electromagnetic field. Frame dragging always occurs in the linearized gravitational field; $\boldsymbol{E}_g$ and $\boldsymbol{B}_g$ are two aspects of the same gravitational manifestation in spacetime. It's only because frame-dragging effects are so small in our everyday experience that we're not aware of it—like the Coriolis force on Earth, which is not ordinarily noticeable.

18.6 SLOWLY ROTATING SOURCE

We now calculate the gravitomagnetic field of a body rotating with constant angular speed, where the velocity $\boldsymbol{v}$ at any point in the object is nonrelativitic, $v \ll c$. The first step is to obtain the vector potential, $\boldsymbol{A}_g$. Combining Eqs. (18.31) and (18.18), we have the multipole expansion,

$$\boldsymbol{A}_g(\boldsymbol{x}) = -\frac{4G}{rc^2}\left[\int \boldsymbol{J}(\boldsymbol{y})\mathrm{d}^3y + \frac{x_k}{r^2}\int \boldsymbol{J}(\boldsymbol{y})y^k\mathrm{d}^3y + \cdots\right]\,. \tag{18.36}$$

The first integral on the right of Eq. (18.36) vanishes due to conservation of mass. Using a variant of the trick used in Section 18.3 ("sub-trick"):

$$\int J^j \mathrm{d}^3y = \int J^i \frac{\partial y^j}{\partial y^i}\mathrm{d}^3y = \int \mathrm{d}S_i J^i y^j - \int y^j \partial_i J^i \mathrm{d}^3y = 0\,, \tag{18.37}$$

where we've integrated by parts. The remaining integrals each vanish because $J^i = 0$ on the surface and because of the conservation law for steady currents, $\partial_i J^i = 0$.

To evaluate the second integral in Eq. (18.36), express $J^i y^k$ as the sum of symmetric and antisymmetric parts, $J^i y^k = \frac{1}{2}(J^i y^k + J^k y^i) + \frac{1}{2}(J^i y^k - J^k y^i)$. By Eq. (18.25), the integral over

[12] The Coriolis force (acceleration seen in a noninertial, rotating frame), $\boldsymbol{\Omega} \times \boldsymbol{v}$, is related to $\boldsymbol{v} \times \boldsymbol{B}_g$ through the correspondence $\boldsymbol{B}_g \propto \boldsymbol{L} \propto \boldsymbol{\Omega}$.

the symmetric part vanishes *for a stationary source* (time derivative vanishes). Thus, $\int y^k J^i \mathrm{d}^3y = \frac{1}{2}\int (y^k J^i - y^i J^k)\mathrm{d}^3y$. Define the angular momentum of the mass source, $\boldsymbol{L} \equiv \int (\boldsymbol{y} \times \boldsymbol{J})\mathrm{d}^3y$. The vector $\boldsymbol{L}$ is independent of the choice of origin because $\int \boldsymbol{J}(\boldsymbol{y})\mathrm{d}^3y = 0$. Let's work out $(\boldsymbol{L} \times \boldsymbol{x})^i = \epsilon^{i\ k}_{\ j} L^j x_k = \epsilon^{i\ k}_{\ j} \epsilon^j_{\ lm} x_k \int y^l J^m \mathrm{d}^3y = (\delta^i_m \delta^k_l - \delta^i_l \delta^k_m) x_k \int y^l J^m \mathrm{d}^3y = x_k \int (y^k J^i - y^i J^k)\mathrm{d}^3y$. We then have the remarkable result:

$$x_k \int y^k \boldsymbol{J}(\boldsymbol{y})\mathrm{d}^3y = \frac{1}{2}\boldsymbol{L} \times \boldsymbol{x} \,. \tag{18.38}$$

Combining Eqs. (18.36)–(18.38), we have the gravitational vector potential at lowest order in a multipole expansion,

$$\boldsymbol{A}_g(\boldsymbol{x}) = -\frac{2G}{c^2 r^3}(\boldsymbol{L} \times \boldsymbol{x}) \,. \tag{18.39}$$

Equation (18.39) is exact for a spherically symmetric source. It's then straightforward to show that $\boldsymbol{B}_g = \boldsymbol{\nabla} \times \boldsymbol{A}_g$ can be written

$$\boldsymbol{B}_g = -\frac{2G}{c^2 r^3}\left(\frac{3\boldsymbol{L} \cdot \boldsymbol{r}}{r^2}\boldsymbol{r} - \boldsymbol{L}\right) , \tag{18.40}$$

what should be a familiar expression.[13] Combining Eqs. (18.39) and (18.30), the spacetime metric associated with a slowly rotating mass is

$$g_{\mu\nu} = \begin{pmatrix} -1 - 2\Phi/c^2 & A_1/c & A_2/c & A_3/c \\ A_1/c & 1 - 2\Phi/c^2 & 0 & 0 \\ A_2/c & 0 & 1 - 2\Phi/c^2 & 0 \\ A_3/c & 0 & 0 & 1 - 2\Phi/c^2 \end{pmatrix} . \tag{18.41}$$

18.7 SPIN PRECESSION II: THE LENSE-THIRRING EFFECT

We now examine spin precession in the gravitoelectromagnetic field. Let's consider some orders of magnitude apropos of the GPB experiment. For Earth, $\Phi/c^2 \sim 10^{-9}$ (Table 1.2). The magnitude of A_g can be estimated by taking $L = I\omega$ (I is Earth's moment of inertia, ω its angular frequency). Using Eq. (18.39), $A_g/c \sim (G\omega M)/c^3 \approx 10^{-15}$, where M is Earth's mass. With these numbers we can, referring to Eq. (18.41), ignore terms like $g_{0i}g_{0j}$, Φg_{0i}, etc. The inverse of Eq. (18.41) is in principle a full matrix. However, given the small magnitude of the off-diagonal elements, we can take as $g^{\mu\nu}$:

$$g^{\mu\nu} = \begin{pmatrix} -1 + 2\Phi/c^2 & A_1/c & A_2/c & A_3/c \\ A_1/c & 1 + 2\Phi/c^2 & 0 & 0 \\ A_2/c & 0 & 1 + 2\Phi/c^2 & 0 \\ A_3/c & 0 & 0 & 1 + 2\Phi/c^2 \end{pmatrix} .$$

The equation of motion for a spin S^μ in free fall is $\mathrm{d}S^\mu/\mathrm{d}\tau = -\Gamma^\mu_{\alpha\beta} S^\alpha U^\beta$, Eq. (17.78). It will be convenient in what follows to work with the covariant spin, S_μ, in which case the equation of motion is[14] $\mathrm{d}S_\mu/\mathrm{d}\tau = \Gamma^\lambda_{\mu\beta} S_\lambda U^\beta$. Multiply by $\mathrm{d}t/\mathrm{d}\tau$ and we have

$$\frac{\mathrm{d}S_\mu}{\mathrm{d}t} = \Gamma^\lambda_{\mu\beta} S_\lambda V^\beta , \tag{18.42}$$

[13] If you're keeping track, Eq. (18.40) is "off" by a factor of -2 as compared with the standard expression for the $\boldsymbol{B}$-field produced by a magnetic dipole, when the correspondence for μ_0 in Eq. (18.33) is used. The minus sign is because masses attract. The factor of two is because in magnetostatics the magnetic moment is defined with a factor of $\frac{1}{2}$.

[14] Use $\mathrm{D}S_\mu = U^\alpha \nabla_\alpha S_\mu = 0$ and Eq. (14.13).

where $V^\alpha \equiv \mathrm{d}x^\alpha/\mathrm{d}t = (c, \boldsymbol{v})$. From the spatial parts of Eq. (18.42),

$$\frac{\mathrm{d}S_i}{\mathrm{d}t} = \Gamma^0_{i0}S_0 c + \Gamma^0_{ij}S_0 v^j + \Gamma^j_{i0}S_j c + \Gamma^j_{ik}S_j v^k \,. \tag{18.43}$$

We can eliminate S_0 from Eq. (18.43) using $S_\mu U^\mu = 0$, Eq. (12.53). Thus,

$$S_0 = -\frac{U^i}{U^0}S_i = -\frac{1}{c}\boldsymbol{v}\cdot\boldsymbol{S} \,. \tag{18.44}$$

The second term in Eq. (18.43) is negligible in magnitude compared to the other terms. The equation of motion at lowest order is therefore

$$\frac{\mathrm{d}S_i}{\mathrm{d}t} = -\Gamma^0_{i0}(\boldsymbol{v}\cdot\boldsymbol{S}) + c\Gamma^j_{i0}S_j + \Gamma^j_{ik}S_j v^k \,. \tag{18.45}$$

Using the entries in Table 18.2, it's straightforward to show that

$$\begin{aligned}\Gamma^j_{ik}S_j v^k &= \frac{1}{c^2}\left[(\boldsymbol{S}\cdot\nabla\Phi)v_i - (\boldsymbol{S}\cdot\boldsymbol{v})\partial_i\Phi - (\boldsymbol{v}\cdot\nabla\Phi)S_i\right] \\ c\Gamma^j_{i0}S_j &= \frac{1}{2}\left(\boldsymbol{S}\times\nabla\times\boldsymbol{A}_g\right)_i \,.\end{aligned} \tag{18.46}$$

Combining Eqs. (18.46) and (18.45), we have the vector equation

$$\frac{\mathrm{d}\boldsymbol{S}}{\mathrm{d}t} = -\frac{2}{c^2}\nabla\Phi(\boldsymbol{v}\cdot\boldsymbol{S}) + \frac{\boldsymbol{v}}{c^2}(\boldsymbol{S}\cdot\nabla\Phi) - \frac{1}{c^2}(\boldsymbol{v}\cdot\nabla\Phi)\boldsymbol{S} + \frac{1}{2}\left(\boldsymbol{S}\times\boldsymbol{B}_g\right) \,, \tag{18.47}$$

where we've used $\boldsymbol{B}_g = \nabla\times\boldsymbol{A}_g$. The velocity-dependent terms lead to geodetic precession, while the remaining term gives rise to the Lense-Thirring effect.

Equation (18.47) is a formidable expression. It can be simplified with the help of an invariant. In free fall, $S_\mu S^\mu$ is constant. Hence,

$$\begin{aligned}g^{\mu\nu}S_\mu S_\nu &= g^{00}(S_0)^2 + 2g^{0i}S_0 S_i + g^{ij}S_i S_j \\ &= g^{00}\frac{(\boldsymbol{v}\cdot\boldsymbol{S})^2}{c^2} - 2\frac{(\boldsymbol{v}\cdot\boldsymbol{S})(\boldsymbol{A}_g\cdot\boldsymbol{S})}{c^2} + g^{11}S^2 = \text{constant} \,,\end{aligned}$$

where we've used Eq. (18.44) to eliminate the time component S_0. The middle term is negligible in comparison with the other terms (orbital speed $\beta \approx 2\times 10^{-5}$); thus, we have an approximate invariant involving only the spatial parts of the spin vector (the spin in its rest frame)

$$S^2\left(1+\frac{2\Phi}{c^2}\right) - \frac{(\boldsymbol{v}\cdot\boldsymbol{S})^2}{c^2} = \text{constant} \,. \tag{18.48}$$

Equation (18.48) suggests that through a suitable change of variables we might find a single quantity, call it $\boldsymbol{\Sigma}$, that approximately captures the constant in Eq. (18.48). Let [74, p234]

$$\boldsymbol{S} \equiv \left(1-\frac{\Phi}{c^2}\right)\boldsymbol{\Sigma} + \frac{1}{2c^2}\boldsymbol{v}(\boldsymbol{v}\cdot\boldsymbol{\Sigma}) \,. \tag{18.49}$$

Combining Eqs. (18.49) and (18.48), we find

$$S^2\left(1+\frac{2\Phi}{c^2}\right) - \frac{(\boldsymbol{v}\cdot\boldsymbol{S})^2}{c^2} = \Sigma^2 + \text{higher order terms} = \text{constant} \,,$$

so that to lowest order $\Sigma^2 = $ constant. The inverse of Eq. (18.49) is, to lowest order,

$$\boldsymbol{\Sigma} = \left(1+\frac{\Phi}{c^2}\right)\boldsymbol{S} - \frac{1}{2c^2}\boldsymbol{v}\left(\boldsymbol{v}\cdot\boldsymbol{S}\right) \,. \tag{18.50}$$

The significance of this development is that because $\Sigma^2 \approx$ constant, we expect $\boldsymbol{\Sigma}$ to precess. Take the time derivative of $\boldsymbol{\Sigma}$; from Eq. (18.50)

$$\begin{aligned}\frac{\mathrm{d}\boldsymbol{\Sigma}}{\mathrm{d}t} &= \frac{\mathrm{d}\boldsymbol{S}}{\mathrm{d}t} + \frac{\boldsymbol{S}}{c^2}\frac{\mathrm{d}\Phi}{\mathrm{d}t} - \frac{\boldsymbol{v}}{2c^2}\left(\boldsymbol{S}\cdot\frac{\mathrm{d}\boldsymbol{v}}{\mathrm{d}t}\right) - \frac{\boldsymbol{v}\cdot\boldsymbol{S}}{2c^2}\frac{\mathrm{d}\boldsymbol{v}}{\mathrm{d}t} \\ &= \frac{\mathrm{d}\boldsymbol{S}}{\mathrm{d}t} + \frac{\boldsymbol{S}}{c^2}(\boldsymbol{v}\cdot\boldsymbol{\nabla}\Phi) + \frac{\boldsymbol{v}}{2c^2}(\boldsymbol{S}\cdot\boldsymbol{\nabla}\Phi) + \frac{(\boldsymbol{v}\cdot\boldsymbol{S})}{2c^2}\boldsymbol{\nabla}\Phi\,,\end{aligned} \tag{18.51}$$

where we've ignored $\boldsymbol{v}\cdot\mathrm{d}\boldsymbol{S}/\mathrm{d}t$ and $\Phi\mathrm{d}\boldsymbol{S}/\mathrm{d}t$ as being higher order, and we've used $\mathrm{d}\Phi/\mathrm{d}t = \boldsymbol{v}\cdot\boldsymbol{\nabla}\Phi$ and $\mathrm{d}\boldsymbol{v}/\mathrm{d}t = -\boldsymbol{\nabla}\Phi$. Combining Eqs. (18.47) and (18.51),

$$\frac{\mathrm{d}\boldsymbol{\Sigma}}{\mathrm{d}t} = \frac{1}{2}\boldsymbol{S}\times\boldsymbol{B}_g - \frac{3}{2c^2}\left(\boldsymbol{\nabla}\Phi(\boldsymbol{v}\cdot\boldsymbol{S}) - \boldsymbol{v}(\boldsymbol{S}\cdot\boldsymbol{\nabla}\Phi)\right) = \boldsymbol{S}\times\left[\frac{1}{2}\boldsymbol{B}_g - \frac{3}{2c^2}(\boldsymbol{\nabla}\Phi\times\boldsymbol{v})\right].$$

The terms in parentheses are equivalent to $\boldsymbol{S}\times(\boldsymbol{\nabla}\Phi\times\boldsymbol{v})$ (“*BAC-CAB*”). At this order of approximation we can replace $\boldsymbol{S}$ by $\boldsymbol{\Sigma}$. Thus, at lowest order in small quantities the equation of motion for a spin in free fall in the gravitoelectromagnetic field is

$$\frac{\mathrm{d}\boldsymbol{\Sigma}}{\mathrm{d}t} = \boldsymbol{\Sigma}\times\boldsymbol{\Omega}\,,$$

where

$$\boldsymbol{\Omega} \equiv \frac{1}{2}\boldsymbol{B}_g - \frac{3}{2c^2}(\boldsymbol{\nabla}\Phi\times\boldsymbol{v}) \equiv \boldsymbol{\Omega}_{\text{LT}} + \boldsymbol{\Omega}_{\text{geodetic}}\,. \tag{18.52}$$

The quantity $\boldsymbol{\Sigma}$ precesses around the direction of $\boldsymbol{\Omega}$ at the rate $|\boldsymbol{\Omega}|$ with no change in magnitude.

In Section 17.9 we calculated, for a *spherically symmetric* non-rotating source, the angle through which a spin precesses in one period of a circular orbit as $3\pi GM/(c^2 r)$, Eq. (17.83). Here we have obtained, from linearized gravity, a general expression for the spin precession rate where such assumptions are not made. If there was no rotation of the source we'd have $\boldsymbol{B}_g = 0$, Eq. (18.40). Setting $\boldsymbol{B}_g = 0$ in Eq. (18.52), we have the geodetic precession rate

$$\boldsymbol{\Omega}_{\text{geodetic}} = -\frac{3}{2c^2}(\boldsymbol{\nabla}\Phi\times\boldsymbol{v})\,. \tag{18.53}$$

Let's check the prediction of Eq. (18.53) for a spherically symmetric source and a circular orbit, $\boldsymbol{\nabla}\Phi = (GM/r^2)\hat{\boldsymbol{r}}$ and $\boldsymbol{v} = r\dot{\phi}\hat{\boldsymbol{\phi}}$. In that case, the geodetic precession rate is

$$|\boldsymbol{\Omega}_{\text{geodetic}}| = \frac{3GM}{2c^2 r}\dot{\phi}\,. \tag{18.54}$$

Equation (18.54) agrees with Eq. (17.83) when we set $\dot{\phi}\,\times$ one orbital period $= 2\pi$. As noted in Section 17.9, the GPB experiment (which used a highly circular orbit) measured the geodetic precession to within 0.3% of the prediction of GR.

The Lense-Thirring precession is caused by the rotation of the source. Using Eq. (18.40),

$$\boldsymbol{\Omega}_{\text{LT}} = \frac{1}{2}\boldsymbol{B}_g = -\frac{GI\omega}{c^2 r^3}\left(3(\hat{\boldsymbol{z}}\cdot\hat{\boldsymbol{r}})\hat{\boldsymbol{r}} - \hat{\boldsymbol{z}}\right),$$

where $\boldsymbol{L} = I\boldsymbol{\omega} = I\omega\hat{\boldsymbol{z}}$ and $\boldsymbol{r} = r\hat{\boldsymbol{r}}$. With the GPB experiment in mind, the direction and magnitude of $\boldsymbol{\Omega}_{\text{LT}}$ changes over the course of the satellite orbit, owing to the term $(\hat{\boldsymbol{z}}\cdot\hat{\boldsymbol{r}})\hat{\boldsymbol{r}}$. We can easily find the *average* value of $\boldsymbol{\Omega}_{\text{LT}}$ over an orbital period. In the plane of the orbit let $\hat{\boldsymbol{r}}(t) = \hat{\boldsymbol{z}}\cos(\alpha t) + \hat{\boldsymbol{x}}\sin(\alpha t)$, where α is the orbital angular frequency. It's then straightforward to show that the time average over an orbital period yields $\langle(\hat{\boldsymbol{z}}\cdot\hat{\boldsymbol{r}})\hat{\boldsymbol{r}}\rangle = \frac{1}{2}\hat{\boldsymbol{z}}$. The average precession rate is thus

$$\langle\boldsymbol{\Omega}_{\text{LT}}\rangle = -\frac{GI\omega}{2c^2 r^3}\hat{\boldsymbol{z}}\,.$$

When all parameters pertaining to the GPB experiment are taken into account, GR predicts a precession rate of -39.2 mas/year,[71] whereas the measured value was -37.2 ± 7.2 mas/year, a 5% agreement with GR. There are proposals to measure the effect to greater accuracy.[75]

The orbit for the GPB satellite was chosen with skill. For the geodetic effect, precession occurs about the direction perpendicular to the orbital plane, Eq. (18.53). To maximize geodetic precession, the spin $\boldsymbol{S}$ of the gyroscope should be *in* the orbital plane. For the Lense-Thirring effect, precession occurs about the direction of $\boldsymbol{B}_g$. For an equatorial orbit, $\boldsymbol{B}_g$ is perpendicular to the orbital plane; in this case, the geodetic and Lense-Thirring effects align, rendering separate measurements of each effect impossible. For a polar orbit, however, $\boldsymbol{B}_g$ is largely oriented along the orbit: $\boldsymbol{\Omega}_{\text{geodetic}}$ and $\boldsymbol{\Omega}_{\text{LT}}$ are orthogonal, making it possible to measure the two types of precession in the same experiment.

18.8 GRAVITATIONAL WAVES

In *source-free* regions Eq. (18.12) reduces to the homogeneous wave equation,

$$\partial_\mu \partial^\mu \psi_{\rho\nu} = 0 \,. \tag{18.55}$$

GR thus predicts propagating "ripples" in spacetime curvature, *gravitational waves*, that travel at the speed of light.[15] One way to view Einstein's equation is in terms of the *rigidity* of spacetime, considered as a medium. The quantity G/c^4 has the dimension of $[\text{force}]^{-1}$. Rewriting the field equation: Stress $= \left(c^4/(8\pi G)\right) \times$ curvature—it takes $\approx 5 \times 10^{42}$ N m^{-2} to produce a curvature of 1 m^{-2}. *Spacetime can be considered an extremely rigid medium* that supports propagating deformations traveling at the speed of light.[16] In the following we consider the simplest type of solution to the wave equation: monochromatic plane waves.[17]

18.8.1 Plane-wave solutions

How to solve differential equations? Guess. Here's our intelligent guess:

$$\psi_{\rho\nu} = A_{\rho\nu} \exp(\mathrm{i} k_\mu x^\mu) \,, \tag{18.56}$$

where $k_\mu = (-\omega/c, \boldsymbol{k})$ is the four-wavevector, Eq. (5.61), and $A_{\rho\nu}$ is a 4×4 symmetric[18] matrix of constant amplitudes, the *polarization tensor.* Equation (18.56) satisfies Eq. (18.55) when ω and $k = |\boldsymbol{k}|$ are related by the linear dispersion relation $\omega = ck$. Thus, k_μ is a null vector with respect to the Lorentz metric, $k_\mu k^\mu = 0$. We can write $k_\mu = (\omega/c)(-1, \hat{\boldsymbol{n}})$, where $\hat{\boldsymbol{n}}$, a unit vector, signifies the direction of propagation.

The gauge condition Eq. (18.11) implies, for plane-wave solutions:

$$k_\nu A^{\rho\nu} = 0 \,. \tag{18.57}$$

Equation (18.57) represents four equations of constraint among the 10 independent elements of $[A^{\rho\nu}]$; it's a system of four equations in 10 unknowns. Thus, we would seemingly have six independent parameters to choose among the $A^{\rho\nu}$. We expect, however, gravitational waves to be described by *two* independent parameters, not six. Under coordinate transformations of the type of Eq. (18.6), we know from Eq. (18.10) that the Lorenz condition, Eq. (18.11), and the wave equation (18.55) are unchanged when the ξ^μ are *harmonic functions*, those that satisfy the wave equation

$$\partial_\nu \partial^\nu \xi^\mu = 0 \,. \tag{18.58}$$

---

[15] Small perturbations of the spacetime metric pertain to the left side of Einstein's equation $G_{\rho\nu} = \kappa T_{\rho\nu}$; gravitational waves are propagating disturbances in the *curvature* of spacetime.

[16] Sounds superficially like the ether frame that is relativity disposed of. Spacetime also supports the propagation of electromagnetic fields. The spacetime of GR is not static and *evolves* in response to matter-energy-momentum.

[17] More complicated wave forms can be synthesized from plane waves through Fourier integrals.

[18] The matrix with elements $A_{\rho\nu}$ is symmetric because the metric tensor is symmetric.

The 10 functions $\psi_{\mu\nu}$ are therefore constrained by eight supplementary conditions, Eqs. (18.11) and (18.58), leaving the metric of gravitational waves to depend on only two independent parameters. Together with the four constraint equations implied by Eq. (18.57), *we're free to choose four of the amplitude parameters* $A_{\rho\nu}$—the simplest choice is to set them to zero! Let's do that.

18.8.2 Transverse traceless gauge

A useful gauge is obtained by requiring the trace be zero:

$$\psi \equiv \psi^{\alpha}_{\alpha} = 0 \,. \tag{18.59}$$

If $\psi = 0$, then from Eq. (18.4) $h = 0$, where h is the trace of $h_{\mu\nu}$, the "original" perturbation to the metric, $g_{\mu\nu} = \eta_{\mu\nu} + h_{\mu\nu}$. If $\psi = 0$, then $h_{\mu\nu} = \psi_{\mu\nu}$, Eq. (18.4). Equation (18.59) places one equation of constraint on the amplitudes ($A^{\alpha}_{\alpha} = 0$) not provided by Eq. (18.57). For the other three "picks," we choose

$$A^{0i} = 0 \,. \tag{18.60}$$

Equations (18.59) and (18.60) represent four conditions we're allowed to impose on the amplitudes $A^{\rho\nu}$. Equation (18.60) combined with Eq. (18.57) implies $A^{00} = 0$. At this point the polarization matrix has the form

$$[A^{\rho\nu}] = \begin{pmatrix} 0 & 0 & 0 & 0 \\ 0 & A^{11} & A^{12} & A^{13} \\ 0 & A^{12} & A^{22} & A^{23} \\ 0 & A^{13} & A^{23} & A^{33} \end{pmatrix}$$

such that $\sum_i A^{ii} = 0$. The gauge condition Eq. (18.57) implies $k_i A^{ji} = 0$, or that

$$\boldsymbol{k} \cdot \boldsymbol{A}^{(j)} = 0 \,, \qquad (j = 1, 2, 3) \tag{18.61}$$

where $\boldsymbol{A}^{(j)}$ is the vector comprised of the row A^{ji}. Equation (18.61) is the *transversality condition.*

To be definite, assume the wave travels in the $+z$-direction. Then, $\boldsymbol{k} = (0, 0, \omega/c)$. Under these conditions, transversality requires $A^{i3} = A^{3i} = 0$, implying, along with the traceless condition, that

$$[A^{\rho\nu}] = \begin{pmatrix} 0 & 0 & 0 & 0 \\ 0 & a & b & 0 \\ 0 & b & -a & 0 \\ 0 & 0 & 0 & 0 \end{pmatrix} , \tag{18.62}$$

i.e., the state of the gravitational wave is characterized by two independent parameters (a, b). Equation (18.62) is referred to as the *transverse traceless gauge* (or TT gauge for short).

18.8.3 Polarization states

The spacetime metric associated with gravitational plane waves in the TT gauge is found by combining Eqs. (18.62), (18.56), and (18.13):

$$\begin{aligned} (\mathrm{d}s)^2 = &-(c\mathrm{d}t)^2 + (\mathrm{d}z)^2 \\ &+ [1 + a\cos(kz - \omega t)]\,(\mathrm{d}x)^2 + 2b\cos(kz - \omega t)\mathrm{d}x\mathrm{d}y + [1 - a\cos(kz - \omega t)]\,(\mathrm{d}y)^2 \,, \end{aligned} \tag{18.63}$$

where we've used the real part of Eq. (18.56). Thus, there are two independent polarization states: $b = 0$, referred to as the $+$ (plus) polarization, and $a = 0$, referred to as the $\times$ (cross) polarization. Both polarization states are pulsating ellipses, oriented at 45° to each other. A general gravitational wave is a superposition of the two polarization states. That the two states are at 45° to each other contrasts the difference between the gravitational field (represented by a second-rank symmetric tensor $g_{\alpha\beta}$) and the electromagnetic field (represented by a four-vector A_{μ}), which has polarization states of plane waves at 90° to each other.

18.8.4 Gravitational wave interferometer

A schematic depiction of a gravitational wave detector is shown in Fig. 18.3. Consider a $(+)$-wave

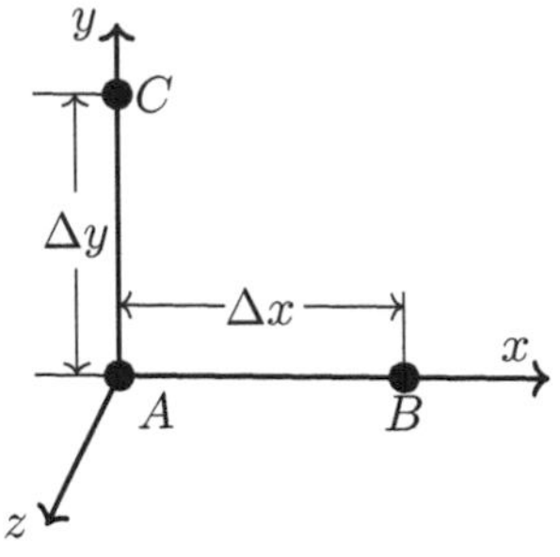

Figure 18.3 Idealized gravitational wave detector, wave propagating in z-direction.

propagating in the z-direction. In the $z = 0$ plane, object A is at the origin, with B situated on the x-axis at coordinate separation Δx and C on the y-axis at coordinate separation Δy. From Eq. (18.63), the proper distances ($\mathrm{d}t = 0$) from A to objects B and C are

$$\begin{aligned}\Delta s_{AB} &= \sqrt{g_{11}}\Delta x = \sqrt{1 + a\cos\omega t}\Delta x \approx \left(1 + \frac{1}{2}a\cos\omega t\right)\Delta x \\ \Delta s_{AC} &= \sqrt{g_{22}}\Delta y = \sqrt{1 - a\cos\omega t}\Delta y \approx \left(1 - \frac{1}{2}a\cos\omega t\right)\Delta y\,.\end{aligned}$$

The distances between (A, C) and (A, B) therefore oscillate with the frequency of the gravitational wave! Figure 18.3 is in the form of the arms of a Michelson interferometer, which, if B and C were mirrors, could be used to show constructive interference when the difference in the lengths of the arms differs by an integer multiple of the wavelength λ of the electromagnetic radiation used in the experiment (not shown in Fig. 18.3). In 2016, the LIGO experiment (Laser Interferometer Gravitational-Wave Observatory) announced the first observation of gravitational waves.[76] Gravitational waves with frequency between 35 and 250 Hz were detected, which was attributed to a pair of black holes merging into a single black hole. Since then, detection of other gravitational waves have been reported. We do not have the necessary space to review the LIGO experiment in any further detail. Suffice to say that with the GPB and LIGO experiments, 100-year-old predictions of GR have at last been validated, and, with the observation of gravitational waves, a new type of astronomy is made possible, in which gravitational waves provide information about the dynamics of massive objects at far-off locations in the universe.

18.9 ENERGY-MOMENTUM OF GRAVITATION

The metric perturbation field $\psi_{\mu\nu}(x)$ has properties in common with the electromagnetic field (Table 18.1). On the basis of these analogies we'd expect radiative excitations of the gravitational field that transport energy and momentum. Such expectations, however, beg the question: What is the energy of the gravitational field?

In 1905 Einstein taught that mass is equivalent to energy, $E = mc^2$; in 1911 he showed that energy has inertia, E/c^2, and in 1916 he showed that energy-momentum is the source of the gravitational field, generalizing the Newtonian model that mass is the source of gravity. Thus, mass is energy, energy is mass, and energy generates gravity. Does the gravitational field, however—curved spacetime, have an energy of its own? If so, then gravity generates itself—gravity gravitates! Einstein's equation is nonlinear: The source of the gravitational field, energy-momentum density, *is determined by the field itself.* In contrast, charge, the source of the electromagnetic field, is unaffected by the field.[19] Electromagnetic waves carry energy, but not charge; gravitational waves in carrying energy in effect carry mass. Is it possible to specifically isolate an energy of the gravitational field?

What is energy? In other branches of physics it's always possible to define a quantity for an isolated system, call it energy, that's constant. Allowing the system to interact with its environment, balance equations express conservation of energy between system and environment. When there are different *forms* of energy, it's possible to convert one form into another. Energy-momentum of the electromagnetic field is not conserved when the field couples to the four-current, Eq. (9.48), but energy-momentum of the combined system of field and matter *is* conserved—expressed as a continuity equation of the combined energy-momentum tensors, $\partial_\nu(\theta_F^{\mu\nu}+\theta_M^{\mu\nu})=0$ (Section 9.5.2). Let's review what is meant when we say a quantity is conserved in spacetime. We showed in Chapter 8 that $\partial_\nu J^\nu=0$ implies zero net flux in spacetime, Eq. (8.14), $\int_{\partial V} J^\alpha \mathrm{d}\Sigma_\alpha = Q(\Sigma_2)-Q(\Sigma_1)=0$, where ∂V is a closed hypersurface, Σ_2 and Σ_1 are spacelike hypersurfaces (see Fig. 8.2), and $Q(\Sigma)\equiv\int_\Sigma J^0\mathrm{d}^3y$. The quantity $Q(\Sigma)$ is independent of Σ: The amount of charge in a spacelike hypersurface is conserved in time. What about curved spacetime? The conclusion remains the same. The covariant divergence can be related to the partial derivative, $\sqrt{|g|}\nabla_\mu J^\mu=\partial_\mu(\sqrt{|g|}J^\mu)$, Eq. (14.35). If $\nabla_\mu J^\mu=0$, we have, integrating Eq. (14.35) and using Gauss's theorem, Eq. (5.90), that the net flux vanishes through a closed hypersurface, $\int_{\partial V} J^\mu \mathrm{d}\Sigma_\mu=0$.

What about tensor quantities? We know that if the action integral associated with matter fields[20] is invariant under the coordinate transformation Eq. (18.6), then $\nabla_\nu T^{\mu\nu}=0$ (Section 15.4.2). If $\nabla_\nu T^{\mu\nu}=0$, then $\nabla_\nu T_\mu^{\ \nu}=0$, which is more convenient here. Using Eqs. (14.14) and (14.35),

$$\begin{aligned}\nabla_\nu T_\mu^{\ \nu} =&\partial_\nu T_\mu^{\ \nu}+\Gamma^\nu_{\nu\lambda}T_\mu^{\ \lambda}-\Gamma^\lambda_{\nu\mu}T_\lambda^{\ \nu}=\frac{1}{\sqrt{|g|}}\partial_\nu\left(\sqrt{|g|}T_\mu^{\ \nu}\right)-\Gamma^\lambda_{\nu\mu}T_\lambda^{\ \nu}\\ =&\frac{1}{\sqrt{|g|}}\partial_\nu\left(\sqrt{|g|}T_\mu^{\ \nu}\right)-\frac{1}{2}T^{\rho\nu}\partial_\mu g_{\nu\rho}\,,\end{aligned}\tag{18.64}$$

where the last identity is the result of Exercise 18.2. If $\nabla_\nu T_\mu^{\ \nu}=0$, then integrating Eq. (18.64) and applying Gauss's theorem,

$$\int_{\partial V} T_\mu^{\ \nu}\mathrm{d}\Sigma_\nu = Q_\mu(\Sigma_2)-Q_\mu(\Sigma_1)=\frac{1}{2}\int_V T^{\rho\nu}\partial_\mu g_{\nu\rho}\sqrt{|g|}\mathrm{d}^4x\,,\tag{18.65}$$

where $Q_\mu(\Sigma)\equiv\int_\Sigma T_\mu^{\ 0}\mathrm{d}^3y$. Thus, $\nabla_\nu T^{\mu\nu}=0$ *does not imply zero net flux in spacetime* and $Q_\mu(\Sigma)$ is not independent of Σ, i.e., *in curved spacetime, energy-momentum of matter fields is not conserved.* Quantities obeying the continuity equation $\partial_\nu T^{\mu\nu}=0$ are locally conserved; $\nabla_\nu T^{\mu\nu}=0$ is not a conservation law.[21]

If we take that "gravity gravitates" characterizes the nonlinear nature of Einstein's field equation (that any form of energy-momentum acts as a source of the gravitational field, including that of the gravitational field), what is discarded in linearizing the theory? The linearized field equation is consistent[22] with $\partial_\nu T^{\rho\nu}=0$, implying that energy-momentum of matter fields is conserved. From Eq. (18.65) we see that *nonconservation* of matter energy-momentum is associated with $\partial_\mu g_{\nu\rho}\neq 0$, i.e.,

[19]The charge of an electron is not altered by an interaction with the electric field. Because there are two *kinds* of charge, however, positive and negative, as opposed to one kind of mass, an electric field can affect the *net* charge density at a point, $\rho(x)$, as in the self-consistent charge distribution in plasmas or semiconductors. In such cases the Poisson equation is nonlinear, $\epsilon_0\nabla^2\phi_e=-\rho(\phi_e(x))$: The charge distribution occurs in response to the same field that charges generate.

[20]When $T_{\mu\nu}$ is defined as in Eq. (15.43).

[21]Landau and Lifshitz state, referring to $\nabla_\nu T_\mu^{\ \nu}=0$:[77, p280] "In this form, however, this equation does not generally express any conservation law whatever. This is related to the fact that in a gravitational field the four-momentum of matter alone must not be conserved, but rather the four-momentum of matter plus gravitational field; the latter is not included in the expression for $T_\mu^{\ \nu}$."

[22]Raise indices (ρ,ν) in Eq. (18.12), contract with ∂_ν, implying $\partial_\nu T^{\rho\nu}=0$ because of the gauge condition, Eq. (18.11). We've relied on the gauge condition to make this statement, but the theory is gauge invariant. We saw in Section 8.10 that gauge invariance of the electromagnetic field is associated with charge conservation.

when there is a gravitational field, when spacetime is curved. Equation (18.65) is a "work-energy theorem": The difference in matter energy-momentum at two times (spacelike hypersurfaces, Σ_1, Σ_2) is due to the coupling of the (matter) source of the field with the gravitational field, $T^{\rho\nu}\partial_\mu g_{\nu\rho}$, over the volume of spacetime enclosed between Σ_1 and Σ_2. *The linearized theory ignores interactions between the source and the gravitational field.* The source-field coupling occurs implicitly in the full field equation; no need to put it in explicitly. The effect of the source-field interaction is reflected in the equation $\nabla_\nu T^{\mu\nu} = 0$, where the covariant derivative depends on the metric, $g_{\alpha\beta}$. We have a self-consistent loop in GR: $T^{\mu\nu}$ determines $G^{\mu\nu}$, but $T^{\mu\nu}$ must satisfy $\nabla_\nu T^{\mu\nu} = 0$, which depends on $g_{\alpha\beta}$, which is found from $G^{\mu\nu}$ (see Fig. 1.13).

The fact that energy-momentum of matter is not conserved should not come as a surprise if gravitational waves transport energy-momentum away from sources.[23] It might be supposed that gravitational energy would be radiated in the *linear* theory, where, just as in electrodynamics, the radiated energy would be supplied by the agency producing the acceleration of matter.[24] Such a scenario would not conserve energy-momentum, because in GR we have to include *all* sources of energy-momentum, including that which produces accelerating masses.

Returning to the question, we'd like a way to identify the energy of the gravitational field. To do so, we must construct an energy-momentum tensor for the gravitational field. A way to do that can be found in the Lagrangian formulation of Einstein's equation, Section 15.4. As this is an involved calculation, the reader uninterested in the details should skip to Eq. (18.72).

We found from Eq. (15.41), which we reproduce here, that for the action of the gravitational field (set $\Lambda = 0$—it's easy to put back)

$$\delta S = \int_V \frac{\delta \mathscr{L}_g}{\delta g^{\mu\nu}} \delta g^{\mu\nu} \mathrm{d}^4 x = \int_V \left[R_{\alpha\beta} - \frac{1}{2} R g_{\alpha\beta} \right] \delta g^{\alpha\beta} \sqrt{-g} \mathrm{d}^4 x \,, \tag{15.41}$$

where $\mathscr{L}_g = \sqrt{-g}R$ (with $\Lambda = 0$). The integrands of the integrals in Eq. (15.41) must agree (up to the divergence of a quantity which vanishes at the boundary of V):

$$R_{\mu\nu} - \frac{1}{2} R g_{\mu\nu} = \frac{1}{\sqrt{-g}} \frac{\delta \mathscr{L}_g}{\delta g^{\mu\nu}} = \frac{1}{\sqrt{-g}} \left[\frac{\partial \mathscr{L}_g}{\partial g^{\mu\nu}} - \partial_\lambda \left(\frac{\partial \mathscr{L}_g}{\partial(\partial_\lambda g^{\mu\nu})} \right) \right] , \tag{18.66}$$

where we've used the functional derivative,[25] Eq. (D.63). It's shown in Exercise 15.7 that an equivalent Lagrangian for the gravitational field is given by:

$$\mathcal{L} = \sqrt{-g} g^{\rho\nu} \left(\Gamma^\mu_{\alpha\rho} \Gamma^\alpha_{\mu\nu} - \Gamma^\mu_{\nu\rho} \Gamma^\alpha_{\mu\alpha} \right) . \tag{18.67}$$

The advantage of this form of the Lagrangian (as opposed to $\mathscr{L}_g = \sqrt{-g}R$) is that it involves the metric tensor and only its first derivatives, what we require on the right side of Eq. (18.66). We invoke Einstein's equation by setting the left side of Eq. (18.66) equal to $\kappa T_{\mu\nu}$, and we use the Lagrangian $\mathcal{L}$:

$$\sqrt{-g}\kappa T_{\mu\nu} = \frac{\partial \mathcal{L}}{\partial g^{\mu\nu}} - \partial_\lambda \left(\frac{\partial \mathcal{L}}{\partial(\partial_\lambda g^{\mu\nu})} \right) . \tag{18.68}$$

We now make use of Eq. (18.64), which, because $\nabla_\nu T_\mu^{\ \nu} = 0$, implies that

$$\begin{aligned} \partial_\nu \left(\sqrt{-g} T_\mu^{\ \nu} \right) &= \frac{1}{2} (\partial_\mu g_{\nu\rho}) \sqrt{-g} T^{\rho\nu} = -\frac{1}{2} \left(\partial_\mu g^{\nu\rho} \right) \sqrt{-g} T_{\rho\nu} \\ &= -\frac{1}{2\kappa} \left(\partial_\mu g^{\nu\rho} \right) \left[\frac{\partial \mathcal{L}}{\partial g^{\rho\nu}} - \partial_\lambda \left(\frac{\partial \mathcal{L}}{\partial(\partial_\lambda g^{\rho\nu})} \right) \right] , \end{aligned} \tag{18.69}$$

[23]And we know that gravitational waves *do* transport energy: Something has to move the mirrors in the LIGO experiment.

[24]We see from Eq. (18.27) that the perturbation to the metric goes as the *second* time derivative of the mass moment.

[25]The functional derivative in the form of Eq. (D.63) assumes that the Lagrangian is a function only of the fields and its first derivatives, whereas the Ricci scalar field (Hilbert Lagrangian) is a function of first and second derivatives of the metric. Enter the results of Exercise 15.7.

where we've used the result of Exercise 18.3 and Eq. (18.68). To make progress, we note that, using the chain rule

$$\partial_\mu \mathcal{L} = \frac{\partial \mathcal{L}}{\partial g^{\rho\nu}} \partial_\mu g^{\rho\nu} + \frac{\partial \mathcal{L}}{\partial(\partial_\lambda g^{\rho\nu})} \partial_\mu \partial_\lambda g^{\rho\nu} \, . \tag{18.70}$$

Combine Eq. (18.70) with the right side of Eq. (18.69). It can be shown:

$$(\partial_\mu g^{\nu\rho}) \left[\frac{\partial \mathcal{L}}{\partial g^{\rho\nu}} - \partial_\lambda \left(\frac{\partial \mathcal{L}}{\partial(\partial_\lambda g^{\rho\nu})} \right) \right] = \partial_\lambda \left(\mathcal{L} \delta^\lambda_\mu - (\partial_\mu g^{\rho\nu}) \frac{\partial \mathcal{L}}{\partial(\partial_\lambda g^{\rho\nu})} \right) .$$

Define

$$\sqrt{-g} t_\mu^{\ \nu} \equiv \frac{1}{2\kappa} \left(\mathcal{L} \delta^\nu_\mu - (\partial_\mu g^{\rho\lambda}) \frac{\partial \mathscr{L}}{\partial(\partial_\nu g^{\rho\lambda})} \right) . \tag{18.71}$$

The net effect of these manipulations is that by combining Eqs. (18.69) and (18.71),

$$\partial_\nu \left[\sqrt{-g} \left(T_\mu^{\ \nu} + t_\mu^{\ \nu} \right) \right] = 0 \, . \tag{18.72}$$

Thus, $\nabla_\nu T_\mu^{\ \nu} = 0$ is equivalent to Eq. (18.72). We've therefore found a local conservation law for the energy-momentum of matter fields, represented by $T_\mu^{\ \nu}$, combined with that for the gravitational field, conventionally denoted $t_\mu^{\ \nu}$. That's the good news. The bad news is that $t_\mu^{\ \nu}$ *is not a tensor*! Because a coordinate system can always be found where at a given point $\partial_\mu g_{\nu\rho} = 0$ and all Christoffel symbols vanish (Section 14.4.4), $t_\mu^{\ \nu}$ cannot be a tensor. A bona fide tensor, if it vanishes in one coordinate system, vanishes in all coordinate systems. Because gravity can be "waved away" locally (equivalence principle), there cannot be an energy-momentum tensor for the gravitational field valid in all reference frames. These considerations imply that *gravitational energy is not localizable*,[26] something that exists at a point; rather, the interaction between matter and gravitational fields occurs over an extended region of spacetime, what we see as an integration in Eq. (18.65). The energy of the electromagnetic field, for example (and other physical fields), is determined by the fields themselves, at a point. The gravitational field is the *geometry* of spacetime, and it's useful to conceive of an interaction between matter and geometry. The part of the metric tensor that can't be made to vanish is associated with second and higher-order derivatives of the metric. These terms become significant over sufficiently extended regions of spacetime.

The quantity $t_\mu^{\ \nu}$, introduced by Einstein in 1916, is known as the *energy-momentum pseudotensor*, a special use of the word unrelated to customary usage.[27,28] The pseudotensor was criticized by Levi-Civita, Schrödinger, and others; the early objections have been summarized by Pauli.[79, p176] Einstein's reply was that even though one cannot attach physical meaning to the values of $t^{\mu\nu}$ (because it's impossible to localize energy-momentum of the gravitational field in a covariant way), the integral expressions

$$J^\mu \equiv \int_\Sigma (T^{\mu 0} + t^{\mu 0}) \sqrt{-g} \mathrm{d}^3 y \tag{18.73}$$

do have physical meaning in prescribed circumstances (Σ is any hypersurface for which $x^0 = 0$). Einstein showed that the quantities J^μ in Eq. (18.73) are 1) independent of the choice of coordinates in systems for which $g_{\mu\nu} \to \eta_{\mu\nu}$ "outside" the system, what we have previously called asymptotically flat, and 2) transform as a four-vector under *linear* coordinate transformations. We can therefore speak of the energy-momentum of the gravitational field in systems that are isolated by a *far-field* zone, far removed from the sources of the field.

---

[26]The *nonlocalizability* of energy is a generic issue with fields. Only the total integral of the Poynting vector over a closed surface has meaning, $\oint \boldsymbol{S} \cdot \mathrm{d}\boldsymbol{a}$; one can add to $\boldsymbol{S}$ any quantity that integrates to zero over a closed surface.

[27]Pseudotensors are objects that transform as tensors for positive Jacobian determinant J, but with an extra minus sign for $J < 0$ (Section 5.9). Apparently Eddington was the first to use the word *pseudotensor* in this context.[78, p134] The two terms together $T_\mu^{\ \nu} + t_\mu^{\ \nu}$ are referred to as the *Einstein complex*.

[28]That $t_\mu^{\ \nu}$ is not a tensor was explicitly called out by Einstein:[9, p147] "It is to be noticed that t_σ^α is not a tensor . . . ".

Equation (18.72) does not possess a unique solution. For any $t^{\mu\nu}$ that satisfies Eq. (18.72) we can add to it the divergence of a third-rank quantity $\psi^{\mu[\nu\lambda]}$ that's antisymmetric in the second and third indices and still satisfy Eq. (18.72) (see Section 9.4). For this reason one finds in the literature a variety of possible forms of $t^{\mu\nu}$. That defined by Eq. (18.71) is not symmetric.[29] A symmetric version of the pseudotensor was found by Landau and Lifshitz,[77, p282] but we won't write it down—it's a lengthy expression! Our purpose here is not to provide an exhaustive survey of pseudotensors.

SUMMARY

By working to first order in a small perturbation $\psi_{\mu\nu}(x)$ to the Lorentz metric, Einstein's field equation reduces to an inhomogeneous wave equation, Eq. (18.12). Once $\psi_{\mu\nu}(x)$ is known, by solving the wave equation, we have an approximate metric tensor that represents a solution to the field equation at lowest order in the perturbation; see Eq. (18.13). The perturbation field $\psi_{\mu\nu}(x)$ has many properties in common with the electromagnetic field (see Table 18.1), most notably that it can be expressed as a retarded field, Eq. (18.15),

$$\psi_{\rho\nu}(t,\boldsymbol{x}) = \frac{4G}{c^4}\int \frac{T_{\rho\nu}(t-|\boldsymbol{x}-\boldsymbol{y}|/c,\boldsymbol{y})}{|\boldsymbol{x}-\boldsymbol{y}|}\mathrm{d}^3y\,, \tag{18.15}$$

where the integral is over the sources evaluated at the retarded time. By approximating Eq. (18.15), we have a systematic way of developing approximate solutions. While there are a few exact solutions of Einstein's equation,[30] the method of this chapter provides a scheme for developing approximate solutions based on perturbations to flat spacetime.

- For a static source with $T_{00} = \rho c^2$ and $T_{\mu\nu} = 0$ otherwise,

  $$\psi_{00}(\boldsymbol{x}) = \frac{4G}{c^2}\int \frac{\rho(\boldsymbol{y})}{|\boldsymbol{x}-\boldsymbol{y}|}\mathrm{d}^3y = -\frac{4}{c^2}\Phi(\boldsymbol{x})\,,$$

  where Φ is the Newtonian gravitational potential. The spacetime geometry is described by Eq. (18.17),

  $$(\mathrm{d}s)^2 = -\left(1+\frac{2}{c^2}\Phi\right)(c\mathrm{d}t)^2 + \left(1-\frac{2}{c^2}\Phi\right)(\mathrm{d}\sigma)^2\,,$$

  where $(\mathrm{d}\sigma)^2$ is the spatial line element. While this metric is in the form of the Schwarzschild metric for weak gravity, it's more general than the Schwarzschild metric. The potential Φ is the solution to the Poisson equation for any static mass distribution, not necessarily spherically symmetric. Moreover, the metric is not a vacuum solution; it is not restricted to the exterior of the mass distribution.

- In the far-field approximation, when the observation point is far removed from the source (in relation to its size), and the source particles have nonrelativistic speeds, we can approximate $\psi_{\mu\nu}$ as the first term in a multipole expansion,

  $$\psi_{\mu\nu}(t,\boldsymbol{x}) = \frac{4G}{c^4 r}\int T_{\mu\nu}(t-r/c,\boldsymbol{y})\mathrm{d}^3y\,,$$

  where $r = |\boldsymbol{x}|$. In this case $\psi_{00}(t,r) = (4G/(c^2r))\,[M]$, $\psi_{0i}(t,r) = -(4G/(c^3r))\,[P^i]$, and $\psi_{ij} = (2G/(c^6r))\,[\ddot{I}^{ij}]$, where the brackets indicate quantities evaluated at the retarded time, $t_r = t - r/c$. Here Mc^2 is the total energy of the source, P^i is its total momentum in the ith direction, and $I^{ij} = \int y^i y^j T^{00}\mathrm{d}^3y$ is the second mass moment of the source.

---

[29] And of course one cannot speak of symmetry or antisymmetry of mixed tensors—Section 5.7. One has to either raise or lower an index.

[30] We've considered the Schwarzschild solution, but there are other exact solutions, notably the Kerr and the Reissner-Nordström metrics, not treated in this book.

- For stationary sources we do not have to consider time retardation and

$$\psi_{\mu\nu}(\boldsymbol{x}) = \frac{4G}{c^4}\int \frac{T_{\mu\nu}(\boldsymbol{y})}{|\boldsymbol{x}-\boldsymbol{y}|}\mathrm{d}^3y\,.$$

When pressure can be neglected, the energy-momentum tensor is such that $T^{0i} \sim T^{00}\beta^i$ and $T^{ij} \sim T^{00}\beta^i\beta^j$. For nonrelativistic speeds of the source particles, $\beta^i \ll 1$, and, neglecting terms second order in β, there are only four independent elements, $T^{0\alpha}$, enabling a useful analogy with electromagnetism. When these approximations apply,

$$(\mathrm{d}s)^2 = -\left(1+\frac{2}{c^2}\Phi\right)(c\mathrm{d}t)^2 + \frac{2}{c}\mathrm{d}x^0 A_i \mathrm{d}x^i + \left(1-\frac{2}{c^2}\Phi\right)(\mathrm{d}\sigma)^2\,,$$

where A_i is a vector potential produced by a mass current, akin to the magnetic vector potential produced by charge currents. The gravitational four-potential, $A_g^\mu = (4\Phi/c, \boldsymbol{A}_g)$, where

$$\boldsymbol{A}_g = -\frac{4G}{c^2}\int \frac{\boldsymbol{J}(\boldsymbol{y})}{|\boldsymbol{x}-\boldsymbol{y}|}\mathrm{d}^3y\,,$$

with $\boldsymbol{J} = \rho\boldsymbol{v}$ the mass current density. There is an analogy between linearized gravity for stationary, nonrelativistic sources and electro- and magnetostatics. The gravitoelectric field $\boldsymbol{E}_g = -\boldsymbol{\nabla}\Phi$ and the gravitomagnetic field $\boldsymbol{B}_g = \boldsymbol{\nabla}\times\boldsymbol{A}_g$. The geodesic for a free particle in the spacetime of a nonrelativistic stationary source is the analog of the Lorentz acceleration, $\boldsymbol{a} = \boldsymbol{E}_g + \boldsymbol{v}\times\boldsymbol{B}_g$. The gravitomagnetic field "drags" the reference frame, causing it to effectively rotate—a free particle appears as if it was observed from a rotating reference frame because $\boldsymbol{v}\times\boldsymbol{B}_g$ is the analog of the Coriolis force.

- For a slowly rotating source, the gravitomagnetic field is in the dipole approximation

$$\boldsymbol{B}_g = -\frac{2G}{c^2r^3}\left(\frac{3\boldsymbol{L}\cdot\boldsymbol{r}}{r^2} - \boldsymbol{L}\right)\,,$$

where $\boldsymbol{L}$ is the angular momentum of the source.

- A spin in the gravitoelectromagnetic field precesses about the direction

$$\boldsymbol{\Omega} = \frac{1}{2}\boldsymbol{B}_g - \frac{3}{2c^2}\left(\boldsymbol{\nabla}\Phi\times\boldsymbol{v}\right)\,,$$

where $\boldsymbol{v}$ is the orbital velocity. The velocity-dependent term is the geodetic precession effect. Spin precession about $\boldsymbol{B}_g$ is the Lense-Thirring effect, measured in the GPB experiment.

- In source-free regions of spacetime, GR predicts radiative solutions to the linearized field equation, gravitational waves, observed in the LIGO experiment.

EXERCISES

18.1 Show that the linearized Riemann tensor Eq. (18.14) is invariant under the infinitesimal coordinate transformation of the type in Eq. (18.6). Verify that the symmetries of the Riemann tensor are exhibited by Eq. (18.14).

18.2 Show that $\Gamma^\lambda_{\nu\mu}T_\lambda{}^\nu = \frac{1}{2}T^{\rho\nu}\partial_\mu g_{\nu\rho}$. Use the symmetry of $g_{\alpha\beta}$ and $T^{\mu\nu}$.

18.3 Show that $(\partial_\mu g_{\nu\rho})T^{\rho\nu} = -\left(\partial_\mu g^{\alpha\beta}\right)T_{\alpha\beta}$. Start with $g_{\nu\rho} = g_{\alpha\nu}g_{\beta\rho}g^{\alpha\beta}$.

CHAPTER 19

Relativistic cosmology

WE now turn to the largest spacetime structure—the universe. Having delved into GR and the extent to which it's been tested on the scale of the solar system, we can ask what it has to say about the universe—the view of spacetime provided by GR must be believable if we're to have confidence in its predictions for cosmology. Over cosmological distances, matter and energy are governed by the *weakest* force, gravity.[1]

19.1 THE COSMOLOGICAL PRINCIPLE

Look out at the night sky. It would be natural to assume the universe is a vast array of independent stars, yet that's not in accord with what we find—stars gather into gravitationally bound systems called galaxies, and there are *vast* numbers of galaxies.[2] One might think that the universe is a huge array of independent galaxies. That too is not in accord with what we find, that galaxies are interacting objects, grouping into galactic clusters and ultimately superclusters, the largest known structures. *Is there a length scale beyond all structure*—what could be termed *cosmological distances*—at which the universe has the same appearance from every point?

The *cosmological principle* is that at sufficiently large length scales the universe *looks the same* regardless of location and direction from which it's viewed. There are two ideas here: *homogeneity*—no unique location in the universe, and *isotropy*—no preferred direction. While these concepts are logically distinct, homogeneity is implied if isotropy holds at every point. In Chapter 4 we used homogeneity and isotropy of *spacetime* to derive the Lorentz transformation. In cosmology we apply these ideas to *space only* because we have in the Big Bang an origin in time.[3] The cosmological principle sounds good, but is it true? Homogeneity and isotropy clearly do not hold on the scale of the solar system, nor on the scale of the galaxy; it's only substantially beyond the scale of intergalactic distances that these features become apparent. Homogeneity is found [81] on length scales greater than $\sim 80h^{-1}$ Mpc.[4] Galaxies, of size ≈ 100 kpc, are mere points on such cosmic length scales. In a first application of GR to cosmology, we presume the validity of the cosmological principle.

[1]The weak and strong forces are short ranged. The electromagnetic force, while long ranged, is *screened* by charges of opposite sign, and the universe is, to the best of our knowledge, electrically neutral.[80] The electromagnetic force does not contribute to the long-range evolution of the universe. That leaves gravity.

[2]It is only since the 1920s that the concept of galaxies existing outside our own was decisively settled.

[3]The *Big-Bang Model* is supported by the redshift-distance relation, *Hubble's law*, implying an expanding universe. Run the movie backwards, and there was a singular time when the universe was considerably smaller, denser, and hotter than at present. An early hot universe is supported by the cosmic microwave background (CMB) radiation permeating the universe, a thermal relic having the Planck frequency spectrum, and the observed abundances of the light elements.

[4]A *parsec* (pc) is the distance from which the astronomical unit (distance from Earth to Sun) subtends an angle of one arcsecond, ≈ 3.26 ly. The value of the Hubble constant is quoted as $H = 100h$ (km/s) Mpc. Recent findings place $h \sim 0.7$.

19.2 A COORDINATE SYSTEM FOR COSMOLOGY

By the cosmological principle (and the evidence supporting it), the universe becomes homogeneous and isotropic at sufficiently large length scales. Over such distances the universe demonstrates a level of simplicity (as simple as it's going to get) that actually *restricts* the type of coordinate system we can use for its description. If the universe is to look the same everywhere and in every direction, the *curvature* of space must be constant, and there are just three types of geometry for which this is possible, as we show. In a sense GR, the ultimate in a relative theory of space and time, begins to *revert* to an absolute theory when we get to the cosmological realm. At every locality there is now a preferred reference frame, that in which the expansion of the universe is isotropic. And if the universe appears the same everywhere, there must be a universal *cosmic time*, the same for all observers at rest relative to the frame in which the universe everywhere has isotropic expansion.

A "moment of time" in GR is specified by a spacelike hypersurface (SH), and spacetime can be partitioned into a one-parameter family, labeled by t, of nonintersecting SHs.[5] Of course there's no unique way to partition spacetime into nonintersecting hypersurfaces; for cosmology we choose SHs having constant curvature (to be consistent with the cosmological principle). Let coordinates be established by a set of *freely falling particles*. Each particle carries a clock and has spatial coordinates x^i, *taken as the coordinates when its clock reads* $t = 0$. This reference frame is known as the *comoving coordinate system*. The coordinates x^i in the comoving frame form a spatial grid that *keep their values fixed*.[6]

At each point of a SH there is a local Lorentz frame whose surface of simultaneity coincides locally with the hypersurface and which is locally spanned by basis vectors $\boldsymbol{e}_i$ orthogonal to the local time direction, $\boldsymbol{e}_0$. The infinitesimal spacetime separation can therefore be written

$$(\mathrm{d}s)^2 = -(c\mathrm{d}t)^2 + g_{ij}\mathrm{d}x^i\mathrm{d}x^j \ , \tag{19.1}$$

because $g_{0i} = \boldsymbol{e}_0 \cdot \boldsymbol{e}_i = 0$. Along a worldline in the comoving coordinates $\mathrm{d}x^i = 0$. Thus, $(\mathrm{d}s)^2 = -(c\mathrm{d}\tau)^2 = -(c\mathrm{d}t)^2$; the parameter used to label SHs is the proper time, $t = \tau$. Let $x^\mu(\tau)$ denote the worldline of a free-falling particle in comoving coordinates

$$x^0 = c\tau \qquad x^i = \text{constant} \, . \tag{19.2}$$

The four-velocity—tangent to the worldline—is then $U^\mu = (c, 0, 0, 0)$. The worldline prescribed by Eq. (19.2) will be a geodesic, $\mathrm{d}^2x^\mu/\mathrm{d}\tau^2 + \Gamma^\mu_{\alpha\beta}U^\alpha U^\beta = 0$, if we can show that $\Gamma^\mu_{00} = 0$. From Eq. (14.23), $\Gamma^\mu_{00} = \frac{1}{2}g^{\mu\rho}(2\partial_0 g_{0\rho} - \partial_\rho g_{00})$, which vanishes because $g_{0i} = 0$ and $g_{00} = -1$.

19.3 SPACES OF CONSTANT CURVATURE

The cosmological principle singles out a class of SHs, those that "look the same" at every point. To comply with the cosmological principle we assume that (at cosmological distances) space is

[5]The concept of congruence, a manifold-filling family of nonintersecting curves, admits higher-dimensional generalizations that for lack of space we omitted in Chapter 13. A manifold can be decomposed into a family of nonintersecting surfaces (higher-dimensional submanifolds).

[6]The earliest occurrence I have found of the term comoving coordinates is a single sentence in Tolman:[82] "For the purposes of the investigation it will be simplest to use a set of comoving coordinates such that the spatial components are determined by a network of meshes drawn so as to connect neighboring particles and allowed to move therewith." The complex of ideas associated with the comoving coordinate system is also known as *Weyl's postulate*. From Weyl:[41, Section 34] "In the three-dimensional space $x_0 = 0$ surrounding O we may mark off a region $\boldsymbol{R}$, such that, in it, $(\mathrm{d}s)^2$ remains definitely positive. Through every point of this region we draw the geodetic world-line which is orthogonal to that region, and which has a time-like direction. These lines will cover singly a certain four-dimensional neighborhood of O. We now introduce new coordinates which will coincide with the previous ones in the three-dimensional space $\boldsymbol{R}$, for which we now assign the coordinates x_0, x_1, x_2, x_3 to the point P at which we arrive, if we go from the point $P_0 = (x_1, x_2, x_3)$ in $\boldsymbol{R}$ along the orthogonal geodetic line passing through it, so far that the proper time of the arc traversed, P_0P, is equal to x_0." You should be able to parse Weyl's words to see the comoving coordinate system as we have defined it.

maximally symmetric in the sense of Section 14.8—that a maximally symmetric space, one that has all the symmetries it *can* have, is characterized by a *constant* curvature scalar, R. The Ricci tensor for a maximally symmetric, three-dimensional space is, from Eq. (14.113),

$$R_{ij} = \frac{R}{3} g_{ij} \,. \tag{19.3}$$

As we'll see, because there are exactly three choices for the sign of R (positive, negative, or zero), there are three types of coordinate system that embody Eq. (19.3).

What is the metric tensor for such spaces? A three-dimensional maximally symmetric space has six symmetries (Killing vectors),[7] three translations and three rotations—homogeneity and isotropy. Just as with the Schwarzschild metric, Eq. (16.1), the line element for a spherically symmetric space has the form

$$(\mathrm{d}\sigma)^2 = B(r)(\mathrm{d}r)^2 + r^2\left((\mathrm{d}\theta)^2 + \sin^2\theta(\mathrm{d}\phi)^2\right) , \tag{19.4}$$

where $B(r)$ is an unknown function.[8] The form of Eq. (19.4) takes care of the isotropy requirement, what about homogeneity? It can be shown that *a space that's isotropic about every point is homogeneous*.[74, p379] The origin of the coordinate r in Eq. (19.4) is thus arbitrary; we can choose any point as the origin because all points are equivalent.

The "recipe" of GR would have us at this point obtain the Christoffel symbols associated with the metric specified by Eq. (19.4). But we've already done that: In Chapter 16 we worked out the Christoffel symbols for the Schwarzschild metric. By setting $A = 0$ in Table 16.1, the Christoffel symbols associated with Eq. (19.4) are given in Table 19.1. We've also already worked out the

Table 19.1 Nonzero Christoffel symbols for isotropic three-space.

$\Gamma^r_{rr} = \dfrac{B'}{2B}$	$\Gamma^\theta_{\phi\phi} = -\sin\theta\cos\theta$
$\Gamma^r_{\theta\theta} = -\dfrac{r}{B}$	$\Gamma^\phi_{\phi\theta} = \cot\theta$
$\Gamma^r_{\phi\phi} = \Gamma^r_{\theta\theta}\sin^2\theta$	$\Gamma^\theta_{\theta r} = \Gamma^\phi_{\phi r} = r^{-1}$

elements of the Ricci tensor. From Eqs. (16.4)–(16.6) with $A = 0$,

$$R_{rr} = \frac{B'}{Br} \qquad R_{\theta\theta} = 1 - \frac{1}{B} + \frac{rB'}{2B^2} \qquad R_{\phi\phi} = R_{\theta\theta}\sin^2\theta \,. \tag{19.5}$$

Equating Eqs. (19.5) and (19.3), we obtain two differential equations for B:

$$R_{rr} = \frac{B'}{Br} = \frac{R}{3}g_{rr} = \frac{R}{3}B \qquad R_{\theta\theta} = 1 - \frac{1}{B} + \frac{rB'}{2B^2} = \frac{R}{3}g_{\theta\theta} = \frac{R}{3}r^2 \,. \tag{19.6}$$

The equation for $R_{\phi\phi}$ reduces to that for $R_{\theta\theta}$. We find from Eq. (19.6) (no integration required!):

$$B(r) = \frac{1}{1 - Kr^2} \,, \tag{19.7}$$

where $K \equiv R/6$. The line element for a maximally symmetric three-space thus has the form

$$(\mathrm{d}\sigma)^2 = \frac{(\mathrm{d}r)^2}{1 - Kr^2} + r^2\left((\mathrm{d}\theta)^2 + \sin^2\theta(\mathrm{d}\phi)^2\right) , \tag{19.8}$$

where K is a constant.

[7] The maximum number of Killing vectors an n-dimensional space can have is $n(n+1)/2$ (Section 14.7).

[8] We can't simply "borrow" the function $B(r)$ from the Schwarzschild metric, which satisfies the vacuum equation in four-dimensional spacetime, $R_{\mu\nu} = 0$. The Ricci tensor for a maximally symmetric space is of the form of Eq. (19.3).

19.4 FRIEDMANN-ROBERTSON-WALKER SPACETIME

A *spacetime* separation that builds in 1) the cosmological principle and 2) the expansion of the universe is: $(\mathrm{d}s)^2 = -(c\mathrm{d}t)^2 + S^2(t)(\mathrm{d}\sigma)^2$, where $S(t)$ is a time-dependent *scale factor* (it will be determined through Einstein's equation) with $(\mathrm{d}\sigma)^2$ given by Eq. (19.8). Homogeneity requires that $S(t)$ is a function of time only, not space. The scale factor provides the ever changing *distance* between galaxies, galaxies that in the comoving frame have constant spatial coordinates.[9] The *Friedmann-Robertson-Walker* (FRW) metric is

$$(\mathrm{d}s)^2 = -(c\mathrm{d}t)^2 + S^2(t)\left[\frac{(\mathrm{d}r)^2}{1-Kr^2} + r^2((\mathrm{d}\theta)^2 + \sin^2\theta(\mathrm{d}\phi)^2)\right] . \tag{19.9}$$

Let's rescale the radial coordinate in Eq. (19.9) so that it absorbs the magnitude of K (for $K \neq 0$). Let $\overline{r} \equiv \sqrt{|K|}r$. With this substitution,

$$(\mathrm{d}s)^2 = -(c\mathrm{d}t)^2 + \frac{S^2(t)}{|K|}\left[\frac{(\mathrm{d}\overline{r})^2}{1-K\overline{r}^2/|K|} + \overline{r}^2((\mathrm{d}\theta)^2 + \sin^2\theta(\mathrm{d}\phi)^2)\right] .$$

Now let $k \equiv K/|K| = (1, 0, -1)$ according to whether $K > 0$, $K = 0$, or $K < 0$. Likewise rescale $S(t)$. Let

$$R(t) \equiv \begin{cases} \dfrac{S(t)}{\sqrt{|K|}} & \text{if } K \neq 0 \\ S(t) & \text{if } K = 0 . \end{cases}$$

Erasing the "bar" on $\overline{r}$, we have the FRW metric as it's usually written

$$(\mathrm{d}s)^2 = -(c\mathrm{d}t)^2 + R^2(t)\left[\frac{(\mathrm{d}r)^2}{1-kr^2} + r^2((\mathrm{d}\theta)^2 + \sin^2\theta(\mathrm{d}\phi)^2)\right] . \tag{19.10}$$

In Eq. (19.10), *the standard spacetime metric for cosmology*, the coordinate r is dimensionless; $R(t)$ carries the dimension of length. There are three possible types of geometry for the universe characterized by $k = (1, 0, -1)$. Einstein's field equation will give us differential equations for $R(t)$; the value of k cannot be determined through theory and is inferred experimentally. The Christoffel symbols associated with the FRW metric are listed in Table 19.2.

The elements of the Ricci tensor can be derived using Eq. (15.28) and Table 19.2. It's found that $R_{\mu\nu} = 0$ for $\mu \neq \nu$. The diagonal terms are

$$\begin{aligned} R_{tt} &= -\frac{3}{c^2}\frac{\ddot{R}}{R} & R_{rr} &= \frac{1}{c^2(1-kr^2)}\left(R\ddot{R} + 2\dot{R}^2 + 2kc^2\right) \\ R_{\theta\theta} &= \frac{r^2}{c^2}\left(R\ddot{R} + 2\dot{R}^2 + 2kc^2\right) & R_{\phi\phi} &= R_{\theta\theta}\sin^2\theta . \end{aligned} \tag{19.11}$$

Continuing with the "checklist" of GR,[10] let's get the Ricci scalar. First, we need to raise an index: $R^\mu_\nu = g^{\mu\alpha}R_{\alpha\nu}$. Because the metric Eq. (19.10) is diagonal, this is particularly simple: $R^\mu_\nu = g^{\mu\mu}R_{\mu\nu}$ (no sum). We have

$$\begin{aligned} R^t_t &= g^{tt}R_{tt} = \frac{3}{c^2}\frac{\ddot{R}}{R} & R^r_r &= g^{rr}R_{rr} = \frac{1}{c^2R^2}\left(R\ddot{R} + 2\dot{R}^2 + 2kc^2\right) \\ R^\theta_\theta &= g^{\theta\theta}R_{\theta\theta} = R^r_r & R^\phi_\phi &= g^{\phi\phi}R_{\phi\phi} = R^\theta_\theta = R^r_r . \end{aligned} \tag{19.12}$$

---

[9]Imagine three galaxies at a given time, on a given SH. The same galaxies appear later (on a different SH) with the same coordinates, yet with a different distance between them, where the distance is obtained from the metric.

[10]Which, in case you've forgotten is $g_{\mu\nu} \to \Gamma^\alpha_{\lambda\rho} \to R_{\mu\nu} \to R^\alpha_\alpha \to G_{\mu\nu}$.

Table 19.2 Christoffel symbols for the Friedmann-Robertson-Walker metric.

$\alpha = t$	$\alpha = r$
$\Gamma^t_{tt} = \Gamma^t_{t\theta} = \Gamma^t_{t\phi} = \Gamma^t_{tr} = 0$	$\Gamma^r_{r\theta} = \Gamma^r_{r\phi} = \Gamma^r_{tt} = 0$
$\Gamma^t_{rr} = \dfrac{R\dot{R}}{c(1-kr^2)}$	$\Gamma^r_{rr} = \dfrac{kr}{1-kr^2}$
$\Gamma^t_{\theta\theta} = \dfrac{r^2 R\dot{R}}{c}$	$\Gamma^r_{rt} = \dfrac{\dot{R}}{cR}$
$\Gamma^t_{\phi\phi} = \Gamma^t_{\theta\theta}\sin^2\theta$	$\Gamma^r_{\theta\theta} = -r(1-kr^2)$
	$\Gamma^r_{\phi\phi} = \Gamma^r_{\theta\theta}\sin^2\theta$
$\alpha = \theta$	$\alpha = \phi$
$\Gamma^\theta_{\theta\theta} = \Gamma^\theta_{tt} = \Gamma^\theta_{rr} = \Gamma^\theta_{\theta\phi} = 0$	$\Gamma^\phi_{\phi\phi} = \Gamma^\phi_{tt} = \Gamma^\phi_{rr} = \Gamma^\phi_{\theta\theta} = 0$
$\Gamma^\theta_{\theta t} = \dfrac{\dot{R}}{Rc}$	$\Gamma^\phi_{\phi t} = \dfrac{\dot{R}}{Rc}$
$\Gamma^\theta_{\phi\phi} = -\sin\theta\cos\theta$	$\Gamma^\phi_{\phi\theta} = \cot\theta$
$\Gamma^\theta_{\theta r} = r^{-1}$	$\Gamma^\phi_{\phi r} = r^{-1}$

The fact that $R^r_r = R^\theta_\theta = R^\phi_\phi$ should be expected for a 3-space with constant curvature. The Ricci scalar $R^\alpha_\alpha = R^t_t + 3R^r_r$ is then

$$R^\alpha_\alpha = \frac{6}{c^2R^2}\left(R\ddot{R} + \dot{R}^2 + kc^2\right) . \tag{19.13}$$

The elements of the Einstein tensor are, from $G_{\mu\nu} = R_{\mu\nu} - \frac{1}{2}g_{\mu\nu}R^\alpha_\alpha$,

$$G_{tt} = \frac{3}{c^2R^2}\left(\dot{R}^2 + kc^2\right) \qquad G_{rr} = -\frac{1}{c^2R^2}\left(2R\ddot{R} + \dot{R}^2 + kc^2\right) g_{rr}$$
$$G_{\theta\theta} = -\frac{1}{c^2R^2}\left(2R\ddot{R} + \dot{R}^2 + kc^2\right) g_{\theta\theta} \qquad G_{\phi\phi} = \sin^2\theta G_{\theta\theta} . \tag{19.14}$$

The spatial elements all have the form $G_{ij} = -g_{ij}(2R\ddot{R}+\dot{R}^2+kc^2)/(cR)^2$. We'll use these results to formulate Einstein's equation for cosmology. First, however, we examine the possible geometries implied by the spatial part of the FRW metric.

19.5 SPATIAL GEOMETRIES

The FRW metric pertains to spacetimes (one for each value of k) that are spatially homogeneous and isotropic at each instant of time. What types of geometry are associated with the three values of k? For fixed t, θ, and ϕ, the radial distance is proportional to the integral

$$\chi \equiv \int \frac{\mathrm{d}r}{\sqrt{1-kr^2}} = \begin{cases} \sin^{-1} r & k=1 \\ r & k=0 \\ \sinh^{-1} r & k=-1 \end{cases} . \tag{19.15}$$

Define the function

$$S_k(\chi) \equiv \begin{cases} \sin\chi & k=1 \\ \chi & k=0 \\ \sinh\chi & k=-1 \end{cases} , \tag{19.16}$$

so that the spatial part of Eq. (19.10) is (at a fixed time, on a SH)

$$(\mathrm{d}\sigma)^2 = R_0^2 \left[(\mathrm{d}\chi)^2 + S_k^2(\chi)\left((\mathrm{d}\theta)^2 + \sin^2\theta(\mathrm{d}\phi)^2\right)\right] , \tag{19.17}$$

where $R_0 \equiv R(t = t_0)$. Writing the metric in terms of χ removes the coordinate singularity at $r = 1$ for $k = 1$. Let's examine the geometries described by Eq. (19.17) for $k = 1, 0, -1$.

19.5.1 k = +1: Positive curvature, closed space (S^3)

For $k = 1$, Eq. (19.17) becomes

$$(\mathrm{d}\sigma)^2 = R_0^2 \left[(\mathrm{d}\chi)^2 + \sin^2\chi\left((\mathrm{d}\theta)^2 + \sin^2\theta(\mathrm{d}\phi)^2\right)\right] . \qquad (k = 1) \tag{19.18}$$

Equation (19.18) is the metric for a space spanned by the coordinates (χ, θ, ϕ), which describes *the distance on the surface of a three-dimensional sphere*, S^3. Mathematically, S^3 is covered by the coordinate ranges $0 \le \chi \le \pi, 0 \le \theta \le \pi$, and $0 \le \phi < 2\pi$ (χ, θ, ϕ are *hyperspherical coordinates*). We can gain an understanding of the geometry described by Eq. (19.18) by embedding it in a four-dimensional geometry (see Section 16.5). Define Euclidean coordinates

$$w = R_0 \cos\chi \qquad x = R_0 \sin\chi \sin\theta \cos\phi \qquad y = R_0 \sin\chi \sin\theta \sin\phi \qquad z = R_0 \sin\chi \cos\theta .$$

An embedding is possible because

$$(\mathrm{d}w)^2 + (\mathrm{d}x)^2 + (\mathrm{d}y)^2 + (\mathrm{d}z)^2 = R_0^2 \left[(\mathrm{d}\chi)^2 + \sin^2\chi\left((\mathrm{d}\theta)^2 + \sin^2\theta(\mathrm{d}\phi)^2\right)\right] .$$

We thus have a four-dimensional space ($\mathbb{R}^4$) such that when its coordinates (w, x, y, z) are restricted to the three-sphere specified by $w^2 + x^2 + y^2 + z^2 = R_0^2$, the distance relation on that surface matches the spatial part of the FRW metric for $k = 1$. Our customary picture of a sphere, $x^2 + y^2 + z^2 = R^2$, is the *two-sphere*, S^2. For S^2 there is a constraint among three Euclidean coordinates that produces a surface covered by two independent coordinates, the usual angles (θ, ϕ). Consider $z = R\cos\theta$, $x = R\sin\theta\cos\phi$, $y = R\sin\theta\sin\phi$; $(\mathrm{d}x)^2 + (\mathrm{d}y)^2 + (\mathrm{d}z)^2 = R^2[(\mathrm{d}\theta)^2 + \sin^2\theta(\mathrm{d}\phi)^2]$ so that we have an embedding of S^2 in $\mathbb{R}^3$.

Back to cosmology, the spatial geometry for $k = 1$ is a three-sphere covered by the *three* angles, (χ, θ, ϕ). The area element for S^3 (letting χ, θ, ϕ vary) is $R_0^3 \mathrm{d}\chi(\sin\chi \mathrm{d}\theta)(\sin\chi\sin\theta \mathrm{d}\phi)$, with the *hyperarea* (three-dimensional "area" of S^3 embedded in $\mathbb{R}^4$)

$$A_3 = R_0^3 \int_0^{2\pi} \mathrm{d}\phi \int_0^{\pi} \sin\theta \mathrm{d}\theta \int_0^{\pi} \sin^2\chi \mathrm{d}\chi = 2\pi^2 R_0^3 . \tag{19.19}$$

We can thus refer to R_0 as the present radius of the universe if it turns out that the universe is described by $k = 1$. Because A_3 in Eq. (19.19) is finite, the geometry associated with $k = 1$ is said to be *closed*. The use of an embedding space lends insight into geometries that are not easily visualized, but the picture provided should not be taken literally. The three-dimensional volume given by Eq. (19.19) is the totality of all the space that exists at any one time (if $k = 1$); it should not be viewed as embedded in a higher-dimensional space.

To get a feeling for what's implied by curvature, consider the simpler example of a two-sphere of radius a. Imagine an ant starting at the top of the sphere that travels a distance $D = a\theta$ when it crawls in a "straight line" (geodesic) to a point with polar angle θ (see Fig. 19.1). If the ant then crawls along a circle at a constant distance D from the top of the sphere, it travels the circumference $C(D) = 2\pi a \sin\theta = 2\pi a \sin(D/a)$. For small values of D/a,

$$C(D) = 2\pi a \sin(D/a) = 2\pi D\left(1 - \frac{D^2}{6a^2} + \cdots\right) .$$

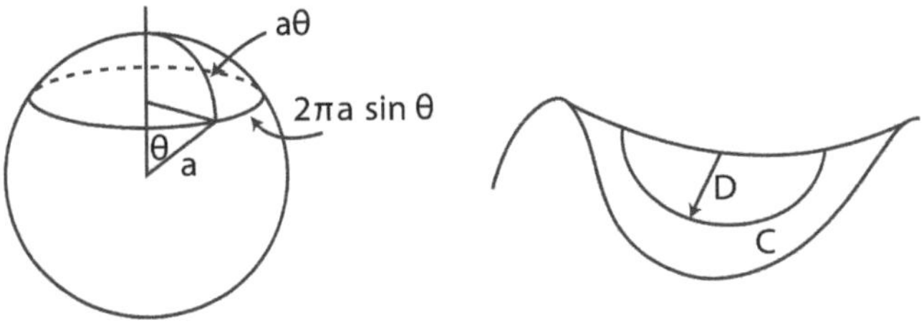

Figure 19.1 A sphere has positive curvature, a saddle has negative curvature.

The circumference of a circle of radius D *on a sphere* is thus not $2\pi D$. Likewise, the area enclosed by a circle of radius D, $A(D) = 2\pi a^2(1-\cos\theta) = 2\pi a^2(1-\cos(D/a))$. For small values of D/a,

$$A(D) = \pi D^2 \left(1 - \frac{D^2}{12a^2} + \cdots\right) .$$

Thus, measurements made on the sphere are such that $C/D < 2\pi$ and $A/(D^2) < \pi$. Geometries with $C/(2\pi D) < 1$ are said to have *positive curvature*; geometries with $C/(2\pi D) > 1$ are said to have *negative curvature*,[11] such as at a saddle point (Fig. 19.1). Curvature can thus be characterized in terms of the "deficit" between the area or the circumference from their Euclidean values. The Ricci scalar field is a number assigned to every point of a manifold that's proportional to the amount by which the local volume of a geodesic ball deviates from the volume of a ball in Euclidean space of the same dimension.

19.5.2 k = 0: Flat space ($\mathbb{R}^3$)

For $k = 0$, set $r = R_0\chi$ and Eq. (19.17) becomes

$$(\mathrm{d}\sigma)^2 = (\mathrm{d}r)^2 + r^2((\mathrm{d}\theta)^2 + \sin^2\theta(\mathrm{d}\phi)^2) . \qquad (k=0) \tag{19.20}$$

Equation (19.20) is the metric for $\mathbb{R}^3$ covered by coordinates $r \geq 0$, $0 \leq \theta \leq \pi$, $0 \leq \phi < 2\pi$. The volume of this space is infinite, and hence is an *open* geometry.

19.5.3 k = −1: Negative curvature, open space (hyperbolic)

For $k = -1$, Eq. (19.17) becomes

$$(\mathrm{d}\sigma)^2 = R_0^2\left[(\mathrm{d}\chi)^2 + \sinh^2\chi\left((\mathrm{d}\theta)^2 + \sin^2\theta(\mathrm{d}\phi)^2\right)\right] , \qquad (k=-1) \tag{19.21}$$

the metric for a three-dimensional space covered by coordinates (χ, θ, ϕ) with ranges $\chi \geq 0$, $0 \leq \theta \leq \pi$, $0 \leq \phi < 2\pi$. The geometry described by Eq. (19.21) cannot be embedded in a four-dimensional Euclidean space. It can, however, be embedded in an abstract four-dimensional Minkowskian space (not spacetime) with metric signature $(-+++)$. Consider the coordinates

$$w = R_0\cosh\chi \qquad x = R_0\sinh\chi\sin\theta\cos\phi \qquad y = R_0\sinh\chi\sin\theta\sin\phi \qquad z = R_0\sinh\chi\cos\theta .$$

As can be shown, the distance relations match

$$-(\mathrm{d}w)^2 + (\mathrm{d}x)^2 + (\mathrm{d}y)^2 + (\mathrm{d}z)^2 = R_0^2\left[(\mathrm{d}\chi)^2 + \sinh^2\chi\left((\mathrm{d}\theta)^2 + \sin^2\theta(\mathrm{d}\phi)^2\right)\right] .$$

The area element of the hypersurface is $R_0^3\sinh^2\chi d\chi\sin\theta d\theta d\phi$, and hence the hyperarea is infinite—an open geometry. The surface specified by the coordinates of the embedding space is given by $w^2 - (x^2+y^2+z^2) = R_0^2$. The embedding space is flat (as is Minkowski space), but the hypersurface has negative curvature.

[11] In Section 12.3.4 we noted that, in a rotating reference frame, space has a negative curvature.

19.6 THE FRIEDMANN EQUATIONS

We now determine the scale factor $R(t)$ so that the FRW metric satisfies the Einstein field equation. For this purpose, we use the energy-momentum tensor for a perfect fluid, Eq. (15.59). Thus, we assume that on cosmological length scales the motions of galaxies can be modeled as the motion of a fluid. By combining $T_{\mu\nu} = (\rho + P/c^2)U_\mu U_\nu + P g_{\mu\nu}$ with the four-velocity in the comoving frame, $U_\mu = (-c, 0, 0, 0)$, we have $T_{00} = \rho c^2$, $T_{0i} = 0$, and $T_{ij} = P g_{ij}$. Now combine Einstein's equation, $G_{\mu\nu} + \Lambda g_{\mu\nu} = \kappa T_{\mu\nu}$, with $G_{\mu\nu}$ in the comoving frame, Eq. (19.14). We find for the time component

$$\left(\frac{\dot{R}}{R}\right)^2 + \frac{kc^2}{R^2} = \frac{8\pi G}{3}\rho + \frac{1}{3}\Lambda c^2 \tag{19.22}$$

and for the three spatial components

$$\frac{1}{R^2}\left(2R\ddot{R} + \dot{R}^2 + kc^2\right) = -\frac{8\pi G}{c^2}P + \Lambda c^2 \,. \tag{19.23}$$

It's not surprising that the spatial parts of the Einstein equation are identical, given the high symmetry built in through the cosmological principle.

By subtracting Eq. (19.22) from Eq. (19.23) we find

$$\frac{\ddot{R}}{R} = -\frac{4\pi G}{3}\left(\rho + 3\frac{P}{c^2}\right) + \frac{1}{3}\Lambda c^2 \,. \tag{19.24}$$

Equations (19.22) and (19.24) are two differential equations for $R(t)$. They are referred to as the *Friedmann equations*. Sometimes Eq. (19.22) is referred to as *the* Friedmann equation, with Eq. (19.24) the *acceleration equation*. Note that P occurs in the acceleration equation, the geometry parameter k occurs in Eq. (19.22), and Λ occurs in both.

We showed in Section 10.5 that projecting the conservation laws $\partial_\nu T^{\mu\nu} = 0$ onto and orthogonal to U^μ, using $T^{\mu\nu}$ for the perfect fluid, results in Eqs. (10.27) and (10.28), the covariant equations of hydrodynamics in flat spacetime. By letting $\partial_\mu \to \nabla_\mu$ and $\eta^{\mu\nu} \to g^{\mu\nu}$ in Eqs. (10.27) and (10.28), we have the analogous projections of $\nabla_\nu T^{\mu\nu} = 0$,

$$\left(\rho + \frac{P}{c^2}\right)\nabla_\nu U^\nu + U^\nu \partial_\nu \rho = 0 \tag{19.25}$$

and

$$\left(\rho + \frac{P}{c^2}\right)U^\nu \nabla_\nu U^\mu + \left(g^{\mu\nu} + \frac{1}{c^2}U^\mu U^\nu\right)\partial_\nu P = 0 \,. \tag{19.26}$$

Equation (19.26) is Euler's equation in GR. Let's evaluate these equations in the comoving frame, with $U^\mu = (c, 0, 0, 0)$. In that case the time component of Eq. (19.26) reduces to $0 = 0$. The spatial components reduce to $\partial^i P = 0$, i.e., no pressure gradients, implying that P can only be a function of time. Turning to Eq. (19.25), use $\nabla_\nu U^\nu = \partial_\nu(\sqrt{|g|}U^\nu)/\sqrt{|g|}$. The determinant of the FRW metric is $\sqrt{|g|} = R^3(t)r^2 \sin\theta/\sqrt{1 - kr^2}$. In the spacetime of cosmology, therefore, free-fall worldlines *diverge* because of the expanding scale factor

$$\nabla_\nu U^\nu = \frac{1}{\sqrt{|g|}}c\partial_0(\sqrt{|g|}) = \frac{1}{R^3}\partial_t R^3 = 3\frac{\dot{R}}{R} \,. \tag{19.27}$$

Combining Eqs. (19.27) and (19.25), in the comoving frame Eq. (19.25) reduces to

$$\dot{\rho} + 3(\rho + P/c^2)\frac{\dot{R}}{R} = 0 \,. \tag{19.28}$$

Equation (19.28) is a statement of energy conservation (it comes from the projection of $\nabla_\nu T^{\mu\nu} = 0$ onto the timelike direction). It can be derived directly, however. Differentiate Eq. (19.25) with respect to time and make use of Eq. (19.24); the result is Eq. (19.28). It's "nice," however, to see Eq. (19.28) emerge from $\nabla_\nu T^{\mu\nu} = 0$.

19.7 NEWTONIAN COSMOLOGY

What do the Friedmann equations tell us? To get a handle on their meaning, let's try to derive them from a Newtonian perspective. Consider a "universe" consisting of a small mass m at the edge of a fixed mass M that's uniformly distributed throughout a sphere of time-dependent radius $R(t)$ (see Fig. 19.2). The mass density $\rho(t)$ therefore varies in time, with $M = 4\pi\rho(t)R^3(t)/3$. From

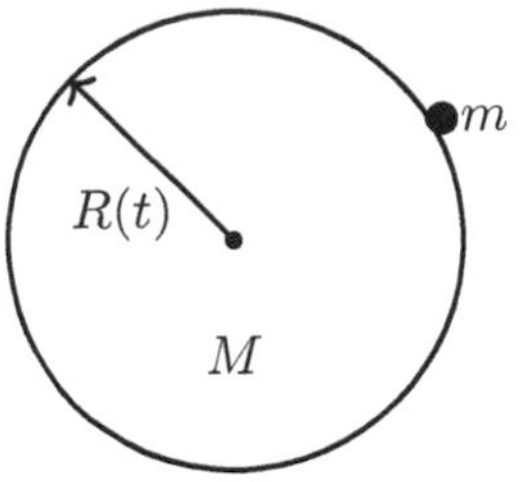

Figure 19.2 Mass M of radius $R(t)$ (the universe) expanding against mass m.

Newton's law of gravity and second law of motion, we have that $R(t)$ varies in time as

$$\ddot{R} = -\frac{GM}{R^2} = -\frac{4\pi G}{3}\rho R\,, \tag{19.29}$$

which is not altogether different from the Friedmann acceleration equation, (19.24). Note that $\ddot{R} < 0$, always. Comparing Eqs. (19.29) and (19.24), what we obtain from Newtonian theory obviously doesn't include Λ, which is new to GR, and we "miss" the pressure term, which can be attributed to the lack of "$E = mc^2$" in pre-relativistic mechanics. Pressure has dimensions of energy density; P/c^2 is an equivalent mass density that should be added to ρ. For matter under normal conditions $\rho c^2 \gg P$ and P can be ignored. One could object that pressure doesn't occur in a single-particle theory (Newton's second law); it enters the cosmological equations from the energy-momentum tensor of a *fluid*. We can group the term containing Λ in Eq. (19.24) into an equivalent pressure. Write Eq. (19.24) as

$$\frac{\ddot{R}}{R} = -\frac{4\pi G}{3}\left(\rho + 3\frac{\widetilde{P}}{c^2}\right)$$

where

$$\widetilde{P} \equiv P - \frac{2}{3}\frac{\Lambda}{\kappa}\,, \tag{19.30}$$

with $\kappa = 8\pi G/c^4$. The effect of Λ (for $\Lambda > 0$) is to *reduce* the effective pressure; *the cosmological constant is associated with a negative pressure* (for $\Lambda > 0$).

We can derive the Newtonian version of Eq. (19.22) by multiplying Eq. (19.29) by $\dot{R}$. We find

$$\frac{1}{2}\frac{\mathrm{d}}{\mathrm{d}t}\left(\dot{R}^2\right) = -GM\frac{\dot{R}}{R^2} = \frac{\mathrm{d}}{\mathrm{d}t}\left(\frac{GM}{R}\right) \implies \frac{\mathrm{d}}{\mathrm{d}t}\left(\frac{1}{2}\dot{R}^2 - \frac{GM}{R}\right) = 0\,,$$

and thus

$$\frac{1}{2}\dot{R}^2 - \frac{GM}{R} = E\,, \tag{19.31}$$

where E is a constant with dimensions of energy per mass. Equation (19.31) can be put in the form of the Friedmann equation:

$$\left(\frac{\dot{R}}{R}\right)^2 = \frac{8\pi G}{3}\rho + \frac{2E}{R^2}\,. \tag{19.32}$$

Comparing Eqs. (19.32) with (19.22), we can identify $kc^2 = -2E$. In the Friedmann equation the geometry parameter k has precisely the values $k = 1, 0, -1$. In the Newtonian version k is related to the energy of the universe. For $E > 0$, the right side of Eq. (19.31) is always positive, implying that $\dot{R} > 0$ always. Positive energy (associated with $k = -1$) corresponds to an *unbound* universe that expands indefinitely. For $E < 0$, the right side of Eq. (19.31) starts out positive (for $R \to 0$) but goes to zero at the maximum scale factor $R_{\text{max}} = GM/|E|$ (energy is all potential), at which point, because $\ddot{R} < 0$, the universe starts to contract. Negative gravitational energy, associated with $k = 1$, a closed geometry, corresponds to a *bound* universe. The third possibility, $E = 0$ ($k = 0$) corresponds to an expanding universe with $\dot{R} \to 0$ as $t \to \infty$ and $\rho \to 0$. The factor of Λ in the Friedmann equation can be absorbed into an equivalent mass density. Write Eq. (19.22) as

$$\left(\frac{\dot{R}}{R}\right)^2 + \frac{kc^2}{R^2} = \frac{8\pi G}{3}\widetilde{\rho},$$

where

$$\widetilde{\rho} \equiv \rho + \frac{\Lambda}{\kappa c^2}. \tag{19.33}$$

The effect of Λ is to *increase* the effective mass density; *the cosmological constant is associated with a positive energy density* (for $\Lambda > 0$).

19.8 COSMOLOGICAL REDSHIFT

With Eqs. (19.22), (19.24), and (19.28) *we have the standard equations of cosmology*. They are not independent of each other. Which ones we use depends on the problem at hand. To implement the Friedmann equations we have to make assumptions about the energy density of the universe, specifically how pressure is related to the density. Before taking up cosmological models, let's see what can be inferred without explicit knowledge of $R(t)$. We could have done so *before* deriving the Friedmann equations. It could be objected, however, that we should not discuss what can be inferred independent of the form of $R(t)$ without having first presented a path by which $R(t)$ can be found, so that we know it exists. Now that we know that, what can we learn?

Take our galaxy to be located at $r = 0$. Another galaxy having coordinate r is then at the distance, using the FRW metric, Eq. (19.10),

$$D(t) = R(t)\int_0^r \frac{\mathrm{d}r}{\sqrt{1-kr^2}} = \chi(r)R(t), \tag{19.34}$$

where χ is defined in Eq. (19.15). The distance $R(t)\chi(r)$ is the *proper distance*—the distance that could be measured by laying a tape measure between galaxies, if such a procedure could be done at one time. At a later time the galaxy is at the distance $D(t+\Delta t) = R(t+\Delta t)\chi(r)$ (in the comoving frame the coordinate r is constant). We observe the galaxy to have the recessional velocity,

$$v = \frac{\mathrm{d}D}{\mathrm{d}t} = \chi\dot{R} = \frac{\dot{R}}{R}(R\chi) = \frac{\dot{R}}{R}D, \tag{19.35}$$

which is precisely the form of Hubble's law, with the Hubble "constant" (parameter) given by

$$H(t) \equiv \frac{\dot{R}(t)}{R(t)}. \tag{19.36}$$

Hubble's law emerges as a consequence of the comoving reference frame description.[12]

[12]This shouldn't come as a surprise—an isotropic expansion was built into the FRW metric from the outset.

Photons propagate along lines of null separation, $(\mathrm{d}s)^2 = 0$. From Eq. (19.10), light travels in such a way that $(c\mathrm{d}t)^2 = R^2(t)(\mathrm{d}r)^2/(1 - kr^2)$, or such that

$$c\frac{\mathrm{d}t}{R(t)} = \frac{\mathrm{d}r}{\sqrt{1 - kr^2}} . \tag{19.37}$$

Now consider two light signals emitted at coordinate r_1 at times t_1 and $t_1 + \Delta t_1$ that reach r_2 at times t_2 and $t_2 + \Delta t_2$. Then, from Eq. (19.37)

$$c\int_{t_1}^{t_2} \frac{\mathrm{d}t}{R(t)} = \int_{r_1}^{r_2} \frac{\mathrm{d}r}{\sqrt{1 - kr^2}} = c\int_{t_1+\Delta t_1}^{t_2+\Delta t_2} \frac{\mathrm{d}t}{R(t)} . \tag{19.38}$$

Comparing the integrals on the two sides of Eq. (19.38), we have, schematically,

$$\int_{t_1+\Delta t_1}^{t_2+\Delta t_2} = \int_{t_1+\Delta t_1}^{t_2} + \int_{t_2}^{t_2+\Delta t_2} = \int_{t_1}^{t_2} = \int_{t_1}^{t_1+\Delta t_1} + \int_{t_1+\Delta t_1}^{t_2} ,$$

so that

$$\int_{t_2}^{t_2+\Delta t_2} \frac{\mathrm{d}t}{R(t)} = \int_{t_1}^{t_1+\Delta t_1} \frac{\mathrm{d}t}{R(t)} . \tag{19.39}$$

Assuming that $R(t)$ varies slowly over Δt, we have from Eq. (19.39)

$$\frac{\Delta t_2}{R(t_2)} = \frac{\Delta t_1}{R(t_1)} \implies \frac{\Delta t_2}{\Delta t_1} = \frac{R(t_2)}{R(t_1)} . \tag{19.40}$$

Let Δt be the time between the emission or reception of two successive wave crests, with $c\Delta t = \lambda$; Eq. (19.40) implies

$$\frac{\lambda_o}{\lambda_e} = \frac{R(t_o)}{R(t_e)} , \tag{19.41}$$

where "o" and "e" refer to observed and emitted. The *redshift parameter* is related to the change in scale factor between emission and absorption

$$z \equiv \frac{\lambda_o - \lambda_e}{\lambda_e} = \frac{R(t_o)}{R(t_e)} - 1 . \tag{19.42}$$

Clearly,

$$\lambda_o = \lambda_e(1 + z) . \tag{19.43}$$

Equation (19.43) indicates that the wavelength becomes *stretched* during the propagation of the photon between the times of emission and reception due to the evolution of $R(t)$ in that time. The larger the value of z, the more the scale factor has increased since the light was emitted. Larger z values thus indicate a longer propagation *distance* as well as time. The redshift parameter specified by Eq. (19.43) is the *cosmological redshift* due to the isotropic, homogeneous expansion of space. The cosmological redshift differs conceptually from the Doppler shift. In a Doppler shift, emitter and receiver are in relative motion in the same IRF. In the cosmological redshift, emitter and receiver are in their own IRFs at different points of the SH of the comoving coordinate system. The relative motion in the latter case is that due to the change of the scale factor $R(t)$. The Doppler and cosmological redshifts are distinct from the gravitational redshift.

19.9 THE EINSTEIN UNIVERSE

Are static solutions to the Friedmann equations possible? Let's try it with $\Lambda = 0$. With $\Lambda = 0$ and $\dot{R} = 0$ in Eq. (19.22) we have $kc^2 = 8\pi G\rho R^2/3$. Because all quantities are positive, we conclude that $k = 1$. This looks good: a static spherical universe. If, however, we then set $\ddot{R} = 0$ in Eq. (19.24), it can be satisfied (for $\Lambda = 0$) only if the pressure is *negative*, with $P = -3\rho c^2$. A negative pressure is difficult to understand. As we'll see, a negative pressure on the cosmological scale has a *repulsive* gravitational effect, and something like a repulsive component to gravity would be necessary to keep the universe from imploding on itself.

Already then we've encountered the prospect of a negative pressure by assuming $\Lambda = 0$. Can we get a static solution with $\Lambda \neq 0$? Set $\dot{R} = \ddot{R} = 0$ in Eqs. (19.22) and (19.24),

$$\frac{kc^2}{R^2} = \frac{8\pi G}{3}\rho + \frac{1}{3}\Lambda c^2 \qquad 0 = -\frac{4\pi G}{3}(\rho + 3p/c^2) + \frac{1}{3}\Lambda c^2 \, .$$

We can again set $k = 1$ (all quantities are positive if $\Lambda > 0$). If we subtract these equations and set $P = 0$ ($P \ll \rho c^2$) we find the radius of the *Einstein universe* (a static spherical universe)

$$R_E = \frac{c}{\sqrt{4\pi G\rho}} \, . \tag{19.44}$$

Substituting $4\pi G\rho = c^2/R_E^2$ into the equations above, we obtain *the* value of Λ,

$$\Lambda_E = \frac{1}{R_E^2} \, . \tag{19.45}$$

So far, so good: a static, spherical universe achieved by introducing Λ, which is indeed cosmological because it's related to the size of the universe. One small problem, however: The solution is not stable! By setting $4\pi G\rho = c^2/R_E^2$ and $\Lambda = 1/R_E^2$ in Eq. (19.22), we find:

$$\dot{R}^2 = c^2 \left(\frac{R^2}{R_E^2} - 1 \right) \, . \tag{19.46}$$

If a small perturbation causes R to deviate from R_E, then from Eq. (19.46) $\dot{R} \neq 0$. The Einstein static solution is therefore unstable. The parameter Λ *does not fix the problem it was introduced to fix*, namely the evolution of the universe.

19.10 THE DE SITTER UNIVERSE

Imagine a universe that's empty of all matter and radiation, but *does* have the cosmological constant as part of its workings. How would such a universe evolve? This is known as the *de Sitter universe*. From the Friedmann equations, (19.22) and (19.24), with $\rho = P = 0$

$$\ddot{R} = \frac{1}{3}\Lambda c^2 R \qquad \dot{R}^2 + kc^2 = \frac{1}{3}\Lambda c^2 R^2 \, . \tag{19.47}$$

The solutions to the acceleration equation are given by $R(t) = C_1 e^{\alpha t} + C_2 e^{-\alpha t}$ where C_1, C_2 are constants and $\alpha \equiv \sqrt{\Lambda c^2/3}$. Note that $\sqrt{\Lambda c^2}$ has the dimension of (time)$^{-1}$. For the other Friedmann equation, a nonlinear differential equation, C_1 and C_2 must be chosen appropriately.[13] The second equation is satisfied when $4\alpha^2 C_1 C_2 = kc^2$, or if $C_1 C_2 = 3k/(4\Lambda)$. If $k = 0$, either C_1 or C_2 must vanish. Setting $C_2 = 0$ leads to an exponentially growing solution $R(t) = e^{\alpha t}$.

[13] This is a good time to note how the solution of a nonlinear differential equation differs from the solutions of linear differential equations. With linear differential equations, we're accustomed to finding *families* of solutions; with nonlinear differential equations, we're fortunate to find just *one* solution.

If $k = 1$, then $C_1 = C_2 = \sqrt{3/(4\Lambda)}$ is a solution with $R(t) = \sqrt{3/\Lambda}\cosh\alpha t$. For $k = -1$, $R(t) = \sqrt{3/\Lambda}\sinh\alpha t$ is a solution.

An "empty" universe thus evolves! Of course, it's not empty: It has the energy density associated with the cosmological constant, and we'll show that Λ acts as a repulsive form of a gravity. The de Sitter universe evolves on its own, without the "feedback" provided by matter that we've come to expect in GR. As the universe expands, the density of matter and radiation decline to the point where they become negligible, and the acceleration established by Λ dominates. The ultimate fate of the universe apparently is to become the de Sitter universe.

19.11 DARK ENERGY

Let's continue with the idea of a universe without matter and radiation. Start with Einstein's equation $G_{\mu\nu}+\Lambda g_{\mu\nu} = \kappa T_{\mu\nu}$ and set to zero $T_{\mu\nu}$ associated with conventional matter fields. Write Einstein's equation *as if* Λ represents the energy-momentum of a new substance, referred to as *dark energy*:

$$G_{\mu\nu} = -\Lambda g_{\mu\nu} \equiv \kappa T^{\Lambda}_{\mu\nu} \,. \tag{19.48}$$

Use the form of $T_{\mu\nu}$ for perfect fluids,[14] $T_{\mu\nu} = P g_{\mu\nu} + (\rho + P/c^2)U_\mu U_\nu$. *Demand* that the right side of Eq. (19.48), $-\Lambda g_{\mu\nu}$, be written in the form of an equivalent stress-energy tensor:

$$-\Lambda g_{\mu\nu} \leftrightarrow \kappa\left(P_\Lambda g_{\mu\nu} + (\rho_\Lambda + P_\Lambda/c^2)U_\mu U_\nu\right) \,. \tag{19.49}$$

To make (19.49) into an equality, require that $\kappa P_\Lambda \equiv -\Lambda$ and $\rho_\Lambda \equiv -P_\Lambda/c^2$. We are thus led to *define* the pressure and energy density of the Λ "fluid,"

$$P_\Lambda = -\frac{\Lambda}{\kappa} = -\frac{c^4}{8\pi G}\Lambda \qquad \rho_\Lambda c^2 = -P_\Lambda = \frac{c^4}{8\pi G}\Lambda \,. \tag{19.50}$$

If $\Lambda > 0$, then $P_\Lambda < 0$ and $\rho_\Lambda c^2 > 0$.

How to interpret negative pressure? Let's do a simple calculation. Take the time derivative of $\rho c^2 R^3$:

$$\frac{\mathrm{d}}{\mathrm{d}t}(\rho c^2 R^3) = 3\rho c^2 R^2 \dot{R} + c^2 R^3 \dot{\rho} = 3\rho c^2 R^2 \dot{R} - 3c^2(\rho + P/c^2)\dot{R}R^2 = -3P\dot{R}R^2 = -P\frac{\mathrm{d}}{\mathrm{d}t}R^3 \,, \tag{19.51}$$

where we've used Eq. (19.28). Equation (19.51), $\mathrm{d}(\rho c^2 R^3) = -P\mathrm{d}(R^3)$, has the form of the first law of thermodynamics, $\mathrm{d}U = -P\mathrm{d}V$. The entropy term that would normally be present, $\mathrm{d}U = T\mathrm{d}S - P\mathrm{d}V$, is absent which is not surprising—the orderly motion of galaxies on geodesics (inherent in the perfect fluid model) produces no change in entropy.[15] Because the energy density $\rho_\Lambda c^2$ is *constant*, expanding the volume by ΔV *increases* the energy by $\Delta U = \rho_\Lambda c^2 \Delta V$. Work is therefore done *on* the system because $\Delta U > 0$. The work done in an adiabatic, constant pressure process is $-P\Delta V$. The pressure of a substance having a constant, positive energy density must be negative!

Einstein's equation can be written (Exercise 15.1) $R_{\mu\nu} = \kappa(T_{\mu\nu} - g_{\mu\nu}T/2) + \Lambda g_{\mu\nu}$. For weak gravity, $R_{00} \approx \nabla^2\Phi/c^2$, Eq. (15.25), and from Eq. (15.24), $T_{00} - \frac{1}{2}g_{00}T = \frac{1}{2}(\rho c^2 + 3P)$. The Poisson equation associated with Λ is therefore:

$$\nabla^2\Phi = 4\pi G(\rho + 3P/c^2) - \Lambda c^2 \,. \tag{19.52}$$

Ignoring the pressure term, the solution to Eq. (19.52) outside a spherical mass M is

$$g = -\boldsymbol{\nabla}\Phi = -\frac{GM}{r}\hat{\boldsymbol{r}} + \frac{1}{3}\Lambda c^2 r\hat{\boldsymbol{r}} \,. \tag{19.53}$$

[14] What we've used to establish the Friedmann equations.

[15] One could say that in a reversible adiabatic expansion $\mathrm{d}S = 0$—this is the universe, from where outside the system would heat flow?

The cosmological term is thus associated with a repulsive form of gravity, one that *increases* in magnitude with r.

To make headway with the Friedmann equations, one must specify an *equation of state*, the connection between the pressure and density, $P = P(\rho)$. Because pressure and *energy* density have the same dimension, in cosmology the equation of state is written simply as $P = w\rho c^2$, where w is a dimensionless constant. From Eq. (19.50), $P_\Lambda = -\rho_\Lambda c^2$, and thus $w = -1$ for dark energy. For electromagnetic energy $P = \frac{1}{3}u$, where u is the energy density, and hence $w = \frac{1}{3}$ for radiation.[16] For "cold" matter (nonrelativistic) $P \ll \rho c^2$, and we approximate the equation of state for matter simply as $w = 0$. Thus, we work with $w = 0, \frac{1}{3}, -1$ for matter, radiation, and dark energy.

From Eq. (19.51), with $P = w\rho c^2$,

$$\mathrm{d}(\rho c^2 R^3) = -3w\rho c^2 R^2 \mathrm{d}R\,. \tag{19.54}$$

A solution to Eq. (19.54) can be had by guessing $\rho \propto R^\alpha$, where α is an unknown exponent. With this substitution, Eq. (19.54) becomes $\mathrm{d}(R^{3+\alpha}) = -3wR^{2+\alpha}\mathrm{d}R$, implying $\alpha = -3(1+w)$. The density therefore evolves with $R(t)$ as $\rho(t) \propto R^{-3(1+w)}$, or $\rho(t)R^{3(1+w)}(t) = $ constant. The constant can be evaluated using the present-dat values of these quantities. Thus,

$$\rho(t) = \rho_0 \left(\frac{R_0}{R(t)}\right)^{3(1+w)}\,. \tag{19.55}$$

Clearly, for matter, radiation, and dark energy, respectively,

$$\rho_M(t) = \rho_{M,0}\left(\frac{R_0}{R(t)}\right)^3 \qquad \rho_R(t) = \rho_{R,0}\left(\frac{R_0}{R(t)}\right)^4 \qquad \rho_\Lambda(t) = \rho_{\Lambda,0}\,. \tag{19.56}$$

We discussed in Section 15.4.2 that the same constant κ occurs in the Lagrangian $\mathscr{L} = (R - 2\Lambda)\sqrt{-g} + 2\kappa\mathscr{L}_m$, no matter the form of $\mathscr{L}_m$, implying a *universality* of the coupling of energy to gravity. Either we can treat Λ as something added on to Einstein's equation, as in Eq. (15.26), or we can treat Λ as representing a *new form of energy*, with its own stress-energy tensor, as in Eq. (19.48). For many purposes it's convenient to move Λ from the left side of Einstein's equation to the right in the form of a stress-energy tensor, $\kappa T^\Lambda_{\mu\nu} = -\Lambda g_{\mu\nu}$. We're then "entitled" to write Einstein's equation in its original form, $G_{\mu\nu} = \kappa T_{\mu\nu}$, but with the total stress-energy tensor as

$$T_{\mu\nu} = T^M_{\mu\nu} + T^R_{\mu\nu} + T^\Lambda_{\mu\nu}\,, \tag{19.57}$$

where each of the terms matter, radiation, and Λ can be modeled as a perfect fluid. The total stress-energy tensor can then be written (for $i = M, R, \Lambda$)

$$T_{\mu\nu} = \sum_i T^{(i)}_{\mu\nu} = \left(\sum_i P_i\right) g_{\mu\nu} + \left(\sum_i \rho_i + \frac{1}{c^2}\sum_i P_i\right) U_\mu U_\nu \tag{19.58}$$

$$= \left(\sum_i w_i \rho_i c^2\right) g_{\mu\nu} + \left(\sum_i \rho_i(1+w_i)\right) U_\mu U_\nu = \left(\frac{1}{3}\rho_R c^2 - \rho_\Lambda c^2\right) g_{\mu\nu} + \left(\rho_M + \frac{4}{3}\rho_R\right) U_\mu U_\nu\,.$$

Note that the energy density $T_{00} = (\rho_M + \rho_R + \rho_\Lambda)c^2$.

---

[16] This formula is for *cavity radiation*—electromagnetic energy in thermal equilibrium with its surroundings (the walls of a cavity) at temperature T—shown in books on thermodynamics; might we suggest [72]. Is the universe a big cavity? The CMB is found to be unpolarized and highly isotropic—what we'd expect from the thermodynamic theory of radiation. The spectral energy density of the CMB is found to occur as a Planck distribution with temperature $T = 2.726$ K. In the early stages of the universe (not covered in this book), photons are scattered by charged particles, particularly free electrons, providing an efficient mechanism for establishing thermal equilibrium; with free charges about the universe has "opaque walls"—just what's required for cavity radiation! In an adiabatic expansion of a photon gas, two things happen: 1) $VT^3 = $ constant, and thus the temperature drops as the universe expands, and 2) the spectral distribution is maintained (Wien's law), explaining the Planck spectrum today. In cosmology, electromagnetic radiation is modeled as a fluid with the equation state $P = \frac{1}{3}u = \frac{1}{3}\rho_R c^2$ where $\rho_R = u/c^2$.

19.12 THE FRIEDMANN EVOLUTION EQUATION

We start with the Friedmann equation (19.22), written now as

$$\frac{\dot{R}^2}{R^2} + \frac{kc^2}{R^2} = \frac{8\pi G}{3}\left(\rho_M + \rho_R + \rho_\Lambda\right) . \tag{19.59}$$

Because from Eq. (19.36) $\dot{R}/R = H(t)$, Eq. (19.59) is equivalent to

$$1 + \frac{kc^2}{H^2R^2} = \frac{8\pi G}{3H^2}\left(\rho_M + \rho_R + \rho_\Lambda\right) . \tag{19.60}$$

Define the *critical density*

$$\rho_C \equiv \frac{3H^2}{8\pi G} . \tag{19.61}$$

Note that ρ_C evolves in time along with the Hubble parameter. At the present time, $\rho_{C,0} = 3H_0^2/(8\pi G) = 1.878h^2 \times 10^{-26}$ kg m^{-3}. Next define the dimensionless (although time-dependent) *density ratios*

$$\Omega_M \equiv \frac{\rho_M}{\rho_C} \qquad \Omega_R \equiv \frac{\rho_R}{\rho_C} \qquad \Omega_\Lambda \equiv \frac{\rho_\Lambda}{\rho_C} . \tag{19.62}$$

With these definitions, Eq. (19.60) becomes

$$\frac{kc^2}{H^2R^2} = \Omega_M + \Omega_R + \Omega_\Lambda - 1 . \tag{19.63}$$

The numerical value of $\Omega_M + \Omega_R + \Omega_\Lambda - 1$ therefore tells us the curvature index k!

$$\text{If } \Omega_M + \Omega_R + \Omega_\Lambda \begin{cases} > 1 & \Rightarrow \text{the universe is closed} \\ = 1 & \Rightarrow \text{the universe is flat} \\ < 1 & \Rightarrow \text{the universe is open} \end{cases} . \tag{19.64}$$

If we rewrite Eq. (19.63) as $1 = \Omega_M + \Omega_R + \Omega_\Lambda - kc^2/(H^2R^2)$, we see that although the density ratios Ω_i, H, and R are separately time dependent, their sum is independent of time and hence the value of k cannot change. *The geometry of the universe is therefore fixed.*

If we now combine the Friedmann equation in the form of Eq. (19.59) with Eq. (19.56),

$$\dot{R}^2 + kc^2 = \frac{8\pi G}{3}R_0^2\left(\rho_{M,0}\frac{R_0}{R} + \rho_{R,0}\frac{R_0^2}{R^2} + \rho_{\Lambda,0}\frac{R^2}{R_0^2}\right) , \tag{19.65}$$

suggesting we work with a dimensionless scale factor $a(t) \equiv R(t)/R_0$. Thus, from Eq. (19.65)

$$\dot{a}^2 = \frac{8\pi G}{3}\left(\frac{\rho_{M,0}}{a} + \frac{\rho_{R,0}}{a^2} + \rho_{\Lambda,0}a^2\right) - \frac{kc^2}{R_0^2} . \tag{19.66}$$

We can massage this equation a bit, by multiplying and dividing by H_0^2. We obtain

$$\dot{a}^2 = H_0^2\left(\frac{\Omega_{M,0}}{a} + \frac{\Omega_{R,0}}{a^2} + \Omega_{\Lambda,0}a^2 - \Omega_{k,0}\right) , \tag{19.67}$$

where $\Omega_{i,0} \equiv \rho_{i,0}/\rho_{C,0}$ and $\Omega_{k,0} \equiv kc^2/(R_0^2H_0^2)$. From (33.21), $\Omega_{k,0} = \Omega_{M,0} + \Omega_{R,0} + \Omega_{\Lambda,0} - 1$. Equation (19.67), which is the Friedmann equation in disguised form, is a differential equation for the future of the universe! How's that for hubris? The future of the universe in a single equation. Note that the future is prescribed by the present values of the cosmological parameters.

Present-day cosmology is concerned with determining the values of $(H_0, \Omega_{M,0}, \Omega_{R,0}, \Omega_{\Lambda,0})$. We do not have the space here to discuss the measurements that lead to our knowledge of these numbers—this concludes our brief look at cosmology. Cosmological models are obtained from Eq. (19.67) under various scenarios concerning the parameters involved.

SUMMARY

- An "instant of time" in GR is a spacelike hypersurface. In the comoving reference frame the spatial coordinates are constant on SHs. In this frame the four-velocity $U^\mu = (c, 0, 0, 0)$. Free particles move along geodesics with the four-velocity orthogonal to SHs. By the cosmological principle, we want SHs to "look the same" everywhere, and hence we choose them as maximally symmetric spaces. The metric for such a three-dimensional space has the form

$$(\mathrm{d}\sigma)^2 = \frac{(\mathrm{d}r)^2}{1 - Kr^2} + r^2((\mathrm{d}\theta)^2 + \sin\theta(\mathrm{d}\phi)^2) \, .$$

  where K is a constant.

- The Friedmann-Robertson-Walker (FRW) model is a metric for relativistic cosmology with $(\mathrm{d}s)^2 = -(c\mathrm{d}t)^2 + S^2(t)(\mathrm{d}\sigma)^2$, where $S(t)$ is an unknown time-dependent scale factor. The FRW metric builds in the cosmological principle through the use of the line element of the maximally symmetric three-space, $(d\sigma)^2$, and the observed expansion of the universe through the scale factor $S(t)$. Because the curvature K is a constant (cosmological principle), it is either positive, negative, or zero. There are thus three possible types of spatial geometry for the universe, each of constant curvature. The radial coordinate and $S(t)$ can be rescaled so that they absorb the magnitude of K. The rescaled line element is

$$(\mathrm{d}s)^2 = -(c\mathrm{d}t)^2 + R^2(t)\left[\frac{(\mathrm{d}r)^2}{1 - kr^2} + r^2((\mathrm{d}\theta)^2 + \sin\theta(\mathrm{d}\phi)^2)\right] ,$$

  where $k = (1, 0, -1)$. This is the standard metric of relativistic cosmology.

- The spatial geometry associated with $k = 1$ has positive curvature, finite volume (closed geometry), and can be seen as a three-dimensional sphere embedded in four-dimensional Euclidean space. The geometry for $k = 0$ is flat three-dimensional Euclidean space. The geometry for $k = -1$ has negative curvature, and can be seen as a hyperboloid embedded in a four-dimensional Minkowskian space (not spacetime) with metric signature $(- + ++)$. This is an open space (infinite volume). The spatial geometry of the universe is an experimental issue.

- The Friedmann equations are differential equations for the scale factor $R(t)$ that result when the FRW metric is combined with Einstein's equation and $T_{\mu\nu}$ for the perfect fluid.

- Hubble's law emerges naturally from the Friedmann model, with the Hubble "constant" (parameter) given by $H = \dot{R}/R$. The Friedmann model also naturally accounts for the cosmological redshift.

Invariance of the wave equation

WE examine how the wave equation $\partial^2/\partial x^2 - (1/c^2)\partial^2/\partial t^2$ transforms first under the general transformation equations $x' = x'(x,t)$ and $t' = t'(x,t)$, and then ask what linear transformation equations preserve its form.

By the rules of calculus,

$$\frac{\partial}{\partial x} = \frac{\partial x'}{\partial x}\frac{\partial}{\partial x'} + \frac{\partial t'}{\partial x}\frac{\partial}{\partial t'} \qquad \frac{\partial}{\partial t} = \frac{\partial x'}{\partial t}\frac{\partial}{\partial x'} + \frac{\partial t'}{\partial t}\frac{\partial}{\partial t'} .$$

The second derivative is surprisingly complicated:

$$\begin{aligned}
\frac{\partial^2}{\partial x^2} &= \frac{\partial}{\partial x}\left(\frac{\partial x'}{\partial x}\frac{\partial}{\partial x'} + \frac{\partial t'}{\partial x}\frac{\partial}{\partial t'}\right) = \frac{\partial^2 x'}{\partial x^2}\frac{\partial}{\partial x'} + \frac{\partial x'}{\partial x}\frac{\partial}{\partial x}\frac{\partial}{\partial x'} + \frac{\partial^2 t'}{\partial x^2}\frac{\partial}{\partial t'} + \frac{\partial t'}{\partial x}\frac{\partial}{\partial x}\frac{\partial}{\partial t'} \\
&= \left(\frac{\partial^2 x'}{\partial x^2}\frac{\partial}{\partial x'} + \frac{\partial^2 t'}{\partial x^2}\frac{\partial}{\partial t'}\right) + \frac{\partial x'}{\partial x}\left[\frac{\partial x'}{\partial x}\frac{\partial}{\partial x'} + \frac{\partial t'}{\partial x}\frac{\partial}{\partial t'}\right]\frac{\partial}{\partial x'} + \frac{\partial t'}{\partial x}\left[\frac{\partial x'}{\partial x}\frac{\partial}{\partial x'} + \frac{\partial t'}{\partial x}\frac{\partial}{\partial t'}\right]\frac{\partial}{\partial t'} \\
&= \left(\frac{\partial^2 x'}{\partial x^2}\frac{\partial}{\partial x'} + \frac{\partial^2 t'}{\partial x^2}\frac{\partial}{\partial t'}\right) + \left(\frac{\partial x'}{\partial x}\right)^2\frac{\partial^2}{\partial x'^2} + 2\frac{\partial x'}{\partial x}\frac{\partial t'}{\partial x}\frac{\partial^2}{\partial t'\partial x'} + \left(\frac{\partial t'}{\partial x}\right)^2\frac{\partial^2}{\partial t'^2} .
\end{aligned}$$

A similar equation holds for $\partial^2/\partial t^2$—let $x \to t$; I won't write it down. Now assemble the wave equation in one system, and see how it compares with that in the other.

$$\begin{aligned}
\frac{\partial^2}{\partial x^2} - \frac{1}{c^2}\frac{\partial^2}{\partial t^2} =& \left[\left(\frac{\partial x'}{\partial x}\right)^2 - \frac{1}{c^2}\left(\frac{\partial x'}{\partial t}\right)^2\right]\frac{\partial^2}{\partial x'^2} - \frac{1}{c^2}\left[\left(\frac{\partial t'}{\partial t}\right)^2 - c^2\left(\frac{\partial t'}{\partial x}\right)^2\right]\frac{\partial^2}{\partial t'^2} \\
&+ \left(\frac{\partial^2 x'}{\partial x^2} - \frac{1}{c^2}\frac{\partial^2 x'}{\partial t^2}\right)\frac{\partial}{\partial x'} + \left(\frac{\partial^2 t'}{\partial x^2} - \frac{1}{c^2}\frac{\partial^2 t'}{\partial t^2}\right)\frac{\partial}{\partial t'} \\
&+ 2\left(\frac{\partial t'}{\partial x}\frac{\partial x'}{\partial x} - \frac{1}{c^2}\frac{\partial x'}{\partial t}\frac{\partial t'}{\partial t}\right)\frac{\partial^2}{\partial x'\partial t'} .
\end{aligned} \tag{A.1}$$

Now restrict transformations to be *linear*, apropos of SR, where all second derivatives vanish. Equation (A.1) simplifies:

$$\begin{aligned}
\frac{\partial^2}{\partial x^2} - \frac{1}{c^2}\frac{\partial^2}{\partial t^2} =& \left[\left(\frac{\partial x'}{\partial x}\right)^2 - \frac{1}{c^2}\left(\frac{\partial x'}{\partial t}\right)^2\right]\frac{\partial^2}{\partial x'^2} - \frac{1}{c^2}\left[\left(\frac{\partial t'}{\partial t}\right)^2 - c^2\left(\frac{\partial t'}{\partial x}\right)^2\right]\frac{\partial^2}{\partial t'^2} \\
&+ 2\left(\frac{\partial t'}{\partial x}\frac{\partial x'}{\partial x} - \frac{1}{c^2}\frac{\partial x'}{\partial t}\frac{\partial t'}{\partial t}\right)\frac{\partial^2}{\partial x'\partial t'} .
\end{aligned} \tag{A.2}$$

Galilean transformation

Under the Galilean transformation, for two frames in relative motion with speed v along their common x-axis, $x' = x - vt$ and $t' = t$. Thus, $\partial x'/\partial x = 1$, $\partial x'/\partial t = -v$, $\partial t'/\partial t = 1$, and $\partial t'/\partial x = 0$, and the wave equation transforms, from Eq. (A.2),

$$\frac{\partial^2}{\partial x^2} - \frac{1}{c^2}\frac{\partial^2}{\partial t^2} = (1 - v^2/c^2)\frac{\partial^2}{\partial x'^2} - \frac{1}{c^2}\frac{\partial^2}{\partial t'^2} + \frac{2v}{c^2}\frac{\partial^2}{\partial x' \partial t'} \,. \tag{A.3}$$

The form of the wave equation is not preserved by the Galilean transformation.

Lorentz transformation

What linear spacetime coordinate transformation *does* preserve the wave equation? A linear transformation that preserves the origin of coordinates can be written as a matrix equation,

$$\begin{pmatrix} t' \\ x' \end{pmatrix} = \begin{pmatrix} a_{11} & a_{12} \\ a_{21} & a_{22} \end{pmatrix} \begin{pmatrix} t \\ x \end{pmatrix} \,. \tag{A.4}$$

Thus, $\partial t'/\partial t = a_{11}$, $\partial t'/\partial x = a_{12}$, $\partial x'/\partial t = a_{21}$, and $\partial x'/\partial x = a_{22}$. From Eq. (A.2), the wave equation will be invariant under Eq. (A.4) if

$$\begin{aligned}
\left(\frac{\partial x'}{\partial x}\right)^2 - \frac{1}{c^2}\left(\frac{\partial x'}{\partial t}\right)^2 &= a_{22}^2 - \frac{1}{c^2}a_{21}^2 = 1 \\
\left(\frac{\partial t'}{\partial t}\right)^2 - c^2\left(\frac{\partial t'}{\partial x}\right)^2 &= a_{11}^2 - c^2 a_{12}^2 = 1 \\
\frac{\partial t'}{\partial x}\frac{\partial x'}{\partial x} - \frac{1}{c^2}\frac{\partial x'}{\partial t}\frac{\partial t'}{\partial t} &= a_{12}a_{22} - \frac{1}{c^2}a_{21}a_{11} = 0 \,.
\end{aligned}$$

These are three equations in four unknowns. The first can be solved identically by setting $a_{22} = \cosh\phi$ and $a_{21} = c\sinh\phi$. Likewise, the second equation can be solved by setting $a_{11} = \cosh\theta$ and $ca_{12} = \sinh\theta$. The third equation then demands that $\sinh\theta\cosh\phi - \sinh\phi\cosh\theta = \sinh(\theta - \phi) = 0$, or $\phi = \theta$ for real values of these parameters. Equation (A.4) can thus be written in terms of a single parameter, θ,

$$\begin{pmatrix} t' \\ x' \end{pmatrix} = \begin{pmatrix} \cosh\theta & \sinh\theta/c \\ c\sinh\theta & \cosh\theta \end{pmatrix} \begin{pmatrix} t \\ x \end{pmatrix} \,.$$

This equation can be brought into symmetrical form by multiplying by c:

$$\begin{pmatrix} ct' \\ x' \end{pmatrix} = \begin{pmatrix} \cosh\theta & \sinh\theta \\ \sinh\theta & \cosh\theta \end{pmatrix} \begin{pmatrix} ct \\ x \end{pmatrix} = \cosh\theta \begin{pmatrix} 1 & \tanh\theta \\ \tanh\theta & 1 \end{pmatrix} \begin{pmatrix} ct \\ x \end{pmatrix} \,. \tag{A.5}$$

Noting the identity $\cosh\theta = (1 - \tanh^2\theta)^{-1/2}$, Eq. (A.5) is fully specified by $\tanh\theta$.

For frames in uniform relative motion, the origin of one frame, $x' = 0$, is equivalent to $x = vt$ in the other frame. We can therefore identify $\tanh\theta = -\beta$, where $\beta \equiv v/c$. Equation (A.5) is then equivalent to

$$\begin{pmatrix} ct' \\ x' \end{pmatrix} = \gamma \begin{pmatrix} 1 & -\beta \\ -\beta & 1 \end{pmatrix} \begin{pmatrix} ct \\ x \end{pmatrix} \,, \tag{A.6}$$

where $\gamma \equiv 1/\sqrt{1 - \beta^2}$. Equation (A.6) is the Lorentz transformation in its most elementary form. Equation (A.5) is the *hyperbolic form* of the Lorentz transformation.

The Doppler effect

We review the Doppler effect from the classical and relativistic perspectives.

Classical Doppler effect

A source of electromagnetic radiation emits a wave at time $t = 0$. After a time t, the wave has traveled a distance ct; point A in Fig. B.1. The source is in motion with speed $v \ll c$ to the right.

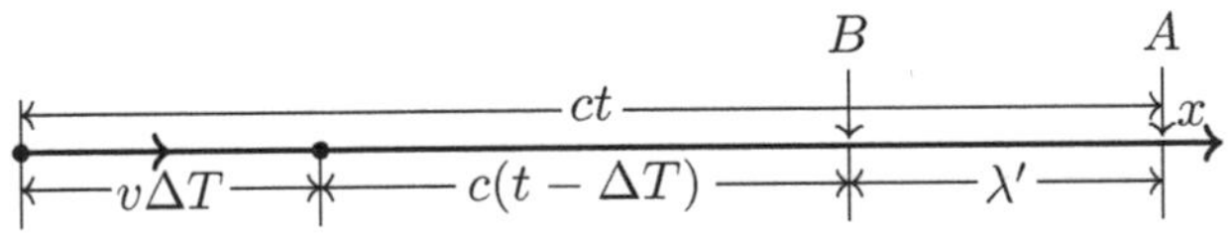

Figure B.1 Classical Doppler effect; source depicted as a black dot.

Let the frequency that the source emits radiation be f_e. After a time $\Delta T = f_e^{-1}$ the source has traveled a distance $v\Delta T$, whereupon it emits another signal. At time t, the second wave has traveled a distance $c(t - \Delta T)$; point B in Fig. B.1. The observed wavelength λ' is the distance between successive wavefronts, $\lambda' = ct - (v\Delta T + c(t - \Delta T)) = (c - v)\Delta T$. The time between reception of successive wavefronts is λ'/c, implying that the frequency of the received signal is $f_o = c/\lambda'$. The relation between the emitted frequency f_e and the observed frequency f_o is thus

$$f_o = \frac{c}{\lambda'} = \frac{c}{(c - v)\Delta T} = \frac{c}{c - v} f_e = \frac{1}{1 - \beta} f_e \,, \qquad \text{(this derivation)} \qquad \text{(B.1)}$$

where $\beta \equiv v/c$. Equation (B.1) would be the classical expression for the Doppler shift, except for a sign convention. In deriving Eq. (B.1) we've considered speed as positive for the emitter moving toward the receiver, as in Fig. B.1. The convention, however, is the opposite: Speed is considered positive when the receiver moves *away* from the source. Thus, let $\beta \to -\beta$ in Eq. (B.1). The classical Doppler factor is

$$f_o = \frac{1}{1 + \beta} f_e \,. \qquad (\beta > 0 \text{ for receiver moving away from source}) \qquad \text{(B.2)}$$

With Eq. (B.2), the pitch of the ambulance is higher when it moves toward you—in that case $\beta < 0$ for source approaching receiver. For acoustical waves, there *is* a "preferred reference frame."

Relativistic Doppler effect

There's a flaw in the above derivation. We shouldn't use $\Delta T = f_e^{-1}$, which originates in the frame of the *source*, to compute the distance traveled $v\Delta T$ in the frame of the *observer*. The time ΔT is the time between emission of photons *in the source frame*, the *proper time*. The time measured in the observer frame is $\gamma\Delta T$ (time dilation). By letting $\Delta T \to \gamma\Delta T$ in the above analysis, we obtain the relativistic Doppler effect:

$$f_o = \frac{1}{\gamma}\frac{1}{1+\beta}f_e = f_e\sqrt{\frac{1-\beta}{1+\beta}}\,. \tag{B.3}$$

Equation (B.3) can be derived by drawing Fig. B.1 as a spacetime diagram, shown in Fig. B.2. The source emits a signal, which in the frame of the receiver, travels a distance ct in time t; point A

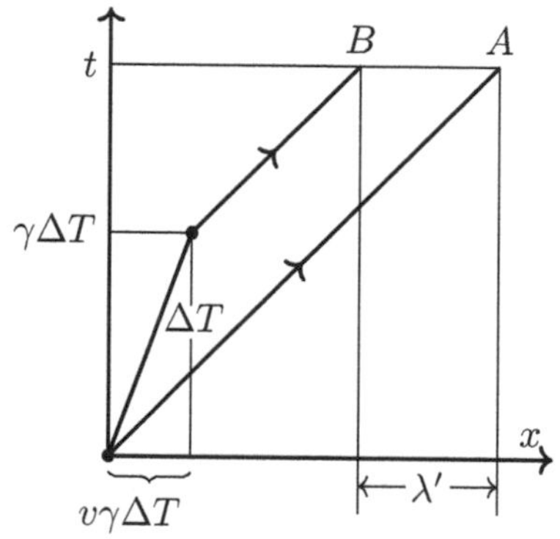

Figure B.2 Relativistic Doppler effect; source depicted as a black dot.

in Fig. B.2. After time ΔT *in the frame of the source*, it emits another signal. In the receiver frame, the time for the second emission event occurs at $\gamma\Delta T$, and therefore the distance the source travels is $v\gamma\Delta T$. In the receiver frame, the second photon has made it to point B in Fig. B.2 at time t. The wavelength in the receiver frame is

$$\lambda' = ct - [v\gamma\Delta T + c(t - \gamma\Delta T)] = \gamma\Delta T(c - v)\,. \tag{B.4}$$

Repeating the same analysis as in the previous derivation,

$$f_o = \frac{c}{\lambda'} = \frac{c}{\gamma\Delta T(c-v)} = \frac{1}{\gamma}\frac{1}{1-\beta}f_e\,, \tag{B.5}$$

where $f_e = (\Delta T)^{-1}$ is the proper frequency. Let $\beta \to -\beta$ (sign convention) in Eq. (B.5), and we have Eq. (B.3).

A high-precision experiment is required to distinguish the relativistic from the classical Doppler effects. Equations (B.2) and (B.3) differ only at $O(\beta^2)$: $\sqrt{(1-\beta)(1+\beta)^{-1}} \approx 1 - \beta + \frac{1}{2}\beta^2$ and $(1+\beta)^{-1} \approx 1 - \beta + \beta^2$. The Ives-Stilwell experiment (1938) confirmed to high accuracy the Doppler shift predicted by SR.[14]

APPENDIX C

Topics in linear algebra

THE theory of relativity draws upon properties of linear spaces that may not be part of the traditional mathematical preparation of physics students. In this appendix we review topics in linear algebra that are used throughout the book. Appropriate references for this material would be Gelfand [24] or Halmos.[28]

C.1 VECTOR SPACE

The traditional definition of vector is a quantity with direction and magnitude. Not all quantities with direction and magnitude, however, are vectors. Rotations through finite angles have magnitude (angle of rotation) and direction (axis of rotation), but do not constitute vectors because they do not *add* like vectors. The *prototype* vector is the displacement vector. Anything called vector must have the essential attributes of the prototype. Referring to Fig. C.1, vectors must combine like physical

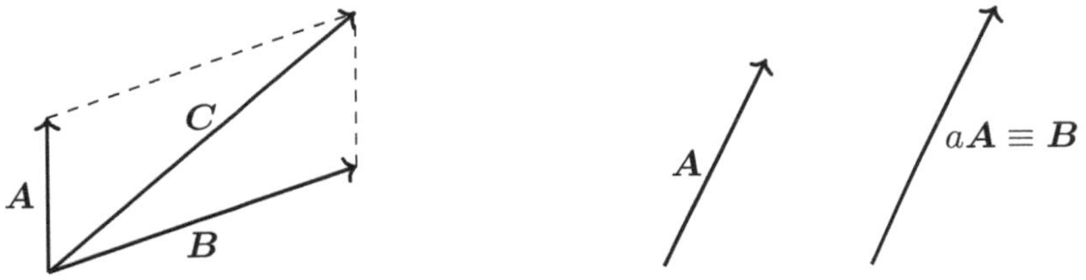

Figure C.1 Prototype properties of vectors: Vector addition and scalar multiplication.

displacements, with $\boldsymbol{C} = \boldsymbol{A} + \boldsymbol{B} = \boldsymbol{B} + \boldsymbol{A}$, and we should be able to multiply vectors by numbers to get new vectors, $\boldsymbol{B} = a\boldsymbol{A}$, where $a > 1$.

A *vector space* is a set $V = \{\phi, \chi, \psi, \ldots\}$ of objects called vectors having two main properties: The sum of vectors is a vector, $\psi + \phi = \chi$, and each vector when multiplied by a number (scalar) a is a vector, $a\phi = \psi$. The scalar is chosen from another set, the *field* $\mathbb{F}$—a special use of the word field—usually $\mathbb{R}$ or $\mathbb{C}$, real or complex numbers.[1] If a is a real (complex) number, V is referred to as a real (complex) vector space. Vector spaces are collections of objects that abstract our experience with elementary vectors, $\boldsymbol{C} = \boldsymbol{A} + \boldsymbol{B}$ and $\boldsymbol{B} = a\boldsymbol{A}$.

The full definition of vector space is that for any $\psi, \phi, \chi \in V$, the rules of addition are satisfied:

- $\psi + \phi = \phi + \psi$
- $\psi + (\phi + \chi) = (\psi + \phi) + \chi$
- there is a unique zero vector 0 such that $\psi + 0 = \psi$
- for every vector ψ there is an opposite vector $-\psi$ such that $\psi + (-\psi) = 0$;

[1]We use the special "blackboard bold" symbols: $\mathbb{R}$ denotes the set of real numbers and $\mathbb{C}$ the set of complex numbers.

as well as that of scalar multiplication: $(a, b \in \mathbb{F})$

- $a(\psi + \phi) = a\psi + a\phi$
- $(a + b)\psi = a\psi + b\psi$
- $(ab\psi) = a(b\psi)$
- $1\psi = \psi$.

We use the *abstract* symbols ϕ, χ, ψ because the rules for vector-space membership apply to a variety of mathematical objects.

C.2 EXAMPLES OF VECTOR SPACES

The concept of vector space is quite robust. As the following examples show, diverse mathematical objects meet the requirements of a vector space.

(1) The set of all n-tuples of real numbers $\psi \equiv (x_1, \cdots, x_n)$ is called n-dimensional Euclidean space, $\mathbb{R}^n$. To show that n-tuples form a vector space, the operations of vector addition and scalar multiplication must be specified. These operations are defined *componentwise*, with $\psi + \phi \equiv (x_1 + y_1, \cdots, x_n + y_n)$ and $a\psi \equiv (ax_1, \cdots, ax_n)$. The scalars for $\mathbb{R}^n$ must be real.

(2) The set of all *infinite sequences* of numbers $\psi = (x_1, \cdots, x_k, \cdots)$ having the property that $\sum_{k=1}^{\infty} |x_k|^2$ is finite, with addition and scalar multiplication defined componentwise, is a *sequence space*, l^2. Convergent infinite series do not naturally come to mind as "vectors," but they satisfy the requirements of a vector space. Minkowski's inequality, $(\sum_{k=1}^{n} |x_k + y_k|^2)^{1/2} \leq (\sum_{k=1}^{n} |x_k|^2)^{1/2} + (\sum_{k=1}^{n} |y_k|^2)^{1/2}$, guarantees that the sum of two elements of l^2 is in l^2.

(3) The set of all continuous functions of a real variable x, with addition and scalar multiplication defined pointwise, $(\psi + \phi)(x) = \psi(x) + \phi(x)$, and $(a\psi)(x) = a\psi(x)$, a *function space*.

(4) The set of all *square-integrable functions* $\psi(x)$ of a real variable x for which $\int |\psi(x)|^2 \, dx$ is finite (for specified limits of integration), with addition and scalar multiplication defined as in the previous example, is a vector space known as L^2.

C.3 DIMENSION AND BASIS: LINEAR INDEPENDENCE

The definition of vector space tells us nothing about the *dimension* of the space. The key notion for that purpose is *linear independence*. A set of vectors $\psi_1, \cdots, \psi_n$ is said to be linearly independent if $\sum_{k=1}^{n} a_k \psi_k = 0$ holds only for the trivial case $a_1 = \cdots = a_n = 0$. Otherwise, the set is *linearly dependent*. Linear independence means that *every nontrivial linear combination of vectors is different from zero*. Thus, no member of a linearly independent set can be expressed as a linear combination of the other vectors in the set. A vector space is n-dimensional if it contains n linearly independent vectors, but not $n+1$. A set of vectors $\psi_1, \cdots, \psi_n$ is said to *span the space* if any vector in the space can be expressed as a linear combination, $\psi = \sum_{k=1}^{n} a_k \psi_k$. A set of vectors $\psi_1, \cdots, \psi_n$ is a *basis* if it spans the space and is linearly independent. A vector space is n-dimensional if and only if it has a basis of n vectors. The numbers $a_1, \cdots, a_n$ in the linear combination $\psi = \sum_{k=1}^{n} a_k \psi_k$ are the *components* of ψ with respect to the basis. The components are *unique* with respect to a given basis; this follows from the linear independence of the basis. The expansion coefficients change, however, if we change the basis.

C.4 INNER PRODUCT

The inner product is a *mapping* $g : V \times V \to \mathbb{F}$. The notation $\times$ denotes the *Cartesian product*, which for sets C and D is a new set $C \times D$, the set of all *ordered pairs* (c, d) where $c \in C$ and $d \in D$. The notation $g : V \times V \to \mathbb{F}$ informs us that g associates members of the set $V \times V$ with elements of $\mathbb{F}$. An *inner product space* is a vector space having the additional structure of an inner product. The standard inner product used in quantum mechanics (for L^2) has the properties:

- $g(\phi, \psi) = g(\psi, \phi)^*$, conjugate symmetry
- $g(\phi, a\psi + b\chi) = ag(\phi, \psi) + bg(\phi, \chi)$, linearity in the second argument
- $g(\phi, \phi) \geq 0$. positive definite, equality if $\phi = 0$

The first two requirements imply $g(a\phi, \psi) = a^* g(\phi, \psi)$. For Minkowski space, however, a real space with an indefinite metric, the inner product is defined such that:

- $g(\phi, \psi) = g(\psi, \phi)$, symmetric
- $g(a\phi + \chi, \psi) = ag(\phi, \psi) + g(\chi, \psi)$, bilinear
- if $g(\phi, \psi) = 0$ for all ψ, then $\phi = 0$. nondegenerate

The last requirement guarantees that the inverse of the metric tensor exists. MS is not a vector space *per se*, but rather is an inner product space. MS has one timelike and three spacelike dimensions, concepts that require the Lorentz invariance of the spacetime separation, which is what specifies the inner product on MS. MS comes to the party equipped with an inner product; it doesn't have to be postulated.

C.5 DUAL VECTOR SPACE

We introduced the dual basis in Chapter 5 without first explaining the *dual space* for which the dual basis is a basis. A *linear transformation* is a mapping $f : V \to W$ between vector spaces V, W (having the same set of scalars), with the property $f(a\phi + b\psi) = af(\phi) + bf(\psi)$. For $W = \mathbb{R}$, f is referred to as a *linear functional*, $f : V \to \mathbb{R}$. The key idea is that *linear functionals form a vector space of their own*: $(f_1 + f_2)\psi \equiv f_1\psi + f_2\psi$ and $(af)\psi \equiv af(\psi)$. The set of all linear functionals on V is a vector space called the *dual space*, V^* (the asterisk has nothing to do with complex conjugation). The elements of V^* are called *dual vectors*.

If V^* is a vector space, it must have a basis. We defined the dual basis in Chapter 5 through the requirement that it be orthogonal to the coordinate basis, Eq. (5.10). We now make that more precise. For vectors $\{\boldsymbol{e}_j\}_{j=1}^n$ of the coordinate basis for an n-dimensional vector space V, the basis of V^*, $\{\boldsymbol{e}^k\}_{k=1}^n$, is defined through the requirement

$$\boldsymbol{e}^k(\boldsymbol{e}_j) = \delta_j^k \, . \tag{C.1}$$

We're changing notation here, away from the dot product in Eq. (5.10), to one that indicates more generally $\boldsymbol{e}^k$ acting on $\boldsymbol{e}_j$; $\boldsymbol{e}^k$, being an element of V^*, is a linear functional that maps vectors onto numbers (here zero or one). The dimensions of V^* and V are the same; V and V^* are *isomorphic* (a one-to-one correspondence between the elements of the sets). An arbitrary linear functional $\boldsymbol{\omega} \in V^*$ can be expressed as a linear combination, $\boldsymbol{\omega} = \omega_j \boldsymbol{e}^j$. For $\boldsymbol{v} \in V$, with $\boldsymbol{v} = v^k \boldsymbol{e}_k$, $\boldsymbol{\omega}(\boldsymbol{v}) = \omega_j v^k \boldsymbol{e}^j(\boldsymbol{e}_k) = \omega_j v^k \delta_k^j = \omega_j v^j$. If $\boldsymbol{v}$ is arranged as column vector, then $\boldsymbol{\omega}$ as a row vector is a linear functional under matrix multiplication. The dual basis specified by Eq. (C.1) is *unique*. Because every vector $\boldsymbol{u} = u^j \boldsymbol{e}_j \in V$ is uniquely determined by its components $(u^1, \cdots, u^n)$ with respect to the basis $\{\boldsymbol{e}_j\}$, there is a *unique* linear functional in V^* given by $\boldsymbol{\tau} = \tau_k \boldsymbol{e}^k$, where $\tau_k \equiv u^k$. There is an *isomorphism* between V and V^*, denoted $V \cong V^*$.

Example. In Dirac notation, functions $\psi(x)$, $\phi(x)$, elements of L^2, are denoted $|\psi\rangle$ and $|\phi\rangle$. The inner product in L^2 between $|\phi\rangle$ and $|\psi\rangle$ is denoted $\langle\psi|\phi\rangle$, shorthand for $\langle\psi|\phi\rangle \equiv \int \psi^*(x)\phi(x)\mathrm{d}x$. In a slight abuse of notation, the linear functional $\langle\psi|$ is defined such that $\langle\psi|\,(|\phi\rangle) = \langle\psi|\phi\rangle$. The linear functional associated with $\psi(x)$ is $\int \psi^*(x) \cdot \mathrm{d}x$, where $\cdot$ denotes a placeholder.

C.6 THE NATURAL PAIRING

The isomorphism between V and V^* depends on the basis. There is a basis-independent way of associating duals with vectors, the *natural pairing*. To bring this out, we need to prove a result about bilinear maps. Let $b : W \times V \to \mathbb{F}$ be a bilinear map.[2] For fixed $w_0 \in W$, $b(w_0, \cdot)$ is linear in V, $b(w_0, \cdot) : V \to \mathbb{F}$ and hence $b(w_0, \cdot) \in V^*$, so that b determines a map $W \to V^*$. Similarly, for fixed $v_0 \in V$, $b(\cdot, v_0)$ is a linear map $b(\cdot, v_0) : W \to \mathbb{F}$, so that $b(\cdot, v_0) \in W^*$ and b determines a map $V \to W^*$. A bilinear map b thus effects the isomorphisms $W \cong V^*$ and $V \cong W^*$. Now consider that the process of evaluating $\boldsymbol{\omega}(\boldsymbol{v})$, $\boldsymbol{\omega} \in V^*$ and $\boldsymbol{v} \in V$, is a function $f : V^* \times V \to \mathbb{F}$ such that $f(\boldsymbol{\omega}, \boldsymbol{v}) = \boldsymbol{\omega}(\boldsymbol{v})$ is a bilinear map. For fixed $\boldsymbol{\omega}$, $f : V \to \mathbb{F}$, and for fixed $\boldsymbol{v}$, $f : V^* \to \mathbb{F}$. The space V is thus isomorphic to the dual of V^*, $V \cong (V^*)^*$, known as the *double dual space*, V^{**}, the space of all linear mappings $V^* \to \mathbb{F}$. The natural isomorphism between V and V^{**} is achieved as follows.[3] For each vector $\boldsymbol{v} \in V$, associate the map in V^{**} on $\boldsymbol{\omega} \in V^*$ *to have the same value* as $\boldsymbol{\omega}(\boldsymbol{v})$. Note that we haven't given a symbol to denote the mappings in V^{**} that map $\boldsymbol{\omega} \in V^*$ onto $\mathbb{F}$. We are *identifying* these mappings with the vectors in V, $\boldsymbol{v} : V^* \to \mathbb{F}$ such that $\boldsymbol{v}(\boldsymbol{\omega}) \equiv \boldsymbol{\omega}(\boldsymbol{v})$. The identification will be used throughout this book,

$$\boldsymbol{v}(\boldsymbol{\omega}) = \boldsymbol{\omega}(\boldsymbol{v}) \, . \tag{C.2}$$

C.7 DERIVATIVE OF A DETERMINANT

An $n \times n$ matrix M with elements M_{ij} has determinant m obtained from its expansion in minors,

$$m \equiv \det M = \sum_{j=1}^{n} M_{ij} \left(-1\right)^{i+j} A_{ij} \tag{C.3}$$

where i labels any row and the cofactor A_{ij} is the determinant of the matrix obtained by striking the ith row and j^{th} column. The inverse M^{-1}, defined by $\sum_{j=1}^{n} \left(M^{-1}\right)_{ij} M_{jk} = \delta_{ik}$, has elements

$$\left(M^{-1}\right)_{ij} = m^{-1}(-1)^{i+j} A_{ji} \, . \tag{C.4}$$

Equation (C.3) facilitates taking the derivative $\partial m / \partial M_{ij}$, because the cofactor does not depend on M_{ij}: $\partial m / \partial M_{ij} = (-1)^{i+j} A_{ij}$. From Eq. (C.4), however, $A_{ij} = (-1)^{i+j} m \left(M^{-1}\right)_{ji}$ and thus

$$\frac{\partial m}{\partial M_{ij}} = m \left(M^{-1}\right)_{ji} \, . \tag{C.5}$$

The total derivative of the determinant with respect to the matrix elements is

$$\mathrm{d}m \equiv \sum_{i=1}^{n} \sum_{j=1}^{n} \frac{\partial m}{\partial M_{ij}} \mathrm{d}M_{ij} \, . \tag{C.6}$$

Combining Eqs. (C.5) and (C.6), we have the desired result (note the placement of indices)

$$\frac{1}{m} \mathrm{d}m = \sum_{i=1}^{n} \sum_{j=1}^{n} \left(M^{-1}\right)_{ji} \mathrm{d}M_{ij} \, . \tag{C.7}$$

Equation (C.7) is used in Chapter 14.

---

[2] Bilinearity is defined in Section 5.5.1.

[3] "Natural" means that the isomorphism depends only on V and V^* and not on the choice of bases in these spaces.

Topics in classical mechanics

WE review aspects of classical mechanics that are used throughout the book.

D.1 EULER-LAGRANGE EQUATION

The calculus of variations is concerned with finding the extremum of a *functional*, a mapping between *functions* and real numbers. A functional $J[y]$ (J acts on function $y(x)$ to produce a number) is often given in the form of a definite integral,

$$J[y] \equiv \int_{x_1}^{x_2} F(y(x))\mathrm{d}x \ ,$$

where the function F and the endpoints x_1, x_2 are known. The task is to find the *extremal function* $y(x)$ that makes $J[y]$ attain an extremum. Often F involves not just $y(x)$ but also the derivative $y' = \mathrm{d}y/\mathrm{d}x$, in which case F is denoted $F(y, y')$. We show that $J[y]$ attains an extremum when $y(x)$ satisfies Eq. (D.12), the Euler-Lagrange equation.

An example of how the type of problem posed by the calculus of variations could arise, consider, for fixed points (a, b) in the Euclidean plane, the path of shortest distance between them. The question can be formulated as which function $y(x)$ extremizes the integral[1]

$$J = \int_a^b \mathrm{d}s = \int_a^b \sqrt{(\mathrm{d}x)^2 + (\mathrm{d}y)^2} = \int_{x_a}^{x_b} \mathrm{d}x\sqrt{1 + (\mathrm{d}y/\mathrm{d}x)^2} = \int_{x_a}^{x_b} \mathrm{d}x\sqrt{1 + (y')^2} \ . \tag{D.1}$$

Clearly, $F(y') = \sqrt{1 + (y')^2}$. Pick a function $y(x)$, take its derivative, substitute into Eq. (D.1) and do the integral. Which function results in the smallest value of J?

The problem sounds fantastically general, too general perhaps to make progress. Plan on success, however: *Assume* an extremal function $y(x)$ exists. Determine $y(x)$ through the requirement that small *functional variations* away from the extremal function produce even smaller changes in the value of J. To do this, introduce a class of varied functions,

$$Y(x, \alpha) \equiv y(x) + \alpha\eta(x) \ , \tag{D.2}$$

[1] *Extremize* appears not to be an official word of English. We're using it to indicate attaining the extremum in some quantity, its minimum or maximum.

where α is a number and $\eta(x)$ is any function that vanishes at the endpoints, $\eta(x_1) = \eta(x_2) = 0$, but is otherwise arbitrary. The functional can then be parameterized (note the change in notation),

$$J(\alpha) = \int_{x_1}^{x_2} F\left(Y(x,\alpha), Y'(x,\alpha)\right) \mathrm{d}x \ , \tag{D.3}$$

where by assumption $J(\alpha = 0)$ is the extremum value. We expect that $J(\alpha)$ is a smooth function of α and is such that

$$\left.\frac{\partial J}{\partial \alpha}\right|_{\alpha=0} = 0 \ , \tag{D.4}$$

i.e., *to first order in small values of α there are only second-order changes in $J(\alpha)$*. When Eq. (D.4) is satisfied, J is said to be *stationary* with respect to small values of α; Eq. (D.4) is the *stationarity condition*. We show that Eq. (D.4) can be achieved for arbitrary $\eta(x)$.

Differentiate Eq. (D.3) (the limits of integration are independent of α):

$$\frac{\partial J}{\partial \alpha} = \int_{x_1}^{x_2} \frac{\partial}{\partial \alpha} F\left(Y(\alpha,x), Y'(\alpha,x)\right) \mathrm{d}x = \int_{x_1}^{x_2} \left(\frac{\partial F}{\partial y}\frac{\partial Y}{\partial \alpha} + \frac{\partial F}{\partial y'}\frac{\partial Y'}{\partial \alpha} \right) \mathrm{d}x \ . \tag{D.5}$$

The *infinitesimal variation*, $\delta y(x)$, is the variation in functional form *at a point* x,

$$\delta y(x) \equiv Y(x, \mathrm{d}\alpha) - y(x) = \frac{\partial Y}{\partial \alpha}\mathrm{d}\alpha = \eta(x)\mathrm{d}\alpha \ , \tag{D.6}$$

where we've used Eq. (D.2). The infinitesimal variation of $y'(x)$ is similarly defined

$$\delta y'(x) = \delta\left(\frac{\mathrm{d}y}{\mathrm{d}x}\right) \equiv Y'(x, \mathrm{d}\alpha) - y'(x) = \frac{\partial Y'}{\partial \alpha}\mathrm{d}\alpha = \eta'(x)\mathrm{d}\alpha = \frac{\mathrm{d}}{\mathrm{d}x}\left(\delta y\right) \ . \tag{D.7}$$

The variation of the derivative equals the derivative of the variation. Thus, *$\delta y'$ is not independent of δy*. The change in J under the variation is $\delta J \equiv (\partial J/\partial \alpha)\mathrm{d}\alpha$, and thus the stationarity condition is expressed by writing $\delta J = 0$. Multiplying Eq. (D.5) by $\mathrm{d}\alpha$ we have, using Eqs. (D.6) and (D.7):

$$\delta J = \int_{x_1}^{x_2} \left(\frac{\partial F}{\partial y}\delta y(x) + \frac{\partial F}{\partial y'}\delta y'(x) \right) \mathrm{d}x \ . \tag{D.8}$$

Equation (D.7) allows the second term in Eq. (D.8) to be integrated by parts:

$$\begin{aligned} \int_{x_1}^{x_2} \frac{\partial F}{\partial y'}\delta y' \mathrm{d}x &= \int_{x_1}^{x_2} \frac{\partial F}{\partial y'}\frac{\mathrm{d}}{\mathrm{d}x}(\delta y)\mathrm{d}x = \left.\frac{\partial F}{\partial y'}\delta y(x)\right|_{x_1}^{x_2} - \int_{x_1}^{x_2} \frac{\mathrm{d}}{\mathrm{d}x}\left(\frac{\partial F}{\partial y'}\right)\delta y(x)\mathrm{d}x \\ &= -\int_{x_1}^{x_2} \frac{\mathrm{d}}{\mathrm{d}x}\left(\frac{\partial F}{\partial y'}\right)\delta y(x)\mathrm{d}x \ , \end{aligned} \tag{D.9}$$

where the integrated part vanishes because $\delta y(x)$ vanishes at the endpoints. Combining Eq. (D.9) with Eq. (D.8), the change of J under the variation $\delta y(x)$ is

$$\delta J = \int_{x_1}^{x_2} \left[\frac{\partial F}{\partial y} - \frac{\mathrm{d}}{\mathrm{d}x}\left(\frac{\partial F}{\partial y'}\right) \right] \delta y(x)\mathrm{d}x \equiv \int_{x_1}^{x_2} \frac{\delta F}{\delta y}\delta y(x)\mathrm{d}x \ . \tag{D.10}$$

The terms in square brackets are known as the *functional derivative* of $F(y, y')$ or its *variational derivative*,

$$\frac{\delta F}{\delta y(x)} \equiv \frac{\partial F}{\partial y} - \frac{\mathrm{d}}{\mathrm{d}x}\left(\frac{\partial F}{\partial y'}\right) \ . \tag{D.11}$$

We want Eq. (D.10) to vanish for *any* variation $\delta y(x)$. The only way that can happen is if the functional derivative of F vanishes identically,

$$\frac{\delta F}{\delta y} = \frac{\partial F}{\partial y} - \frac{\mathrm{d}}{\mathrm{d}x}\left(\frac{\partial F}{\partial y'}\right) = 0\,. \tag{D.12}$$

Equation (D.12), the *Euler-Lagrange equation*, solves the problem of the calculus of variations by producing a differential equation to be satisfied by the extremal function,[2] $y(x)$.

The generalization to functionals of several functions is straightforward. The goal is to find the set of functions $\{y_i(x)\}_{i=1}^n$ such that

$$\delta J = \delta \int_{x_1}^{x_2} F\left(y_1(x),\cdots,y_n(x);y_1'(x),\cdots,y_n'(x)\right)\mathrm{d}x = 0\,.$$

Following the method set out above, one obtains the generalization of Eq. (D.10)

$$\delta J = \int_{x_1}^{x_2} \sum_{i=1}^{n} \frac{\delta F}{\delta y_i(x)}\delta y_i(x)\mathrm{d}x\,. \tag{D.13}$$

If the variations δy_i can be performed *independently*, the vanishing of Eq. (D.13) requires the vanishing of each of the terms in the integrand,

$$\frac{\delta F}{\delta y_i(x)} = \frac{\partial F}{\partial y_i} - \frac{\mathrm{d}}{\mathrm{d}x}\left(\frac{\partial F}{\partial y_i'}\right) = 0\,, \qquad (i=1,\cdots,n) \tag{D.14}$$

providing a differential equation to be satisfied by each of the functions $y_i(x)$.

D.2 HAMILTON'S PRINCIPLE

The problem posed by Newtonian dynamics of finding the trajectory $x(t)$ for given forces can be formulated as a problem in the calculus of variations. We seek a function $F = F(T,V)$ such that the condition

$$\delta \int_{t_1}^{t_2} F(T(\dot{x}),V(x))\mathrm{d}t = 0$$

is equivalent to the equation of motion, where a "clean" separation between $x(t)$ and $\dot{x}(t)$ has been assumed, with $T = T(\dot{x})$ and $V = V(x)$ the kinetic and potential energy functions of the particle. Substitute $F(T,V)$ into Eq. (D.12):

$$\frac{\partial F}{\partial V}\frac{\partial V}{\partial x} - \frac{\mathrm{d}}{\mathrm{d}t}\left(\frac{\partial F}{\partial T}\frac{\partial T}{\partial \dot{x}}\right) = 0\,. \tag{D.15}$$

Equation (D.15) reproduces Newton's second law ($\dot{p} = -\partial V/\partial x$) if we identify $p \equiv \partial T/\partial \dot{x}$ and if F is such that $\partial F/\partial T = -\partial F/\partial V =$ a nonzero constant. Take $F = T - V \equiv L$, where F has been renamed L, the *Lagrangian function*.

The Lagrangian can be extended to a point mass having n *independent degrees of freedom* described by *generalized coordinates* q^i, $i = 1,\cdots,n$ (the use of superscripts is deliberate; see Section 1.3). The space spanned by these coordinates is known as *configuration space*. The distance between infinitesimally separated points in configuration space is expressed in terms of the metric tensor, $(\mathrm{d}s)^2 = g_{ij}\mathrm{d}q^i\mathrm{d}q^j$ (the metric tensor is explained in Chapter 5, as well as the summation

[2] We've relied on a theorem: If $\int_{x_1}^{x_2} \eta(x)\phi(x)\mathrm{d}x = 0$ for all functions $\eta(x)$ which vanish on the boundary and are continuous together with their first two derivatives, then $\phi(x) = 0$ identically.[83, p185]

convention). Introducing time as a parameter,[3] the kinetic energy $T = \frac{1}{2}m(\mathrm{d}s/\mathrm{d}t)^2 = \frac{1}{2}mg_{ij}\dot{q}^i\dot{q}^j$. The Lagrangian function is then

$$L(q^k, \dot{q}^k) = \tfrac{1}{2}mg_{ij}\dot{q}^i\dot{q}^j - V(q^j)\,. \tag{D.16}$$

The time derivatives $\dot{q}^i$ are called *generalized velocities*. The point with all this *generalized* business is that generalized coordinates need not have the dimension of length—they're whatever variables (angles, for example) it takes to describe the geometric configuration of the mass, possibly taking into account constraints on the motion. The kinetic energy T (wherein the metric tensor appears) is a *kinematic* quantity, descriptive of the geometry of motion, while the potential energy V is a *dynamic* quantity, expressing the *coupling* of a particle to its environment.

Hamilton's principle is that of all the paths $q^i(t)$ a point mass *could* take through configuration space between fixed endpoints, the actual path is the one for which the *action integral* is stationary,

$$\delta S \equiv \delta \int_{t_1}^{t_2} L(q^k, \dot{q}^k)\mathrm{d}t = 0\,. \tag{D.17}$$

The time development of each coordinate q^i follows from Eq. (D.14),

$$\frac{\delta L}{\delta q^k} = \frac{\partial L}{\partial q^k} - \frac{\mathrm{d}}{\mathrm{d}t}\left(\frac{\partial L}{\partial \dot{q}^k}\right) = 0\,. \qquad (k = 1, \cdots, n) \tag{D.18}$$

Along the path of stationary action, the Lagrangian is *variationally constant* ($\delta L/\delta q^k = 0$), invariant under small functional variations δq^k *taken for the motion as a whole* between fixed endpoints.

Now, the metric tensor can itself be a function of coordinates, $g_{ij} = g_{ij}(q^k)$, a *tensor field*, e.g., the metric tensor for the spherical coordinate system, Eq. (5.5). The kinetic energy is then more generally a function of generalized velocities *and* coordinates, $T = T(\dot{q}^i, q^j)$. Using Eq. (D.16):

$$\frac{\partial L}{\partial q^k} = \frac{\partial T}{\partial q^k} - \frac{\partial V}{\partial q^k} = \frac{1}{2}m\frac{\partial g_{ij}}{\partial q^k}\dot{q}^i\dot{q}^j - \frac{\partial V}{\partial q^k}\,. \tag{D.19}$$

The time derivative in Eq. (D.18) must take into account the position dependence of the metric tensor. Starting from Eq. (D.16),

$$\frac{\mathrm{d}}{\mathrm{d}t}\left(\frac{\partial L}{\partial \dot{q}^k}\right) = mg_{kj}\ddot{q}^j + m\dot{q}^j\dot{q}^l\frac{\partial g_{kj}}{\partial q^l} = mg_{kj}\ddot{q}^j + \frac{m}{2}\dot{q}^j\dot{q}^l\left(\frac{\partial g_{kj}}{\partial q^l} + \frac{\partial g_{kl}}{\partial q^j}\right)\,, \tag{D.20}$$

where the second step is a cosmetic operation to make the expression symmetric in the indices (j, l) (which we're free to do because (j, l) are dummy indices and the metric tensor is symmetric in its indices). Combining Eqs. (D.20) and (D.19) with Eq. (D.18), we have the equation of motion

$$g_{kj}\ddot{q}^j + \frac{1}{2}\left[\frac{\partial g_{kj}}{\partial q^l} + \frac{\partial g_{kl}}{\partial q^j} - \frac{\partial g_{lj}}{\partial q^k}\right]\dot{q}^l\dot{q}^j = -\frac{1}{m}\frac{\partial V}{\partial q^k}\,. \tag{D.21}$$

Equation (D.21) can be simplified with the inverse metric tensor g^{ij}, defined by $g^{lk}g_{kj} = \delta^l_j$, Eq. (5.18). Multiplying Eq. (D.21) by g^{ik} and summing over an index, the path of stationary action is described by a set of coupled second-order differential equations:

$$\ddot{q}^i + \Gamma^i_{jl}\dot{q}^l\dot{q}^j = -\frac{1}{m}g^{ik}\frac{\partial V}{\partial q^k}\,; \qquad (i = 1, \cdots, n) \tag{D.22}$$

[3] In pre-relativistic physics it's assumed that all paths in configuration space can be parameterized by the same time. That assumption must be addressed in relativistic applications of the calculus of variations.

where the terms

$$\Gamma^i_{jl} \equiv \frac{1}{2} g^{ik} \left[\frac{\partial g_{kj}}{\partial q^l} + \frac{\partial g_{kl}}{\partial q^j} - \frac{\partial g_{lj}}{\partial q^k} \right] \tag{D.23}$$

are the *Christoffel symbols*, which involve derivatives of the metric tensor. The Christoffel symbols play an important role in GR.

Equation (D.22) is Newton's second law $a^i = F^i/m$, as it appears in an arbitrary coordinate system. It becomes for $V = 0$ the differential equation for a *geodesic curve* (Chapter 14), the shortest distance between two points. How do we know that? Consider that the arc length can be written[4]

$$\int \mathrm{d}s = \int \frac{\mathrm{d}s}{\mathrm{d}t}\mathrm{d}t = \int \sqrt{g_{ij}\dot{q}^i\dot{q}^j}\,\mathrm{d}t\,. \tag{D.24}$$

The requirement $\delta \int \mathrm{d}s = 0$ leads to Eq. (D.22) with $V = 0$. *The path of extremal action for a free particle is a geodesic curve.* Is it obvious we have a minimum and not a maximum? Equation (D.12) specifies the condition for a functional to attain an extremum, not whether it's a minimum or a maximum. For the extremal path to represent the minimum of a functional, the *Legendre condition* [83, p214] must be satisfied that the matrix of second derivatives $\partial^2 F/\partial\dot{q}^i\partial\dot{q}^j$ be positive definite, which for Lagrangians in the form of Eq. (D.16) is the requirement that the metric be positive definite.

D.3 HAMILTONIAN EQUATIONS OF MOTION

Starting from Eq. (D.16), we define the *canonical momentum* p_k associated with the coordinate q^k:

$$p_k \equiv \frac{\partial L}{\partial \dot{q}^k} = m g_{kj}\dot{q}^j \equiv m\dot{q}_k\,, \tag{D.25}$$

where the use of subscripts is deliberate (see Chapter 5 for index manipulations). For conservative forces there is no difference between the canonical momentum and the *kinetic momentum*, $\partial T/\partial \dot{q}^k$. The distinction is useful if we allow velocity-dependent forces, $V = V(q^k, \dot{q}^k)$.

The *Hamiltonian function* is the *Legendre transformation* of the Lagrangian function:

$$H \equiv p_k\dot{q}^k - L(q^j, \dot{q}^j)\,. \tag{D.26}$$

The Legendre transformation is an equivalent way of specifying *convex functions*, not in terms of point-wise values (as usual), but in terms of its tangents at a point.[5][84, p61] The Legendre transformation shifts emphasis away from the dependence of a function on one variable, here the generalized velocities $\dot{q}^k$, in favor of the slope of the function,[6] here the generalized momenta, p_k. The Hamiltonian is a function of canonical momenta and generalized coordinates, $H = H(p_k, q^k)$, as can be seen from the differential of Eq. (D.26):

$$\delta H = p_k\delta\dot{q}^k + \dot{q}^k\delta p_k - \frac{\partial L}{\partial q^k}\delta q^k - \frac{\partial L}{\partial \dot{q}^k}\delta\dot{q}^k = \dot{q}^k\delta p_k - \dot{p}_k\delta q^k\,,$$

where we've used Eqs. (D.18) and (D.25), and that variations obey the product rule: $\delta(f(x)g(x)) = f(x)\delta g(x) + g(x)\delta f(x)$, which follows from Eq. (D.6). By inverting Eq. (D.25), $\dot{q}^l = g^{lk}p_k/m$, we find from Eq. (D.26) that

$$H(p_k, q^k) = \frac{1}{2m}p_kp^k + V(q^j)\,. \tag{D.27}$$

[4]Equation (D.24) assumes that $g_{ij}\dot{q}_i\dot{q}_j > 0$, which is true for a positive-definite metric or for spacelike separated events. For timelike separated events, $\sqrt{g_{ij}\dot{q}_i\dot{q}_j}$ is replaced with $\sqrt{-g_{ij}\dot{q}_i\dot{q}_j}$; Eq. (4.14).

[5]The Lagrangian in the form of Eq. (D.16) is a convex function of the generalized velocities if the metric tensor is positive definite and if $m > 0$. A function f is concave if $(-f)$ is convex.

[6]An example from thermodynamics is the shift away from the dependence of the internal energy function $U = U(S, V)$ on entropy S (and volume V), to an equivalent function, its Legendre transformation, the Helmholtz free energy $F \equiv U - TS$, which involves the slope of $U(S, V)$, $\partial U/\partial S = T$ (absolute temperature), and hence $F = F(T, V)$.

Differentiate Eq. (D.26) with respect to p_k,

$$\frac{\partial H}{\partial p_k} = \dot{q}^k , \qquad (k = 1, \cdots, n) \tag{D.28}$$

and again with respect to q^k,

$$\frac{\partial H}{\partial q^k} = -\frac{\partial L}{\partial q^k} = -\dot{p}_k , \qquad (k = 1, \cdots, n) \tag{D.29}$$

where the latter equality follows from Lagrange's equations. Equations (D.28) and (D.29) are the *Hamiltonian equations of motion*. They are an equivalent formulation of classical mechanics: Rather than n second-order differential equations, as in Eq. (D.22), there are $2n$ first-order differential equations to describe the motion.

The Legendre transformation is *self-dual*: the transformation applied twice reproduces the original function. With that noted, the Lagrangian is the Legendre transformation of the Hamiltonian,

$$L = p_k \dot{q}^k - H(p_k, q^k) . \tag{D.30}$$

By varying Eq. (D.30) with respect to p_k,

$$\delta L = \left(\dot{q}^k - \frac{\partial H}{\partial p_k} \right) \delta p_k = 0 , \tag{D.31}$$

which follows from Eq. (D.28). Equation (D.31) implies that *the canonical momenta are independent of q^k and $\dot{q}^k$*. Arbitrary variations of the set of quantities $\{p_k\}$ have, by Eq. (D.31), no influence on the variation of L, implying that such variations (in p_k) have no influence on the action integral of L over time. *The canonical momenta $\{p_k\}$ are a second set of independent variables*, which they must be to have $2n$ first-order differential equations to describe the dynamics of the system.

The Hamiltonian is a constant of the motion, as can be shown by taking the total time derivative of $H(p_k, q^k)$:

$$\frac{\mathrm{d}H}{\mathrm{d}t} = \frac{\partial H}{\partial p_k}\dot{p}_k + \frac{\partial H}{\partial q^k}\dot{q}^k = -\frac{\partial H}{\partial p_k}\frac{\partial H}{\partial q^k} + \frac{\partial H}{\partial q^k}\frac{\partial H}{\partial p_k} = 0 , \tag{D.32}$$

where we have used Eqs. (D.28) and (D.29). That constant can often be identified with the energy, but not always.

D.4 D'ALEMBERT'S PRINCIPLE

We noted in Section 1.6.3 that d'Alembert's principle implies Hamilton's principle; we now show that. D'Alembert's principle is that $\boldsymbol{F} - m\boldsymbol{a} = 0$: An object in motion is *as if* it's in static equilibrium between applied forces $\boldsymbol{F}$ and the inertial reaction force $-m\boldsymbol{a}$. The *principle of virtual work* is that the equilibrium configuration of a mechanical system is such that the work done by all forces under *virtual displacements*[7] $\delta\boldsymbol{r}$ about that configuration vanishes. Thus, $(\boldsymbol{F} - m\boldsymbol{a}) \cdot \delta\boldsymbol{r} = 0$. Multiply by $\mathrm{d}t$ and integrate:

$$0 = \int_{t_1}^{t_2} \left(\boldsymbol{F} - \frac{\mathrm{d}}{\mathrm{d}t}(m\boldsymbol{v}) \right) \cdot \delta\boldsymbol{r}\mathrm{d}t = \int_{t_1}^{t_2} \boldsymbol{F} \cdot \delta\boldsymbol{r}\mathrm{d}t - \int_{t_1}^{t_2} \frac{\mathrm{d}}{\mathrm{d}t}(m\boldsymbol{v}) \cdot \delta\boldsymbol{r}\mathrm{d}t . \tag{D.33}$$

Assume that $\boldsymbol{F}$ is derivable from a scalar function V; thus $\boldsymbol{F} \cdot \delta\boldsymbol{r} = -\delta V$. The first integral on the right side of Eq. (D.33) can therefore be written

$$\int_{t_1}^{t_2} \boldsymbol{F} \cdot \delta\boldsymbol{r}\mathrm{d}t = -\int_{t_1}^{t_2} \delta V \mathrm{d}t = -\delta \int_{t_1}^{t_2} V \mathrm{d}t , \tag{D.34}$$

[7]A virtual displacement, denoted $\delta\boldsymbol{r}$, is a kind of "mathematical experiment" of *possible* variations in the coordinates of a system, consistent with any constraints, *at a fixed time*; it's not the *actual* displacement of the system $\mathrm{d}\boldsymbol{r}$ in the time $\mathrm{d}t$.

where we've used that *the variation of a definite integral equals the definite integral of the variation*: (referring to Eq. (D.6))

$$\int_a^b \delta y(x)\mathrm{d}x = \int_a^b \left(Y(x,\mathrm{d}\alpha) - y(x)\right)\mathrm{d}x = \int_a^b Y(x,\mathrm{d}\alpha)\mathrm{d}x - \int_a^b y(x)\mathrm{d}x \equiv \delta \int_a^b y(x)\mathrm{d}x \,.$$

The second integral on the right side of Eq. (D.33) is set up to integrate by parts:

$$-\int_{t_1}^{t_2} \frac{\mathrm{d}}{\mathrm{d}t}(m\boldsymbol{v})\boldsymbol{\cdot}\delta\boldsymbol{r}\mathrm{d}t = \int_{t_1}^{t_2} m\boldsymbol{v}\boldsymbol{\cdot}\delta\dot{\boldsymbol{r}}\mathrm{d}t = \frac{1}{2}\int_{t_1}^{t_2}\delta(m\boldsymbol{v}\boldsymbol{\cdot}\boldsymbol{v})\mathrm{d}t = \int_{t_1}^{t_2}\delta T\mathrm{d}t = \delta\int_{t_1}^{t_2} T\mathrm{d}t \,, \tag{D.35}$$

where the first equality follows from Eq. (D.9) and in the final equality we've used that the operations of integration and variation commute. Combining Eqs. (D.34) and (D.35) with Eq. (D.33), we have Hamilton's principle, $\delta\int(T-V)\mathrm{d}t = 0$. D'Alembert's principle holds at each instant of time; Hamilton's principle involves the integral over time of the motion taken as a whole.

D.5 CHARGED PARTICLE

We consider a particle coupled to electric and magnetic fields $\boldsymbol{E}$ and $\boldsymbol{B}$. This case is atypical in that the Lorentz force is not conservative,[8] yet—as we show—it can be derived from a scalar function $V(q^k, \dot{q}^k)$ such that

$$F_k = q\left(\boldsymbol{E} + \boldsymbol{v}\times\boldsymbol{B}\right)_k = -\frac{\partial V}{\partial q^k} + \frac{\mathrm{d}}{\mathrm{d}t}\left(\frac{\partial V}{\partial \dot{q}^k}\right) \equiv -\frac{\delta V}{\delta q^k}\,, \tag{D.36}$$

where q is the charge, $\boldsymbol{v}$ is the velocity, F_k is the component of the force in the system of generalized coordinates (the *generalized force*). *The Lagrange equations have the structure of d'Alembert's principle*, that the applied force is balanced by the inertial force:

$$\frac{\delta L}{\delta q^k} = \frac{\delta T}{\delta q^k} - \frac{\delta V}{\delta q^k} = 0 \implies F_k - \left(-\frac{\delta T}{\delta q^k}\right) = 0\,,$$

where the acceleration terms (reaction force) are given by

$$-\frac{\delta T}{\delta q^k} = \frac{\mathrm{d}}{\mathrm{d}t}\left(\frac{\partial T}{\partial \dot{q}^k}\right) - \frac{\partial T}{\partial q^k}\,.$$

With $T = \frac{m}{2}g_{ij}\dot{q}^i\dot{q}^j$, the terms implied by $\delta T/\delta q^k$ appear on the left side of Eq. (D.22) involving the Christoffel symbols.

The homogeneous Maxwell equations can be "solved" by introducing scalar and vector potential functions ϕ and $\boldsymbol{A}$ such that $\boldsymbol{B} = \boldsymbol{\nabla}\times\boldsymbol{A}$ and $\boldsymbol{E} = -\boldsymbol{\nabla}\phi - \partial\boldsymbol{A}/\partial t$. An equivalent expression for the Lorentz force is therefore

$$\boldsymbol{F} = q\left(-\boldsymbol{\nabla}\phi - \frac{\partial\boldsymbol{A}}{\partial t} + \boldsymbol{v}\times\boldsymbol{\nabla}\times\boldsymbol{A}\right)\,. \tag{D.37}$$

Let's massage the term in Eq. (D.37) involving $\boldsymbol{v}$:

$$\begin{aligned}\left(\boldsymbol{v}\times\boldsymbol{\nabla}\times\boldsymbol{A}\right)_i &= \epsilon_{ijk}v^j\left(\boldsymbol{\nabla}\times\boldsymbol{A}\right)^k = \epsilon_{ijk}\epsilon^{klm}v^j\partial_l A_m = \left(\delta_i^l\delta_j^m - \delta_j^l\delta_i^m\right)v^j\partial_l A_m \\ &= v^m\partial_i A_m - v^l\partial_l A_i \equiv v^m\partial_i A_m - \left(\boldsymbol{v}\cdot\boldsymbol{\nabla}\right)A_i\,,\end{aligned} \tag{D.38}$$

[8]Is it true that $\boldsymbol{\nabla}\times(\boldsymbol{E}+\boldsymbol{v}\times\boldsymbol{B}) = 0$?

where we've used the properties of the Levi-Civita symbol (Section 5.8.2). Now, the vector potential is a function of space and time, $\boldsymbol{A} = \boldsymbol{A}(\boldsymbol{r}, t)$, and thus its total time derivative can be written

$$\frac{\mathrm{d}\boldsymbol{A}}{\mathrm{d}t} = \frac{\partial \boldsymbol{A}}{\partial t} + (\boldsymbol{v} \cdot \boldsymbol{\nabla}) \boldsymbol{A} . \tag{D.39}$$

Combining Eqs. (D.38) and (D.39), we have the (not particularly pretty) identity

$$\left(\boldsymbol{v} \times \boldsymbol{\nabla} \times \boldsymbol{A} - \frac{\partial \boldsymbol{A}}{\partial t} \right)_i = v^m \partial_i A_m - \frac{\mathrm{d}A_i}{\mathrm{d}t} .$$

An equivalent way of writing the components of the Lorentz force is therefore

$$q \left(-\boldsymbol{\nabla}\phi - \frac{\partial \boldsymbol{A}}{\partial t} + \boldsymbol{v} \times \boldsymbol{\nabla} \times \boldsymbol{A} \right)_k = q \left(-\partial_k \phi + v^m \partial_k A_m - \frac{\mathrm{d}A_k}{\mathrm{d}t} \right) . \tag{D.40}$$

The virtue of this (somewhat tortured) manipulation is that Eq. (D.40) can be written in the form of Eq. (D.36) if we take the potential energy function to be $V \equiv q\phi - q\boldsymbol{v} \cdot \boldsymbol{A}$ (show this). The Lagrangian for a particle coupled to the electromagnetic field is thus

$$L = T - q\phi + q\boldsymbol{v} \cdot \boldsymbol{A} . \tag{D.41}$$

The canonical momentum obtained from Eq. (D.41) is $\boldsymbol{p} = m\boldsymbol{v} + q\boldsymbol{A}$. The Hamiltonian function for a charged particle follows from Eq. (D.27),

$$H = \frac{1}{2m} (p_i - qA_i) \left(p^i - qA^i \right) + q\phi . \tag{D.42}$$

D.6 INVARIANCE UNDER COORDINATE TRANSFORMATIONS

The methods of analytical mechanics afford considerable flexibility in choosing coordinates. In particular, generalized coordinates allow constraints in the configuration of the system to be taken into account. The number of possible systems of coordinates (with the same number of degrees of freedom) is unlimited. Apropos of our study of relativity, to what extent are the equations of motion preserved under changes of coordinates? As we now show, *the form of the Lagrange and Hamilton equations is invariant under coordinate transformations in configuration space.*

Any set $\{Q^k\}_{k=1}^n$ of n independent, analytic functions of the coordinates $\{q^i\}$ provides another set of coordinates to describe the system (Section 5.1.4), a new set of numbers to attach to the same points in configuration space,

$$Q^k = Q^k(q^1, \cdots, q^n) . \qquad (k = 1, \cdots, n) \tag{D.43}$$

Time is not a coordinate in classical mechanics, as it is in the theory of relativity; we restrict ourselves to coordinate transformations not explicitly involving the time. The transformation given by Eq. (D.43) is by assumption invertible, with $q^j = q^j(Q^1, \cdots, Q^n)$, $(j = 1, \cdots, n)$. The generalized velocities then transform as

$$\dot{q}^j = \frac{\partial q^j}{\partial Q^k} \dot{Q}^k , \tag{D.44}$$

i.e., the generalized velocities in one coordinate system, $\dot{q}^j$, are linear combinations of those in another, $\dot{Q}^k$, multiplied by functions solely of the Q^i, $\partial q^j / \partial Q^k$. The quantities $\{\partial q^j / \partial Q^k\}$ play the role of the Jacobian matrices in tensor analysis. Note that Eq. (D.44) implies the relation

$$\frac{\partial \dot{q}^j}{\partial \dot{Q}^k} = \frac{\partial q^j}{\partial Q^k} . \tag{D.45}$$

(The reciprocal to Eq. (D.44) is $\dot{Q}^k = \left(\partial Q^k/\partial q^j\right)\dot{q}^j$, with $\partial \dot{Q}^k/\partial \dot{q}^j = \partial Q^k/\partial q^j$.) A similar expression holds for the four-velocity in SR, $U^\mu = \mathrm{d}x^\mu/\mathrm{d}\tau$, Eq. (7.6); using its transformation property as a contravariant four-vector, Eq. (5.35),

$$\frac{\partial U^{\mu'}}{\partial U^\nu} = A^{\mu'}_\nu = \frac{\partial x^{\mu'}}{\partial x^\nu} . \tag{D.46}$$

Equation (D.46) is possible because there is an observer-independent time, the proper time $\mathrm{d}\tau$; Eq. (D.45) is possible because of the absolute time of classical mechanics, $\mathrm{d}t$.

The Lagrangian function in the "old" coordinates, $L(q^k, \dot{q}^k)$, *defines* through Eq. (D.44) and the inverse of Eq. (D.43) the Lagrangian $\widetilde{L}(Q^j, \dot{Q}^j)$ in the new coordinate system:

$$L(q^k, \dot{q}^k) \equiv \widetilde{L}(Q^j, \dot{Q}^j) . \tag{D.47}$$

The value of a scalar field is invariant under coordinate transformations, but the *form* of the function of the transformed coordinates may change (Section 5.1). The form invariance of the Lagrange equations is then straightforward to demonstrate. From Eq. (D.18) and the chain rule,

$$0 = \frac{\delta L}{\delta q^k} = \frac{\partial Q^j}{\partial q^k}\frac{\delta \widetilde{L}}{\delta Q^j} , \qquad (k = 1, \cdots n) \tag{D.48}$$

where we've used Eq. (D.47). We therefore have[9]

$$\frac{\delta \widetilde{L}}{\delta Q^j} \equiv \frac{\partial \widetilde{L}}{\partial Q^j} - \frac{\mathrm{d}}{\mathrm{d}t}\left(\frac{\partial \widetilde{L}}{\partial \dot{Q}^j}\right) = 0 . \qquad (j = 1, \cdots, n) \tag{D.49}$$

The form of the Lagrange equations is independent of the choice of coordinates so long as they are connected by invertible coordinate transformations. The Lagrange equations (and Hamilton's equations, as we'll show) are thus *covariant equations of motion under point transformations*. Equation (D.18) (or Eq. (D.49)) is a necessary and sufficient condition for the action integral to be stationary. Finding the extremum of the action integral is thus independent of any special coordinate system; we can adopt whatever coordinates are best suited to the problem. Because the Lagrangian is an invariant of the transformation Eq. (D.47), we arrive at the important conclusion: *The value of the action integral S is invariant.*

The canonical momentum in the new coordinates is defined analogously to its definition in the original coordinates, Eq. (D.25):

$$P_k \equiv \frac{\partial \widetilde{L}}{\partial \dot{Q}^k} = \frac{\partial \dot{q}^j}{\partial \dot{Q}^k}\frac{\partial L}{\partial \dot{q}^j} = \frac{\partial q^j}{\partial Q^k} p_j , \tag{D.50}$$

where we've used Eq. (D.47) and the chain rule in the second equality, and Eq. (D.45) in the last. The canonical momenta P_k therefore transform as the components of a vector, just as do the velocities, Eq. (D.44). Note the invariance of the scalar product:

$$p_j \dot{q}^j = P_k \frac{\partial Q^k}{\partial q^j}\frac{\partial q^j}{\partial Q^l}\dot{Q}^l = P_k \dot{Q}^l \delta^k_l = P_k \dot{Q}^k , \tag{D.51}$$

where we've used the inverse of Eq. (D.50) and Eq. (D.44).

The "old" Hamiltonian defines, through these transformation equations, its form for the new coordinates and momenta:

$$H(p_k, q^k) \equiv \widetilde{H}(P_j, Q^j) . \tag{D.52}$$

[9] Because of the linear independence of the coordinate transformations, $\partial Q^j/\partial q^k$.

(Equivalently, Eq. (D.52) follows from the Legendre transform, $\widetilde{H} = P_k\dot{Q}^k - \widetilde{L}(Q,\dot{Q})$, together with Eq. (D.47) and Eq. (D.51).) The form invariance of Hamilton's equations then follows using the steps by which Eqs. (D.28) and (D.29) were derived:

$$\frac{\partial \widetilde{H}}{\partial P_k} = \dot{Q}^k \qquad \frac{\partial \widetilde{H}}{\partial Q^k} = -\frac{\partial \widetilde{L}}{\partial Q^k} = -\dot{P}_k \,.$$

D.7 LAGRANGIAN DENSITY

We now consider the dynamics of a *continuous system*. To do so, we examine a system of discrete masses connected by springs and show how it transitions to a continuous elastic medium in the limit as the separation between masses vanishes. Figure D.1 shows a segment of a one-dimensional array of identical masses m connected by identical springs k at equilibrium locations $x = na$, where n is an integer and a is the equilibrium spring length. The figure shows a configuration of the instantaneous *displacements* ϕ_n of the masses away from their equilibrium positions.

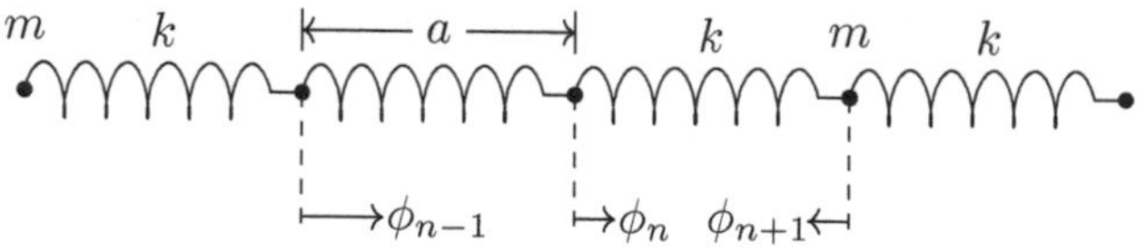

Figure D.1 System of coupled masses and springs.

The Lagrangian for the system of masses and springs is

$$L = \frac{1}{2}\sum_n \left[m\dot{\phi}_n^2 - k\left(\phi_n - \phi_{n-1}\right)^2\right] \,. \tag{D.53}$$

The equations of motion for the displacements $\{\phi_n\}$ are, from Eq. (D.18), a coupled set of differential-difference equations

$$m\ddot{\phi}_n = -k\left(2\phi_n - \phi_{n-1} - \phi_{n+1}\right) \,. \tag{D.54}$$

At this point, one could Fourier transform Eq. (D.54) to get the normal-mode frequencies. But, as they say, that's not important now.

Rewrite Eq. (D.53),

$$L = \frac{1}{2}\sum_n a\left[\frac{m}{a}\dot{\phi}_n^2 - ka\left(\frac{\phi_n - \phi_{n-1}}{a}\right)^2\right] \equiv \sum_n aL_n \,. \tag{D.55}$$

Now let $a \to 0$ (while the number of masses $N \to \infty$) such that there is a uniform mass per length $m/a \to \rho$ and a string tension $ka \to T$. We also generalize the notion of displacement, away from its definition at discrete locations $x = na$, to a continuous *function*, $\phi(x = na, t) \to \phi(x,t)$ as $a \to 0$. The *displacement field* $\phi(x,t)$ is the elastic distortion of the medium at location x at time t. We can also think of the process $a \to 0$ on the right side of Eq. (D.55) as replacing a with $\mathrm{d}x$, and converting the sum to an integral,

$$L \to \frac{1}{2}\int \mathrm{d}x \left[\rho\left(\frac{\partial\phi}{\partial t}\right)^2 - T\left(\frac{\partial\phi}{\partial x}\right)^2\right] \equiv \int \mathscr{L}\mathrm{d}x \,, \tag{D.56}$$

where we've replaced $(\phi_n - \phi_{n-1})/a$ with $\partial\phi/\partial x$ as $a \to 0$ and we've defined the *Lagrangian density* for a one-dimensional elastic medium,

$$\mathscr{L} = \frac{1}{2}\left[\rho\left(\frac{\partial\phi}{\partial t}\right)^2 - T\left(\frac{\partial\phi}{\partial x}\right)^2\right] \,. \tag{D.57}$$

Whenever we have a continuous system we work with a Lagrangian density, $L = \int \mathscr{L} \mathrm{d}x$. In three dimensions, $L = \int \mathscr{L} \mathrm{d}x\mathrm{d}y\mathrm{d}z$. The dimension of $\mathscr{L}$ in d dimensions is energy/(length)^d.

What are the generalized coordinates for a continuous system? It's *not* the position coordinate x and it's useful to keep in mind the process by which we arrived at the Lagrangian density. The discrete index n "becomes" the position coordinate x in the transition from a discrete to a continuous system. The generalized coordinates that specify the *configuration* of the system, the displacements $\{\phi_n\}$ in the discrete case, become the displacement *field* $\phi(x)$ for the continuous system. It might seem then that there is only one generalized coordinate for the continuous system, the field $\phi(x, t)$, but that's misleading: The field ϕ represents the displacement at a non-denumerably *infinite* number of locations, those labeled by the variable x.

The Lagrangian density is a function of the field and its space and time derivatives, $\mathscr{L} = \mathscr{L}(\phi, \partial_t\phi, \partial_x\phi, \partial_y\phi, \partial_z\phi)$, which we can abbreviate as $\mathscr{L} = \mathscr{L}(\phi, \partial_\mu\phi)$. The action integral $S = \int L\mathrm{d}t$, Eq. (D.17), becomes in terms of the Lagrangian density, $S = \int \mathscr{L}\mathrm{d}x\mathrm{d}y\mathrm{d}z\mathrm{d}t$. If $\mathscr{L}$ is Lorentz invariant, then so is S owing to the Lorentz invariance of the volume element in Minkowski space (Section 5.2). Hamilton's principle for fields is then

$$\delta S = \delta \int \mathscr{L}(\phi, \partial_\mu\phi)\mathrm{d}x\mathrm{d}y\mathrm{d}z\mathrm{d}t = 0 \,, \tag{D.58}$$

where the variations indicated in Eq. (D.58) are with respect to the *fields* for fixed x, y, z, t. The analogs of Eqs. (D.6) and (D.7) are

$$\delta\phi = \frac{\partial\phi}{\partial\alpha}\mathrm{d}\alpha \qquad \delta(\partial_\mu\phi) = \frac{\partial}{\partial\alpha}\frac{\partial\phi}{\partial x^\mu}\mathrm{d}\alpha = \frac{\partial}{\partial x^\mu}\frac{\partial\phi}{\partial\alpha}\mathrm{d}\alpha = \partial_\mu\delta\phi \,. \tag{D.59}$$

The analog of Eq. (D.8) is

$$\delta S = \int \left[\frac{\partial\mathscr{L}}{\partial\phi}\delta\phi + \frac{\partial\mathscr{L}}{\partial(\partial_\mu\phi)}\delta(\partial_\mu\phi)\right]\mathrm{d}x\mathrm{d}y\mathrm{d}z\mathrm{d}t \,. \tag{D.60}$$

Just as in Eq. (D.9), the second term in Eq. (D.60) can be integrated by parts using Eq. (D.59),

$$\begin{aligned}\int \frac{\partial\mathscr{L}}{\partial(\partial_\mu\phi)}(\partial_\mu\delta\phi)\mathrm{d}x\mathrm{d}y\mathrm{d}z\mathrm{d}t &= \int \mathrm{d}x\mathrm{d}z\mathrm{d}t\left[\frac{\partial\mathscr{L}}{\partial(\partial_y\phi)}\delta\phi\bigg|_{\text{surface}} - \int\frac{\partial}{\partial y}\left(\frac{\partial\mathscr{L}}{\partial(\partial_y\phi)}\right)\delta\phi\mathrm{d}y\right] \\ &= -\int \partial_\mu\left(\frac{\partial\mathscr{L}}{\partial(\partial_\mu\phi)}\right)\delta\phi\mathrm{d}x\mathrm{d}y\mathrm{d}z\mathrm{d}t \,,\end{aligned} \tag{D.61}$$

where we have shown explicitly the integration by parts for $\mu = y$, but there will be similar terms for $\mu = x, z, t$. The variation $\delta\phi$ is taken to vanish at the hypersurfaces bounding the system in spacetime. Combining Eq. (D.61) with Eq. (D.60), Eq. (D.58) is equivalent to

$$\delta S = \int \left[\frac{\partial\mathscr{L}}{\partial\phi} - \partial_\mu\left(\frac{\partial\mathscr{L}}{\partial(\partial_\mu\phi)}\right)\right]\delta\phi\mathrm{d}x\mathrm{d}y\mathrm{d}z\mathrm{d}t = 0 \,. \tag{D.62}$$

The integral can vanish for arbitrary $\delta\phi$ only if the functional derivative vanishes (the analog of Eq. (D.12))

$$\frac{\delta\mathscr{L}}{\delta\phi} \equiv \frac{\partial\mathscr{L}}{\partial\phi} - \partial_\mu\left(\frac{\partial\mathscr{L}}{\partial(\partial_\mu\phi)}\right) = 0 \,. \tag{D.63}$$

Equation (D.63) is the Euler-Lagrange equation for fields.

It's often the case that a system supports *several* fields $\{\phi_i\}_{i=1}^n$ at each point in spacetime, with $\mathscr{L} = \mathscr{L}(\phi_i, \partial_\mu\phi_i)$. Stationarity of the action is then expressed as in Eq. (D.13),

$$\delta S = \int \sum_i \frac{\delta\mathscr{L}}{\delta\phi_i}\delta\phi_i\mathrm{d}t\mathrm{d}x\mathrm{d}y\mathrm{d}z = 0 \,. \tag{D.64}$$

As in Eq. (D.14), for independent variations $\delta\phi_i$, stationarity requires

$$\partial_\mu \left(\frac{\partial \mathscr{L}}{\partial(\partial_\mu \phi_i)} \right) = \frac{\partial \mathscr{L}}{\partial \phi_i} \qquad i = 1, \cdots, n\,. \tag{D.65}$$

What equation of motion is obtained from the Lagrangian density given by Eq. (D.57), which we can write as

$$\mathscr{L} = \frac{1}{2}\left[\rho\left(\partial_0\phi\right)^2 - T\left(\partial_1\phi\right)^2\right]? \tag{D.66}$$

Using Eq. (D.66), we have the derivatives

$$\frac{\partial \mathscr{L}}{\partial \phi} = 0 \qquad \frac{\partial \mathscr{L}}{\partial(\partial_0\phi)} = \rho\partial_0\phi \qquad \frac{\partial \mathscr{L}}{\partial(\partial_1\phi)} = -T\partial_1\phi\,. \tag{D.67}$$

From Eq. (D.63), we find, using Eq. (D.67),

$$\partial_0\left(\frac{\partial \mathscr{L}}{\partial(\partial_0\phi)}\right) + \partial_1\left(\frac{\partial \mathscr{L}}{\partial(\partial_1\phi)}\right) = \rho\frac{\partial^2\phi}{\partial t^2} - T\frac{\partial^2\phi}{\partial x^2} = 0\,, \tag{D.68}$$

which is the wave equation for a one-dimensional elastic medium. From the equation of motion for the discrete system, Eq. (D.54), if we let $a \to 0$ presumably we should obtain the wave equation. Starting from Eq. (D.54), divide by a,

$$\frac{m}{a}\ddot{\phi}_n = -ka\left(\frac{2\phi_n - \phi_{n-1} - \phi_{n+1}}{a^2}\right)\,. \tag{D.69}$$

In the limit as $a \to 0$ we have

$$\rho\ddot{\phi} = T\frac{\partial^2\phi}{\partial x^2}\,, \tag{D.70}$$

the same as Eq. (D.68), where we recognize the terms in parentheses in Eq. (D.69) as the finite-difference approximation to the second derivative.

The Hamiltonian for continuous systems is the analog of Eq. (D.26) :

$$H = \int \mathrm{d}^3x \left(\frac{\partial \mathscr{L}}{\partial\dot{\phi}}\dot{\phi} - \mathscr{L}\right) \equiv \int \mathrm{d}^3x \mathscr{H}\,, \tag{D.71}$$

where the *Hamiltonian density*

$$\mathscr{H} \equiv \frac{\partial \mathscr{L}}{\partial\dot{\phi}}\dot{\phi} - \mathscr{L} = \partial_0\phi\frac{\partial \mathscr{L}}{\partial(\partial_0\phi)} - \mathscr{L}\,. \tag{D.72}$$

APPENDIX E

Photon and particle orbits

WE gather several results useful in analyzing the integrals that give the orbits of photons and particles in Schwarzschild spacetime.

E.1 PHOTON SCATTERING ANGLE

The scattering angle $\Delta\phi$ in a photon trajectory is given by Eq. (17.49), which we reproduce here

$$\Delta\phi = 2\int_0^{u_1} du \left[\frac{1}{b^2} - u^2 + r_S u^3\right]^{-1/2}, \tag{E.1}$$

where b is the impact parameter and u_1 is the smallest positive factor of the cubic polynomial in the integrand. The cubic can be factored: $r_S u^3 - u^2 + 1/b^2 = r_S(u-u_1)(u-u_2)(u-u_3)$ where $u_3 \le 0 \le u_1 \le u_2$ are

$$\begin{aligned} u_1 &= \frac{1}{3r_S}\left(1 - \cos(\theta/3) + \sqrt{3}\sin(\theta/3)\right) \qquad u_2 = \frac{1}{3r_S}\left(1 + 2\cos(\theta/3)\right) \\ u_3 &= \frac{1}{3r_S}\left(1 - \cos(\theta/3) - \sqrt{3}\sin(\theta/3)\right), \end{aligned} \tag{E.2}$$

with θ given by

$$\cos\theta = 1 - \frac{27}{2}\left(\frac{r_S}{b}\right)^2. \tag{E.3}$$

For $b \to \infty$ ($\mathscr{E} \to 0$), $\theta \to 0$, and no scattering occurs. As b decreases to the point where a circular orbit is possible, $4b^2 \to 27r_S^2$, $\theta \to \pi$. For $0 \le \theta \le \pi$, $0 \le u_1 \le 2/(3r_S)$, $2/(3r_S) \le u_2 \le 1/r_S$, and $-1/(3r_S) \le u_3 \le 0$. The factor u_3 is unphysical; it does not represent a positive distance. The factor u_1 is associated with the turning point in the orbit. As $\theta \to \pi$, $u_1 = u_2 = 2/(3r_S)$ and the integral diverges: A circular orbit represents an infinite scattering angle.

Equation (E.1) is an elliptic integral. In the notation of Abramowitz and Stegun,[73, p597]

$$\Delta\phi = \frac{2}{\sqrt{r_S}}\frac{2}{\sqrt{u_2 - u_3}}F(\psi\backslash\alpha) = \frac{4\sqrt{3}}{\sqrt{3\cos\theta/3 + \sqrt{3}\sin\theta/3}}F(\psi\backslash\alpha),$$

where $F(\psi\backslash\alpha)$ is an elliptic integral of the first kind, $F(\psi\backslash\alpha) \equiv \int_0^\psi \frac{d\lambda}{\sqrt{(1-\sin^2\alpha\sin^2\lambda)}}$, with

$$\cos^2\psi = \frac{-u_3\,(u_2-u_1)}{u_2\;(u_1-u_3)} = \frac{\sin\theta/3+\sin 2\theta/3-\sqrt{3}(\cos\theta/3-\cos 2\theta/3)}{2\sin(\theta/3)(1+2\cos\theta/3)}$$

$$\sin^2\alpha = \frac{u_1-u_3}{u_2-u_3} = \frac{2\tan(\theta/3)}{\sqrt{3}+\tan(\theta/3)} .$$

As $b \to \infty$, $\theta \to 0$; consequently $\alpha \to 0$ and $\psi \to \pi/4$. In this limit $\Delta\phi \to \pi$, as it should (scattering angle $\Delta = 0$). As $4b^2 \to 27r_S^2$ (circular orbit), $\theta \to \pi$, with the result that $\alpha \to \pi/2$ and $\psi \to \pi/2$; hence in this limit $\Delta\phi \to 4F(\psi \to \pi/2\backslash\alpha \to \pi/2)$, which diverges.

In Section 17.4 a perturbative treatment of Eq. (E.1) is given for $r_S << b$. If we let $x \equiv r_S/b$, then for $x \ll 1$ we find from an analysis of Eq. (E.3) that

$$\theta = 3\sqrt{3}x\left(1+\frac{9}{8}x^2+O(x^4)\right) . \tag{E.4}$$

From (E.2), expanding u_1 and u_3 to third order in θ, we have

$$u_{1,3} = \pm\frac{\sqrt{3}}{9r_S}\theta\left(1\pm\frac{\sqrt{3}}{18}\theta-\frac{1}{54}\theta^2+O(\theta^3)\right) , \tag{E.5}$$

where the upper (lower) sign refers to u_1 (u_3). Combining Eqs. (E.4) and (E.5), we find

$$u_{1,3} = \pm\frac{1}{b}\left(1\pm\frac{1}{2}x+\frac{5}{8}x^2+O(x^3)\right) . \tag{E.6}$$

From (E.2), u_2 is given by

$$u_2 = \frac{1}{r_S}(1-x^2+O(x^4)) . \tag{E.7}$$

E.2 BOUND ORBITS

The angle through one period of a bound orbit is given by Eq. (17.60), which we reproduce here:

$$\Delta\phi = 2\int_{u_3}^{u_2} du\left[\frac{2\mathscr{E}}{l^2}+\frac{2GM}{l^2}u-u^2+\frac{2GM}{c^2}u^3\right]^{-1/2} . \tag{E.8}$$

The quantities u_3, u_2 are the smaller two of the three factors of the cubic polynomial,

$$\frac{2\mathscr{E}}{l^2}+\frac{2GM}{l^2}u-u^2+r_Su^3 \equiv r_S(u-u_3)(u-u_2)(u-u_1) , \tag{E.9}$$

where generally we have $u_3 \le u_2 \le u_1$ and $u_3 < u_1$. They are given by

$$\begin{aligned} u_1 &= \frac{1}{3r_S}\left(1+2\sqrt{1-3/\xi^2}\cos\theta/3\right) \\ u_2 &= \frac{1}{3r_S}\left(1-\sqrt{1-3/\xi^2}(\cos\theta/3-\sqrt{3}\sin\theta/3)\right) \\ u_3 &= \frac{1}{3r_S}\left(1-\sqrt{1-3/\xi^2}(\cos\theta/3+\sqrt{3}\sin\theta/3)\right) , \end{aligned} \tag{E.10}$$

where $\xi \equiv lc/(2GM)$ and

$$\cos\theta = \frac{1}{(1-3/\xi^2)^{3/2}}\left(1 - \frac{9}{2\xi^2} - \frac{27\mathscr{E}}{\xi^2 c^2}\right) . \tag{E.11}$$

For these quantities to be real, $\xi > \sqrt{3}$, consistent with what we found in Section 17.2.

From (E.10), $u_3 \le u_2$, with equality occurring for $\theta = 0$. The factors u_3, u_2 become equal for a stable circular orbit with $\mathscr{E}_{\text{sco}} \equiv c^2[y_-(y_- - 6)]/(54\xi^2)$ (Eqs. (17.17) and (17.16)), where Eq. (E.11) indicates $\cos\theta = 1$. Thus, $\theta = 0$ corresponds to a stable circular orbit, for any $\xi > \sqrt{3}$. From (E.10), $u_2 \le u_1$, with equality occurring for $\theta = \pi$. The factors u_1, u_2 coincide at the *unstable* circular orbit, where $\mathscr{E}_{\text{uco}} \equiv c^2[y_+(y_+ - 6)]/(54\xi^2)$ implies $\cos\theta = -1$ for any $\xi > \sqrt{3}$. As can be shown, $\mathscr{E}_{\text{uco}} \le 0$ only for $\xi \le 2$. Thus, for $\sqrt{3} < \xi \le 2$, $0 \le \theta \le \pi$. For $\xi > 2$, however, $0 \le \theta \le \theta_{\text{max}}$, with $\theta_{\text{max}} < \pi$ given by Eq. (E.11) for $\mathscr{E} = 0$.

For the purposes of a perturbative treatment of Eq. (E.8) (Section 17.5) we need to analyze the factors of (E.9) for large ξ. And to do that, we need to characterize the angle θ in this limit. It's convenient to write $\mathscr{E} = \mathscr{E}_c + \delta\mathscr{E}$, where $\mathscr{E}_c$ is the energy of the stable circular orbit, implying that $\delta\mathscr{E} > 0$. Because $\cos\theta = 1$ for $\mathscr{E} = \mathscr{E}_c$, we have from Eq. (E.11) that

$$\cos\theta = 1 - \frac{27}{\xi^2 c^2}\frac{\delta\mathscr{E}}{(1-3/\xi^2)^{3/2}} . \tag{E.12}$$

Approximating $\cos\theta \approx 1 - \frac{1}{2}\theta^2$,

$$\theta^2 \approx \frac{54}{\xi^2 c^2}\frac{\delta\mathscr{E}}{(1-3/\xi^2)^{3/2}} . \tag{E.13}$$

For large c, $\theta \approx 1/c^2$. To first order in $1/c^2$, we have from (E.10)

$$\begin{aligned} u_1 &= \frac{1}{r_S}\left(1 - \frac{1}{\xi^2} + O(1/c^4)\right) \\ \frac{u_2+u_3}{u_1} &= \frac{1}{\xi^2} + O(1/c^4) . \end{aligned} \tag{E.14}$$

Bibliography

[1] E.P. Wigner. The unreasonable effectiveness of mathematics in the natural sciences. *Communications on Pure and Applied Mathematics*, 13:1–14, 1960.

[2] A. Forsee. *Albert Einstein: Theoretical Physicist*. Macmillan, 1963.

[3] Galileo. *Dialogue Concerning the Two Chief World Systems*. University of California Press, 1967.

[4] I. Newton. *The Principia: Mathematical Principles of Natural Philosophy*. University of California Press, 1999.

[5] A. Einstein. *The Meaning of Relativity*. Princeton University Press, 1955.

[6] M.R. Ayers. *George Berkeley: Philosophical Works*. Tuttle Publishing, 1993.

[7] E. Mach. *Science of Mechanics*. Open Court Publishing, 1960.

[8] O. Veblen. *Invariants of Quadratic Differential Forms*. Cambridge University Press, 1927.

[9] H.A. Lorentz, A. Einstein, H. Minkowski, and H. Weyl. *The Principle of Relativity*. Dover Publications, 1952.

[10] C. Lanczos. *The Variational Principles of Mechanics*. University of Toronto Press, 1949.

[11] M.K. Munitz. *Theories of the Universe*. The Free Press, 1957.

[12] H. Bondi. *Relativity and Common Sense*. Dover Publications, 1980.

[13] R.P. Feynman. *The Feynman Lectures on Physics, Volume III*. Addison-Wesley, 1965.

[14] H.E. Ives and G.R. Stilwell. An experimental study of the rate of a moving clock. *Journal of the Optical Society of America*, 28:215–226, 1938.

[15] A.A. Michelson and E.W. Morley. On the relative motion of the earth and the luminiferous ether. *American Journal of Science*, 34:333–345, 1887.

[16] R.J. Kennedy and E.M. Thorndike. Experimental establishment of the relativity of time. *Physical Review*, 42:400–418, 1932.

[17] A. Brillet and J.L. Hall. Improved laser test of the isotropy of space. *Physical Review Letters*, 42:549–552, 1979.

[18] S. Hermann et al. Rotating optical cavity experiment testing Lorentz invariance at the 10^{-17} level. *Physical Review D*, 80:105011, 2009.

[19] W. Bertozzi. Speed and kinetic energy of relativistic electrons. *American Journal of Physics*, 32:551–555, 1964.

[20] Z.G.T. Guiragossian et al. Relative velocity measurements of electrons and gamma rays at 15 GeV. *Phys. Rev. Lett.*, 34:335, 1975.

[21] D. Hasselkamp, E. Mondry, and A. Scharmann. Direct observation of the transversal Doppler-shift. *Zeitschrift für Physik A*, 289:151–155, 1979.

[22] M.D. Weir, J. Hess, and G.B. Thomas. *Thomas' Calculus*. Pearson, 2010.

[23] O. Veblen and J.H.C. Whitehead. *The Foundations of Differential Geometry*. Cambridge University Press, 1932.

[24] I.M. Gelfand. *Lectures on Linear Algebra*. Interscience Publishers, 1961.

[25] J.A. Schouten. *Tensor Analysis for Physicists*. Dover Publications, 1989.

[26] G. Weinreich. *Geometrical Vectors*. University of Chicago Press, 1998.

[27] C.W. Norman. *Undergraduate Algebra*. Oxford University Press, 1986.

[28] P.R. Halmos. *Finite-Dimensional Vector Spaces*. Van Nostrand, 1958.

[29] R.L. Bishop and S.I. Goldberg. *Tensor Analysis on Manifolds*. Dover, 1980.

[30] A.C. Aitken. *Determinants and Matrices*. Oliver and Boyd, 1964.

[31] H.S.M. Coxeter. *Regular Polytopes*. Dover Publications, 1973.

[32] R.P. Feynman. The development of the space-time view of quantum electrodynamics. *Science*, 153:699–708, 1966.

[33] L.H. Thomas. Motion of the spinning electron. *Nature*, 117:514, 1926.

[34] E.P. Wigner. On unitary representations of the Lorentz group. *Ann. Math.*, 40:149, 1939.

[35] E.M. Purcell. *Electricity and Magnetism*. McGraw-Hill, 1965.

[36] J.C. Maxwell. *A Treatise on Electricity and Magnetism, Vol I, Third Edition*. Oxford University Press, 1892.

[37] R.M. Wald. *General Relativity*. University of Chicago Press, 1984.

[38] S.W. Hawking and G.F.R. Ellis. *The Large Scale Structure of Space-Time*. Cambridge University Press, 1973.

[39] B.E. Schaefer. Severe limits on variations of the speed of light with frequency. *Physical Review Letters*, 82:4964–4966, 1999.

[40] F. Gross. *Relativistic Quantum Mechanics and Field Theory*. John Wiley, 1993.

[41] H. Weyl. *Space-Time-Matter*. Dover Publications, 1952.

[42] Galileo. *Two New Sciences*. University of Wisconsin Press, 1974.

[43] S. Schlamminger, K.-Y. Choi, T.A. Wagner, et al. Test of the equivalence principle using a rotating torsion balance. *Phys. Rev. Lett.*, 100:041101, 2008.

[44] C.W.F. Everitt et al. The Gravity Probe B test of general relativity. *Class. Quantum Grav.*, 32:224001, 2015.

[45] K. Nordvedt. Equivalence principle for massive bodies. *Physical Review*, 169:1014–1025, 1968.

[46] J.G. Williams, S.G. Turyshev, and D.H. Boggs. Lunar laser ranging tests of the equivalence principle with the earth and moon. *International Journal of Modern Physics D*, 18:1129–1175, 2009.

[47] R.V. Pound and G.A. Rebka. Apparent weight of photons. *Phys. Rev. Lett.*, 4:337, 1959.

[48] R.V. Pound and J.L. Snider. Effect of gravity on nuclear resonance. *Phys. Rev. Lett.*, 13:539, 1964.

[49] A.A. Michelson and H.G. Gale. The effect of the earth's rotation on the velocity of light. *The Astrophysical Journal*, 61:140, 1925.

[50] V. Bargmann, L. Michel, and V.L. Telegdi. Precession of the polarization of particles moving in a homogeneous electromagnetic field. *Phys. Rev. Lett.*, 2:435, 1959.

[51] R. Geroch. Lecture notes on differential geometry. Unpublished, Univerity of Chicago, 1972.

[52] S. Kobayashi and K. Nomizu. *Foundations of Differential Geometry: Volume I.* John Wiley, 1963.

[53] R. Geroch. Space-time structure from a global viewpoint. In R.K. Sachs, editor, *General Relativity and Cosmology*, pages 71–103. Academic Press, 1971.

[54] H. Flanders. *Differential Forms*. Academic Press, 1963.

[55] M. Göckeler and T. Schücker. *Differential Geometry, Gauge Theories, and Gravity*. Cambridge University Press, 1987.

[56] R.L. Bishop and R.J. Crittenden. *Geometry of Manifolds*. Academic Press, 1964.

[57] M. Spivak. *Calculus on Manifolds*. W.A. Benjamin, 1965.

[58] D. Greenspan. *Theory and Solutions of Ordinary Differential Equations*. Macmillan, 1960.

[59] E.A. Coddington and N. Levinson. *The Theory of Ordinary Differential Equations*. McGraw-Hill, 1955.

[60] L.P. Eisenhart. *Riemannian Geometry*. Princeton University Press, 1997.

[61] F. De Felice and C.J.S. Clarke. *Relativity on Curved Manifolds*. Cambridge University Press, 1990.

[62] M.D. Kruskal. Maximal extension of Schwarzschild metric. *Phys. Rev.*, 119:1743–1745, 1960.

[63] G. Szekeres. On the singularities of a Riemannian manifold. *Publicationes Mathematicae Debrecen*, 7:285–301, 1960.

[64] D.E. Lebach et al. Measurements of the solar gravitational deflection of radio waves using very-long-baseline interferometry. *Phys. Rev. Lett.*, 75:1439, 1995.

[65] P.K. Seidelmann. *Explanatory Supplement to the Astronomical Almanac*. University Science Books, 1992.

[66] E.V. Pitjeva. EPM ephemerides and relativity. In S.A. Kiloner, P.K. Seidelmann, and M.H. Soffel, editors, *Proceedings IAU Symposium 261, Relativity in Fundamental Astronomy*, page 170. Cambridge University Press, 2010.

[67] I.I. Shapiro et al. Fourth test of general relativity: New radar result. *Phys. Rev. Lett.*, pages 1132–1135, 1971.

[68] C.M. Will. *Theory and Experiment in Gravitational Physics*. Cambridge University Press, 1993.

[69] B. Bertotti, L. Iess, and P. Tortura. A test of general relativity using radio links with the Cassini spacecraft. *Nature*, 425:374–376, 2003.

[70] N. Ashby. Relativity and the Global Positioning System. *Physics Today*, 55:41, 2002.

[71] C.W.F. Everitt et al. Gravity Probe B: Final results of a space experiment to test general relativity. *Phys. Rev. Lett.*, 106:221101, 2011.

[72] J.H. Luscombe. *Thermodynamics*. CRC Press, 2018.

[73] M. Abramowitz and I.A. Stegun. *Handbook of Mathematical Functions*. United States Department of Commerce, 1964.

[74] S. Weinberg. *Gravitation and Cosmology*. John Wiley, 1972.

[75] I. Ciufolini et al. Towards a one percent measurement of frame dragging by spin with satellite laser ranging to LAGEOS, LAGEOS 2 and LARES and GRACE gravity models. *Space Science Reviews*, 148:71, 2009.

[76] B.P. Abbott et al. Observation of gravitational waves from a binary black hole merger. *Phys. Rev. Lett.*, 116:061102, 2016.

[77] L.D. Landau and E.M. Lifshitz. *The Classical Theory of Fields*. Pergamon Press, 1975.

[78] A.S. Eddington. *The Mathematical Theory of Relativity*. Cambridge University Press, 1930.

[79] W. Pauli. *Theory of Relativity*. Pergamon Press, 1958.

[80] E. Masso and F. Rota. Primordial helium production in a charged universe. *Physics Letters B*, 545:221–225, 2005.

[81] M.I. Scrimgeour et al. The WiggleZ Dark Energy Survey: the transition to large-scale cosmic homogeneity. *Mon. Not. R. Astron. Soc.*, 425:116–134, 2012.

[82] R.C. Tolman. Effect of inhomogeneity on cosmological models. *Proceedings of the National Academy of Science*, 20:169–176, 1934.

[83] R. Courant and D. Hilbert. *Methods of Mathematical Physics, Volume I*. Interscience Publishers, 1953.

[84] V.I. Arnold. *Mathematical Methods of Classical Mechanics*. Springer-Verlag, 1989.

Index

Milton Keynes UK
Ingram Content Group UK Ltd.
UKHW051944071024
449327UK00026B/2163